FUNDAMENTALS
OF
ABSTRACT ALGEBRA

INTERNATIONAL SERIES IN PURE & APPLIED MATHEMATICS

SCHAUM'S SOLVER PROBLEMS SERIES

Each title in this series is a complete and expert source of solved problems with solutions worked out in step-by-step detail.

Related titles on the current list include:

3000 Solved Problems in Calculus
2500 Solved Problems in Differential
  Equations
2000 Solved Problems in Discrete
  Mathematics
3000 Solved Problems in Linear Algebra
2000 Solved Problems in Numerical
  Analysis
3000 Solved Problems in Precalculus

Bob Miller's Math Helpers
Bob Miller's Calc I Helper
Bob Miller's Calc II Helper
Bob Miller's Precalc Helper

McGraw-Hill Paperbacks
Arithmetic and Algebra...Again
How to Solve Word Problems in Algebra
Mind Over Math

Available at most college bookstores, or for a complete list of titles and prices, write to:

Schaum Division
The McGraw-Hill Companies, Inc.
11 West 19th Street
New York, NY 10011

To
Sadhana Malik
Patricia Mordeson
Monisha Sen

FUNDAMENTALS

OF

ABSTRACT ALGEBRA

D. S. Malik
Creighton University

John M. Mordeson
Creighton University

M. K. Sen
Calcutta University

The McGraw-Hill Companies, Inc.

New York St. Louis San Francisco Auckland Bogotá Caracas
Lisbon London Madrid Mexico City Milan Montreal
New Delhi San Juan Singapore Sydney Tokyo Toronto

McGraw-Hill

A Division of The McGraw-Hill Companies

FUNDAMENTALS OF ABSTRACT ALGEBRA

This book is printed on acid-free paper.

1 2 3 4 5 6 7 8 9 0   FGR   FGR   9 0 9 8 7 6

ISBN 0-07-040035-0

The editors were Karen M. Minette and John M. Morriss;
the production supervisor was Kathryn Porzio.
The jacket was designed by Robin Hoffmann.
The photo editor was Elyse Rieder.
Quebecor Printing/Fairfield was printer and binder.

Library of Congress Cataloging-in-Publication data
Malik, D. S.
   Fundamentals of abstract algebra / D.S. Malik, John M. Mordeson,
M.K. Sen.
      p.   cm. — (International series in pure and applied
mathematics)
   Includes bibliographical references (p.   -   ) and index.
   ISBN 0-07-040035-0
   1. Algebra, Abstract.   I. Mordeson, John M.   II. Sen, M. K.
III. Title.   IV. Series.
QA162.M346   1997
512' .02—dc20   96-33312

http://www.mhcollege.com

Contents

PREFACE

This book is intended for a one-year undergraduate course in abstract algebra. Its design is such that the book can also be used for a one-semester course. The book contains more material than normally would be taught in a one-year course. This should give the teacher flexibility with respect to the selection of the content and the level at which the book is to be used. We give a rigorous treatment of the fundamentals of abstract algebra with numerous examples to illustrate the concepts. It usually takes students some time to become comfortable with the seeming abstractness of modern algebra. Hence we begin at a leisurely pace paying great attention to the clarity of our proofs. The only real prerequisite for the course is the appropriate mathematical maturity of the students. Although the material found in calculus is independent of that of abstract algebra, a year of calculus is typically given as a prerequisite. Since many of the examples in algebra comes from matrices, we assume that the reader has some basic knowledge of matrix theory. The book should prepare the student for higher level mathematics courses and computer science courses. We have many problems of varying difficulty appearing after each section. We occasionally leave as an exercise the verification of a certain point in a proof. However, we do not rely on exercises to introduce concepts which will be needed later on in the text.

Topics are introduced that have never appeared in this type of textbook. They include Gröbner basis, rings of matrices, and Noetherian and Artinian rings. Another distinguishing feature of the book is the Worked-Out Exercises which appear after every section. These Worked-Out Exercises provide not only techniques of problem solving, but also supply additional information to enhance the level of knowledge of the reader. For example, in Chapter 7, we illustrate several techniques that are very effective in determining the Sylow subgroups of a group, whether the group is simple or not, and in determining the structure of a group. In Chapter 9, we give numerous examples and show how to determine different Abelian groups of a given order. We also show how to find the elementary divisors, the torsion coefficients, and the betti number of a finitely generated Abelian group. In Chapter 15, we give an algorithmic procedure to find the greatest common divisor and illustrate it in full detail.

We also illustrate how to show whether an element is prime and/or irreducible. In Chapter 24, we give numerous examples to show how to determine the Galois group and the intermediate fields of a Galois field extension. Of course, each section is followed by problems of varying difficulty for the reader to further master the subject. The reader should study the Worked-Out Exercises that are marked with $\diamond$ along with the chapter. Those not marked with $\diamond$ may be skipped during the first reading. Sprinkled throughout the book are comments dealing with the historical development of abstract algebra.

This book has been class-tested at Creighton University and at the University of Calcutta. During preparation of the manuscript, we used an approach which would help students who need a text to pass different types of aptitude tests in algebra.

In Chapter 1, the necessary ideas of sets, relations, functions, and binary operations are presented. We recommend that the chapter be gone through quickly in order to provide enough time to cover essential topics from abstract algebra. The students can refer back to material omitted on the first pass, as needed. For example, Zorn's lemma may be omitted on the first reading. It is not needed until Chapter 17.

Chapters 2 through 6 contain basic results on group theory. Most of the material in these chapters should be covered in the first semester. Chapters 10 through 14 contain basic results on ring theory. Most of the results in these chapters should also be covered in the first semester.

The second semester should cover Chapters 15 through 17. These chapters deal with Euclidean domains, unique factorization domains, and prime and maximal ideals. Students should now be well prepared to study field theory in the remaining part of the semester. Those who have not had a course on linear algebra should spend some time on vector spaces in Chapter 20. The students should finish the semester with Chapter 21 and as much of Chapter 22 through 24 as possible. There is plenty of material remaining from which special topics may be chosen.

We have included chapters on coding theory and Gröbner bases so that the student can gain some appreciation of the applications of abstract algebra. The chapter on coding theory contains enough material to allow the student to see applications of groups, ideals, and fields. We present a chapter on Gröbner bases because of its currency. It can be a first step into the area of computational algebra. The chapter also provides important applications of commutative algebra.

We would like to thank Professor James K. Deveney of Virginia Commonwealth University and his abstract algebra class for their valuable suggestions. We express our sincere gratitude to Fr. Michael Proterra, Dean, Creighton College of Arts and Sciences, for making possible Dr. Sen's visit during 1992–1993. We would like to thank Dr. Mark J. Wierman for showing us many important features of LaTex which were very helpful in preparing this manuscript in its present form and also for drawing all diagrams in the book. In addition, we express our sincere thanks to Dr. T.K. Mukherjee, Dr. S. Ganguly, and Dr. S.R. Lopéz-Permouth for their critical comments. We are very thankful to our families for their constant support and encouragement throughout this project. We would like to give special thanks to Shelly Malik, who constantly inquired about the manuscript and counted each chapter every time the manuscript was printed. Finally we would like to thank Karen Minette of McGraw-Hill for making this project a success.

We welcome any comments concerning the text. The comments may be forwarded to the following e-mail addresses: malik@bluejay.creighton.edu or mordes@bluejay.creighton.edu

D. S. Malik
John N. Mordeson
M. K. Sen

Chapter Dependency Diagram

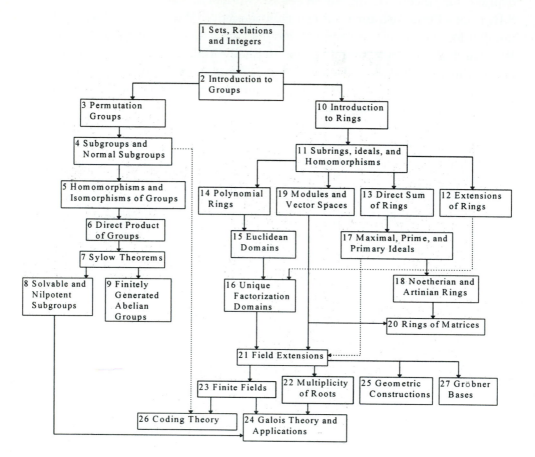

LIST OF SYMBOLS

$\in$	belongs to
$\notin$	does not belong to
$\subseteq$	subset
$\subset$	proper subset
$\supseteq$	contains
$\supset$	properly contains
$\triangle$	symmetric difference
$A \backslash B$	set difference
(a, b)	ordered pair
A'	complement of a set A
$\mathbf{N}$	set of positive integers
$\mathbf{Z}$	set of integers
$\mathbf{Z}^{\#}$	set of nonnegative integers
$\mathbf{Q}$	set of rational numbers
$\mathbf{Q}^{+}$	set of positive rational numbers
$\mathbf{Q}^{*}$	set of nonzero rational numbers
$\mathbf{R}$	set of real numbers
$\mathbf{R}^{+}$	set of positive real numbers
$\mathbf{R}^{*}$	set of nonzero real numbers
$\mathbf{C}$	set of complex numbers
$\mathbf{C}^{*}$	set of nonzero complex numbers
$\mathcal{P}(S)$	power set of the set S
$\cup$	union of sets
$\cap$	intersection of sets
$\binom{n}{i}$	number of combinations of n objects taken i at a time
$n!$	n factorial
$a \mid b$	a divides b
$a \nmid b$	a does not divide b
$\gcd(a, b)$	greatest common divisor of a and b
$\operatorname{lcm}(a, b)$	least common multiple of a and b
$\sum_{i=1}^{n} a_i$	$a_1 + a_2 + \cdots + a_n$
$\sum_{a \in S} a$	sum of all elements of S
$\equiv_n$	congruence modulo n
$f : A \to B$	f is a function from a set A into a set B
$f(x)$	image of x under f
$\mathcal{D}(f)$	domain of f

$\mathcal{I}(f)$	image of f
$g \circ f$	composition of mappings g and f
f^{-1}	inverse of a mapping f
I_n	$I_n = \{1, 2, \ldots, n\}$
$f(A)$	$f(A) = \{f(a) \mid a \in A\}$, A is a set contained in the domain of the function f
$f^{-1}(B)$	$f^{-1}(B) = \{x \in X \mid f(x) \in B\}$, where $f : X \to Y$ and $B \subseteq Y$
$\circ$	composition
Π	product
$M_n(R)$	set of all $n \times n$ matrices over R
$|X|$	number of elements in a set X
$|G|$	order of the group G
$\circ(a)$	order of an element a
$\mathbf{Z}_n$	set of integers modulo n
$Z(G)$	center of the group G
G/H	quotient group
aH, Ha	left, right coset of a in H
aHa^{-1}	$aHa^{-1} = \{aha^{-1} \mid h \in H\}$
$[G : H]$	index of the subgroup H in G
K_4	Klein 4-group
S_n	symmetric group on n symbols
A_n	alternating group on n symbols
D_n	dihedral group of degree n
$\langle S \rangle$	the subgroup generated by S
$\langle a \rangle$	the subgroup generated by a
$\oplus$	direct sum
$N(H)$	Normalizer of H
$C(a)$	centralizer of a
Ker f	kernel of f
$\simeq$	isomorphism
Aut(G)	set of all automorphisms of the group G
Inn(G)	set of all inner automorphisms of the group G
G_a	stabilizer of a or isotropy group of a
$Cl(a)$	conjugacy class of a
G'	commutator subgroup of the group G
$G[n]$	set of all $x \in G$ with $nx = 0$, G is a group
nG	$nG = \{nx \mid x \in G\}$
$C(R)$	center of the ring R
$Q_{\mathbf{R}}$	real quaternions
$\mathbf{Z}[\sqrt{n}]$	$\mathbf{Z}[\sqrt{n}] = \{a + b\sqrt{n} \mid a, b \in \mathbf{Z}\}$, n is a fixed positive integer
$\mathbf{Z}[i]$	$\mathbf{Z}[i] = \{a + bi \mid a, b \in \mathbf{Z}\}$
$\mathbf{Z}[i\sqrt{n}]$	$\mathbf{Z}[i\sqrt{n}] = \{a + bi\sqrt{n} \mid a, b \in \mathbf{Z}\}$, n is a fixed positive integer
$\mathbf{Q}[\sqrt{n}]$	$\mathbf{Q}[\sqrt{n}] = \{a + b\sqrt{n} \mid a, b \in \mathbf{Q}\}$, n is a fixed positive integer
$\mathbf{Q}[i]$	$\mathbf{Q}[i] = \{a + bi \mid a, b \in \mathbf{Q}\}$

$\mathbf{Q}[i\sqrt{n}]$	$\mathbf{Q}[i\sqrt{n}] = \{a + bi\sqrt{n} \mid a, b \in \mathbf{Q}\}$, n is a fixed positive integer
$\langle a \rangle_l$	the left ideal generated by a
$\langle a \rangle_r$	the right ideal generated by a
$\langle a \rangle$	the ideal generated by a
R/I	quotient ring
$Q(R)$	quotient field of the ring R
$R[x]$	polynomial ring in x
$\deg f(x)$	degree of the polynomial $f(x)$
$R[x_1, x_2, \ldots, x_n]$	polynomial ring in n indeterminates
$\sqrt{I}$	radical of an ideal I
$\mathrm{rad}R$	Jacobson radical of a ring R
F/K	field extension
$K(C)$	smallest subfield containing the subfield K and the subset C of a field
$[F : K]$	degree of the field F over the field K
$\mathrm{GF}(n)$	Galois field of n elements
$G(F/K)$	Galois group of the field F over the field K
F_G	fixed field of the group G
$\Phi_n(x)$	nth cyclotomic polynomial
P_F	plane of the field F
B^n	set of all binary n-tuples
x^α	$x_1^{\alpha_1} \cdots x_n^{\alpha_n}$
K^n	affine space over the field K
$I(V)$	ideal of the variety V
$\succ$	total ordering
$\succ_l$	lexicographic order
$\succ_{grl}$	graded lexicographic order
$\succ_{grel}$	graded reverse lexicographic order
$\mathrm{multideg}f$	multidegree of the polynomial f
$\mathrm{LC}(f)$	leading coefficient of the polynomial f
$\mathrm{LM}(f)$	leading monomial of the polynomial f
$\mathrm{LT}(f)$	leading term of the polynomial f
∎	end of proof

Chapter 1

Sets, Relations, and Integers

The purpose of this introductory chapter is mainly to review briefly some familiar properties of sets, functions, and number theory. Although most of these properties are familiar to the reader, there are certain concepts and results which are basic to the understanding of the body of the text.

This chapter is also used to set down the conventions and notations to be used throughout the book. Sets will always be denoted by capital letters. For example, we use the notation $\mathbf{N}$ for the set of positive integers, $\mathbf{Z}$ for the set of integers, $\mathbf{Z}^{\#}$ for the set of nonnegative integers, $\mathbf{E}$ for the set of even integers, $\mathbf{Q}$ for the set of rational numbers, $\mathbf{Q}^{+}$ for the set of positive rational numbers, $\mathbf{Q}^{*}$ for the set of nonzero rational numbers, $\mathbf{R}$ for the set of real numbers, $\mathbf{R}^{+}$ for the set of positive real numbers, $\mathbf{R}^{*}$ for the set of nonzero real numbers, $\mathbf{C}$ for the set of complex numbers, and $\mathbf{C}^{*}$ for the set of nonzero complex numbers.

1.1 Sets

We will not attempt to give an axiomatic treatment of set theory. Rather we use an intuitive approach to the subject. Consequently, we think of a **set** as some given collection of objects. A set S with only a finite number of elements is called a **finite** set; otherwise S is called an **infinite** set. We let $|S|$ denote the number of elements of S. We quite often denote a finite set by a listing of its elements within braces. For example, $\{1, 2, 3\}$ is the set consisting of the objects $1, 2, 3$. This technique is sometimes used for infinite sets. For instance, the set of positive integers $\mathbf{N}$ may be denoted by $\{1, 2, 3, \ldots\}$.

Given a set S, we use the notation $x \in S$ and $x \notin S$ to mean x is a member of S and x is not a member of S, respectively. For the set $S = \{1, 2, 3\}$, we have $1 \in S$ and $4 \notin S$.

A set A is said to be a **subset** of a set S if every element of A is an element of S. In this case, we write $A \subseteq S$ and say that A is contained in S. If $A \subseteq S$, but $A \neq S$, then we write $A \subset S$ and say that A is properly contained in S or that A is a **proper subset** of S. As an example, we have $\{1,2,3\} \subseteq \{1,2,3\}$ and $\{1,2\} \subset \{1,2,3\}$.

Let A and B be sets. If every member of A is a member of B and every member of B is a member of A, then we say that A and B are the **same** or **equal**. In this case, we write $A = B$. It is immediate that $A = B$ if and only if $A \subseteq B$ and $B \subseteq A$. Thus, we have the following theorem.

Theorem 1.1.1 *Let A and B be sets. Then $A = B$ if and only if $A \subseteq B$ and $B \subseteq A$.* ∎

The **null set** or **empty set** is the set with no elements. We usually denote the empty set by ϕ. For any set A, we have $\phi \subseteq A$. The later inclusion follows vacuously. That is, every element of ϕ is an element of A since ϕ has no elements.

We also describe sets in the following manner. Given a set S, the notation

$$A = \{x \mid x \in S, P(x)\}$$

or

$$A = \{x \in S \mid P(x)\}$$

means that A is the set of all elements x of S such that x satisfies the property P. For example, $\mathbf{N} = \{x \mid x \in \mathbf{Z}, x > 0\}$.

We can combine sets in several ways.

Definition 1.1.2 *The **union** of two sets A and B, written $A \cup B$, is defined to be the set*

$$A \cup B = \{x \mid x \in A \text{ or } x \in B\}.$$

In the above definition, we mean x is a member of A or x is a member of B or x is a member of both A and B.

Definition 1.1.3 *The **intersection** of two sets A and B, written $A \cap B$, is defined to be the set*

$$A \cap B = \{x \mid x \in A \text{ and } x \in B\}.$$

Here x is an element of $A \cap B$ if and only if x is a member of A and at the same time x is a member of B.

Let A and B be sets. By the definition of the union of sets, every element of A is an element of $A \cup B$. That is, $A \subseteq A \cup B$. Similarly, every element of B is also an element of $A \cup B$ and so $B \subseteq A \cup B$. Also, by the definition of the intersection of sets, every element of $A \cap B$ is an element of A and also an element of B. Hence, $A \cap B \subseteq A$ and $A \cap B \subseteq B$. We record these results in the following theorem.

Theorem 1.1.4 *Let A and B be sets. Then the following statements hold:*
   (i) $A \subseteq A \cup B$ and $B \subseteq A \cup B$.
   (ii) $A \cap B \subseteq A$ and $A \cap B \subseteq B$. ∎

The union and intersection of two sets A and B is described pictorially in the following diagrams. The shaded area represents the set in question.

   AUB   A∩B

Two sets A and B are said to be **disjoint** if $A \cap B = \phi$.

Example 1.1.5 *Let A be the set $\{1, 2, 3, 4\}$ and B be the set $\{3, 4, 5, 6\}$. Then*

$$A \cup B = \{1, 2, 3, 4, 5, 6\}$$

and $A \cap B = \{3, 4\}$. If C is the set $\{5, 6\}$, then

$$A \cup C = \{1, 2, 3, 4, 5, 6\}$$

while $A \cap C = \phi$.

Now that the union and intersection have been defined for two sets, these operations can be similarly defined for any finite number of sets. That is, suppose that $A_1, A_2, \ldots, A_n$ are n sets. The union of $A_1, A_2, \ldots, A_n$, denoted by $\cup_{i=1}^{n} A_i$ or $A_1 \cup A_2 \cup \cdots \cup A_n$, is the set of all elements x such that x is an element of some A_i, where $1 \leq i \leq n$. The intersection of $A_1, A_2, \ldots, A_n$, denoted by $\cap_{i=1}^{n} A_i$ or $A_1 \cap A_2 \cap \cdots \cap A_n$, is the set of all elements x such that $x \in A_i$ for all i, $1 \leq i \leq n$.

We say that a set I is an **index set** for a collection of sets $\mathcal{A}$ if for any $\alpha \in I$, there exists a set $A_\alpha \in \mathcal{A}$ and $\mathcal{A} = \{A_\alpha \mid \alpha \in I\}$. I can be any nonempty set, finite or infinite.

The **union** of the sets A_α, $\alpha \in I$, is defined to be the set $\{x \mid x \in A_\alpha$ for at least one $\alpha \in I\}$ and is denoted by $\cup_{\alpha \in I} A_\alpha$. The **intersection** of the sets A_α, $\alpha \in I$, is defined to be the set $\{x \mid x \in A_\alpha$ for all $\alpha \in I\}$ and is denoted by $\cap_{\alpha \in I} A_\alpha$.

Definition 1.1.6 *Given two sets A and B, the **relative complement** of B in A, denoted by the set difference $A \backslash B$, is the set*

$$A \backslash B = \{x \mid x \in A, \ but \ x \notin B\}.$$

The following diagram describes the set difference of two sets.

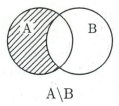

A\B

Example 1.1.7 *Let* $A = \{1, 2, 3, 4\}$ *and* $B = \{3, 4, 5, 6\}$. *Then* $A\backslash B = \{1, 2\}$.

We now define a concept which is a building block for all of mathematics, namely, the concept of an ordered pair.

Definition 1.1.8 *Let* A *and* B *be nonempty sets and* $x \in A$, $y \in B$.
  *(i) The **ordered pair** (x, y) is defined to be the set $\{\{x\}, \{x, y\}\}$.*
  *(ii) The **Cartesian cross product (Cartesian product)** of* A *and* B, *written* $A \times B$, *is defined to be the set*

$$A \times B = \{(x, y) \mid x \in A, y \in B\}.$$

Let (x, y), $(z, w) \in A \times B$. We claim that $(x, y) = (z, w)$ if and only if $x = z$ and $y = w$. First suppose that $x = z$ and $y = w$. Then $\{\{x\}, \{x, y\}\} = \{\{z\}, \{z, w\}\}$ and so $(x, y) = (z, w)$. Now suppose that $(x, y) = (z, w)$. Then

$$\{\{x\}, \{x, y\}\} = \{\{z\}, \{z, w\}\}.$$

Since $\{x\} \in \{\{x\}, \{x, y\}\}$, it follows that $\{x\} \in \{\{z\}, \{z, w\}\}$. This implies that $\{x\} = \{z\}$ or $\{x\} = \{z, w\}$. If $\{x\} = \{z\}$, then we must have $\{x, y\} = \{z, w\}$. From this, it follows that $x = z$ and $y = w$. If $\{x\} = \{z, w\}$, then we must have $\{x, y\} = \{z\}$. This implies that $x = z = w$ and $x = y = z$. Thus, in this case, $x = y = z = w$. This establishes our claim.

It now follows that if A has m elements and B has n elements, then $A \times B$ has mn elements.

Example 1.1.9 *Let* $A = \{1, 2, 3\}$ *and* $B = \{3, 4\}$. *Then*

$$A \times B = \{(1, 3), (1, 4), (2, 3), (2, 4), (3, 3), (3, 4)\}.$$

For the set $\mathbf{R}$ *of real numbers, the Cartesian product* $\mathbf{R} \times \mathbf{R}$ *is merely the Euclidean plane.*

Definition 1.1.10 *For any set* X, *the **power set** of* X, *written* $\mathcal{P}(X)$, *is defined to be the set* $\{A \mid A$ *is a subset of* $X\}$.

Example 1.1.11 *Let $X = \{1, 2, 3\}$. Then*

$$\mathcal{P}(X) = \{\phi, \{1\}, \{2\}, \{3\}, \{1, 2\}, \{1, 3\}, \{2, 3\}, \{1, 2, 3\}\}.$$

Here $\mathcal{P}(X)$ has 2^3 elements.

Remark 1.1.12 *Let P and Q be statements. Throughout the text we will encounter questions in which we will be asked to show that P if and only if Q; that is, show that statement P is true if and only if statement Q is true. In situations like this, we first assume that statement P is true and show that statement Q is true. Then we assume that statement Q is true and show that statement P is true. The statement P if and only if Q is also equivalent to the statement: if P, then Q, and if Q, then P. For example, see Worked-Out Exercise 1, below.*

1.1.1 Worked-Out Exercises

◇ **Exercise 1** Prove for sets A and B that $A \subseteq B$ if and only if $A \cup B = B$.

Solution: First suppose $A \subseteq B$. We now show that $A \cup B = B$. Let x be any element of $A \cup B$. Then either $x \in A$ or $x \in B$. This implies that $x \in B$ since $A \subseteq B$. Thus, we find that every element of $A \cup B$ is an element of B and so $A \cup B \subseteq B$. Also, $B \subseteq A \cup B$ by Theorem 1.1.4(i). Hence, $A \cup B = B$.

Conversely, suppose $A \cup B = B$. Now by Theorem 1.1.4(i), $A \subseteq A \cup B$. Since $A \cup B = B$, it now follows that $A \subseteq B$.

◇ **Exercise 2** For a subset A of a set S, let A' denote the subset $S \backslash A$. A' is called the **complement** of A in S. Let A and B be subsets of S. Prove that $(A \cap B)' = A' \cup B'$, **DeMorgan's law**.

Solution: First we show that $(A \cap B)' \subseteq A' \cup B'$. Then we show that $A' \cup B' \subseteq (A \cap B)'$. The result then follows by Theorem 1.1.1.

Let x be any element of $(A \cap B)'$. Now $(A \cap B)' = S \backslash (A \cap B)$ and so $x \in S$ and $x \notin A \cap B$. Also, $x \notin A \cap B$ implies that either $x \notin A$ or $x \notin B$. If $x \in S$ and $x \notin A$, then $x \in A'$, and if $x \in S$ and $x \notin B$, then $x \in B'$. Thus, either $x \in A'$ or $x \in B'$, i.e., $x \in A' \cup B'$. Hence, $(A \cap B)' \subseteq A' \cup B'$.

Let us now show that $A' \cup B' \subseteq (A \cap B)'$. Suppose x is any element of $A' \cup B'$. Then either $x \in A'$ or $x \in B'$. Suppose $x \in A'$, then $x \in S$ and $x \notin A$. Since $A \cap B \subseteq A$ and $x \notin A$, we must have $x \notin A \cap B$. This implies that $x \in (A \cap B)'$. Similarly, we can show that if $x \in B'$, then $x \notin A \cap B$, i.e., $x \in (A \cap B)'$. Hence, $A' \cup B' \subseteq (A \cap B)'$. Consequently, $(A \cap B)' = A' \cup B'$.

◇ **Exercise 3** Let A, B, and C be sets. Prove that

$$A \cap (B \cup C) = (A \cap B) \cup (A \cap C).$$

Solution: As in the previous exercise, we first show that $A \cap (B \cup C) \subseteq (A \cap B) \cup (A \cap C)$ and then $(A \cap B) \cup (A \cap C) \subseteq A \cap (B \cup C)$. The result then follows by Theorem 1.1.1.

Let x be any element of $A \cap (B \cup C)$. Then $x \in A$ and $x \in B \cup C$. Thus, $x \in A$ and $x \in B$ or $x \in C$. If $x \in A$ and $x \in B$, then $x \in A \cap B$, and if $x \in A$ and $x \in C$, then $x \in A \cap C$. Therefore, $x \in A \cap B$ or $x \in A \cap C$. Hence, $x \in (A \cap B) \cup (A \cap C)$. This shows that $A \cap (B \cup C) \subseteq (A \cap B) \cup (A \cap C)$.

Let us now show that $(A \cap B) \cup (A \cap C) \subseteq A \cap (B \cup C)$. Suppose x is any element of $(A \cap B) \cup (A \cap C)$. Then $x \in A \cap B$ or $x \in A \cap C$. Suppose $x \in A \cap B$, then $x \in A$ and $x \in B$. Since $B \subseteq B \cup C$, we have $x \in B \cup C$. Thus, $x \in A$ and $x \in B \cup C$ and so $x \in A \cap (B \cup C)$. Similarly, if $x \in A$ and $x \in C$, then $x \in A \cap (B \cup C)$. Hence, $(A \cap B) \cup (A \cap C) \subseteq A \cap (B \cup C)$. Consequently, $A \cap (B \cup C) = (A \cap B) \cup (A \cap C)$.

1.1.2 Exercises

1. Let $A = \{x, y, z\}$ and $B = \{y, w\}$. Determine each of the following sets: $A \cup B$, $A \cap B$, $A \backslash B$, $B \backslash A$, $A \times B$, and $\mathcal{P}(A)$.

2. Prove for sets A and B that $A \subseteq B$ if and only if $A \cap B = A$.

3. Prove for sets A, B, and C that

   (i) $A \cup B = B \cup A$ and $A \cap B = B \cap A$,

   (ii) $(A \cup B) \cup C = A \cup (B \cup C)$ and $(A \cap B) \cap C = A \cap (B \cap C)$,

   (iii) $A \cup (B \cap C) = (A \cup B) \cap (A \cup C)$,

   (iv) $A \cup (A \cap B) = A$,

   (v) $A \cap (A \cup B) = A$.

4. If a set S has 12 elements, how many elements does $\mathcal{P}(S)$ have? How many of these are properly contained in S?

5. For subsets A and B of a set S, prove **DeMorgan's law**:

$$(A \cup B)' = A' \cap B'.$$

6. The **symmetric difference** of two sets A and B is the set

$$A \triangle B = (A \cup B) \backslash (A \cap B).$$

   (i) If $A = \{a, b, c\}$ and $B = \{b, c, d, e\}$, find $A \triangle B$.

   (ii) Show that $A \triangle B = (A \backslash B) \cup (B \backslash A)$.

7. Let A and B be finite subsets of a set S. Show that

   (i) if $A \cap B = \phi$, then $|A \cup B| = |A| + |B|$,

   (ii) $|A \backslash B| = |A| - |A \cap B|$,

   (iii) $|A \cup B| = |A| + |B| - |A \cap B|$.

8. In each of the following exercises, write the proof if the statement is true; otherwise give a counterexample. The sets A, B, and C are subsets of a set U.

   (i) $A \cap (B \backslash C) = (A \cap B) \backslash (A \cap C)$.

   (ii) $A \backslash (B \cup C) = (A \backslash B) \cup C$.

   (iii) $(A \backslash B)' = (B \backslash A)'$.

   (iv) $A \times (B \cup C) = (A \times B) \cup (A \times C)$.

   (v) $A \triangle C = B \triangle C$ implies $A = B$.

1.2   Integers

Throughout abstract algebra, the set of **integers** provides a source of examples. In fact, many algebraic abstractions come from the integers. An axiomatic development of the integers is not given in this text. Instead, certain basic properties of integers are taken for granted. For example, if n and m are integers with $n < m$, then there exists a positive integer $t \in \mathbf{Z}$ such that $m = n + t$. In this section, we review and prove some important properties of the integers.

The proofs of many results of algebra depend on the following basic principle of the integers.

Principle of Well-Ordering: Every nonempty subset of $\mathbf{Z}^{\#}$ has a smallest (least) element, i.e., if $\phi \neq S \subseteq \mathbf{Z}^{\#}$, then there exists $x \in S$ such that $x \leq y$ for all $y \in S$.

Let S be a subset of $\mathbf{Z}^{\#}$. Suppose that S has the following properties:
(i) $n_0 \in S$, i.e., there exists an element $n_0 \in S$.
(ii) For all $n \geq n_0$, $n \in \mathbf{Z}^{\#}$, if $n \in S$, then $n + 1 \in S$.
We show that the set of all integers greater than or equal to n_0 is a subset of S, i.e.,

$$\{n \in \mathbf{Z}^{\#} \mid n \geq n_0\} \subseteq S.$$

Let T denote the set $\{n \in \mathbf{Z}^{\#} \mid n \geq n_0\}$. We wish to show that $T \subseteq S$. On the contrary, suppose $T \not\subseteq S$. Then there exists $a \in T$ such that $a \notin S$. Let T_1 be the set of all elements of T that are not in S, i.e., $T_1 = T \backslash S$. Since $a \in T$ and $a \notin S$, we have $a \in T_1$. Thus, T_1 is a nonempty subset of $\mathbf{Z}^{\#}$. Hence, by the principle of well-ordering, T_1 has a smallest element m, say. Then $m \in T$ and $m \notin S$. Since $m \in T$, $m \geq n_0$. If $m = n_0$, then $m \in S$, a contradiction. Thus, $m > n_0$. This implies that $m - 1 \geq n_0$ and so $m - 1 \in T$. Now $m - 1 \notin T_1$ since m is the smallest element of T_1. Since $m - 1 \in T$ and $m - 1 \notin T_1$, we must have $m - 1 \in S$. But then by (ii), $m = (m - 1) + 1 \in S$, which is a contradiction. Hence, $T \subseteq S$.

Thus, from the principle of well-ordering, we deduce another important property of integers. This property is known as the principle of mathematical induction. We thus have the following theorem.

Theorem 1.2.1 (Principle of Mathematical Induction) *Let $S \subseteq \mathbf{Z}^{\#}$. Let $n_0 \in S$. Suppose S satisfies either of the following conditions.*
(i) For all $n \geq n_0$, $n \in \mathbf{Z}^{\#}$, if $n \in S$, then $n + 1 \in S$.
(ii) For all $m < n$, $n \in \mathbf{Z}^{\#}$, if $m \in S$, then $n \in S$.
Then
$$\{n \in \mathbf{Z}^{\#} \mid n \geq n_0\} \subseteq S. \ \blacksquare$$

We proved, above, Theorem 1.2.1, when S satisfies (i). We leave it for the reader to prove Theorem 1.2.1 if S satisfies (ii).

We have seen the following mathematical statement in a college algebra or in a calculus course.
$$1 + 2 + \cdots + n = \frac{n(n + 1)}{2}, \quad n \geq 1.$$

We now show how this statement can be proved using the principle of mathematical induction. Let $S(n)$ denote the above mathematical statement, i.e.,
$$S(n): \quad 1 + 2 + \cdots + n = \frac{n(n + 1)}{2}, \quad n \geq 1.$$

This statement will be true if the left-hand side of the statement is equal to the right-hand side. Let
$$S = \{n \in \mathbf{Z}^{\#} \mid S(n) \text{ is true}\}.$$

That is, S is the set of all nonnegative integers n for which the statement $S(n)$ is true. We will show that S is the set of all positive integers. Now
$$1 = \frac{1 \cdot (1 + 1)}{2},$$

i.e., $S(1)$ is true. Hence, $1 \in S$. Let n be an integer such that $n \geq 1$ and suppose $S(n)$ is true, i.e., $n \in S$. We now show that $S(n + 1)$ is true. Now
$$S(n + 1): \quad 1 + 2 + \cdots + n + (n + 1) = \frac{(n + 1)(n + 2)}{2}.$$

Consider the left-hand side.

$$1 + 2 + \cdots + n + (n+1) = \frac{n(n+1)}{2} + (n+1) \text{ (since } S(n) \text{ is true)}$$
$$= \frac{(n+1)(n+2)}{2}.$$

Hence, the left-hand side is equal to the right-hand side and so $S(n+1)$ is true. Thus, $n+1 \in S$. Hence, by the principle of mathematical induction, $S = \{n \in \mathbf{Z}^{\#} \mid n \geq 1\}$. This proves our claim, which in turn shows that

$$1 + 2 + \cdots + n = \frac{n(n+1)}{2}$$

is true for all positive integers n.

Sometimes we use the word induction for the principle of mathematical induction.

A proof by the principle of mathematical induction consists of three steps.

Step 1: Show that $n_0 \in S$, i.e., the statement $S(n_0)$ is true for some $n_0 \in \mathbf{Z}^{\#}$.

Step 2: Write the induction hypothesis: n is an integer such that $n \geq n_0$ and $n \in S$, i.e., $S(n)$ is true for some integer n such that $n \geq n_0$ (or k is an integer such that $n_0 \leq k \leq n$ and $S(k)$ is true).

Step 3: Show that $n+1 \in S$, i.e., $S(n+1)$ is true.

Example 1.2.2 *In this example, we show that $2n + 1 \leq 2^n$ for all $n \geq 3$.*
Let $S(n)$ be the statement:

$$S(n): \quad 2n + 1 \leq 2^n, \quad n \geq 3.$$

Since we want to show that $S(n)$ is true for all $n \geq 3$, as the first step of our induction, we must verify that $S(3)$ is true. Let $n = 3$. Now $2n+1 = 2 \cdot 3 + 1 = 7$ and $2^n = 2^3 = 8$. Thus, for $n = 3$, $2n + 1 \leq 2^n$. This shows that $S(3)$ is true. Suppose that $2n + 1 \leq 2^n$ for some $n \geq 3$, i.e., $S(n)$ is true for some $n \geq 3$. Consider $S(n+1)$,

$$S(n+1): \quad 2(n+1) + 1 \leq 2^{n+1}.$$

Let us evaluate the left-hand side of $S(n+1)$. We have

$$
\begin{aligned}
2(n+1) + 1 &= 2n + 2 + 1 \\
&= (2n + 1) + 2 \\
&\leq 2^n + 2 \qquad \text{since } S(n) \text{ is true} \\
&\leq 2^n + 2^n \qquad (\text{since } n \geq 3, \ 2 \leq 2^n) \\
&= 2^{n+1}.
\end{aligned}
$$

Thus, $S(n+1)$ is true. Hence, by the principle of mathematical induction, $2n + 1 \leq 2^n$ for all $n \geq 3$.

The principle of mathematical induction is a very useful tool in mathematics. We will make use of this result throughout the text.

We now prove the following important properties of integers with the help of the principle of well-ordering.

Theorem 1.2.3 (Division Algorithm) *Let x, $y \in \mathbf{Z}$ with $y \neq 0$. Then there exist unique integers q and r such that $x = qy + r$, $0 \leq r < |y|$.*

Proof. Let us first assume $y > 0$. Then $y \geq 1$. Consider the set

$$S = \{x - uy \mid u \in \mathbf{Z}, x - uy \geq 0\}.$$

Since $y \geq 1$, we have $x - (-|x|)y = x + |x|y \geq 0$ so that $x - (-|x|)y \in S$. Thus, S is a nonempty set of nonnegative integers. Hence, by the principle of well-ordering, S must have a smallest element, say, r. Since $r \in S$, we have $r \geq 0$ and $r = x - qy$ for some $q \in \mathbf{Z}$. Then $x = qy + r$. We must show that $r < |y|$. Suppose on the contrary that $r \geq |y| = y$. Then

$$x - (q+1)y = (x - qy) - y = r - y \geq 0$$

so that $r - y \in S$, a contradiction since r is the smallest nonnegative integer in S and $r - y < r$. Hence, it must be the case that $r < |y|$. This proves the theorem in case $y > 0$.

Suppose now that $y < 0$. Then $|y| > 0$. Thus, there exist integers q', r such that $x = q'|y| + r$, $0 \leq r < |y|$ by the above argument. Since $y < 0$, $|y| = -y$. Hence, $x = -q'y + r$. Let $q = -q'$. Then $x = qy + r$, $0 \leq r < |y|$, the desired conclusion.

The uniqueness of q and r remains to be shown. Suppose there are integers q', r' such that

$$x = qy + r = q'y + r',$$

$0 \leq r' < |y|$, $0 \leq r < |y|$. Then

$$r' - r = (q - q')y.$$

Thus,

$$|r' - r| = |q - q'| \, |y|.$$

Now $-|y| < -r \leq 0$ and $0 \leq r' < |y|$. Therefore, if we add these inequalities, we obtain

$$-|y| < r' - r < |y|,$$

or $|r' - r| < |y|$. Hence, we have

$$0 \leq |q - q'| < 1.$$

Since $q - q'$ is an integer, we must have $0 = |q - q'|$. It now also follows that $|r - r'| = 0$. Thus, $q - q' = 0$ and $r - r' = 0$ or $q = q'$ and $r = r'$. Consequently, q and r are unique. ■

In Theorem 1.2.3, the integer q is called the **quotient** of x and y on dividing x by y and the integer r is called the **remainder** of x and y on dividing x by y.

The following corollary is a special case of Theorem 1.2.3.

Corollary 1.2.4 *For any two integers x and y with $y > 0$, there exist unique integers q and r such that $x = qy + r$, where $0 \leq r < y$.*

Proof.   By Theorem 1.2.3, there exist unique integers q and r such that $x = qy + r$, where $0 \leq r < |y|$. Since $y > 0$, $|y| = y$. Hence, $x = qy + r$, where $0 \leq r < y$. ■

Definition 1.2.5 *Let $x, y \in \mathbf{Z}$ with $x \neq 0$. Then x is said to **divide** y or x is a **divisor** (or **factor**) of y, written $x|y$, provided there exists $q \in \mathbf{Z}$ such that $y = qx$. When x does not divide y, we sometimes write $x \nmid y$.*

Let x, y, z be integers with $x \neq 0$. Suppose $x|y$ and $x|z$. Then for all integers s and t, $x|(sy + tz)$. We ask the reader to prove this fact in Exercise 5(iii) (page 19).

Definition 1.2.6 *Let $x, y \in \mathbf{Z}$. A nonzero integer c is called a **common divisor** of x and y if $c|x$ and $c|y$.*

Definition 1.2.7 *A nonzero integer d is called a **greatest common divisor** **(gcd)** of the integers x and y if*
  (i) $d|x$ and $d|y$,
  (ii) for all $c \in \mathbf{Z}$ if $c|x$ and $c|y$, then $c|d$.

Let d and d' be two greatest common divisors of integers x and y. Then $d|d'$ and $d'|d$. Hence, there exist integers u and v such that $d' = du$ and $d = d'v$. Therefore, $d = duv$, which implies that $uv = 1$ since $d \neq 0$. Thus, either $u = v = 1$ or $u = v = -1$. Hence, $d' = \pm d$. It now follows that two different gcd's of x and y differ in their sign. Of the two gcd's of x and y, the positive one is denoted by $\gcd(x, y)$. For example, 2 and -2 are the greatest common divisors of 4 and 6. Hence, $2 = \gcd(4, 6)$.

In the next theorem, we show that the gcd always exists for any two nonzero integers.

Theorem 1.2.8 *Let $x, y \in \mathbf{Z}$ with either $x \neq 0$ or $y \neq 0$. Then x and y have a positive greatest common divisor d. Moreover, there exist elements $s, t \in \mathbf{Z}$ such that $d = sx + ty$.*

Proof.   Let

$$S = \{mx + ny \mid m, n \in \mathbf{Z}, mx + ny > 0\}.$$

Suppose $x \neq 0$. Then

$$|x| = \begin{cases} x & \text{if } x > 0 \\ -x & \text{if } x < 0 \end{cases}$$

$$= \begin{cases} 1x + 0y & \text{if } x > 0 \\ (-1)x + 0y & \text{if } x < 0. \end{cases}$$

Hence, $|x| \in S$ and so $S \neq \phi$. By the well-ordering principle, S contains a smallest positive integer, say, d. We now show that d is the greatest common divisor of x and y.

Since $d \in S$, there exist $s, t \in \mathbf{Z}$ such that $d = sx + ty$. First we show that $d|x$ and $d|y$. Since $d \neq 0$, by the division algorithm (Theorem 1.2.3), there exist integers q and r such that

$$x = dq + r,$$

where $0 \leq r < |d| = d$. Thus,

$$\begin{aligned} r &= x - dq \\ &= x - (sx + ty)q \quad \text{(substituting for } d) \\ &= (1 - qs)x + (-qt)y. \end{aligned}$$

Suppose $r > 0$. Then $r \in S$, which is a contradiction since d is the smallest element of S and $r < d$. Thus, $r = 0$. This implies that $x = dq$ and so $d|x$. Similarly, $d|y$. Hence, d satisfies (i) of Definition 1.2.7. Suppose $c|x$ and $c|y$ for some integer c. Then $c|(sx + ty)$ by Exercise 5(iii) (page 19), i.e., $c|d$. Thus, d satisfies (ii) of Definition 1.2.7. Consequently, $d = \gcd(x, y)$. ∎

Let x and y be nonzero integers. By Theorem 1.2.8, $\gcd(x, y)$ exists and if $d = \gcd(x, y)$, then there exist integers s and t such that $d = sx + ty$. The integers s and t in the representation $d = sx + ty$ are not unique. For example, let $x = 45$ and $y = 126$. Then $\gcd(x, y) = 9$, and $9 = 3 \cdot 45 + (-1) \cdot 126 = 129 \cdot 45 + (-46) \cdot 126$.

The proof of Theorem 1.2.8 does not indicate how to find $\gcd(x, y)$ or the integers s, t. In the following, we indicate how these integers can be found.

Let $x, y \in \mathbf{Z}$ with $y \neq 0$. By the division algorithm, there exist $q_1, r_1 \in \mathbf{Z}$ such that

$$x = q_1 y + r_1, \qquad 0 \leq r_1 < |y|.$$

If $r_1 \neq 0$, then by the division algorithm, there exist $q_2, r_2 \in \mathbf{Z}$ such that

$$y = q_2 r_1 + r_2, \qquad 0 \leq r_2 < r_1.$$

If $r_2 \neq 0$, then again by the division algorithm, there exist $q_3, r_3 \in \mathbf{Z}$ such that

$$r_1 = q_3 r_2 + r_3, \qquad 0 \leq r_3 < r_2.$$

Since $r_1 > r_2 > r_3 \geq 0$, we must in a finite number of steps find integers q_n, q_{n+1}, and $r_n > 0$ such that

$$\begin{aligned} r_{n-2} &= q_n r_{n-1} + r_n, \quad 0 < r_n < r_{n-1} \\ r_{n-1} &= q_{n+1} r_n + 0. \end{aligned}$$

We assert that r_n (the last nonzero remainder) is the greatest common divisor of x and y. Now $r_n | r_{n-1}$. Since $r_n | r_n$, $r_n | r_{n-1}$, and $r_{n-2} = q_n r_{n-1} + r_n$, we have $r_n | r_{n-2}$ by Exercise 5(iii) (page 19). Working our way back in this fashion, we have $r_n | r_1$ and $r_n | r_2$. Thus, $r_n | y$ since $y = q_2 r_1 + r_2$. Since $r_n | y$, $r_n | r_1$, and $x = q_1 y + r_1$, we have $r_n | x$. Hence, r_n is a common divisor of x and y. Now if c is any common divisor of x and y, then we see that $c | r_1$. Since $c | y$ and $c | r_1$, $c | r_2$. Continuing, we finally obtain $c | r_n$. Thus, $r_n = \gcd(x, y)$.

We now find $s, t \in \mathbf{Z}$ such that $\gcd(x, y) = sx + ty$ as follows:

$$\begin{aligned} r_n &= r_{n-2} + r_{n-1}(-q_n) \\ &= r_{n-2} + [r_{n-3} + r_{n-2}(-q_{n-1})](-q_n) \\ &= r_{n-3}(-q_n) + r_{n-2}(1 + q_{n-1}q_n) \qquad \text{(simplifying).} \end{aligned}$$

We now substitute $r_{n-4} + r_{n-3}(-q_{n-2})$ for r_{n-2}. We repeat this "back" substitution process until we reach $r_n = sx + ty$ for some integers s and t.

We illustrate the above procedure for finding the gcd and integers s and t with the help of the following example.

Example 1.2.9 *Consider the integers 45 and 126. Now*

$$\begin{aligned} 126 &= 2 \cdot 45 + 36 \\ 45 &= 1 \cdot 36 + 9 \\ 36 &= 4 \cdot 9 + 0 \end{aligned}$$

Thus, $9 = \gcd(45, 126)$. *Also,*

$$\begin{aligned} 9 &= 45 - 1 \cdot 36 \\ &= 45 - 1 \cdot [126 - 2 \cdot 45] \\ &= 3 \cdot 45 + (-1) \cdot 126. \end{aligned}$$

Here $s = 3$ *and* $t = -1$.

We now define prime integers and study their basic properties.

Definition 1.2.10 *(i) An integer $p > 1$ is called **prime** if the only divisors of p are ± 1 and $\pm p$.*

*(ii) Two integers x and y are called **relatively prime** if $\gcd(x, y) = 1$.*

The following theorem gives a necessary and sufficient condition for two nonzero integers to be relatively prime.

Theorem 1.2.11 *Let x and y be nonzero integers. Then x and y are relatively prime if and only if there exist $s, t \in \mathbf{Z}$ such that $1 = sx + ty$.*

Proof. Let x and y be relatively prime. Then $\gcd(x, y) = 1$. By Theorem 1.2.8, there exist integers s and t such that $1 = sx + ty$.

Conversely, suppose $1 = sx + ty$ for some pair of integers s, t. Let $d = \gcd(x, y)$. Then $d|x$ and $d|y$ and so $d|(sx + ty)$ (by Exercise 5(iii) (page 19)) or $d|1$. Since d is a positive integer and $d|1$, $d = 1$. Thus, $\gcd(x, y) = 1$ and so x and y are relatively prime. ∎

Theorem 1.2.12 *Let x, y, $z \in \mathbf{Z}$ with $x \neq 0$. If $x|yz$ and x, y are relatively prime, then $x|z$.*

Proof. Since x and y are relatively prime, there exist $s, t \in \mathbf{Z}$ such that $1 = sx + ty$ by Theorem 1.2.11. Thus, $z = sxz + tyz$. Now $x|x$ and by hypothesis $x|yz$. Thus, $x|(sxz + tyz)$ by Exercise 5(iii) (page 19) and so $x|z$. ∎

Corollary 1.2.13 *Let $x, y, p \in \mathbf{Z}$ with p a prime. If $p|xy$, then either $p|x$ or $p|y$.*

Proof. If $p|x$, then we have the desired result. Suppose that p does not divide x. Since the only positive divisors of p are 1 and p, we must have that p and x are relatively prime. Thus, $p|y$ by Theorem 1.2.12. ∎

The following corollary is a generalization of Corollary 1.2.13.

Corollary 1.2.14 *Let $x_1, x_2, \ldots, x_n, p \in \mathbf{Z}$ with p a prime. If*

$$p|x_1 x_2 \cdots x_n,$$

then $p|x_i$ for some i, $1 \leq i \leq n$.

Proof. The proof follows by Corollary 1.2.13 and induction. ∎

Consider the integer 24. We can write $24 = 2^3 \cdot 3$. That is, 24 can be written as product of prime powers. Similarly, $49500 = 2^2 \cdot 3^2 \cdot 5^3 \cdot 11$. In the next theorem, called the fundamental theorem of arithmetic, we prove that any positive integer can be written as product of prime powers.

Theorem 1.2.15 (Fundamental Theorem of Arithmetic) *Any integer* n *> 1 has a unique factorization (up to order)*

$$n = p_1^{e_1} p_2^{e_2} \cdots p_s^{e_s}, \tag{1.1}$$

where $p_1, p_2, \ldots, p_s$ *are distinct primes and* $e_1, e_2, \ldots, e_s$ *are positive integers.*

Proof.   First we show that any integer $n > 1$ has a factorization like Eq. (1.1) and then we show the uniqueness of the factorization.

We show the existence of the factorization by induction. If $n = 2$, then clearly n has the above factorization as a product of prime powers. Make the induction hypothesis that any integer k such that $2 \le k < n$ has a factorization like Eq. (1.1). If n is prime, then n already has the above factorization as a product of prime powers, namely n itself. If n is not prime, then $n = xy$ for integers x, y, with $1 < x < n$ and $1 < y < n$. By the induction hypothesis, there exist primes $q_1, q_2, \ldots, q_k, q_1', q_2', \ldots, q_t'$ and positive integers $e_1, e_2, \ldots,$ $e_k, e_1', e_2', \ldots, e_t'$ such that $q_1, q_2, \ldots, q_k$ are distinct primes, $q_1', q_2', \ldots, q_t'$ are distinct primes and

$$x = q_1^{e_1} q_2^{e_2} \cdots q_k^{e_k}$$
$$y = q_1'^{e_1'} q_2'^{e_2'} \cdots q_t'^{e_t'}.$$

Thus,

$$n = q_1^{e_1} q_2^{e_2} \cdots q_k^{e_k} q_1'^{e_1'} q_2'^{e_2'} \cdots q_t'^{e_t'},$$

i.e., n can be factored as a product of prime powers. If $q_i = q_j'$ for some i and j, then we replace $q_i^{e_i} q_j'^{e_j'}$ by $q_i^{e_i+e_j'}$. It now follows that $n = p_1^{e_1} p_2^{e_2} \cdots p_s^{e_s}$, where $p_1, p_2, \ldots, p_s$ are distinct primes and $e_1, e_2, \ldots, e_s$ are positive integers. Hence, by induction, any integer $n > 1$ has a factorization like (1.1).

We now prove the uniqueness property by induction also. If $n = 2$, then clearly n has a unique factorization as a product of prime powers. Suppose the uniqueness property holds for all integers k such that $2 \le k < n$. Let

$$n = p_1^{e_1} p_2^{e_2} \cdots p_s^{e_s} = q_1^{c_1} q_2^{c_2} \cdots q_t^{c_t} \tag{1.2}$$

be two factorizations of n into a product of prime powers. Suppose n is prime. Then in Eq. (1.2), we must have $s = t = 1$ and $e_1 = 1 = c_1$ since the only positive divisors of n are 1 and n itself. This implies that $n = p_1 = q_1$ and so the factorization is unique.

Suppose n is not a prime. Now $p_1 | n$ and

$$\frac{n}{p_1} = p_1^{e_1-1} p_2^{e_2} \cdots p_s^{e_s}$$

is an integer. If $s = 1$, then $n = p_1^{e_1}$ and since n is not a prime, we have $e_1 > 1$. Hence, $\frac{n}{p_1} = p_1^{e_1-1} \ge 2$. If $s > 1$, then $\frac{n}{p_1} = p_1^{e_1-1} p_2^{e_2} \cdots p_s^{e_s} \ge 2$. Thus, in either case, $\frac{n}{p_1}$ is an integer ≥ 2. Now $p_1 | n$ implies that $p_1 | q_1^{c_1} q_2^{c_2} \cdots q_t^{c_t}$ and so by Corollary 1.2.14, $p_1 | q_i^{c_i}$ for some i. By reordering the q_i if necessary, we can assume that $i = 1$. Thus, $p_1 | q_1^{c_1}$ and so by Corollary 1.2.14, $p_1 | q_1$. Since p_1 and q_1 are primes, $p_1 = q_1$. Thus,

$$\frac{n}{p_1} = p_1^{e_1 - 1} p_2^{e_2} \cdots p_s^{e_s} = p_1^{c_1 - 1} q_2^{c_2} \cdots q_t^{c_t}. \tag{1.3}$$

Now $e_1 - 1 = 0$ if and only if $c_1 - 1 = 0$. For suppose $e_1 - 1 = 0$ and $c_1 - 1 > 0$. Then $\frac{n}{p_1} = p_2^{e_2} \cdots p_s^{e_s}$ implies that $p_1 \nmid \frac{n}{p_1}$ and $\frac{n}{p_1} = p_1^{c_1 - 1} q_2^{c_2} \cdots q_t^{c_t}$ implies that $p_1 | \frac{n}{p_1}$, which is of course impossible. We can get a similar contradiction if we assume $e_1 - 1 > 0$ and $c_1 - 1 = 0$.

Now $\frac{n}{p_1}$ is an integer and $2 \leq \frac{n}{p_1} < n$. Hence, by the induction hypothesis, we obtain from Eq. (1.3) that $s = t$, and $p_1 = q_1, \ldots, p_s = q_s$ (without worrying about the order), and $e_1 - 1 = c_1 - 1$, $e_2 = c_2, \ldots, e_s = c_s$. Hence, by induction, we have the desired uniqueness property. ∎

Corollary 1.2.16 *Any integer $n < -1$ has a unique factorization (up to order)*

$$n = (-1) p_1^{e_1} p_2^{e_2} \cdots p_s^{e_s},$$

where $p_1, p_2, \ldots, p_s$ are distinct primes and $e_1, e_2, \ldots, e_s$ are positive integers.

Proof. Since $n < -1$, $-n > 1$. Hence, by Theorem 1.2.15, $-n$ has a unique factorization (up to order)

$$-n = p_1^{e_1} p_2^{e_2} \cdots p_s^{e_s},$$

where $p_1, p_2, \ldots, p_s$ are distinct primes and $e_1, e_2, \ldots, e_s$ are positive integers. Thus,

$$n = (-1) p_1^{e_1} p_2^{e_2} \cdots p_s^{e_s},$$

where $p_1, \ldots, p_s$ are distinct primes and $e_1, \ldots, e_s$ are positive integers. ∎

Theorem 1.2.15 says that any positive integer greater than 1 can be written as a product of prime powers. Now we pose the obvious question: How many prime numbers are there? This is answered by the following theorem due to Euclid.

Theorem 1.2.17 (Euclid) *There are an infinite number of primes.*

Proof.   Let $p_1, p_2, \ldots, p_n$ be a finite number of distinct primes. Set $x = p_1 p_2 \cdots p_n + 1$. Since p_i does not divide 1, p_i does not divide x, $i = 1, 2, \ldots, n$. By the fundamental theorem of arithmetic, it follows that there is some prime p such that $p \mid x$. Thus, p is distinct from $p_1, p_2, \ldots, p_n$ so that we have $n + 1$ distinct primes. That is, for any finite set of primes we can always find one more. Thus, there must be an infinite number of primes. ∎

We close this section with the following definition. There are a few places in the text where we will be making use of it.

Definition 1.2.18 *Let n be a positive integer. Let $\phi(n)$ denote the number of positive integers m such that $m \leq n$ and $\gcd(m, n) = 1$, i.e.,*

$$\phi(n) = |\{m \in \mathbf{N} \mid m \leq n \text{ and } \gcd(m, n) = 1\}|.$$

*$\phi(n)$ is called the **Euler ϕ-function**.*

Clearly $\phi(2) = 1$, $\phi(3) = 2$, $\phi(4) = 2$. Since 1, 5, 7, 11 are the only positive integers less than 12 and relatively prime to 12, $\phi(12) = 4$.

Let $\{a_1, \ldots, a_n\} \subseteq \mathbf{Z}$. We use the notation $\sum_{i=1}^{n} a_i$ to denote the sum of $a_1, \ldots, a_n$, i.e.,

$$\sum_{i=1}^{n} a_i = a_1 + \cdots + a_n.$$

If S is any finite subset of $\mathbf{Z}$, then $\sum_{a \in S} a$ denotes the sum of all elements of S. For example, if $S = \{2, 4, 7\}$, then $\sum_{a \in S} a = 2 + 4 + 7 = 13$.

1.2.1   Worked-Out Exercises

◇ **Exercise 1** By the principle of mathematical induction, prove that

$$3^{2n+1} + (-1)^n 2 \equiv 0 (\bmod\ 5)$$

for all positive integers n. (For integers a and b, $a \equiv b (\bmod\ 5)$ means 5 divides $a - b$.)

Solution: Let $S(n)$ be the statement

$$S(n): \qquad 3^{2n+1} + (-1)^n 2 \equiv 0 (\bmod\ 5), \qquad n \geq 1.$$

We wish to show that $S(n)$ is true for all positive integers. We first must verify that $S(1)$ is true as the first step of our induction. Let $n = 1$. Then

$$3^{2n+1} + (-1)^n 2 = 3^{2+1} + (-1)2 = 27 - 2 = 25 \equiv 0 (\bmod\ 5).$$

Thus, $S(1)$ is true. Now suppose that $S(n)$ is true for some positive integer n, i.e., $3^{2n+1} + (-1)^n 2 \equiv 0 \pmod 5$ for some integer $n \geq 1$. We now show that

$$S(n+1): \qquad 3^{2(n+1)+1} + (-1)^{n+1} 2 \equiv 0 \pmod 5$$

is true. Now

$$
\begin{aligned}
3^{2(n+1)+1} + (-1)^{n+1} 2 &= 3^{2n+1} \cdot 3^2 - (-1)^n 2 \\
&= 9(3^{2n+1} + (-1)^n 2) - (-1)^n 18 - (-1)^n 2 \\
&= 9(3^{2n+1} + (-1)^n 2) - (-1)^n 20.
\end{aligned}
$$

Since $3^{2n+1} + (-1)^n 2 \equiv 0 \pmod 5$ and $20 \equiv 0 \pmod 5$, it follows that $3^{2(n+1)+1} + (-1)^{n+1} 2 \equiv 0 \pmod 5$. This shows that $S(n+1)$ is true. Hence, by the principle of mathematical induction, $3^{2n+1} + (-1)^n 2 \equiv 0 \pmod 5$ for all positive integers n.

◇ **Exercise 2** Let a and b be integers such that $\gcd(a, 4) = 2$ and $\gcd(b, 4) = 2$. Prove that $\gcd(a + b, 4) = 4$.

Solution: Since $\gcd(a, 4) = 2$, $2|a$, but 4 does not divide a. Therefore, $a = 2x$ for some integer x such that $\gcd(2, x) = 1$. Similarly, $b = 2y$ for some integer y such that $\gcd(2, y) = 1$. Thus, x and y are both odd integers. This implies that $x + y$ is an even integer and so $x + y = 2n$ for some integer n. Now $a + b = 2(x + y) = 4n$. Hence, $\gcd(a + b, 4) = \gcd(4n, 4) = 4$.

◇ **Exercise 3** Let a, b, and c be integers such that $\gcd(a, c) = \gcd(b, c) = 1$. Prove that $\gcd(ab, c) = 1$.

Solution: If $c = 0$, then $\gcd(a, 0) = \gcd(b, 0) = 1$ implies that $a = \pm 1$ and $b = \pm 1$. Thus, $\gcd(ab, c) = \gcd(\pm 1, 0) = 1$. Suppose now $c \neq 0$. By Theorem 1.2.8, $\gcd(ab, c)$ exists. Let $d = \gcd(ab, c)$. Also, by Theorem 1.2.8, there exist integers x_1, y_1, x_2, y_2 such that $1 = ax_1 + cy_1$, $1 = bx_2 + cy_2$. Thus, $(ax_1)(bx_2) = (1 - cy_1)(1 - cy_2) = 1 - cy_1 - cy_2 + cy_1 cy_2$. Hence, $1 = (ab)x_1 x_2 + c(y_1 + y_2 - cy_1 y_2)$. Thus, any common divisor of ab and c is also a divisor of 1. Hence, $d|1$. Since $d > 0$, $d = 1$.

Exercise 4 Let $a, b \in \mathbf{Z}$ with either $a \neq 0$ or $b \neq 0$. Prove that for any integer c,

$$\gcd(a, b) = \gcd(a, -b) = \gcd(a, b + ac).$$

Solution: Suppose $a \neq 0$. Then $\gcd(a, b)$, $\gcd(a, -b)$ and $\gcd(a, b + ac)$ exist. Let $d = \gcd(a, b)$. Then there exist integers x and y such that $d = ax + by = ax + (-b)(-y)$. Thus, any common divisor of a and $-b$ is also a divisor of d. Hence, $\gcd(a, -b)|d$. Similarly, $d|\gcd(a, -b)$. Since $\gcd(a, b)$ and $\gcd(a, -b)$ are positive, $\gcd(a, b) = \gcd(a, -b)$.

Let $e = \gcd(a, b + ac)$. Then there exist integers p and q such that $e = ap + (b + ac)q = ap + bq + acq = a(p + cq) + bq$. Since $d|a$ and $d|b$, $d|e$. Also, $d = ax + by = ax + (b + ac)y - acy = a(x - cy) + (b + ac)y$. Since $e|a$ and $e|b + ac$, $e|d$. Hence, $e = d$.

◇ **Exercise 5** Find integers x and y such that $512x + 320y = 64$.

Solution:

$$\begin{aligned} 512 &= 320 \cdot 1 + 192 \\ 320 &= 192 \cdot 1 + 128 \\ 192 &= 128 \cdot 1 + 64 \\ 128 &= 64 \cdot 2 + 0. \end{aligned}$$

Thus, $64 = 192 - 128 = 192 - (320 - 192) = 192 \cdot 2 + 320 \cdot (-1) = (512 - 320) \cdot 2 + 320 \cdot (-1) = 512 \cdot 2 + 320 \cdot (-3)$. Hence, $x = 2$ and $y = -3$.

1.2.2 Exercises

1. Determine $\gcd(90, 252)$. Find integers s and t such that

$$\gcd(90, 252) = s \cdot 90 + t \cdot 252.$$

2. Find integers s and t such that $\gcd(963, 652) = s \cdot 963 + t \cdot 652$.

3. Find integers s and t such that $657s + 963t = 9$.

4. Use the principle of mathematical induction to prove the following.

   (i) $1^2 + 2^2 + 3^2 + \cdots + n^2 = \frac{n(n+1)(2n+1)}{6}$, $n = 1, 2, \ldots$.

   (ii) $7^n - 1$ is divisible by 6 for all $n \in \mathbf{Z}^{\#}$.

   (iii) $6 \cdot 7^n - 2 \cdot 3^n$ is divisible by 4 for all $n \in \mathbf{Z}^{\#}$.

   (iv) $5^{2n} + 3$ is divisible by 4 for all $n \in \mathbf{Z}^{\#}$.

   (v) $n < 2^n$ for all $n \in \mathbf{Z}^{\#}$.

   (vi) $2^n \geq n^2$, $n = 4, 5, \ldots$.

   (vii) $n! \geq 3^n$, $n = 7, 8, \ldots$.

5. Let a, b, and c be three integers such that $a \neq 0$. Prove the following:

   (i) If $a|b$, then $a|bc$ for all $c \in \mathbf{Z}$.

   (ii) If $b \neq 0$, $a|b$ and $b|c$, then $a|c$.

   (iii) If $a|b$ and $a|c$, then $a|(bx + cy)$ for all $x, y \in \mathbf{Z}$.

   (iv) If a, b are positive integers such that $a|b$, then $a \leq b$.

   (v) If $b \neq 0$, $a|b$, and $b|a$, then $a = \pm b$.

6. Let a, b, and c be integers. Prove that if $ac \neq 0$ and $ac|bc$, then $a|b$.

7. Let a, b, c, and d be integers such that $a \neq 0$ and $b \neq 0$. Prove that if $a|c$ and $b|d$, then $ab|cd$.

8. Let p be a prime integer, m, n integers and r a positive integer. Suppose $p^r|mn$ and $p \nmid m$. Show that $p^r|n$.

9. Let a and b be integers and $\gcd(a, b) = d$. If $a = dm$ and $b = dn$, prove that $\gcd(m, n) = 1$.

10. Let a, b, and c be positive integers. Prove that $\gcd(ab, ac) = a \gcd(b, c)$.

11. Prove that if $\gcd(x, y) = \gcd(x, z) = 1$, then $\gcd(x, yz) = 1$ for all $x, y, z \in$ **N**.

12. Prove that if $\gcd(x, y) = 1, x|z$, and $y|z$, then $xy|z$ for all $x, y, z \in$ **N**.

13. Let $a, b \in$ **N**. Show that $\gcd(a, b) = \gcd(a, a + b)$.

14. Prove that $\gcd(a, b) = 1$ for any two positive consecutive integers a and b.

15. Let x and y be nonzero integers. The **least common multiple** of x and y, written $\mathrm{lcm}(x, y)$, is defined to be a positive integer m such that

    (i) $x|m$ and $y|m$ and

    (ii) if $x|c$ and $y|c$, then $m|c$.

    Prove that $\mathrm{lcm}(x, y)$ exists and is unique.

16. Let x and y be nonzero integers. Prove that $\mathrm{lcm}(x, y) \cdot \gcd(x, y) = |xy|$.

17. Let x and y be nonzero integers. Show that $\mathrm{lcm}(x, y) = |xy|$ if and only if $\gcd(x, y) = 1$.

18. Show that there are infinitely many prime integers of the form $6n - 1$, $n \geq 1$.

19. Let S be a set with n elements, $n \geq 1$. Show by mathematical induction that $|\mathcal{P}(S)| = 2^n$.

20. Determine whether the following assertions are true or false. If true, then prove it, and if false give a counterexample.

    (i) If p is a prime such that $p|a^5$, then $p|a$, where a is an integer.

    (ii) If p is a prime such that $p|(a^2 + b^2)$ and $p|a$, then $p|b$, where a and b are integers.

(iii) For any integer a, $\gcd(a, a+3) = 1$ or 3.

(iv) If $\gcd(a, 6) = 3$ and $\gcd(b, 6) = 3$, then $\gcd(a+b, 6) = 6$, where a and b are integers.

(v) If $\gcd(b, c) = 1$ and $a|b$, then $\gcd(a, c) = 1$.

1.3   Relations

Some describe or define mathematics as the study of relations. Since a relation is a set of ordered pairs, we get our first glimpse of the fundamental importance of the concept of an ordered pair.

Definition 1.3.1 *A **binary relation** or simply a **relation** R from a set A into a set B is a subset of $A \times B$.*

Let R be a relation from a set A into a set B. If $(x, y) \in R$, we write xRy or $R(x) = y$. If xRy, then sometimes we say that x is related to y (or y is in relation with x) with respect to R or simply x is related to y. If $A = B$, then we speak of a binary relation on A.

Example 1.3.2 *Let A denote the names of all states in the USA and $B = \mathbf{Z}$. With each state a in A associate an integer n which denotes the number of people in that state in the year 1996. Then $R = \{(a, n) \mid a \in A$ and n is the number of people in state a in 1996$\}$ is a subset of $A \times \mathbf{Z}$. Thus, R defines a relation from A into $\mathbf{Z}$.*

Example 1.3.3 *Consider the set of integers $\mathbf{Z}$. Let R be the set of all ordered pairs (m, n) of integers such that $m < n$, i.e.,*

$$R = \{(m, n) \in \mathbf{Z} \times \mathbf{Z} \mid m < n\}.$$

Then R is a binary relation on $\mathbf{Z}$.

Let R be a relation from a set A into a set B. By looking at the elements of R, we can find out which elements of A are related to elements of B with respect to R. The elements of A that are related to elements of B form a subset of A, called the **domain** of R, and the elements of B that are in relation with elements of A form a subset of B, called the **range** of R. More formally, we have the following definition.

Definition 1.3.4 *Let R be a relation from a set A into a set B. Then the **domain** of R, denoted by $\mathcal{D}(R)$, is defined to be the set*

$$\{x \mid x \in A \text{ and there exists } y \in B \text{ such that } (x, y) \in R\}.$$

*The **range** or **image** of R, denoted by $\mathcal{I}(R)$, is defined to be the set*

$$\{y \mid y \in B \text{ and there exists } x \in A \text{ such that } (x, y) \in R\}.$$

Example 1.3.5 *Let $A = \{4, 5, 7, 8, 9\}$ and $B = \{16, 18, 20, 22\}$. Define $R \subseteq A \times B$ by*

$$R = \{(4, 16), (4, 20), (5, 20), (8, 16), (9, 18)\}.$$

Then R is a relation from A into B. Here $(a, b) \in R$ if and only if a divides b, where $a \in A$ and $b \in B$. Note that for the domain of R, we have $\mathcal{D}(R) = \{4, 5, 8, 9\}$ and for the range of R, we have $\mathcal{I}(R) = \{16, 18, 20\}$.

Example 1.3.6 *Let $S = \{(x, y) \mid x, y \in \mathbf{R}, x^2 + y^2 = 1, y > 0\}$. Then S is a binary relation on $\mathbf{R}$. S is the set of points in the Euclidean plane constituting the semicircle lying above the x-axis with center $(0, 0)$ and radius 1.*

Definition 1.3.7 *Let R be a binary relation on a set A. Then R is called*
*(i) **reflexive** if for all $x \in A$, xRx,*
*(ii) **symmetric** if for all $x, y \in A$, xRy implies yRx,*
*(iii) **transitive** if for all $x, y, z \in A$, xRy and yRz imply xRz.*

Definition 1.3.8 *A binary relation E on a set A is called an **equivalence relation** on A if E is reflexive, symmetric, and transitive.*

The important concept of an equivalence relation is due to Gauss. We will use this concept repeatedly throughout the text.

Example 1.3.9 *Let $A = \{1, 2, 3, 4, 5, 6\}$ and $E = \{(1, 1), (2, 2), (3, 3), (4, 4), (5, 5), (6, 6), (2, 3), (3, 2)\}$. Then E is an equivalence relation on A.*

Example 1.3.10 *(i) Let L denote the set of all straight lines in the Euclidean plane and E be the relation on L defined by for all $l_1, l_2 \in L$, $(l_1, l_2) \in E$ if and only if l_1 and l_2 are parallel. Then E is an equivalence relation on L.*
(ii) Let L be defined as in (i) and P be the relation defined on L by for all $l_1, l_2 \in L$, $(l_1, l_2) \in P$ if and only if l_1 and l_2 are perpendicular. Let l be a line in L. Since l cannot be perpendicular to itself, $(l, l) \notin P$. Hence, P is not reflexive and so P is not an equivalence relation on L. Also, P is not transitive.

Example 1.3.11 *Let n be a fixed positive integer in $\mathbf{Z}$. Define the relation $\equiv_n$ on $\mathbf{Z}$ by for all $x, y \in \mathbf{Z}$, $x \equiv_n y$ if and only if $n \mid (x - y)$, i.e., $x - y = nk$ for some $k \in \mathbf{Z}$. We now show that $\equiv_n$ is an equivalence relation on $\mathbf{Z}$.*
(i) For all $x \in \mathbf{Z}$, $x - x = 0 = 0n$. Hence, for all $x \in \mathbf{Z}$, $x \equiv_n x$. Thus, $\equiv_n$ is reflexive.
(ii) Let $x, y \in \mathbf{Z}$. Suppose $x \equiv_n y$. Then there exists $q \in \mathbf{Z}$ such that $qn = x - y$. Thus, $(-q)n = y - x$ and so $n \mid (y - x)$, i.e., $y \equiv_n x$. Hence, $\equiv_n$ is symmetric.
(iii) Let $x, y, z \in \mathbf{Z}$. Suppose $x \equiv_n y$ and $y \equiv_n z$. Then there exist $q, r \in \mathbf{Z}$ such that $qn = x - y$ and $rn = y - z$. Thus, $(q + r)n = x - z$ and $q + r \in \mathbf{Z}$. This implies that $x \equiv_n z$. Hence, $\equiv_n$ is transitive.
Consequently, $\equiv_n$ is an equivalence relation on $\mathbf{Z}$.

The equivalence relation, $\equiv_n$, as defined in Example 1.3.11 is called **congruence modulo** n. (Another commonly used notation for $x \equiv_n y$ is $x \equiv y \pmod{n}$.)

Definition 1.3.12 *Let E be an equivalence relation on a set A. For all $x \in A$, let $[x]$ denote the set*

$$[x] = \{y \in A \mid yEx\}.$$

The set $[x]$ is called the **equivalence class** (with respect to E) determined by x.

In the following theorem, we prove some basic properties of equivalence classes.

Theorem 1.3.13 *Let E be an equivalence relation on the set A. Then*
   (i) for all $x \in A$, $[x] \neq \phi$,
   (ii) if $y \in [x]$, then $[x] = [y]$, where $x, y \in A$,
   (iii) for all $x, y \in A$, either $[x] = [y]$ or $[x] \cap [y] = \phi$,
   (iv) $A = \cup_{x \in A}[x]$, i.e., A is the union of all equivalence classes with respect to E.

Proof.   (i) Let $x \in A$. Since E is reflexive, xEx. Hence, $x \in [x]$ and so $[x] \neq \phi$.
   (ii) Let $y \in [x]$. Then yEx and by the symmetric property of E, xEy. In order to show that $[x] = [y]$, we will show that $[x] \subseteq [y]$ and $[y] \subseteq [x]$. The result then will follow by Theorem 1.1.1. Let $u \in [y]$. Then uEy. Since uEy and yEx, the transitivity of E implies that uEx. Hence, $u \in [x]$. Thus, $[y] \subseteq [x]$. Now let $u \in [x]$. Then uEx. Since uEx and xEy, uEy by transitivity and so $u \in [y]$. Hence, $[x] \subseteq [y]$. Consequently, $[x] = [y]$.
   (iii) Let $x, y \in A$. Suppose $[x] \cap [y] \neq \phi$. Then there exists $u \in [x] \cap [y]$. Thus, $u \in [x]$ and $u \in [y]$, i.e., uEx and uEy. Since E is symmetric and uEy, we have yEu. Now yEu and uEx and so by the transitivity of E, yEx. This implies that $y \in [x]$. Hence, by (ii), $[y] = [x]$.
   (iv) Let $x \in A$. Then $x \in [x] \subseteq \cup_{x \in A}[x]$. Thus, $A \subseteq \cup_{x \in A}[x]$. Also, $\cup_{x \in A}[x] \subseteq A$. Hence, $A = \cup_{x \in A}[x]$. ∎

One of the main objectives of this section is to study the relationship between an equivalence relation and a partition of a set. We now focus our attention to partitions. We begin with the following definition.

Definition 1.3.14 *Let A be a set and $\mathcal{P}$ be a collection of nonempty subsets of A. Then $\mathcal{P}$ is called a **partition** of A if the following properties are satisfied:*
   (i) for all $B, C \in \mathcal{P}$, either $B = C$ or $B \cap C = \phi$.
   (ii) $A = \cup_{B \in \mathcal{P}} B$.

In other words, if $\mathcal{P}$ is a partition of A, then (i) $B \subseteq A$ for all $B \in \mathcal{P}$, i.e., every element of $\mathcal{P}$ is a subset of A, (ii) distinct elements of $\mathcal{P}$ are either equal or disjoint, and (iii) the union of the members of $\mathcal{P}$ is A.

Example 1.3.15 *(i) Let $A = \{1, 2, 3, 4, 5, 6\}$. Let $A_1 = \{1\}$, $A_2 = \{2, 4, 6\}$, and $A_3 = \{3, 5\}$. Now $A = A_1 \cup A_2 \cup A_3$, $A_1 \cap A_2 = \phi$, $A_1 \cap A_3 = \phi$, and $A_2 \cap A_3 = \phi$. Hence, $\mathcal{P} = \{A_1, A_2, A_3\}$ is a partition of A.*

(ii) Consider $\mathbf{Z}$. Let A be the set of all even integers and B be the set of all odd integers. Then $A \cap B = \phi$ and $A \cup B = \mathbf{Z}$. Thus, $\{A, B\}$ is a partition of $\mathbf{Z}$.

The following theorem is immediate from Theorem 1.3.13.

Theorem 1.3.16 *Let E be an equivalence relation on the set A. Then*

$$\mathcal{P} = \{[x] \mid x \in A\}$$

is a partition of A. ∎

Example 1.3.17 *Consider the equivalence relation $\equiv_n$ on $\mathbf{Z}$ as defined in Example 1.3.11. Let $\mathbf{Z}_n = \{[x] \mid x \in \mathbf{Z}\}$. By Theorem 1.3.16, $\mathbf{Z}_n$ is a partition of $\mathbf{Z}$. Suppose $n = 6$. We claim that*

$$\mathbf{Z}_6 = \{[0], [1], [2], [3], [4], [5]\}$$

and

$$[i] = \{0 + i, \pm 6 + i, \pm 12 + i, \ldots\} = \{6q + i \mid q \in \mathbf{Z}\} \text{ for all } i \in \mathbf{Z}.$$

Let $0 \leq n < m < 6$. Suppose $[n] = [m]$. Then $m \in [n]$ and so $6 \mid (m - n)$. This is a contradiction since $0 < m - n < 6$. Hence, the equivalence classes $[0], [1], [2], [3], [4], [5]$ are distinct. We now show that these are the only distinct equivalence classes.

Let k be any integer. By the division algorithm, $k = 6q + r$ for some integers q and r such that $0 \leq r < 6$. Thus, $k - r = 6q$ and so $6 \mid (k - r)$. This implies that $k \equiv_6 r$ and so $[k] = [r]$. Since $0 \leq r < 6$ we have $[r] \in \{[0], [1], [2], [3], [4], [5]\}$ and so $[k] \in \{[0], [1], [2], [3], [4], [5]\}$. This proves our first claim.

Let $i \in \mathbf{Z}$. Then $x \in [i]$ if and only if $6 \mid (x - i)$ if and only if $6q = x - i$ for some $q \in \mathbf{Z}$ if and only if $x = 6q + i$ for some $q \in \mathbf{Z}$. This proves our second claim. It now follows that for all $i = 0, 1, \ldots, 5$, $[i] = [6q + i]$ for all $q \in \mathbf{Z}$. Hence,

$$\begin{aligned}
&\text{for } i = 0, \; [0] = [6] = [12] = \cdots = [-6] = [-12] = \cdots; \\
&\text{for } i = 1, \; [1] = [7] = [13] = \cdots = [-5] = [-11] = \cdots; \\
&\text{for } i = 2, \; [2] = [8] = [14] = \cdots = [-4] = [-10] = \cdots; \\
&\text{for } i = 3, \; [3] = [9] = [15] = \cdots = [-3] = [-9] = \cdots; \\
&\text{for } i = 4, \; [4] = [10] = [16] = \cdots = [-2] = [-8] = \cdots; \\
&\text{for } i = 5, \; [5] = [11] = [17] = \cdots = [-1] = [-7] = \cdots.
\end{aligned}$$

By Theorem 1.3.16, given an equivalence relation E on a set A, the set of all equivalence classes forms a partition of A. We now prove that corresponding to any partition, we can associate an equivalence relation.

Theorem 1.3.18 *Let $\mathcal{P}$ be a partition of the set A. Define a relation E on A by for all $x, y \in A$, xEy if there exists $B \in \mathcal{P}$ such that $x, y \in B$. Then E is an equivalence relation on A and the equivalence classes are precisely the elements of $\mathcal{P}$.*

Proof.   Note that if two elements x and y of A are related, i.e., xEy, then x and y must belong to the same member of $\mathcal{P}$. Also, if $B \in \mathcal{P}$, then any two elements of B are related, i.e., xEy for all $x, y \in B$. We now prove the result.

Since $\mathcal{P}$ is a partition of A, $A = \cup_{B \in \mathcal{P}}B$. First we show that E is reflexive. Let x be any element of A. Then there exists $B \in \mathcal{P}$ such that $x \in B$. Since $x, x \in B$, we have xEx. Hence, E is reflexive. We now show that E is symmetric. Let xEy. Then $x, y \in B$ for some $B \in \mathcal{P}$. Thus, $y, x \in B$ and so yEx. Hence, E is symmetric. We now establish the transitivity of E. Let $x, y, z \in A$. Suppose xEy and yEz. Then $x, y \in B$ and $y, z \in C$ for some B, $C \in \mathcal{P}$. Since $y \in B \cap C$, $B \cap C \neq \phi$. Also, since $\mathcal{P}$ is a partition and $B \cap C \neq \phi$, we have $B = C$ so that $x, z \in B$. Hence, xEz. This shows that E is transitive. Consequently, E is an equivalence relation.

We now show that the equivalence classes determined by E are precisely the elements of $\mathcal{P}$. Let $x \in A$. Consider the equivalence class $[x]$. Since $A = \cup_{B \in \mathcal{P}}B$, there exists $B \in \mathcal{P}$ such that $x \in B$. We claim that $[x] = B$. Let $u \in [x]$. Then uEx and so $u \in B$ since $x \in B$. Thus, $[x] \subseteq B$. Also, since $x \in B$, we have yEx for all $y \in B$ and so $y \in [x]$ for all $y \in B$. This implies that $B \subseteq [x]$. Hence, $[x] = B$. Finally, note that if $C \in \mathcal{P}$, then $C = [u]$ for all $u \in C$. Thus, the equivalence classes are precisely the elements of $\mathcal{P}$. ∎

The relation E in Theorem 1.3.18 is called the **equivalence relation** on A **induced by the partition** $\mathcal{P}$.

New relations can be constructed from existing relations. For example, given relations R and S from a set A into a set B, we can form relations $R \cap S$, $R \cup S$, $R \backslash S$, $(A \times B) \backslash R$ in a natural way. In all these relations, the domain and range of the relations under consideration are subsets of A and B, respectively. Now given a relation R from a set A into a set B and a relation S from B into a set C, there is a relation from A into C that arises in a natural way as follows: Let us denote the new relation by T. Suppose $(a, b) \in R$ and $(b, c) \in S$. Then we make $(a, c) \in T$. Every element of T is constructed in this way. That is, $(a, c) \in T$ for some $a \in A$ and $c \in C$ if and only if there exists $b \in B$ such that $(a, b) \in R$ and $(b, c) \in S$. This relation T is called the **composition** of R and S and is denoted by $S \circ R$. Note that to form the composition of R and S, we must have the domain of S and the range of R to be subsets of the same set. More formally we have the following definition.

Definition 1.3.19 *Let R be a relation from a set A into a set B and S be a relation from B into a set C. The **composition** of R and S, denoted by $S \circ R$, is the relation from A into C defined by*

$$x(S \circ R)y \text{ if there exists } z \in B \text{ such that } xRz \text{ and } zSy$$

for all $x \in A, y \in C$.

Let R be a relation on a set A. Recursively, we define a relation R^n, $n \in \mathbf{N}$, as follows:

$$\begin{aligned} R^1 &= R \\ R^n &= R \circ R^{n-1} \text{ if } n > 1. \end{aligned}$$

Definition 1.3.20 *Let R be a relation from a set A into a set B. The **inverse** of R, denoted by R^{-1}, is the relation from B into A defined by*

$$xR^{-1}y \text{ if } yRx$$

for all $x \in B$, $y \in A$.

The following theorem gives a necessary and sufficient condition for a binary relation to be an equivalence relation.

Theorem 1.3.21 *Let R be a relation on a set A. Then R is an equivalence relation on A if and only if*
(i) $\triangle \subseteq R$, where $\triangle = \{(x, x) \mid x \in A\}$,
(ii) $R = R^{-1}$, and
(iii) $R \circ R \subseteq R$.

Proof.   Suppose R is an equivalence relation. Let $(x, x) \in \triangle$, where $x \in A$. Since R is reflexive, $(x, x) \in R$. Hence, $\triangle \subseteq R$, i.e., (i) holds. Let $(x, y) \in R$. Since R is symmetric, $(y, x) \in R$. Thus, by the definition of R^{-1}, $(x, y) \in R^{-1}$. Hence, $R \subseteq R^{-1}$. On the other hand, let $(x, y) \in R^{-1}$. Then $(y, x) \in R$. Therefore, by the symmetric property, $(x, y) \in R$. Hence, $R^{-1} \subseteq R$. Thus, $R = R^{-1}$, i.e., (ii) holds. We now prove (iii). Let $(x, y) \in R \circ R$. Then there exists $z \in A$ such that $(x, z) \in R$ and $(z, y) \in R$. Since R is transitive, $(x, y) \in R$. Thus, $R \circ R \subseteq R$, i.e., (iii) holds.
    Conversely, suppose that (i), (ii), and (iii) hold for R. For all $x \in A$, $(x, x) \in \triangle \subseteq R$. Thus, R is reflexive. Next, we show that R is symmetric. Let $(x, y) \in R$. Then by (ii), $(x, y) \in R^{-1}$. This implies that $(y, x) \in R$. Hence, R is symmetric. For the transitivity of R, let $(x, z) \in R$ and $(z, y) \in R$. Then $(x, y) \in R \circ R$ by the definition of composition of relations. Since $R \circ R \subseteq R$, $(x, y) \in R$. Hence, R is transitive. Consequently, R is an equivalence relation. ∎

1.3.1 Worked-Out Exercises

◇ **Exercise 1** In $\mathbf{Z}_{10}$, which of the following equivalence classes are equal: $[2]$, $[-5]$, $[5]$, $[-8]$, $[12]$, $[15]$, $[-3]$, $[7]$, $[22]$?

Solution: We note that $[2] = [2+10] = [12]$, $[-8] = [-8+10] = [2]$, $[12] = [12+10] = [22]$, $[-5] = [-5+10] = [5] = [5+10] = [15]$ and $[-3] = [-3+10] = [7]$. Also, $[2] \neq [5]$, $[2] \neq [7]$ and $[5] \neq [7]$. Hence, it now follows that $[2] = [12] = [-8] = [22]$, $[-5] = [5] = [15]$ and $[-3] = [7]$.

Exercise 2 Let R be a reflexive and transitive relation on a set S. Prove that $R \cap R^{-1}$ is an equivalence relation.

Solution: Since $(x, x) \in R$ for all $x \in S$, $(x, x) \in R^{-1}$ for all $x \in S$. Thus, $(x, x) \in R \cap R^{-1}$ for all $x \in S$. Hence, $R \cap R^{-1}$ is reflexive. Let $(x, y) \in R \cap R^{-1}$. Then $(x, y) \in R$ and $(x, y) \in R^{-1}$. Thus, $(y, x) \in R^{-1}$ and $(y, x) \in R$. Therefore, $(y, x) \in R \cap R^{-1}$. Hence, $R \cap R^{-1}$ is symmetric. Now suppose that $(x, y), (y, z) \in R \cap R^{-1}$. Then $(x, y), (y, z) \in R$ and $(x, y), (y, z) \in R^{-1}$. Since R is transitive, $(x, z) \in R$. Now since $(x, y), (y, z) \in R^{-1}$, $(y, x), (z, y) \in R$. Since R is transitive, $(z, x) \in R$ and so $(x, z) \in R^{-1}$. Thus, $(x, z) \in R \cap R^{-1}$. Hence, $R \cap R^{-1}$ is transitive. We have thus proved that $R \cap R^{-1}$ is reflexive, symmetric, and transitive and hence $R \cap R^{-1}$ is an equivalence relation.

◇ **Exercise 3** Give an example of an equivalence relation on the set $S = \{1, 2, 3, 4, 5, 6, 7, 8\}$ such that R has exactly four equivalence classes.

Solution: $R = \{(1, 1), (2, 2), (3, 3), (4, 4), (5, 5), (6, 6), (7, 7), (8, 8), (1, 2), (2, 1), (3, 4), (4, 3), (5, 6), (6, 5), (7, 8), (8, 7)\}$. The equivalence classes are $[1] = [2]$, $[3] = [4]$, $[5] = [6]$, and $[7] = [8]$.

Exercise 4 Let R_1 and R_2 be two symmetric relations on a set S. Prove that $R_1 \circ R_2$ is symmetric if and only if $R_1 \circ R_2 = R_2 \circ R_1$.

Solution: Suppose $R_1 \circ R_2$ is symmetric. Let (x, y) be any element of $R_1 \circ R_2$. Then $(y, x) \in R_1 \circ R_2$ since $R_1 \circ R_2$ is symmetric. Thus, there exists $z \in S$ such that $(y, z) \in R_2$ and $(z, x) \in R_1$ by the definition of composition of relations. Since R_1 and R_2 are symmetric, $(z, y) \in R_2$ and $(x, z) \in R_1$. Hence, $(x, y) \in R_2 \circ R_1$. Thus, $R_1 \circ R_2 \subseteq R_2 \circ R_1$. Similarly, $R_2 \circ R_1 \subseteq R_1 \circ R_2$. Hence, $R_1 \circ R_2 = R_2 \circ R_1$.

Conversely, suppose that $R_1 \circ R_2 = R_2 \circ R_1$. Let $(x, y) \in R_1 \circ R_2$. Then $(x, y) \in R_2 \circ R_1$. Thus, there exists $z \in S$ such that $(x, z) \in R_1$ and $(z, y) \in R_2$. Since R_1 and R_2 are symmetric, $(z, x) \in R_1$ and $(y, z) \in R_2$. Hence, $(y, x) \in R_2 \circ R_1 = R_1 \circ R_2$. Thus, $R_1 \circ R_2$ is symmetric.

◇ **Exercise 5** Let $A = \{1, 2, 3, 4, 5\}$ and $R = \{(1, 1), (2, 2), (3, 3), (4, 4), (5, 5), (1, 2), (2, 1), (4, 5), (5, 4)\}$. Show that R is an equivalence relation.

Solution: Let $B = \{1, 2\}$, $C = \{3\}$, and $D = \{4, 5\}$. Let $\mathcal{P} = \{B, C, D\}$. Then $\mathcal{P}$ is a partition of A. Also, note that if $x, y \in A$, then $(x, y) \in R$ if and only if $x, y \in X$ for some $X \in \mathcal{P}$, i.e., the relation R is induced by the partition $\mathcal{P}$. Hence, R is an equivalence relation on A by Theorem 1.3.18.

◇ **Exercise 6** Let $X = \{1, 2, 3, 4, 5, 6, 7\}$. Then

$$\mathcal{P} = \{\{1, 3, 5\}, \{2, 6\}, \{4, 7\}\}$$

is a partition of X. List the elements of the corresponding equivalence relation R on X induced by $\mathcal{P}$.

Solution: $R = \{(a, b) \in X \times X \mid a \text{ and } b \text{ both belong to the same element of } \mathcal{P}\}$. Then $R = \{(1, 1), (2, 2), (3, 3), (4, 4), (5, 5), (6, 6), (7, 7), (1, 3), (3, 1), (1, 5), (5, 1), (3, 5), (5, 3), (2, 6), (6, 2), (4, 7), (7, 4)\}$.

Exercise 7 Let R be a relation on a set S. Prove that the following conditions are equivalent.

(i) R is an equivalence relation on S.

(ii) R is reflexive and for all $a, b, c \in S$, if aRb and bRc, then cRa.

Solution: (i)$\Rightarrow$(ii): Suppose R is an equivalence relation on S. Then R is reflexive. Let $a, b, c \in S$. Suppose aRb and bRc. The transitive property of R implies that aRc. Hence, cRa since R is symmetric.

(ii)$\Rightarrow$(i): Since R is given to be reflexive, to show that R is an equivalence relation, we only need to check that R is symmetric and transitive. For symmetry, suppose aRb. Since R is reflexive, we have aRa. Now since we have aRa and aRb, bRa by hypothesis. This shows that R is symmetric. To show that R is transitive, suppose aRb and bRc. Then by the hypothesis, cRa. Since we have shown that R is symmetric, cRa implies that aRc. Hence, R is transitive. Consequently, R is an equivalence relation on S.

1.3.2 Exercises

1. Let R be a relation on the set $A = \{1, 2, 3, 4, 5, 6, 7\}$ defined by $R = \{(a, b) \in A \times A \mid 4 \text{ divides } a - b\}$.

   (i) List the elements of R.

   (ii) Find the domain of R.

   (iii) Find the range of R.

(iv) Find the elements of R^{-1}.

(v) Find the domain of R^{-1}.

(vi) Find the range of R^{-1}.

2. Let R be a relation on the set $A = \{1, 2, 3, 4, 5, 6\}$ defined by $R = \{(a, b) \in A \times A \mid a + b \leq 9\}$.

(i) List the elements of R.

(ii) Is $\Delta \subseteq R$, where $\Delta = \{(x, x) \mid x \in A\}$?

(iii) Is $R = R^{-1}$?

(iv) Is $R \circ R \subseteq R$?

3. Which of the following relations E are equivalence relations on the set of integers $\mathbf{Z}$?

(i) xEy if and only if $x - y$ is an even integer.

(ii) xEy if and only if $x - y$ is an odd integer.

(iii) xEy if and only if $x \leq y$.

(iv) xEy if and only if x divides y.

(v) xEy if and only if $x^2 = y^2$.

(vi) xEy if and only if $|x| = |y|$.

(vii) xEy if and only if $|x - y| \leq 2$.

4. Let $R = \{(a, b) \mid a, b \in \mathbf{Q} \text{ and } a - b \in \mathbf{Z}\}$. Prove that R is an equivalence relation on $\mathbf{Q}$.

5. Let $A = \{1, 2, 3, 4, 5, 6, 7, 8\}$. Define a relation R on A by

$$aRb \text{ if and only if } 3 \text{ divides } a - b$$

for all $a, b \in A$. Show that R is an equivalence relation on A. Find the equivalence classes $[1]$, $[2]$, $[3]$, and $[4]$.

6. Let R be an equivalence relation on a set A. Find the domain and range of R.

7. Find all equivalence relations on the set $S = \{a, b, c\}$.

8. In $\mathbf{Z}_6$, which of the following equivalence classes are equal: $[-1]$, $[2]$, $[8]$, $[5]$, $[-2]$, $[11]$, $[23]$?

9. Let $x, y \in \mathbf{Z}$ be such that $x \equiv_n y$, where $n \in \mathbf{N}$. Show that for all $z \in \mathbf{Z}$,

(i) $x + z \equiv_n y + z$,

(ii) $xz \equiv_n yz$

10. Let $x, y, z, w \in \mathbf{Z}$ and n be a positive integer. Suppose that $x \equiv_n y$ and $z \equiv_n w$. Show that $x + z \equiv_n y + w$ and $xz \equiv_n yw$.

11. Let n be a positive integer and $[x], [y] \in \mathbf{Z}_n$. Show that the following conditions are equivalent.

(i) $[x] = [y]$.

(ii) $x - y = nr$ for some integer r.

(iii) $n \mid (x - y)$.

12. (**Chinese Remainder Theorem**) Let m and n be positive integers such that $\gcd(m, n) = 1$. Prove that for any integers a and b, the congruences $x \equiv_m a$ and $x \equiv_n b$ have a common solution in $\mathbf{Z}$. Furthermore, if u and v are two solutions of these congruences, prove that $u \equiv_{mn} v$.

13. Define relations R_1, R_2, R_3 such that R_1 is reflexive and symmetric but not transitive, R_2 is reflexive and transitive but not symmetric, and R_3 is symmetric and transitive but not reflexive.

14. Prove that the intersection of two equivalence relations on a set S is an equivalence relation on S.

15. Let R be a relation on a set A. Define $T(R) = R \cup R^{-1} \cup \{(x, x) \mid x \in A\}$. Show that $T(R)$ is reflexive and symmetric.

16. Let R be a relation on a set S. Set $R^\infty = R \cup R^2 \cup R^3 \cup \cdots$. Prove the following:

(i) R^∞ is a transitive relation on S.

(ii) If T is a transitive relation on A such that $R \subseteq T$, then $R^\infty \subseteq T$.

(R^∞ is called the **transitive closure** of R.)

17. Let R_1 and R_2 be symmetric relations on a set S such that $R_1 \circ R_2 \subseteq R_2 \circ R_1$. Prove that $R_2 \circ R_1$ is symmetric and $R_1 \circ R_2 = R_2 \circ R_1$.

18. Let R_1 and R_2 be equivalence relations on a set S such that $R_1 \circ R_2 = R_2 \circ R_1$. Prove that $R_1 \circ R_2$ is an equivalence relation.

19. Let R_1 and R_2 be relations on a set S. Determine whether each statement is true or false. If the statement is false, give a counterexample.

(i) If R_1 and R_2 are reflexive, then $R_1 \circ R_2$ is reflexive.

(ii) If R_1 and R_2 are transitive, then $R_1 \circ R_2$ is transitive.

(iii) If R_1 and R_2 are symmetric, then $R_1 \circ R_2$ is symmetric.

(iv) If R_1 is transitive, then R_1^{-1} is transitive.

(v) If R_1 is reflexive and transitive, then $R_1 \circ R_1$ is transitive.

1.4 Partially Ordered Sets

In the previous section, we defined binary relations and studied their basic properties. More specifically, we looked at equivalence relations and showed that equivalence relations and partitions are closely related. In this section, we will consider binary relations which are reflexive, are transitive, and satisfy a new property, called antisymmetric. We begin with the following definition.

Definition 1.4.1 *A relation R on a set S is called a **partial order** on S if it satisfies the following conditions:*

(i) *$(a, a) \in R$ for all $a \in S$ (i.e., R is reflexive).*

(ii) *For all $a, b \in S$ if $(a, b) \in R$ and $(b, a) \in R$, then $a = b$ (i.e., R is **antisymmetric**).*

(iii) *For all $a, b, c \in S$, if $(a, b) \in R$ and $(b, c) \in R$, then $(a, c) \in R$ (i.e., R is transitive).*

In other words, a reflexive, antisymmetric, and transitive relation on a set S is called a partial order on S.

Example 1.4.2 *Let R be the relation on $\mathbf{Z}$ defined by $R = \{(a, b) \in \mathbf{Z} \times \mathbf{Z} \mid a - b \leq 0\}$. We show that R is a partial order on $\mathbf{Z}$.*

First note that $a - a = 0 \leq 0$ for all $a \in \mathbf{Z}$. Thus, $(a, a) \in R$ for all $a \in \mathbf{Z}$ and so R is reflexive. For antisymmetry, let $(a, b), (b, a) \in R$. Then $a - b \leq 0$, i.e., $a \leq b$ and $b - a \leq 0$, i.e., $b \leq a$. This implies that $a = b$. Thus, R is antisymmetric. Finally, we show that R is transitive. Let $(a, b), (b, c) \in R$. Then $a - b \leq 0$ and $b - c \leq 0$. Thus, $a \leq b$ and $b \leq c$. This implies that $a \leq c$ and so $a - c \leq 0$. Hence, $(a, c) \in R$. Thus, R is transitive. Consequently, R is a partial order on $\mathbf{Z}$.

Example 1.4.3 *Let R be the relation on $\mathbf{N}$ defined by $R = \{(a, b) \in \mathbf{N} \times \mathbf{N} \mid a$ divides b in $\mathbf{N}\}$. Then R is a partial order on $\mathbf{N}$.*

As in the previous example, we show that R is reflexive, antisymmetric, and transitive.

Reflexive: Let $a \in \mathbf{N}$. Since $a = 1a$, we have $a|a$ and so $(a, a) \in R$. Thus, R is reflexive.

Antisymmetric: Let $(a, b), (b, a) \in R$. Then $a|b$ and $b|a$. Thus, $b = ad$ and $a = bc$ for some positive integers c and d. Therefore, $a = bc = adc$ and so $1 = cd$. Since c and d are positive integers and $cd = 1$, it follows that $c = d = 1$. Hence, $a = b$. Thus, R is antisymmetric.

Transitive: Let $(a, b), (b, c) \in R$. Then $a|b$ and $b|c$ in $\mathbf{N}$. Thus, $b = an$ and $c = bm$ for some positive integers m and n. This implies that $c = bm = anm$ and since m and n are positive integers, nm is a positive integer. Thus, $a|c$ in $\mathbf{N}$ and so $(a, c) \in R$. Hence, R is transitive.

Consequently, R is a partial order on $\mathbf{N}$.

Example 1.4.4 *Consider the relation* $R = \{(a,b) \in \mathbf{Z} \times \mathbf{Z} \mid a$ *divides* b *in* $\mathbf{Z}\}$ *on* $\mathbf{Z}$. *As in the previous example, we can show that* R *is reflexive and transitive. Since* $6 = (-1)(-6)$ *and* $-6 = (-1)6$, $(6, -6) \in R$ *and* $(-6, 6) \in R$, *but* $6 \neq -6$. *Thus,* R *is not antisymmetric, proving that* R *is not a partial order on* $\mathbf{Z}$.

Example 1.4.5 *Let* S *be a set and* $\mathcal{P}(S)$ *the power set of* S. *Let* R *be a relation on* $\mathcal{P}(S)$ *given by* $R = \{(A, B) \in \mathcal{P}(S) \times \mathcal{P}(S) \mid A \subseteq B\}$. *We show that* R *is a partial order on* $\mathcal{P}(S)$. *Since* $A \subseteq A$ *for all* $A \in \mathcal{P}(S)$, *we find that* $(A, A) \in R$ *for all* $A \in \mathcal{P}(S)$. *This shows that* R *is reflexive. For antisymmetry, let* $(A, B), (B, A) \in R$. *Then by the definition of* R, $A \subseteq B$ *and* $B \subseteq A$ *and so* $A = B$. *Thus,* R *is antisymmetric. To show that* R *is transitive, let* $(A, B), (B, C) \in R$. *Then* $A \subseteq B$ *and* $B \subseteq C$ *and so* $A \subseteq C$. *Thus,* $(A, C) \in R$. *Hence,* R *is transitive. Consequently,* R *is a partial order on* $\mathcal{P}(S)$.

A partial order on a set S is usually denoted by $\leq$. Instead of writing $(a, b) \in \leq$, from now on we shall write $a \leq b$.

Definition 1.4.6 *A set* S *together with a partial order is called a* **partially ordered set (poset)**.

If S is a partially ordered set with partial order $\leq$, then we write $(S, \leq)$.

In Example 1.4.2, R is a partial order. This relation is the usual "less than or equal to" relation on $\mathbf{Z}$. In Example 1.4.3, R is a partial order. We call this relation the **divisibility relation** on $\mathbf{N}$. Hence, $\mathbf{N}$ together with the divisibility relation is a poset. From Example 1.4.4, we find that $\mathbf{Z}$ together with the divisibility relation is not a poset. The partial order in Example 1.4.5 is known as **set inclusion relation**. $\mathcal{P}(S)$ together with set inclusion relation is a poset.

Let S be a poset and $a, b \in S$. If either $a \leq b$ or $b \leq a$, then we say that a and b are **comparable**.

Definition 1.4.7 *A partially ordered set* $(S, \leq)$ *is called a* **linearly ordered set** *or a* **chain** *if for all* $x, y \in S$ *either* $x \leq y$ *or* $y \leq x$.

Thus, a linearly ordered set or a chain is a poset in which any two elements are comparable.

Example 1.4.8 *(i)* $\mathbf{Z}$ *together with the usual "less than or equal to" (Example 1.4.2) relation is a chain.*

(ii) $\mathbf{N}$ *with the divisibility relation (Example 1.4.3) is not a chain because neither* 3 *divides* 5 *nor* 5 *divides* 3, *i.e.,* 3 *and* 5 *are not comparable.*

(iii) Let S *be a set with more than one element. Then* $\mathcal{P}(S)$ *together with the set inclusion relation (Example 1.4.5) is not a chain since if* a *and* b *are distinct elements of* S, *then neither* $\{a\}$ *is a subset of* $\{b\}$ *nor* $\{b\}$ *is a subset of* $\{a\}$, *i.e.,* $\{a\}$ *and* $\{b\}$ *are not comparable.*

Definition 1.4.9 *Let $(S, \leq)$ be a poset and $\{a, b\}$ be a subset of S. An element $c \in S$ is called an **upper bound** of $\{a, b\}$ if $a \leq c$ and $b \leq c$.*

*An element $d \in S$ is called a **least upper bound (lub)** of $\{a, b\}$ if*
(i) d is an upper bound of $\{a, b\}$ and
(ii) if $c \in S$ is an upper bound of $\{a, b\}$, then $d \leq c$.

Example 1.4.10 *(i) Consider the set $\mathbf{N}$ together with the divisibility relation (Example 1.4.3). For all $a, b \in \mathbf{N}$, $a \leq b$ if and only if a divides b. Now for the subset $\{4, 6\}$, $12, 24, 36$ are all upper bounds of $\{4, 6\}$. However, 12 is the least upper bound of $\{4, 6\}$.*

(ii) Consider the set $\mathbf{Z}$ together with the usual "less than or equal to" relation (Example 1.4.2). For the subset $\{4, 6\}$, $6, 7, 8, \ldots$ are all upper bounds of $\{4, 6\}$. However, 6 is the least upper bound of $\{4, 6\}$.

(iii) Let $S = \{1, 2, 3, 4\}$. Let $\leq$ denote the set inclusion relation (Example 1.4.5). Then $(\mathcal{P}(S), \leq)$ is a poset. Let $A = \{1, 2\}$ and $B = \{1, 4\}$. Then $A \cup B = \{1, 2, 4\}$ is the least upper bound of $\{A, B\}$.

Remark 1.4.11 *(i) In a poset $(S, \leq)$, a subset $\{a, b\}$ of S may not have an upper bound.*

(ii) In a poset $(S, \leq)$, a subset $\{a, b\}$ of S may have more than one upper bound.

(iii) In a poset $(S, \leq)$, a subset $\{a, b\}$ of S may not have a lub.

(iv) In a poset $(S, \leq)$, if a subset $\{a, b\}$ of S has a lub, then this lub is unique.

We leave the verification of (i), (ii), and (iii) as an exercise and verify (iv). Let $c, d \in S$ be two lubs of $\{a, b\}$. Then c and d are upper bounds of $\{a, b\}$. Since c is a lub of $\{a, b\}$ and d is an upper bound of $\{a, b\}$, $c \leq d$. Similarly, $d \leq c$. Hence, $c = d$.

Notation: The lub of $\{a, b\}$ in $(S, \leq)$, if it exists, is denoted by $a \vee b$.

Definition 1.4.12 *Let $(S, \leq)$ be a poset and $\{a, b\}$ be a subset of S. An element $c \in S$ is called a **lower bound** of $\{a, b\}$ if $c \leq a$ and $c \leq b$. An element $d \in S$ is called a **greatest lower bound (glb)** of $\{a, b\}$ if*
(i) d is a lower bound of $\{a, b\}$ and
(ii) if $c \in S$ is a lower bound of $\{a, b\}$, then $c \leq d$.

Remark 1.4.13 *(i) In a poset $(S, \leq)$, a subset $\{a, b\}$ of S may not have a lower bound.*

(ii) In a poset $(S, \leq)$, a subset $\{a, b\}$ of S may have more than one lower bound.

(iii) In a poset $(S, \leq)$, a subset $\{a, b\}$ of S may not have a glb.

(iv) In a poset $(S, \leq)$, if a subset $\{a, b\}$ of S has a glb, then this glb is unique.

Notation: The glb of $\{a, b\}$ in $(S, \leq)$, if it exists, is denoted by $a \wedge b$.

A useful device in the study of posets is the poset diagram. Let $(S, \leq)$ be a poset and $x, y \in S$. We say that y **covers** x, denoted by $y \succ x$, if $x \leq y$, $x \neq y$, and there are no elements $z \in S$ such that $x \leq z \leq y$, $x \neq z$, $z \neq y$. We represent the elements of S by the elements themselves in the plane such that if $x \leq y$, then y occurs above x, and we connect x with y by a line segment if and only if y covers x. The resulting diagram is called the poset diagram of $(S, \leq)$.

Example 1.4.14 *Let* $S = \{1, 2, 3\}$. *Then*

$$\mathcal{P}(S) = \{\phi, \{1\}, \{2\}, \{3\}, \{1, 2\}, \{2, 3\}, \{1, 3\}, S\}.$$

Now $(\mathcal{P}(S), \leq)$ *is a poset, where* $\leq$ *denotes the set inclusion relation. The poset diagram of* $(\mathcal{P}(S), \leq)$ *is given below.*

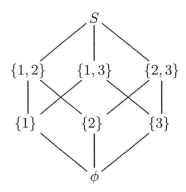

Definition 1.4.15 *Let* $(S, \leq)$ *be a poset. An element* $u \in S$ *is called a **maximal (minimal)** element of* S *if there is no element* $v \in S$ *such that* $u \leq v$ $(v \leq u)$ *and* $u \neq v$.

Example 1.4.16 *Let* $S = \{1, 2, 3\}$ *and* T *be the set of all proper nonempty subsets of* S. *Now* $(T, \leq)$ *is a poset, where* $\leq$ *is the set inclusion relation. In this poset* $\{1\}, \{2\}$, *and* $\{3\}$ *are minimal elements and* $\{1, 2\}, \{1, 3\}, \{2, 3\}$ *are maximal elements.*

Next, we state the following fundamental axiom of set theory. There are several places in this text, where we will use it very effectively.

Zorn's Lemma: If every chain in a poset $(S, \leq)$ has an upper bound in S, then S contains a maximal element.

We have seen several examples of posets in which lub (glb) need not exist. Next, we study those posets for which lub (glb) exists.

Definition 1.4.17 *A poset* $(L, \leq)$ *is called a* **lattice** *if* $a \wedge b$ *and* $a \vee b$ *exist in* L *for all* $a, b \in L$.

Example 1.4.18 *Let* $L = [0, 1] = \{x \in \mathbf{R} \mid 0 \leq x \leq 1\}$. *Then* $(L, \leq)$ *is a poset,* *where* $\leq$ *denotes the usual "less than or equal to" relation. Let* $a, b \in [0, 1]$. *Now* $\max\{a, b\} \in L$ *and* $\min\{a, b\} \in L$. *It is easy to see that* $\max\{a, b\}$ *is the lub of* $\{a, b\}$ *and* $\min\{a, b\}$ *is the glb of* $\{a, b\}$. *For example,* $\max\{.2, .3\} = .3 = .2 \vee .3$ *and* $\min\{.2, .3\} = .2 = .2 \wedge .3$. *Hence,* $(L, \leq)$ *is a lattice.*

Example 1.4.19 *Let* S *be a set. Then* $(\mathcal{P}(S), \leq)$ *is a poset, where* $\leq$ *is the set inclusion relation. For* $A, B \in \mathcal{P}(S)$, *we can show that* $A \vee B = A \cup B$ *and* $A \wedge B = A \cap B$. *Hence,* $(\mathcal{P}(S), \leq)$ *is a lattice.*

In the following theorem, we collect several useful properties of a lattice.

Theorem 1.4.20 *Let* $(L, \leq)$ *be a lattice and* $a, b, c \in L$. *Then*
(L1) $a \vee b = b \vee a$, $a \wedge b = b \wedge a$ *(commutative laws),*
(L2) $a \vee (b \vee c) = (a \vee b) \vee c$, $a \wedge (b \wedge c) = (a \wedge b) \wedge c$ *(associative laws),*
(L3) $a \vee a = a$, $a \wedge a = a$ *(idempotent laws),*
(L4) $a \vee (a \wedge b) = a$, $a \wedge (a \vee b) = a$ *(absorption laws).*

Proof. (L1) $a \vee b = $ lub of $\{a, b\} = $ lub of $\{b, a\} = b \vee a$. Note that the proof follows from the fact that the set $\{a, b\}$ is the same as the set $\{b, a\}$.

We leave the remainder of the proof to the exercises except for L4.

(L4) Now $a \leq a$ and $a \wedge b \leq a$. Hence, a is an upper bound of $\{a, a \wedge b\}$. Thus, by the definition of least upper bound, $a \vee (a \wedge b) \leq a$. Since $a \vee (a \wedge b)$ is the lub of $\{a, a \wedge b\}$, $a \leq a \vee (a \wedge b)$. Hence, $a = a \vee (a \wedge b)$ since $\leq$ is antisymmetric. ∎

The proof of the following result is left as an exercise.

Theorem 1.4.21 *Let* $(S, \leq)$ *be a poset and* $a, b \in S$. *Then the following conditions are equivalent.*
(i) $a \leq b$.
(ii) $a \vee b = b$.
(iii) $a \wedge b = a$.

Definition 1.4.22 *A lattice* $(L, \leq)$ *is called a* **modular** *lattice if for all* $a, b, c \in L$, $a \leq c$ *implies*

$$a \vee (b \wedge c) = (a \vee b) \wedge c.$$

The lattices defined in Examples 1.4.18 and 1.4.19 are modular lattices.

Example 1.4.23 *Consider the lattice given by the following diagram*

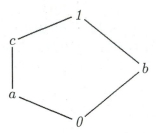

    Here $a \leq c$, but $a \vee (b \wedge c) = a \vee 0 = a \neq c = (a \vee b) \wedge c$. Hence, this lattice is not modular.

Definition 1.4.24 *A lattice $(L, \leq)$ is called* **distributive** *if it satisfies*
    (D1) $a \wedge (b \vee c) = (a \wedge b) \vee (a \wedge c)$
for all $a, b, c \in L$.

    The lattices defined in Examples 1.4.18 and 1.4.19 are distributive lattices.

Theorem 1.4.25 *A lattice $(L, \leq)$ is distributive if and only if*
    (D2) $a \vee (b \wedge c) = (a \vee b) \wedge (a \vee c)$
for all $a, b, c \in L$.

Proof.   Suppose $(L, \leq)$ is distributive. Let $a, b, c \in L$. Then

$$
\begin{aligned}
(a \vee b) \wedge (a \vee c) &= ((a \vee b) \wedge a) \vee ((a \vee b) \wedge c) &&\text{by D1} \\
&= (a \wedge (a \vee b)) \vee ((a \vee b) \wedge c) &&\text{by L1} \\
&= a \vee ((a \vee b) \wedge c) &&\text{by L4} \\
&= a \vee (c \wedge (a \vee b)) &&\text{by L1} \\
&= a \vee ((c \wedge a) \vee (c \wedge b)) &&\text{by D1} \\
&= (a \vee (c \wedge a)) \vee (c \wedge b) &&\text{by L2} \\
&= (a \vee (c \wedge a)) \vee (b \wedge c) &&\text{by L1} \\
&= a \vee (b \wedge c) &&\text{by L4.}
\end{aligned}
$$

Hence, $a \vee (b \wedge c) = (a \vee b) \wedge (a \vee c)$. Similarly, D2$\Rightarrow$D1. ∎

Theorem 1.4.26 *Every distributive lattice is a modular lattice.*

Proof.   Let $(L, \leq)$ be a distributive lattice and $a, b, c \in L$ be such that $a \leq c$. Then $a \vee (b \wedge c) = (a \vee b) \wedge (a \vee c) = (a \vee b) \wedge c$. Hence, $(L, \leq)$ is a modular lattice. ∎

    Theorem 1.4.26 says that every distributive lattice is a modular lattice. However, the converse of this result is not true, as shown by the following example.

Example 1.4.27 *Consider the lattice given by the following poset*

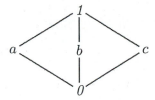

This is a modular lattice, but not a distributive lattice since $a \vee (b \wedge c) = a \vee 0 = a \neq 1 = (a \vee b) \wedge (a \vee c)$.

Theorem 1.4.28 *In a distributive lattice* $(L, \leq)$,

$$a \wedge b = a \wedge c \text{ and } a \vee b = a \vee c \text{ imply that } b = c$$

for all $a, b, c \in L$.

Proof.   Now $b = b \wedge (a \vee b) = b \wedge (a \vee c) = (b \wedge a) \vee (b \wedge c) = (a \wedge c) \vee (b \wedge c) = (c \wedge a) \vee (c \wedge b) = c \wedge (a \vee b) = c \wedge (a \vee c) = c.$ ∎

1.4.1   Worked-Out Exercises

◇ **Exercise 1** Suppose that in a poset $(P, \leq)$, $a \wedge b$, $b \wedge c$, and $a \wedge (b \wedge c)$ exist, where $a, b, c \in P$. Show that $(a \wedge b) \wedge c$ exists and $a \wedge (b \wedge c) = (a \wedge b) \wedge c$.

Solution: Now $a \wedge (b \wedge c) \leq a$, $a \wedge (b \wedge c) \leq b \wedge c$, $b \wedge c \leq b$, and $b \wedge c \leq c$. Hence, $a \wedge (b \wedge c)$ is a lower bound of a, b. Since $a \wedge b$ exists, we find that $a \wedge (b \wedge c) \leq a \wedge b$. Also, $a \wedge (b \wedge c) \leq c$. Hence, $a \wedge (b \wedge c)$ is a lower bound of $\{a \wedge b, c\}$. Let d be a lower bound of $\{a \wedge b, c\}$. Then $d \leq a \wedge b$ and $d \leq c$. Thus, $d \leq a$, $d \leq b$, and $d \leq c$. Since $b \wedge c$ exists, $d \leq b \wedge c$. Also, $a \wedge (b \wedge c)$ exists. Hence, $d \leq a \wedge (b \wedge c)$. Thus, $a \wedge (b \wedge c)$ is the glb of $\{a \wedge b, c\}$. Consequently, $(a \wedge b) \wedge c$ exists and $a \wedge (b \wedge c) = (a \wedge b) \wedge c$.

Exercise 2 Show that every chain is a distributive lattice.

Solution: Let $(L, \leq)$ be a chain and $a, b, c \in L$. Since L is a chain, either $a \leq b$ or $b \leq a$. If $a \leq b$, then $a \vee b = b$ and $a \wedge b = a$. If $b \leq a$, then $a \vee b = a$ and $a \wedge b = b$. Hence, for any two elements $a, b \in L$, $a \wedge b$ and $a \vee b$ exist in L. Suppose $a \leq b$.
    Case 1: $b \leq c$.
    Now $a \wedge (b \vee c) = a \wedge c = a$ and $(a \wedge b) \vee (a \wedge c) = a \vee a = a$. Hence, we have $a \wedge (b \vee c) = (a \wedge b) \vee (a \wedge c)$.
    Case 2: $c \leq b$.
    Subcase 2a: $a \leq c$.

In this case, we have $a \leq c \leq b$. Now $a \wedge (b \vee c) = a \wedge b = a$ and $(a \wedge b) \vee (a \wedge c) = a \vee a = a$. Hence, $a \wedge (b \vee c) = (a \wedge b) \vee (a \wedge c)$.

Subcase 2b: $c \leq a$.

In this case, we have $c \leq a \leq b$. Now $a \wedge (b \vee c) = a \wedge b = a$ and $(a \wedge b) \vee (a \wedge c) = a \vee c = a$. Hence, $a \wedge (b \vee c) = (a \wedge b) \vee (a \wedge c)$.

Similarly, if $b \leq a$, then $a \wedge (b \vee c) = (a \wedge b) \vee (a \wedge c)$.

Exercise 3 In a lattice $(L, \leq)$, prove that $(a \wedge b) \vee (a \wedge c) \leq a \wedge (b \vee (a \wedge c))$ for all $a, b, c \in L$.

Solution: $a \wedge b \leq a$, $a \wedge c \leq a$. Hence, $(a \wedge b) \vee (a \wedge c) \leq a$. Again $a \wedge b \leq b$ implies $(a \wedge b) \vee (a \wedge c) \leq b \vee (a \wedge c)$. Thus, we find that $(a \wedge b) \vee (a \wedge c)$ is a lower bound of $\{a, b \vee (a \wedge c)\}$. But $a \wedge (b \vee (a \wedge c))$ is the glb of $\{a, b \vee (a \wedge c)\}$. Hence, $(a \wedge b) \vee (a \wedge c) \leq a \wedge (b \vee (a \wedge c))$.

Exercise 4 Prove that a lattice $(L, \leq)$ is modular if and only if $(a \wedge b) \vee (a \wedge c) = a \wedge (b \vee (a \wedge c))$ for all $a, b, c \in L$.

Solution: Suppose $(L, \leq)$ is modular. Then

$$
\begin{aligned}
(a \wedge b) \vee (a \wedge c) &= (a \wedge c) \vee (a \wedge b) \\
&= (a \wedge c) \vee (b \wedge a) \\
&= ((a \wedge c) \vee b) \wedge a \qquad \text{(by modularity since } a \wedge c \leq a) \\
&= a \wedge (b \vee (a \wedge c)).
\end{aligned}
$$

Conversely, suppose that $(a \wedge b) \vee (a \wedge c) = a \wedge (b \vee (a \wedge c))$ for all $a, b, c \in L$. Let $a, b, c \in L$ be such that $a \leq c$. Then $a \wedge c = a$. Now $(c \wedge b) \vee (c \wedge a) = c \wedge (b \vee (a \wedge c))$. Hence, $(c \wedge b) \vee a = c \wedge (b \vee a)$, i.e., $a \vee (b \wedge c) = (a \vee b) \wedge c$.

1.4.2 Exercises

1. Draw the poset diagram for each of the following posets.

   (i) $(\{a \mid a$ is a positive divisor of $20\}, \leq)$, where $\leq$ denotes the divisibility relation.

   (ii) $(\mathbf{N}, \leq)$, where $\leq$ denotes the natural order relation.

   (iii) $(\mathcal{P}(S), \leq)$, $S = \{1, 2, 3, 4\}$, where $\leq$ denotes the set inclusion relation.

   (iv) $(\mathcal{P}(S) \backslash \{\phi\}, \leq)$, $S = \{1, 2, 3\}$, where $\leq$ denotes the set inclusion relation.

2. Give an example of a relation R which is antisymmetric, but not reflexive.

3. Give an example of a poset $(P, \leq)$ such that P has two elements a and b for which $a \wedge b$ does not exist.

4. Show that $(\mathbf{R}, \leq)$ is not a poset, where $a \leq b$ means that $b = ad$ for some $d \in \mathbf{R}$.

5. Let $\leq_1$ and $\leq_2$ be two partial orders on a set S. Is $\leq_1 \cap \leq_2$ a partial order on S?

6. Let $(A, \leq_1)$ and $(B, \leq_2)$ be two posets. Prove that $(A \times B, \leq)$ is a poset, where $(a, b) \leq (c, d)$ if and only if $a \leq_1 c$ and $b \leq_2 d$.

7. Let $(P, \leq)$ be a poset and $a, b, c \in P$.

   (i) If $a \vee b$, $b \vee c$, and $a \vee (b \vee c)$ exist, show that $(a\vee b) \vee c$ exists and $a \vee (b \vee c) = (a\vee b) \vee c$.

   (ii) If $a \vee b$ exists, prove that $a \vee (a \vee b)$ exists and $a \vee b = a \vee (a \vee b)$.

8. Which of the following posets are lattices?

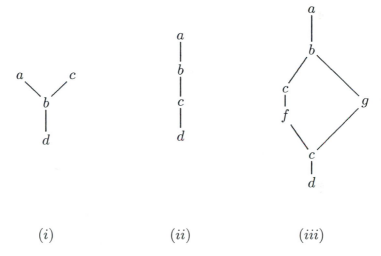

| (i) | (ii) | (iii) |

9. Let $D(40)$ denote the set of all positive divisors of 40. Consider the lattice

$$(D(40), \leq),$$

where $\leq$ denotes the divisibility relation. Find $4 \wedge (8 \vee 10)$ and $(2 \vee (2 \wedge 8)) \vee 20$.

10. In a lattice $(L, \leq)$, prove the following.

   (i) $a \vee (b \wedge c) \leq (a \vee b) \wedge (a \vee c)$,

   (ii) $(a \wedge b) \vee (a \wedge c) \leq a \wedge (b \vee c)$,

   (iii) $(a \wedge b) \vee (b \wedge c) \vee (c \wedge a) \leq (a \vee b) \wedge (b \vee c) \wedge (c \vee a)$,

   (iv) if $a \leq c$, then $a \vee (b \wedge c) \leq (a \vee b) \wedge c$,

   for all $a, b, c \in L$.

11. In a modular lattice $(L, \leq)$, prove that for all $a, b, c \in L$, $a \leq c$, $a \wedge b = c \wedge b$, and $a \vee b = c \vee b$ imply that $a = c$.

12. Prove that a lattice $(L, \leq)$ is distributive if and only if for all $a, b, c \in L$,

$$(a \wedge b) \vee (b \wedge c) \vee (c \wedge a) = (a \vee b) \wedge (b \vee c) \wedge (c \vee a).$$

13. Determine whether the following assertions are true or false. If true prove the result; and if false give a counterexample.

(i) The relation $R = \{(a, b) \in \mathbf{Z} \times \mathbf{Z} \mid |a - b| \leq 1\}$ is a partial order on $\mathbf{Z}$.

(ii) The relation $R = \{(a, b) \in \mathbf{Z} \times \mathbf{Z} \mid |a| \leq |b|\}$ is a partial order on $\mathbf{Z}$.

(iii) The relation $R = \{(a, b) \in S \times S \mid a$ divides b in $\mathbf{N}\}$ is a partial order on $S = \{1, 2, 3, 4, 6, 12\}$.

1.5   Functions

Like sets, functions play a central role in mathematics. Readers may already be familiar with the notion of a function either through a college algebra or a calculus course. In these courses, functions were usually real valued. Throughout the text we will encounter functions which do not have to be real valued. Functions help us study the relationship between various algebraic structures. In this section, we review some of their basic properties. Roughly speaking, a function is a special type of correspondence between elements of one set and those of another set. More precisely, a function is a particular set of ordered pairs.

Definition 1.5.1 *Let A and B be nonempty sets. A relation f from A into B is called a **function** (or **mapping**) from A into B if*
  (i) $\mathcal{D}(f) = A$ and
  (ii) for all $(x, y), (x', y') \in f$, $x = x'$ implies $y = y'$.
  *When (ii) is satisfied by a relation f, we say that f is **well defined** or **single-valued**.*

We use the notation $f : A \rightarrow B$ to denote a function f from a set A into a set B. For $(x, y) \in f$, we usually write $f(x) = y$ and say that y is the **image** of x under f and x is a **preimage** of y under f.

Leibniz seems to be the first to have used the word "function" to stand for any quantity related to a curve. Clairant (1734) originated the notation $f(x)$ and Euler made extensive use of it. Dirichlet is responsible for the current definition of a function.

Let us now explain the above definition. Suppose $f : A \to B$. Then f is a subset of $A \times B$ such that for all $x \in A$, there exists a unique $y \in B$ such that $(x, y) \in f$. Hence, we like to think of a function as a rule which associates to each element x of A exactly one element y of B. In order to show that a relation f from A into B is a function, we first show that the domain of f is A and next we show that f well defined or single-valued, i.e., if $x = y$ in A, then $f(x) = f(y)$ in B for all $x, y \in A$.

We now consider some examples of relations, some of which are functions and some of which are not.

Example 1.5.2 Let f be the subset of $\mathbf{Z} \times \mathbf{Z}$ defined by

$$f = \{(n, 2n + 3) \mid n \in \mathbf{Z}\}.$$

Then $\mathcal{D}(f) = \{n \mid n \in \mathbf{Z}\} = \mathbf{Z}$. We now show that f is well defined. Let $n, m \in \mathbf{Z}$. Suppose $n = m$. Then $2n + 3 = 2m + 3$, i.e., $f(n) = f(m)$. Therefore, f is well defined. Hence, f satisfies (i) and (ii) of Definition 1.5.1 and so f is a function.

Example 1.5.3 Let $A = \{1, 2, 3, 4\}$ and $B = \{a, b, c\}$. Let f be the subset of $A \times B$ defined by

$$f = \{(1, a), (2, b), (3, c), (4, b)\}.$$

First note that $\mathcal{D}(f) = \{1, 2, 3, 4\} = A$ and so f satisfies (i) of Definition 1.5.1. From the definition of f, it is immediate that for all $x \in A$, there exists a unique $y \in B$ such that $(x, y) \in f$. Therefore, f is well defined and so f satisfies (ii) of Definition 1.5.1. Hence, f is a function.

Example 1.5.4 Let f be the subset of $\mathbf{Q} \times \mathbf{Z}$ defined by

$$f = \{(\tfrac{p}{q}, p) \mid p, q \in \mathbf{Z}, \ q \neq 0\}.$$

First we note that $\mathcal{D}(f) = \{\tfrac{p}{q} \mid p, q \in \mathbf{Z}, \ q \neq 0\} = \mathbf{Q}$. Thus, f satisfies (i) of Definition 1.5.1. Now $(\tfrac{2}{3}, 2) \in f$, $(\tfrac{4}{6}, 4) \in f$ and $\tfrac{2}{3} = \tfrac{4}{6}$. But $f(\tfrac{2}{3}) = 2 \neq 4 = f(\tfrac{4}{6})$. Thus, f is not well defined. Hence, f is not a function from $\mathbf{Q}$ into $\mathbf{Z}$.

Example 1.5.5 Let f be the subset of $\mathbf{Z} \times \mathbf{Z}$ defined by

$$f = \{(mn, m + n) \mid m, n \in \mathbf{Z}\}.$$

First we show that f satisfies (i) of Definition 1.5.1. Let x be any element of $\mathbf{Z}$. Then we can write $x = x \cdot 1$. Hence, $(x, x + 1) = (x \cdot 1, x + 1) \in f$. This implies that $x \in \mathcal{D}(f)$. Thus, $\mathbf{Z} \subseteq \mathcal{D}(f)$. However, $\mathcal{D}(f) \subseteq \mathbf{Z}$ and so $\mathcal{D}(f) = \mathbf{Z}$. Thus, f satisfies (i) of Definition 1.5.1. Now $4 \in \mathbf{Z}$ and $4 = 4 \cdot 1 = 2 \cdot 2$. Thus, $(4 \cdot 1, 4 + 1) \in f$ and $(2 \cdot 2, 2 + 2) \in f$. Hence, we find that $4 \cdot 1 = 2 \cdot 2$ and $f(4 \cdot 1) = 5 \neq 4 = f(2 \cdot 2)$. This implies that f is not well defined, i.e., f does not satisfy (ii) of Definition 1.5.1. Hence, f is not a function from $\mathbf{Z}$ into $\mathbf{Z}$.

We now explore the meaning of equality of two functions.

Let $f : A \to B$ and $g : A \to B$ be two functions. Then f and g are subsets of $A \times B$. Suppose $f = g$. Let x be any element of A. Then $(x, f(x)) \in f = g$. Also, $(x, g(x)) \in g$. Since g is a function and $(x, f(x)), (x, g(x)) \in g$, we must have $g(x) = f(x)$. Conversely, assume that $g(x) = f(x)$ for all $x \in A$. Let $(x, y) \in f$. Then $y = f(x) = g(x)$. Thus, $(x, y) \in g$. This implies that $f \subseteq g$. Similarly, we can show that $g \subseteq f$. It now follows that $f = g$. Thus, two functions $f : A \to B$ and $g : A \to B$ are **equal** if and only if $f(x) = g(x)$ for all $x \in A$.

Example 1.5.6 *Let $f : \mathbf{Z} \to \mathbf{Z}^{\#}$ and $g : \mathbf{Z} \to \mathbf{Z}^{\#}$ be defined by $f = \{(n, n^2) \mid n \in \mathbf{Z}\}$ and $g = \{(n, |n|^2) \mid n \in \mathbf{Z}\}$. Now for all $n \in \mathbf{Z}$,*

$$f(n) = n^2 = |n|^2 = g(n).$$

Hence, $f = g$.

Definition 1.5.7 *Let f be a function from a set A into a set B. Then*
*(i) f is called **one-one** if for all $x, x' \in A$, $f(x) = f(x')$ implies $x = x'$.*
*(ii) f is called **onto** B (or f **maps** A **onto** B) if $\mathcal{I}(f) = B$.*

We note that if $f : A \to B$, then $\mathcal{I}(f) = B$ if and only if for all $y \in B$, there exists $x \in A$ such that $f(x) = y$. In other words, $\mathcal{I}(f) = B$ if and only if every element of B has a preimage. We also note that f is one-one if and only if every element of B has at most one preimage.

Let A be a nonempty set. The function $i_A : A \to A$ defined by $i_A(x) = x$ for all $x \in A$ is a one-one function of A onto A. i_A is called the **identity map** on A.

Example 1.5.8 *Consider the relation f from $\mathbf{Z}$ into $\mathbf{Z}$ defined by*

$$f(n) = n^2$$

for all $n \in \mathbf{Z}$. Now $\mathcal{D}(f) = \mathbf{Z}$. Also, if $n = n'$, then $n^2 = (n')^2$, i.e., $f(n) = f(n')$. Hence, f is well defined. Thus, f is a function. Now $f(1) = 1 = f(-1)$ and $1 \neq -1$. This implies that f is not one-one. Now for all $n \in \mathbf{Z}$, $f(n)$ is a nonnegative integer. This shows that a negative integer has no preimage. Hence, f is not onto $\mathbf{Z}$. Note that f is onto $\{0, 1, 4, 9, \ldots\}$.

Example 1.5.9 *Consider the relation f from $\mathbf{Z}$ into $\mathbf{Z}$ defined by for all $n \in \mathbf{Z}$, $f(n) = 2n$. As in the previous examples, we can show that f is a function. Let $n, n' \in \mathbf{Z}$ and suppose that $f(n) = f(n')$. Then $2n = 2n'$, i.e., $n = n'$. Hence, f is a one-one function. Since for all $n \in \mathbf{Z}$, $f(n)$ is an even integer, we see that an odd integer has no preimage. Thus, f is not onto $\mathbf{Z}$. However, we note that f is onto $\mathbf{E}$.*

Definition 1.5.10 *Let A, B, and C be nonempty sets and $f : A \to B$ and $g : B \to C$. The **composition** $\circ$ of f and g, written $g \circ f$, is the relation from A into C defined as follows:*

$$g \circ f \;=\; \{(x, z) \mid x \in A, z \in C, \text{ there exists } y \in B \\ \text{ such that } f(x) = y \text{ and } g(y) = z\}.$$

Let $f : A \to B$ and $g : B \to C$ and $(x, z) \in g \circ f$, i.e., $(g \circ f)(x) = z$. Then by the definition of composition of functions, there exists $y \in B$ such that $f(x) = y$ and $g(y) = z$. Now

$$z = g(y) = g(f(x)).$$

Hence, $(g \circ f)(x) = g(f(x))$.

In the following, we describe some properties of composition of functions.

Theorem 1.5.11 *Suppose that $f : A \to B$ and $g : B \to C$. Then*
(i) $g \circ f : A \to C$, i.e., $g \circ f$ is a function from A into C.
(ii) If f and g are one-one, then $g \circ f$ is one-one.
(iii) If f is onto B and g is onto C, then $g \circ f$ is onto C.

Proof.   (i) Let $x \in A$. Since f is a function and $x \in A$, there exists $y \in B$ such that $f(x) = y$. Now since g is a function and $y \in B$, there exists $z \in C$ such that $g(y) = z$. Thus, $(g \circ f)(x) = g(f(x)) = g(y) = z$, i.e., $(x, z) \in g \circ f$. Hence, $x \in \mathcal{D}(g \circ f)$. This shows that $A \subseteq \mathcal{D}(g \circ f)$. But $\mathcal{D}(g \circ f) \subseteq A$ and so $\mathcal{D}(g \circ f) = A$. Next, we show that $g \circ f$ is well defined.
Suppose that $(x, z) \in g \circ f$, $(x_1, z_1) \in g \circ f$ and $x = x_1$, where $x, x_1 \in A$ and $z, z_1 \in C$. By the definition of composition of functions, there exist $y, y_1 \in B$ such that $f(x) = y$, $g(y) = z$, $f(x_1) = y_1$ and $g(y_1) = z_1$. Since f is a function and $x = x_1$, we have $y = y_1$. Similarly, since g is a function and $y = y_1$, we have $z = z_1$. Thus, $g \circ f$ is well defined. Hence, $g \circ f$ is a function from A into C.
(ii) Let $x, x' \in A$. Suppose $(g \circ f)(x) = (g \circ f)(x')$. Then $g(f(x)) = g(f(x'))$. Since g is one-one, $f(x) = f(x')$. Since f is one-one, $x = x'$. Thus, $g \circ f$ is one-one.
(iii) Let $z \in C$. Then there exists $y \in B$ such that $g(y) = z$ since g is onto C. Since f is onto B, there exists $x \in A$ such that $f(x) = y$. Thus, $(g \circ f)(x) = g(f(x)) = g(y) = z$. Hence, $g \circ f$ is onto C. ∎

Example 1.5.12 *Consider the function $f : \mathbf{Z} \to \mathbf{Z}$ and $g : \mathbf{Z} \to \mathbf{E}$, where $f(n) = n^2$ and $g(n) = 2n$ for all $n \in \mathbf{Z}$. Then $g \circ f : \mathbf{Z} \to \mathbf{E}$ and $(g \circ f)(n) = g(f(n)) = g(n^2) = 2n^2$.*

Theorem 1.5.13 *Let $f : A \to B$, $g : B \to C$, and $h : C \to D$. Then*

$$h \circ (g \circ f) = (h \circ g) \circ f.$$

That is, composition of functions is associative.

Proof.   First note that $h \circ (g \circ f) : A \to D$ and $(h \circ g) \circ f : A \to D$. Let $x \in A$. Then

$$[h \circ (g \circ f)](x) = h((g \circ f)(x)) = h(g(f(x))) = (h \circ g)(f(x)) = [(h \circ g) \circ f](x).$$

Thus, by the equality of two functions, $h \circ (g \circ f) = (h \circ g) \circ f$. ■

Let A be a set and $f : A \to A$. Recursively, we define

$$\begin{aligned} f^1(x) &= f(x) \\ f^{n+1}(x) &= (f \circ f^n)(x) \end{aligned}$$

for all $x \in A$, $n \in \mathbf{N}$.

Let A and B be sets. A and B are said to be **equipollent**, written $A \sim B$, if there exists a one-one function from A onto B, i.e., the elements of A and B are in **one-one correspondence**.

From Theorem 1.5.11, it follows that $\sim$ is an equivalence relation. If $A \sim B$, then sometimes we write $|A| = |B|$. It is immediate that if A and B are finite sets, then $|A| = |B|$ if and only if A and B have the same number of elements.

The following lemma, which follows from Theorem 1.5.11(ii), is of independent interest. We give a direct proof of this result.

Lemma 1.5.14 *Let A be a set and $f : A \to A$ be a one-one function. Then $f^n : A \to A$ is a one-one function for all integers $n \geq 1$.*

Proof.   Suppose there exists $n > 1$ such that f^n is not one-one. Let $m > 1$ be the smallest positive integer such that f^m is not one-one. Then there exist x, $y \in A$ such that $x \neq y$ and $f^m(x) = f^m(y)$. But then $f(f^{m-1}(x)) = f(f^{m-1}(y))$ and hence $f^{m-1}(x) = f^{m-1}(y)$ since f is one-one. Now since m is the smallest positive integer such that f^m is not one-one, f^{m-1} is one-one. Hence, $x = y$, which is a contradiction. Thus, f^n is one-one for all $n \geq 1$. ■

That one-one functions on a finite set are onto is proved next.

Theorem 1.5.15 *Let A be a finite set. If $f : A \to A$ is one-one, then f is onto A.*

Proof.   Let $y \in A$. Now $f^n(y) \in A$ for all $n \geq 1$. Hence,

$$\{y, f(y), f^2(y), \ldots\} \subseteq A.$$

Since A is finite, all elements of the set $\{y, f(y), f^2(y), \ldots\}$ cannot be distinct. Thus, there exist positive integers s and t such that $s > t$ and $f^s(y) = f^t(y)$. Then $f^t(f^{s-t}(y)) = f^t(y)$. Hence, $f^{s-t}(y) = y$ since by Lemma 1.5.14, f^t is one-one. Let $x = f^{s-t-1}(y) \in A$. Then $f(x) = y$. Hence, f is onto A. ■

Definition 1.5.16 *Let A and B be sets and $f : A \to B$.*
    *(i) f is called **left invertible** if there exists $g : B \to A$ such that*

$$g \circ f = i_A.$$

    *(ii) f is called **right invertible** if there exists $h : B \to A$ such that*

$$f \circ h = i_B.$$

    A function $f : A \to B$ is called **invertible** if f is both left and right invertible.

Example 1.5.17 *Let $f : \mathbf{Z} \to \mathbf{Z}$ and $g : \mathbf{Z} \to \mathbf{Z}$ be as defined below.*

$$f(n) = 3n$$

$$g(n) = \begin{cases} \frac{n}{3} & \text{if } n \text{ is a multiple of } 3 \\ 0 & \text{if } n \text{ is not a multiple of } 3 \end{cases}$$

for all $n \in \mathbf{Z}$. Now

$$\begin{aligned} (f \circ g)(n) &= f(g(n)) \\ &= \begin{cases} n & \text{if } n \text{ is a multiple of } 3 \\ 0 & \text{if } n \text{ is not a multiple of } 3. \end{cases} \end{aligned}$$

Hence, $f \circ g \neq i_{\mathbf{Z}}$. But $(g \circ f)(n) = g(f(n)) = g(3n) = n$ for all $n \in \mathbf{Z}$. Thus, $g \circ f = i_{\mathbf{Z}}$. Hence, g is a left inverse of f.

    Often we are required to find a left (right) inverse of a function. However, not every function has a left (right) inverse. Thus, before we attempt to find a left (right) inverse of a function, it would be helpful to know if a given function has a left (right) inverse or not. The following theorem is very useful in determining whether a function is left (right) invertible or invertible.

Theorem 1.5.18 *Let A and B be sets and $f : A \to B$. Then the following assertions hold.*
    (i) f is one-one if and only if f is left invertible.
    (ii) f is onto B if and only if f is right invertible.
    (iii) f is one-one and onto B if and only if f is invertible.

Proof. (i) Suppose f is left invertible. Then there exists $g : B \to A$ such that $g \circ f = i_A$. Let $x, y \in A$ be such that $f(x) = f(y)$. Then $g(f(x)) = g(f(y))$ or $(g \circ f)(x) = (g \circ f)(y)$. Hence, $i_A(x) = i_A(y)$, i.e., $x = y$. Thus, f is one-one.

Conversely, suppose f is one-one. Then for $y \in B$, either y has no preimage or there exists a unique $x_y \in A$ such that $f(x_y) = y$. Fix $x \in A$. Define $g : B \to A$ by

$$g(y) = \begin{cases} x & \text{if } y \text{ has no preimage under } f \\ x_y & \text{if } y \text{ has a preimage under } f \text{ and } f(x_y) = y \end{cases}$$

for all $y \in B$. By the definition of g, $\mathcal{D}(g) = B$. To show g is well defined, suppose $y, y' \in B$ and $y = y'$. Then either both y and y' have no preimages or there exist unique $x_y, x_{y'} \in A$ such that $f(x_y) = y$ and $f(x_{y'}) = y'$. Suppose both y and y' have no preimages. Then $g(y) = x = g(y')$. Now suppose there exist unique $x_y, x_{y'} \in A$ such that $f(x_y) = y$ and $f(x_{y'}) = y'$. Thus, $g(y) = x_y$ and $g(y') = x'_y$. Since $y = y'$, we have $f(x_y) = f(x_{y'})$. Since f is one-one, $x_y = x_{y'}$ and so $g(y) = g(y')$. We have thus shown that g is well defined and so g is a function. We now show that $g \circ f = i_A$. Let $u \in A$ and suppose $f(u) = v$ for some $v \in B$. Then by the definition of g, $g(v) = u$. Thus,

$$(g \circ f)(u) = g(f(u)) = g(v) = u = i_A(u).$$

Hence, $g \circ f = i_A$.

(ii) Suppose f is right invertible. Then there exists $g : B \to A$ such that $f \circ g = i_B$. Let $y \in B$. Let $x = g(y) \in A$. Now $y = i_B(y) = (f \circ g)(y) = f(g(y)) = f(x)$. Hence, f is onto B.

Conversely, suppose f is onto B. Let $y \in B$. Since f is onto, there exists $x \in A$ such that $f(x) = y$. Let $A_y = \{x \in A \mid f(x) = y\}$. Then $A_y \neq \phi$. Choose $x_y \in A_y$ for all $y \in B$. Define $h : B \to A$ such that $h(y) = x_y$ for all $y \in B$. Then h is a function. Let $y \in B$. Then $(f \circ h)(y) = f(h(y)) = f(x_y) = y = i_B(y)$. Hence, $f \circ h = i_B$ and so f is right invertible.

(iii) The result here follows from (i) and (ii). ∎

Let $f : A \to B$ be invertible. Let g be a left inverse of f and h be a right inverse of f. Then $g \circ f = i_A$ and $f \circ h = i_B$. Now $g = g \circ i_B = g \circ (f \circ h) = (g \circ f) \circ h = i_A \circ h = h$. Thus, if f is invertible, then left and right inverses of f are the same. This also proves that the inverse of a function, if it exists, is unique.

If f is an invertible function, then the inverse of f is denoted by f^{-1}.

Let $f : A \to B$ and $A' \subseteq A$. Then f induces a function from A' into B in a natural way as defined next.

Definition 1.5.19 *Let $f : A \to B$ and A' be a nonempty subset of A. The* **restriction of f to A'**, *written $f|_{A'}$, is defined to be*

$$f|_{A'} = \{(x', f(x')) \mid x' \in A'\}.$$

We see that $f|_{A'}$ is really the function f except that we are considering f on a smaller domain.

Definition 1.5.20 *Let $f : A' \to B$ and A be a set containing A'. A function $g : A \to B$ is called an **extension** of f to A if $g|_{A'} = f$.*

Example 1.5.21 *Consider the function $f : \mathbf{E} \to \mathbf{Z}$ and $g : \mathbf{Z} \to \mathbf{Z}$, where $f(2n) = 2n + 1$ and $g(n) = n+1$ for all $n \in \mathbf{Z}$. Then g is an extension of f to $\mathbf{Z}$ and f is the restriction of g to $\mathbf{E}$. Let the function $h : \mathbf{Z} \to \mathbf{Z}$ be defined by for all $m \in \mathbf{Z}$, $h(m) = m + 1$ if $m \in \mathbf{E}$ and $h(m) = m$ if $m \notin \mathbf{E}$. Then h is an extension of f to $\mathbf{Z}$. However, $h \neq g$. Thus, a function may have more than one extension.*

In Section 1.1, we defined the Cartesian cross product, $A \times B$, of two sets A and B. We now extend this notion to a family of sets $\{A_\alpha \mid \alpha \in I\}$, where I is an index set. First let us make the following observation: Suppose $I = \{1, 2\}$. Let S be the set of all functions $f : I \to A \cup B$ such that $f(1) \in A$ and $f(2) \in B$. Then every function $f \in S$ defines an ordered pair $(f(1), f(2)) \in A \times B$. Conversely, given $x \in A$ and $y \in B$, define $f \in S$ by $f(1) = x$ and $f(2) = y$. Then the ordered pair (x, y) defines a function $f \in S$. Hence, there is a one-one correspondence between the elements of S and $A \times B$. We now define the Cartesian product of $\{A_\alpha \mid \alpha \in I\}$.

Let $\{A_\alpha \mid \alpha \in I\}$ be a family of sets. The **Cartesian (cross) product** of $\{A_\alpha \mid \alpha \in I\}$, denoted by $\prod_{\alpha \in I} A_\alpha$, is defined to be the set

$$\{f \mid f : I \to \cup_{\alpha \in I} A_\alpha \text{ and } f(\alpha) \in A_\alpha \text{ for all } \alpha \in I\}.$$

Let $f \in \prod_{\alpha \in I} A_\alpha$. Then $f(\alpha) \in A_\alpha$ for all $\alpha \in I$. Let us write $f(\alpha) = x_\alpha$ for all $\alpha \in I$. We usually write $(x_\alpha)_{\alpha \in I}$ for f, i.e., a typical member of $\prod_{\alpha \in I} A_\alpha$ is denoted by $(x_\alpha)_{\alpha \in I}$, where $x_\alpha \in A_\alpha$ for all $\alpha \in I$.

Suppose $I = \{1, 2, \ldots, n\}$ is a finite set. Then the Cartesian product $\prod_{i \in I} A_\alpha$, is denoted by $A_1 \times A_2 \times \cdots \times A_n$. A typical member of $A_1 \times A_2 \times \cdots \times A_n$ is denoted by $(x_1, x_2, \ldots, x_n)$, $x_i \in A_i$ for all $i = 1, 2, \ldots, n$. The elements of $A_1 \times A_2 \times \cdots \times A_n$ are called **ordered n-tuples**. For two elements $(x_1, x_2, \ldots, x_n), (y_1, y_2, \ldots, y_n) \in A_1 \times A_2 \times \cdots \times A_n$, $(x_1, x_2, \ldots, x_n) = (y_1, y_2, \ldots, y_n)$ if and only if $x_i = y_i$ for all i.

1.5.1   Worked-Out Exercises

◇ **Exercise 1** Determine which of the following mappings $f : \mathbf{R} \to \mathbf{R}$ are one-one and which are onto $\mathbf{R}$:

(i) $f(x) = x + 4$,

(ii) $f(x) = x^2$

for all $x \in \mathbf{R}$.

Solution: (i) Let $x, y \in \mathbf{R}$. Suppose $f(x) = f(y)$. Then $x + 4 = y + 4$ or $x = y$. Hence, f is one-one. Now f is onto $\mathbf{R}$ if and only if for all $y \in \mathbf{R}$ there exists $x \in \mathbf{R}$ such that $f(x) = y$. Let $y \in \mathbf{R}$. If $f(x) = y$, then $x + 4 = y$ or $x = y - 4$. Also, $y - 4 \in \mathbf{R}$. Thus, we can take x to be $y - 4$. Now $f(y - 4) = y - 4 + 4 = y$. Hence, f is onto $\mathbf{R}$.

(ii) We note that $f(x)$ is a nonnegative real number for all $x \in \mathbf{R}$. This means that negative real numbers have no preimages. In particular, for all $x \in \mathbf{R}$, $f(x) = x^2 \neq -1$. Hence, f is not onto $\mathbf{R}$. Also, $f(-1) = 1 = f(1)$ and $-1 \neq 1$. Thus, f is not one-one. Thus, f is neither one-one nor onto $\mathbf{R}$.

$\diamond$ **Exercise 2** (i) Let $f : \mathbf{Z} \to \mathbf{Z}$ be a mapping defined by

$$f(x) = \begin{cases} x \text{ if } x \text{ is even} \\ 2x + 1 \text{ if } x \text{ is odd} \end{cases}$$

for all $x \in \mathbf{Z}$. Find a left inverse of f if one exists.

(ii) Let $f : \mathbf{Z} \to \mathbf{Z}$ be the mapping defined by $f(x) = |x| + x$ for all $x \in \mathbf{Z}$. Find a right inverse of f if one exists.

Solution: (i) By Theorem 1.5.18, f has a left inverse if and only if f is one-one. Before we attempt to find a left inverse of f, let us first check whether f is one-one or not. Let $x, y \in \mathbf{R}$ and $f(x) = f(y)$. By the definition of f, $f(x)$ is even if x is even and $f(x)$ is odd if x is odd. Thus, since $f(x) = f(y)$, we have both x and y are either even or odd. If x and y are both even then $f(x) = x$ and $f(y) = y$ and so $x = y$. Suppose x and y are odd. Then $f(x) = 2x + 1$ and $f(y) = 2y + 1$. Then $2x + 1 = 2y + 1$ or $x = y$. Hence, f is one-one and so f has a left inverse. Thus, there exists a function $g : \mathbf{Z} \to \mathbf{Z}$ such that $g \circ f = i_{\mathbf{Z}}$. Let $x \in \mathbf{Z}$. Suppose x is even. Now $x = i_{\mathbf{Z}}(x) = (g \circ f)(x) = g(f(x)) = g(x)$. This means $g(x) = x$ when x is even. Now suppose x is odd. Then $x = i_{\mathbf{Z}}(x) = (g \circ f)(x) = g(f(x)) = g(2x + 1)$. Put $t = 2x + 1$. Then $x = \frac{t-1}{2}$. This shows that $g(x) = \frac{x-1}{2}$ if x is odd. Thus, our choice of g is

$$g(x) = \begin{cases} x & \text{if } x \text{ is even} \\ \frac{x-1}{2} & \text{if } x \text{ is odd.} \end{cases}$$

(ii) Note that $f(x) = |x| + x \geq 0$ for all $x \in \mathbf{Z}$. This shows that negative integers do not belong to $\mathcal{I}(f)$. In particular, $f(x) \neq -1$ for all $x \in \mathbf{Z}$. Thus, f is not onto $\mathbf{Z}$ and so f does not have a right inverse.

$\diamond$ **Exercise 3** Let X and Y be nonempty sets and $f : X \to Y$. If $T \subseteq X$, then $f(T)$ denotes the set $\{f(x) \mid x \in T\}$. $f(T)$ is called the **image** of T under f. Prove that f is one-one if and only if

$$f(A \cap B) = f(A) \cap f(B)$$

for all nonempty subsets A and B of X.

Solution: Suppose that f is one-one. Let A and B be nonempty subsets of X. Let $y \in f(A \cap B)$. Then $y = f(x)$ for some $x \in A \cap B$. Hence, $y \in f(A) \cap f(B)$. Thus, $f(A \cap B) \subseteq f(A) \cap f(B)$. Now let $y \in f(A) \cap f(B)$. Then $y \in f(A)$ and $y \in f(B)$. Thus, $y = f(a)$ for some $a \in A$ and $y = f(b)$ for some $b \in B$. Since f is one-one and $f(a) = f(b)$, we find that $a = b$. Thus, $y \in f(A \cap B)$. Hence, $f(A) \cap f(B) \subseteq f(A \cap B)$. Consequently, $f(A \cap B) = f(A) \cap f(B)$.

Conversely, suppose that $f(A \cap B) = f(A) \cap f(B)$ for all subsets A and B of X. Suppose f is not one-one. Then there exist $x, y \in X$ such that $f(x) = f(y)$ and $x \neq y$. Let $A = \{x\}$ and $B = \{y\}$. Since $A \cap B = \phi$, $f(A \cap B) = \phi$. However, $f(A) \cap f(B) = \{f(x)\} \neq \phi$. Thus, $f(A \cap B) \neq f(A) \cap f(B)$, a contradiction. Hence, f is one-one.

◇ **Exercise 4** Let A be a nonempty set and E be an equivalence relation on A. Let $B = \{[x] \mid x \in A\}$, i.e., B is the set of all equivalence classes with respect to E. Prove that there exists a function f from A onto B. The set B is usually denoted by A/E and is called the **quotient set** of A determined by E.

Solution: Define $f : A \to B$ by $f(x) = [x]$ for all $x \in A$. By the definition of f, $\mathcal{D}(f) = A$. Let $x, y \in A$. Suppose $x = y$. Then $[x] = [y]$ and so $f(x) = f(y)$. Thus, f is well defined. Let $[a] \in B$. Then $a \in A$ and $f(a) = [a]$. Hence, f is onto B.

Exercise 5 Let $S = \{x \in \mathbf{R} \mid -1 < x < 1\}$. Show that $\mathbf{R} \sim S$.

Solution: Define $f : \mathbf{R} \to S$ by

$$f(x) = \frac{x}{1 + |x|}$$

for all $x \in \mathbf{R}$. Let $x \in \mathbf{R}$. Then $-|x| \leq x \leq |x|$, $-1 - |x| < -|x|$, and $|x| \leq 1 + |x|$. Hence, $-1 - |x| < x < 1 + |x|$. Thus, $-1 < \frac{x}{1+|x|} < 1$ and so $-1 < f(x) < 1$. This shows that $f(x) \in S$. Let $x, y \in \mathbf{R}$ and $f(x) = f(y)$. Then $\frac{x}{1+|x|} = \frac{y}{1+|y|}$. Thus, $\frac{|x|}{1+|x|} = \frac{|y|}{1+|y|}$. This implies that $|x| + |x||y| = |y| + |x||y|$ and so $|x| = |y|$. Now $\frac{x}{1+|x|} = \frac{y}{1+|y|}$ implies that $x \geq 0$ if and only if $y \geq 0$. Therefore, since $|x| = |y|$, $x = y$. Thus, f is one-one.

Now let $z \in \mathbf{R}$ and $-1 < z < 1$. If $0 \leq z < 1$, then

$$f\left(\frac{z}{1-z}\right) = \frac{\frac{z}{1-z}}{1 + \left|\frac{z}{1-z}\right|} = \frac{\frac{z}{1-z}}{1 + \frac{z}{1-z}} = z.$$

If $-1 < z < 0$, then

$$f\left(\frac{z}{1+z}\right) = \frac{\frac{z}{1+z}}{1 + \left|\frac{z}{1+z}\right|} = \frac{\frac{z}{1+z}}{1 + \frac{-z}{1+z}} = z.$$

Hence, f is onto $\mathbf{R}$. Consequently, $\mathbf{R} \sim S$.

1.5.2 Exercises

1. Determine which of the following mappings $f : \mathbf{R} \to \mathbf{R}$ are one-one and which are onto $\mathbf{R}$:

   (i) $f(x) = x + 1$,

   (ii) $f(x) = x^3$,

   (iii) $f(x) = |x| + x$

   for all $x \in \mathbf{R}$.

2. Consider the function $f = \{(x, x^2) \mid x \in S\}$ of $S = \{-3, -2, -1, 0, 1, 2, 3\}$ into $\mathbf{Z}$. Is f one-one? Is f onto $\mathbf{Z}$?

3. Let $f : \mathbf{R}^+ \to \mathbf{R}^+$ and $g : \mathbf{R}^+ \to \mathbf{R}^+$ be functions defined by $f(x) = \sqrt{x}$ and $g(x) = 3x + 1$ for all $x \in \mathbf{R}^+$, where $\mathbf{R}^+$ is the set of all positive real numbers. Find $f \circ g$ and $g \circ f$. Is $f \circ g = g \circ f$?

4. Let $f : \mathbf{Q}^+ \to \mathbf{R}$ and $g : \mathbf{R} \to \mathbf{R}$ be defined by $f(x) = 1 + \frac{1}{x}$ for all $x \in \mathbf{Q}^+$ and $g(x) = x + 1$ for all $x \in \mathbf{R}$, where $\mathbf{Q}^+$ is the set of all positive rational numbers. Find $g \circ f$.

5. For each of the mappings $f : \mathbf{Z} \to \mathbf{Z}$ given below, find a left inverse of f whenever one exists.

   (i) $f(x) = x + 2$,

   (ii) $f(x) = 2x$,

   (iii) $f(x) = \begin{cases} \frac{x}{2} & \text{if } x \text{ is even} \\ 5 & \text{if } x \text{ is odd} \end{cases}$

   for all $x \in \mathbf{Z}$.

6. For each of the mappings $f : \mathbf{Z} \to \mathbf{Z}$ given below, find a right inverse of f whenever one exists.

   (i) $f(x) = x - 3$,

   (ii) $f(x) = 2x$,

   (iii) $f(x) = \begin{cases} x & \text{if } x \text{ is even} \\ x + 1 & \text{if } x \text{ is odd} \end{cases}$

   for all $x \in \mathbf{Z}$.

7. Let $A = \{1, 2, 3\}$. List all one-one functions from A onto A.

8. Let $A = \{1, 2, \ldots, n\}$. Show that the number of one-one functions of A onto A is $n!$

9. Let $f : A \to B$ be a function. Define a relation R on A by for all $a, b \in A$, aRb if and only if $f(a) = f(b)$. Show that R is an equivalence relation.

10. Given $f : X \to Y$ and $A, B \subseteq X$, prove that

    (i) $f(A \cup B) = f(A) \cup f(B)$,

    (ii) $f(A \cap B) \subseteq f(A) \cap f(B)$,

    (iii) $f(A \backslash B) \subseteq f(A) \backslash f(B)$ if f is one-one.

11. Given $f : X \to Y$. Let $S \subseteq Y$. Define $f^{-1}(S) = \{x \in X \mid f(x) \in S\}$. Let $A, B \subseteq Y$. Prove that

    (i) $f^{-1}(A \cup B) = f^{-1}(A) \cup f^{-1}(B)$,

    (ii) $f^{-1}(A \cap B) = f^{-1}(A) \cap f^{-1}(B)$,

    (iii) $f^{-1}(A \backslash B) = f^{-1}(A) \backslash f^{-1}(B)$.

12. Let $f : A \to B$. Let f^* be the inverse relation, i.e.,

$$f^* = \{(y, x) \in B \times A \mid f(x) = y\}.$$

    (i) Show by an example that f^* need not be a function.

    (ii) Show that f^* is a function from $\mathcal{I}(f)$ into A if and only if f is one-one.

    (iii) Show that f^* is a function from B into A if and only if f is one-one and onto B.

    (iv) Show that if f^* is a function from B into A, then $f^{-1} = f^*$.

13. Show that $\mathbf{Z} \sim \mathbf{E}$, where $\mathbf{E}$ is the set of all even integers.

14. Let $A = \{x \in \mathbf{R} \mid 0 \le x \le 1\}$ and $B = \{x \in \mathbf{R} \mid 5 \le x \le 8\}$. Show that $f : A \to B$ defined by $f(x) = 5 + (8 - 5)x$ is a one-one function from A onto B.

15. (i) Show that $\mathbf{Z}$ and $3\mathbf{Z}$ are equipollent.

    (ii) Show that $5\mathbf{Z}$ and $7\mathbf{Z}$ are equipollent.

16. Let $S = \{x \in \mathbf{R} \mid 0 < x < 1\}$. Show that $\mathbf{R}^+ \sim S$.

17. (**Schröder-Bernstein**) Let A and B be sets. If $A \sim Y$ for some subset Y of B and $B \sim X$ for some subset X of A, prove that $A \sim B$.

18. Find a one-one mapping from $\mathbf{R}$ onto $\mathbf{R}^+$.

19. Is $\mathbf{Z} \sim \mathbf{Q}$?

20. Let $A = \{x \in \mathbf{R} \mid 0 \le x \le 1\}$ and $B = \{x \in \mathbf{R} \mid 0 < x < 1\}$. Is it true that $A \sim B$?

21. For each of the following statements, write the proof if the statement is true, otherwise give a counterexample.

(i) A function $f : A \to B$ is one-one if and only if $g \circ f = h \circ f$ for all functions $g, h : B \to A$.

(ii) A function $f : A \to B$ is one-one if and only if for all subsets C of A, $f(A \backslash C) \supseteq B \backslash f(C)$.

1.6   Binary Operations

The concept of a binary operation is very important in abstract algebra. Throughout the text we will be concerned with sets together with one or more binary operations. In this section, we define binary operations and examine their basic properties.

Definition 1.6.1 *Let S be a nonempty set. A **binary operation** on S is a function from $S \times S$ into S.*

For any ordered pair (x, y) of elements $x, y \in S$, a binary operation assigns a third member of S. For example, $+$ is a binary operation on $\mathbf{Z}$ which assigns 3 to the pair $(2, 1)$.

If $*$ is a binary operation on S, we write $x * y$ for $*(x, y)$, where $x, y \in S$. Since the image of $*$ is a subset of S, we say S is **closed under** $*$.

$\mathbf{Z}$ is closed under $+$ since if we add two integers we obtain an integer. Since $2, 5 \in \mathbf{N}$ and $2 - 5 = -3 \notin \mathbf{N}$, we see that $-$ (subtraction) is not a binary operation of $\mathbf{N}$ and we say that $\mathbf{N}$ is not closed under $-$.

Definition 1.6.2 *A **mathematical system** is an ordered $(n + 1)$-tuple $(S, *_1, \ldots, *_n)$, where S is a nonempty set and $*_i$ is a binary operation on S, $i = 1, 2, \ldots, n$. S is called the underlying set of the system.*

Definition 1.6.3 *Let $(S, *)$ be a mathematical system. Then*
*(i) $*$ is called **associative** if for all $x, y, z \in S$, $x * (y * z) = (x * y) * z$.*
*(ii) $*$ is called **commutative** if for all $x, y \in S$, $x * y = y * x$.*

Example 1.6.4 *Consider the mathematical system $(\mathbf{Z}, +)$. Since addition of integers is both associative and commutative, $+$ is both associative and commutative.*

Example 1.6.5 *Let A be a nonempty set. Let S be the set of all functions on A, i.e.,*

$$S = \{f \mid f : A \to A\}.$$

Since composition of functions is a function (Theorem 1.5.11), $(S, \circ)$ is a mathematical system. By Theorem 1.5.13, $\circ$ is associative.

Example 1.6.6 *Let $M_2(\mathbf{R})$ be the set of all 2×2 matrices over $\mathbf{R}$, i.e.,*

$$M_2(\mathbf{R}) = \left\{ \begin{bmatrix} a & b \\ c & d \end{bmatrix} \mid a, b, c, d \in \mathbf{R} \right\}.$$

Let $+$ denote the usual addition of matrices and $\cdot$ denote the usual multiplication of matrices. Since addition (multiplication) of 2×2 matrices over $\mathbf{R}$ is a 2×2 matrix over $\mathbf{R}$, it follows that $+$ $(\cdot)$ is a binary operation on $M_2(\mathbf{R})$. Hence, $(M_2(\mathbf{R}), +, \cdot)$ is a mathematical system. Note that $+$ is both associative and commutative and $\cdot$ is associative, but not commutative.

The following is an example of a mathematical system for which the binary operation is neither associative nor commutative.

Example 1.6.7 *Consider the mathematical system $(\mathbf{Z}, -)$, where $-$ denotes the binary operation of subtraction on $\mathbf{Z}$. Then $3 - (2 - 1) = 2 \neq 0 = (3 - 2) - 1$ and so $-$ is not associative. Also, since $3 - 2 \neq 2 - 3$, $-$ is not commutative.*

A convenient way to define a binary operation on a finite set S is by means of an operation or multiplication table. For example, let $S = \{a, b, c\}$. Define $*$ on S by the following operation table.

$*$	a	b	c
a	c	b	a
b	a	a	a
c	b	b	b

To determine the element of S assigned to $a * b$, we look at the intersection of the row labeled by a and the column headed by b. We see that $a * b = b$. Note that $b * a = a$.

Definition 1.6.8 *Let $(S, *)$ be a mathematical system. An element $e \in S$ is called an **identity** of $(S, *)$ if for all $x \in S$,*

$$e * x = x = x * e.$$

Example 1.6.9 *Let $S = \{e, a, b\}$. Define $*$ on S by the following multiplication table*

$*$	e	a	b
e	e	a	b
a	a	a	a
b	b	a	a

We note that $e * a = a = a * e$, $e * b = b = b * e$ and $e * e = e = e * e$. Thus, e is an identity of $(S, *)$.

Example 1.6.10 *(i) In Example 1.6.5, i_A is an identity element of $(S, \circ)$.*

(ii) In Example 1.6.6, $\begin{bmatrix} 0 & 0 \\ 0 & 0 \end{bmatrix}$ *is an identity element for the mathematical system $(M_2(\mathbf{R}), +)$ and* $\begin{bmatrix} 1 & 0 \\ 0 & 1 \end{bmatrix}$ *is an identity element for the mathematical system $(M_2(\mathbf{R}), \cdot)$.*

Theorem 1.6.11 *An identity element (if it exists) of a mathematical system $(S, *)$ is unique.*

Proof.   Let e, f be identities of $(S, *)$. Since e is identity, $e * a = e$ for all $a \in S$. Substituting f for a, we get

$$e * f = e. \tag{1.4}$$

Now f is identity and so $a * f = f$ for all $a \in S$. Substituting e for a we get

$$e * f = f. \tag{1.5}$$

From Eqs. (1.4) and (1.5), we get $e = f$. Hence, an identity element (if it exists) is unique. ■

1.6.1   Worked-Out Exercises

◇ **Exercise 1** Which of the following are associative binary operations?

  (i) $(\mathbf{Z}, *)$, where $x * y = (x + y) - (x \cdot y)$ for all $x, y \in \mathbf{Z}$.

  (ii) $(\mathbf{R}, *)$, where $x * y = \max(x, y)$ for all $x, y \in \mathbf{R}$.

  (iii) $(\mathbf{R}, *)$, where $x * y = |x + y|$ for all $x, y \in \mathbf{R}$.

Solution: (i) $(x * y) * z = ((x + y) - (x \cdot y)) * z = (x + y) - (x \cdot y) + z - ((x + y) - (x \cdot y)) \cdot z = x + y + z - x \cdot y - x \cdot z - y \cdot z + x \cdot y \cdot z$. Similarly, $x * (y * z) = x + y + z - x \cdot y - x \cdot z - y \cdot z + x \cdot y \cdot z$. Thus, $(x * y) * z = x * (y * z)$. Hence, $*$ is associative.

  (ii) $(x * y) * z = \max(x, y) * z = \max(\max(x, y), z) = \max(x, y, z) = \max(x, \max(y, z)) = x * \max(y, z) = x * (y * z)$. Thus, $*$ is associative.

  (iii) $(2 * (-3)) * 6 = |2 + (-3)| * 6 = 1 * 6 = |1 + 6| = 7$ and $2 * ((-3) * 6) = 2 * (|(-3) + 6|) = 2 * 3 = |2 + 3| = 5$. Hence, $(2 * (-3)) * 6 \neq 2 * ((-3) * 6)$ and so $*$ is not associative.

1.6.2 Exercises

1. Which of the following are associative binary operations?

(i) $(\mathbf{N}, *)$, where $x * y = x^y$ for all $x, y \in \mathbf{N}$.

(ii) $(\mathbf{Z}, *)$, where $x * y = x + y + 1$ for all $x, y \in \mathbf{Z}$.

(iii) $(\mathbf{N}, *)$, where $x * y = \gcd(x, y)$ for all $x, y \in \mathbf{N}$.

(iv) $(\mathbf{N}, *)$, where $x * y = \operatorname{lcm}(x, y)$ for all $x, y \in \mathbf{N}$.

(v) $(\mathbf{R}, *)$, where $x * y = \min(x, y)$ for all $x, y \in \mathbf{R}$.

(vi) $(\mathbf{R}, *)$, where $x * y = |x| + |y|$ for all $x, y \in \mathbf{R}$.

2. In Exercise 1, which of the operations are commutative?

3. In Exercise 1, which mathematical systems have an identity?

Carl Friedrich Gauss (1777–1855) was born on April 30, 1777, in Brunswick, Germany. Gauss is considered to be one of the last mathematicians to know everything in his subject.

Gauss's genius was revealed at a very early age. He was able to do long calculations in his head. He rediscovered the law of quadratic reciprocity, related the arithmetic-geometric mean to infinite series expansion, and conjectured the prime number theorem. Before the age of twenty, he showed that a regular polygon of seventeen sides was constructible with ruler and compass—an unsolved problem since Greek times. At the age of twenty, he published the first proof of the fundamental theorem of algebra. He completed his Ph.D. at the University of Helmstedt, under the supervision of Pfaff, when he was twenty-two .

In 1801, Gauss published his monumental book on number theory, *Disquisitiones Arithmeticae*. In his *Disquisitiones*, Gauss summarized previous work in a systematic way and solved some of the most difficult outstanding questions. He introduced the notion of congruence of integers modulo an integer ($a \equiv b \mod(c)$) and extensively studied $\mathbf{Z}_n$ and obtained many of its important properties. He is credited for coining the term **complex number** and the notation i for $\sqrt{-1}$. He showed that $\mathbf{Z}[i]$ is a unique factorization domain. In his honor, $\mathbf{Z}[i]$ is called the ring of Gaussian integers. *Disquisitiones* laid the foundations of algebraic number theory. Leopold Kronecker said, "It is really astonishing to think a single man of such young years was able to bring to light such a wealth of results, and above all, to present such a profound and well-organized treatment of an entirely new discipline."

Besides being a mathematician he was also a physicist and an astronomer. In January 1801, a new planet was briefly observed, which the astronomers were unable to locate later. Gauss calculated the position of the planet by using a more accurate orbit theory than the usual circular approximation. Gauss used a theory based on the ellipse. At the end of the year the planet was discovered at the precise location he predicted. The methods he developed are still in use. They include the theory of least squares.

He was appointed director of the observatory at Göttingen and remained there for forty years. Gauss disliked teaching and preferred his job at the observatory. He usually rejected students who sought his guidance. However, he did accept students such as Dedekind, Dirichlet, Eisenstein, Riemann, and Kummer, who themselves became famous mathematicians. Gauss died on February 23, 1855. As E.T. Bell has said,"He lives everywhere in mathematics."

Chapter 2

Introduction to Groups

There are four major sources from which group theory evolved, namely, classical algebra, number theory, geometry, and analysis. Classical algebra originated in 1770 with J.L. Lagrange's work on polynomial equations. His work appeared in a memoir entitled, "Réflexions sur la résolution algébrique des équations." C.F. Gauss is considered the originator of number theory with his work, "*Disquistiones Arithmeticae*," which was published in 1801. F. Klein's lecture in 1872, "A Comparative Review of Recent Researches in Geometry," dealt with the classification of geometry as the study of invariants under groups of transformations. The impact of his lecture was so strong as to allow Klein to be considered as the originator of this source of group theory. The originators of the analysis source are S. Lie (1874) and H. Poincaré and F. Klein (1876).

2.1   Elementary Properties of Groups

In this chapter, and in fact in the remainder of the text, we will be concerned with mathematical systems. These systems are composed of a nonempty set together with binary operations defined on this set so that certain properties hold. From these properties, results concerning these systems are derived. This axiomatic approach to abstract algebra unifies diverse examples and also strips away nonessential ideas.

Although noted for his geometry, Euclid inspired the use of the axiomatic method, which has proved so indispensable in mathematics. His axiomatic approach also affected philosophy, where in the 17th century Baruch Spinoza laid down (in The *Ethics*) an axiomatic system from which he was able to prove the existence of God. His proof, of course, depended on his axioms. His proof lost its conviction with the emergence of noneuclidean geometries whose axioms were as logical and practical as Euclid's.

We will be primarily concerned with mathematical systems called groups in this chapter. The theory of groups is one of the oldest branches of abstract algebra. The first effective use of groups was in the early nineteenth century by A. Cauchy and E. Galois. They used groups to describe the effect of permutations of roots of a polynomial equation. Their use of groups was not based on an axiomatic approach. In 1854, A. Cayley gave the first postulates for a group. However, his definition was lost sight of. Kronecker again set down the axioms for an Abelian group in 1870. H. Weber gave the definition for finite groups (in 1882) and the definition for infinite groups in 1883.

As previously mentioned, the notion of a group arose from the study of one-one functions on the set of roots of a polynomial equation. We have seen that the set S of all one-one functions from a set X onto itself satisfies the following properties:

(i) Composition of functions, $\circ$, is a binary operation on S.

(ii) For all $f, g, h \in S$, $f \circ (g \circ h) = (f \circ g) \circ h$.

(iii) There exists $i \in S$ such that $f \circ i = f = i \circ f$ for all $f \in S$.

(iv) For all $f \in S$ there exists an element $f^{-1} \in S$ such that $f \circ f^{-1} = i = f^{-1} \circ f$.

These properties lead us to the definition of an abstract group.

Definition 2.1.1 *A **group** is an ordered pair $(G, *)$, where G is a nonempty set and $*$ is a binary operation on G such that the following properties hold:*

*(G1) For all $a, b, c \in G$, $a * (b * c) = (a * b) * c$ (**associative law**).*

*(G2) There exists $e \in G$ such that for all $a \in G$, $a*e = a = e*a$ (**existence of an identity**).*

*(G3) For all $a \in G$, there exists $b \in G$ such that $a * b = e = b*a$ (**existence of an inverse**).*

Thus, a group is a mathematical system $(G, *)$ satisfying axioms G1 to G3.

In what follows, we will see several examples of groups. However, let us first observe the following important properties of groups.

Theorem 2.1.2 *Let $(G, *)$ be a group.*

(i) There exists a unique element $e \in G$ such that $e * a = a = a * e$ for all $a \in G$.

(ii) For all $a \in G$, there exists a unique $b \in G$ such that $a * b = e = b * a$.

Proof.   (i) By G2, there exists $e \in G$ such that $e * a = a = a * e$ for all $a \in G$. Since $(G, *)$ is a mathematical system, e is unique by Theorem 1.6.11.

(ii) Let $a \in G$. By G3, there exists $b \in G$ such that $a*b = e = b*a$. Suppose there exists $c \in G$ such that $a * c = e = c * a$. We show that $b = c$. Now

$$
\begin{aligned}
b &= b * e \\
&= b * (a * c) && \text{(substituting } e = a * c) \\
&= (b * a) * c && \text{(using the associativity of } *) \\
&= e * c && \text{(since } b * a = e) \\
&= c.
\end{aligned}
$$

Thus, b is unique. ∎

The unique element $e \in G$ that satisfies G2 is called the **identity** element of the group $(G, *)$. Let $a \in G$. Then the unique element $b \in G$ that satisfies G3 is called the **inverse** of a and is denoted by a^{-1}.

If a group $(G, *)$ has the property that $a * b = b * a$ for all $a, b \in G$, then $(G, *)$ is called a **commutative** or **Abelian** group. A group $(G, *)$ is called **noncommutative** if it is not commutative.

Example 2.1.3 *Consider $\mathbf{Z}$, the set of integers, together with the binary operation $+$, where $+$ is the usual addition. We know that $+$ is associative. Now $0 \in \mathbf{Z}$ and for all $a \in \mathbf{Z}$, $a + 0 = a = 0 + a$ and so 0 is the identity. Also, for all $a \in \mathbf{Z}$, $-a \in \mathbf{Z}$ and $a + (-a) = 0 = (-a) + a$. That is, $-a$ is the inverse of a. Hence, it now follows that $(\mathbf{Z}, +)$ is a group. Since $a + b = b + a$ for all $a, b \in \mathbf{Z}$, $+$ is commutative. Thus, $(\mathbf{Z}, +)$ is a commutative group.*

Similarly, we can show that $(\mathbf{Q}, +)$, $(\mathbf{R}, +)$, $(\mathbf{C}, +)$, $(\mathbf{Q}\backslash\{0\}, \cdot)$, $(\mathbf{R}\backslash\{0\}, \cdot)$, $(\mathbf{C}\backslash\{0\}, \cdot)$ are all examples of commutative groups, where $+$ is the usual addition and $\cdot$ is the usual multiplication. Note that for each of the groups $(\mathbf{Q}\backslash\{0\}, \cdot)$, $(\mathbf{R}\backslash\{0\}, \cdot)$, $(\mathbf{C}\backslash\{0\}, \cdot)$ the identity element is 1.

Example 2.1.4 *Let a be any fixed integer. Let $G = \{na \mid n \in \mathbf{Z}\}$. Then $(G, +)$ is a commutative group, where $+$ is the usual addition of integers. Note that $0 = 0 \cdot a$ and $-(na) = (-n)a$ are members of G.*

Gauss's work yielded many new directions of research in Abelian groups. The next two examples are due to Gauss.

Example 2.1.5 *Consider $\mathbf{Z}_n$ (Examples 1.3.11 and 1.3.17). Define $+_n$ on $\mathbf{Z}_n$ by*

$$[a] +_n [b] = [a + b]$$

for all $[a], [b] \in \mathbf{Z}_n$. We show that $(\mathbf{Z}_n, +_n)$ is a commutative group.

We first prove that $+_n$ is a binary operation. Let $[a], [b], [c], [d] \in \mathbf{Z}_n$. Suppose $[a] = [c]$ and $[b] = [d]$. Then $n|(a-c)$ and $n|(b-d)$, i.e., there exist integers s and t such that $ns = a - c$ and $nt = b - d$. Hence, $n(s+t) = ((a+b) - (c+d))$ and so $n|((a + b) - (c + d))$. This implies that $a + b \equiv_n c + d$. Therefore,

$[a+b] = [c+d]$. As a result $+_n$ is well defined and so $+_n$ is a binary operation. For all $[a], [b], [c] \in \mathbf{Z}_n$, $([a] +_n [b]) +_n [c] = [a + b] +_n [c] = [(a + b) + c] = [a + (b + c)] = [a] +_n [b + c] = [a] +_n ([b] +_n [c])$. Hence, $+_n$ is associative. Now $[0] \in \mathbf{Z}_n$ and for all $[a] \in \mathbf{Z}_n$,

$$[a] +_n [0] = [a + 0] = [a] = [0 + a] = [0] +_n [a].$$

This shows that $[0]$ is the identity element. Also, for all $[a] \in \mathbf{Z}_n$, $[-a] \in \mathbf{Z}_n$ and

$$[a] +_n [-a] = [a - a] = [0] = [-a + a] = [-a] +_n [a].$$

Thus, $[-a]$ is the inverse of $[a]$. Finally, for all $[a], [b] \in \mathbf{Z}_n$

$$[a] +_n [b] = [a + b] = [b + a] = [b] +_n [a]$$

and so $+_n$ is commutative. Hence, $(\mathbf{Z}_n, +_n)$ is a commutative group.

Example 2.1.6 *Consider $\mathbf{Z}_n$ (Examples 1.3.11 and 1.3.17). Define $\cdot_n$ on $\mathbf{Z}_n$ by*

$$[a] \cdot_n [b] = [ab]$$

for all $[a], [b] \in \mathbf{Z}_n$. With the help of a little calculation as in Example 2.1.5, we can show that $\cdot_n$ is a binary operation on $\mathbf{Z}_n$ and $\cdot_n$ is associative. Now $[1] \in \mathbf{Z}_n$ and for all $[a] \in \mathbf{Z}_n$,

$$[a] \cdot_n [1] = [a \cdot 1] = [a] = [1 \cdot a] = [1] \cdot_n [a].$$

This implies that $[1]$ is the identity element. We now show that if $[a] \in \mathbf{Z}_n$ and $[a] \neq [0]$, then $[a]$ has an inverse if and only if $\gcd(a, n) = 1$.

   Let $[a] \in \mathbf{Z}_n$ and $[a] \neq [0]$. Suppose $\gcd(a, n) = 1$. Then there exist $b, r \in \mathbf{Z}$ such that $ab + nr = 1$ by Theorem 1.2.11, i.e., $ab - 1 = nr$. This implies that $[ab] = [1]$ or $[a] \cdot_n [b] = [1]$. Since $ab = ba$, we also have $[b] \cdot_n [a] = [ba] = [ab] = [1]$. Thus, there exists $[b] \in \mathbf{Z}_n$ such that $[a][b] = [1] = [b][a]$ and so $[a]$ has an inverse. Conversely, suppose $[a] \in \mathbf{Z}_n$, $[a] \neq [0]$ and $[a]$ has an inverse. Then there exists $[b] \in \mathbf{Z}_n$ such that $[a][b] = [1]$. This implies that $n|(ab - 1)$ (by Exercise 11, page 30) and so $ab - 1 = nr$ for some $r \in \mathbf{Z}$. Thus, $ab + nr = 1$ and hence by Theorem 1.2.11, $\gcd(a, n) = 1$. This proves our claim.

   Thus, we see that in general, not every element of $\mathbf{Z}_n \backslash \{[0]\}$ has an inverse. For example if $n = 6$, then the only elements of $\mathbf{Z}_6$ that have inverses are $[1]$, $[3]$ and $[5]$. Hence, in general $(\mathbf{Z}_n \backslash \{[0]\}, \cdot_n)$ is not a group.

   Let U_n be the set of all elements of $\mathbf{Z}_n \backslash \{[0]\}$ that have an inverse in $(\mathbf{Z}_n \backslash \{[0]\}, \cdot_n)$, i.e.,

$$U_n = \{[a] \in \mathbf{Z}_n \backslash \{[0]\} \,|\, \gcd(a, n) = 1\}.$$

We ask the reader to verify in Exercise 10 (page 78) that $(U_n, \cdot_n)$ is a group. Note that for $n = 8$, $U_8 = \{[1], [3], [5], [7]\}$ and for $n = 7$,

$$U_7 = \{[1], [2], [3], [4], [5], [6]\} = \mathbf{Z}_7 \backslash \{[0]\}.$$

Example 2.1.7 *Let*

$$\mathbf{Q}[\sqrt{2}] = \{a + b\sqrt{2} \mid a, b \in \mathbf{Q}\}.$$

Then $(\mathbf{Q}[\sqrt{2}], +)$ *and* $(\mathbf{Q}[\sqrt{2}]\backslash\{0\}, \cdot)$ *are commutative groups, where* $+$ *is the usual addition and* $\cdot$ *is the usual multiplication. The identity of* $(\mathbf{Q}[\sqrt{2}], +)$ *is* $0 + 0\sqrt{2}$ *and the inverse of* $a + b\sqrt{2}$ *is* $-a + (-b)\sqrt{2}$. *The identity of* $(\mathbf{Q}[\sqrt{2}]\backslash\{0\}, \cdot)$ *is* $1 = 1 + 0\sqrt{2}$ *and the inverse of* $a + b\sqrt{2} \neq 0$ *is* $\frac{a}{a^2 - 2b^2} - \frac{b}{a^2 - 2b^2}\sqrt{2}$.

Example 2.1.8 *Let* $\mathcal{P}(X)$ *be the power set of a set* X. *Consider the operation* Δ *(symmetric difference, Exercise 6, page 6) on* $\mathcal{P}(X)$. *Then for all* A, B $\in \mathcal{P}(X)$,

$$A \Delta B = (A \backslash B) \cup (B \backslash A).$$

$(\mathcal{P}(X), \Delta)$ *is a commutative group. The empty set* ϕ *is the identity of* $(\mathcal{P}(X), \Delta)$ *and every element of* $\mathcal{P}(X)$ *is its own inverse. We warn the reader that verification of the associative law is tedious.*

Example 2.1.9 *Let* X *be a set and* S_X *the set of all one-one functions of* X *onto* X. *Since* i_X, *the identity function on* X, *is one-one and onto* X, $i_X \in S_X$. *Thus,* $S_X \neq \phi$. *Let* $f, g \in S_X$. *Then* $f \circ g$ *is a one-one function of* X *onto* X *by Theorem 1.5.11. Hence,* $f \circ g \in S_X$. *By Theorem 1.5.13,* $\circ$ *is associative. Also, for all* $f \in S_X$, $f^{-1} \in S_X$ *and* $f \circ f^{-1} = i_X = f^{-1} \circ f$. *Consequently,* $(S_X, \circ)$ *is a group. However,* $(S_X, \circ)$ *is not necessarily commutative. For example, let* $X = \{a, b, c\}$. *Let* $f, g \in S_X$ *be defined by* $f(a) = b$, $f(b) = a$, $f(c) = c$, $g(a) = b$, $g(b) = c$, $g(c) = a$. *Then* $(f \circ g)(b) = f(g(b)) = f(c) = c$ *and* $(g \circ f)(b) = g(f(b)) = g(a) = b$. *Hence,* $f \circ g \neq g \circ f$. *Thus,* $(S_X, \circ)$ *is not commutative.*

Example 2.1.10 *Let* $GL(2, \mathbf{R}) = \left\{ \begin{bmatrix} a & b \\ c & d \end{bmatrix} \mid a, b, c, d \in \mathbf{R}, ad - bc \neq 0 \right\}$.
Define a binary operation $*$ *on* $GL(2, \mathbf{R})$ *by*

$$\begin{bmatrix} a & b \\ c & d \end{bmatrix} * \begin{bmatrix} u & v \\ w & s \end{bmatrix} = \begin{bmatrix} au + bw & av + bs \\ cu + dw & cv + ds \end{bmatrix}$$

for all $\begin{bmatrix} a & b \\ c & d \end{bmatrix}, \begin{bmatrix} u & v \\ w & s \end{bmatrix} \in GL(2, \mathbf{R})$. *This binary operation is the usual matrix multiplication. Since matrix multiplication is associative, we have* $*$ *is associative. The element* $\begin{bmatrix} 1 & 0 \\ 0 & 1 \end{bmatrix} \in GL(2, \mathbf{R})$ *and is the identity element of*

GL(2, **R**). *Let* $\begin{bmatrix} a & b \\ c & d \end{bmatrix} \in GL(2,\mathbf{R})$. *Then* $ad - bc \neq 0$. *Consider the matrix*

$\begin{bmatrix} \frac{d}{ad-bc} & \frac{-b}{ad-bc} \\ \frac{-c}{ad-bc} & \frac{a}{ad-bc} \end{bmatrix}$. *Since*

$$\frac{d}{ad-bc} \cdot \frac{a}{ad-bc} - \frac{-b}{ad-bc} \cdot \frac{-c}{ad-bc} = \frac{1}{ad-bc} \neq 0,$$

we have

$$\begin{bmatrix} \frac{d}{ad-bc} & \frac{-b}{ad-bc} \\ \frac{-c}{ad-bc} & \frac{a}{ad-bc} \end{bmatrix} \in GL(2,\mathbf{R}).$$

Now

$$\begin{bmatrix} a & b \\ c & d \end{bmatrix} * \begin{bmatrix} \frac{d}{ad-bc} & \frac{-b}{ad-bc} \\ \frac{-c}{ad-bc} & \frac{a}{ad-bc} \end{bmatrix} = \begin{bmatrix} 1 & 0 \\ 0 & 1 \end{bmatrix}$$

and

$$\begin{bmatrix} \frac{d}{ad-bc} & \frac{-b}{ad-bc} \\ \frac{-c}{ad-bc} & \frac{a}{ad-bc} \end{bmatrix} * \begin{bmatrix} a & b \\ c & d \end{bmatrix} = \begin{bmatrix} 1 & 0 \\ 0 & 1 \end{bmatrix}.$$

Thus, $\begin{bmatrix} \frac{d}{ad-bc} & \frac{-b}{ad-bc} \\ \frac{-c}{ad-bc} & \frac{a}{ad-bc} \end{bmatrix}$ *is the inverse of* $\begin{bmatrix} a & b \\ c & d \end{bmatrix}$. *Hence,* $(GL(2,\mathbf{R}),*)$ *is a group. Now*

$$\begin{bmatrix} 1 & 1 \\ 0 & 1 \end{bmatrix}, \begin{bmatrix} 1 & 0 \\ 1 & 1 \end{bmatrix} \in GL(2,\mathbf{R})$$

and

$$\begin{bmatrix} 1 & 1 \\ 0 & 1 \end{bmatrix} * \begin{bmatrix} 1 & 0 \\ 1 & 1 \end{bmatrix} = \begin{bmatrix} 2 & 1 \\ 1 & 1 \end{bmatrix} \neq \begin{bmatrix} 1 & 1 \\ 1 & 2 \end{bmatrix} = \begin{bmatrix} 1 & 0 \\ 1 & 1 \end{bmatrix} * \begin{bmatrix} 1 & 1 \\ 0 & 1 \end{bmatrix}.$$

Hence, $(GL(2,\mathbf{R}),*)$ *is a noncommutative group.*

The group in Example 2.1.10 is known as the **general linear group of degree 2**.

We now prove some elementary properties of a group in the following theorem.

Theorem 2.1.11 *Let* $(G,*)$ *be a group.*

(i) $(a^{-1})^{-1} = a$ *for all* $a \in G$.

(ii) $(a*b)^{-1} = b^{-1} * a^{-1}$ *for all* $a, b \in G$.

(iii) *(**Cancellation Law**) For all* $a, b, c \in G$, *if either* $a*c = b*c$ *or* $c*a = c*b$, *then* $a = b$.

(iv) *For all* $a, b \in G$, *the equations* $a * x = b$ *and* $y * a = b$ *have unique solutions in* G *for* x *and* y.

Proof.   (i) Let $a \in G$. Then $a^{-1} * a = e = a * a^{-1}$ and so a is an inverse of
a^{-1}. Since the inverse of an element is unique in a group (Theorem 2.1.2) and
since $(a^{-1})^{-1}$ denotes the inverse of a^{-1}, it follows that $a = (a^{-1})^{-1}$.

(ii) Let $a, b \in G$. Then

$$
\begin{aligned}
(a * b) * (b^{-1} * a^{-1}) &= ((a * b) * b^{-1}) * a^{-1} \\
&= (a * (b * b^{-1})) * a^{-1} \\
&= (a * e) * a^{-1} \\
&= a * a^{-1} \\
&= e.
\end{aligned}
$$

Similarly, $(b^{-1} * a^{-1}) * (a * b) = e$. Hence, $b^{-1} * a^{-1}$ is an inverse of $a * b$. Since
the inverse of an element is unique in a group and since $(a * b)^{-1}$ denotes the
inverse of $a * b$, it follows that $(a * b)^{-1} = b^{-1} * a^{-1}$.

(iii) Let $a, b, c \in G$. Suppose $a * c = b * c$. Now $(a * c) * c^{-1} = (b * c) * c^{-1}$
implies that $a * (c * c^{-1}) = b * (c * c^{-1})$. Hence, $a * e = b * e$ or $a = b$. Similarly,
if $c * a = c * b$, then $a = b$.

(iv) Let $a, b \in G$. First we consider the equation $a * x = b$. Now $a^{-1} * b \in G$.
Substituting $a^{-1} * b$ for x in the equation $a * x = b$, we obtain

$$
a * (a^{-1} * b) = (a * a^{-1}) * b = e * b = b.
$$

Thus, $a^{-1} * b$ is a solution of the equation $a * x = b$. We now establish the
uniqueness of the solution.   Suppose c is any solution of $a * x = b$. Then
$a * c = b$. Hence,

$$
\begin{aligned}
c &= e * c & \\
&= (a^{-1} * a) * c & (\text{since } a^{-1} * a = e) \\
&= a^{-1} * (a * c) & (\text{since } * \text{ is associative}) \\
&= a^{-1} * b & (\text{since } a * c = b).
\end{aligned}
$$

This yields the uniqueness of the solution.   Similar arguments hold for the
equation $y * a = b$. ∎

Corollary 2.1.12 *Let $(G, *)$ be a group and $a \in G$. If $a * a = a$, then $a = e$.* ∎

Proof.   Since $a = a * a$, we have $a * a = a * e$. By the cancellation law, $a = e$.

Corollary 2.1.13 *In a multiplication table for a group $(G, *)$, each element
appears exactly once in each row and exactly once in each column.*

Proof. Let $b \in G$ be such that b occurs twice in the row marked by $a \in G$. Then there exists $u, v \in G$ with $u \neq v$ such that $a * u = b$ and $a * v = b$. Thus, the equation $a * x = b$ has two distinct solutions, u and v. This is a contradiction to Theorem 2.1.11(iv) since the equation $a * x = b$ has a unique solution for x. A similar argument for columns can be used. ∎

Let $(G, *)$ be a group and $a, b, c \in G$. Then by the associative law, $a*(b*c) = (a*b)*c$. Hence, we can define $a*b*c = a*(b*c) = (a*b)*c$. Let $a, b, c, d \in G$. Then $(a * b * c) * d = (a * (b * c)) * d = a * ((b * c) * d)) = a * (b * (c * d)) = (a * b) * (c * d) = ((a * b) * c) * d$. Thus, there is more than one way of inserting parentheses in the expression $a * b * c * d$ to produce a "meaningful product" of a, b, c, d (in this order). We now extend this notion to any finite number of elements.

Definition 2.1.14 *Let $(G, *)$ be a group and $a_1, a_2, \ldots, a_n \in G$ be n elements of G (not necessarily distinct). The **meaningful product** of $a_1, a_2, \ldots, a_n$ (in this order) is defined as follows: If $n = 1$, then the meaningful product is a_1. If $n > 1$, then the meaningful product of $a_1, a_2, \ldots, a_n$ is any product of the form*

$$(a_1 * \cdots * a_m) * (a_{m+1} * \cdots * a_n),$$

where $1 \leq m < n$ and $(a_1* \cdots * a_m)$ and $(a_{m+1}* \cdots * a_n)$ are meaningful products of m and $n - m$ elements, respectively.

Definition 2.1.15 *Let $(G, *)$ be a group and $a_1, a_2, \ldots, a_n \in G$, $n \geq 1$. The **standard product** of $a_1, a_2, \ldots, a_n$ denoted by $a_1* a_2* \cdots * a_n$ is defined recursively as*

$$a_1 = a_1$$
$$a_1 * a_2 * \cdots * a_n = (a_1 * a_2 * \cdots * a_{n-1}) * a_n \text{ if } n > 1.$$

In the next theorem, we establish the equality between any meaningful product and standard product.

Theorem 2.1.16 *Let $(G, *)$ be a group and $a_1, a_2, \ldots, a_n \in G$, $n \geq 1$. Then all possible meaningful products of $a_1, a_2, \ldots, a_n$ (in this order) are equal to the standard product of $a_1, a_2, \ldots, a_n$ (in this order).*

Proof. We prove the result by induction. If $n = 1$, then a_1 is the only meaningful product of a_1, which is equal to the standard product a_1 of a_1. Thus, the result is true if $n = 1$. Suppose that the theorem is true for all integers m such that $1 \leq m < n$. Let $a_1, a_2, \ldots, a_n \in G$. Let $(a_1* \cdots * a_t) * (a_{t+1}* \cdots * a_n)$ be a meaningful product of $a_1, a_2, \ldots, a_n$ (in this order). Now $t < n$ and $n-t < n$. If $t = n-1$, then $(a_1 * a_2 * \cdots * a_t) * a_{t+1} = a_1 * a_2 * \cdots * a_t * a_{t+1}$. Suppose $t < n-1$.

Then $(a_1 * \cdots * a_t) * (a_{t+1} * \cdots * a_n) = (a_1 * \cdots * a_t) * ((a_{t+1} * \cdots * a_{n-1}) * a_n) = ((a_1 * \cdots * a_t) * (a_{t+1} * \cdots * a_{n-1})) * a_n = (a_1 * a_2 * \cdots * a_{n-1}) * a_n = a_1 * \cdots * a_n$ since by the induction hypothesis $(a_1 * \cdots * a_t) * (a_{t+1} * \cdots * a_{n-1}) = a_1 * a_2 * \cdots * a_{n-1}$. Hence, the result is true for n. The result now follows by induction. ∎

We have seen several examples of groups. In order to show that a given set with a given binary operation is a group, we need to verify G1 to G3 of Definition 2.1.1. However, it would be helpful if we had some criteria that could be used to show whether a given set with a binary operation is a group or not instead of verifying all the properties G1–G3 explicitly. Partly for this reason we define what a semigroup is. Following the examples, we develop some results that can be used to test whether a given set with a binary operation is a group or not.

Definition 2.1.17 *A **semigroup** is an ordered pair $(S, *)$, where S is a nonempty set and $*$ is an associative binary operation on S.*

Thus, a semigroup is a mathematical system with one binary operation such that the binary operation is associative. We note that every group $(G, *)$ is a semigroup.

A semigroup $(S, *)$ is **commutative** if $*$ is commutative, i.e., $a * b = b * a$ for all $a, b \in S$. A semigroup $(S, *)$ which is not commutative is called **non-commutative.**

Let $(S, *)$ be a semigroup. We say that $(S, *)$ is with identity if the mathematical system $(S, *)$ has an identity. An element $a \in S$ is called **idempotent** if $a * a = a$.

Example 2.1.18 *Consider **N**, the set of positive integers. We know that addition of positive integers is again a positive integer. Thus, $+$ is a binary operation on **N**. We also know that $+$ is associative and commutative. Thus, $(\mathbf{N}, +)$ is a commutative semigroup.*

Example 2.1.19 *Let X be a nonempty set and S the set of all functions $f : X \to X$. If $\circ$ denotes the composition of functions, then $(S, \circ)$ is a semigroup with identity. The associativity of $\circ$ follows from Theorem 1.5.13. When X has two or more elements, the semigroup $(S, \circ)$ is noncommutative. For example, let $X = \{a, b\}$. Let $g, h \in S$ be defined by $g(a) = b$, $g(b) = b$, $h(a) = b$, $h(b) = a$. Then $(g \circ h)(a) = b \neq a = (h \circ g)(a)$. Therefore, $g \circ h \neq h \circ g$. Let $f \in S$ be defined by $f(a) = a$ and $f(b) = a$. Now $(f \circ g)(x) = f(g(x)) = a = f(h(x)) = (f \circ h)(x)$ for all $x \in G$. Hence, $f \circ g = f \circ h$. But $g \neq h$. This shows that the cancellation laws do not hold in S. Thus, $(S, \circ)$ is not a group.*

Example 2.1.20 *Let X be a set with two or more elements and S' the set of all functions $f : X \to X$ which are not one-one. Then $(S', \circ)$ is a noncommutative semigroup without identity.*

Example 2.1.21 *Let X be a set and $\mathcal{P}(X)$ the power set of X. Then $(\mathcal{P}(X), \cup)$ and $(\mathcal{P}(X), \cap)$ are commutative semigroups with identity. The identity of $(\mathcal{P}(X), \cup)$ is ϕ and the identity of $(\mathcal{P}(X), \cap)$ is X.*

The following three theorems give necessary and sufficient conditions for a semigroup to be a group.

Theorem 2.1.22 *A semigroup $(S, *)$ is a group if and only if*
(i) there exists $e \in S$ such that $e * a = a$ for all $a \in S$ and
(ii) for all $a \in S$ there exists $b \in S$ such that $b * a = e$.

Proof. Suppose $(S, *)$ is a semigroup that satisfies (i) and (ii). Let a be any element of S. Then there exists $b \in S$ such that $b * a = e$ by (ii). For $b \in S$, there exists $c \in S$ such that $c * b = e$ by (ii). Now

$$a = e * a = (c * b) * a = c * (b * a) = c * e$$

and

$$a * b = (c * e) * b = c * (e * b) = c * b = e.$$

Hence, $a * b = e = b * a$. Also,

$$a * e = a * (b * a) = (a * b) * a = e * a = a.$$

Thus, $a * e = a = e * a$. This shows that e is the identity element of S. Now since $a * b = e = b * a$, we have $b = a^{-1}$. Therefore, $(S, *)$ is a group. The converse follows from the definition of a group. ∎

Theorem 2.1.23 *A semigroup $(S, *)$ is a group if and only if for all $a, b \in S$ the equations $a * x = b$ and $y * a = b$ have solutions in S for x and y.*

Proof. Suppose the given equations have solutions in S. Let $a \in S$. Consider the equation $y * a = a$. By our assumption, $y * a = a$ has a solution $u \in S$, say. Then $u * a = a$. Let b be any element of S. Consider the equation $a * x = b$. Again by our assumption, $a * x = b$ has a solution in S. Let $c \in S$ be a solution of $a * x = b$. Then $a * c = b$. Now

$$
\begin{aligned}
u * b &= u * (a * c) & \text{(since } b = a * c) \\
&= (u * a) * c & \text{(since } * \text{ is asociative)} \\
&= a * c & \text{(since } u * a = a) \\
&= b.
\end{aligned}
$$

Since b was an arbitrary element of S, we find that $u * b = b$ for all $b \in S$. Thus, $(S, *)$ satisfies (i) of Theorem 2.1.22. Consider the equation $y * a = u$. Let $d \in S$ be a solution of $y * a = u$. Then $d * a = u$. This shows that $(S, *)$ satisfies (ii) of Theorem 2.1.22. Hence, $(S, *)$ is a group by Theorem 2.1.22.
    The converse follows by Theorem 2.1.11(iv). ■

Theorem 2.1.24 *A finite semigroup* $(S, *)$ *is a group if and only if* $(S, *)$ *satisfies the cancellation laws (i.e.,* $a * c = b * c$ *implies* $a = b$ *and* $c * a = c * b$ *implies* $a = b$ *for all* $a, b, c \in S$).

Proof.   Let $(S, *)$ be a finite semigroup satisfying the cancellation laws. Let $a, b \in S$. Consider the equation $a * x = b$. We show that this equation has a solution in S. Let us write $S = \{a_1, a_2, \ldots, a_n\}$, where the a_i's are all distinct elements of S. Since S is a semigroup, $a * a_i \in S$ for all $i = 1, 2, \ldots, n$. Thus, $\{a * a_1, a * a_2, \ldots, a * a_n\} \subseteq S$. Suppose $a * a_i = a * a_j$ for some $i \neq j$. Then by the cancellation law we have $a_i = a_j$, which is a contradiction since $a_i \neq a_j$. Hence, all elements in $\{a * a_1, a * a_2, \ldots, a * a_n\}$ are distinct. Thus, $S = \{a * a_1, a * a_2, \ldots, a * a_n\}$. Let $b \in S$. Then $b = a * a_k$ for some $a_k \in S$. Therefore, the equation $a * x = b$ has a solution in S. Similarly, we can show that the equation $y * a = b$ has a solution in S. Hence, by Theorem 2.1.23, $(S, *)$ is a group. The converse follows by Theorem 2.1.11(iii). ■

    Let $(G, *)$ be a group, $a \in G$, and $n \in \mathbf{Z}$. We now define the **integral power** a^n of a as follows:

$$
\begin{aligned}
a^0 &= e \\
a^n &= a * a^{n-1} \text{ if } n > 0 \\
a^n &= (a^{-1})^{-n} \text{ if } n < 0.
\end{aligned}
$$

    Note that $a^n = (a^{-n})^{-1}$ if $n < 0$. In the exercises at the end of this section, we ask the reader to verify certain basic properties of integral powers. It should be pointed out that when we use additive notation for the binary operation $*$, we speak of multiples of an element a of the group $(G, +)$, which are defined as follows:

$$
\begin{aligned}
0a &= 0, \text{ where the 0 on the right-hand side denotes the identity of the} \\
&\quad \text{group } (G, +) \text{ and the 0 on the left-hand side denotes the integer 0.} \\
na &= a + (n - 1)a \quad \text{if } n > 0 \\
na &= (-n)(-a) \quad \text{if } n < 0.
\end{aligned}
$$

For example, in $(\mathbf{Z}_6, +_6)$, $2[3] = [3] +_6 [3] = [6] = [0]$. By the notation na, we do not mean n and a multiplied together since no multiplicative operation between elements of $\mathbf{Z}$ and G has been defined.

Definition 2.1.25 *A group* $(G, *)$ *is called a **finite group** if* G *has only a finite number of elements. The **order**, written* $|G|$, *of a group* $(G, *)$ *is the number of elements of* G.

Example 2.1.5 shows that for every positive integer n, there is a commutative group of order n.

The groups in Examples 2.1.5 and 2.1.6 are finite groups.

A group with an infinite number of elements is referred to as an **infinite group**. Klein and Lie's use of groups in geometry influenced the turn from finite groups to infinite groups.

The groups in Examples 2.1.3, 2.1.4, and 2.1.7 are infinite groups.

Let G be a finite group and $a \in G$. Now $a^2 = a * a \in G$ and by induction, we can show that $a^m \in G$ for all $m \geq 1$. Thus, $\{a, a^2, \ldots, a^m, \ldots\} \subseteq G$. Since G is finite, all elements of the set $\{a, a^2, \ldots, a^m, \ldots\}$ cannot be distinct. Hence, $a^k = a^l$ for some positive integers k, l, $k > l$. This implies that $a^{k-l} = e$. Let us write $n = k - l$. Therefore, $a^n = e$ for some positive integer n. Also, if G is an infinite group and $a \in G$, then it may still be possible that $a^n = e$ for some positive integer n. This leads us to the following definition.

Definition 2.1.26 *Let* $(G, *)$ *be a group and* $a \in G$. *If there exists a positive integer* n *such that* $a^n = e$, *then the smallest such positive integer is called the **order** of* a. *If no such positive integer* n *exists, then we say that* a *is of **infinite order**.*

We denote the order of an element a of a group $(G, *)$ by $o(a)$.

The concept of the order of an element is very important in group theory. We shall see in later chapters how effectively information about the order of an element of a group reveals the nature of the group and in several instances leads us to determine the structure of the group itself.

Example 2.1.27 *Consider the group* $(\mathbf{Z}_6, +_6)$. $\mathbf{Z}_6$ *has order* 6. *The elements* $[0], [1], [2], [3], [4], [5]$ *have orders* 1, 6, 3, 2, 3, 6, *respectively. For example* $2[3] = [3] +_6 [3] = [6] = [0]$ *and* 2 *is the smallest positive integer* n *such that* $n[3] = [0]$.

Let G be a group and $a \in G$. If $o(a)$ is infinite, then by the definition of the order of an element it follows that $o(a^k)$ is also infinite for all $k \geq 1$, i.e., the order of every positive power of a is also infinite. If $o(a)$ is finite, then the next theorem tells us how to compute the order of various powers of a.

Theorem 2.1.28 *Let* $(G, *)$ *be a group and* a *be an element of* G *such that* $o(a) = n$.

(i) *If* $a^m = e$ *for some positive integer* m, *then* n *divides* m.

(ii) For every positive integer t,

$$o(a^t) = \frac{n}{\gcd(t, n)}.$$

Proof.   (i) By the division algorithm, there exist $p, q \in \mathbf{Z}$ such that $m = nq + r$, where $0 \le r < n$. Now $a^r = a^{m-nq} = a^m * a^{-nq} = a^m * (a^n)^{-q} = e * (e)^{-q} = e$. Since n is the smallest positive integer such that $a^n = e$ and $a^r = e$, it follows that $r = 0$. Thus, $m = nq$. This implies that n divides m.

(ii) Let $o(a^t) = k$. Then $a^{kt} = e$. By (i), n divides kt. Thus, there exists $r \in \mathbf{Z}$ such that $kt = nr$. Let $\gcd(t, n) = d$. Then there exist integers u and v such that $t = du$ and $n = dv$ and $\gcd(u, v) = 1$ by Exercise 9 (page 20). Now $kt = nr$ implies that $kdu = dvr$. Hence, $ku = rv$. Thus, v divides ku. Since $\gcd(u, v) = 1$, v divides k. Thus, $\frac{n}{d}$ divides k. Now $(a^t)^{\frac{n}{d}} = a^{\frac{nt}{d}} = a^{\frac{ndu}{d}} = a^{nu} = (a^n)^u = e^u = e$. Since $o(a^t) = k$, k divides $\frac{n}{d}$. Since k and $\frac{n}{d}$ are positive integers, $k = \frac{n}{d}$. Hence, $o(a^t) = k = \frac{n}{d} = \frac{n}{\gcd(t,n)}$. $\blacksquare$

A group $(G, *)$ is called a **torsion group** if every element of G is of finite order. If every nonidentity element of G is of infinite order, then G is called a **torsion-free group.**

The group of Example 2.1.27 is a torsion group. The groups $(\mathbf{R}, +)$, $(\mathbf{R}^+, \cdot)$, $(\mathbf{Q}^+, \cdot)$ are torsion-free groups. The group $(\mathbf{R}\backslash\{0\}, \cdot)$ is neither a torsion group nor a torsion-free group, since -1 is of order 2 and all other nonidentity elements are of infinite order.

We close this chapter with the following example. The ideas set forth in this example are due to Klein.

Example 2.1.29 *Imagine a square having its sides parallel to the axes of a coordinate system and its center at the origin.*

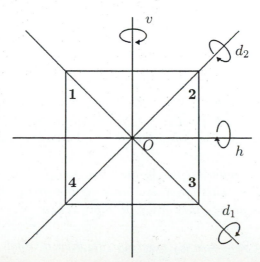

We label the vertices as in the figure and we allow the following rigid motions of the square: clockwise rotations of the square about the center and through angles of 90°, 180°, 270°, 360°, say, r_{90}, r_{180}, r_{270}, r_{360}, respectively; reflections h and v about the horizontal and vertical axes; reflections d_1, d_2 about the diagonals. The following figures should prove helpful.

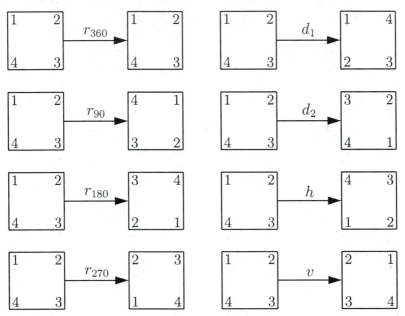

A multiplication $*$ on two rigid motions can be defined by performing two such motions in succession. For example, $r_{90}*h$ is determined by first performing motion h and then the motion r_{90}. We see that $r_{90} * h = d_1$. The complete multiplication table for the operation $*$ follows.

$*$	r_{360}	r_{90}	r_{180}	r_{270}	h	v	d_1	d_2
r_{360}	r_{360}	r_{90}	r_{180}	r_{270}	h	v	d_1	d_2
r_{90}	r_{90}	r_{180}	r_{270}	r_{360}	d_1	d_2	v	h
r_{180}	r_{180}	r_{270}	r_{360}	r_{90}	v	h	d_2	d_1
r_{270}	r_{270}	r_{360}	r_{90}	r_{180}	d_2	d_1	h	v
h	h	d_2	v	d_1	r_{360}	r_{180}	r_{270}	r_{90}
v	v	d_1	h	d_2	r_{180}	r_{360}	r_{90}	r_{270}
d_1	d_1	h	d_2	v	r_{90}	r_{270}	r_{360}	r_{180}
d_2	d_2	v	d_1	h	r_{270}	r_{90}	r_{180}	r_{360}

$$(2.1)$$

We leave it for the reader to verify that the set of rigid motions is a group under the operation $*$. This group is known as the **group of symmetries of the square**. Let us denote this group by Sym. Then

$$Sym = \{r_{360}, r_{90}, r_{180}, r_{270}, h, v, d_1, d_2\}.$$

Since $h * r_{270} = d_1 \neq d_2 = r_{270} * h$, we see that the Sym is noncommutative. We also note that r_{360} is the identity element.

Let us now determine the order of the elements. Consider r_{90}. Now $r_{90}^2 = r_{90} * r_{90} = r_{180}$, $r_{90}^3 = r_{90}^2 * r_{90} = r_{270}$, and $r_{90}^4 = r_{90}^3 * r_{90} = r_{360}$. Thus, $o(r_{90}) = 4$. Similarly, $o(r_{180}) = o(r_{90}^2) = \frac{4}{gcd(4,2)}$ (by Theorem 2.1.28) $= \frac{4}{2} = 2$, $o(r_{270}) = 4$, $o(h) = 2$, $o(v) = 2$, $o(d_1) = 2$, and $o(d_2) = 2$.

Let us write $\alpha = r_{90}$ and $\beta = d_2$. Then $\alpha^2 = r_{180}$, $\alpha^3 = r_{270}$, $\alpha^4 = r_{360}$, $\beta * \alpha = v$, $\beta * \alpha^2 = d_1$, and $\beta * \alpha^3 = h$. Also, note that $\beta * \alpha = \alpha^{-1} * \beta = \alpha^3 * \beta$. Thus, we see that

$$Sym = \{e, \alpha, \alpha^2, \alpha^3, \beta, \beta * \alpha, \beta * \alpha^2, \beta * \alpha^3\}.$$

Finally, we make the following observations. Consider r_{90}. We can think of r_{90} as a one-one function of $\{1, 2, 3, 4\}$ onto $\{1, 2, 3, 4\}$ by defining $r_{90}(1) = 2$, $r_{90}(2) = 3$, $r_{90}(3) = 4$, $r_{90}(4) = 1$. In a similar manner, we can consider other rigid motions of the square as one-one functions of $\{1, 2, 3, 4\}$ onto $\{1, 2, 3, 4\}$.

A fundamental phenomenon of nature is that of symmetry. A figure or an object is said to have a symmetry if a rotation, a translation, an inversion, a minor reflection, or a combination of these operations leaves the figure or object indistinguishable from its original position. The 1890s saw the first application of group theory to the natural and physical sciences. An important application of group theory was to crystallography. Groups were used to give a theoretical classification of the different kinds of symmetry arrangements possible within crystalline matter 20 years before experimental means were available for analyzing the crystals themselves.

Group theory is used in quantum mechanics. It is used to study the atom's internal structure. In the 1950s, a new generation of particle accelerators produced a variety of subatomic particles. Group theory was used to predict the existence of a tenth nucleon in a tenfold symmetry scheme of nucleons of which nine particles had already been detected. In 1964, the tracks of Omega-Minus, the tenth nucleon, were identified.

2.1.1   Worked-Out Exercises

◊ **Exercise 1** Let $G = \{a \in \mathbf{R} \mid -1 < a < 1\}$. Define a binary operation $*$ on G by

$$a * b = \frac{a + b}{1 + ab}$$

for all $a, b \in G$. Show that $(G, *)$ is a group.

Solution: Note that $-1 < x < 1$ if and only if $x^2 < 1$ for all $x \in \mathbf{R}$. Let $a, b \in G$. First we show that $a * b \in G$. Now $a^2 < 1$ and $b^2 < 1$. Thus,

$(1 - a^2)(1 - b^2) > 0$. This implies that $1 - a^2 - b^2 + a^2 b^2 > 0$. Now $(1 + ab)^2 - (a + b)^2 = 1 + a^2 b^2 + 2ab - a^2 - b^2 - 2ab = 1 - a^2 - b^2 + a^2 b^2 > 0$ and so $(\frac{a+b}{1+ab})^2 < 1$. Therefore, $a * b \in G$. Hence, G is closed under $*$. We now show that $*$ is well defined. Let $a, b, c, d \in G$ and $(a, b) = (c, d)$. Then $a = c$ and $b = d$. Thus,

$$a * b = \frac{a + b}{1 + ab} = \frac{c + d}{1 + cd} = c * d$$

and so $*$ is well defined. To show that $*$ is associative, let $a, b, c \in G$. Now

$$(a * b) * c = \frac{a + b}{1 + ab} * c = \frac{\frac{a+b}{1+ab} + c}{1 + (\frac{a+b}{1+ab})c} = \frac{a + b + c + abc}{1 + ab + ac + bc}.$$

Similarly,

$$a * (b * c) = \frac{a + b + c + abc}{1 + ab + ac + bc}.$$

Therefore, $(a * b) * c = a * (b * c)$ and so $*$ is associative. Hence, we have shown that $(G, *)$ is a semigroup. Now $0 \in G$ and

$$0 * a = \frac{0 + a}{1 + 0a} = a \quad \text{for all } a \in G.$$

This shows that $(G, *)$ satisfies (i) of Theorem 2.1.22. Let $a \in G$. Then $-a \in G$ and

$$(-a) * a = \frac{-a + a}{1 + (-a)a} = 0.$$

Thus, $(G, *)$ satisfies (ii) of Theorem 2.1.22. Consequently, by Theorem 2.1.22, $(G, *)$ is a group.

◇ **Exercise 2** Let $G = \{(a, b) \mid a, b \in \mathbf{R}, a \neq 0\} = \mathbf{R}\backslash\{0\} \times \mathbf{R}$. Define a binary operation $*$ on G by

$$(a, b) * (c, d) = (ac, b + d)$$

for all $(a, b), (c, d) \in G$. Show that

(i) $(G, *)$ is a group,

(ii) G has exactly one element of order 2,

(iii) G has no elements of order 3.

Solution: (i) As in Worked-Out Exercise 1, we show that $(G, *)$ satisfies the conditions of Theorem 2.1.22. Let $(a, b), (c, d) \in G$. Then $a \neq 0$ and $c \neq 0$ and so $ac \neq 0$. Thus, $(a, b) * (c, d) = (ac, b + d) \in G$. Hence, G is closed under $*$. It is a direct computation to verify that $*$ is well defined and associative, so we ask the reader to do the verification. Now $(1, 0) \in G$ and

$$(1, 0) * (a, b) = (1a, 0 + b) = (a, b) \quad \text{for all } (a, b) \in G$$

and so $(G, *)$ satisfies (i) of Theorem 2.1.22. Let $(a, b) \in G$. Then $a \neq 0$ and so $\frac{1}{a} \in \mathbf{R}$ and $\frac{1}{a} \neq 0$. Thus, $(\frac{1}{a}, -b) \in G$ and

$$(\frac{1}{a}, -b) * (a, b) = (\frac{1}{a}a, -b + b) = (1, 0).$$

Therefore, $(G, *)$ satisfies (ii) of Theorem 2.1.22. Hence, $(G, *)$ is a group by Theorem 2.1.22.

(ii) First note that $(-1, 0) \in G$ and $(-1, 0) * (-1, 0) = (1, 0)$. Thus, $(-1, 0)$ is of order 2. We now show that this is the only element of order 2 by showing that if (a, b) is any other element of G of order 2, then $(a, b) = (-1, 0)$. Let $(a, b) \in G$ be an element of order 2. Then $(a, b) * (a, b) = (1, 0)$ implies that $(a^2, b + b) = (1, 0)$. Therefore, $a^2 = 1$ and $b = 0$. Now $a^2 = 1$ implies that $a = \pm 1$. If $a = 1$, then $(a, b) = (1, 0)$, which is a contradiction since $(1, 0)$ is of order 1. Hence, $a = -1$ and so $(a, b) = (-1, 0)$. Thus, $(-1, 0)$ is the only element of order 2.

(iii) Suppose that (a, b) is an element of order 3. Then $(a, b) * (a, b) * (a, b) = (1, 0)$. This implies that $(a^3, 3b) = (1, 0)$. Thus, $a^3 = 1$ and $b = 0$. Now $a^3 = 1$ implies that $a = 1$. Hence, $(a, b) = (1, 0)$. But $(1, 0)$ is of order 1. Consequently, G has no element of order 3.

◇ **Exercise 3** Let G be the set of all rational numbers except -1. Show that $(G, *)$ is a group where

$$a * b = a + b + ab$$

for all $a, b \in G$.

Solution: As in Worked-Out Exercise 1, we show that $(G, *)$ satisfies the conditions of Theorem 2.1.22. Our first step is to show that $*$ is well defined. Let $a, b, c, d \in G$ and $(a, b) = (c, d)$. Then $a = c$ and $b = d$. Thus, $a * b = a + b + ab = c + d + cd = c * d$ and so $*$ is well defined. Let $a, b \in G$. Then $a \neq -1$ and $b \neq -1$. We now show that $a * b \in G$ by showing that $a * b \neq -1$ and $a * b$ is a rational number. Suppose $a * b = a + b + ab = -1$. Then $(a + 1)(b + 1) = 0$. Hence, either $(a + 1) = 0$ or $(b + 1) = 0$ and so either $a = -1$ or $b = -1$, which is a contradiction. Therefore, $a * b \neq -1$. Since addition and multiplication of rational numbers is a rational number, it follows that $a * b$ is a rational number. Hence, $a * b \in G$. Thus, $*$ is a binary operation on G. Let $a, b, c \in G$. Then

$$
\begin{aligned}
(a * b) * c &= (a + b + ab) * c \\
&= a + b + ab + c + ac + bc + abc \\
&= a + (b + c + bc) + a(b + c + bc) \\
&= a + b * c + a(b * c) \\
&= a * (b * c).
\end{aligned}
$$

This shows that $*$ is associative. Thus, $(G, *)$ is a semigroup. Now $0 \in G$ and $0*a = 0+a+0\cdot a = a$ for all $a \in G$. Hence, $(G, *)$ satisfies (i) of Theorem 2.1.22. Now for all $a \in G$, $a+1 \neq 0$. Note that $-\frac{a}{a+1} \neq -1$. Therefore, $-\frac{a}{a+1} \in G$ and

$$-\frac{a}{a+1} * a = -\frac{a}{a+1} + a + (-\frac{a}{a+1})a = \frac{-a+a+a^2-a^2}{a+1} = 0.$$

This implies that $(G, *)$ satisfies (ii) of Theorem 2.1.22. Hence, by Theorem 2.1.22, $(G, *)$ is a group.

$\diamond$ **Exercise 4** Let G be a group and $x \in G$. Suppose $o(x) = mn$, where m and n are relatively prime. Show that there exist $y, z \in G$ such that $x = y * z = z * y$ and $o(y) = m$ and $o(z) = n$.

Solution: Since $\gcd(m, n) = 1$ there exist $s, t \in \mathbf{Z}$ such that $1 = ms + nt$. Now $x = x^{ms+nt} = x^{ms} * x^{nt}$. Let $y = x^{nt}$ and $z = x^{ms}$. Then $x = y * z = z * y$. Now $y^m = (x^{nt})^m = x^{mnt} = e$. Hence, $o(y)$ divides m. Similarly, $o(z)$ divides n. Suppose $o(y) = m_1$ and $o(z) = n_1$. It is an easy exercise to verify that $(y*z)^l = y^l * z^l$ for all positive integers l. Thus, $x^{m_1 n_1} = (y * z)^{m_1 n_1} = y^{m_1 n_1} * z^{m_1 n_1} = e * e = e$. Hence, $mn|m_1 n_1$. But since $m_1|m$ and $n_1|n$, we must have $m = m_1$ and $n = n_1$.

$\diamond$ **Exercise 5** Let $(G, *)$ be a group of even order. Show that there exists $a \in G$ such that $a \neq e$, $a^2 = e$.

Solution: Let $A = \{g \in G \mid g \neq g^{-1}\} \subseteq G$. Then $e \notin A$. If $g \in A$, then $g^{-1} \in A$, i.e., elements of A occurs in pairs. Therefore, the number of elements in A is even. This implies that the number of elements in $\{e\} \cup A$ is odd. Since the number of elements in G is even and $\{e\} \cup A \subseteq G$, there exists $a \in G$ such that $a \notin \{e\} \cup A$. But then $a \neq e$ and $a \notin A$. Hence, there exists $a \in G$ such that $a \neq e$ and $a = a^{-1}$ or $a^2 = e$.

$\diamond$ **Exercise 6** Let $(G, *)$ be a group and $a, b \in G$. Suppose that $a*b = b*a^{-1}$ and $b * a = a * b^{-1}$. Show that $a^4 = b^4 = e$.

Solution: Since $a*b = b*a^{-1}$, $a = b*a^{-1}*b^{-1}$. Similarly, $b = a*b^{-1}*a^{-1}$. Thus, $b*a = a*b^{-1} = (b*a^{-1}*b^{-1})*b^{-1} = b*a^{-1}*b^{-2}$. Multiply both sides of the equation $b*a = b*a^{-1}*b^{-2}$ by b^{-1} to get $a = a^{-1}*b^{-2}$. This implies that $a^2 = b^{-2}$. Hence, $a^4 = a^2 * a^2 = a^2 * b^{-2} = a*(a*b^{-1})*b^{-1} = a*(b*a)*b^{-1} = (a*b)*a*b^{-1} = (b*a^{-1})*a*b^{-1} = b*(a^{-1}*a)*b^{-1} = b*e*b^{-1} = e$. Also, $b^4 = a^{-4} = e$.

Exercise 7 Let $(G, *)$ be a group and $a, b \in G$. Suppose that $a * b^n = b^{n+1} * a$ and $b * a^n = a^{n+1} * b$ for some $n \in \mathbf{N}$. Show that $a = b = e$.

Solution: Multiply both sides of the equation $a * b^n = b^{n+1} * a$ by b^{-n} to get $a = b^{n+1} * a * b^{-n}$. Thus, $a^2 = a * a = a* b^{n+1} * a * b^{-n} = (a* b^n) * b * a * b^{-n} = (b^{n+1} * a) * b * a * b^{-n} = b^{n+1} * (a * b) * a * b^{-n}$. Now $a^3 = a * a^2 = a * (b^{n+1} * (a * b) * a * b^{-n}) = (a * b^n) * b * (a * b) * a * b^{-n} = (b^{n+1} * a) * b * (a * b) * a * b^{-n} = b^{n+1} * (a * b)^2 * a * b^{-n}$. Hence, we see that we could use induction to obtain

$$a^n = b^{n+1} * (a * b)^{n-1} * a * b^{-n} \tag{2.2}$$

for all $n \in \mathbf{N}$. Also,

$$
\begin{aligned}
b * a^n &= a^{n+1} * b \\
&= a * a^n * b \\
&= a * (b^{n+1} * (a * b)^{n-1} * a * b^{-n}) * b \\
&= a * b^{n+1} * (a * b)^{n-1} * a * b^{1-n} \\
&= (a * b^n) * b * (a * b)^{n-1} * a * b^{1-n} \\
&= (b^{n+1} * a) * b * (a * b)^{n-1} * a * b^{1-n} \\
&= b^{n+1} * (a * b)^n * a * b^{1-n},
\end{aligned}
$$

which implies that

$$a^n = b^n * (a * b)^n * a * b^{1-n}. \tag{2.3}$$

From Eqs. (2.2) and (2.3),

$$b^{n+1} * (a * b)^{n-1} * a * b^{-n} = b^n * (a * b)^n * a * b^{1-n},$$

which implies that

$$b * (a * b)^{n-1} * a = (a * b)^n * a * b = (a * b)^{n+1}.$$

Thus,

$$
\begin{aligned}
(a * b)^{n+1} &= b * (a * b)^{n-1} * a \\
&= b* \underbrace{((a * b) * \cdots * (a * b))}_{n-1 \text{ times}} *a \\
&= \underbrace{(b * a) * \cdots * (b * a)}_{n \text{ times}} \\
&= (b * a)^n.
\end{aligned}
\tag{2.4}
$$

Interchange the role of a and b to get

$$(b * a)^{n+1} = (a * b)^n. \tag{2.5}$$

Hence, $(a * b)^n = (b * a)^{n+1} = (b * a)^n * (b * a) = (a * b)^{n+1} * (b * a)$ and so $e = (a * b) * (b * a)$, which implies that

$$a^2 = b^{-2}. \tag{2.6}$$

Now
$$b * a^n = b * a^2 * a^{n-2} = b * b^{-2} * a^{n-2} = b^{-1} * a^{n-2} \qquad (2.7)$$

and
$$a^{n+1} * b = a^{n-1} * a^2 * b = a^{n-1} * b^{-2} * b = a^{n-1} * b^{-1}. \qquad (2.8)$$

Thus, from Eqs. (2.7) and (2.8) it follows that $b^{-1} * a^{n-2} = a^{n-1} * b^{-1}$ and so
$$a^{n-1} = b^{-1} * a^{n-2} * b = (b^{-1} * a * b)^{n-2}. \qquad (2.9)$$

Now $b * a^n = a^{n+1} * b$ implies that
$$a^n = (b^{-1} * a * b)^{n+1}. \qquad (2.10)$$

Hence, $a^n = (b^{-1}*a*b)^{n+1} = (b^{-1}*a*b)^{n-2}*(b^{-1}*a*b)^3 = a^{n-1}*(b^{-1}*a*b)^3$, which implies that $a = (b^{-1}*a*b)^3 = b^{-1}*a^3*b$. Thus, $a^3*b = b*a$. Therefore, $b * a = a^3 * b = a* a^2 * b = a * b^{-2} * b = a * b^{-1}$ by Eq. (2.6). That is, we have
$$b * a = a * b^{-1}. \qquad (2.11)$$

Similarly,
$$a * b = b * a^{-1}. \qquad (2.12)$$

Now $a * b = b * a^{-1}$ implies that $a * b * a = b$. Thus, $b = a * b * a = a * a * b^{-1}$ [by Eq. (2.11)]. Hence,
$$a^2 = b^2.$$

Suppose n is even. Then $a^2 = b^2$ implies that $a^n = b^n$. Hence, $a * b^n = b^{n+1} * a$ implies that $a^{n+1} = a^n * b * a$ and so $b = e$. Similarly, $a = e$. Suppose n is odd. Let $n = 2k + 1$. Then $a^{2k} = b^{2k}$. Now $a * b^n = b^{n+1} * a \Rightarrow a * b^{2k+1} = b^{2k+2} * a \Rightarrow a * a^{2k} * b = a^{2k+2} * a$. Thus, $b = a^2 = b^2$. Hence, $b = e$. Similarly, $a = e$.

Exercise 8 (Hays) Let $(S, *)$ be a semigroup. Show that S is a group if and only if for all $a \in S$ there exists a unique $b \in S$ such that $a * b * a = a$.

Solution: Suppose for all $a \in S$, there exists a unique $b \in S$ such that $a * b * a = a$. Let $a \in S$. Then there exists $b \in S$ such that $a * b * a = a$. Thus, $a * b * a * b = a * b$ and so $(a * b)^2 = a * b$. Hence, S has an idempotent element. If $(S, *)$ is to be a group, then it can have only one idempotent (Corollary 2.1.12), namely, the identity element. Therefore, first we show that S has only one idempotent.

Suppose e and f are two idempotents in S. Since $e * f \in S$, there exists a unique g such that $(e * f) * g * (e * f) = e * f$. Now $(e * f) * (g * e) * (e * f) = (e * f) * g * (e * e) * f = (e * f) * g * e^2 * f = (e * f) * g * (e * f) = e * f$. Since g is unique such that $(e * f) * g * (e * f) = (e * f)$, it follows that $g * e = g$. Similarly, since $(e * f) * (f * g) * (e * f) = (e * f) * g * (e * f) = e * f$, the uniqueness of g implies that $f * g = g$. Also, $(e * f) * (g * (e * f) * g) * (e * f) = ((e * f) * g * (e * f)) * g * (e * f) = (e * f) * g * (e * f)$. Again, the uniqueness of g implies that $g * (e * f) * g = g$. Hence, $g^2 = g * g = (g * e) * (f * g) = g*(e*f)*g = g$. Thus, g is an idempotent. Now $g = g*g*g$ and $g*(e*f)*g = g$. Hence, by the uniqueness of the middle element $g = e * f$. Therefore, $e * f$ is an idempotent. Now $(e * f) * f * (e * f) = (e * (f * f)) * (e * f) = (e * f) * (e * f) = e * f$ and similarly $(e * f) * e * (e * f) = e * f$. By the uniqueness of the middle element, it follows that $e = f$. Hence, S has a unique idempotent element.

Let e be the idempotent element of S. Let $a \in S$. Then there exists $b \in S$ such that $a * b * a = a$, which implies that $(a * b)^2 = a * b$. Hence, $a * b = e$. Also, $a * b * a = a$ implies that $b * a * b * a = b * a$. Thus, $b * a$ is an idempotent. Hence, $b * a = e$. Also, $a * b * a = a$ together with $a * b = e = b * a$ implies that $e * a = a = a * e$. Therefore, e is the identity element. Since $a * b = e = b * a$, b is an inverse of a. Consequently, $(S, *)$ is a group.

Conversely, suppose $(S, *)$ is a group. Let $a \in S$. Note that $a * a^{-1} * a = a$. This shows the existence of an element $b \in S$ such that $a * b * a = a$, namely, $b = a^{-1}$. To show the uniqueness, suppose there exist $b, c \in S$ such that $a * b * a = a$ and $a * c * a = a$. Then $a * b * a = a * c * a$ and by the cancellation laws, $b = c$. Thus, b is unique such that $a * b * a = a$.

2.1.2   Exercises

1. Which of the following mathematical systems are semigroups? Which are groups?

   (i) $(\mathbf{N}, *)$, where $a * b = a$ for all $a, b \in \mathbf{N}$.

   (ii) $(\mathbf{Z}, *)$, where $a * b = a - b$ for all $a, b \in \mathbf{Z}$.

   (iii) $(\mathbf{R}, *)$, where $a * b = |a|b$ for all $a, b \in \mathbf{R}$.

   (iv) $(\mathbf{R}, *)$, where $a * b = a + b + 1$ for all $a, b \in \mathbf{R}$.

   (v) $(\mathbf{R}, *)$, where $a * b = a + b - ab$ for all $a, b \in \mathbf{R}$.

   (vi) $(\mathbf{Q}, *)$, where $a * b = \frac{ab}{2}$ for all $a, b \in \mathbf{Q}$.

   (vii) $(G, *)$, where

   $$G = \left\{ \begin{bmatrix} a & b \\ -b & a \end{bmatrix} \mid \begin{bmatrix} a & b \\ -b & a \end{bmatrix} \neq \begin{bmatrix} 0 & 0 \\ 0 & 0 \end{bmatrix} \text{ and } a, b \in \mathbf{R} \right\}$$

   and $*$ is the usual matrix multiplication.

   (viii) $(G, *)$, where G is the set of all matrices of the following form over $\mathbf{Z}$

   $$\begin{bmatrix} 1 & a & b \\ 0 & 1 & c \\ 0 & 0 & 1 \end{bmatrix}$$

and $*$ is the usual matrix multiplication.

2. Let $G = \{(a, b) \mid a, b \in \mathbf{R}, b \neq 0\}$. Define a binary operation $*$ on G by $(a, b) * (c, d) = (a + bc, bd)$ for all $(a, b), (c, d) \in G$. Show that $(G, *)$ is a noncommutative group.

3. Let $G = \left\{ \begin{bmatrix} a & b \\ c & d \end{bmatrix} \mid a, b, c, d \in \mathbf{R}, \ ad - bc = 1 \right\}$. Show that G is a group under usual matrix multiplication. (This group is usually denoted by $SL(2, \mathbf{R})$ and is called the **special linear group of degree 2**.)

4. Let $G = \left\{ \begin{bmatrix} 1 & n \\ 0 & 1 \end{bmatrix} \mid n \in \mathbf{Z} \right\}$. Show that $(G, *)$ is a commutative group, where $*$ denotes the usual matrix multiplication. Also, show that $(G, *)$ is torsion-free.

5. In $\mathbf{Z}_{14}$, find the smallest positive integer n such that $n[6] = [0]$.

6. Find an element $[b] \in \mathbf{Z}_9$ such that $[8] \cdot_9 [b] = [1]$. Does $[b] \in U_9$?

7. In U_{24}, find the smallest positive integer n such that $[7]^n = [1]$.

8. Describe U_6, U_9, U_{12}, U_{24} of Example 2.1.6.

9. Let p be a prime. Show that $U_p = \mathbf{Z}_p \backslash \{[0]\}$.

10. Let $U_n = \{[a] \in \mathbf{Z}_n \backslash \{[0]\} \mid \gcd(a, n) = 1\}$. Show that $(U_n, \cdot_n)$ is a group, where $\cdot_n$ is multiplication modulo n.

11. Show that $U_n = \{[a] \in \mathbf{Z}_n \backslash \{[0]\} \mid$ additive order of $[a] = n \}$.

12. Let $(G, *)$ be a group and $a, b \in G$. Suppose that $a^2 = e$ and $a * b^4 * a = b^7$. Show that $b^{33} = e$.

13. Let $(G, *)$ be a group and $a, b \in G$. Suppose that $a^{-1} * b^2 * a = b^3$ and $b^{-1} * a^2 * b = a^3$. Show that $a = b = e$.

14. Let $(G, *)$ be a group. If $a, b \in G$ are such that $a^4 = e$ and $a^2 * b = b * a$, show that $a = e$.

15. Let $(G, *)$ be a group and $x, a, b \in G$. Let $c = x * a * x^{-1}$ and $d = x * b * x^{-1}$. Show that $a * b = b * a$ if and only if $c * d = d * c$.

16. Let $(G, *)$ be a group such that $a^2 = e$ for all $a \in G$. Show that G is commutative.

17. Prove that a group $(G, *)$ is commutative if and only if $(a * b)^{-1} = a^{-1} * b^{-1}$ for all $a, b \in G$.

18. Let $(G, *)$ be a group. Prove that if $(a * b)^2 = a^2 * b^2$ for all $a, b \in G$, then $(G, *)$ is commutative.

19. Prove that a group $(G, *)$ is commutative if and only if for all $a, b \in G$, $(a * b)^n = a^n * b^n$ for any three consecutive integers n.

20. Let $(G, *)$ be a group. If G has only two elements, prove that G is commutative.

21. Let $(G, *)$ be a group and $a, b, c \in G$. Find an element $x \in G$ such that $a * x * b = c$. Is x unique?

22. Let $(G, *)$ be a group and $a, b \in G$. Show that $(a * b * a^{-1})^n = a * b^n * a^{-1}$ for all integers n.

23. Let $(G, *)$ be a finite group and $a \in G$. Show that there exists $n \in \mathbf{N}$ such that $a^n = e$.

24. If $(G, *)$ is a group and $a_1, \ldots, a_n \in G$, prove that $(a_1 * \cdots * a_n)^{-1} = a_n^{-1} * \cdots * a_1^{-1}$.

25. Let $(G, *)$ and $(H, \cdot)$ be groups. Define the operation $\star$ on $G \times H = \{(a, b) \mid a \in G, b \in H\}$ by $(a, b) \star (c, d) = (a * c, b \cdot d)$. Prove that $(G \times H, \star)$ is a group. If $(G, *)$ and $(H, \cdot)$ are commutative, prove that $(G \times H, \star)$ is commutative. The group $(G \times H, \star)$ is called the **direct product** of G and H.

26. Let $(G, *)$ be a finite group and $a \in G$. Show that $o(a) \leq |G|$.

27. Let $(G, *)$ be a group and $a, b \in G$.
    (i) Show that a and a^{-1} have the same order.
    (ii) Show that a and $b * a * b^{-1}$ have the same order.
    (iii) Show that $a * b$ and $b * a$ have the same order.

28. Let $(G, *)$ be a group and $a, b \in G$.
    (i) Suppose that $a * b = b^5 * a^3$. Show that $o(b * a^{-1}) = o(b^5 * a) = o(b^3 * a^3)$.
    (ii) Generalize (i) to arbitrary powers of a and b.

29. Let $(G, *)$ be a group, $a \in G$ and $o(a) = n$. Let $1 \leq p \leq n$ be such that p and n are relatively prime. Show that $o(a^p) = n$.

30. Let $(G, *)$ be a group, $a \in G$, and $o(a) = p$, where p is a prime.
    (i) Show that $o(a^k) = p$ for all $1 \leq k < p$.
    (ii) Show that for all $m \in \mathbf{N}$, either $a^m = e$ or $o(a^m) = p$.

31. Let $(G, *)$ be a group and $a \in G$. Suppose that $o(a) = n$ and $n = mk$ for some $m, k \in \mathbf{Z}$. What is $o(a^k)$?

32. (i) Let $(G, *)$ be a group, $a, b \in G$, $o(a) = n$, $o(b) = m$, $\gcd(m, n) = 1$, and $a * b = b * a$. Show that $o(a * b) = mn$.

    (ii) Let $(G, *)$ be a group, $a_i \in G$, $o(a_i) = n_i$, $1 \le i \le m$. Suppose $\gcd(n_i, n_j) = 1$ and $a_i a_j = a_j a_i$ for all i and j. Let $x = a_1 * a_2 * \cdots * a_m$. Show that $o(x) = n_1 n_2 \cdots n_m$.

33. Let $(G, *)$ be a group and $x \in G$. Suppose $o(x) = n = n_1 n_2 \cdots n_k$, where for all $i \ne j$, n_i and n_j are relatively prime. Show that there exists $x_i \in G$ such that $o(x_i) = n_i$ for all $i = 1, 2, \ldots, k$, $x = x_1 * x_2 * \cdots * x_k$ and $x_i * x_j = x_j * x_i$ for all i and j.

34. Let $G = \{(a, b) \mid a, b \in \mathbf{R}, a \ne 0\}$. Then G is a group under the binary operation $(a, b) * (c, d) = (ac, bc + d)$ for all $(a, b), (c, d) \in G$. Show that G has infinitely many elements of order 2, but G has no element of order 3.

35. Let $a, b \in \text{Sym}$. As remarked in Example 2.1.29, every rigid motion of the square can be considered a one-one function of $\{1, 2, 3, 4\}$ onto itself. Consider $a * b$ as a function. Show that $a * b = a \circ b$, where $*$ represents the binary operation of rigid motions of the square and $\circ$ is the composition of functions.

36. Let $(S, *)$ be a finite semigroup. Prove that there exists $a \in S$ such that $a^2 = a$.

37. Let $(G, *)$ be a finite semigroup with identity. Prove that $(G, *)$ is a group if and only if G has only one element a such that $a^2 = a$.

38. Prove that a semigroup $(S, *)$ is a group if and only if $a * S = S$ and $S * a = S$ for all $a \in S$, where $a * S = \{a * s \mid s \in S\}$ and $S * a = \{s * a \mid s \in S\}$.

39. Prove that a semigroup $(S, *)$ is a group if and only if

    (i) there exists $e \in S$ such that $a * e = a$ for all $a \in S$, and

    (ii) for all $a \in S$ there exists $b \in S$ such that $a * b = e$.

40. Rewrite the statements and proofs of the theorems in this chapter using additive notation.

41. Let $(G, *)$ be a group, $a, b \in G$ and $m, n \in \mathbf{Z}$. Prove that

    (i) $a^n * a^m = a^{n+m} = a^m * a^n$,

    (ii) $(a^n)^m = a^{nm}$,

(iii) $a^{-n} = (a^n)^{-1}$,

(iv) $e^n = e$,

(v) $(a * b)^n = a^n * b^n$, if $(G, *)$ is commutative.

42. Write the proof if the following statements are true; otherwise, give a counterexample.

(i) Let $T(S)$ be the set of all functions on $S = \{1, 2, 3\}$. $T(S)$ is a group under composition of functions.

(ii) $M_2(\mathbf{R}) = \left\{ \begin{bmatrix} a & b \\ c & d \end{bmatrix} \middle| a, b, c, d \in \mathbf{R} \right\}$ is a group under usual matrix multiplication.

(iii) Every group of four elements is commutative.

(iv) A group has only one idempotent element.

(v) A semigroup with only one idempotent is a group.

(vi) If a semigroup S satisfies the cancellation laws, then S is a group.

Niels Henrik Abel (1802–1829) was born on August 5, 1802, in Finnöy, Norway. He was the second of six children. Abel and his brothers received their first education from their father.

At the age of 13, Abel along with his older brother, was sent to the Cathedral school in Christiania (Oslo). In 1817, his mathematics teacher was Bernt Michael Holmbë, who was seven years older that Abel. Holmbë recognized Abel's talent and started giving him special problems and recommended special books outside the curriculum. Abel and Holmbë read the calculus text of Euler and the work of Lagrange and Laplace. Soon Abel became familiar with most of the important mathematical literature.

Abel's father died when he was 18 years old and the responsibility of supporting the family fell on his shoulders. He gave private lessons and did odd jobs. However, he continued to carry out his mathematical research.

Abel, in his last year of school, attacked the problem of the solvability of the quintic equation, a problem that had been unsettled since the sixteenth century. Abel thought that he had solved the problem and submitted his work for publication. Unable to find an error and understand his arguments, he was asked by the editor to illustrate his method. In 1824, during the process of illustration he discovered an error. This discovery led Abel to a proof that no such solution exists. He also worked on elliptic functions and in essence revolutionized the theory of elliptic functions.

He traveled to Paris and Berlin in order to find a teaching position. Then poverty took its toll, and Abel died from tuberculosis on April 6, 1829. Two days later a letter from Crelle reached his address, conveying the news of his appointment to the professorship of mathematics at the University of Berlin.

Abel is honored by such terms as Abelian group and Abelian function.

Chapter 3

Permutation Groups

Permutation groups is one of the specialized theories of groups which arose from the source, classical algebra, in the evolution of group theory.

3.1   Permutation Groups

As stated earlier, there are four major sources from which abstract group theory evolved. Mathematicians' interest in finding formulas to solve polynomial equations by means of radicals led some mathematicians to the study of permutations of the roots of rational functions. Lagrange, Rufini, and Cauchy were among the earlier mathematicians to work with permutation groups. However, it was Cauchy whose systematic study of permutation groups (between 1815 and 1845) is believed, by some, to be the origin of abstract group theory. Many of the concepts and major results in this chapter are due to Cauchy.

    We begin our study of permutation groups by defining what a permutation is.

Definition 3.1.1 *Let X be a nonempty set. A **permutation** π of X is a one-one function from X onto X.*

Definition 3.1.2 *A group $(G, *)$ is called a **permutation** group on a nonempty set X if the elements of G are permutations of X and the operation $*$ is the composition of two functions.*

Example 3.1.3 *Let X be any nonempty set and S_X be the set of all one-one functions from X onto X, as defined in Example 2.1.9. Then $(S_X, \circ)$ is a group as we have shown in Example 2.1.9, where $\circ$ is the composition of functions. Hence, $(S_X, \circ)$ is a permutation group.*

    In this chapter, and in fact in this text, our study of permutation groups will focus on permutation groups on finite sets, i.e., X is a finite set.

Before we consider more examples of permutation groups, let us fix some notation which will be useful when working with permutations.

Let $I_n = \{1, 2, \ldots, n\}$, $n \geq 1$. Let π be a permutation on I_n. Then

$$\pi = \{(1, \pi(1)), (2, \pi(2)), \ldots, (n, \pi(n))\}.$$

(Recall that a function $f : A \to A$ is a subset of $A \times A$.) It is sometimes convenient to describe a permutation by means of the following notational device:

$$\pi = \begin{pmatrix} 1 & 2 & 3 & \cdots & n \\ \pi(1) & \pi(2) & \pi(3) & \cdots & \pi(n) \end{pmatrix}.$$

This notation is due to Cauchy and is called the **two-row notation**. In the upper row, we list all the elements of I_n and in the lower row under each element $i \in I_n$, we write the image of the element, i.e., $\pi(i)$.

Example 3.1.4 *Let $n = 4$ and π be the permutation on I_4 defined by $\pi(1) = 2$, $\pi(2) = 4$, $\pi(3) = 3$, and $\pi(4) = 1$. Then using the two-row notation we can write*

$$\pi = \begin{pmatrix} 1 & 2 & 3 & 4 \\ 2 & 4 & 3 & 1 \end{pmatrix}.$$

As we shall see, the two-row notation of permutations is quite convenient while doing computations such as determining the composition of permutations.

Let $n = 7$ and π and σ be two permutations on I_7 defined by

$$\pi = \begin{pmatrix} 1 & 2 & 3 & 4 & 5 & 6 & 7 \\ 1 & 3 & 4 & 6 & 7 & 2 & 5 \end{pmatrix}$$

and

$$\sigma = \begin{pmatrix} 1 & 2 & 3 & 4 & 5 & 6 & 7 \\ 2 & 5 & 3 & 1 & 7 & 6 & 4 \end{pmatrix}.$$

Let us compute $\pi \circ \sigma$. Now by definition, $(\pi \circ \sigma)(i) = \pi(\sigma(i))$ for all $i \in I_7$. Thus,

$$(\pi \circ \sigma)(1) = \pi(\sigma(1)) = \pi(2) = 3,$$

$$(\pi \circ \sigma)(2) = \pi(\sigma(2)) = \pi(5) = 7$$

and so on. From this, it is clear that when determining, say, $(\pi \circ \sigma)(1)$, we start with σ and finish with π and read as follows: 1 goes to 2 (under σ) and 2 goes to 3 (under π) and so 1 goes to 3 (under $\pi \circ \sigma$). We can exhibit this in the following form:

$$
\begin{array}{lll}
1 \xrightarrow{\sigma} 2 \xrightarrow{\pi} 3 & \qquad & 1 \xrightarrow{\pi \circ \sigma} 3 \\
2 \xrightarrow{\sigma} 5 \xrightarrow{\pi} 7 & & 2 \xrightarrow{\pi \circ \sigma} 7 \\
3 \xrightarrow{\sigma} 3 \xrightarrow{\pi} 4 & & 3 \xrightarrow{\pi \circ \sigma} 4 \\
4 \xrightarrow{\sigma} 1 \xrightarrow{\pi} 1 & & 4 \xrightarrow{\pi \circ \sigma} 1 \\
5 \xrightarrow{\sigma} 7 \xrightarrow{\pi} 5 & & 5 \xrightarrow{\pi \circ \sigma} 5 \\
6 \xrightarrow{\sigma} 6 \xrightarrow{\pi} 2 & & 6 \xrightarrow{\pi \circ \sigma} 2 \\
7 \xrightarrow{\sigma} 4 \xrightarrow{\pi} 6 & & 7 \xrightarrow{\pi \circ \sigma} 6.
\end{array}
$$

Thus,

$$
\pi \circ \sigma = \begin{pmatrix} 1 & 2 & 3 & 4 & 5 & 6 & 7 \\ 3 & 7 & 4 & 1 & 5 & 2 & 6 \end{pmatrix}.
$$

Example 3.1.5 *Let $n = 6$ and α and β be permutations on I_6 defined by*

$$
\alpha = \begin{pmatrix} 1 & 2 & 3 & 4 & 5 & 6 \\ 3 & 1 & 4 & 6 & 5 & 2 \end{pmatrix}
$$

and

$$
\beta = \begin{pmatrix} 1 & 2 & 3 & 4 & 5 & 6 \\ 1 & 3 & 5 & 4 & 2 & 6 \end{pmatrix}.
$$

Let us first determine $\alpha \circ \beta$. Now $1 \xrightarrow{\beta} 1 \xrightarrow{\alpha} 3$, i.e., $1 \xrightarrow{\alpha \circ \beta} 3$. Similarly, $2 \xrightarrow{\alpha \circ \beta} 4$, $3 \xrightarrow{\alpha \circ \beta} 5$, $4 \xrightarrow{\alpha \circ \beta} 6$, $5 \xrightarrow{\alpha \circ \beta} 1$, $6 \xrightarrow{\alpha \circ \beta} 2$. Thus,

$$
\alpha \circ \beta = \begin{pmatrix} 1 & 2 & 3 & 4 & 5 & 6 \\ 3 & 4 & 5 & 6 & 1 & 2 \end{pmatrix}.
$$

Similarly, for $\beta \circ \alpha$; $1 \xrightarrow{\alpha} 3 \xrightarrow{\beta} 5$, i.e., $1 \xrightarrow{\beta \circ \alpha} 5$ and so on. In this case, we start with α and finish with β. Note that

$$
\beta \circ \alpha = \begin{pmatrix} 1 & 2 & 3 & 4 & 5 & 6 \\ 5 & 1 & 4 & 6 & 2 & 3 \end{pmatrix}.
$$

We note that $\alpha \circ \beta \neq \beta \circ \alpha$.

Let S_n denote the set of all permutations on I_n, $n \geq 1$.

Example 3.1.6 *In this example, we describe S_3, i.e., the set of all permutations on $I_3 = \{1, 2, 3\}$. From Exercise 8 (page 50), we know that the number of one-one functions of I_3 onto I_3 is $3! = 6$. Thus, $|S_3| = 6$. Let e denote the identity permutation on I_3, i.e., $e = \begin{pmatrix} 1 & 2 & 3 \\ 1 & 2 & 3 \end{pmatrix}$. Let α_1 be a nonidentity permutation on I_3. Let us see some of the choices for α_1. Suppose $\alpha_1(1) = 1$.*

If $\alpha_1(2) = 2$, then we must have $\alpha_1(3) = 3$ since α_1 is a permutation. In this case, we see that $\alpha_1 = e$, a contradiction. Thus, we must have $\alpha_1(2) = 3$ and $\alpha_1(3) = 2$, i.e., $\alpha_1 = \begin{pmatrix} 1 & 2 & 3 \\ 1 & 3 & 2 \end{pmatrix}$. In a similar manner, we can show that the other four permutations on I_3 are $\alpha_2 = \begin{pmatrix} 1 & 2 & 3 \\ 2 & 1 & 3 \end{pmatrix}$, $\alpha_3 = \begin{pmatrix} 1 & 2 & 3 \\ 3 & 2 & 1 \end{pmatrix}$, $\alpha_4 = \begin{pmatrix} 1 & 2 & 3 \\ 2 & 3 & 1 \end{pmatrix}$, and $\alpha_5 = \begin{pmatrix} 1 & 2 & 3 \\ 3 & 1 & 2 \end{pmatrix}$. Thus,

$$S_3 = \{e, \alpha_1, \alpha_2, \alpha_3, \alpha_4, \alpha_5\}.$$

Let us denote α_2 by α and α_4 by β. We ask the reader to check that $\beta^2 = \alpha_5$, $\alpha \circ \beta = \alpha_1$, and $\alpha \circ \beta^2 = \alpha_3$. Hence, we can write

$$S_3 = \{e, \beta, \beta^2, \alpha, \alpha \circ \beta, \alpha \circ \beta^2\}.$$

Since $(S_3, \circ)$ is also a group, we ask the reader to show that $o(\alpha) = 2$ and $o(\beta) = 3$ by showing that $\alpha^2 = e$ and $\beta^2 \neq e$, but $\beta^3 = e$.

In the previous example, the permutation group $(S_3, \circ)$ consisted of all permutations on the set I_3. Next, we give an example of a permutation group that does not contain all permutations on a given set.

Example 3.1.7 Let $n = 4$ and consider $I_4 = \{1, 2, 3, 4\}$. Recall that in Example 2.1.29, we remarked that rigid motions of the square can be viewed as permutations on I_4. Let S be the set of all permutations that corresponds to the rigid motions of the square. We will use the same notation for the permutations, i.e., r_{90} is the permutation $\begin{pmatrix} 1 & 2 & 3 & 4 \\ 2 & 3 & 4 & 1 \end{pmatrix}$, r_{360} is the identity permutation, etc. By Exercise 35 (page 80), it follows that the multiplication table of $(S, \circ)$ is the same as the multiplication table of the group $(Sym, *)$. Now composition of functions is associative and from the multiplication table, it follows that S is closed under $\circ$, r_{360} is the identity of $(S, \circ)$, and every element of S has an inverse. Thus, $(S, \circ)$ is a group. Hence, the group of symmetries of a square can be thought of as a permutation group on I_4.

The following theorem describes some basic properties of S_n.

Theorem 3.1.8 (i) $(S_n, \circ)$ is a group for any positive integer $n \geq 1$.
(ii) If $n \geq 3$, then $(S_n, \circ)$ is noncommutative.
(iii) $|S_n| = n!$

Proof.   (i) We have already noted that the set of all one-one functions of any nonempty set onto itself forms a group under composition of functions in Example 2.1.9. Thus, $(S_n, \circ)$ is a group for any positive integer $n \geq 1$.

(ii) Let $n \geq 3$. Let $\alpha, \beta \in S_n$ be defined by

$$\alpha = \begin{pmatrix} 1 & 2 & 3 & 4 & \cdots & n \\ 1 & 3 & 2 & 4 & \cdots & n \end{pmatrix} \text{ and } \beta = \begin{pmatrix} 1 & 2 & 3 & 4 & \cdots & n \\ 3 & 2 & 1 & 4 & \cdots & n \end{pmatrix}.$$

Now

$$\alpha \circ \beta = \begin{pmatrix} 1 & 2 & 3 & 4 & \cdots & n \\ 2 & 3 & 1 & 4 & \cdots & n \end{pmatrix}$$

and

$$\beta \circ \alpha = \begin{pmatrix} 1 & 2 & 3 & 4 & \cdots & n \\ 3 & 1 & 2 & 4 & \cdots & n \end{pmatrix}.$$

Thus, $(\alpha \circ \beta)(1) = 2 \neq 3 = (\beta \circ \alpha)(1)$. Hence, $\alpha \circ \beta \neq \beta \circ \alpha$ and so S_n is noncommutative.

(iii) This follows from Exercise 8 (page 50).

Definition 3.1.9 *The group* $(S_n, \circ)$ *is called the* **symmetric group on** I_n.

Consider the permutation $\pi = \begin{pmatrix} 1 & 2 & \cdots & n \\ \pi(1) & \pi(2) & \cdots & \pi(n) \end{pmatrix}$. If $\pi(i) = i$, then we drop the column $\begin{matrix} i \\ \pi(i) \end{matrix}$. For example, $\alpha = \begin{pmatrix} 1 & 2 & 3 & 4 \\ 1 & 4 & 3 & 2 \end{pmatrix}$ is denoted by $\begin{pmatrix} 2 & 4 \\ 4 & 2 \end{pmatrix}$.

Definition 3.1.10 *Let* π *be an element of* S_n. *Then* π *is called a* **k-cycle**, *written* $(i_1 \ i_2 \cdots i_k)$, *if*

$$\pi = \begin{pmatrix} i_1 & i_2 & \cdots & i_{k-1} & i_k \\ i_2 & i_3 & \cdots & i_k & i_1 \end{pmatrix},$$

i.e., $\pi(i_j) = i_{j+1}$, $j = 1, 2, \ldots, k-1$, $\pi(i_k) = i_1$, *and* $\pi(a) = a$ *for any other element of* I_n.

Note that if $\pi = (i_1 i_2 \cdots i_k)$, then

$$\begin{aligned} \pi &= (i_1 i_2 \cdots i_k) \\ &= (i_2 i_3 \cdots i_k i_1) \\ &\vdots \\ &= (i_j i_{j+1} \cdots i_k i_1 \cdots i_{j-1}). \end{aligned}$$

A k-cycle is called a **transposition** when $k = 2$.

We know that in Example 3.1.7, the permutation r_{90} is a 4-cycle and d_2 is a 2-cycle. We write

$$r_{90} = (1\ 2\ 3\ 4)$$

and

$$d_2 = (1\ 3).$$

The identity of S_n is sometimes denoted by (1) or e.

Example 3.1.11 *Using the cycle notation, we can write*

$$S_3 = \{e, (1\ 2), (1\ 3), (2\ 3), (1\ 2\ 3), (1\ 3\ 2)\}.$$

We now note some of the properties of the group $(S_3, \circ)$.

(i) $(S_3, \circ)$ is a noncommutative group of order 6 by Theorem 3.1.8.

(ii) S_3 contains two elements of order 3; for $(1\ 2\ 3) \circ (1\ 2\ 3) = (1\ 3\ 2) \neq e$ and $(1\ 2\ 3) \circ (1\ 2\ 3) \circ (1\ 2\ 3) = e$. Hence, the order of $(1\ 2\ 3)$ is 3. Similarly, the order of $(1\ 3\ 2)$ is 3. The order of $(1\ 2), (1\ 3),$ and $(2\ 3)$ is 2 since $(1\ 2) \circ (1\ 2) = e$, $(1\ 3) \circ (1\ 3) = e$, and $(2\ 3) \circ (2\ 3) = e$.

(iii) In S_3, the product of distinct elements of order 2 is an element of order 3. $(1\ 2) \circ (2\ 3) = (1\ 2\ 3)$, $(1\ 3) \circ (1\ 2) = (1\ 2\ 3)$, $(1\ 2) \circ (1\ 3) = (1\ 3\ 2)$, $(2\ 3) \circ (1\ 2) = (1\ 3\ 2)$, $(1\ 3) \circ (2\ 3) = (1\ 3\ 2)$, and $(2\ 3) \circ (1\ 3) = (1\ 2\ 3)$.

Definition 3.1.12 *Let $\alpha, \beta \in S_n$. Then α and β are called **conjugate** if there exists $\gamma \in S_n$ such that*

$$\gamma \circ \alpha \circ \gamma^{-1} = \beta.$$

The following theorem shows how to compute the conjugate of a cycle.

Theorem 3.1.13 *Let $\pi = (i_1 i_2 \cdots i_l) \in S_n$ be a cycle. Then for all $\alpha \in S_n$,*

$$\alpha \circ \pi \circ \alpha^{-1} = (\alpha(i_1)\ \alpha(i_2)\ \cdots\ \alpha(i_l)).$$

Proof.   Since $\alpha \in S_n$, α is a one-one mapping of I_n onto I_n. Thus, the elements $\alpha(1), \ldots, \alpha(n) \in I_n$ are all distinct and so $I_n = \{\alpha(1), \alpha(2), \ldots, \alpha(n)\}$. Let r be any integer such that $1 \leq r < l$. Then

$$
\begin{aligned}
(\alpha \circ \pi \circ \alpha^{-1})(\alpha(i_r)) &= \alpha(\pi(\alpha^{-1}(\alpha(i_r)))) \\
&= \alpha(\pi(i_r)) \\
&= \alpha(i_{r+1}).
\end{aligned}
$$

Also, $(\alpha \circ \pi \circ \alpha^{-1})(\alpha(i_l)) = \alpha(\pi(\alpha^{-1}(\alpha(i_l)))) = \alpha(\pi(i_l)) = \alpha(i_1)$. Now let $a \in I_n$ be such that $a \neq \alpha(i_r)$ for all r, $1 \leq r \leq l$. Then $\alpha^{-1}(a) \in I_n$ and $\alpha^{-1}(a) \neq i_r$ for all r, $1 \leq r \leq l$, and so $\pi(\alpha^{-1}(a)) = \alpha^{-1}(a)$. Thus,

$$
\begin{aligned}
(\alpha \circ \pi \circ \alpha^{-1})(a) &= \alpha(\pi(\alpha^{-1}(a))) \\
&= \alpha(\alpha^{-1}(a)) \\
&= a.
\end{aligned}
$$

It now follows that $\alpha \circ \pi \circ \alpha^{-1} = (\alpha(i_1)\ \alpha(i_2)\ \cdots\ \alpha(i_l))$. ∎

Definition 3.1.14 *Let* $\pi_1, \pi_2, \ldots, \pi_k \in S_n$. *Then* $\pi_1, \pi_2, \ldots, \pi_k$ *are called **disjoint** if for all* i, $1 \le i \le k$ *and for all* $a \in I_n$, $\pi_i(a) \ne a$ *implies* $\pi_j(a) = a$ *for all* $j \ne i$, $1 \le j \le k$.

In other words, $\pi_1, \pi_2, \ldots, \pi_k \in S_n$ are disjoint if for all $1 \le i \le k$ and for all $a \in I_n$, if π_i moves a, then all other permutations π_j must fix a, i.e., $\pi_j(a) = a$ for all $j \ne i$, $1 \le j \le k$.

Let π and λ be disjoint permutations on I_n. Let $a \in S$ be such that $\pi(a) \ne a$. Then $\lambda(a) = a$. Let $\pi(a) = b$. Then $(\pi \circ \lambda)(a) = \pi(\lambda(a)) = \pi(a) = b$. Also, $(\lambda \circ \pi)(a) = \lambda(\pi(a)) = \lambda(b)$. If $\pi(b) = b$, then $\pi(b) = b = \pi(a)$ and so $a = b$. Thus, $\pi(a) = b = a$, a contradiction. Hence, $\pi(b) \ne b$ and so $\lambda(b) = b$. Thus, $(\lambda \circ \pi)(a) = \lambda(\pi(a)) = \lambda(b) = b$. Hence, $(\pi \circ \lambda)(a) = (\lambda \circ \pi)(a)$. Suppose $\pi(a) = a$. If $\lambda(a) = a$, then $(\pi \circ \lambda)(a) = a = (\lambda \circ \pi)(a)$. Suppose $\lambda(a) \ne a$. By a similar argument as before, $(\pi \circ \lambda)(a) = (\lambda \circ \pi)(a)$. Therefore, $\pi \circ \lambda = \lambda \circ \pi$. Consequently, if π and λ are disjoint permutations, then they commute.

Consider $\pi = \begin{pmatrix} 1 & 2 & 3 & 4 & 5 & 6 & 7 & 8 \\ 2 & 5 & 1 & 8 & 3 & 7 & 6 & 4 \end{pmatrix} \in S_n$. Then $\pi = (1\ 2\ 5\ 3) \circ (4\ 8) \circ (6\ 7)$ can be written as a product of disjoint cycles. This leads us to the following theorem.

Theorem 3.1.15 *Any nonidentity permutation* π *of* S_n $(n \ge 2)$ *can be uniquely expressed (up to the order of the factors) as a product of disjoint cycles, where each cycle is of length at least 2.*

Proof.   We prove the result by induction on n. Suppose $n = 2$. Now $|S_2| = 2$ and the nonidentity element of S_2 is $\alpha = \begin{pmatrix} 1 & 2 \\ 2 & 1 \end{pmatrix}$. Now $\alpha = (1\ 2)$, i.e., α is a cycle. Thus, the theorem is true for $n = 2$. Suppose $n > 2$ and the theorem is true for all S_k such that $2 \le k < n$. Let π be a nonidentity element of S_n. Now $\pi^i(1) \in I_n$ for all integers i, $i \ge 1$. Therefore, $\{\pi(1), \pi^2(1), \ldots, \pi^i(1), \ldots\} \subseteq I_n$. Since I_n is a finite set, we must have $\pi^l(1) = \pi^m(1)$ for some integers l and m such that $l > m \ge 1$. This implies that $\pi^{l-m}(1) = 1$. Let us write $j = l - m$. Then $j > 0$ and $\pi^j(1) = 1$. Let i be the smallest positive integer such that $\pi^i(1) = 1$. Let
$$A = \{1, \pi(1), \pi^2(1), \ldots, \pi^{i-1}(1)\}.$$

Then all elements of the set A are distinct. Let $\tau \in S_n$ be the permutation defined by
$$\tau = (1\ \pi(1)\ \pi^2(1)\ \cdots\ \pi^{i-1}(1)),$$

i.e., τ is a cycle. Let $B = I_n \backslash A$. If $B = \phi$, then π is a cycle. Suppose $B \ne \phi$. Let $\sigma = \pi|_B$. If σ is the identity, then π is a cycle. Suppose that σ is not the identity. Now by the induction hypothesis, σ is a product of disjoint cycles on B, say, $\sigma = \sigma_1 \circ \sigma_2 \circ \cdots \circ \sigma_r$. Now for $1 \leq i \leq r$, define π_i by

$$\pi_i(a) = \begin{cases} \sigma_i(a) \text{ if } a \in B \\ a \text{ if } a \notin B. \end{cases}$$

Then $\pi_1, \pi_2, \ldots, \pi_r$ and τ are disjoint cycles in S_n. It is easy to see that $\pi = \pi_1 \circ \pi_2 \circ \cdots \circ \pi_r \circ \tau$. Thus, π is a product of disjoint cycles.

To prove the uniqueness, let $\pi = \pi_1 \circ \pi_2 \circ \cdots \circ \pi_r = \mu_1 \circ \mu_2 \circ \cdots \circ \mu_s$, a product of r disjoint cycles and also a product of s disjoint cycles, respectively. We show that every π_i is equal to some μ_j and every μ_k is equal to some π_t. Consider π_i, $1 \leq i \leq r$. Suppose $\pi_i = (i_1 i_2 \ldots i_l)$. Then $\pi(i_1) \neq i_1$. This implies that i_1 is moved by some μ_l. By the disjointness of the cycles, there exists unique μ_j, $1 \leq j \leq s$, such that i_1 appears as an element in μ_j. By reordering, if necessary, we may write $\mu_j = (i_1 \ c_2 \ \ldots \ c_m)$. Now

$$
\begin{array}{ccccccccc}
i_2 & = & \pi_i(i_1) & = & \pi(i_1) & = & \mu_j(i_1) & = & c_2 \\
i_3 & = & \pi_i(i_2) & = & \pi(i_2) & = & \pi(c_2) & = & \mu_j(c_2) & = & c_3 \\
\cdot & & \cdot & & \cdot & & \cdot & & \cdot & & \cdot \\
\cdot & & \cdot & & \cdot & & \cdot & & \cdot & & \cdot \\
\cdot & & \cdot & & \cdot & & \cdot & & \cdot & & \cdot \\
i_l & = & \pi_i(i_{l-1}) & = & \pi(i_{l-1}) & = & \pi(c_{l-1}) & = & \mu_j(c_{l-1}) & = & c_l.
\end{array}
$$

If $l < m$, then $i_1 = \pi_i(i_l) = \pi(i_l) = \pi(c_l) = \mu_j(c_l) = c_{l+1}$, a contradiction. Thus, $l = m$. Hence, $\pi_i = \mu_j$ for some j, $1 \leq j \leq s$. Similarly, every $\mu_k = \pi_t$ for some t, $1 \leq t \leq r$. ∎

Corollary 3.1.16 *Let $n \geq 2$. Any permutation π of S_n can be expressed as a product of transpositions.*

Proof.   In view of the preceding theorem, it suffices to show that every k-cycle can be expressed as a product of transpositions. This fact is immediate from the following equations:

$$e = (1) = (1\ 2) \circ (1\ 2)$$

and for $k \geq 2$

$$(i_1\ i_2\ \cdots\ i_k) = (i_1\ i_k) \circ (i_1\ i_{k-1}) \circ \cdots \circ (i_1\ i_2),$$

where $\{i_1, i_2 \ldots, i_k\} \subseteq I_n$. ∎

Let $\pi \in S_n$. Since S_n is a finite group, we know that $o(\pi)$ is finite. Thus, in order to find the order of π, we need to compute $\pi, \pi^2, \pi^3, \ldots$, until we find the first positive integer k such that $\pi^k = e$. Finding such a positive integer could be a tedious task. However, we can effectively make use of the decomposition of π as a product of disjoint cycles, compute the order of each cycle, which is nothing but the length of the cycle (Exercise 17, page 97) and from the order of the cycles deduce the order of π. We ask the reader to consider this problem in Exercise 18 (page 97).

Theorem 3.1.15 tells us that any permutation $\alpha \in S_n$, $n \geq 2$, can be written as a product of disjoint cycles. However, the theorem does not tell us how to find the disjoint cycles in the decomposition of α. Next, we illustrate how to find these cycles.

Let π be a permutation on I_n, $n \geq 2$. In order to express π as a product of disjoint cycles, first consider $1, \pi(1), \pi^2(1), \pi^3(1), \ldots$ and find the smallest positive integer r such that $\pi^r(1) = 1$. Let

$$\sigma_1 = (1 \; \pi(1) \; \pi^2(1) \; \cdots \; \pi^{r-1}(1)).$$

Then σ_1 is a cycle of length r. Let i be the first element of I_n not appearing in σ_1. Now consider $i, \pi(i), \pi^2(i), \pi^3(i), \ldots$ and find the smallest positive integer s such that $\pi^{s-1}(i) = i$. Let

$$\sigma_2 = (i \; \pi(i) \; \pi^2(i) \; \cdots \; \pi^{s-1}(i)).$$

Then σ_2 is a cycle of length s. Now

$$\{1, \pi(1), \pi^2(1), \ldots, \pi^{r-1}(1)\} \cap \{i, \pi(i), \pi^2(i), \ldots, \pi^{s-1}(i)\} = \phi,$$

for if $j \in \{1, \pi(1), \pi^2(1), \ldots, \pi^{r-1}(1)\} \cap \{i, \pi(i), \pi^2(i), \ldots, \pi^{s-1}(i)\}$, then $j = \pi^p(i)$ for some p, $1 \leq p < r$, and $j = \pi^k(1)$ for some k, $1 \leq k < s$. Thus, $\{1, \pi(1), \pi^2(1), \ldots, \pi^{r-1}(1)\} = \{i, \pi(i), \pi^2(i), \ldots, \pi^{s-1}(i)\}$, which is a contradiction. Hence, σ_1 and σ_2 are disjoint cycles. If $\{1, \pi(1), \pi^2(1), \ldots, \pi^{r-1}(1)\} \cup \{i, \pi(i), \pi^2(i), \ldots, \pi^{s-1}(i)\} \neq I_n$, then consider the first element of I_n not appearing in $\{1, \pi(1), \pi^2(1), \ldots, \pi^{r-1}(1)\} \cup \{i, \pi(i), \pi^2(i), \ldots, \pi^{s-1}(i)\}$ and continue the above process to construct the cycle σ_3. Since I_n is finite, the above process must stop with some cycle σ_m. Then $\pi = \sigma_1 \circ \sigma_2 \circ \cdots \circ \sigma_m$.

We illustrate the above procedure with the help of the following example.

Example 3.1.17 *Consider the permutation*

$$\pi = \begin{pmatrix} 1 & 2 & 3 & 4 & 5 & 6 & 7 \\ 6 & 3 & 5 & 2 & 4 & 7 & 1 \end{pmatrix}$$

on I_7. Here $\pi(1) = 6$, $\pi^2(1) = \pi(6) = 7$, and $\pi^3(1) = \pi(7) = 1$. That is, $1 \xrightarrow{\pi} 6 \xrightarrow{\pi} 7 \xrightarrow{\pi} 1$. Hence, $\sigma_1 = (1 \; 6 \; 7)$ is a 3-cycle. Now 2 is the first element of I_7 not appearing in $(1 \; 6 \; 7)$. Also, $\pi(2) = 3$, $\pi^2(2) = \pi(3) = 5$, $\pi^3(2) = \pi(5) = 4$, and $\pi^4(2) = \pi(4) = 2$. That is, $2 \xrightarrow{\pi} 3 \xrightarrow{\pi} 5 \xrightarrow{\pi} 4 \xrightarrow{\pi} 2$. Hence, $\sigma_2 = (2 \; 3 \; 5 \; 4)$ is a cycle of length 4. Now σ_1 and σ_2 are disjoint and $\pi = \sigma_1 \circ \sigma_2$.

While writing a permutation as a product of disjoint cycles, it is customary not to write cycles of length one in the product. Thus, if some element of I_n does not appear in any of the cycles, then it is assumed to be fixed. For example, if $\pi = (1\ 2\ 5) \circ (4\ 6) \in S_7$, then since 3 and 7 neither appear in $(1\ 2\ 5)$ nor in $(4\ 6)$, they are fixed, i.e., $\pi(3) = 3$ and $\pi(7) = 7$.

Given a permutation $\pi \in S_n$, $n \geq 2$, we can write π as a product of disjoint cycles. We can also write π as a product of transpositions. However, the representation of π as a product of transposition need not be unique. For example, $(1\ 2\ 3) = (1\ 3) \circ (1\ 2) = (2\ 1) \circ (2\ 3)$. Also, $(1\ 3) = (1\ 2) \circ (1\ 3) \circ (2\ 3)$. That is, $(1\ 3)$ can be written as a product of one transposition or as a product of three transpositions. However, we will show that the number of transpositions in any representation of a permutation is either even or odd, but not both. We now proceed to prove this result.

Consider the formal product

$$\mathcal{X} = \prod_{1 \leq i < j \leq n}(a_i - a_j) = \begin{matrix}(a_1 - a_2)(a_1 - a_3)\cdots(a_1 - a_n) \\ (a_2 - a_3)\cdots(a_2 - a_n) \\ \vdots \\ (a_{n-1} - a_n).\end{matrix}$$

If $n = 4$, then $\mathcal{X} = (a_1 - a_2)(a_1 - a_3)(a_1 - a_4)(a_2 - a_3)(a_2 - a_4)(a_3 - a_4)$. For any permutation $\pi \in S_n$, let

$$\pi(\mathcal{X}) = \prod_{1 \leq i < j \leq n}(a_{\pi(i)} - a_{\pi(j)}).$$

Let us first examine $\sigma(\mathcal{X})$ for any transposition $\sigma \in S_n$.

Lemma 3.1.18 *Let $n \geq 2$. Let $\sigma = (i\ j) \in S_n$, $i < j$, be a transposition. Then $\sigma(\mathcal{X}) = -\mathcal{X}$.*

Proof. First consider the factor $(a_i - a_j)$ in the product $\mathcal{X}$. The corresponding factor in $\sigma(\mathcal{X})$ is $a_{\sigma(i)} - a_{\sigma(j)}$. Now

$$a_{\sigma(i)} - a_{\sigma(j)} = a_j - a_i = -(a_i - a_j).$$

Next, consider the factor $a_k - a_l$, where both k and l are neither equal to i nor equal to j. The corresponding factor in $\sigma(\mathcal{X})$ is $a_{\sigma(k)} - a_{\sigma(l)}$ and

$$a_{\sigma(k)} - a_{\sigma(l)} = a_k - a_l.$$

Thus, the factor $a_k - a_l$ remains unaltered. Now consider the factor $a_k - a_l$, where either k or l (but not both) is equal to i or j. Let $1 \leq t \leq n$. Suppose

$t < i < j$. We have the pair of factors $(a_t - a_i)$ and $(a_t - a_j)$ in the product $\mathcal{X}$. The corresponding factors in $\sigma(\mathcal{X})$ are $a_{\sigma(t)} - a_{\sigma(i)}$ and $a_{\sigma(t)} - a_{\sigma(j)}$ and

$$(a_{\sigma(t)} - a_{\sigma(i)})(a_{\sigma(t)} - a_{\sigma(j)}) = (a_t - a_j)(a_t - a_i) = (a_t - a_i)(a_t - a_j).$$

Therefore, the product $(a_t - a_i)(a_t - a_j)$ remains unchanged. Now suppose $i < t < j$. Then we have the pair of factors $(a_i - a_t)$ and $(a_t - a_j)$ in the product $\mathcal{X}$. The corresponding factors in $\sigma(\mathcal{X})$ are $a_{\sigma(i)} - a_{\sigma(t)}$ and $a_{\sigma(t)} - a_{\sigma(j)}$ and

$$(a_{\sigma(i)} - a_{\sigma(t)})(a_{\sigma(t)} - a_{\sigma(j)}) = (a_j - a_t)(a_t - a_i) = (a_i - a_t)(a_t - a_j).$$

Hence, the product $(a_i - a_t)(a_t - a_j)$ remains unaltered. Finally, let $i < j < t$. Then we have the pair of factors $(a_i - a_t)$ and $(a_j - a_t)$ in the product $\mathcal{X}$. The corresponding factors in $\sigma(\mathcal{X})$ are $a_{\sigma(i)} - a_{\sigma(t)}$ and $a_{\sigma(j)} - a_{\sigma(t)}$ and

$$(a_{\sigma(i)} - a_{\sigma(t)})(a_{\sigma(j)} - a_{\sigma(t)}) = (a_j - a_t)(a_i - a_t) = (a_i - a_t)(a_j - a_t).$$

Therefore, the product $(a_i - a_t)(a_j - a_t)$ remains unaltered. Thus, all factors other than $a_i - a_j$ and $a_k - a_l$, where both k and l are neither equal to i nor equal to j, can be paired so that the product of factors under σ remains unaltered. Hence, it now follows that $\sigma(\mathcal{X}) = -\mathcal{X}$. ∎

Theorem 3.1.19 *Let $n \geq 2$. Let $\pi \in S_n$. Suppose*

$$\pi = \sigma_1 \circ \sigma_2 \circ \cdots \circ \sigma_r = \tau_1 \circ \tau_2 \circ \cdots \circ \tau_s,$$

where σ_i, $\tau_j \in S_n$ are transpositions, $i = 1, 2, \ldots, r$, and $j = 1, 2, \ldots, s$. Then both r and s are either even or odd.

Proof. By Lemma 3.1.18, $\sigma_i(\mathcal{X}) = -\mathcal{X}$ and $\tau_j(\mathcal{X}) = -\mathcal{X}$ for all $i = 1, 2, \ldots, r$, and $j = 1, 2, \ldots, s$. First we compute $(\sigma_1 \circ \sigma_2 \circ \cdots \circ \sigma_r)(\mathcal{X})$. Now

$$\begin{aligned}(\sigma_1 \circ \sigma_2 \circ \cdots \circ \sigma_r)(\mathcal{X}) &= \sigma_1(\sigma_2(\cdots(\sigma_r(\mathcal{X})))) \\ &= (-1)^r \mathcal{X}.\end{aligned}$$

Similarly, $(\tau_1 \circ \tau_2 \circ \cdots \circ \tau_s)(\mathcal{X}) = (-1)^s \mathcal{X}$. Hence, $(-1)^r = (-1)^s$. Thus, both r and s are either even of odd. ∎

By the above theorem, if $\pi \in S_n$, then π can be written as a product of either an even or an odd number of transpositions, but not both. This leads us to the following definition.

Definition 3.1.20 *Let $\pi \in S_n$. If π is a product of an even number of transpositions, then π is called an* **even permutation**; *otherwise π is called an* **odd permutation**.

Corollary 3.1.21 *Let $\pi \in S_n$ be a k-cycle. Then π is an even permutation if and only if k is odd.*

Proof. Let $\pi = (1\ 2\ \cdots\ k)$. Then $\pi = (1\ k) \circ (1\ k-1) \circ \cdots \circ (1\ 2)$, i.e., π is a product of $k-1$ transposition. If π is an even permutation then $k-1$ is even and so k is odd. On the other hand, if k is odd, then $k-1$ is even and so π is an even permutation. This completes the proof. ∎

Let A_n denote the subset of S_n consisting of all even permutations, $n \geq 2$.

Theorem 3.1.22 *For $n \geq 2$, the pair $(A_n, \circ)$ is a group, called the **alternating group on** I_n.*

Proof. Since $e = (1\ 2) \circ (1\ 2)$, $e \in A_n$. Thus, $A_n \neq \phi$. A product $\pi_1 \circ \pi_2$ is even if and only if π_1 and π_2 are both even or both odd by Theorem 3.1.19. Therefore, A_n is closed under $\circ$. If $\pi \in A_n$, then $\pi \circ \pi^{-1} = e$ is even and hence $\pi^{-1} \in A_n$. Hence, $(A_n, \circ)$ is a group. ∎

Cauchy recognized many important properties of A_n. Among others, he proved the following theorem.

Theorem 3.1.23 *Every element in A_n is a product of 3-cycles, $n \geq 3$.*

Proof. Let $\pi \in A_n$. Then $\pi = \sigma_1 \circ \sigma_2 \circ \cdots \circ \sigma_r$, where σ_i is a transposition, $1 \leq i \leq r$, and r is even. Now for any transposition $(a\ b)$,

$$(a\ b) = (1\ a) \circ (1\ b) \circ (1\ a).$$

Thus,

$$\pi = (1\ i_1) \circ (1\ i_2) \circ \cdots \circ (1\ i_m)$$

where m is even. Since $(1\ i_1) \circ (1\ i_2) = (1\ i_2\ i_1)$, it follows that π is a product of 3-cycles. ∎

3.1.1 Worked-Out Exercises

◇ **Exercise 1** Prove that two cycles in S_n are conjugate if and only if they have the same length.

Solution: Let $\alpha = (i_1 i_2 \cdots i_r)$ and $\beta = (j_1 j_2 \cdots j_s)$ be two cycles in S_n. First suppose that α and β are conjugate. Then $\beta = \sigma^{-1} \circ \alpha \circ \sigma$ for some $\sigma \in S_n$. Since σ is onto and $i_l \in I_n$, there exists k_l such that $\sigma(k_l) = i_l$ for all $l = 1, 2, \ldots, r$. Now

$$\begin{aligned}(j_1 j_2 \cdots j_s) &= (\sigma^{-1}(i_1)\sigma^{-1}(i_2) \cdots \sigma^{-1}(i_r)) \quad (\text{ by Theorem 3.1.13})\\ &= (k_1 k_2 \cdots k_r).\end{aligned}$$

Hence, $s = r$ and so α and β are of the same length.

Conversely, let $\alpha = (i_1 i_2 \cdots i_r)$ and $\beta = (j_1 j_2 \cdots j_r)$ be two cycles in S_n of the same length. Let $\sigma = \begin{pmatrix} i_1 & i_2 & \cdots & i_r \\ j_1 & j_2 & \cdots & j_r \end{pmatrix}$, i.e., $\sigma(i_l) = j_l$ for all $l = 1, 2, \ldots, r$, and $\sigma(a) = a$ for all $a \in I_n \setminus \{i_1, i_2, \ldots, i_r\}$. Then $\sigma \in S_n$. Now

$$\sigma^{-1} \circ \beta \circ \sigma = (\sigma^{-1}(j_1) \sigma^{-1}(j_2) \cdots \sigma^{-1}(j_r)) = (i_1 i_2 \cdots i_r) = \alpha.$$

$\Diamond$ **Exercise 2** Express the permutation

$$\sigma = \begin{pmatrix} 1 & 2 & 3 & 4 & 5 & 6 & 7 & 8 \\ 2 & 3 & 8 & 5 & 6 & 4 & 7 & 1 \end{pmatrix}$$

on I_8 as a product of disjoint cycles and then as a product of transposition. Is σ an even permutation?

Solution: We have $\sigma(1) = 2$, $\sigma^2(1) = \sigma(2) = 3$, $\sigma^3(1) = \sigma(3) = 8$, and $\sigma^4(1) = \sigma(8) = 1$. Thus, $(1\ 2\ 3\ 8)$ is a cycle. Now 4 is the first element of I_8 not appearing in $(1\ 2\ 3\ 8)$. We have $\sigma(4) = 5$, $\sigma^2(4) = \sigma(5) = 6$, and $\sigma^3(4) = \sigma(6) = 4$. Hence, $(4\ 5\ 6)$ is also a cycle in σ. Next, 7 is the first element of I_8 not appearing in $(1\ 2\ 3\ 8)$ and $(4\ 5\ 6)$. Now $\sigma(7) = 7$. Since all the elements of I_8 appear in one of the cycles $(1\ 2\ 3\ 8)$, $(4\ 5\ 6)$, and (7), we have $\sigma = (1\ 2\ 3\ 8) \circ (4\ 5\ 6)$. Now $(1\ 2\ 3\ 8) = (1\ 8) \circ (1\ 3) \circ (1\ 2)$ and $(4\ 5\ 6) = (4\ 6) \circ (4\ 5)$. Thus, $\sigma = (1\ 8) \circ (1\ 3) \circ (1\ 2) \circ (4\ 6) \circ (4\ 5)$. Since σ is a product of five transpositions, σ is not an even permutation.

$\Diamond$ **Exercise 3** Write all elements of S_4. Show that S_4 has no elements of order ≥ 5.

Solution: Let $\sigma \in S_4$ and $\sigma = \sigma_1 \circ \sigma_2 \circ \cdots \circ \sigma_k$, a product of disjoint cycles. Since S_4 is a permutation group on I_4, $k \leq 2$. If $k = 1$, then σ is a 2-cycle, 3-cycle, or 4-cycle. If $k = 2$, then σ is a product of two disjoint transpositions. The number of distinct cycles of length 2 is 6, the number of distinct cycles of length 3 is 8, and the number of distinct cycles of length 4 is 6. Hence, $S_4 = \{e,$ $(1\ 2), (1\ 3), (1\ 4), (2\ 3), (2\ 4), (3\ 4), (1\ 2\ 3), (1\ 3\ 2), (2\ 3\ 4), (2\ 4\ 3), (1\ 3\ 4),$ $(1\ 4\ 3), (1\ 2\ 4), (1\ 4\ 2), (1\ 2\ 3\ 4), (1\ 3\ 2\ 4), (1\ 4\ 2\ 3), (1\ 2\ 4\ 3), (1\ 3\ 4\ 2), (1\ 4\ 3\ 2), (1\ 2) \circ (3\ 4), (1\ 4) \circ (3\ 2), (1\ 3) \circ (2\ 4)\}$.

Since each 2-cycle is of order 2, each 3-cycle is of order 3, each 4-cycle is of order 4, and the order of the product of two disjoint 2-cycles is 2, S_4 has no element of order ≥ 5.

$\Diamond$ **Exercise 4** Find the order of $(1\ 2\ 3\ 4) \circ (5\ 6\ 7)$ in S_7.

Solution: $o(1\ 2\ 3\ 4) = 4$, $o(5\ 6\ 7) = 3$. Now $(1\ 2\ 3\ 4)$ and $(5\ 6\ 7)$ are disjoint. Hence, $(1\ 2\ 3\ 4) \circ (5\ 6\ 7) = (5\ 6\ 7) \circ (1\ 2\ 3\ 4)$. If a and b are two elements of a group G such that $o(a) = m$, $o(b) = n$, and $\gcd(m, n) = 1$, then $o(ab) = mn$. Using this result, we find that the order of $(1\ 2\ 3\ 4) \circ (5\ 6\ 7)$ is 12.

◇ **Exercise 5** Find the order of $(1\ 2\ 3\ 4) \circ (5\ 6)$ in S_6.

Solution: $o(1\ 2\ 3\ 4) = 4$, $o(5\ 6) = 2$. Now $(1\ 2\ 3\ 4)$ and $(5\ 6)$ are disjoint and so they commute. Thus, $((1\ 2\ 3\ 4) \circ (5\ 6))^4 = e$. Now $((1\ 2\ 3\ 4) \circ (5\ 6))^1 \neq e$, $((1\ 2\ 3\ 4) \circ (5\ 6))^2 = (1\ 2\ 3\ 4)^2 \circ (5\ 6)^2 = (1\ 2\ 3\ 4)^2 \neq e$. If $((1\ 2\ 3\ 4) \circ (5\ 6))^3 = e$, then the order of $(1\ 2\ 3\ 4) \circ (5\ 6)$ will be 3 and 3 divides 4, a contradiction. Hence, the order of $(1\ 2\ 3\ 4) \circ (5\ 6)$ is 4.

3.1.2 Exercises

1. Express the following permutations as (i) a product of disjoint cycles and (ii) a product of transpositions:

$$\begin{pmatrix} 1 & 2 & 3 & 4 & 5 & 6 \\ 3 & 5 & 4 & 1 & 6 & 2 \end{pmatrix}, \begin{pmatrix} 1 & 2 & 3 & 4 & 5 & 6 \\ 3 & 2 & 1 & 5 & 4 & 6 \end{pmatrix}.$$

2. Let $\alpha = (1\ 2\ 5\ 7)$ and $\beta = (2\ 4\ 6) \in S_7$. Find $\alpha \circ \beta \circ \alpha^{-1}$.

3. Let $\alpha = (1\ 3\ 5\ 7)$ and $\beta = (2\ 4\ 8) \circ (1\ 3\ 6) \in S_8$. Find $\alpha \circ \beta \circ \alpha^{-1}$.

4. Let $\alpha = (1\ 3) \circ (5\ 8)$ and $\beta = (2\ 3\ 6\ 7) \in S_8$. Find $\alpha \circ \beta \circ \alpha^{-1}$.

5. Let $\alpha = (2\ 5\ 9) \circ (1\ 3\ 6)$ and $\beta = (1\ 5\ 7) \circ (2\ 4\ 6\ 9) \in S_9$. Find $\alpha \circ \beta \circ \alpha^{-1}$.

6. Let $(1\ 3\ 5\ 7)$ and $(2\ 3\ 6\ 8) \in S_8$. Find $\alpha \in S_8$ such that $\alpha \circ (1\ 3\ 5\ 7) \circ \alpha^{-1} = (2\ 3\ 6\ 8)$.

7. If $\alpha = (1\ 2\ 3\ 4\ 5\ 6)$, show that $\alpha = (1\ 6) \circ (1\ 5) \circ (1\ 4) \circ (1\ 3) \circ (1\ 2)$.

8. Find the order of $(1\ 2\ 3) \circ (4\ 5)$ in S_5.

9. Prove that $(1\ 2 \cdots n - 1\ n)^{-1} = (n\ n - 1 \cdots 2\ 1)$.

10. Prove that every transposition is its own inverse.

11. Prove that the symmetric group on two symbols $(S_2, \circ)$ is commutative.

12. Let $\alpha = (a_1\ a_2 \cdots a_k) \in S_n$ be a k-cycle. Show that

$$\alpha^2 = \begin{cases} (a_1\ a_3 \cdots a_{2m-1}) \circ (a_2\ a_4\ a_6 \cdots a_{2m}) & \text{if } k = 2m, \text{ i.e., } k \text{ is even} \\ (a_1\ a_3 \cdots a_{2m+1}\ a_2\ a_4 \cdots a_{2m}) & \text{if } k = 2m + 1, \text{ i.e., } k \text{ is odd.} \end{cases}$$

13. Determine A_4.

14. Let $\alpha, \beta \in S_n$. Show that $\alpha^{-1} \circ \beta^{-1} \circ \alpha \circ \beta \in A_n$.

15. Prove that $|A_n| = \frac{n!}{2}$.

16. Show that the number of distinct cycles of length r in S_n is $\frac{1}{r}\frac{n!}{(n-r)!}$.

17. Let $n \geq 2$ and $\sigma \in S_n$ be a cycle. Show that σ is a k-cycle if and only if $\circ(\sigma) = k$.

18. Let $\sigma \in S_n$ and $\sigma = \sigma_1 \circ \sigma_2 \circ \cdots \circ \sigma_k$ be a product of disjoint cycles. Suppose $\circ(\sigma_i) = n_i$, $i = 1, 2, \ldots, n$. Show that $\circ(\sigma) = \text{lcm}(n_1, n_2, \ldots, n_k)$.

19. Let $\alpha \in S_n$ and p be a prime.

(i) Show that $\circ(\alpha) = p$ if and only if either α is a p-cycle or α is a product of disjoint cycles, where each cycle is either of length 1 or length p and at least one cycle is of length p.

(ii) If α is a p-cycle, prove that either $\alpha^m = e$ or α^m is a p-cycle for all $m \in \mathbf{N}$.

20. Let α and $\beta \in S_n$. Let $\alpha = \alpha_1 \circ \alpha_2 \circ \cdots \circ \alpha_k$ and $\beta = \beta_1 \circ \beta_2 \circ \cdots \circ \beta_s$ be a product of disjoint cycles. Let $\text{length}(\alpha_i) = d_i$ and $\text{length}(\beta_j) = m_j$ for all $i = 1, 2, \ldots, k$ and $j = 1, 2, \ldots, s$ and $d_1 \leq d_2 \leq \cdots \leq d_k$ and $m_1 \leq m_2 \leq \cdots \leq m_s$. We say that α and β have the same **cyclic structure** if $k = s$ and $d_i = m_i$ for all $i = 1, 2, \ldots, k$. Prove that α and β have the same cyclic structure if and only if α and β are conjugate.

21. Prove that for $\pi \in S_n$, π is an even permutation if and only if $\pi(\mathcal{X}) = \mathcal{X}$.

22. (i) Let $\alpha = (k\ l)$, $\beta \in S_n$ be two distinct transpositions, $n \geq 3$. Show that there exist transpositions $\mu, \nu \in S_n$ such that $\beta \circ \alpha = \nu \circ \mu$, $\mu(k) = k$ and ν moves k.

(ii) Prove that if the identity permutation $e \in S_n$ can be written as a product of r (≥ 3) transpositions, then e can be written as a product of $r - 2$ transpositions.

(iii) Prove that if $e = \sigma_1 \circ \sigma_2 \circ \cdots \circ \sigma_r \in S_n$ as a product of transpositions, then r is even.

(iv) Use (i), (ii), and (iii) to prove that if $\pi \in S_n$, then π can be written as a product of either an even or an odd number of transpositions, but not both.

Augustin-Louis Cauchy (1789–1857) was born on August 21, 1789, in Paris, France. He received his first education from his father. He was a neighbor of Laplace and Berthollet. Cauchy became acquainted with famous scientists at a young age. Lagrange is said to have warned his father not to show Cauchy any mathematics book before the age of seventeen.

At the age of fifteen, he completed his classic studies with distinction. He became an engineer in 1810, in the Napoleon army. In 1813, he returned to Paris.

In 1811, Cauchy started his mathematical career by solving a problem sent to him by Lagrange on convex polygons. In 1812, he solved Fermat's famous classical problem on polygon numbers. His treatise on the definite integral, which he submitted in 1814 to the French Academy, later became a basis of the theory of complex functions.

In 1816, he was appointed full professor at the École Polytechnique. More theorems and concepts have been named for Cauchy than for any other mathematician. There are sixteen concepts and theorems named for Cauchy in elasticity alone.

He worked on mathematics, mathematical physics, and celestial mechanics. In mathematics, he worked on several areas, such as calculus, complex functions, algebra, differential equations, geometry, and analysis. The notion of continuity used today was invented by Cauchy. He also proved that a continuous function has a zero between two points where the function changes its signs, a result also proved by Bolzano. The first adequate definitions of indefinite integral and definite improper integral are due to Cauchy

In algebra, the notion of the order of an element, a subgroup, and conjugates are found in his papers. He proved the famous Cauchy's theorem for finite groups, that is, if the order of a finite group is divisible by a prime p, then the group has a subgroup of order p. Cauchy's role in shaping the theory of permutation groups is central. He is regarded by some to be the founder of finite group theory. The two-row notation for permutations was introduced by Cauchy. He also defined the product of permutations, inverse permutations, transpositions, and the cyclic notation. He wrote his first paper on this subject in 1815, but did not return to it for nearly thirty years. In 1844, he proved that every permutation is a product of disjoint cycles.

He also did work of fundamental importance in the theory of determinants. His treatise on determinants, published in 1812, contains important results concerning product theorems and the inverse of a matrix.

Cauchy enjoyed teaching. He published more than 800 papers and eight books. He died on May 22, 1857.

Chapter 4

Subgroups and Normal Subgroups

In Chapter 2, we began a discussion of the evolution of group theory. This chapter seems a good place to renew the discussion. It took more than 100 years for the abstract concept of a group to evolve. The evolution followed lines similar to the evolution of other theories. First came the discovery of isolated phenomena, followed by the recognition of features common to all. Then came the search and classification of other instances. Next, general principles emerged. Last, the abstract postulates which define the system were uncovered. A deeper account can be found in Bell.

4.1 Subgroups

In the previous chapter, we saw that for the groups $(A_n, \circ)$ and $(S_n, \circ)$, A_n is a subset of S_n. One can think of many examples, where the underlying set of one group is a subset of the underlying set of another group. This leads us to the concept of a subgroup.

Let $(G, *)$ be a group and H be a nonempty subset of G. Then H is said to be closed under the binary operation $*$ if $a * b \in H$ for all $a, b \in H$.

Suppose H is closed under the binary operation $*$. Then the restriction of $*$ to $H \times H$ is a mapping from $H \times H$ into H. Thus, the binary operation $*$ defined on G induces a binary operation on H. We denote this induced binary operation on H by $*$ also. Thus, $(H, *)$ is a mathematical system. It also follows that $*$ is associative as a binary operation on H, i.e., $a * (b * c) = (a * b) * c$ for all $a, b, c \in H$. If $(H, *)$ is a group, then we call H a subgroup of G. More formally, we have the following definition.

Definition 4.1.1 *Let $(G, *)$ be a group and H be a nonempty subset of G. Then $(H, *)$ is called a **subgroup** of $(G, *)$ if $(H, *)$ is a group.*

Let $(H, *)$ be a subgroup of a group $(G, *)$. Let e_H denote the identity of H and e denote the identity of G. Now $e_H * e_H = e_H = e_H * e$. Hence, by the cancellation property, $e_H = e$. Thus, the identity elements of G and H are the same. Now let $h \in H$. Let h' denote the inverse of h in H and h^{-1} denote the inverse of h in G. Then $h' = h' * e = h' * (h * h^{-1}) = (h' * h) * h^{-1} = e * h^{-1} = h^{-1}$. Thus, the inverse of h in H and the inverse of h in G are the same.

Of course, if $(G, *)$ is a group, then $(\{e\}, *)$ and $(G, *)$ are subgroups of $(G, *)$. These subgroups are called **trivial**.

Example 4.1.2 *Consider the following list of groups.*
(i) $(\{0\}, +), (\mathbf{Z}, +), (\mathbf{Q}, +), (\mathbf{R}, +), (\mathbf{C}, +),$
(ii) $(\{1\}, \cdot), (\mathbf{Q}\backslash\{0\}, \cdot), (\mathbf{R}\backslash\{0\}, \cdot), (\mathbf{C}\backslash\{0\}, \cdot),$
where $+$ is the usual addition operation and $\cdot$ is the usual multiplication operation. Each group is a subgroup of the group listed to its right. For example, $(\mathbf{Z}, +)$ is a subgroup of $(\mathbf{Q}, +), (\mathbf{R}, +),$ and $(\mathbf{C}, +),$ and $(\mathbf{R}\backslash\{0\}, \cdot)$ is a subgroup of $(\mathbf{C}\backslash\{0\}, \cdot).$

In the remainder of the text, we shall generally use the notation G instead of $(G, *)$ for a group and we write ab for $a*b$. We shall refer to ab as the product of a and b. This notation is usually called multiplicative notation.

Readers with some knowledge of linear algebra should notice the similarity with respect to the type of results and order of presentation of those which immediately follow. First comes a result which gives an easy method of determining if a nonempty subset is a substructure. This is followed by the result that the intersection of any collection of substructures is a substructure. Next, comes the definition of a substructure "generated" by a subset. Finally, a theorem describing the substructure generated by a given subset. These ideas appear throughout algebra. We will encounter them again, for example, when we examine ideals of a ring.

Theorem 4.1.3 *Let G be a group and H be a nonempty subset of G. Then H is a subgroup of G if and only if for all $a, b \in H$, $ab^{-1} \in H$.*

Proof.   Suppose H is a subgroup of G. Let $a, b \in H$. Since H is a subgroup, it is a group and so $b^{-1} \in H$. Thus, $ab^{-1} \in H$ since H is closed under the binary operation. Conversely, suppose H is a nonempty subset of G such that $a, b \in H$ implies $ab^{-1} \in H$. Since $H \neq \phi$, there exists $a \in H$. Therefore, $e = aa^{-1} \in H$, i.e., H contains the identity. Now for all $b \in H$, $b^{-1} = eb^{-1} \in H$, i.e., every element of H has an inverse in H. Thus, for all $a, b \in H$, $a, b^{-1} \in H$ and so $ab = a(b^{-1})^{-1} \in H$, i.e., H is closed under the binary operation. From the statements preceding Definition 4.1.1, associativity holds for H. Hence, H is a group and so H is subgroup of G. ∎

In order to see whether a certain nonempty subset of a given group is a subgroup or not, we can use Theorem 4.1.3.

Corollary 4.1.4 *Let G be a group and H be a finite nonempty subset of G. Then H is a subgroup of G if and only if for all $a, b \in H$, $ab \in H$.*

Proof. If H is a subgroup, then for all $a, b \in H$, $ab \in H$. Conversely, suppose that for all $a, b \in H$, $ab \in H$. Let $h \in H$. Then $h, h^2, \ldots, h^n, \ldots \in H$ and so $\{h, h^2, \ldots, h^n, \ldots\} \subseteq H$. Since H is finite, all elements of $\{h, h^2, \ldots, h^n, \ldots\}$ cannot be distinct. Thus, there exist integers r and s such that $0 \leq r < s$ and $h^r = h^s$. Hence, $e = h^{s-r} \in H$. Now $s - r \geq 1$. Thus, $e = hh^{s-r-1}$ implies that $h^{-1} = h^{s-r-1} \in H$. Let $a, b \in H$. Then $a, b^{-1} \in H$ and so $ab^{-1} \in H$ by the hypothesis. Thus, by Theorem 4.1.3, H is a subgroup. ∎

Theorem 4.1.5 *Let G be a group and $Z(G) = \{b \in G \mid ab = ba \text{ for all } a \in G\}$. Then $Z(G)$ is a commutative subgroup of G. $Z(G)$ is called the **center** of G.*

Proof. Since $ae = a = ea$ for all $a \in G$, $e \in Z(G)$ and so $Z(G) \neq \phi$. Let $a, b \in Z(G)$. Then $bc = cb$ for all $c \in G$. From this, it follows that $cb^{-1} = b^{-1}c$ for all $c \in G$ and so $b^{-1} \in Z(G)$. Now $(ab^{-1})c = a(b^{-1}c) = a(cb^{-1}) = (ac)b^{-1} = (ca)b^{-1} = c(ab^{-1})$ for all $c \in G$ and so $ab^{-1} \in Z(G)$. Hence by Theorem 4.1.3, $Z(G)$ is a subgroup of G. That $Z(G)$ is commutative follows by the definition of $Z(G)$. ∎

In the remainder of this section, we will see how new subgroups arise from existing subgroups of a group.

Theorem 4.1.6 *Let G be a group and $\{H_\alpha \mid \alpha \in I\}$ be any nonempty collection of subgroups of G. Then $\cap_{\alpha \in I} H_\alpha$ is a subgroup of G.*

Proof. Since each H_α is a subgroup, $e \in H_\alpha$ for all $\alpha \in I$. Hence, $e \in \cap_{\alpha \in I} H_\alpha$ and so $\cap_{\alpha \in I} H_\alpha \neq \phi$. Let $a, b \in \cap_{\alpha \in I} H_\alpha$. Then $a, b \in H_\alpha$ for all $\alpha \in I$. Thus, $ab^{-1} \in H_\alpha$ for all $\alpha \in I$ since each H_α is a subgroup and so $ab^{-1} \in \cap_{\alpha \in I} H_\alpha$. Consequently, $\cap_{\alpha \in I} H_\alpha$ is a subgroup of G by Theorem 4.1.3. ∎

Definition 4.1.7 *Let G be a group and S be a subset of G. Let*

$$\mathcal{S} = \{H \mid H \text{ is a subgroup of } G \text{ and } S \subseteq H\}.$$

Define

$$= \cap_{H \in \mathcal{S}} H,$$

*i.e., $\langle S \rangle$ is the intersection of all subgroups H of G such that $S \subseteq H$. Then the subgroup $\langle S \rangle$ of G is called the **subgroup generated by** S. If $G = \langle S \rangle$, then S is called a set of **generators** for G.*

If either $S = \phi$ or $S = \{e\}$, then $\langle S \rangle = \{e\}$. Also, $\langle G \rangle = G$.

We now proceed to obtain a characterization of a subgroup generated by a nonempty subset in terms of the elements of the group.

Let $\mathcal{S} = \{H \mid H$ is a subgroup of G and $S \subseteq H\}$, where $S \neq \phi$. Then $(\mathcal{S}, \leq)$ is a partially ordered set, where $\leq$ denotes the set inclusion relation. In this poset, $\langle S \rangle$ is the least element. Hence, $\langle S \rangle$ is the smallest subgroup of G which contains S. Since $\langle S \rangle$ is a subgroup of G, we must have for any $s_1, \ldots, s_n \in S$, the product $s_1^{e_1} \cdots s_n^{e_n} \in \langle S \rangle$, where $e_i = \pm 1$ for $i = 1, 2, \ldots, n$. Thus, if A denotes the set $\{s_1^{e_1} \cdots s_n^{e_n} \mid s_i \in S, e_i = \pm 1, i = 1, 2, \ldots, n; n = 1, 2, \ldots\}$, then $A \subseteq \langle S \rangle$. Note that if $s \in S$, then $e = ss^{-1} \in A$. In the following theorem, we show that $A = \langle S \rangle$. Therefore, S does "generate" $\langle S \rangle$ in the sense of multiplying elements of S or their inverses together to build up the smallest subgroup containing S.

Theorem 4.1.8 *Let S be a nonempty subset of a group G. Then*

$$\langle S \rangle = \{s_1^{e_1} \cdots s_n^{e_n} \mid s_i \in S, \ e_i = \pm 1, \ i = 1, 2, \ldots, n; \ n = 1, 2, \ldots\}.$$

Proof.   Let

$$A = \{s_1^{e_1} \cdots s_n^{e_n} \mid s_i \in S, e_i = \pm 1, i = 1, 2, \ldots, n; \ n = 1, 2, \ldots\}.$$

We have already noted that $A \subseteq \langle S \rangle$. We show that $\langle S \rangle \subseteq A$ by showing that A is a subgroup of G containing S. (Recall that $\langle S \rangle$ is the smallest subgroup of G containing S.) Let $s \in S$. Then $s = s^1 \in A$ and so $S \subseteq A$. Let $s_1^{f_1} \cdots s_m^{f_m}$, $t_1^{g_1} \cdots t_q^{g_q} \in A$. Then

$$(s_1^{f_1} \cdots s_m^{f_m})(t_1^{g_1} \cdots t_q^{g_q})^{-1} = s_1^{f_1} \cdots s_m^{f_m} t_q^{-g_q} \cdots t_1^{-g_1} \in A.$$

Thus, A is a subgroup of G by Theorem 4.1.3. Hence, $\langle S \rangle \subseteq A$. ∎

For $a \in G$, we use the notation $\langle a \rangle$ rather than $\langle \{a\} \rangle$ to denote the subgroup of G generated by $\{a\}$.

Corollary 4.1.9 *Let G be a group and $a \in G$. Then $\langle a \rangle = \{a^n \mid n \in \mathbf{Z}\}$.*

Proof.   By Theorem 4.1.8, we have $\langle a \rangle = \{a^{e_1} \cdots a^{e_m} \mid e_i = \pm 1, i = 1, 2, \ldots, m; m = 1, 2, \ldots\} = \{a^{e_1 + \cdots + e_m} \mid e_i = \pm 1, i = 1, 2, \ldots, m; m = 1, 2, \ldots\} = \{a^n \mid n \in \mathbf{Z}\}$. ∎

In additive notation, we would have $\langle a \rangle = \{na \mid n \in \mathbf{Z}\}$.

Let $n \geq 3$. In Chapter 3, we proved that every element of A_n is a product of 3-cycles (Theorem 3.1.23). In the following theorem, we conclude that A_n is generated by the set of all 3-cycles.

Theorem 4.1.10 *Let $n \geq 3$. Then A_n is generated by the set of all 3-cycles.*

Proof. Since a 3-cycle is an even permutation, every 3-cycle is in A_n. By Theorem 3.1.23, every element of A_n is a product of 3-cycles. Hence, A_n is generated by the set of all 3-cycles. ∎

We now turn our attention to the product of subgroups.

Definition 4.1.11 *Let H and K be nonempty subsets of a group G. The product of H and K is defined to be the set*

$$HK = \{hk \mid h \in H, k \in K\}.$$

Let H_1, H_2, ..., H_n be nonempty subsets of a group G. We define the product, $H_1 H_2 \cdots H_n$, of H_1, H_2, ..., H_n to be the set

$$H_1 H_2 \cdots H_n = \{h_1 h_2 \cdots h_n \mid h_i \in H_i, \ i = 1, 2, \ldots, n\}.$$

Example 4.1.12 *Consider the group of symmetries of the square. Let $H = \{r_{360}, d_1\}$ and $K = \{r_{360}, h\}$. Then H and K are subgroups of G. Now*

$$HK = \{r_{360} r_{360}, r_{360} h, d_1 r_{360}, d_1 h\} = \{r_{360}, h, d_1, r_{90}\}.$$

Since $hd_1 = r_{270} \notin HK$, HK is not closed under the binary operation. Hence, HK is not a subgroup of the symmetries of the square. Also, note that

$$KH = \{r_{360} r_{360}, r_{360} d_1, h r_{360}, h d_1\} = \{r_{360}, d_1, h, r_{270}\},$$

and

$$\langle H \cup K \rangle = \{r_{360}, r_{90}, r_{180}, r_{270}, h, v, d_1, d_2\}.$$

Example 4.1.12 shows that in general the product of subgroups need not be a subgroup. In the following theorem, we give a necessary and sufficient condition for the product of subgroups to be a subgroup.

Theorem 4.1.13 *Let H and K be subgroups of a group G. Then HK is a subgroup of G if and only if $HK = KH$.*

Proof. Suppose HK is a subgroup of G. Let $kh \in KH$, where $h \in H$ and $k \in K$. Now $h = he \in HK$ and $k = ek \in HK$. Since HK is a subgroup, it follows that $kh \in HK$. Hence, $KH \subseteq HK$. On the other hand, let $hk \in HK$. Then $(hk)^{-1} \in HK$ and so $(hk)^{-1} = h_1 k_1$ for some $h_1 \in H$ and $k_1 \in K$. Thus, $hk = (h_1 k_1)^{-1} = k_1^{-1} h_1^{-1} \in KH$. This implies that $HK \subseteq KH$. Hence, $HK = KH$.

Conversely, suppose $HK = KH$. Let $h_1k_1, h_2k_2 \in HK$. Now $k_2^{-1}h_2^{-1} \in KH = HK$. This implies that $k_2^{-1}h_2^{-1} = h_3k_3$ for some $h_3 \in H$ and $k_3 \in K$. Similarly, $k_1h_3 = h_4k_4$ for some $h_4 \in H$ and $k_4 \in K$. Thus,

$$
\begin{aligned}
(h_1k_1)(h_2k_2)^{-1} &= h_1k_1k_2^{-1}h_2^{-1} \\
&= h_1k_1h_3k_3 \\
&= h_1h_4k_4k_3 \in HK.
\end{aligned}
$$

Hence, HK is a subgroup of G by Theorem 4.1.3. ∎

Corollary 4.1.14 *If H and K are subgroups of a commutative group G, then HK is a subgroup of G.*

Proof.   Since G is commutative, $HK = KH$. The result now follows by Theorem 4.1.13 ∎

The following theorem gives another necessary and sufficient condition for a product of subgroups to be a subgroup.

Theorem 4.1.15 *Let H and K be subgroups of a group G. Then HK is a subgroup of G if and only if $HK = \langle H \cup K \rangle$.*

Proof.   First suppose that HK is a subgroup of G. Let $h \in H$. Then $h = he \in HK$. Thus, $H \subseteq HK$. Similarly, $K \subseteq HK$. Hence, $H \cup K \subseteq HK$. Since $\langle H \cup K \rangle$ is the smallest subgroup of G containing $H \cup K$, it follows that $\langle H \cup K \rangle \subseteq HK$. Let $hk \in HK$, where $h \in H$ and $k \in K$. Since $H \subseteq \langle H \cup K \rangle$ and $K \subseteq \langle H \cup K \rangle$, we have $h, k \in \langle H \cup K \rangle$. Thus, $hk \in \langle H \cup K \rangle$. This implies that $HK \subseteq \langle H \cup K \rangle$. Hence, $HK = \langle H \cup K \rangle$. The converse is immediate since $\langle H \cup K \rangle$ is a subgroup and $HK = \langle H \cup K \rangle$. ∎

Let G be a group. We denote by $S(G)$ the set of all subgroups of G.

Theorem 4.1.16 *Let G be a group. Then $(S(G), \leq)$ is a lattice, where $\leq$ is set inclusion relation.*

Proof.   Proceeding as in Example 1.4.5, we can show that the set inclusion relation is a partial order on $S(G)$. We now show that for all $A, B \in S(G)$, $A \vee B, A \wedge B \in S(G)$. Let $A, B \in S(G)$. By Theorem 4.1.6, $A \cap B \in S(G)$ and by the definition of $S(G)$, $\langle A \cup B \rangle \in S(G)$. Now $A, B \subseteq \langle A \cup B \rangle$ and so $\langle A \cup B \rangle$ is an upper bound of A and B. Let $C \in S(G)$ be such that $A \subseteq C$ and $B \subseteq C$. Then $A \cup B \subseteq C$ and so $\langle A \cup B \rangle \subseteq C$. Thus, $\langle A \cup B \rangle$ is the least upper bound of A and B, i.e., $A \vee B = \langle A \cup B \rangle$. Hence, $A \vee B \in S(G)$. Next, we show that $A \wedge B = A \cap B$, i.e., $A \cap B$ is the greatest lower of A and B.

Now $A \cap B \subseteq A$ and $A \cap B \subseteq B$ and so $A \cap B$ is a lower bound of A and B. Let $D \in S(G)$, $D \subseteq A$, and $D \subseteq B$. Then $D \subseteq A \cap B$ and so $A \cap B$ is the greatest lower bound of A and B, i.e., $A \wedge B = A \cap B$. Therefore, $A \wedge B \in S(G)$. Consequently, $(S(G), \leq)$ is a lattice. ∎

The lattice $(S(G), \leq)$ in Theorem 4.1.16 is called the **subgroup lattice** of the group G. Let $(\mathcal{T}, \leq)$ be a sublattice of $(S(G), \leq)$, i.e., $\mathcal{T} \subseteq S(G)$ and $(\mathcal{T}, \leq)$ is a lattice. The poset diagram of $(\mathcal{T}, \leq)$ is called the **lattice diagram**. This lattice diagram will be useful in studying the interrelations among the subgroups of a group. Consider the following example.

Example 4.1.17 *(i) Let $G = \{1, -1, i, -i\}$. Then $(G, *)$ is a group, where $*$ is the usual multiplication of complex numbers. Let*

$$S = \{\{1\},\ \{1, -1\},\ G\}.$$

The lattice diagram of S is:

$$
\begin{array}{c}
G \\
| \\
\{1, -1\} \\
| \\
\{1\}
\end{array}
$$

(ii) Let $G = \{(1,1),\ (1,-1),\ (-1,1),\ (-1,-1)\}$. Then $(G, *)$ is a group, where $*$ is defined by $(a, b) * (c, d) = (ac, bd)$ for all (a, b), $(c, d) \in G$, where the multiplication ac and bd take place in the integers. Let $E = \{(1,1)\}$, $H_1 = \{(1,1),\ (1,-1)\}$, $H_2 = \{(1,1),\ (-1,1)\}$, and $H_3 = \{(1,1),\ (-1,-1)\}$. Let $S = \{E,\ H_1,\ H_2,\ H_3,\ G\}$. The lattice diagram of S is:

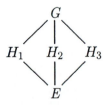

We see from these examples that a lattice diagram gives a visual picture of how subgroups of a given group are related.

Next, we consider an example of a group generated by two elements. We list several properties of the group. We ask the reader to verify these properties. We will study these types of groups in more detail in later chapters.

Example 4.1.18 *Let* $G = \langle a, b \rangle$, *where* $a^3 = e$, $b^2 = e$, *and* $(ab)^2 = e$. *Then*
(i) $ab = ba^{-1}$, $ba = a^{-1}b$, *and* $a^2b = ba$.
(ii) *G is not commutative since* $ab \neq ba$.
(iii) $ba^s = a^{-s}b$ *for all positive integers* s.
(iv) *By* (i) *and* (iii)

$$a^r b^i a^s b^j = \begin{cases} a^{r+s} b^j & if \ i = 0 \\ a^{r-s} b^{i+j} & if \ i = 1. \end{cases}$$

(v) *Since* $a^3 = e = b^2$, *every element of G is of the form* $a^r b^i$, $0 \leq r < 3$, $i = 0, 1$ *by* (iv).
(vi) $G = \{e, a, b, ab, a^2, a^2b\}$. *Thus,* $|G| = 6$.
(vii) $\circ(a) = 3 = \circ(a^2)$, $\circ(b) = \circ(ab) = \circ(a^2b) = 2$.
(viii) *The only subgroups of G are* $\{e\}$, $\langle a \rangle = \langle a^2 \rangle$, $\langle b \rangle$, $\langle ab \rangle$, $\langle a^2b \rangle$, *and* G.

G is called a **dihedral group** *of degree* 3 *and is denoted by* D_3. *In general, a dihedral group*[1] *of degree n is* $D_n = \langle a, b \rangle$, *where* $(ab)^2 = e$, $\circ(a) = n$, *and* $\circ(b) = 2$. *In Chapter 5, we consider a dihedral group of degree 4, D_4, and study this group in detail.*

4.1.1   Worked-Out Exercises

◇ **Exercise 1** Let H be a subgroup of a group G. Let $g \in G$. Prove that

(i) $gHg^{-1} = \{ghg^{-1} \mid h \in H\}$ is a subgroup of G,

(ii) $|gHg^{-1}| = |H|$.

Solution: (i) We first show that $gHg^{-1} \neq \phi$ and then use Theorem 4.1.3. Since $e = geg^{-1} \in gHg^{-1}$, $gHg^{-1} \neq \phi$. Let $gh_1g^{-1}, gh_2g^{-1} \in gHg^{-1}$. Then

$$(gh_1g^{-1})(gh_2g^{-1})^{-1} = gh_1g^{-1}gh_2^{-1}g^{-1} = gh_1h_2^{-1}g^{-1} \in gHg^{-1}.$$

Hence, gHg^{-1} is a subgroup of G.
(ii) Let $g \in G$. To prove that $|gHg^{-1}| = |H|$, we show that there exists a one-one onto function of H onto gHg^{-1}. Define $f : H \rightarrow gHg^{-1}$ by $f(h) = ghg^{-1}$ for all $h \in H$. Let $h, h' \in H$. If $h = h'$, then $ghg^{-1} = gh'g^{-1}$, i.e., f is well defined. Also, $ghg^{-1} \in gHg^{-1}$. Thus, f is a function of H into gHg^{-1}. Suppose $f(h) = f(h')$. Then $ghg^{-1} = gh'g^{-1}$. From this it follows that $h = h'$. This shows that f is one-one. To show f is onto gHg^{-1}, let $a \in gHg^{-1}$. Then $a = gbg^{-1} = f(b)$ for some $b \in H$, namely, $b = g^{-1}ag$. Thus, f is onto gHg^{-1}.

◇ **Exercise 2** Prove that S_n is generated by $\{(1 \ 2), (1 \ 3), (1 \ 4), \ldots, (1 \ n)\}$.

---

[1] We show the existence of such groups in Chapter 7.

Solution: Let π be any permutation in S_n. Then π is a product of trans-positions. Thus, it is sufficient to show that if $(i\ j)$ is any transposition in S_n, $i < j$, then

$$(i\ j) \in \langle (1\ 2), (1\ 3), (1\ 4), \ldots, (1\ n) \rangle\,.$$

This follows from the fact that $(i\ j) = (1\ i) \circ (1\ j) \circ (1\ i)$. Hence, S_n is generated by $\{(1\ 2), (1\ 3), (1\ 4), \ldots, (1\ n)\}$.

◇ **Exercise 3** Find all subgroups of $(\mathbf{Z}, +)$.

Solution: Let H be a subgroup of $\mathbf{Z}$. Suppose $H \neq \{0\}$. Let a be a nonzero element of H. Then $-a \in H$. Since either a or $-a$ is a positive integer, H contains a positive integer. With the help of the principle of well-ordering, we can show that H contains a smallest positive integer. Let a be the smallest positive integer in H. We claim that $H = \{na \mid n \in \mathbf{Z}\}$.

Now $na \in H$ for all $n \in \mathbf{Z}$ and so $\{na \mid n \in \mathbf{Z}\} \subseteq H$. On the other hand, let $b \in H$. By the division algorithm, there exist c and r in $\mathbf{Z}$ such that $b = ca + r$, where $0 \le r < a$. Suppose $r \neq 0$. Then $r = b - ca \in H$. Thus, H contains a positive integer smaller than a, a contradiction. Hence, $r = 0$ and so $b = ca \in \{na \mid n \in \mathbf{Z}\}$. This implies that $H \subseteq \{na \mid n \in \mathbf{Z}\}$. Thus, $H = \{na \mid n \in \mathbf{Z}\}$ for some $a \in \mathbf{Z}$. Also, for all $n \in \mathbf{Z}$, the set $T = \{nm \mid m \in \mathbf{Z}\} = n\mathbf{Z}$ is a subgroup of $\mathbf{Z}$. Hence, $n\mathbf{Z}$, $n = 0, 1, 2, \ldots$ are the subgroups of $\mathbf{Z}$.

4.1.2 Exercises

1. Prove that H is a subgroup of the group G, where

   (i) $H = \{[0], [2], [4], [6], [8], [10]\}$, $G = \mathbf{Z}_{12}$,

   (ii) $H = \{[0], [3], [6], [9]\}$, $G = \mathbf{Z}_{12}$

   and where the group operation under consideration is $+_{12}$.

2. Let $GL(2, \mathbf{R})$ denote the group of all nonsingular 2×2 matrices over $\mathbf{R}$. Show that each of the following sets is a subgroup of $GL(2, \mathbf{R})$.

   (i) $S = \left\{ \begin{bmatrix} a & b \\ c & d \end{bmatrix} \;\middle|\; ad - bc = 1 \right\}$.

   (ii) $S = \left\{ \begin{bmatrix} a & 0 \\ 0 & a \end{bmatrix} \;\middle|\; a \neq 0 \right\}$.

   (iii) $S = \left\{ \begin{bmatrix} a & b \\ -b & a \end{bmatrix} \;\middle|\; \text{either } a \text{ or } b \text{ is nonzero} \right\}$.

   (iv) $S = \left\{ \begin{bmatrix} a & b \\ 0 & d \end{bmatrix} \;\middle|\; ad \neq 0 \right\}$.

(v) $S = \left\{ \begin{bmatrix} a & -b \\ b & a \end{bmatrix} \mid a, b \in \mathbf{R} \text{ and } a^2 + b^2 \neq 0 \right\}$.

3. Show that the set $H = \{a + bi \in \mathbf{C}^* \mid a^2 + b^2 = 1\}$ is a subgroup of $(\mathbf{C}^*, \cdot)$, where $\cdot$ is the multiplication operation of complex numbers.

4. Let $G = \{(a, b) \mid a, b \in \mathbf{R}, b \neq 0\}$. Prove that $(G, *)$ is a noncommutative group under the binary operation $(a, b) * (c, d) = (a + bc, bd)$ for all $(a, b), (c, d) \in G$.

   (i) Let $H = \{(a, b) \in G \mid a = 0\}$. Show that H is a subgroup of G.

   (ii) Let $K = \{(a, b) \in G \mid b > 0\}$. Show that K is a subgroup of G.

   (iii) Let $T = \{(a, b) \in G \mid b = 1\}$. Show that T is a subgroup of G.

   (iv) Find all elements of order 2 in G.

5. In S_3, determine the set $T = \{x \in S_3 \mid x^2 = e\}$. Is T a subgroup of S_3?

6. Determine the subgroup $\langle 4, 6 \rangle$ in $(\mathbf{Z}, +)$.

7. In $(\mathbf{Z}, +)$, determine the subgroup generated by $\{4, 5\}$.

8. List the elements of the following subgroups.

   (i) $\left\langle \begin{pmatrix} 1 & 2 & 3 & 4 \\ 4 & 3 & 2 & 1 \end{pmatrix}, \begin{pmatrix} 1 & 2 & 3 & 4 \\ 2 & 1 & 4 & 3 \end{pmatrix} \right\rangle$ in S_4.

   (ii) $\langle h, v \rangle$ in the symmetries of the square.

9. Let $a = (1\ 2\ 3\ 4)$ and $b = (2\ 4) \in S_4$.

   (i) Find $o(a)$ and $o(b)$.

   (ii) Show that $ba = a^3 b = a^{-1} b$.

   (iii) Find $H = \langle a, b \rangle$ in S_4.

   (iv) Find $|H|$.

10. Let G be a group generated by a, b such that $o(b) = 2$, $o(a) = 6$, and $(ab)^2 = e$. Show that

   (i) $aba = b$,

   (ii) $(a^2 b)^2 = e$,

   (iii) $ba^2 b = a^4$,

   (iv) $ba^3 b = a^3$.

11. Let G be a group. Prove that a nonempty subset H of G is a subgroup if and only if for all $a, b \in H$, $ab \in H$ and $a^{-1} \in H$.

12. Let G be a commutative group. Show that the set H of all elements of finite order is a subgroup of G.

13. Let G be a group and $a \in G$. Show that if a is the only element of order n in G, then $a \in Z(G)$.

14. Show that $Z(S_n) = \{e\}$ for all $n \geq 3$.

15. Let G be a group and $a \in G$. Let $C(a) = \{b \in G \mid ba = ab\}$. Prove that $C(a)$ is a subgroup of G and that $Z(G) = \cap_{a \in G} C(a)$. $C(a)$ is called the **centralizer** of a in G.

16. Prove that a group G cannot be written as the union of two proper subgroups.

17. Let G be a group and H be a nonempty subset of G.

    (i) Show that if H is a subgroup of G, then $HH = H$.

    (ii) If H is finite and $HH \subseteq H$, prove that H is a subgroup of G.

    (iii) Give an example of a group G and a nonempty subset H of G such that $HH \subseteq H$, but H is not a subgroup of G.

18. Let H be a subgroup of a group G. Prove that $\langle H \rangle = H$.

19. If A and B are subgroups of a group G, prove that $A \cup B$ is a subgroup of G if and only if $A \subseteq B$ or $B \subseteq A$. If C is also a subgroup of G, does a similar necessary and sufficient condition hold for $A \cup B \cup C$ to be a subgroup of G?

20. Let G be a commutative group. If a and b are two distinct elements of G such that $o(a) = 2 = o(b)$, show that $|\langle a, b \rangle| = 4$.

21. (i) Prove that S_n is generated by $\{(1\ 2), (1\ 2\ 3 \cdots n)\}$.

    (ii) Prove that S_n is generated by $\{(1\ 2), (2\ 3), (3\ 4), \ldots, (n-1\ n)\}$.

22. Show that $(\mathbf{Q}, +)$ is not finitely generated.

23. Let G be a group. Prove that if G is finite, then G has finitely many subgroups.

24. Does there exist an infinite group with only a finite number of subgroups?

25. For the following statements, write the proof if the statement is true; otherwise, give a counterexample.

    (i) All nontrivial subgroups of $(\mathbf{Z}, +)$ are infinite groups.

(ii) If A, B, and C are subgroups of a group G such that $A \cup B \subseteq C$, then $ABC \subseteq C$.

(iii) If G is a noncommutative group, then $Z(G) = \{e\}$.

(iv) Let G be a group. If H is a nonempty subset of G such that $a^{-1} \in H$ for all $a \in H$, then H is a subgroup of G.

(v) There exists a proper subgroup A of $(\mathbf{Z}, +)$ such that A contains both $2\mathbf{Z}$ and $3\mathbf{Z}$.

(vi) If H is a subgroup of $(\mathbf{Q}, +)$ such that $\mathbf{Z} \subset H$, then $H = \mathbf{Q}$.

(vii) If H is a subgroup of $(\mathbf{Q}^*, \cdot)$ such that $\mathbf{Z} \backslash \{0\} \subseteq H$, then $H = \mathbf{Q}^*$.

4.2 Cyclic Groups

In the previous section, we introduced the notion of a subgroup generated by a set. Groups that are generated by a single element, called cyclic groups, are of special importance. As we shall see throughout the text, these groups play an important role in studying the structure of a group. In fact, all of Chapter 9 revolves around these groups. Cyclic groups are easier to study than any other group. They have special properties, some of which we will discover in this section.

Definition 4.2.1 *A group G is called a **cyclic group** if there exists $a \in G$ such that*

$$G = \langle a \rangle .$$

We recall that $\langle a \rangle$ in Definition 4.2.1 is the set $\{a^n \mid n \in \mathbf{Z}\}$ (Corollary 4.1.9).

Let $G = \langle a \rangle$ be a cyclic group and $b, c \in G$. Then $b = a^n$ and $c = a^m$ for some $n, m \in \mathbf{Z}$. Now $bc = a^n a^m = a^{n+m} = a^{m+n} = a^m a^n = cb$. This shows that G is commutative. Hence, every cyclic group is commutative.

Example 4.2.2 (*i*) $(\mathbf{Z}, +)$ *is a cyclic group since* $\mathbf{Z} = \langle 1 \rangle$.
(*ii*) $(\{na \mid n \in \mathbf{Z}\}, +)$ *(Example 2.1.4) is a cyclic group, where a is any fixed element of $\mathbf{Z}$.*
(*iii*) $(\mathbf{Z}_n, +_n)$ *is a cyclic group since* $\mathbf{Z}_n = \langle [1] \rangle$.

Example 4.2.3 *Let a be a symbol and n a positive integer. Define $*$ by means of the following operation table.*

$*$	a^0	a^1	a^2	$\cdots$	a^{n-2}	a^{n-1}
a^0	a^0	a^1	a^2	$\cdots$	a^{n-2}	a^{n-1}
a^1	a^1	a^2	a^3	$\cdots$	a^{n-1}.	a^0
a^2	a^2	a^3	a^4	$\cdots$	a^0	a^1
$\vdots$	$\vdots$	$\vdots$	$\vdots$	$\vdots$	$\vdots$	$\vdots$
a^{n-2}	a^{n-2}	a^{n-1}	a^0	$\cdots$	a^{n-4}	a^{n-3}
a^{n-1}	a^{n-1}	a^0	a^1	$\cdots$	a^{n-3}	a^{n-2}

Then $(\{a^0, a^1, \ldots, a^{n-1}\}, *)$ is a cyclic group generated by a^1.

Example 4.2.4 *Consider the set $G = \{e, a, b, c\}$. Define $*$ on G by means of the following operation table.*

$*$	e	a	b	c
e	e	a	b	c
a	a	e	c	b
b	b	c	e	a
c	c	b	a	e

From the multiplication table, it follows that $(G, *)$ is a commutative group. However, G is not a cyclic group since

$$\langle e \rangle = \{e\}, \langle a \rangle = \{e, a\}, \langle b \rangle = \{e, b\}, \text{ and } \langle c \rangle = \{e, c\}$$

*and each of these subgroups is properly contained in G. G is known as the **Klein 4-group**.*

The next theorem gives the exact description of a finite cyclic group.

Theorem 4.2.5 *Let $\langle a \rangle$ be a finite cyclic group of order n. Then $\langle a \rangle = \{e, a, a^2, \ldots, a^{n-1}\}$.*

Proof. By Corollary 4.1.9, $\langle a \rangle = \{a^i \mid i \in \mathbf{Z}\}$. Since $\langle a \rangle$ is finite, there exist $i, j \in \mathbf{Z}$ $(j > i)$ such that $a^i = a^j$. Thus, $a^{j-i} = e$ and $j - i$ is positive. Let m be the smallest positive integer such that $a^m = e$. Then for all integers i, j such that $0 \le i < j < m$, $a^i \ne a^j$ otherwise $a^{j-i} = e$ for some $0 \le i < j < m$, which contradicts the minimality of m. Hence, the elements of the set $S = \{e, a, a^2, \ldots, a^{m-1}\}$ are distinct. Clearly $S \subseteq \langle a \rangle$. Let $a^k \in \langle a \rangle$. By the division algorithm, there exist integers q, r such that $k = qm + r$, $0 \le r < m$. Thus, $a^k = a^{qm+r} = (a^m)^q a^r = ea^r = a^r \in S$. Therefore, $\langle a \rangle \subseteq S$. Thus, $S = \langle a \rangle$. Since the elements of S are distinct and $\langle a \rangle$ has order n, it must be the case that $m = n$. ■

The following corollaries are immediate from the proof of Theorem 4.2.5. We omit the proofs.

Corollary 4.2.6 *Let* $\langle a \rangle$ *be a finite cyclic group. Then* $o(a) = |\langle a \rangle|$. ■

Corollary 4.2.7 *A finite group* G *is a cyclic group if and only if there exists an element* $a \in G$ *such that* $o(a) = |G|$. ■

As stated in the beginning of this section, cyclic groups have special properties. We now proceed to discover some of these properties. Subgroups of a cyclic group are themselves cyclic; this is proved in the next theorem.

Theorem 4.2.8 *Every subgroup of a cyclic group is cyclic.*

Proof. Let H be a subgroup of a cyclic group $G = \langle a \rangle$. If $H = \{e\}$, then $H = \langle e \rangle$ and so H is cyclic. Suppose $\{e\} \subset H$. Then there exists $b \in H$ such that $b \neq e$. Since $b \in G$, we have $b = a^m$ for some integer m. Thus, $m \neq 0$ since $b \neq e$. Since H is a group, $a^{-m} = b^{-1} \in H$. Now either m or $-m$ is positive. Therefore, H contains at least one element which is a positive power of a. Let n be the smallest positive integer such that $a^n \in H$. We now show that $H = \langle a^n \rangle$.

Since $a^n \in H$, we must have $\langle a^n \rangle \subseteq H$. Let $h \in H$. Then $h = a^k$ for some integer k. By the division algorithm, there exist integers q, r such that $k = nq + r$, $0 \leq r < n$. Since a^n and $a^k \in H$, we have $a^r = a^{k-nq} = a^k(a^n)^{-q} \in H$. However, if $r > 0$, we contradict the minimality of n. Therefore, $r = 0$ so that $a^k = (a^n)^q \in \langle a^n \rangle$. Hence, $H \subseteq \langle a^n \rangle$ and so $H = \langle a^n \rangle$. Thus, H is cyclic. ■

Corollary 4.2.9 *Let* $G = \langle a \rangle$ *be a cyclic group of order* m, $m > 1$, *and* H *be a proper subgroup of* G. *Then* $H = \langle a^k \rangle$ *for some integer* k *such that* k *divides* m *and* $k > 1$. *Furthermore,* $|H|$ *divides* m.

Proof. If $H = \{e\}$, then $H = \langle a^m \rangle$. Suppose that $H \neq \{e\}$. Let k be the smallest positive integer such that $a^k \in H$. Then $H = \langle a^k \rangle$. Now there exist integers q and r such that $m = qk + r$, where $0 \leq r < k$, and

$$a^r = a^{m-qk} = a^m a^{-qk} = a^{-qk} = ((a^k)^{-1})^q \in H.$$

The minimality of k implies that $r = 0$. Hence, $m = qk$ and so k divides m. Since $H \neq G$, $k > 1$. Next, we show that $|H|$ divides m. By Theorem 2.1.28(ii), $o(a^k) = \frac{m}{\gcd(m,k)} = \frac{m}{k} = q$. As a result Corollary 4.2.6 implies that

$$|H| = o(a^k) = q.$$

Since $m = qk$, we have $q|m$, i.e., $|H|$ divides m. ∎

By Corollary 4.2.9, if G is a finite cyclic group and H is a subgroup of G, then $|H|$ divides $|G|$. This is a special case of a more general result, called Lagrange's theorem, which we will prove in the next section.

Let $G = \langle a \rangle$ be an infinite cyclic group. Then $\circ(a)$ is infinite and this implies that $\circ(a^k)$ is infinite for any nonzero integer k. Thus, the order of any nonidentity element of G is infinite. Let H be a nontrivial subgroup of G. Then H is cyclic. Let $H = \langle b \rangle$. Then $b \neq e$ and $b \in G$ and so $\circ(b)$ is infinite. This in turn shows that $|H|$ is infinite. Thus, every nontrivial subgroup of an infinite cyclic group is infinite.

Now let $G = \langle a \rangle$ be a finite cyclic group of order n and H be a proper subgroup of G. Then by Corollary 4.2.9, $|H|$ divides $|G|$. If $H = \{e\}$, then $|H| = 1$ and if $H = G$, then $|H| = |G|$ and so $|H|$ divides $|G|$. Thus, the order of every subgroup of G divides the order of G. The following theorem shows that the converse of this result is also true for finite cyclic groups.

Theorem 4.2.10 *Let G be a finite cyclic group of order m. Then for every positive divisor d of m, there exists a unique subgroup of G of order d.*

Proof. Let $G = \langle a \rangle$ and d be a positive divisor of m. Since $d|m$, there exists $k \in \mathbf{Z}$ such that $m = kd$. Now $a^k \in G$ and by Theorem 2.1.28(ii),

$$\circ(a^k) = \frac{\circ(a)}{\gcd(k, m)} = \frac{m}{k} = d.$$

Let $H = \langle a^k \rangle$. Then $|H| = \circ(a^k) = d$. Thus, G has a subgroup of order d. Next, we establish that H is unique.

Let K be a subgroup of order d. Let t be the smallest positive integer such that $a^t \in K$. Then $K = \langle a^t \rangle$. Since K is of order d, $\circ(a^t) = d$ by Corollary 4.2.6. But $\circ(a^t) = \frac{m}{\gcd(t,m)}$ by Theorem 2.1.28(ii). Hence, $d = \frac{m}{\gcd(t,m)}$, which implies that $\gcd(t, m) = \frac{m}{d} = k$. This shows that $k|t$. Let $t = kl$ for some $l \in \mathbf{Z}$. Now $a^t = a^{kl} = (a^k)^l \in H$. Hence, $K \subseteq H$. Since $|K| = |H|$ and H and K are finite, we have $H = K$. Thus, there exists a unique subgroup of order d. ∎

4.2.1 Worked-Out Exercises

◇ **Exercise 1** $(\mathbf{Q}, +)$ is not cyclic.

Solution: Suppose $\mathbf{Q}$ is cyclic. Then $\mathbf{Q} = \left\langle \frac{p}{q} \right\rangle$ for some $\frac{p}{q} \in \mathbf{Q}$, where p and q are relatively prime. Since $\frac{p}{2q} \in \mathbf{Q}$, there exists $n \in \mathbf{Z}$, $n \neq 0$ such that $\frac{p}{2q} = n\frac{p}{q}$ by Corollary 4.1.9. This implies that $\frac{1}{2} = n \in \mathbf{Z}$, which is a contradiction. Thus, $\mathbf{Q}$ is not cyclic.

Exercise 2 Let G be a group such that $|G| = mn$, $m > 1$, $n > 1$. Show that G has a nontrivial subgroup.

Solution: First suppose that G is cyclic. Let $G = \langle a \rangle$. Then $o(a) = mn$. Clearly $o(a^m) = n$. Let $H = \langle a^m \rangle$. Then H is a nontrivial subgroup of G. Now suppose that G is not cyclic. Then for all $a \in G$, $o(a) < mn$ by Exercise 26 (page 79). Let $e \neq a \in G$ and let $H = \langle a \rangle$. Then H is a nontrivial subgroup of G.

$\Diamond$ **Exercise 3** Let G be an infinite cyclic group generated by a. Show that

(i) $a^r = a^t$ if and only if $r = t$, where $r, t \in \mathbf{Z}$,

(ii) G has exactly two generators.

Solution: (i) Suppose $a^r = a^t$ and $r \neq t$. Let $r > t$. Then $a^{r-t} = e$. Thus, $o(a)$ is finite, say, $o(a) = n$. Then $G = \{e, a, \ldots, a^{n-1}\}$, which is a contradiction since G is an infinite group. The converse is straightforward.

(ii) Let $G = \langle b \rangle$ for some $b \in G$. Since $a \in G = \langle b \rangle$ and $b \in G = \langle a \rangle$, $a = b^r$ and $b = a^t$ for some $r, t \in \mathbf{Z}$. Thus, $a = b^r = (a^t)^r = a^{rt}$. Hence, by (i), $rt = 1$. This implies that either $r = 1 = t$ or $r = -1 = t$. Thus, either $b = a$ or $b = a^{-1}$. Now from (i), $a \neq a^{-1}$. Therefore, G has exactly two generators.

$\Diamond$ **Exercise 4** (i) Let $G = \langle a \rangle$ be a finite cyclic group of order n. Show that a^k is a generator of G if and only if $\gcd(k, n) = 1$, where k is a positive integer.

(ii) Find all generators of $\mathbf{Z}_{10}$.

Solution: (i) Suppose a^k is a generator of G. Since $|G| = n$, $o(a^k) = n$. But $o(a^k) = \frac{n}{\gcd(k,n)}$. Hence, $\frac{n}{\gcd(k,n)} = n$. Thus, $\gcd(k, n) = 1$. Conversely, suppose that $\gcd(k, n) = 1$. Then $o(a^k) = \frac{n}{\gcd(k,n)} = n$. Hence, $\left| \langle a^k \rangle \right| = n$. Since $\langle a^k \rangle \subseteq G$ and $|G| = n$, $G = \langle a^k \rangle$.

(ii) Now $\mathbf{Z}_{10} = \langle [1] \rangle$ and $|\mathbf{Z}_{10}| = 10$. By (i), $k[1]$ is a generator if and only if $\gcd(k, 10) = 1$, where $1 \leq k \leq 10$. Now if $k = 1, 3, 7,$ or 9, then $\gcd(k, 10) = 1$. Thus, the generators of $\mathbf{Z}_{10}$ are $1[1] = [1]$, $3[1] = [3]$, $7[1] = [7]$ and $9[1] = [9]$.

4.2.2 Exercises

1. Let $G = \langle a \rangle$ be a cyclic group of order 30. Determine the following subgroups.

(i) $\langle a^5 \rangle$.

(ii) $\langle a^2 \rangle$.

2. Let G be a cyclic group of order 30. Find the number of elements of order 6 in G and also find the number of elements of order 5 in G.

3. Prove that 1 and -1 are the only generators of $\mathbf{Z}$.

4. (i) Show that $(\mathbf{R}, +)$ is not cyclic.

   (ii) Show that $(\mathbf{Q}^*, \cdot)$ is not cyclic.

   (iii) Show that $(\mathbf{R}^*, \cdot)$ is not cyclic.

5. If G is a cyclic group of order n, show that the number of generators of G is $\phi(n)$, where ϕ is the Euler ϕ-function.

6. Show that every proper subgroup of S_3 is cyclic.

7. Give an example of a noncyclic Abelian group all of whose proper subgroups are cyclic.

8. Let G be a group. Suppose that G has at most two nontrivial subgroups. Show that G is cyclic.

9. Let G be a finite group. Show that if G has exactly one nontrivial subgroup, then order of G is p^2 for some prime p.

10. Let G be a noncommutative group. Show that G has a nontrivial subgroup.

11. Give an example of an infinite group which contains a nontrivial finite cyclic group.

12. Show that there are cyclic subgroups of order $1, 2, 3$, and 4 in S_4, but S_4 does not contain any cyclic subgroup of order ≥ 5.

13. For the following statements, write the proof if the statement is true; otherwise, give a counterexample.

    (i) For every positive integer n, there exists a cyclic group of order n.

    (ii) Every proper subgroup of A_4 is cyclic.

    (iii) A_3 is a cyclic group.

    (iv) A_4 is a cyclic group.

    (v) All proper subgroups of $(\mathbf{R}, +)$ are cyclic.

4.3   Lagrange's Theorem

In the last section, we noted that the order of a subgroup of a finite cyclic group divides the order of the group (Corollary 4.2.9). We also remarked that this is a special case of a general result, called Lagrange's theorem, i.e., the order of a subgroup of a finite group divides the order of the group. Lagrange proved this result in 1770, long before the creation of group theory, while working on the permutations of the roots of a polynomial equation. Lagrange's theorem is a basic theorem of finite group theory and is considered by some to be the most important result in finite group theory. In this section, we prove this result. We begin with the following definition.

Definition 4.3.1 *Let H be a subgroup of a group G and $a \in G$. The sets $aH = \{ah \mid h \in H\}$ and $Ha = \{ha \mid h \in H\}$ are called the **left** and **right** cosets of H in G, respectively. The element a is called a representative of aH and Ha.*

If G is commutative, then of course $aH = Ha$. Observe that $eH = H = He$ and that $a = ae \in aH$ and $a = ea \in Ha$.

Example 4.3.2 *Consider the symmetric group S_3 (Example 3.1.6). Then*

$$H = \left\{ e, \begin{pmatrix} 1 & 2 & 3 \\ 2 & 3 & 1 \end{pmatrix}, \begin{pmatrix} 1 & 2 & 3 \\ 3 & 1 & 2 \end{pmatrix} \right\}$$

and

$$H' = \left\{ e, \begin{pmatrix} 1 & 2 & 3 \\ 1 & 3 & 2 \end{pmatrix} \right\}$$

are subgroups of S_3. We now compute the left and right cosets of H in S_3. The left cosets of H in S_3 are

$$\begin{pmatrix} 1 & 2 & 3 \\ 1 & 2 & 3 \end{pmatrix} H = \begin{pmatrix} 1 & 2 & 3 \\ 2 & 3 & 1 \end{pmatrix} H = \begin{pmatrix} 1 & 2 & 3 \\ 3 & 1 & 2 \end{pmatrix} H = H$$

and

$$\begin{pmatrix} 1 & 2 & 3 \\ 1 & 3 & 2 \end{pmatrix} H = \begin{pmatrix} 1 & 2 & 3 \\ 3 & 2 & 1 \end{pmatrix} H = \begin{pmatrix} 1 & 2 & 3 \\ 2 & 1 & 3 \end{pmatrix} H =$$
$$\left\{ \begin{pmatrix} 1 & 2 & 3 \\ 1 & 3 & 2 \end{pmatrix}, \begin{pmatrix} 1 & 2 & 3 \\ 2 & 1 & 3 \end{pmatrix}, \begin{pmatrix} 1 & 2 & 3 \\ 3 & 2 & 1 \end{pmatrix} \right\}$$

and the right cosets of H in S_3 are

$$H \begin{pmatrix} 1 & 2 & 3 \\ 1 & 2 & 3 \end{pmatrix} = H \begin{pmatrix} 1 & 2 & 3 \\ 2 & 3 & 1 \end{pmatrix} = H \begin{pmatrix} 1 & 2 & 3 \\ 3 & 1 & 2 \end{pmatrix} = H$$

and

$$H\begin{pmatrix} 1 & 2 & 3 \\ 1 & 3 & 2 \end{pmatrix} = H\begin{pmatrix} 1 & 2 & 3 \\ 3 & 2 & 1 \end{pmatrix} = H\begin{pmatrix} 1 & 2 & 3 \\ 2 & 1 & 3 \end{pmatrix} =$$
$$\left\{ \begin{pmatrix} 1 & 2 & 3 \\ 1 & 3 & 2 \end{pmatrix}, \begin{pmatrix} 1 & 2 & 3 \\ 2 & 1 & 3 \end{pmatrix}, \begin{pmatrix} 1 & 2 & 3 \\ 3 & 2 & 1 \end{pmatrix} \right\}.$$

Thus, for all $a \in S_3$, $aH = Ha$.

Next, we compute the left and right cosets of H' in S_3. The left cosets of H' in S_3 are

$$\begin{pmatrix} 1 & 2 & 3 \\ 1 & 2 & 3 \end{pmatrix} H' = \begin{pmatrix} 1 & 2 & 3 \\ 1 & 3 & 2 \end{pmatrix} H' = H',$$

$$\begin{pmatrix} 1 & 2 & 3 \\ 3 & 2 & 1 \end{pmatrix} H' = \begin{pmatrix} 1 & 2 & 3 \\ 3 & 1 & 2 \end{pmatrix} H' = \left\{ \begin{pmatrix} 1 & 2 & 3 \\ 3 & 2 & 1 \end{pmatrix}, \begin{pmatrix} 1 & 2 & 3 \\ 3 & 1 & 2 \end{pmatrix} \right\},$$

and

$$\begin{pmatrix} 1 & 2 & 3 \\ 2 & 1 & 3 \end{pmatrix} H' = \begin{pmatrix} 1 & 2 & 3 \\ 2 & 3 & 1 \end{pmatrix} H' = \left\{ \begin{pmatrix} 1 & 2 & 3 \\ 2 & 1 & 3 \end{pmatrix}, \begin{pmatrix} 1 & 2 & 3 \\ 2 & 3 & 1 \end{pmatrix} \right\}$$

and the right cosets of H' in S_3 are

$$H'\begin{pmatrix} 1 & 2 & 3 \\ 1 & 2 & 3 \end{pmatrix} = H'\begin{pmatrix} 1 & 2 & 3 \\ 1 & 3 & 2 \end{pmatrix} = H',$$

$$H'\begin{pmatrix} 1 & 2 & 3 \\ 3 & 2 & 1 \end{pmatrix} = H'\begin{pmatrix} 1 & 2 & 3 \\ 2 & 3 & 1 \end{pmatrix} = \left\{ \begin{pmatrix} 1 & 2 & 3 \\ 3 & 2 & 1 \end{pmatrix}, \begin{pmatrix} 1 & 2 & 3 \\ 2 & 3 & 1 \end{pmatrix} \right\},$$

and

$$H'\begin{pmatrix} 1 & 2 & 3 \\ 2 & 1 & 3 \end{pmatrix} = H'\begin{pmatrix} 1 & 2 & 3 \\ 3 & 1 & 2 \end{pmatrix} = \left\{ \begin{pmatrix} 1 & 2 & 3 \\ 2 & 1 & 3 \end{pmatrix}, \begin{pmatrix} 1 & 2 & 3 \\ 3 & 1 & 2 \end{pmatrix} \right\}.$$

We see that

$$\begin{pmatrix} 1 & 2 & 3 \\ 3 & 1 & 2 \end{pmatrix} H' \neq H'\begin{pmatrix} 1 & 2 & 3 \\ 3 & 1 & 2 \end{pmatrix}.$$

Thus, the left and right cosets of H' in S_3 are not the same.

There are some interesting phenomena happening in the above example. We see that all left and right cosets of H in S_3 have the same number of elements, namely, 3; that there are the same number of distinct left cosets of H in S_3 as of right cosets, namely, 2; that the set of all left cosets and the set of all right cosets form partitions of S_3; and, finally, that $3 \cdot 2$ equals the order of S_3. Similar statements hold for the subgroup H'. We show, in the results to follow, that these phenomena hold in general.

In the next few theorems, we prove some properties of left and right cosets of a subgroup which will eventually lead us to the proof of Lagrange's theorem. The following theorem tells us when two left (right) cosets are equal. It is a result that is used often in the study of groups.

Theorem 4.3.3 *Let H be a subgroup of a group G and $a, b \in G$. Then*
    (i) $aH = bH$ if and only if $b^{-1}a \in H$.
    (ii) $Ha = Hb$ if and only if $ab^{-1} \in H$.

Proof.   (i) Suppose $aH = bH$. Since $a \in aH$ and $aH = bH$, there exists $h' \in H$ such that $a = bh'$. This implies that $b^{-1}a = h' \in H$.

Conversely, suppose $b^{-1}a \in H$. Then there exists $h' \in H$ such that $b^{-1}a = h'$, i.e., $a = bh'$. Let $ah \in aH$. Then $ah = bh'h \in bH$. This implies that $aH \subseteq bH$. Next, we show that $bH \subseteq aH$. Now $b^{-1}a = h'$ implies that $ah'^{-1} = b$. Let $bh \in bH$. Then $bh = ah'^{-1}h \in aH$. Hence, $bH \subseteq aH$. Consequently, $aH = bH$.

(ii) The proof is similar to (i). We leave it as an exercise. ∎

Theorem 4.3.4 *Let H be a subgroup of a group G. Then for all $a, b \in G$, either $aH = bH$ or $aH \cap bH = \phi$ (i.e., two left cosets are either equal or they are disjoint).*

Proof.   Let $a, b \in G$. Suppose that $aH \cap bH \neq \phi$. We wish to show that $aH = bH$. Since $aH \cap bH \neq \phi$, there exists $c \in aH \cap bH$. Hence, $c \in aH$ and $c \in bH$ and so there exist $h_1, h_2 \in H$ such that $c = ah_1$ and $c = bh_2$. Thus, $ah_1 = bh_2$ and from this, it follows that $b^{-1}a = h_2h_1^{-1}$. Therefore, $b^{-1}a \in H$. By Theorem 4.3.3(i), $aH = bH$. ∎

Corollary 4.3.5 *Let H be a subgroup of a group G. Then $\{aH \mid a \in G\}$ forms a partition of G.*

Proof.   Let $\mathcal{P} = \{aH \mid a \in G\}$, i.e., $\mathcal{P}$ is the set of all left cosets of H in G. By Theorem 4.3.4, for all $aH, bH \in \mathcal{P}$, either $aH = bH$ or $aH \cap bH = \phi$. Thus, $\mathcal{P}$ satisfies (i) of Definition 1.3.14. Since $aH \subseteq G$ for all $a \in G$, $\cup_{aH \in \mathcal{P}} aH \subseteq G$. If $a \in G$, then $a \in aH \subseteq \cup_{aH \in \mathcal{P}} aH$. Therefore, $G \subseteq \cup_{aH \in \mathcal{P}} aH$. Hence, $G = \cup_{aH \in \mathcal{P}} aH$. This shows that $\mathcal{P}$ satisfies (ii) of Definition 1.3.14. Consequently, $\mathcal{P}$ is a partition of G. ∎

Theorem 4.3.6 *Let H be a subgroup of a group G. Then the elements of H are in one-one correspondence with the elements of any left (right) coset of H in G.*

Proof.   Let a be any element of G and aH be a left coset of H in G. To show that the elements of H are in one-one correspondence with the elements of aH, we show that there exists a one-one function of H onto aH. Define $f : H \rightarrow aH$ by $f(h) = ah$ for all $h \in H$. Let $h, h_1 \in H$. If $h = h_1$, then $ah = ah_1$, i.e., $f(h) = f(h_1)$. Hence, f is well defined. Suppose $f(h) = f(h_1)$. Then $ah = ah_1$ and this implies that $h = h_1$. Thus, f is a one-one function. To show f is onto aH, let $ah \in aH$, where $h \in H$. Then $ah = f(h)$. Hence, f maps H onto aH. Similarly, we can show that the elements of H are in one-one correspondence with the elements of Ha. ∎

The following corollary is immediate from Theorem 4.3.6.

Corollary 4.3.7 *Let H be a subgroup of a group G. Then for all $a \in G$,* $|H| = |aH| = |Ha|$. ∎

The next theorem says that there are the same number of left cosets as right cosets.

Theorem 4.3.8 *Let H be a subgroup of a group G. Then there is a one-one correspondence of the set of all left cosets of H in G onto the set of all right cosets of H in G.*

Proof.   Let $\mathcal{L} = \{aH \mid a \in G\}$ be the set of all left cosets of H in G and $\mathcal{R} = \{Ha \mid a \in G\}$ be the set of all right cosets of H in G. To establish a one-one correspondence between the elements of $\mathcal{L}$ and $\mathcal{R}$, we need to show the existence of a one-one function of $\mathcal{L}$ onto $\mathcal{R}$.
       Define $f : \mathcal{L} \rightarrow \mathcal{R}$ by

$$f(aH) = Ha^{-1}$$

for all $aH \in \mathcal{L}$. First note that $Ha^{-1} \in \mathcal{R}$ for all $a \in G$. Let $aH, bH \in \mathcal{L}$. Suppose $aH = bH$. Then by Theorem 4.3.3(i), $b^{-1}a \in H$. This implies that $b^{-1}(a^{-1})^{-1} = b^{-1}a \in H$ and so by Theorem 4.3.3(ii), $Hb^{-1} = Ha^{-1}$. Thus, $f(bH) = f(aH)$. Hence, f is well defined. To show f is one-one, suppose $f(aH) = f(bH)$. Then $Ha^{-1} = Hb^{-1}$ and so $a^{-1}(b^{-1})^{-1} \in H$ by Theorem 4.3.3(ii), i.e., $a^{-1}b \in H$. Therefore, $b^{-1}a = (a^{-1}b)^{-1} \in H$ and so $aH = bH$. Hence, f is one-one. Since for all $Ha \in \mathcal{R}$, $Ha = H(a^{-1})^{-1} = f(a^{-1}H)$ and $a^{-1}H \in \mathcal{L}$, it follows that f is onto $\mathcal{R}$. Thus, f is a one-one function from $\mathcal{L}$ onto $\mathcal{R}$. ∎

Definition 4.3.9 *Let H be a subgroup of a group G. Then the number of distinct left (or right) cosets, written $[G : H]$, of H in G is called the **index** of H in G.*

By Theorem 4.3.8, the number of left cosets and the number of right cosets of a subgroup H of a group G are the same. Thus, $[G : H]$ is well defined.

If G is finite, then of course $[G : H]$ is finite. The following example is one, where G is infinite and $[G : H]$ is finite.

Example 4.3.10 *Let n be a fixed positive integer. Consider the cyclic subgroup $(\langle n \rangle, +)$ of $(\mathbf{Z}, +)$. Let $k + \langle n \rangle$ be a left coset of $\langle n \rangle$ in $\mathbf{Z}$. By the division algorithm, there exist integers q and r such that $k = qn + r$, where $0 \leq r < n$. Then $k - r = qn \in \langle n \rangle$ and so $k + \langle n \rangle = r + \langle n \rangle$ by Theorem 4.3.3. Suppose $i + \langle n \rangle = j + \langle n \rangle$, where $0 \leq i, j < n$. Then $i - j \in \langle n \rangle$ by Theorem 4.3.3. This implies that $n | (i - j)$ and so we must have $i - j = 0$ or $i = j$ since $0 \leq i, j < n$. Thus, the distinct left cosets of $\langle n \rangle$ in $\mathbf{Z}$ are $0 + \langle n \rangle, 1 + \langle n \rangle, \ldots, n - 1 + \langle n \rangle$.*

We are now ready to prove Lagrange's theorem. It is interesting to note that Lagrange proved the result for the symmetric group S_n. Some credit Galois for proving the result in general.

Theorem 4.3.11 (Lagrange) *Let H be a subgroup of a finite group G. Then the order of H divides the order of G. In particular,*

$$|G| = [G : H]|H|.$$

Proof. Since G is a finite group, the number of left cosets of H in G is finite. Let $\{a_1 H, a_2 H, \ldots, a_r H\}$ be the set of all distinct left cosets of H in G. Then by Corollary 4.3.5, $G = \cup_{i=1}^r a_i H$ and $a_i H \cap a_j H = \phi$ for all $i \neq j$, $1 \leq i, j \leq r$. Hence, $[G : H] = r$ and

$$|G| = |a_1 H| + |a_2 H| + \cdots + |a_r H|.$$

By Corollary 4.3.7, $|H| = |a_i H|$ for all i, $1 \leq i \leq r$. Therefore,

$$
\begin{aligned}
|G| &= \underbrace{|H| + |H| + \cdots + |H|}_{r \text{ times}} \\
&= r|H| \\
&= [G : H]|H|.
\end{aligned}
$$

Thus, the order of H divides the order of G. ∎

Corollary 4.3.12 *Let G be a group of finite order n. Then the order of any element a of G divides n and $a^n = e$.*

Proof.   Let $a \in G$ and $\circ(a) = k$. Let $H = \langle a \rangle$. Then by Corollary 4.2.6, $|H| = |\langle a \rangle| = \circ(a) = k$. Hence, by Theorem 4.3.11, k divides n. Thus, there exists $q \in \mathbf{Z}$ such that $n = kq$. Hence, $a^n = a^{kq} = (a^k)^q = e^q = e$. ∎

Let G be a finite group of order n and $a \in G$. Then $\circ(a)$ divides n by Corollary 4.3.12. Thus, to find $\circ(a)$, we only need to check a^k, where k is a positive divisor of n. For example, consider $\mathbf{Z}_{20}$ and $[6] \in \mathbf{Z}_{20}$. Now $|\mathbf{Z}_{20}| = 20$ and $1, 2, 4, 5, 10,$ and 20 are the only positive divisors of 20. Now $1[6] = [6] \neq [0]$, $2[6] = [12] \neq [0]$, $4[6] = [24] = [4] \neq [0]$, $5[6] = [30] = [10] \neq [0]$, and $10[6] = [60] = [0]$. Thus, $\circ([6]) = 10$. Hence, the above corollary can be used to find the order of an element in a finite group.

Corollary 4.3.13 *Let G be a group of prime order. Then G is cyclic.*

Proof.   Since $|G| \geq 2$, there exists $a \in G$ such that $a \neq e$. Let $H = \langle a \rangle$. Then $\{e\} \subset H$ and $|H|$ divides $|G|$. But $|G|$ is prime and so $|H| = |G|$. Since $H \subseteq G$ and $|H| = |G|$, it follows that $G = H$. Therefore, G is cyclic. ∎

G.H. Hardy (1877–1947) believed that no result of number theory would have a practical application. However, number theoretic results have recently been applied to cryptography, the study of secret codes. The following is such a result. It is known as Fermat's little theorem.

Theorem 4.3.14 (Fermat) *Let p be a prime integer and a be an integer such that p does not divide a. Then p divides $a^{p-1} - 1$, i.e.,*

$$a^{p-1} \equiv_p 1.$$

Proof.   Let $U_p = \mathbf{Z}_p \backslash \{0\}$. Then by Exercise 10 (page 78), U_p is a group. Also, by Exercise 9 (page 78), $|U_p| = p - 1$. Let a be an integer such that p does not divide a. Then $[a]$ is a nonzero element of $\mathbf{Z}_p$ and so $[a] \in U_p$. Thus, by Corollary 4.3.12, $[a]^{p-1} = [1]$, i.e., $[a^{p-1}] = [1]$. Hence, $a^{p-1} \equiv_p 1$ by Exercise 11 (page 30). ∎

Let H and K be subgroups of a group G. If either H or K is infinite, then, of course, HK is infinite. Suppose H and K are both finite. We know that HK need not be a subgroup of G. Thus, $|HK|$ need not divide $|G|$. However, with the help of Lagrange's theorem, we can determine $|HK|$. This is a very useful result and we will use it very effectively in this text. In the next theorem, we determine $|HK|$ when H and K are both finite.

Theorem 4.3.15 *Let H and K be finite subgroups of a group G. Then*

$$|HK| = \frac{|H|\,|K|}{|H \cap K|}.$$

Proof. Let us write $A = H \cap K$. Since H and K are subgroups of G, A is a subgroup of G and since $A \subseteq H$, A is also a subgroup of H. By Lagrange's theorem, $|A|$ divides $|H|$. Let $n = \frac{|H|}{|A|}$. Then $[H : A] = n$ and so A has n distinct left cosets in H. Let $\{x_1 A, x_2 A, \ldots, x_n A\}$ be the set of all distinct left cosets of A in H. Then $H = \cup_{i=1}^{n} x_i A$. Since $A \subseteq K$, it follows that

$$HK = (\cup_{i=1}^{n} x_i A)K = \cup_{i=1}^{n} x_i K.$$

We now show that $x_i K \cap x_j K = \phi$ if $i \neq j$. Suppose $x_i K \cap x_j K \neq \phi$ for some $i \neq j$. Then $x_j K = x_i K$. Thus, $x_i^{-1} x_j \in K$. Since $x_i^{-1} x_j \in H$, $x_i^{-1} x_j \in A$ and so $x_j A = x_i A$. This contradicts the assumption that $x_1 A, \ldots, x_n A$ are all distinct left cosets. Hence, $x_1 K, \ldots, x_n K$ are distinct left cosets of K. Also, $|K| = |x_i K|$ by Corollary 4.3.7 for all $i = 1, 2, \ldots, n$. Thus,

$$
\begin{aligned}
|HK| &= |x_1 K| + \cdots + |x_n K| \\
&= \underbrace{|K| + \cdots + |K|}_{n \text{ times}} \\
&= n |K| \\
&= \frac{|H||K|}{|A|} \\
&= \frac{|H||K|}{|H \cap K|}. \blacksquare
\end{aligned}
$$

The following corollary is an immediate consequence of the above theorem.

Corollary 4.3.16 *Let H and K be finite subgroups of a group G such that $H \cap K = \{e\}$. Then*

$$|HK| = |H||K|. \blacksquare$$

4.3.1 Worked-Out Exercises

$\diamond$ **Exercise 1** Let H be a subgroup of a group G. Show that for all $a \in G$, $aH = H$ if and only if $a \in H$.

Solution: Let $a \in G$. Suppose $aH = H$. Then $a = ae \in aH = H$. Conversely, suppose that $a \in H$. Now for any $h \in H$, $ah \in H$. Hence, $aH \subseteq H$. Let $h \in H$. Then $a^{-1}h \in H$. Thus, $h = a(a^{-1}h) \in aH$. Therefore, $H \subseteq aH$, proving that $aH = H$.

$\diamond$ **Exercise 2** Let G be a noncyclic group of order p^2, p a prime integer. Show that the order of each nonidentity element is p.

Solution: Let $g \in G$ and $g \neq e$. Now $o(g)$ divides $|G| = p^2$. Hence, $o(g) = 1, p$ or p^2. Since $g \neq e$, $o(g) \neq 1$. If $o(g) = p^2$, then G contains an element g such that $o(g) = |G|$ and this implies that G is cyclic, which contradicts the hypothesis. Hence, $o(g) = p$.

Exercise 3 Let $G = \{a, b, c, d\}$ be a group. Complete the following Cayley table for this group.

	a	b	c	d
a				
b				
c			b	
d		b		

Solution: From the table, $c^2 = b$ and $db = b$. Now $db = b$ implies that $d = e$, the identity element of G. Since $c^2 = b \neq d$, $o(c) \neq 2$. Hence, $o(c) = 4$. Thus, G is a cyclic group generated by c. Then $G = \{e, c, c^2, c^3\}$. Since $d = e$ and $c^2 = b$, it follows that $c^3 = a$. Hence, the Cayley table is

	a	b	c	d
a	b	c	d	a
b	c	d	a	b
c	d	a	b	c
d	a	b	c	d

Exercise 4 Let G be a finite nontrivial group. Suppose for all $x \in G$, there exists $y \in G$ such that $x = y^2$. Prove that the order of G is odd and conversely.

Solution: Suppose G is of odd order. Then $|G| = 2n + 1$ for some positive integer n and for all $x \in G$, $x^{2n+1} = e$. Now $x^{2n+1} = e$ implies $x = x^{-2n} = (x^{-n})^2 = y^2$, where $y = x^{-n}$. Conversely, suppose $|G|$ is not odd. Let $|G| = 2n$ and $x \in G$. Then there exists $y \in G$ such that $x = y^2$. Hence, $x^n = y^{2n} = e$. Thus, for all $x \in G$, $x^n = e$. Suppose n is odd, say, $n = 2m + 1$. Then $x^{2m+1} = e$ for all $x \in G$. By Worked-Out Exercise 5 (page 74), there exists $z \in G$ such that $z \neq e$ and $z^2 = e$ since $|G|$ is even. Hence, $e = z^{2m+1} = zz^{2m} = z(z^2)^m = ze = z$, which is a contradiction. So n is even, say, $n = 2m$. Then $x^{2m} = e$ for all $x \in G$. As before, we can show that $x^m = e$ for all $x \in G$ and m is even. Continuing in this way, we can conclude that $x^2 = e$ for all $x \in G$. Let $x \in G$. Then there exists $y \in G$ such that $x = y^2$. Therefore, $x = e$. Thus, $|G| = 1$, which is a contradiction. Consequently, G is of odd order.

$\diamond$ **Exercise 5** Let G be a group such that $|G| > 1$. Prove that G has only the trivial subgroups if and only if $|G|$ is prime.

Solution: Let $|G| = p$, p a prime. Let H be a subgroup of G. Then $|H|$ divides $|G|$. This implies that $|H| = 1$ or p. Thus, $H = \{e\}$ or $H = G$. Conversely, suppose that G has only the trivial subgroups. Let $a \in G$ be such that $a \neq e$. Now $\langle a \rangle = \{a^n \mid a \in \mathbf{Z}\}$ is a cyclic subgroup of G and $\langle a \rangle \neq \{e\}$. Therefore, $G = \langle a \rangle$. If G is infinite, then $a^r \neq a^s$ for all $r, s \in \mathbf{Z}$, $r \neq s$. Hence,

$\{a^{2n} \mid n \in \mathbf{Z}\}$ is a nontrivial subgroup of G, which is a contradiction. Thus, $|G|$ is a finite cyclic group of order, say, $m > 1$. Suppose m is not prime. Then $m = rs$ for some $r, s \in \mathbf{Z}$, $1 < r, s < m$. Since $r \mid |G|$ and G is cyclic, G has a cyclic subgroup H of order r. This contradicts the assumption that G has only the trivial subgroups. Hence, $|G|$ is prime.

◇ **Exercise 6** Let G be a group of order p^n, p a prime. Show that G contains an element of order p.

Solution: Let $a \in G$, $a \neq e$. Then $H = \langle a \rangle$ is a cyclic subgroup of G. Now $|H|$ divides $|G| = p^n$. Thus, $|H| = p^m$ for some $m \in \mathbf{Z}$, $0 < m \leq n$. Now H is a cyclic group of order p^m. Hence, for every divisor d of p^m, there exists a subgroup of order d. So for p, there exists a subgroup T of H such that $|T| = p$. By Corollary 4.3.13, there exists $b \in T$ such that $T = \langle b \rangle$ and b is of order p. Hence, G contains an element of order p.

Exercise 7 Let G be a finite commutative group such that G contains two distinct elements of order 2. Show that $|G|$ is a multiple of 4. Also, show that this result need not be true if G is not commutative.

Solution: Let a and b be two distinct elements of order 2. Let $H = \{e, a\}$ and $K = \{e, b\}$. Now H and K are subgroups of G. Since G is commutative, $HK = \{e, a, b, ab\}$ is a subgroup of G of order 4. Now $|HK| = 4$ divides $|G|$. Thus, $|G|$ is a multiple of 4.

The symmetric group S_3 is noncommutative, $(1\ 2)$ and $(1\ 3)$ are elements of S_3, and each is of order 2. But 4 does not divide $|S_3| = 6$.

Exercise 8 Find all subgroups of S_3 and draw the lattice diagram of the subgroup lattice of S_3.

Solution: $S_3 = \{e, (1\ 2), (1\ 3), (2\ 3), (1\ 2\ 3), (1\ 3\ 2)\}$. $\circ(1\ 2) = 2$, $\circ(1\ 3) = 2$, $\circ(2\ 3) = 2$, $\circ(1\ 2\ 3) = 3$, and $\circ(1\ 3\ 2) = 3$. Now $\{e\}$, $\{e, (1\ 2)\}$, $\{e, (1\ 3)\}$, $\{e, (2\ 3)\}$, $\{e, (1\ 2\ 3), (1\ 3\ 2)\}$, and S_3 are subgroups of S_3. Let H be a subgroup of S_3. Now $|H|$ divides $|G|$. Thus, $|H| = 1, 2, 3$, or 6. If $|H| = 1$, then $H = \{e\}$. If $|H| = 6$, then $H = S_3$. If $|H| = 2$, then H is a cyclic group of order 2. Hence, H is one of $\{e, (1\ 2)\}$, $\{e, (1\ 3)\}$, $\{e, (2\ 3)\}$. Suppose $|H| = 3$. Then by Lagrange's theorem, H has no subgroup of order 2. Thus, $(1\ 2), (1\ 3), (2\ 3) \notin H$. Therefore, $e, (1\ 2\ 3), (1\ 3\ 2) \in H$. Also, $\{e, (1\ 2\ 3), (1\ 3\ 2)\}$ is a subgroup and so $H = \{e, (1\ 2\ 3), (1\ 3\ 2)\}$. Hence, $H_0 = \{e\}$, $H_1 = \{e, (1\ 2)\}$, $H_2 = \{e, (1\ 3)\}$, $H_3 = \{e, (2\ 3)\}$, $H_4 = \{e, (1\ 2\ 3), (1\ 3\ 2)\}$, and S_3 are the only subgroups of S_3.

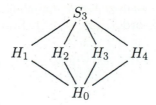

4.3.2 Exercises

1. In S_3,

   (i) find all right cosets of $H = \{e, (2\ 3)\}$,

   (ii) find a subgroup B of G such that $H(1\ 2\ 3)$ is a left coset of B.

2. Find all right cosets of the subgroup $6\mathbf{Z}$ in the group $(\mathbf{Z}, +)$.

3. Let

$$H = \left\{ e, \begin{pmatrix} 1 & 2 & 3 & 4 \\ 4 & 3 & 2 & 1 \end{pmatrix}, \begin{pmatrix} 1 & 2 & 3 & 4 \\ 2 & 1 & 4 & 3 \end{pmatrix}, \begin{pmatrix} 1 & 2 & 3 & 4 \\ 3 & 4 & 1 & 2 \end{pmatrix} \right\},$$

   where e is the identity permutation. Show that H is a subgroup of S_4. List all the left and right cosets of H in S_4.

4. Let H denote the subgroup $\{r_{360}, h\}$ of the group of symmetries of the square. List all the left and right cosets of H in G.

5. Find all subgroups of the Klein 4-group.

6. Find all subgroups of order 4 in S_4.

7. Let $G = \{a, b, c, d\}$ be a group. Complete the following Cayley table for this group.

	d	a	b	c
d	d			
a			c	d
b				
c				

8. Let G be a group and H and K be subgroups of G. Show that $(H \cap K)x = Hx \cap Kx$ for all $x \in G$.

9. Let G be a group and H and K be subgroups of G. Let $a, b \in G$. Show that either $Ha \cap Kb = \phi$ or $Ha \cap Kb = (H \cap K)c$ for some $c \in G$.

10. (Poincaré) Let G be a group and H and K be subgroups of G of finite indices. Show that $H \cap K$ is of finite index.

11. Give an example of a group G and a subgroup H of G such that $aH = bH$, but $Ha \neq Hb$ for some $a, b \in G$.

12. Let G be a group of order pq, where p and q are prime integers. Show that every proper subgroup of G is cyclic.

13. Let H be a subgroup of a group G. Define a relation $\sim$ on G by for all $a, b \in G$, $a \sim b$ if and only if $b^{-1}a \in H$ (i.e., if and only if $aH = bH$). Show that $\sim$ is an equivalence relation on G and the equivalence classes of $\sim$ are the cosets aH, $a \in G$.

14. Let $n > 1$. Show that there exists a proper subgroup H of S_n such that $[S_n : H] \leq n$.

15. Let H and K be subgroups of a finite group G such that $|H| > \sqrt{|G|}$ and $|K| > \sqrt{|G|}$. Show that $|H \cap K| > 1$.

16. Let $|G| = pq$, $(p > q)$, where p and q are distinct primes. Show that G has at most one subgroup of order p.

17. Let G be a group. If a subset A is a left coset of some subgroup of G, show that A is a right coset of some subgroup of G.

18. Let G be a finite group and A and B be subgroups of G such that $A \subseteq B \subseteq G$. Prove that
$$[G : A] = [G : B][B : A].$$

19. Let G be a group such that $|G| < 200$. Suppose G has subgroups of order 25 and 35. Find the order of G.

20. Let G be a group of order 35 and A and B be subgroups of G of order 5 and 7, respectively. Show that $G = AB$.

21. Let A and B be subgroups of a group G. If $|A| = p$, a prime integer, show that either $A \cap B = \{e\}$ or $A \subseteq B$.

22. Let H and K be subgroups of a group G. Define a relation $\sim$ on G by for all $a, b \in G$, $a \sim b$ if and only if $b = hak$ for some $h \in H$ and $k \in K$.

(i) Show that $\sim$ is an equivalence relation on G.

(ii) Let $a \in G$ and $[a]$ denote the equivalence class of a in G. Show that

$$[a] = \{hak \mid h \in H, \ k \in K\} = HaK.$$

The set HaK is called a **double coset** of H and K in G.

(iii) If G is a finite group, prove that

$$|HaK| = \frac{|H||K|}{|H \cap aKa^{-1}|}$$

for all $a \in G$.

23. For the following, if the statement is true, then write the proof. Otherwise justify why the statement is false.

(i) Every left coset of a subgroup of a group is also a right coset.

(ii) The product of two left cosets of a subgroup of a group is also a left coset.

(iii) There may exist a subgroup of order 12 in a group of order 40.

(iv) Let $G = \langle a \rangle$ be a cyclic group of order 30. Then $[G : \langle a^5 \rangle] = 5$.

(v) Every proper subgroup of a group of order p^2 (p a prime) is cyclic.

(vi) Let G be a group. If H is a subgroup of order p and K is a subgroup of order q, where p and q are distinct primes, then $|HK| = pq$.

4.4 Normal Subgroups and Quotient Groups

In the previous section, we saw that a subgroup H of a group G induced two decompositions of G, one by left cosets and another by right cosets. In other words, if H is a subgroup of a group G, then G can be written as a disjoint union of distinct left (right) cosets of H in G. These two decompositions were first recognized by Galois in 1831 in the context of permutation groups. Galois called the decomposition "proper" if the two decompositions coincide, i.e., if left cosets are the same as right cosets. We call such a subgroup normal in our present-day terminology. Normal subgroups are the subject of this section. Galois showed how the solvability of a polynomial equation by means of radicals is related to the concept of a normal subgroup of the group of permutations of the roots and the group, called the quotient group, created by the normal subgroup.

Perhaps the notion of a normal subgroup is one of the most innovative ideas in group theory. I.N. Herstein (1923–1988) remarked about normal subgroups that "It is a tribute to the genius of Galois that he recognized that those subgroups for which the left and right cosets coincide are distinguished ones. Very often in mathematics the crucial problem is to recognize and to discover what are the relevant concepts; once this is accomplished the job may be more than half done."

Later C. Jordan defined normal subgroups without using the term normal as we define it in our present-day terminology.

We shall see in this text that normal subgroups play a crucial role in obtaining structural results of groups. Let us now begin our study of normal subgroups.

Definition 4.4.1 *Let G be a group. A subgroup H of G is said to be a **normal** (or **invariant**) subgroup of G if $aH = Ha$ for all $a \in G$.*

From the definition of a normal subgroup, it follows that for any group G, G and $\{e\}$ are normal subgroups of G.

If H is a normal subgroup of G, this does not always mean that $ah = ha$ for all $h \in H$ and for all $a \in G$ as shown by the following example.

Example 4.4.2 *Recall Example 4.3.2. H is a normal subgroup of S_3. Consider*
$$h = \begin{pmatrix} 1 & 2 & 3 \\ 2 & 3 & 1 \end{pmatrix} \in H. \text{ Then}$$

$$\begin{pmatrix} 1 & 2 & 3 \\ 1 & 3 & 2 \end{pmatrix} \circ h = \begin{pmatrix} 1 & 2 & 3 \\ 3 & 2 & 1 \end{pmatrix}$$

and

$$h \circ \begin{pmatrix} 1 & 2 & 3 \\ 1 & 3 & 2 \end{pmatrix} = \begin{pmatrix} 1 & 2 & 3 \\ 2 & 1 & 3 \end{pmatrix}.$$

Hence,

$$\begin{pmatrix} 1 & 2 & 3 \\ 1 & 3 & 2 \end{pmatrix} \circ h \neq h \circ \begin{pmatrix} 1 & 2 & 3 \\ 1 & 3 & 2 \end{pmatrix},$$

even though

$$\begin{pmatrix} 1 & 2 & 3 \\ 1 & 3 & 2 \end{pmatrix} H = H \begin{pmatrix} 1 & 2 & 3 \\ 1 & 3 & 2 \end{pmatrix}.$$

The following theorem gives a necessary and sufficient condition for a subgroup to be a normal subgroup. For $a \in G$, $\phi \neq H \subseteq G$, let $aHa^{-1} = \{aha^{-1} \mid h \in H\}$.

Theorem 4.4.3 *Let H be a subgroup of a group G. Then H is a normal subgroup of G if and only if for all $a \in G$, $aHa^{-1} \subseteq H$.*

Proof.   First suppose that H is a normal subgroup of G. Let $a \in G$. We now show that $aHa^{-1} \subseteq H$. Let $aha^{-1} \in aHa^{-1}$, where $h \in H$. Since H is a normal subgroup of G, $aH = Ha$. Also, since $ah \in aH$, we have $ah \in Ha$ and so $ah = h'a$ for some $h' \in H$. Thus, $aha^{-1} = h' \in H$. Hence, $aHa^{-1} \subseteq H$.

Conversely, suppose $aHa^{-1} \subseteq H$ for all $a \in G$. Let $a \in G$. We show that $aH = Ha$. Let $ah \in aH$, where $h \in H$. Now $aha^{-1} \in aHa^{-1}$ and so $aha^{-1} \in H$.

Thus, $aha^{-1} = h'$ for some $h' \in H$. This implies that $ah = h'a \in Ha$. Therefore, $aH \subseteq Ha$. Similarly, we can show that $Ha \subseteq aH$. Hence, $aH = Ha$. Consequently, H is a normal subgroup of G. ■

There are several other criteria that can be used to test the normality of a subgroup. We consider some of these criteria in exercises at the end of this section.

The following theorem describes some important properties of normal subgroups.

Theorem 4.4.4 *Let H and K be normal subgroups of a group G. Then*
(i) $H \cap K$ is a normal subgroup of G,
(ii) $HK = KH$ is a normal subgroup of G,
(iii) $\langle H \cup K \rangle = HK$.

Proof. (i) Since the intersection of subgroups is a subgroup, $H \cap K$ is a subgroup of G. Let $g \in G$. Consider $g(H \cap K)g^{-1}$. Let gag^{-1} be any element of $g(H \cap K)g^{-1}$, where $a \in H \cap K$. Since $a \in H \cap K$, we have $a \in H$ and $a \in K$. Hence, $gag^{-1} \in H$ and $gag^{-1} \in K$. Thus, $gag^{-1} \in H \cap K$. This shows that $g(H \cap K)g^{-1} \subseteq H \cap K$. Hence, $H \cap K$ is a normal subgroup by Theorem 4.4.3.

(ii) First we show that $HK = KH$. Let $hk \in HK$, where $h \in H$ and $k \in K$. Since K is a normal subgroup of G and $h \in G$, we have $hK = Kh$. Thus, $hk \in hK = Kh$. Since $Kh \subseteq KH$, we have $hk \in KH$. Hence, $HK \subseteq KH$. Similarly, $KH \subseteq HK$ and so $HK = KH$. Since H and K are subgroups and $HK = KH$, HK is a subgroup of G by Theorem 4.1.13. To show that HK is a normal subgroup, let $g \in G$. Then $gHg^{-1} \subseteq H$ and $gKg^{-1} \subseteq K$ since H and K are normal subgroups. Now

$$
\begin{aligned}
g(HK)g^{-1} &= g(Hg^{-1}gK)g^{-1} \\
&= (gHg^{-1})(gKg^{-1}) \\
&\subseteq HK.
\end{aligned}
$$

Therefore, HK is a normal subgroup of G by Theorem 4.4.3.

(iii) By (ii), HK is a subgroup of G. Hence, by Theorem 4.1.15,

$$ HK = \langle H \cup K \rangle. \ ■ $$

We know that if H and K are subgroups of a group G, then HK need not be a subgroup of G (Example 4.1.12). By the above theorem, if H and K are normal subgroups, then HK is a normal subgroup and hence a subgroup. However, in order to show that HK is a subgroup, we only need either H or K to be a normal subgroup. We consider one of these situations in Exercise 13 (page 137).

In Theorem 4.1.16, we proved that the set of all subgroups of a group G is a lattice. In the next theorem, we prove that the set of all normal subgroups of a group G is a modular lattice.

Corollary 4.4.5 *Let $\mathcal{N}(G)$ denote the set of all normal subgroups of a group G. Then $(\mathcal{N}(G), \leq)$ is a modular lattice, where $\leq$ is the set inclusion relation.*

Proof. Proceeding as in Theorem 4.1.16, we can show that $(\mathcal{N}(G), \leq)$ is a lattice, where $H \wedge K = H \cap K$ and $H \vee K = \langle H \cup K \rangle = HK$ for all $H, K \in \mathcal{N}(G)$. Let $H, K, L \in \mathcal{N}(G)$ be such that $H \leq L$. We now show that $H(K \cap L) = HK \cap L$. Since $H \subseteq HK$ and $H \subseteq L$, we find that $H \subseteq HK \cap L$. Also, $K \cap L \subseteq K \subseteq HK$ and $K \cap L \subseteq L$. As a result $K \cap L \subseteq HK \cap L$, showing that $H(K \cap L) \subseteq HK \cap L$. Let $a \in HK \cap L$. Then $a \in HK$ and $a \in L$. Thus, $a = hk$ for some $h \in H$ and $k \in K$. This implies that $k = h^{-1}a \in L$ and so $k \in K \cap L$. Hence, $a \in H(K \cap L)$, which implies that $HK \cap L \subseteq H(K \cap L)$. Consequently, we must have $H(K \cap L) = HK \cap L$, i.e., $H \vee (K \wedge L) = (H \vee K) \wedge L$. Hence, $(\mathcal{N}(G), \leq)$ is a modular lattice. ∎

We now focus our attention on the study of quotient groups. First, let us consider the following example.

Example 4.4.6 *Consider the subgroup H' of Example 4.3.2. Now H' is not a normal subgroup of S_3. Let S_3/H' be the set of all left cosets of H' in S_3. Now let us try to define a binary operation $*$ on S_3/H'. The natural way would be to define $(\pi_1 H') * (\pi_2 H')$ to be $(\pi_1 \circ \pi_2)H'$. Now*

$$\begin{pmatrix} 1 & 2 & 3 \\ 3 & 2 & 1 \end{pmatrix} H' = \begin{pmatrix} 1 & 2 & 3 \\ 3 & 1 & 2 \end{pmatrix} H'$$

and

$$\begin{pmatrix} 1 & 2 & 3 \\ 2 & 1 & 3 \end{pmatrix} H' = \begin{pmatrix} 1 & 2 & 3 \\ 2 & 3 & 1 \end{pmatrix} H'.$$

However,

$$\left(\begin{pmatrix} 1 & 2 & 3 \\ 2 & 1 & 3 \end{pmatrix} H' \right) * \left(\begin{pmatrix} 1 & 2 & 3 \\ 3 & 2 & 1 \end{pmatrix} H' \right) = \begin{pmatrix} 1 & 2 & 3 \\ 3 & 1 & 2 \end{pmatrix} H'$$

and

$$\left(\begin{pmatrix} 1 & 2 & 3 \\ 2 & 3 & 1 \end{pmatrix} H' \right) * \left(\begin{pmatrix} 1 & 2 & 3 \\ 3 & 1 & 2 \end{pmatrix} H' \right) = \begin{pmatrix} 1 & 2 & 3 \\ 1 & 2 & 3 \end{pmatrix} H'.$$

Since

$$\begin{pmatrix} 1 & 2 & 3 \\ 3 & 1 & 2 \end{pmatrix} H' \neq \begin{pmatrix} 1 & 2 & 3 \\ 1 & 2 & 3 \end{pmatrix} H',$$

* is not well defined. That * is not well defined is due to the fact that H' is not a normal subgroup of S_3.

Theorem 4.4.7 Let H be a normal subgroup of a group G. Denote the set of all left cosets $\{aH \mid a \in G\}$ by G/H and define * on G/H by for all aH, $bH \in G/H$,

$$(aH) * (bH) = abH.$$

Then $(G/H, *)$ is a group.

Proof.   First we show that * is well defined. Let $aH, bH, a'H, b'H \in G/H$ and suppose $(aH, bH) = (a'H, b'H)$. Then $aH = a'H$ and $bH = b'H$. We need to show that $aH * bH = a'H * b'H$ or $abH = a'b'H$. Now $aH = a'H$ and $bH = b'H$ imply that $a = a'h_1$ and $b = b'h_2$ for some $h_1, h_2 \in H$. Thus,

$$\begin{aligned}
(a'b')^{-1}(ab) &= b'^{-1}a'^{-1}ab \\
&= b'^{-1}a'^{-1}a'h_1b'h_2 \\
&= b'^{-1}h_1b'h_2.
\end{aligned}$$

Since H is a normal subgroup and $h_1 \in H$, we have $b'^{-1}h_1b'h_2 = (b'^{-1}h_1b')h_2 \in H$ and so $(a'b')^{-1}(ab) \in H$. Hence, $abH = a'b'H$ by Theorem 4.3.3(i). Thus, * is well defined and so $(G/H, *)$ is a mathematical system.

Next, we show that * is associative. Let $aH, bH, cH \in G/H$. Now $(aH) * [(bH) * (cH)] = (aH) * (bcH) = a(bc)H = (ab)cH = (abH) * (cH) = [(aH) * (bH)] * (cH)$. Hence, * is associative. Now $eH \in G/H$ and

$$(aH) * (eH) = aeH = aH = eaH = (eH) * (aH)$$

for all $aH \in G/H$. Therefore, eH is the identity of G/H. Also, for all $aH \in G/H$, $a^{-1}H \in G/H$ and

$$(aH) * (a^{-1}H) = aa^{-1}H = eH = a^{-1}aH = (a^{-1}H) * (aH).$$

Thus, for all $aH \in G/H$, $a^{-1}H$ is the inverse of aH. Consequently, $(G/H, *)$ is a group. ∎

Definition 4.4.8 Let G be a group and H be a normal subgroup of G. The group G/H is called the **quotient group** of G by H.

Example 4.4.9 Consider the subgroup $(\langle n \rangle, +)$ of the group $(\mathbf{Z}, +)$, where n is a fixed positive integer. Since $\mathbf{Z}$ is commutative, $\langle n \rangle$ is a normal subgroup of $\mathbf{Z}$ (Exercise 15, page 137). Hence, $(\mathbf{Z}/\langle n \rangle, +)$ is a group, where

$$(a + \langle n \rangle) + (b + \langle n \rangle) = (a + b) + \langle n \rangle$$

for all $a + \langle n \rangle, b + \langle n \rangle \in \mathbf{Z}/\langle n \rangle$. In Example 4.3.10, we determined the distinct left cosets of $\langle n \rangle$ in $\mathbf{Z}$. We found that

$$\mathbf{Z}/\langle n \rangle = \{0 + \langle n \rangle, 1 + \langle n \rangle, 2 + \langle n \rangle, \ldots, n - 1 + \langle n \rangle\}.$$

Example 4.4.10 *Consider the normal subgroup H of S_3 of Example 4.4.2. Since $|S_3| = 6$ and $|H| = 3$, $[S_3 : H] = 2$ by Lagrange's theorem. Now $|S_3/H| = [S_3 : H] = 2$ and for all $h \in H$, $hH = H$. Thus, $eH = H$, $(1\ 2\ 3)H = H$ and $(1\ 3\ 2)H = H$. We have shown in Example 4.3.2 that $(2\ 3)H = (1\ 3)H = (1\ 2)H$. Thus,*

$$S_3/H = \{H,\ (2\ 3)H\}.$$

We also note that S_3/H is cyclic and $(2\ 3)H$ is a generator for S_3/H.

Example 4.4.11 *Consider $\mathbf{Z}_8$ and let $H = \{[0], [4]\}$. Then H is a normal subgroup of $\mathbf{Z}_8$. Now $|H| = 2$ and $|\mathbf{Z}_8| = 8$. Thus, $|\mathbf{Z}_8/H| = \frac{|\mathbf{Z}_8|}{|H|} = 4$. Hence, $\mathbf{Z}_8/H$ has four elements. Now*

$$[0] + H = H = [4] + H,$$

$$[1] + H = \{[1],\ [5]\} = [5] + H,$$

$$[2] + H = \{[2],\ [6]\} = [6] + H,$$

and

$$[3] + H = \{[3],\ [7]\} = [7] + H.$$

Hence, $\mathbf{Z}_8/H = \{[0] + H, [1] + H, [2] + H, [3] + H\}$.

Example 4.4.12 *Consider $\mathbf{Z}_4 \times \mathbf{Z}_6$, the direct product of $\mathbf{Z}_4$ and $\mathbf{Z}_6$. Let*

$$H = \langle([0], [1])\rangle = \{([0], [0]), ([0], [1]), ([0], [2]), ([0], [3]), ([0], [4]), ([0], [5])\}.$$

Then H is a subgroup of $\mathbf{Z}_4 \times \mathbf{Z}_6$ and since $\mathbf{Z}_4 \times \mathbf{Z}_6$ is commutative, H is a normal subgroup of $\mathbf{Z}_4 \times \mathbf{Z}_6$. Now $|\mathbf{Z}_4 \times \mathbf{Z}_6| = 24$ and $|H| = 6$. Hence,

$$|(\mathbf{Z}_4 \times \mathbf{Z}_6)/H| = \frac{|\mathbf{Z}_4 \times \mathbf{Z}_6|}{|H|} = 4.$$

Thus, $(\mathbf{Z}_4 \times \mathbf{Z}_6)/H$ has four elements. Since for all $[n] \in \mathbf{Z}_6$, $([0], [n]) \in H$, we have for all $[n] \in \mathbf{Z}_6$, $([0], [n]) + H = H$. Let $([m], [n]) \in \mathbf{Z}_4 \times \mathbf{Z}_6$. Then $([m], [n]) = ([m], [0]) + ([0], [n])$ and from this, it follows that $([m], [n]) + H = ([m], [0]) + H$. Let us now compute $([m], [0]) + H$ for $m = 0, 1, 2, 3$. Now $([0], [0]) + H = H$,

$$([1], [0]) + H = \{([1], [0]), ([1], [1]), ([1], [2]), ([1], [3]), ([1], [4]), ([1], [5])\},$$

$$([2], [0]) + H = \{([2], [0]), ([2], [1]), ([2], [2]), ([2], [3]), ([2], [4]), ([2], [5])\},$$

and

$$([3], [0]) + H = \{([3], [0]), ([3], [1]), ([3], [2]), ([3], [3]), ([3], [4]), ([3], [5])\}.$$

From above, we see that $([0], [0]) + H$, $([1], [0]) + H$, $([2], [0]) + H$, and $([3], [0]) + H$ are all distinct. Hence,

$$(\mathbf{Z}_4 \times \mathbf{Z}_6)/H = \{([0], [0]) + H, ([1], [0]) + H, ([2], [0]) + H, ([3], [0]) + H\}.$$

Groups of the type given in the next definition are building blocks for all groups. They are important because they help to determine the structure of groups. We will discuss this in more detail when we introduce the concept of a composition series of a group (Chapter 8).

Definition 4.4.13 *Let G be a group. Then G is called **simple** if $G \neq \{e\}$ and the only normal subgroups of G are $\{e\}$ and G.*

The only simple commutative groups are given in the next example. We will determine the simple groups of order ≤ 60 (in Section 7.4).

Example 4.4.14 *Let G be a cyclic group of order p, p a prime. Since the only subgroups of G are $\{e\}$ and G, G is simple.*

We now proceed to establish the simplicity of A_n, $n \geq 5$. Thus, there is a large class of simple groups.

Lemma 4.4.15 *Let H be a normal subgroup of A_n, $n \geq 5$. If H contains a 3-cycle, then $H = A_n$.*

Proof. Suppose H contains a 3-cycle, say, $(a\ b\ c) \in H$. Let $(u\ v\ w) \in A_n$ and let $\pi \in S_n$ be such that $\pi(a) = u$, $\pi(b) = v$, and $\pi(c) = w$. Now $\pi \circ (a\ b\ c) \circ \pi^{-1} = (u\ v\ w)$. If $\pi \in A_n$, then $(u\ v\ w) \in H$. Suppose $\pi \notin A_n$. Then π is an odd permutation. Since $n \geq 5$, there exist $d, f \in I_n$ such that d and f are distinct from a, b and c. Then $\pi \circ (d\ f) \in A_n$. Now $(u\ v\ w) = \pi \circ (a\ b\ c) \circ \pi^{-1} = \pi \circ (a\ b\ c) \circ (d\ f) \circ (d\ f)^{-1} \circ \pi^{-1} = \pi \circ (d\ f) \circ (a\ b\ c) \circ (d\ f)^{-1} \circ \pi^{-1} = (\pi \circ (d\ f)) \circ (a\ b\ c) \circ (\pi \circ (d\ f))^{-1} \in H$. Thus, H contains all 3-cycles. Since A_n is generated by the set of all 3-cycles, $H = A_n$. $\blacksquare$

Theorem 4.4.16 *Let H be a normal subgroup of A_n, $n \geq 5$. If H contains a product of two disjoint transpositions, then $H = A_n$.*

Proof. Suppose $(a\ b) \circ (c\ d) \in H$, where $(a\ b)$ and $(c\ d)$ are disjoint transpositions. Let $w \in I_n$ be such that $w \notin \{a, b, c, d\}$. Let $\pi = (c\ d\ w)$. Since π is a 3-cycle, $\pi \in A_n$. Since H is a normal subgroup of A_n, we have $\pi \circ (a\ b) \circ (c\ d) \circ \pi^{-1} \in H$. But

$$\pi \circ (a\ b) \circ (c\ d) \circ \pi^{-1} = (d\ w) \circ (a\ b)$$

and so $(d\ w) \circ (a\ b) \in H$. Since H is a subgroup,

$$(c\ d\ w) = (a\ b) \circ (c\ d) \circ (d\ w) \circ (a\ b) \in H.$$

Hence, H contains a 3-cycle and so by Lemma 4.4.15, $H = A_n$. $\blacksquare$

Theorem 4.4.17 *A_n is simple if $n \geq 5$.*

Proof. Let H be a normal subgroup of A_n and $H \neq \{e\}$. Let $\pi \in H$, $\pi \neq e$ be a permutation that moves the smallest number of elements, say, m. Then $m \geq 3$. We claim that $m = 3$, in which case the result follows by Lemma 4.4.15. Suppose $m > 3$. Write $\pi = \pi_1 \circ \pi_2 \circ \cdots \circ \pi_k$ as a product of disjoint cycles.

Suppose that π_i is a transposition for all $i = 1, 2, \ldots, k$. Then $k \geq 2$. Let $\pi_1 = (a\ b)$ and $\pi_2 = (c\ d)$. Let $f \in I_n$ be such that $f \notin \{a, b, c, d\}$ and let $\sigma = (c\ d\ f)$. Since $\sigma \in A_n$ and H is a normal subgroup of A_n, $\pi' = \pi^{-1} \circ \sigma \circ \pi \circ \sigma^{-1} \in H$. Clearly $\pi'(a) = a$ and $\pi'(b) = b$. If $u \in I_n$ and $u \notin \{a, b, c, d, f\}$ is such that $\pi(u) = u$, then $\pi'(u) = u$. Since $\pi'(f) = c$, $\pi' \neq e$. Thus, $\pi' \in H$, $\pi' \neq e$, and π' moves fewer elements than π, which is a contradiction. Hence, for some i, $1 \leq i \leq k$, π_i is a cycle of length ≥ 3. Since disjoint cycles commute, by renumbering if necessary, we may assume that $i = 1$. Then $\pi_1 = (a\ b\ c\ \cdots)$. If $m = 4$, then π is a cycle of length of 4 and hence an odd permutation, a contradiction. Thus, $m \geq 5$. Hence, π moves at least five elements. Let d, $f \in I_n$ and $d, f \notin \{a, b, c\}$. Let $\sigma = (c\ d\ f)$. As before, $\pi' = \pi^{-1} \circ \sigma \circ \pi \circ \sigma^{-1} \in H$. Since $\pi'(b) = \pi^{-1}(d) \neq b$, $\pi' \neq e$. Now for any $u \notin \{a, b, c, d, f\}$, if $\pi(u) = u$, then $\pi'(u) = u$. Clearly $\pi'(a) = a$. Hence, π' moves fewer elements than π, which is again a contradiction. Hence, $m = 3$. ∎

4.4.1 Worked-Out Exercises

◇ **Exercise 1** Let H be a subgroup of a group G. Then $W = \cap_{g \in G} gHg^{-1}$ is a normal subgroup of G.

Solution: By Worked-Out Exercise 1 (page 106), gHg^{-1} is a subgroup of G for all $g \in G$. Since the intersection of subgroups is a subgroup, W is a subgroup of G. Let $x \in G$, $w \in W$. Then $w \in gHg^{-1}$ for all $g \in G$. We show that $xwx^{-1} \in gHg^{-1}$ for all $g \in G$, which in turn will yield that $xwx^{-1} \in W$. Let $g \in G$.

Let us work our way backward and suppose $xwx^{-1} \in gHg^{-1}$. Then $xwx^{-1} = ghg^{-1}$ for some $h \in H$. Thus, $g^{-1}xw\,x^{-1}g = h \in H$. This implies that

$$(g^{-1}x)w(g^{-1}x)^{-1} \in H.$$

Set $y = x^{-1}g$. Then $g = xy$. Hence, in order to show that $xwx^{-1} \in gHg^{-1}$ for a given $g \in G$, first we need to find $y \in G$ such that $g = xy$. Since $g = x(x^{-1}g)$, we can choose $y = x^{-1}g$.

So there exists $y \in G$ such that $g = xy$. Since $y \in G$, we have $w \in yHy^{-1}$ and so $w = yhy^{-1}$ for some $h \in H$. Therefore, $xwx^{-1} = x(yhy^{-1})x^{-1} = xyhy^{-1}x^{-1} = (xy)h(xy)^{-1} = ghg^{-1} \in gHg^{-1}$. Since $g \in G$ was arbitrary, $xwx^{-1} \in gHg^{-1}$ for all $g \in G$. Consequently, W is a normal subgroup of G.

◇ **Exercise 2** Let H be a subgroup of G.

(i) If $x^2 \in H$ for all $x \in G$, prove that H is a normal subgroup of G and G/H is commutative.

(ii) If $[G : H] = 2$, prove that H is a normal subgroup of G.

Solution: (i) Let $g \in G$ and $h \in H$. Consider ghg^{-1} and note that

$$ghg^{-1} = (gh)^2 h^{-1} g^{-2}.$$

Now $h^{-1} \in H$ and by our hypothesis $(gh)^2, g^{-2} \in H$. This implies that $ghg^{-1} \in H$, which in turn shows that $gHg^{-1} \subseteq H$. Hence, H is a normal subgroup of G. To show that G/H is commutative, let $xH, yH \in G/H$. We wish to show that $xHyH = yHxH$ or $xyH = yxH$ or $(yx)^{-1}(xy) \in H$. Consider $(yx)^{-1}(xy)$. Now

$$(yx)^{-1}(xy) = (x^{-1}y^{-1})(xy) = (x^{-1}y^{-1})^2(yxy^{-1})^2 y^2.$$

Since $a^2 \in H$ for all $a \in G$, it follows that $(x^{-1}y^{-1})^2(yxy^{-1})^2 y^2 \in H$ and so $(yx)^{-1}(xy) \in H$. Thus, G/H is commutative.

(ii) We prove that H is a normal subgroup of G first by showing that $x^2 \in H$ for all $x \in G$ and then by using (i). Suppose there exists $x \in G$ such that $x^2 \notin H$. Then $x \notin H$ and so H and xH are distinct left cosets of H in G. Since $[G : H] = 2$, it follows that $G/H = \{H, xH\}$. Hence, $G = H \cup xH$. This implies that $x^2 \in H \cup xH$. Since $x^2 \notin H$, we must have $x^2 \in xH$. Hence, $x^2 = xh$ for some $h \in H$. But then $x = h \in H$, which is a contradiction. Hence, $x^2 \in H$ for all $x \in G$. By (i), H is a normal subgroup of G.

Exercise 3 Let G be a group such that every cyclic subgroup of G is a normal subgroup of G. Prove that every subgroup of G is a normal subgroup of G.

Solution: Let H be a subgroup of G. Let $g \in G$ and $a \in H$. Then $g^{-1}ag \in \langle a \rangle \subseteq H$. Hence, H is normal in G.

Exercise 4 Let H be a proper subgroup of G such that for all $x, y \in G\backslash H$, $xy \in H$. Prove that H is a normal subgroup of G.

Solution: Let $x \in G\backslash H$. Then $x^{-1} \in G\backslash H$. Let $y \in H$. Then $xy \in G\backslash H$. Thus, $xy, x^{-1} \in G\backslash H$. Hence, $xyx^{-1} \in H$. Therefore, H is a normal subgroup of G.

$\diamond$ **Exercise 5** Let G be a group and $\{N_i \mid i \in \Omega\}$ be a family of proper normal subgroups of G. Suppose $G = \cup_i N_i$ and $N_i \cap N_j = \{e\}$ for $i \neq j$. Prove that G is commutative.

Solution: Let $x, y \in G$. Then there exist i and j such that $x \in N_i$ and $y \in N_j$. If $i \neq j$, then since $N_i \cap N_j = \{e\}$, $xy = yx$ (Exercise 12, page 137). Let $i = j$. Now there exists $z \in G$ such that $z \notin N_i$. Then $zx \notin N_i$. Hence, $zx \in N_l$ for some $l \neq i$ and so $(zx)y = y(zx)$. Thus, $z(xy) = (zx)y = y(zx) = (yz)x = (zy)x = z(yx)$. This implies that $xy = yx$. Consequently, G is commutative.

Exercise 6 Let H be a subgroup of a group G. Suppose that the product of two left cosets of H in G is again a left coset of H in G. Prove that H is a normal subgroup of G.

Solution: Let $g \in G$. Then $gHg^{-1}H = tH$ for some $t \in G$. Thus, $e = geg^{-1}e \in tH$. Hence, $e = th$ for some $h \in H$. Thus, $t = h^{-1} \in H$ so that $tH = H$. Now $gHg^{-1} \subseteq gHg^{-1}H = H$. Therefore, H is a normal subgroup of G.

◇ **Exercise 7** Let G be a group. Show that if $G/Z(G)$ is cyclic, then G is commutative.

Solution: Write $Z = Z(G)$. Let $G/Z = \langle gZ \rangle$. Let $a, b \in G$. Then $aZ, bZ \in G/Z$. Hence, $aZ = g^n Z$ and $bZ = g^m Z$ for some $n, m \in \mathbf{Z}$. Then $a \in g^n Z$ and $b \in g^m Z$. Thus, $a = g^n d$ and $b = g^m h$ for some $d, h \in Z$. Now $ab = g^n d g^m h = g^n g^m dh$ (since $d \in Z$) $= g^{n+m} hd$ (since $h \in Z$) $= g^m g^n hd = g^m h g^n d = ba$. Hence, G is commutative.

4.4.2 Exercises

1. Let
$$H = \left\{ e, \begin{pmatrix} 1 & 2 & 3 & 4 \\ 4 & 3 & 2 & 1 \end{pmatrix}, \begin{pmatrix} 1 & 2 & 3 & 4 \\ 2 & 1 & 4 & 3 \end{pmatrix}, \begin{pmatrix} 1 & 2 & 3 & 4 \\ 3 & 4 & 1 & 2 \end{pmatrix} \right\},$$
   where e is the identity permutation. Determine whether or not H is a normal subgroup of S_4.

2. Let H denote the subgroup $\{r_{360}, h\}$ of the group of symmetries of the square. Determine whether or not H is a normal subgroup of G.

3. Let G be a group and H be a subgroup of G. Show that H is normal if and only if $ghg^{-1} \in H$ for all $g \in G$, $h \in H$.

4. Let G be a group and H be a subgroup of G. If for all $a, b \in G$, $ab \in H$ implies $ba \in H$, prove that H is a normal subgroup of G.

5. Let H be a proper subgroup of a group G and $a \in G$, $a \notin H$. Suppose that for all $b \in G$, either $b \in H$ or $Ha = Hb$. Show that H is a normal subgroup of G.

6. Let G be a group. Prove that $Z(G)$ is a normal subgroup of G.

7. Let G be a group. Let H be a subgroup of G such that $H \subseteq Z(G)$. Show that if G/H is cyclic, then $G = Z(G)$, i.e., G is commutative.

8. Let H and K be subgroups of a group G such that H is a normal subgroup of G. Prove that $H \cap K$ is a normal subgroup of K.

9. Determine the quotient groups of

    (i) $(\mathbf{E}, +)$ in $(\mathbf{Z}, +)$,

    (ii) $(\mathbf{Z}, +)$ in $(\mathbf{Q}, +)$,

    (iii) $(\langle [4] \rangle, +_{12})$ in $(\mathbf{Z}_{12}, +_{12})$.

10. Let H be a normal subgroup of a group G. Prove that if G is commutative, then so is the quotient group G/H.

11. Let H be a nonempty subset of a group G. The set $N(H) = \{a \in G \mid aHa^{-1} = H\}$ is called the **normalizer** of H in G.

    (i) Prove that $N(H)$ is a subgroup of G.

    Suppose H is a subgroup of G.

    (ii) Prove that H is normal in G if and only if $N(H) = G$.

    (iii) Prove that H is normal in $N(H)$.

    (iv) Prove that $N(H)$ is the largest subgroup of G in which H is normal, i.e., if H is normal in a subgroup K of G, then $K \subseteq N(H)$.

12. Let H and K be normal subgroups of a group G. If $H \cap K = \{e\}$, prove that $hk = kh$ for all $h \in H$ and $k \in K$.

13. Let G be a group. Let H be a subgroup of G and K be a normal subgroup of G. Prove that HK is a subgroup of G.

14. Give an example of a noncommutative group in which every subgroup is normal.

15. Show that every subgroup of a commutative group is normal.

16. Let H be a normal subgroup of a group G such that $|H| = 2$. Show that $H \subseteq Z(G)$.

17. Show that if H is the only subgroup of order n in a group G, then H is a normal subgroup of G.

18. Let $K = \{e, (1\ 2) \circ (3\ 4), (1\ 4) \circ (3\ 2), (1\ 3) \circ (2\ 4)\}$.

    (i) Show that K is the only subgroup of order 4 in A_4.

    (ii) Show that K is a normal subgroup of A_4.

19. Show that A_4 has no subgroup of order 6.

20. Find all subgroups of A_4. Draw the subgroup lattice diagram. Is this lattice a modular lattice?

21. Let G be a commutative group. Show that G is simple if and only if G is of prime order.

22. Let G be a group. An equivalence relation ρ on G is called a **congruence relation** if

$$\text{for all } a, b, c \in G, a\rho b \text{ implies that } ca\rho cb \text{ and } ac\rho bc.$$

Let H be a normal subgroup of G. Define the relation ρ_H on G by

$$\text{for all } a, b \in G, a\rho_H b \text{ if and only if } a^{-1}b \in H.$$

Prove that

(i) ρ_H is a congruence relation on G,

(ii) the ρ_H class $a\rho_H = \{b \in G \mid a\rho_H b\}$ is the left coset aH,

(iii) $H = e\rho_H$.

23. Let H be a subgroup of a group G. Define a relation ρ_H on G by $\rho_H = \{(a, b) \in G \times G \mid a^{-1}b \in H\}$. Show that if ρ_H is a congruence relation, then H is a normal subgroup of G.

24. Let ρ be a congruence relation on a group G. Show that there exists a normal subgroup H of G such that $\rho = \{(a, b) \in G \times G \mid a^{-1}b \in H\}$.

25. For the following statements, write the proof if the statement is true; otherwise, give a counterexample.

(i) A subgroup H of a group G is a normal subgroup if and only if every right coset of H is also a left coset.

(ii) If A, B and C are normal subgroups of a group G, then $A(B \cap C)$ is a normal subgroup of G.

(iii) If A is a normal subgroup of a finite group G, then $[G : A] = 2$.

(iv) Every commutative subgroup of a group G is a normal subgroup of G.

(v) If G is a group of order $2p$, p an odd prime, then either G is commutative or G contains a normal subgroup of order p.

Joseph Louis Lagrange (1736–1813) was born on January 25, 1736, in Turin, Italy. He spent the early part of his life in Turin. While there he was involved in carrying out research work in calculus of variations and mechanics.

In 1766, Lagrange was invited by the Prussian king, Frederick II, to fill the position vacated by Euler in Berlin. Frederick the Great proclaimed in his appointment that "the greatest king in Europe" ought to have "the greatest mathematician in Europe." In 1787, after the death of Frederick II, he went to Paris, accepting an invitation from Louis XVI. In 1797, he accepted a position at the newly formed École Polytechnique in Paris. He was made a count by Napoleon and remained at the École Polytechnique till his death. He died on April 10, 1813.

Throughout his life, Lagrange did work of fundamental importance. He made numerous contributions to many branches of mathematics, including number theory, the theory of equations, differential equations, celestial mechanics, and fluid mechanics. In 1770, he proved the famous Lagrange's theorem in group theory.

He is responsible for the work leading to Galois theory. In his paper, "Réflexion sur la théorie algébriques des équations," Lagrange carefully analyzed the various known methods to solve a polynomial equation of degree ≤ 4 by means of radicals. He was interested in finding a general method of solution for polynomials of higher degree. He was unable to find a general solution, but in his paper he introduced several key ideas on the permutations of roots which finally led Abel and Galois to develop the necessary theory to answer the question. Lagrange's work on the solution of polynomial equations is one of the sources from which modern group theory evolved.

Chapter 5

Homomorphisms and Isomorphisms of Groups

One of the main uses of the concept of an isomorphism is the classification of algebraic structures—in particular, groups. Readers with some knowledge of linear algebra may recall that the concept of an isomorphism is used to completely characterize vector spaces with the same field of scalars in terms of a single integer, the dimension of the vector space. Another important use of an isomorphism is the representation of one algebraic structure by means of another. This is done in linear algebra, where it is shown that the vector space of all linear transformations from one finite dimensional vector space into another is isomorphic to a certain vector space of matrices.

5.1   Homomorphisms of Groups

In this section, we consider certain mappings between groups. These mappings will be defined in such a way as to preserve the algebraic structure of the groups involved. More precisely, suppose we are given a function f from a group G into a group G_1, where $*_1$ denotes the operation of G_1. Let $a, b \in G$. Then under f, a corresponds to $f(a)$, b to $f(b)$, and $a * b$ to $f(a * b)$. If f is to preserve the operations of G and G_1, $a * b$ must correspond to $f(a) *_1 f(b)$. Since f is a function, this forces the requirement that $f(a * b) = f(a) *_1 f(b)$.

Definition 5.1.1 *Let* $(G, *)$ *and* $(G_1, *_1)$ *be groups and* f *a function from* G *into* G_1. *Then* f *is called a* **homomorphism** *of* G *into* G_1 *if for all* $a, b \in G$,

$$f(a * b) = f(a) *_1 f(b).$$

Let the identity element of the group G_1 be denoted by e_1.

Define $f : G \to G_1$ by $f(a) = e_1$ for all $a \in G$. Since $f(a * b) = e_1 = e_1 *_1 e_1 = f(a) *_1 f(b)$ for all $a, b \in G$, we find that f is a homomorphism from

G into G_1. This shows that there always exists a homomorphism from a group G into a group G_1. This homomorphism is called the **trivial homomorphism**.

The identity map from G onto G is also a homomorphism.

Before we consider more examples of homomorphisms, let us prove some basic properties of homomorphisms.

Theorem 5.1.2 *Let f be a homomorphism of a group G into a group G_1. Then*

(i) $f(e) = e_1$.

(ii) $f(a^{-1}) = f(a)^{-1}$ for all $a \in G$.

(iii) If H is a subgroup of G, then $f(H) = \{f(h) \mid h \in H\}$ is a subgroup of G_1.

(iv) If H_1 is a subgroup of G_1, then $f^{-1}(H_1) = \{g \in G \mid f(g) \in H_1\}$ is a subgroup of G, and if H_1 is a normal subgroup, then $f^{-1}(H_1)$ is a normal subgroup of G.

(v) If G is commutative, then $f(G)$ is commutative.

(vi) If $a \in G$ is such that $\circ(a) = n$, then $\circ(f(a))$ divides n.

Proof.   (i) Since f is a homomorphism, $f(e)f(e) = f(ee) = f(e) = f(e)e_1$. This implies that $f(e) = e_1$ by the cancellation law.

(ii) Let $a \in G$. Then $f(a)f(a^{-1}) = f(aa^{-1}) = f(e) = e_1$. Similarly, $f(a^{-1})f(a) = e_1$. Since $f(a)$ has a unique inverse, $f(a^{-1}) = f(a)^{-1}$.

(iii) Let H be a subgroup of G. Then $e \in H$ and by (i), $f(e) = e_1$. Thus, $e_1 = f(e) \in f(H)$ and so $f(H) \neq \phi$. Let $f(a), f(b) \in f(H)$, where $a, b \in H$. Since H is a subgroup, $ab^{-1} \in H$. Thus, $f(a)f(b)^{-1} = f(a)f(b^{-1}) = f(ab^{-1}) \in f(H)$. Hence, by Theorem 4.1.3, $f(H)$ is a subgroup of G_1.

(iv) By (i), $e \in f^{-1}(H_1)$ and so $f^{-1}(H_1) \neq \phi$. Let $a, b \in f^{-1}(H_1)$. Then $f(a)$, $f(b) \in H_1$. Hence, $f(ab^{-1}) = f(a)f(b^{-1}) = f(a)f(b)^{-1} \in H_1$ and so $ab^{-1} \in f^{-1}(H_1)$. Thus, by Theorem 4.1.3, $f^{-1}(H_1)$ is a subgroup of G. Suppose H_1 is a normal subgroup of G_1. Let $g \in G$. We now show that $gf^{-1}(H_1)g^{-1} \subseteq f^{-1}(H_1)$. Let $a \in gf^{-1}(H_1)g^{-1}$. Then $a = gbg^{-1}$ for some $b \in f^{-1}(H_1)$. Now $f(a) = f(gbg^{-1}) = f(g)f(b)f(g^{-1}) = f(g)f(b)f(g)^{-1} \in H_1$ since H_1 is a normal subgroup of G_1 and $f(b) \in H_1$. Hence, $a \in f^{-1}(H_1)$ and this shows that $gf^{-1}(H_1)g^{-1} \subseteq f^{-1}(H_1)$. Thus, $f^{-1}(H_1)$ is a normal subgroup of G.

(v) Suppose G is commutative. Let $f(a), f(b) \in f(G)$. Then $f(a)f(b) = f(ab) = f(ba) = f(b)f(a)$. Hence, $f(G)$ is commutative.

(vi) Since $(f(a))^n = f(a^n) = f(e) = e_1$, we have $\circ(f(a))$ divides n by Theorem 2.1.28. ∎

Definition 5.1.3 *Let f be a homomorphism of a group G into a group G_1. The **kernel** of f, written $Ker\ f$, is defined to be the set*

$$Ker\ f = \{a \in G \mid f(a) = e_1\}.$$

By Theorem 5.1.2, $e \in$ Ker f.

Example 5.1.4 *Define the function f from $(\mathbf{Z}, +)$ into $(\mathbf{Z}_n, +_n)$ by $f(a) = [a]$ for all $a \in \mathbf{Z}$. From the definition of f, it follows that f maps $\mathbf{Z}$ onto $\mathbf{Z}_n$. Let $a, b \in \mathbf{Z}$. Then*

$$f(a + b) = [a + b] = [a] +_n [b] = f(a) +_n f(b).$$

Thus, f is a homomorphism of $\mathbf{Z}$ onto $\mathbf{Z}_n$. Now

$$\begin{aligned}
\text{Ker } f &= \{a \in \mathbf{Z} \mid f(a) = [0]\} \\
&= \{a \in \mathbf{Z} \mid [a] = [0]\} \\
&= \{a \in \mathbf{Z} \mid a \text{ is divisible by } n\} \\
&= \{a \in \mathbf{Z} \mid a = qn \text{ for some } q \in \mathbf{Z}\} \\
&= \{qn \mid q \in \mathbf{Z}\}.
\end{aligned}$$

The above example shows that a nontrivial finite group may be an image of an infinite group under a homomorphism. By Theorem 5.1.2(v), a noncommutative group cannot be an image under a homomorphism of a commutative group. In the next example, we show that two finite groups G and G_1 having same number of elements need not have a homomorphism from G onto G_1.

Example 5.1.5 *The groups $\mathbf{Z}_4 \times \mathbf{Z}_4$ and $\mathbf{Z}_8 \times \mathbf{Z}_2$ are commutative and each is of order 16. Suppose there exists a homomorphism f of $\mathbf{Z}_4 \times \mathbf{Z}_4$ onto $\mathbf{Z}_8 \times \mathbf{Z}_2$. Now $a = ([7], [0]) \in \mathbf{Z}_8 \times \mathbf{Z}_2$ and $\circ(a) = 8$. Since f is onto $\mathbf{Z}_8 \times \mathbf{Z}_2$, there exists $b \in \mathbf{Z}_4 \times \mathbf{Z}_4$ such that $f(b) = a$. By Theorem 5.1.2(vi), $\circ(f(b))$ divides $\circ(b)$. Since $\circ(f(b)) = 8$ and $\mathbf{Z}_4 \times \mathbf{Z}_4$ has elements of order $1, 2$, and 4 only, $\circ(f(b))$ cannot divide $\circ(b)$. This is a contradiction. Hence, there does not exist any homomorphism from $\mathbf{Z}_4 \times \mathbf{Z}_4$ onto $\mathbf{Z}_8 \times \mathbf{Z}_2$.*

Definition 5.1.6 *Let G and G_1 be groups. A homomorphism $f : G \to G_1$ is called an **epimorphism** if f is onto G_1 and f is called a **monomorphism** if f is one-one. If there is an epimorphism f from G onto G_1, then G_1 is called a **homomorphic image** of G.*

The homomorphism in Example 5.1.4 is an epimorphism, but not a monomorphism.

Example 5.1.7 *Let $\mathbf{R}^*$ be the group of all nonzero real numbers under multiplication. Define $f : \mathbf{R}^* \to \mathbf{R}^*$ by $f(a) = |a|$. Now $f(ab) = |ab| = |a| |b| = f(a) f(b)$, which implies that f is a homomorphism. Since $f(1) = 1 = f(-1)$ and $1 \neq -1$, f is not one-one. Also, from the definition of f, it follows that f is not onto $\mathbf{R}^*$. Hence, f is neither an epimorphism nor a monomorphism.*

The following theorem gives a necessary and sufficient condition for a homomorphism to be a one-one mapping in terms of its kernel.

Theorem 5.1.8 *Let f be a homomorphism of a group G into a group G_1. Then f is one-one if and only if Ker $f = \{e\}$.*

Proof.   Suppose f is one-one. Let $a \in \text{Ker } f$. Then $f(a) = e_1 = f(e)$ by Theorem 5.1.2(i). Since f is one-one, we must have $a = e$. Hence, $\text{Ker } f = \{e\}$. Conversely, suppose that $\text{Ker } f = \{e\}$. Let $a, b \in G$. Suppose $f(a) = f(b)$. Then

$$f(ab^{-1}) = f(a)f(b^{-1}) = f(a)f(b)^{-1} = e_1.$$

Thus, $ab^{-1} \in \text{Ker } f = \{e\}$ and so $ab^{-1} = e$, i.e., $a = b$. This proves that f is one-one. ∎

Theorem 5.1.9 *Let f be a homomorphism of a group G into a group G_1. Then $\text{Ker } f$ is a normal subgroup of G.*

Proof.   Since $e \in \text{Ker } f$, $\text{Ker } f \neq \phi$. Let $a, b \in \text{Ker } f$. Then $f(ab^{-1}) = f(a)f(b^{-1}) = f(a)f(b)^{-1} = e_1(e_1)^{-1} = e_1 e_1 = e_1$. Thus, $ab^{-1} \in \text{Ker } f$ and hence $\text{Ker } f$ is a subgroup of G by Theorem 4.1.3. Let $a \in G$ and $h \in \text{Ker } f$. Then $f(aha^{-1}) = f(a)f(h)f(a^{-1}) = f(a)f(h)f(a)^{-1} = f(a)e_1 f(a)^{-1} = e_1$. Therefore, $aha^{-1} \in \text{Ker } f$. This proves that $a\text{Ker } fa^{-1} \subseteq \text{Ker } f$. Hence, $\text{Ker } f$ is a normal subgroup of G by Theorem 4.4.3. ∎

Example 5.1.10 *Let $GL(2, \mathbf{R}) = \left\{ \begin{bmatrix} a & b \\ c & d \end{bmatrix} \mid a, b, c, d \in \mathbf{R}, ad - bc \neq 0 \right\}$ be the noncommutative group of Example 2.1.10. Let $\mathbf{R}^*$ be the group of all nonzero real numbers under multiplication. Define $f : GL(2, \mathbf{R}) \to \mathbf{R}^*$ by*

$$f\left(\begin{bmatrix} a & b \\ c & d \end{bmatrix} \right) = ad - bc$$

for all $\begin{bmatrix} a & b \\ c & d \end{bmatrix} \in GL(2, \mathbf{R})$. *Let* $\begin{bmatrix} a & b \\ c & d \end{bmatrix}, \begin{bmatrix} u & v \\ w & s \end{bmatrix} \in GL(2, \mathbf{R})$. *Now*

$$
\begin{aligned}
f\left(\begin{bmatrix} a & b \\ c & d \end{bmatrix} \begin{bmatrix} u & v \\ w & s \end{bmatrix} \right) &= f\left(\begin{bmatrix} au + bw & av + bs \\ cu + dw & cv + ds \end{bmatrix} \right) \\
&= (au + bw)(cv + ds) - (av + bs)(cu + dw) \\
&= (ad - bc)(us - vw) \\
&= f\left(\begin{bmatrix} a & b \\ c & d \end{bmatrix} \right) f\left(\begin{bmatrix} u & v \\ w & s \end{bmatrix} \right).
\end{aligned}
$$

This proves that f is a homomorphism. To show that f is onto $\mathbf{R}^*$, let $a \in \mathbf{R}^*$. Then $\begin{bmatrix} a & 0 \\ 0 & 1 \end{bmatrix} \in GL(2, \mathbf{R})$ and $f\left(\begin{bmatrix} a & 0 \\ 0 & 1 \end{bmatrix} \right) = a$. Hence, f is onto $\mathbf{R}^*$. Since

$f\left(\begin{bmatrix} a & 0 \\ 0 & 1 \end{bmatrix} \right) = a = f\left(\begin{bmatrix} a & 1 \\ 0 & 1 \end{bmatrix} \right)$ *and* $\begin{bmatrix} a & 0 \\ 0 & 1 \end{bmatrix} \neq \begin{bmatrix} a & 1 \\ 0 & 1 \end{bmatrix}$, *$f$ is not one-one.*

The previous example shows that there may exist a homomorphism of a noncommutative group onto a commutative group.

Example 5.1.11 *Consider S_3 and the normal subgroup*

$$H = \left\{ \left(\begin{array}{ccc} 1 & 2 & 3 \\ 1 & 2 & 3 \end{array} \right), \left(\begin{array}{ccc} 1 & 2 & 3 \\ 2 & 3 & 1 \end{array} \right), \left(\begin{array}{ccc} 1 & 2 & 3 \\ 3 & 1 & 2 \end{array} \right) \right\}.$$

Define $f : S_3 \rightarrow S_3/H$ by for all $\pi \in S_3$, $f(\pi) = \pi H$. Then

$$f(\pi \circ \pi') = (\pi \circ \pi')H = (\pi H) \circ (\pi' H) = f(\pi) \circ f(\pi')$$

for all $\pi, \pi' \in S_3$. Hence, f is a homomorphism. Also, $\mathrm{Ker}\, f = \{\alpha \in S_3 \mid \alpha H = H\} = \{\alpha \in S_3 \mid \alpha \in H\} = H$.

In Theorem 5.1.9, we showed that if f is a homomorphism of a group into a group G_1, then $\mathrm{Ker}\, f$ is a normal subgroup of G. In the following theorem, we show that every normal subgroup H of a group induces a homomorphism g of G onto the quotient group G/H such that $\mathrm{Ker}\, g = H$. We note that in Example 5.1.11, the conclusion did not depend on the nature of S_3. The conclusion was made by use of general arguments. This also leads us to the following theorem.

Theorem 5.1.12 *Let H be a normal subgroup of a group G. Define the function g from G onto the quotient group G/H by $g(a) = aH$ for all $a \in G$. Then g is a homomorphism of G onto G/H and $\mathrm{Ker}\, g = H$. (The homomorphism g is called the **natural homomorphism** of G onto G/H.)*

Proof.   From the definition of g, it follows that g is a function from G onto G/H. To show g is a homomorphism, let $a,\ b \in G$. Then $g(ab) = (ab)H = (aH)(bH) = g(a)g(b)$. Hence, g is a homomorphism of G onto G/H. Finally, we show that $\mathrm{Ker}\, g = H$. Now $a \in \mathrm{Ker}\, g$ if and only if $g(a) = eH$ if and only if $aH = eH$ if and only if $e^{-1}a \in H$ if and only if $a \in H$. Thus, $\mathrm{Ker}\, g = H$. ∎

We now define a particular type of homomorphism between groups in order to introduce the important idea of groups being algebraically indistinguishable.

Definition 5.1.13 *A homomorphism f of a group G into a group G_1 is called an **isomorphism** of G onto G_1 if f is one-one and onto G_1. In this case, we write $G \simeq G_1$ and say that G and G_1 are **isomorphic**. An isomorphism of a group G onto G is called an **automorphism**.*

For a group G, $\mathrm{Aut}(G)$, denotes the set of all automorphisms of G.

In the following theorem, we collect some properties of isomorphisms, which will be useful in determining whether given groups are isomorphic or not.

Theorem 5.1.14 *Let f be an isomorphism of a group G onto a group G_1. Then*

(i) $f^{-1} : G_1 \to G$ *is an isomorphism.*

(ii) *G is commutative if and only if G_1 is commutative.*

(iii) *For all $a \in G$, $o(a) = o(f(a))$.*

(iv) *G is a torsion group if and only if G_1 is a torsion group.*

(v) *G is cyclic if and only if G_1 is cyclic.*

Proof. (i) Since f is one-one and onto G_1, f^{-1} is one-one and onto G. Now we only need to verify that f^{-1} is a homomorphism. Let $u, v \in G_1$. Then there exist $a, b \in G$ such that $f(a) = u$ and $f(b) = v$. This implies that $a = f^{-1}(u)$, $b = f^{-1}(v)$, and $uv = f(a)f(b) = f(ab)$. Thus, $f^{-1}(uv) = ab = f^{-1}(u)f^{-1}(v)$ and so f^{-1} is a homomorphism. Hence, f^{-1} is an isomorphism.

(ii) Suppose G is commutative. Let $u, v \in G_1$. Since f is onto G_1, there exist $a, b \in G$ such that $f(a) = u$ and $f(b) = v$. Now $uv = f(a)f(b) = f(ab) = f(ba) = f(b)f(a) = vu$. Thus, G_1 is commutative. Conversely, suppose G_1 is commutative. Let $a, b \in G$. Now $f(ab) = f(a)f(b) = f(b)f(a) = f(ba)$. Since f is one-one, we have $ab = ba$. This proves that G is commutative.

(iii) Let $a \in G$. By induction, it follows that for all positive integers n, $f(a^n) = (f(a))^n$. Since f is one-one, for all $b \in G$, $f(b) = e_1$ if and only if $b = e$. Hence, $a^n = e$ if and only if $(f(a))^n = e_1$. Thus, a is of finite order if and only if $f(a)$ is of finite order. Suppose $o(a) = m$ and $o(f(a)) = n$. Since $a^m = e$, $(f(a))^m = e_1$. By Theorem 2.1.28, n divides m. Also, $(f(a))^n = e_1$ implies that $a^n = e$. Hence, m divides n. Since m and n are both positive integers and m divides n and n divides m, it follows that $m = n$.

(iv) This follows immediately by (iii).

(v) Suppose G is cyclic. Then $G = \langle a \rangle$ for some $a \in G$. Since $f(a) \in G_1$, $\langle f(a) \rangle \subseteq G_1$. Let $b \in G_1$. Since f is onto G_1, there exists $c \in G$ such that $f(c) = b$. Now $c = a^n$ for some $n \in \mathbf{Z}$. Thus, $b = f(c) = f(a^n) = (f(a))^n \in \langle f(a) \rangle$. Hence, $G_1 = \langle f(a) \rangle$ and so G_1 is cyclic. The converse follows since f^{-1} is an isomorphism. ∎

In order to develop a feel for two groups being algebraically indistinguishable, let us consider two sets S and S' such that there is a one-one function f of S onto S'. Then in a set-theoretic sense, S and S' are the same sets "under f". For instance, let A and B be subsets of S. Then $f(A)$ and $f(B)$ are corresponding subsets of S'. Now $f(A \cap B) = f(A) \cap f(B)$ and $f(A \cup B) = f(A) \cup f(B)$; that is, union and intersection are preserved under f. Other purely set-theoretic operations can be seen to be preserved under f also. Now suppose binary operations $*$ and $*'$ are defined on S and S', respectively, so that $(S, *)$ and $(S', *')$ are groups. Now even though S and S' are the same sets "under f," they need not be the same as groups, i.e., f may not preserve operations. We have seen that the requirement for f to preserve operations is that $f(a * b) = f(a) *' f(b)$ for all $a, b \in S$.

We now consider examples of groups that are isomorphic and examples of groups that are not isomorphic.

Example 5.1.15 *Let n be a positive integer. Define f from $\mathbf{Z}_n$ into $\mathbf{Z}/\langle n \rangle$ by for all $[a] \in \mathbf{Z}_n$, $f([a]) = a + \langle n \rangle$. Then $[a] = [b]$ if and only if $n|(a - b)$ if and only if $a - b = nq$ for some $q \in \mathbf{Z}$ if and only if $a - b \in \langle n \rangle$ if and only if $a + \langle n \rangle = b + \langle n \rangle$ if and only if $f([a]) = f([b])$. Therefore, we find that f is a one-one function. From the definition of f, it follows that f maps $\mathbf{Z}_n$ onto $\mathbf{Z}/\langle n \rangle$. Now $f([a] +_n [b]) = f([a + b]) = (a + b) + \langle n \rangle = (a + \langle n \rangle) + (b + \langle n \rangle) = f([a]) + f([b])$. Thus, f is an isomorphism of $\mathbf{Z}_n$ onto $\mathbf{Z}/\langle n \rangle$.*

Example 5.1.16 *Consider the sets $G = \{e, a, b, c\}$ and $G_1 = \{1, -1, i, -i\}$. Define $*$ and $\cdot$ on G and G_1, respectively, by means of the following operation tables.*

$*$	e	a	b	c
e	e	a	b	c
a	a	e	c	b
b	b	c	e	a
c	c	b	a	e

$\cdot$	1	-1	i	$-i$
1	1	-1	i	$-i$
-1	-1	1	$-i$	i
i	i	$-i$	-1	1
$-i$	$-i$	i	1	-1

Now G_1 is a cyclic group generated by i. G is also a group. However, since $aa = e$, $bb = e$, and $cc = e$, no element of G has order 4 and so G is not cyclic. Thus, G and G_1 are not isomorphic.

Example 5.1.17 *Let $(\mathbf{R}, +)$ be the group of real numbers under addition and $(\mathbf{R}^+, \cdot)$ be the group of positive real numbers under multiplication. Define $f : \mathbf{R} \to \mathbf{R}^+$ by $f(a) = e^a$ for all $a \in \mathbf{R}$. Clearly f is well defined. Let $a, b \in \mathbf{R}$. Then $f(a+b) = e^{a+b} = e^a e^b = f(a)f(b)$. Hence, f is a homomorphism. Suppose $f(a) = f(b)$. Then $e^a = e^b$ and so $\log_e e^a = \log_e e^b$. This implies that $a = b$, whence f is one-one. Let $b \in \mathbf{R}^+$. Then $\log_e b \in \mathbf{R}$ and $f(\log_e b) = e^{\log_e b} = b$. Thus, f is onto $\mathbf{R}^+$. Consequently, f is an isomorphism of $(\mathbf{R}, +)$ onto $(\mathbf{R}^+, \cdot)$.*

Example 5.1.18 *Consider the groups $(\mathbf{Z}, +)$ and $(\mathbf{Q}, +)$. By Worked-Out Exercise 1 (page 113), $(\mathbf{Q}, +)$ is not cyclic. Since $(\mathbf{Z}, +)$ is cyclic and $(\mathbf{Q}, +)$ is not cyclic, $(\mathbf{Z}, +)$ is not isomorphic to $(\mathbf{Q}, +)$ by Theorem 5.1.14(v).*

Example 5.1.19 *The group* $(\mathbf{Q}, +)$ *is not isomorphic to* $(\mathbf{Q}^*, \cdot)$ *since every nonidentity element of* $(\mathbf{Q}, +)$ *is of infinite order while* -1 *is a nonidentity element of* $(\mathbf{Q}^*, \cdot)$ *which is of finite order.*

Let us now characterize finite and infinite cyclic groups.

Theorem 5.1.20 *Every finite cyclic group of order n is isomorphic to* $(\mathbf{Z}_n, +_n)$ *and every infinite cyclic group is isomorphic to* $(\mathbf{Z}, +)$.

Proof. Let $(\langle a \rangle, *)$ be a cyclic group of order n. Let $G = \langle a \rangle$. Define the function $f : G \rightarrow \mathbf{Z}_n$ by for all $a^i \in G$, $f(a^i) = [i]$. Now $a^i = a^j$ if and only if $a^{j-i} = e$ if and only if $n | (j - i)$ if and only if $[i] = [j]$ (Exercise 11, page 30) if and only if $f(a^i) = f(a^j)$. Thus, f is a one-one function. Now $f(a^i a^j) = f(a^{i+j}) = [i + j] = [i] +_n [j] = f(a^i) +_n f(a^j)$. Since f is one-one and G and $\mathbf{Z}_n$ are finite with same number of elements, f is onto $\mathbf{Z}_n$. Hence, $G \simeq \mathbf{Z}_n$.

Now let $G = \langle a \rangle$ be an infinite cyclic group. Define the function $f : G \rightarrow \mathbf{Z}$ by $f(a^i) = i$ for all $i \in \mathbf{Z}$. Since $a^i = a^j$ if and only if $a^{i-j} = e$ if and only if $i - j = 0$ (since a is of infinite order) if and only if $i = j$, we have that f is a one-one function of G into $\mathbf{Z}$. From the definition of f, f is onto $\mathbf{Z}$. Now $f(a^i a^j) = f(a^{i+j}) = i + j = f(a^i) + f(a^j)$. Hence, $G \simeq \mathbf{Z}$. ∎

Corollary 5.1.21 *Any two cyclic groups of the same order are isomorphic.* ∎

From the above corollary, it follows that there is only one (up to isomorphism) cyclic group having a prescribed order.

In Example 5.1.16, we saw that there are at least two nonisomorphic groups of order 4. We now show that these are exactly two nonisomorphic groups of order 4.

Let G be a group of order 4 which is not cyclic. (Example 5.1.16 shows that such a group exists.) Then no element of G can have order 4, for if $a \in G$ has order 4, then e, a, a^2, a^3 would be distinct elements of G and thus G would be cyclic, i.e., $G = \langle a \rangle$. This is contrary to the assumption that G is not cyclic. Let $G = \{e, a, b, c\}$. Since the order of every element of G divides the order of G, a, b, and c have order 2. If $ab = a$, then $b = e$, a contradiction. Thus, $ab \neq a$. Similarly, $ab \neq b$. Suppose $ab = e$, then $a(ab) = ae$. Therefore, $b = a$ since $a^2 = e$, a contradiction. Thus, $ab = c$. Similarly, $ba = c$. Hence, $ab = ba$. By similar arguments, we have $ac = b = ca$ and $bc = a = cb$. Thus, we find that G is a commutative group and its operation table is given by the table in Example 5.1.16. Consequently, there is essentially one group of order 4 which is not cyclic. This is the Klein 4-group. Since all cyclic groups of the same orders are isomorphic, we thus have exactly two nonisomorphic groups of order 4, namely, the Klein 4-group and the cyclic group of order 4. We have thus proved the following result.

Theorem 5.1.22 *There are only two groups of order* 4 *(up to isomorphism), a cyclic group of order* 4 *and* K_4 *(Klein 4-group).*

Since every cyclic group is commutative and every group of prime order is cyclic, it follows that that if a group is noncommutative, then it must have order at least 6. Indeed, the symmetric group S_3 is noncommutative and of order 6. Since all cyclic groups of the same order are isomorphic and since every group of prime order is cyclic, there is exactly one group of order 1, 2, 3, 5 (up to isomorphism), respectively. We have seen that there are two nonisomorphic groups of order 4. In the next theorem, we show that there are only two (up to isomorphism) nonisomorphic groups of order 6.

Theorem 5.1.23 *There are only two (up to isomorphism) groups of order* 6.

Proof. The group $\mathbf{Z}_6$ is a cyclic group of order 6 and S_3 is a noncommutative group of order 6. Note that $\mathbf{Z}_6$ is not isomorphic to S_3. To show that there are only two (up to isomorphism) nonisomorphic groups of order 6, we will show that any group of order 6 is isomorphic to either $\mathbf{Z}_6$ or S_3.

Let G be a group of order 6. Since $|G|$ is even, there exists $a \in G$, $a \neq e$ such that $a^2 = e$. If $x^2 = e$ for all $x \in G$, then G is commutative and for any two distinct nonidentity elements a and b, $\{e, a, b, ab\}$ is a subgroup of G. Since $|G| = 6$, G has no subgroups of order 4. Hence, there exists $b \in G$ such that $b^2 \neq e$, i.e., $b \neq e$ and $o(b) \neq 2$. Since $o(b)|6$, $o(b) = 6$ or 3. If $o(b) = 6$, then $G = \langle b \rangle$ is a cyclic group of order 6 and $G \simeq \mathbf{Z}_6$. Suppose G is not cyclic. Then $o(b) = 3$. Let $H = \{e, b, b^2\}$. Then H is a subgroup of G of index 2. Thus, H is a normal subgroup of G. Clearly $a \notin H$. Now $G = H \cup aH$ and $H \cap aH = \phi$. Hence, $G = \{e, b, b^2, a, ab, ab^2\}$. Now $aba^{-1} \in H$ since H is normal and $b \in H$. Therefore, $aba^{-1} = e$ or $aba^{-1} = b$ or $aba^{-1} = b^2$. If $aba^{-1} = e$, then $b = e$, which is a contradiction. If $aba^{-1} = b$, then $ab = ba$. Since $o(a)$ and $o(b)$ are relatively prime and $ab = ba$, $o(ab) = o(a) \cdot o(b) = 6$. Thus, G is cyclic, a contradiction. Hence, $aba^{-1} = b^2$. Thus, $G = \langle a, b \rangle$, where $o(a) = 2$, $o(b) = 3$, and $aba^{-1} = b^2$. It is now easy to see that $G \simeq S_3$. ∎

We conclude this section by proving Cayley's theorem, which says that any group can be realized as a permutation group.

Let a be an element of a group G. Define the function $f_a : G \rightarrow G$ by for all $b \in G$, $f_a(b) = ab$. Then $b = c$ if and only if $ab = ac$ if and only if $f_a(b) = f_a(c)$. Thus, f_a is a one-one function of G into G. For any $b \in G$, $f_a(a^{-1}b) = a(a^{-1}b) = b$. So we find that f_a maps G onto G. Hence, f_a is a permutation of G. Let $F(G) = \{f_a \mid a \in G\}$. Then $F(G)$ is a subset of the set $S(G)$ of all permutations on G. Recall that $(S(G), \circ)$ is a group.

As previously mentioned, early mathematicians worked only with groups of permutations. The following theorem says that every group is isomorphic to a group of permutations of its own elements. In fact, we will show that

$(F(G), \circ)$ is a group which is isomorphic to G. First let us note that for all $b \in G$, $f_{a^{-1}}(b) = a^{-1}b$, and $f_a(a^{-1}b) = b$ implies $(f_a)^{-1}(b) = a^{-1}b$. Thus, $(f_a)^{-1} = f_{a^{-1}}$.

Theorem 5.1.24 (Cayley) *For any group G, $(F(G), \circ)$ is a group and $G \simeq F(G)$.*

Proof.   We first show that $(F(G), \circ)$ is a group. It suffices to show that $F(G)$ is a subgroup of $(S(G), \circ)$. Let $f_a, f_b \in F(G)$. Then $(f_a \circ f_b^{-1})(c) = (f_a \circ f_{b^{-1}})(c) = f_a(f_{b^{-1}}(c)) = f_a(b^{-1}c) = a(b^{-1}c) = (ab^{-1})c = f_{ab^{-1}}(c)$ for all $c \in G$ and so $f_a \circ f_b^{-1} = f_{ab^{-1}} \in F(G)$. Hence, $F(G)$ is a subgroup by Theorem 4.1.3. Define $g : G \to F(G)$ by for all $a \in G$, $g(a) = f_a$. Then $a = b$ if and only if $ac = bc$ for all $c \in G$ if and only if $f_a(c) = f_b(c)$ for all $c \in G$ if and only if $f_a = f_b$ if and only if $g(a) = g(b)$. This proves that g is a one-one function of G into $F(G)$. Clearly g maps G onto $F(G)$. Now $g(ab) = f_{ab}$ and $g(a) \circ g(b) = f_a \circ f_b$. Also, for all $c \in G$, $f_{ab}(c) = (ab)c = a(bc) = f_a(bc) = f_a(f_b(c)) = (f_a \circ f_b)(c)$. Thus, $f_{ab} = f_a \circ f_b$. Hence, $g(ab) = g(a) \circ g(b)$ and so g is an isomorphism. ∎

Cayley's theorem is another example of a representation theorem. However, Cayley realized that the best way of studying general problems in group theory was not necessarily by the use of permutations.

5.1.1   Worked-Out Exercises

◇ **Exercise 1** Let $f : G \to G_1$ be an epimorphism of groups. If H is a normal subgroup of G, then show that $f(H)$ is a normal subgroup of G_1.

Solution: By Theorem 5.1.2, we find that $f(H)$ is a subgroup of G_1. Let $g_1 \in G_1$. Since f is onto G_1, there exists $g \in G$ such that $f(g) = g_1$. Let $a \in g_1 f(H) g_1^{-1} = f(g) f(H) f(g)^{-1}$. Then $a = f(g) f(h) f(g)^{-1} = f(ghg^{-1})$ for some $h \in H$. Since H is a normal subgroup of G, $ghg^{-1} \in H$ and so $a \in f(H)$. Thus, $g_1 f(H) g_1^{-1} \subseteq f(H)$. Hence, $f(H)$ is a normal subgroup of G_1.

◇ **Exercise 2** Let G and H be finite groups such that $\gcd(|G|, |H|) = 1$. Show that the trivial homomorphism is the only homomorphism from G into H.

Solution: Let $f : G \to H$ be a homomorphism and let $a \in G$. We show that every element of G is mapped onto the identity element of H, i.e., $f(a) = e_H$ for all $a \in G$, where e_H denotes the identity element of H. Now $\circ(a)| |G|$ and $\circ(f(a))| |H|$. Also, by Theorem 5.1.2, $\circ(f(a))| \circ (a)$. Hence, $\circ(f(a))| |G|$. Since $|G|$ and $|H|$ are relatively prime, $\circ(f(a)) = 1$, proving $f(a) = e_H$. Thus, f is the trivial homomorphism.

$\Diamond$ **Exercise 3** Show that the group $(\mathbf{Q}, +)$ is not isomorphic to $(\mathbf{Q}/\mathbf{Z}, +)$.

Solution: In $(\mathbf{Q}, +)$, every nonzero element is of infinite order. Let $\frac{p}{q} + \mathbf{Z} \in \mathbf{Q}/\mathbf{Z}$, where $p, q \in \mathbf{Z}$ and $q \neq 0$. Then $q(\frac{p}{q} + \mathbf{Z}) = p + \mathbf{Z} = \mathbf{Z}$. This shows that every element of $\mathbf{Q}/\mathbf{Z}$ is of finite order. Hence, $(\mathbf{Q}, +)$ is not isomorphic to $(\mathbf{Q}/\mathbf{Z}, +)$.

Exercise 4 Show that $\mathbf{R}^*$, the group of all nonzero real numbers under multiplication, is not isomorphic to $\mathbf{C}^*$, the group of all nonzero complex numbers under multiplication.

Solution: In the group $\mathbf{C}^*$, i is an element of order 4. But $\mathbf{R}^*$ does not contain any element of order 4. Hence, by Theorem 5.1.14, $\mathbf{R}^*$ is not isomorphic to $\mathbf{C}^*$.

$\Diamond$ **Exercise 5** Find all homomorphisms from $\mathbf{Z}_6$ into $\mathbf{Z}_4$.

Solution: $\mathbf{Z}_6 = \langle [1] \rangle$. Let $f : \mathbf{Z}_6 \to \mathbf{Z}_4$ be a homomorphism. For any $[a] \in \mathbf{Z}_6$, $f([a]) = af([1])$ shows that f is completely known if $f([1])$ is known. Now $\circ(f([1]))$ divides $\circ([1])$ and 4, i.e., $\circ(f([1]))$ divides 6 and 4. Hence, $\circ(f([1])) = 1$ or 2. Thus, $f([1]) = [0]$ or $[2]$. If $f([1]) = [0]$, then f is the trivial homomorphism which maps every element to $[0]$. On the other hand, $f([1]) = [2]$ implies that $f([a]) = [2a]$ for all $[a] \in \mathbf{Z}_6$. Thus, $f([a] + [b]) = f([a + b]) = [2(a + b)] = [2a + 2b] = [2a] + [2b] = f([a]) + f([b])$, proving that the mapping $f : \mathbf{Z}_6 \to \mathbf{Z}_4$ defined by $f([a]) = [2a]$ for all $[a] \in \mathbf{Z}_6$ is a homomorphism. Hence, there are two homomorphisms from $\mathbf{Z}_6$ into $\mathbf{Z}_4$.

Exercise 6 Let G be a finite commutative group. Let $n \in \mathbf{Z}$ be such that n and $|G|$ are relatively prime. Show that the function $\phi : G \to G$ defined by $\phi(a) = a^n$ for all $a \in G$ is an isomorphism of G onto G.

Solution: Let $a, b \in G$. Now

$$\begin{aligned} \phi(ab) &= (ab)^n \\ &= a^n b^n \quad \text{(since } G \text{ is commutative)} \\ &= \phi(a)\phi(b). \end{aligned}$$

This implies that ϕ is a homomorphism. Let $\phi(a) = \phi(b)$. Then $a^n = b^n$ and so $(ab^{-1})^n = e$. Therefore, $\circ(ab^{-1})$ divides n. Since $\circ(ab^{-1})$ divides $|G|$ and n and $|G|$ are relatively prime, $\circ(ab^{-1}) = 1$. This implies that $ab^{-1} = e$, i.e., $a = b$, proving that ϕ is one-one. Since G is a finite group and ϕ is one-one, ϕ is onto G. Hence, ϕ is an isomorphism of G onto G.

Exercise 7 (i) Let G be a group and $f : G \to G$ be defined by $f(a) = a^n$ for all $a \in G$, where n is a positive integer. Suppose f is an isomorphism. Prove that $a^{n-1} \in Z(G)$ for all $a \in G$.

(ii) Let G be a group and $f : G \to G$ defined by for all $a \in G$, $f(a) = a^3$ be an isomorphism. Prove that G is commutative.

Solution: (i) Let $a, b \in G$. Then $f(a^{-1}ba) = (a^{-1}ba)^n = a^{-1}b^n a$. Thus,

$$a^{-n}b^n a^n = f(a^{-1})f(b)f(a) = f(a^{-1}ba) = a^{-1}b^n a.$$

Hence, $a^{-(n-1)} b^n a^{n-1} = b^n$ or $(a^{-(n-1)}ba^{n-1})^n = b^n$. Thus, $f(a^{-(n-1)}ba^{n-1}) = f(b)$. Since f is one-one, $a^{-(n-1)}ba^{n-1} = b$. Hence, $a^{n-1}b = ba^{n-1}$, proving that $a^{n-1} \in Z(G)$.

(ii) By (i), $a^2 \in Z(G)$ for all $a \in G$. Let $a, b \in G$. Then $f(ab) = (ab)^3 = ab(ab)^2 = a(ab)^2b = aababb = a^2bab^2 = ba^2b^2a = bb^2a^2a = b^3a^3 = f(b)f(a) = f(ba)$. Hence, $ab = ba$ since f is one-one. Thus, G is commutative.

5.1.2 Exercises

1. Determine whether the indicated function f is a homomorphism from the first group into the second group. If f is a homomorphism, determine its kernel.

    (i) $f(a) = a^2$; $(\mathbf{R}^+, \cdot)$, $(\mathbf{R}^+, \cdot)$ for all $a \in \mathbf{R}^+$.

    (ii) $f(a) = 2^a$; $(\mathbf{R}, +)$, $(\mathbf{R}^+, \cdot)$ for all $a \in \mathbf{R}$.

    (iii) $f(a) = |a|$; $(\mathbf{R}\backslash\{0\}, \cdot)$, $(\mathbf{R}^+, \cdot)$ for all $a \in \mathbf{R}\backslash\{0\}$.

    (iv) $f(a) = a + 1$; $(\mathbf{Z}, +)$, $(\mathbf{Z}, +)$ for all $a \in \mathbf{Z}$.

    (v) $f(a) = 2a$; $(\mathbf{Z}, +)$, $(\mathbf{Z}, +)$ for all $a \in \mathbf{Z}$.

2. Find all homomorphisms from $\mathbf{Z}$ into $\mathbf{Z}$. How many homomorphisms are onto?

3. Find all homomorphisms from $\mathbf{Z}$ onto $\mathbf{Z}_6$.

4. Find all homomorphisms from $\mathbf{Z}_8$ into $\mathbf{Z}_{12}$ and from $\mathbf{Z}_{20}$ into $\mathbf{Z}_{10}$.

5. Show that $\mathbf{Q}^*$, the group of all nonzero rational numbers under multiplication, is not isomorphic to $\mathbf{R}^*$, the group of all nonzero real numbers under multiplication.

6. Show that $(\mathbf{Q}, +)$ is not isomorphic to $(\mathbf{R}, +)$.

7. Show that $(\mathbf{Z}, +)$ is not isomorphic to $(\mathbf{R}, +)$.

8. Let G be a group. Define the function $f : G \to G$ by for all $a \in G$, $f(a) = a^{-1}$. Prove that f is a homomorphism if and only if G is commutative.

9. Let $G = \{(a, b) \mid a, b \in \mathbf{R}, b \neq 0\}$. Then $(G, *)$ is a noncommutative group under the binary operation $(a, b) * (c, d) = (a + bc, bd)$ for all $(a, b), (c, d) \in G$. Let $H = \{(a, b) \in G \mid a = 0\}$ and $K = \{(a, b) \in G \mid b > 0\}$. Show that $H \cap K \simeq (\mathbf{R}^+, \cdot)$, where $(\mathbf{R}^+, \cdot)$ is the group of all positive real numbers under multiplication.

10. Let $G = \{a \in \mathbf{R} \mid -1 < a < 1\}$. Show that $(G, *) \simeq (\mathbf{R}, +)$, where the binary operation $*$ on G is defined by

$$a * b = \frac{a + b}{1 + ab}$$

for all $a, b \in G$.

11. (i) Let f be a homomorphism from a cyclic group of order 8 onto a cyclic group of order 4. Determine Ker f.

(ii) Let f be a homomorphism from a cyclic group of order 8 onto a cyclic group of order 2. Determine Ker f.

12. Prove that a homomorphic image of a cyclic group is cyclic.

13. Show that S_3 and $\mathbf{Z}_6$ are not isomorphic groups, but for every proper subgroup A of S_3 there exists a proper subgroup B of $\mathbf{Z}_6$ such that $A \simeq B$.

14. Let G, H, and K be groups. Suppose that the functions $f : G \to H$ and $g : H \to K$ are homomorphisms. Prove that $g \circ f : G \to K$ is also a homomorphism.

15. Let G and H be groups. Define the function $f : G \times H \to G$ by for all $(a, b) \in G \times H$, $f((a, b)) = a$. Prove that f is a homomorphism from $G \times H$ onto G. Determine Ker f.

16. Let $f : G \to H$ be an isomorphism of groups. Prove that $f^{-1} : H \to G$ is also an isomorphism of groups.

17. Let G, H, and K be groups. Prove that

(i) $G \times H \simeq H \times G$.

(ii) If $G \simeq H$ and $H \simeq K$, then $G \simeq K$.

(iii) $G \times (H \times K) \simeq (G \times H) \times K$.

18. Let G and H be groups. Let $f : G \to H$ be a homomorphism of G onto H. Show that if $G = \langle S \rangle$ for some subset S of G, then $H = \langle f(S) \rangle$.

19. Let $f : G \to H$ be an isomorphism of groups. Show that for any integer k and for any $g \in G$, the sets $A = \{a \in G \mid a^k = g\}$ and $B = \{b \in H \mid b^k = f(g)\}$ have the same number of elements.

20. Let G be a simple group and $\psi : S_n \to G$ be an epimorphism for some positive integer n. Prove that $G \simeq S_k$ for some $k \leq n$.

21. Which of the following statements are true? Justify.

    (i) A cyclic group with more than one element may be a homomorphic image of a noncyclic group.

    (ii) There does not exist a nontrivial homomorphism from a group G of order 5 into a group H of order 4.

    (iii) The group $(\mathbf{Z}, +)$ is isomorphic to $(\mathbf{Q}, +)$.

    (iv) There exists a monomorphism from a group of order 20 into a group of order 70.

    (v) There exists an epimorphism of $(\mathbf{R}, +)$ onto $(\mathbf{Z}, +)$.

    (vi) There does not exist any epimorphism of $(\mathbf{Q}, +)$ onto $(\mathbf{Z}, +)$.

    (vii) If f and g are two epimorphisms of a group G onto a group H such that Ker $f =$ Ker g, then $f = g$.

5.2 Isomorphism and Correspondence Theorems

In this section, we continue our study of isomorphisms. Our objective is to prove the fundamental theorem of homomorphisms, the isomorphism theorems, and the correspondence theorem. These theorems show us the relationship between homomorphisms and quotient groups.

Theorem 5.2.1 *Let f be a homomorphism of a group G onto a group G_1, H be a normal subgroup of G such that $H \subseteq$ Ker f, and g be the natural homomorphism of G onto G/H. Then there exists a unique homomorphism h of G/H onto G_1 such that $f = h \circ g$. Furthermore, h is one-one if and only if $H =$ Ker f.*

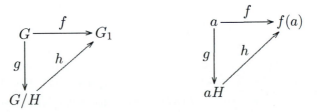

Proof. Define $h : G/H \to G_1$ by

$$h(aH) = f(a)$$

for all $aH \in G/H$.

Now $aH = bH$ implies $b^{-1}a \in H \subseteq$ Ker f and so $f(b^{-1}a) = e_1$ or $f(a) = f(b)$. Hence, $h(aH) = h(bH)$ and so h is well defined. Let $a \in G$. Then $(h \circ g)(a) = h(g(a)) = h(aH) = f(a)$. Therefore, $h \circ g = f$. Since f maps G onto G_1, h must map G/H onto G_1. Now $h((aH)(bH)) = h((ab)H) = f(ab) = f(a)f(b) = h(aH)h(bH)$. Hence, h is a homomorphism of G/H onto G_1 satisfying $f = h \circ g$. To prove the uniqueness part, let us assume $f = h' \circ g$ for some homomorphism h' from G/H onto G_1. Then $h(aH) = f(a) = (h' \circ g)(a) = h'(g(a)) = h'(aH)$ for all $aH \in G/H$ and so $h = h'$. Hence, h is the only homomorphism of G/H onto G_1 such that $f = h \circ g$.

Suppose h is one-one. Let $a \in$ Ker f. Then $f(a) = e_1$ and so $h(aH) = e_1$. Since $h(eH) = e_1$ and h is one-one, $aH = eH$. Thus, $a \in H$ and so Ker $f \subseteq H$. By hypothesis, $H \subseteq$ Ker f and so $H =$ Ker f. Conversely, assume $H =$ Ker f. Suppose $h(aH) = h(bH)$. Then $f(a) = f(b)$ or $f(b^{-1}a) = e_1$. Thus, $b^{-1}a \in$ Ker $f = H$ and so $aH = bH$, proving that h is one-one. ∎

From Theorem 5.2.1, it follows that if $H =$ Ker f, then h is an isomorphism and hence $G/$Ker f is isomorphic to G_1, i.e., every homomorphism of a group G onto a group G_1 induces an isomorphism of $G/$Ker f onto G_1. This result plays a fundamental role in group theory. It is known as **the fundamental theorem of homomorphisms** for groups. This result is also called the first isomorphism theorem for groups. Considering the importance of this theorem, we state it in its general form and also give a direct proof of it.

Theorem 5.2.2 (First Isomorphism Theorem) *Let f be a homomorphism of a group G into a group G_1. Then $f(G)$ is a subgroup of G_1 and*

$$G/\text{Ker } f \simeq f(G).$$

Proof.   By Theorem 5.1.2, $f(G)$ is a subgroup of G_1. Let $H =$ Ker f. Define $h : G/H \to f(G)$ by

$$h(aH) = f(a)$$

for all $aH \in G/H$. Now $aH = bH$ if and only if $b^{-1}a \in H =$ Ker f if and only if $f(b^{-1}a) = e_1$ if and only if $f(b^{-1})f(a) = e_1$ if and only if $f(a) = f(b)$. Thus, h is a one-one function. Let $x \in f(G)$. Then $x = f(b)$ for some $b \in G$. Therefore, $h(bH) = f(b) = x$. This shows that h is onto $f(G)$. Finally, $h(aHbH) = h(abH) = f(ab) = f(a)f(b) = h(aH)h(bH)$ for all $aH, bH \in G/H$, proving that h is a homomorphism. Consequently, $G/$Ker $f \simeq f(G)$. ∎

In the following example we illustrate the first isomorphism theorem.

Example 5.2.3 *Let f be the homomorphism of $(\mathbf{Z}, +)$ onto $(\mathbf{Z}_3, +_3)$ defined by $f(n) = [n]$ for all $n \in \mathbf{Z}$. Let g be the natural homomorphism of $\mathbf{Z}$ onto*

$\mathbf{Z}/\langle 6\rangle$. *Now* $\langle 6\rangle$ *is a normal subgroup of* $\mathbf{Z}$ *and* $\langle 6\rangle \subset \langle 3\rangle = Ker\ f$. *Thus, there exists a homomorphism* h *of* $\mathbf{Z}/\langle 6\rangle$ *onto* $\mathbf{Z}_3$ *such that* $f = h \circ g$. *The homomorphism* h *is defined by* $h(n + \langle 6\rangle) = [n]$.

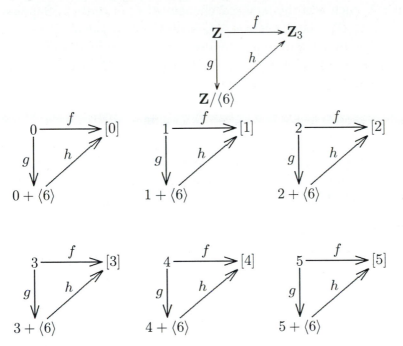

Recall that a group G_1 is called a **homomorphic image** of a group G if there exists a homomorphism of G onto G_1.

From Theorem 5.2.1 and Corollary 5.2.2, we find that for each normal subgroup N of a group G, G/N is a homomorphic image of G, and for each homomorphic image G_1, there exists a normal subgroup N of G such that $G/N \simeq G_1$.

Example 5.2.4 *The group* S_3 *has (up to isomorphism) only three homomorphic images. This follows from the fact that* S_3 *has only three normal subgroups. The homomorphic images are* S_3, $\mathbf{Z}_1$, *and* $\mathbf{Z}_2$ *since* $\{e\}$, S_3, *and* $\{e, (1\ 2\ 3), (1\ 3\ 2)\}$ *are the only normal subgroups of* S_3 *and* $S_3 \simeq S_3/\{e\}$, $\mathbf{Z}_1 \simeq S_3/S_3$, *and* $\mathbf{Z}_2 \simeq S_3/\{e, (1\ 2\ 3), (1\ 3\ 2)\}$.

Theorem 5.2.5 *Let* G_1 *be a homomorphic image of a group* G. *Then the following assertions hold.*

(i) *If* G *is cyclic, then* G_1 *is cyclic.*

(ii) *If* G *is commutative, then* G_1 *is commutative.*

(iii) *If* G_1 *contains an element of order* n *and* $|G|$ *is finite, then* G *contains an element of order* n.

Proof.   (i) Follows by Exercise 12 (page 152).

(ii) Follows by Theorem 5.1.2(v).

(iii) Let $f : G \to G_1$ be an epimorphism and let a' be an element of G_1 of order n. If $n = 1$, then e is the required element of G of order 1. Suppose $n > 1$. Since f is onto G_1, there exists $a \in G$ such that $f(a) = a'$. Now $o(a)$ is finite and by Theorem 5.1.2(v), $o(a')$ divides $o(a)$, i.e., n divides $o(a)$. Let $t \in \mathbf{Z}^+$ be such that $o(a) = nt$. Then $t < o(a)$. Hence, $a^t \neq e$. Now $a^{nt} = e$. Let $b = a^t$. Then $b^n = e$ and by Theorem 2.1.28,

$$o(a^t) = \frac{o(a)}{\gcd(t, o(a))} = \frac{nt}{t} = n. \ \blacksquare$$

Note that the result in Theorem 5.2.5(iii) does not hold if $|G|$ is not finite. For example, $\mathbf{Z}_6$ is a homomorphic image of $\mathbf{Z}$; $\mathbf{Z}_6$ contains an element of order 3, but $\mathbf{Z}$ has no element of order 3.

Theorem 5.2.6 (Second Isomorphism Theorem) *Let H and K be subgroups of a group G with K normal in G. Then*

$$H/(H \cap K) \simeq (HK)/K.$$

Proof.   Define $f : H \to (HK)/K$ by $f(h) = hK$ for all $h \in H$. Now $f(h_1 h_2) = h_1 h_2 K = h_1 K h_2 K = f(h_1)f(h_2)$ for all $h_1, h_2 \in H$, proving that f is a homomorphism. Let $xK \in (HK)/K$. Then $x = hk$ for some $h \in H$ and $k \in K$. Thus, $xK = (hk)K = (hK)(kK) = hK = f(h)$. This proves that f is onto $(HK)/K$ and so $f(H) = (HK)/K$. Hence, by the first isomorphism theorem, it follows that

$$H/\mathrm{Ker}\ f \simeq (HK)/K.$$

To complete the proof, we show that $\mathrm{Ker}\ f = H \cap K$. Now

$$\begin{aligned}
\mathrm{Ker}\ f &= \{h \in H \mid f(h) = \text{ identity element of } HK/K\} \\
&= \{h \in H \mid hK = K\} \\
&= \{h \in H \mid h \in K\} \\
&= H \cap K.
\end{aligned}$$

Consequently, $H/H \cap K \simeq (HK)/K$. $\blacksquare$

We illustrate the second isomorphism theorem with the help of the following example.

Example 5.2.7 *Consider the group $(\mathbf{Z}, +)$ and its subgroups $H = \langle 2 \rangle$ and $K = \langle 3 \rangle$. Then $H + K = \langle 2 \rangle + \langle 3 \rangle = \mathbf{Z}$ and $H \cap K = \langle 6 \rangle$. Theorem 5.2.6 says that*

$$H/(H \cap K) \simeq (H + K)/K,$$

i.e.,

$$\langle 2 \rangle / \langle 6 \rangle \simeq \mathbf{Z} / \langle 3 \rangle .$$

This isomorphism is evident if we notice that $\langle 2 \rangle / \langle 6 \rangle = \{0 + \langle 6 \rangle, \ 2 + \langle 6 \rangle, \ 4 + \langle 6 \rangle\}$ *while* $\mathbf{Z} / \langle 3 \rangle = \{0 + \langle 3 \rangle, \ 1 + \langle 3 \rangle, \ 2 + \langle 3 \rangle\}$. *The mapping*

$$h : \langle 2 \rangle / \langle 6 \rangle \to \mathbf{Z} / \langle 3 \rangle$$

defined by $h : 0 + \langle 6 \rangle \to 0 + \langle 3 \rangle, \ 2 + \langle 6 \rangle \to 2 + \langle 6 \rangle, \ 4 + \langle 6 \rangle \to 1 + \langle 3 \rangle$ *is the desired isomorphism.*

Theorem 5.2.8 *Let* f *be a homomorphism of a group* G *onto a group* G_1, H *be a normal subgroup of* G *such that* $H \supseteq$ *Ker* f, *and* g, g' *be the natural homomorphisms of* G *onto* G/H *and* G_1 *onto* $G_1/f(H)$, *respectively. Then there exists a unique isomorphism* h *of* G/H *onto* $G_1/f(H)$ *such that* $g' \circ f = h \circ g$.

Proof. If we show Ker $g' \circ f = H$, then there exists a unique isomorphism h of G/H onto $G_1/f(H)$ by Theorem 5.2.1. Let $a \in H$. Then $(g' \circ f)(a) = g'(f(a)) =$ the identity of $G_1/f(H)$ since $f(a) \in f(H) = $ Ker g'. Thus, $a \in$ Ker $g' \circ f$ and hence $H \subseteq$ Ker $g' \circ f$. Let $a \in$ Ker $g' \circ f$. Then $g'(f(a)) = $ the identity of $G_1/f(H)$ and so $f(a) \in$ Ker $g' = f(H)$. Therefore, there exists $b \in H$ such that $f(b) = f(a)$ or $f(ab^{-1}) = e_1$. This implies that $ab^{-1} \in$ Ker $f \subseteq H$ and so $a = (ab^{-1})b \in H$. Thus, Ker $g' \circ f \subseteq H$. Hence, Ker $g' \circ f = H$. ∎

Corollary 5.2.9 (Third Isomorphism Theorem) *Let* H_1, H_2 *be normal subgroups of a group* G *such that* $H_1 \subseteq H_2$. *Then*

$$(G/H_1)/(H_2/H_1) \simeq G/H_2.$$

Proof. Make the following substitutions in Theorem 5.2.8: G/H_1 for G_1, H_2 for H, and $(G/H_1)/(H_2/H_1)$ for $G_1/f(H)$, where in this case f is the natural homomorphism of G onto G/H_1. Note that $f(H_2) = H_2/H_1$.

We illustrate the third isomorphism theorem with the help of the following example.

Example 5.2.10 *Consider the group* $(\mathbf{Z}, +)$ *and the subgroups* $\langle 6 \rangle$ *and* $\langle 3 \rangle$ *of* $\mathbf{Z}$. *Then*

$$\mathbf{Z}/\langle 3 \rangle = \{0 + \langle 3 \rangle, 1 + \langle 3 \rangle, 2 + \langle 3 \rangle\}.$$

$$\mathbf{Z}/\langle 6 \rangle = \{0 + \langle 6 \rangle, 1 + \langle 6 \rangle, 2 + \langle 6 \rangle, 3 + \langle 6 \rangle, 4 + \langle 6 \rangle, 5 + \langle 6 \rangle\}.$$

$$\langle 3 \rangle/\langle 6 \rangle = \{0 + \langle 6 \rangle, 3 + \langle 6 \rangle\}.$$

Now,

$$(\mathbf{Z}/\langle 6 \rangle)/(\langle 3 \rangle/\langle 6 \rangle) = \{\bar{0}, \bar{1}, \bar{2}\},$$

where

$$\begin{aligned} \bar{0} &= 0 + \langle 6 \rangle + (\langle 3 \rangle/\langle 6 \rangle) \\ \bar{1} &= 1 + \langle 6 \rangle + (\langle 3 \rangle/\langle 6 \rangle) \\ \bar{2} &= 2 + \langle 6 \rangle + (\langle 3 \rangle/\langle 6 \rangle). \end{aligned}$$

It is now clear that

$$\mathbf{Z}/\langle 3 \rangle \simeq (\mathbf{Z}/\langle 6 \rangle)/(\langle 3 \rangle/\langle 6 \rangle)$$

since both are cyclic groups of order 3 *and of course, by Corollary* 5.2.9.

We can at times determine the subgroups of a group G_1 from a group G whose subgroups are known if there is a homomorphism f of G onto G_1. For if such an f exists, the following result says that the subgroups of G_1 can be determined from the subgroups of G which contain Ker f.

Theorem 5.2.11 (Correspondence Theorem) *Let* f *be a homomorphism of a group* G *onto a group* G_1. *Then* f *induces a one-one inclusion preserving correspondence between the subgroups of* G *containing Ker* f *and the subgroups of* G_1. *In fact, if* H *and* K *are corresponding subgroups of* G *and* G_1, *respectively, then* H *is a normal subgroup of* G *if and only if* K *is a normal subgroup of* G_1.

Proof. Let

$$\mathcal{H} = \{H \mid H \text{ is a subgroup of } G \text{ such that Ker } f \subseteq H\}$$

and

$$\mathcal{K} = \{K \mid K \text{ is a subgroup of } G_1\}.$$

Define $f^* : \mathcal{H} \to \mathcal{K}$ by for all $H \in \mathcal{H}$, $f^*(H) = \{f(h) \mid h \in H\}$. Then $f^*(H) \in \mathcal{K}$ by Theorem 5.1.2. Hence, f^* is a function since f is a function. Let $K \in \mathcal{K}$. Denote the preimage, $f^{-1}(K)$, of K in G by H. Let $a \in$ Ker f. Then $f(a) = e_1 \in K$ and so $a \in f^{-1}(K) = H$. Thus, Ker $f \subseteq H$. Let a,

$b \in H$. Then $f(a)$, $f(b) \in K$ and so $f(ab^{-1}) = f(a)f(b^{-1}) = f(a)f(b)^{-1} \in K$. Therefore, $ab^{-1} \in H$ and so H is a subgroup of G containing Ker f, i.e., $H \in \mathcal{H}$. Hence, f^* maps $\mathcal{H}$ onto $\mathcal{K}$. Let H_1, $H_2 \in \mathcal{H}$. Suppose $f^*(H_1) = f^*(H_2)$. Let $h_1 \in H_1$. Then there exists $h_2 \in H_2$ such that $f(h_1) = f(h_2)$. This implies that $f(h_1 h_2^{-1}) = e_1$ and so $h_1 h_2^{-1} \in$ Ker $f \subseteq H_2$. Hence, $h_1 = (h_1 h_2^{-1})h_2 \in H_2$. Therefore, $H_1 \subseteq H_2$. Similarly, $H_2 \subseteq H_1$. Thus, $H_1 = H_2$ and so f^* is one-one. Clearly $H_1 \subseteq H_2$ if and only if $f^*(H_1) \subseteq f^*(H_2)$. In fact, since f^* is one-one, $H_1 \subset H_2$ if and only if $f^*(H_1) \subset f^*(H_2)$.

Suppose H is a normal subgroup of G such that Ker $f \subseteq H$. Let $K = f^*(H)$. We show that K is a normal subgroup of G. Let $f(a) \in G_1$ and $f(h) \in K$. Now $aha^{-1} \in H$ since H is a normal subgroup of G and so $f(a)f(h)f(a)^{-1} = f(aha^{-1}) \in K$. Hence, K is a normal subgroup of G_1. Let J be a normal subgroup of G_1 and $L \in \mathcal{H}$ be such that $f^*(L) = J$. Let $a \in G$ and $h \in L$. Then $f(aha^{-1}) = f(a)f(h)f(a)^{-1} \in J$ and so $aha^{-1} \in L$. This proves that L is a normal subgroup of G. ∎

Corollary 5.2.12 *Let N be a normal subgroup of a group G. Then every subgroup of G/N is of the form K/N, where K is a subgroup of G that contains N. Also, K/N is a normal subgroup of G/N if and only if K is a normal subgroup of G.*

Proof. Let $g : G \rightarrow G/N$ be the natural homomorphism. If $a \in G$, then $g(a) = aN$. From Theorem 5.2.11, we find that this homomorphism induces a one-one mapping g^* between the subgroups of G which contain Ker $g = N$ and the subgroups of G/N. Let H be a subgroup of G/N. Then there exists a subgroup K of G such that $N \subseteq K$ and $H = g^*(K) = \{g(a) \mid a \in K\} = K/N$. The last part follows from Theorem 5.2.11. ∎

The following example illustrates the correspondence theorem.

Example 5.2.13 *Let f be a homomorphism of $(\mathbf{Z}, +)$ onto $(\mathbf{Z}_{12}, +_{12})$ defined by $f(n) = [n]$ for all $n \in \mathbf{Z}$. Then for $\mathcal{H}$ and $\mathcal{K}$ of Theorem 5.2.11,*

$$\mathcal{H} = \{\langle 12 \rangle, \langle 6 \rangle, \langle 4 \rangle, \langle 3 \rangle, \langle 2 \rangle, \mathbf{Z}\}$$

and

$$\mathcal{K} = \{\langle [0] \rangle, \langle [6] \rangle, \langle [4] \rangle, \langle [3] \rangle, \langle [2] \rangle, \mathbf{Z}_{12}\}.$$

$$f^* : \langle 12 \rangle \rightarrow \langle [0] \rangle, \qquad f^* : \langle 3 \rangle \rightarrow \langle [3] \rangle,$$
$$f^* : \langle 2 \rangle \rightarrow \langle [2] \rangle, \qquad f^* : \langle 6 \rangle \rightarrow \langle [6] \rangle,$$
$$f^* : \langle 4 \rangle \rightarrow \langle [4] \rangle, \qquad f^* : \mathbf{Z} \rightarrow \mathbf{Z}_{12}.$$

The following diagram indicates the one-one inclusion preserving the correspondence property of f^*.

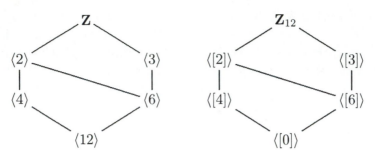

Now $\langle[9]\rangle = \{n[9] \mid n \in \mathbf{Z}\} \subseteq \{n[3] \mid n \in \mathbf{Z}\} = \langle[3]\rangle$. *Also,* $[3] = [27] = 3[9] \in \langle[9]\rangle$. *Therefore,* $\langle[3]\rangle \subseteq \langle[9]\rangle$. *Hence,* $\langle[3]\rangle = \langle[9]\rangle$. *Thus, the subgroup* $\langle 9 \rangle$ *of* $\mathbf{Z}$ *gets mapped to the subgroup* $\langle[3]\rangle$ *of* $\mathbf{Z}_{12}$ *by* f. *However, this does not contradict Theorem 5.2.11 since* $\langle 9 \rangle \not\supseteq \langle 12 \rangle$.

In the remainder of this section, we consider all isomorphisms of a group G onto itself. Recall that $\mathrm{Aut}(G)$ is the set of all automorphisms of G.

Theorem 5.2.14 *Let* G *be a group. Then* $(\mathrm{Aut}(G), \circ)$ *is a group, where* $\circ$ *denotes the composition of functions.*

Proof. Since $i_G \in \mathrm{Aut}(G)$, $\mathrm{Aut}(G) \neq \phi$. Let $f, g \in \mathrm{Aut}(G)$. Then $f \circ g$ is an automorphism by Exercise 14 (page 152) and Theorem 1.5.11. Hence, $f \circ g \in \mathrm{Aut}(G)$. Clearly i_G is the identity of $\mathrm{Aut}(G)$ and f^{-1} is the inverse of f. Also, $\circ$ is associative by Theorem 1.5.13. Consequently, $(\mathrm{Aut}(G), \circ)$ is a group. ∎

Theorem 5.2.15 *Let* G *be a group and* $a \in G$. *Define* $\theta_a : G \to G$ *by* $\theta_a(b) = aba^{-1}$ *for all* $b \in G$. *Then*
  (i) $\theta_a \in \mathrm{Aut}(G)$,
  (ii) $\theta_a \circ \theta_b = \theta_{ab}$ *for all* $a, b \in G$,
  (iii) $(\theta_a)^{-1} = \theta_{a^{-1}}$,
  (iv) for all $\alpha \in \mathrm{Aut}(G)$, $\alpha \circ \theta_a \circ \alpha^{-1} = \theta_{\alpha(a)}$.

Proof. (i) Let $c, d \in G$. Suppose $c = d$. Then $aca^{-1} = ada^{-1}$ or $\theta_a(c) = \theta_a(d)$. Therefore, θ_a is well defined. Now $\theta_a(cd) = a(cd)a^{-1} = (aca^{-1})(ada^{-1}) = \theta_a(c)\theta_a(d)$. This shows that θ_a is a homomorphism. Also, $c = \theta_a(a^{-1}ca)$, proving that θ_a is onto G. Suppose $\theta_a(c) = \theta_a(d)$. Then $aca^{-1} = ada^{-1}$ and so $c = d$. Thus, θ_a is one-one. Consequently, $\theta_a \in \mathrm{Aut}(G)$.
 (ii) Let $a, b \in G$. Then $(\theta_a \circ \theta_b)(c) = \theta_a(\theta_b(c)) = \theta_a(bcb^{-1}) = a(bcb^{-1})a^{-1} = (ab)c(ab)^{-1} = \theta_{ab}(c)$ for all $c \in G$. Hence, $\theta_a \circ \theta_b = \theta_{ab}$.
 (iii) Note that $\theta_a \circ \theta_{a^{-1}} = \theta_{aa^{-1}} = \theta_e = i_G$ and $\theta_{a^{-1}} \circ \theta_a = \theta_{a^{-1}a} = \theta_e = i_G$. Thus, $(\theta_a)^{-1} = \theta_{a^{-1}}$.
 (iv) Let $\alpha \in \mathrm{Aut}(G)$. Now $(\alpha \circ \theta_a \circ \alpha^{-1})(b) = \alpha(\theta_a(\alpha^{-1}(b))) = \alpha(a\alpha^{-1}(b)a^{-1}) = \alpha(a)\alpha(\alpha^{-1}(b))\alpha(a^{-1}) = \alpha(a)b(\alpha(a))^{-1} = \theta_{\alpha(a)}(b)$ for all $b \in G$. Hence,

$$\alpha \circ \theta_a \circ \alpha^{-1} = \theta_{\alpha(a)}. \blacksquare$$

The automorphism θ_a of Theorem 5.2.15 is called an **inner automorphism** of G. We denote by $\text{Inn}(G)$ the set of all inner automorphisms of G.

Theorem 5.2.16 *Let G be a group. Then $\text{Inn}(G)$ is a normal subgroup of $\text{Aut}(G)$.*

Proof. Since $i_G = \theta_e \in \text{Inn}(G)$, $\text{Inn}(G) \neq \phi$. By Theorem 5.2.15(i), $\text{Inn}(G) \subseteq \text{Aut}(G)$. Let $\theta_a, \theta_b \in \text{Inn}(G)$. Then $\theta_a \circ \theta_b^{-1} = \theta_a \circ \theta_{b^{-1}} = \theta_{ab^{-1}} \in \text{Inn}(G)$. Hence, $\text{Inn}(G)$ is a subgroup of $\text{Aut}(G)$ by Theorem 4.1.3. Let $\alpha \in \text{Aut}(G)$. Then by Theorem 5.2.15(iv), $\alpha \circ \theta_a \circ \alpha^{-1} = \theta_{\alpha(a)} \in \text{Inn}(G)$. Hence, $\text{Inn}(G)$ is a normal subgroup of $\text{Aut}(G)$. $\blacksquare$

Theorem 5.2.17 *Let G be a group and H be a subgroup of G. Then*

$$\frac{N(H)}{C(H)} \simeq a \text{ subgroup of } Aut(H),$$

where $N(H) = \{x \in G \mid xHx^{-1} = H\}$ is the normalizer of H and $C(H) = \{x \in G \mid xhx^{-1} = h \text{ for all } h \in H\}$ is the centralizer of H.

Proof. Define $f : N(H) \to \text{Aut}(H)$ by for all $a \in N(H)$,

$$f(a) = \theta_a|_H.$$

Then f is well defined. Let $a_1, a_2 \in N(H)$. Then $f(a_1 a_2) = \theta_{a_1 a_2}|_H = \theta_{a_1}|_H \circ \theta_{a_2}|_H = f(a_1) \circ f(a_2)$. Thus, f is a homomorphism. Now

$$\begin{aligned}
\text{Ker } f &= \{a \in G \mid f(a) = i_H\} \\
&= \{a \in G \mid \theta_a = i_H\} \\
&= \{a \in G \mid \theta_a(b) = i_H(b) \text{ for all } b \in H\} \\
&= \{a \in G \mid aba^{-1} = b \text{ for all } b \in H\} \\
&= \{a \in G \mid ab = ba \text{ for all } b \in H\} \\
&= C(H).
\end{aligned}$$

Thus, by the first isomorphism theorem, we have the desired result. $\blacksquare$

Corollary 5.2.18 *Let G be a group. Then*

$$\frac{G}{Z(G)} \simeq Inn(G).$$

Proof. Let $H = G$ in Theorem 5.2.17. Then we have $N(G) = G$ and $C(G) = Z(G)$. $\blacksquare$

5.2.1 Worked-Out Exercises

◇ **Exercise 1** Find all homomorphic images of the additive group $\mathbf{Z}$.

Solution: Let H be a homomorphic image of $(\mathbf{Z}, +)$. There exists a homomorphism f of $\mathbf{Z}$ onto H. By the first isomorphism theorem, $\mathbf{Z}/\mathrm{Ker}\ f \simeq H$. Since $\mathrm{Ker}\ f$ is a subgroup of $\mathbf{Z}$, $\mathrm{Ker}\ f = n\mathbf{Z}$ for some integer $n \geq 0$. Hence, $H \simeq \mathbf{Z}/n\mathbf{Z}$ for some integer $n \geq 0$. On the other hand, for any $n \geq 0$, $n\mathbf{Z}$ is a subgroup of $\mathbf{Z}$ and since $\mathbf{Z}$ is commutative, $n\mathbf{Z}$ is a normal subgroup of $\mathbf{Z}$. There exists a natural homomorphism f from $\mathbf{Z}$ onto $\mathbf{Z}/n\mathbf{Z}$ given by $f(m) = m + n\mathbf{Z}$ for all $m \in \mathbf{Z}$. This shows that $\mathbf{Z}/n\mathbf{Z}$ is a homomorphic image of $\mathbf{Z}$ for all $n \geq 0$. Consequently, the homomorphic images of $\mathbf{Z}$ are the groups (up to isomorphism) $\mathbf{Z}/n\mathbf{Z}$, $n \geq 0$. Now for $n = 0$, $\mathbf{Z}/n\mathbf{Z} \simeq \mathbf{Z}$ and for $n > 0$, $\mathbf{Z}/n\mathbf{Z} \simeq \mathbf{Z}_n$ (Exercise 2, page 164). Therefore, we conclude that the homomorphic images of $\mathbf{Z}$ are the cyclic groups $\mathbf{Z}$ and $\mathbf{Z}_n$, $n > 0$.

◇ **Exercise 2** If there exists an epimorphism of a finite group G onto the group $\mathbf{Z}_8$, show that G has normal subgroups of index 4 and 2.

Solution: Let $f : G \to \mathbf{Z}_8$ be an epimorphism. Then by the first isomorphism theorem, $G/\mathrm{Ker}\ f \simeq \mathbf{Z}_8$. Hence, $G/\mathrm{Ker}\ f$ is a cyclic group of order 8. Thus, $G/\mathrm{Ker}\ f$ has a normal subgroup H_1 of order 4 and a normal subgroup H_2 of order 2. By the correspondence theorem, there exist normal subgroups N_1 and N_2 of G such that $\mathrm{Ker}\ f \subseteq N_1$, $\mathrm{Ker}\ f \subseteq N_2$, $N_1/\mathrm{Ker}\ f = H_1$, and $N_2/\mathrm{Ker}\ f = H_2$. Thus,

$$8 = |G/\mathrm{Ker}\ f| = [G : \mathrm{Ker}\ f] = [G : N_1][N_1 : \mathrm{Ker}\ f] = [G : N_1]4.$$

This implies that $[G : N_1] = 2$. Similarly, $[G : N_2] = 4$.

◇ **Exercise 3** Show that $4\mathbf{Z}/12\mathbf{Z} \simeq \mathbf{Z}_3$.

Solution: Define $f : 4\mathbf{Z} \to \mathbf{Z}_3$ by $f(4n) = [n]$ for all $4n \in 4\mathbf{Z}$. One can show that f is an epimorphism. Then from the first isomorphism theorem, $4\mathbf{Z}/\mathrm{Ker}\ f \simeq \mathbf{Z}_3$. Now $\mathrm{Ker}\ f = \{4n \in 4\mathbf{Z} \mid f(4n) = [0]\} = \{4n \in 4\mathbf{Z} \mid [n] = [0]\} = 12\mathbf{Z}$.

Exercise 4 Let G be a finite group and f be an automorphism of G such that for all $a \in G$, $f(a) = a$ if and only if $a = e$. Show that for all $g \in G$, there exists $a \in G$ such that $g = a^{-1}f(a)$.

Solution: Let $G = \{a_1, a_2, \ldots, a_n\}$. Let $S = \{a_1^{-1}f(a_1), \ldots, a_n^{-1}f(a_n)\}$. Then $S \subseteq G$. Next, we show that all elements of S are distinct. Now $a_i^{-1}f(a_i) = a_j^{-1}f(a_j)$ if and only if $f(a_i)f(a_j)^{-1} = a_ia_j^{-1}$ if and only if $f(a_ia_j^{-1}) = a_ia_j^{-1}$ if and only if $a_ia_j^{-1} = e$ if and only if $a_i = a_j$. This shows that all elements of S are distinct and so $|S| = n$. Thus, $S = G$. Let $g \in G$. Then $g \in S$. Hence, $g = a^{-1}f(a)$ for some $a \in G$.

Exercise 5 Let G be a finite group and f be an automorphism of G such that for all $a \in G$, $f(a) = a$ if and only if $a = e$. Suppose that $f^2 = i_G$, where i_G denotes the identity map. Prove that G is commutative.

Solution: Let $g \in G$. By Worked-Out Exercise 4, $g = a^{-1}f(a)$ for some $a \in G$. Then $g = i_G(g) = f^2(a^{-1}f(a)) = f(f(a^{-1}f(a))) = f(f(a^{-1})f^2(a)) = f(f(a)^{-1}a) = f(g^{-1})$. This implies that $f(g) = g^{-1}$ for all $g \in G$. Let $a, b \in G$. Then $(ab)^{-1} = f(ab) = f(a)f(b) = a^{-1}b^{-1} = (ba)^{-1}$ and so $ab = ba$. Hence, G is commutative.

◇ **Exercise 6** Let H be a subgroup of index 2 in a finite group G. If the order of H is odd and every element of $G \backslash H$ is of order 2, prove that H is commutative.

Solution: Since $[G : H] = 2$, H is a normal subgroup of G. Now $G = H \cup Hg$, where $g \notin H$. Then $\circ(g) = 2$. Define $f : G \to G$ by for all $a \in G$, $f(a) = gag^{-1}$. Then f is an automorphism of G. Now $f^2(a) = f(f(a)) = f(gag^{-1}) = g(gag^{-1})g^{-1} = g^2ag^{-2} = a$ since $g^2 = e$. Hence, $f^2 = i_G$. Since H is a normal subgroup of G, $f(h) = aha^{-1} \in H$ for all $h \in H$. Thus, f is also an automorphism of H. Let $h \in H$. Suppose $f(h) = h$. Then $ghg^{-1} = h$ or $gh = hg$. Since $gh \notin H$, $\circ(gh) = 2$. Therefore, $h^2 = g^2h^2 = (gh)^2 = e$. Since the order of H is odd, $h^2 = e$ implies that $h = e$. Hence, $f(h) = h$ if and only if $h = e$. Thus, f is an automorphism of H such that $f^2 = i_G$ and $f(h) = h$ if and only if $h = e$. By Worked-Out Exercise 5, H is commutative.

◇ **Exercise 7** Show that $\text{Aut}(\mathbf{Z}_n) \simeq U_n$.

Solution: Define $\alpha : \text{Aut}(\mathbf{Z}_n) \to U_n$ by $\alpha(f) = f([1])$ for all $f \in \text{Aut}(\mathbf{Z}_n)$. Now $mf([1]) = f([m])$. Hence, $f([m]) = [0]$ if and only if m is divisible by n. Thus, $\circ(f([1])) = n$. This implies that $f([1]) \in U_n$ and so α is well defined. Let $f, g \in \text{Aut}(\mathbf{Z}_n)$. Then $\alpha(f \circ g) = (f \circ g)([1]) = f(g([1]))$. Suppose $g([1]) = [k]$. Then $\alpha(f \circ g) = f([k]) = kf([1]) = k[1]f([1]) = [k]f([1]) = f([1])g([1]) = \alpha(f)\alpha(g)$. Hence, α is a homomorphism. Now

$$\begin{aligned} \text{Ker } \alpha &= \{f \in \text{Aut}(\mathbf{Z}_n) \mid \alpha(f) = [1]\} \\ &= \{f \in \text{Aut}(\mathbf{Z}_n) \mid f([1]) = [1]\} \\ &= \{f \in \text{Aut}(\mathbf{Z}_n) \mid f \text{ is the identity map}\}. \end{aligned}$$

Hence, α is a monomorphism. Finally, we show that α is onto U_n. Let $[t] \in U_n$. Then t and n are relatively prime. Define $f : \mathbf{Z}_n \to \mathbf{Z}_n$ by $f([m]) = [mt]$ for all $[m] \in \mathbf{Z}_n$. Let $[r], [s] \in \mathbf{Z}_n$. Suppose $[r] = [s]$. Then $r - s = nq$ for some $q \in \mathbf{Z}$. Thus, $rt - st = nqt$. Hence, $[rt] = [st]$, proving that f is well defined. Clearly f is a homomorphism. Suppose $f([r]) = f([s])$. Then $[rt] = [st]$ and so n divides $rt - st = (r - s)t$. Since t and n are relatively prime, n divides $r - s$. Therefore,

$[r] = [s]$. This implies that f is one-one. Now let $[r] \in \mathbf{Z}_n$. Since $\gcd(n, t) = 1$, there exist $p, q \in \mathbf{Z}$ such that $1 = tp + nq$. Hence, $r = ptr + qnr$. This implies $[r] = [ptr]$. Now $[pr] \in \mathbf{Z}_n$. Thus, $f([pt] = [ptr] = [r]$. We therefore find that f is onto. Hence, $f \in \text{Aut}(\mathbf{Z}_n)$. Now $\alpha(f) = f([1]) = [t]$ shows that α is onto U_n. Thus, α is an isomorphism. Consequently, $\text{Aut}(\mathbf{Z}_n) \simeq U_n$.

5.2.2 Exercises

1. Let $\mathbf{R}^*$ be the multiplicative group of all nonzero real numbers and $T = \{1, -1\}$. Then T is a subgroup of $\mathbf{R}^*$. Prove that the quotient group $\mathbf{R}^*/T$ is isomorphic to the multiplicative group $\mathbf{R}^+$ of positive real numbers.

2. For any positive integer n, prove that $\mathbf{Z}/n\mathbf{Z} \simeq \mathbf{Z}_n$.

3. Show that $8\mathbf{Z}/56\mathbf{Z} \simeq \mathbf{Z}_7$.

4. Let G be a group and A and B be normal subgroups of G such that $A \simeq B$. Show by an example that $G/A \not\simeq G/B$.

5. For any two positive integers m, n such that $\gcd(m, n) = 1$, prove that $m\mathbf{Z}/mn\mathbf{Z} \simeq \mathbf{Z}_n$.

6. Let G be the group of symmetries of the square and K_4 the Klein 4-group.

   Show that the mapping $f : G \to K_4$ defines a homomorphism of G onto K_4, where $f(r_{180}) = f(r_{360}) = e$, $f(r_{90}) = f(r_{270}) = a$, $f(h) = f(v) = b$, $f(d_1) = f(d_2) = c$.

7. In Exercise 6, exhibit the one-one inclusion preserving correspondence between the subgroups of G containing $Z(G)$ and the subgroups of K_4.

8. Let G and K_4 be as in Exercise 6. Let g be the natural homomorphism of G onto $G/Z(G)$, where $Z(G)$ is the center of G. Prove that $Z(G) = \text{Ker } f$ and exhibit the isomorphism h of $G/Z(G)$ onto K_4 such that $f = h \circ g$.

9. Show that $\mathbf{Z}_8$ is not a homomorphic image of $\mathbf{Z}_{15}$.

10. Show that $\mathbf{Z}_9$ is not a homomorphic image of $\mathbf{Z}_3 \times \mathbf{Z}_3$.

11. Show that if there exists an epimorphism from a finite group G onto the group $\mathbf{Z}_{15}$, then G has normal subgroups of indices 5 and 3, respectively.

12. Partition the following collection of groups into subcollections of groups such that any two groups in the same subcollection are isomorphic.

    (i) $(\mathbf{Z}, +)$, (ii) $(\mathbf{Z}_6, +)$, (iii) $(\mathbf{Z}_2, +)$, (iv) S_2, (v) S_6, (vi) $(17\mathbf{Z}, +)$, (vii) $(3\mathbf{Z}, +)$, (vii) $(\mathbf{Q}, +)$, (ix) $(\mathbf{R}, +)$, (x)$(\mathbf{R}^*, \cdot)$, (xi) $(\mathbf{R}^+, \cdot)$, (xii) $(\mathbf{Q}^*, \cdot)$, (xiii) $(\mathbf{C}^*, \cdot)$, (xiv) $(\langle \pi \rangle, \cdot)$, where $\mathbf{R}^*$ denotes the set of nonzero real numbers, $\mathbf{Q}^*$ denotes the set of nonzero rational numbers, $\mathbf{C}^*$ denotes the set of nonzero real numbers, $\mathbf{R}^+$ denotes the set of positive real numbers, and $(\langle \pi \rangle, \cdot)$ is the cyclic subgroup of $(\mathbf{R}^+, \cdot)$ generated by π.

13. Show that

    (i) $\mathrm{Aut}(\mathbf{Z}_5) \simeq \mathbf{Z}_4$.

    (ii) $\mathrm{Aut}(\mathbf{Z}_8) \simeq$ Klein 4-group.

14. Find all automorphisms of the group $\mathbf{Z}_6$.

15. Show that $|\mathrm{Aut}(\mathbf{Z}_p)| = p - 1$, where p is a prime.

16. Prove that $\mathrm{Inn}(S_3) \simeq S_3 \simeq \mathrm{Aut}(S_3)$.

17. Determine $\mathrm{Aut}(S_4)$.

18. Let G be a cyclic group of order n and ϕ be the Euler ϕ-function. Prove that $|\mathrm{Aut}(G)| = \phi(n)$.

19. Let G be a group such that $Z(G) = \{e\}$. Prove that $Z(\mathrm{Aut}(G)) = \{e\}$.

20. Let G be a group and H be a subgroup of G. H is called a **characteristic** subgroup of G if $f(H) \subseteq H$ for all $f \in \mathrm{Aut}(G)$.

    (i) Show that every characteristic subgroup of G is a normal subgroup of G.

    (ii) Give an example of a group G and a subgroup H such that H is a normal subgroup of G, but H is not a characteristic subgroup of G.

    (iii) Show that $Z(G)$ is a characteristic subgroup of G.

    (iv) Let H and K be characteristic subgroups of G. Show that HK and $H \cap K$ are characteristic subgroups of G.

    (v) Let H and K be subgroups of G such that $H \subseteq K$. Show that if K is a normal subgroup of G and H is a characteristic subgroup of G, then H is a normal subgroup of G.

    (vi) Let H and K be subgroups of G such that $H \subseteq K$. Show that if H is a characteristic subgroup of K and K is a characteristic subgroup of G, then H is a characteristic subgroup of G.

    (vii) Suppose G is cyclic. Show that every subgroup of G is a characteristic subgroup of G.

21. Show that the only characteristic subgroups of $(\mathbf{Q}, +)$ are $\{0\}$ and $\mathbf{Q}$.

22. Which of the following statements are true? Justify.

   (i) Any epimorphism of $\mathbf{Z}$ onto $\mathbf{Z}$ is an isomorphism.

   (ii) Any epimorphism of a group G onto G is an isomorphism.

   (iii) The quotient group $4\mathbf{Z}/64\mathbf{Z}$ has five subgroups.

   (iv) $\mathbf{Z}_5$ has five homomorphic images.

   (v) $2\mathbf{Z}/6\mathbf{Z}$ is a subgroup of $\mathbf{Z}/6\mathbf{Z}$.

   (vi) There exist four subgroups of $\mathbf{Z}$ which contain $10\mathbf{Z}$ as a subgroup.

   (vii) Let G and H be two groups, A be a normal subgroup of G, and B be a normal subgroup of H. If $G \simeq H$ and $A \simeq B$, then $G/A \simeq H/B$.

5.3   The Groups D_4 and Q_8

In Section 5.1, we saw that there are two types of groups of order 4 and two types of groups of order 6. In this section, we wish to classify all noncommutative groups of order 8. We will consider finite commutative groups in Chapter 9. First we introduce two groups D_4 and Q_8 and study these groups in detail. The study of these groups will eventually lead us to the classification of noncommutative groups of order 8.

Definition 5.3.1 *A group G is called a **dihedral** group of degree 4 if G is generated by two elements a and b satisfying the relations*

$$o(a) = 4, \quad o\,(b) = 2, \quad and \ ba = a^3b.$$

Example 5.3.2 *Let T be the group of all 2×2 invertible matrices over $\mathbf{R}$ under usual matrix multiplication. Let G be the subgroup of T generated by the matrices*

$$A = \begin{bmatrix} 0 & 1 \\ -1 & 0 \end{bmatrix} \ and \ B = \begin{bmatrix} 0 & 1 \\ 1 & 0 \end{bmatrix}.$$

Then $o(A) = 4$ and $o(B) = 2$. Now

$$BA = \begin{bmatrix} 0 & 1 \\ 1 & 0 \end{bmatrix}\begin{bmatrix} 0 & 1 \\ -1 & 0 \end{bmatrix} = \begin{bmatrix} -1 & 0 \\ 0 & 1 \end{bmatrix}$$

and

$$A^3B = \begin{bmatrix} 0 & -1 \\ 1 & 0 \end{bmatrix}\begin{bmatrix} 0 & 1 \\ 1 & 0 \end{bmatrix} = \begin{bmatrix} -1 & 0 \\ 0 & 1 \end{bmatrix}.$$

Thus, $BA = A^3B$. Hence, G is a dihedral group of degree 4.

Example 5.3.3 *Consider S_4. Let G be the subgroup of S_4 such that G is generated by the permutations*

$$a = (1\ 2\ 3\ 4)\ and\ b = (2\ 4).$$

Then $a^2 = (1\ 3) \circ (2\ 4)$, $a^3 = (1\ 4\ 3\ 2)$, $a^4 = e$, $b^2 = e$, and $b \circ a = (1\ 4) \circ (2\ 3) = a^3 \circ b$. Hence, $\circ(a) = 4$, $\circ(b) = 2$, and $b \circ a = a^3 \circ b$. Thus, G is a dihedral group of degree 4.

The following theorem reveals some interesting properties of D_4. These properties are similar to the properties listed in Example 4.1.18 for D_3.

Theorem 5.3.4 *Let G be a dihedral group of degree 4 generated by the elements a and b such that*

$$\circ(a) = 4,\ \ \circ(b) = 2,\ \ and\ ba = a^3 b.$$

Then the following assertions hold.
(i) Every element of G is of the form $a^i b^j$, $0 \le i < 4$, $0 \le j < 2$.
(ii) G has exactly eight elements, i.e., $|G| = 8$.
(iii) G is a noncommutative group.

Proof.   (i) Since $G = \langle a, b \rangle$,

$$G = \{a^{i_1} b^{j_1} a^{i_2} b^{j_2} \cdots a^{i_n} b^{j_n} \mid i_t, j_t \in \mathbf{Z}, 1 \le t \le n,\ n \in \mathbf{N}\}.$$

Since $ba = a^3 b$, it follows that every element of G is of the form $a^n b^m$, where $n, m \in \mathbf{Z}$. Now $a^4 = e$, $b^2 = e$, $a^{-1} = a^3$, and $b^{-1} = b$. This implies that every element of G is of the form $a^i b^j$, $0 \le i < 4$, $0 \le j < 2$.

(ii) By (i), every element of G is of the form $a^i b^j$, $0 \le i < 4$, $0 \le j < 2$. Thus, $|G| \le 8$. Since $\circ(a) = 4$, it follows that e, a, a^2, a^3 are distinct elements of G. Then $b, ab, a^2 b, a^3 b$ are also distinct elements of G. Also, since $a^{-1} = a^3$, $b^{-1} = b$, and $a \ne b \ne e$,

$$\{e, a, a^2, a^3\} \cap \{b, ab, a^2 b, a^3 b\} = \phi.$$

Thus, $G = \{e, a, a^2, a^3, b, ab, a^2 b, a^3 b\}$. Hence, G has eight elements.
    (iii) Suppose $ab = ba$. Then $ab = a^3 b$. This implies that $a^2 = e$, which is a contradiction. Hence, $ab \ne ba$, proving that G is noncommutative. ∎

It is easy to see that any two dihedral groups of degree 4 are isomorphic. Hence, there exists only one dihedral group (up to isomorphism) of degree 4. We denote a dihedral group of degree 4 by D_4.
    We now describe all subgroups of D_4 and draw the lattice diagram of subgroups of D_4.

In D_4,
$$\circ(a) = 4, \circ(a^2) = 2, \circ(a^3) = 4, \circ(b) = 2,$$
$$(ab)^2 = abab = aa^3bb = e,$$
$$(a^2b)^2 = a^2ba^2b = a^2(a^3b)ab = abab = e,$$
$$(a^3b)^2 = a^3ba^3b = a^3(a^3b)a^2b = a^2ba^2b = e.$$

From this, it follows that $H_1 = \{e, a^2\}$, $H_2 = \{e, b\}$, $H_3 = \{e, ab\}$, $H_4 = \{e, a^2b\}$, and $H_5 = \{e, a^3b\}$ are subgroups of order 2. By Lagrange's theorem, D_4 has no subgroups of order 3, 5, 6, or 7. Now

$$T_1 = \{e, a, a^2, a^3\}$$
$$T_2 = \{e, a^2, b, a^2b\}$$
$$T_3 = \{e, ab, a^2, a^3b\}$$

are subgroups of order 4. We ask the reader to verify that $\{e\}$, H_1, H_2, H_3, H_4, H_5, T_1, T_2, T_3, and D_4 are the only subgroups of D_4. Hence, the lattice diagram of the subgroup lattice of D_4 is the following:

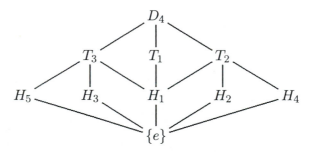

It is interesting to note in D_4 that H_5 is a normal subgroup of T_3 and T_3 is a normal subgroup of D_4, but H_5 is not a normal subgroup of D_4. We also note that every nontrivial subgroup of D_4 is of order 2 or 4. Therefore, every nontrivial subgroup of D_4 is commutative. However, since T_2 is a nontrivial subgroup of D_4 and T_2 is not cyclic, it follows that not every nontrivial subgroup of D_4 is cyclic. Finally, we also note that D_4 is isomorphic to Sym, the group of symmetries of a square (page 69). This follows from Theorem 5.3.4 and the group table of the group of symmetries of the square given on page 70.

Next, we consider Q_8.

Definition 5.3.5 *A group G is called a **quaternion** group if G is generated by two elements a, b satisfying the relation*

$$\circ(a) = 4, \quad a^2 = b^2, \quad and \quad ba = a^3b.$$

Example 5.3.6 *Let T be the group of all 2×2 invertible matrices over $\mathbf{C}$ under usual matrix multiplication. Let G be the subgroup of T generated by the matrices*

$$A = \begin{bmatrix} 0 & 1 \\ -1 & 0 \end{bmatrix} \quad and \quad B = \begin{bmatrix} 0 & i \\ i & 0 \end{bmatrix}.$$

Then $o(A) = 4$ and

$$A^2 = \begin{bmatrix} -1 & 0 \\ 0 & -1 \end{bmatrix} = B^2.$$

Now

$$BA = \begin{bmatrix} 0 & i \\ i & 0 \end{bmatrix} \cdot \begin{bmatrix} 0 & 1 \\ -1 & 0 \end{bmatrix} = \begin{bmatrix} -i & 0 \\ 0 & i \end{bmatrix}$$

and

$$A^3 B = \begin{bmatrix} 0 & -1 \\ 1 & 0 \end{bmatrix} \cdot \begin{bmatrix} 0 & i \\ i & 0 \end{bmatrix} = \begin{bmatrix} -i & 0 \\ 0 & i \end{bmatrix}.$$

Thus, $BA = A^3 B$. Hence, G is a quaternion group.

We leave the proof of the following theorem, which is similar to the proof of Theorem 5.3.4, as an exercise.

Theorem 5.3.7 *Let G be a quaternion group generated by the elements a and b such that*

$$o(a) = 4, \ a^2 = b^2, \ and \ ba = a^3 b.$$

Then the following assertions hold.
(i) Every element of G is of the form $a^i b^j$, $0 \le i < 4$, $0 \le j < 2$.
(ii) G has exactly eight elements, i.e., $|G| = 8$.
(iii) G is a noncommutative group. ∎

It is easy to see that any two quaternion groups are isomorphic. Hence, there exists only one quaternion group (up to isomorphism) and we denote it by Q_8.

Next, we determine all subgroups of Q_8.

Let $Q_8 = \langle a, b \rangle$, where $o(a) = 4$, $a^2 = b^2$, and $ba = a^3 b$. Then

$$Q_8 = \{e, a, a^2, a^3, b, ab, a^2 b, a^3 b\}.$$

In Q_8,

$$o(a) = 4, o(a^2) = 2, o(a^3) = 4, o(b) = 4.$$

Now

$$(ab)^2 = abab = aa^3 bb = b^2 = a^2.$$

Thus, $\circ(ab) = 4$. Also,

$$(a^2b)^2 = a^2ba^2b = a^2(a^3b)ab = a^5bab = abab$$

and

$$(a^3b)^2 = a^3ba^3b = a^3(a^3b)a^2b = a^2ba^2b.$$

Hence, $\circ(a^2b) = 4$ and $\circ(a^3b) = 4$. It now follows that $H_0 = \{e\}$, $H_1 = \{e, a^2\}$, $H_2 = \{e, a, a^2, a^3\}$, $H_3 = \{e, ab, a^2, a^3b\}$, and $H_4 = \{e, b, a^2, a^2b\}$ are subgroups of Q_8. We ask the reader to verify that H_0, H_1, H_2, H_3, H_4, and Q_8 are the only subgroups of Q_8. Thus, the lattice diagram of the subgroup lattice of Q_8 is the following:

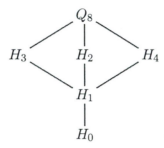

Since $[Q_8 : H_2] = [Q_8 : H_3] = [Q_8 : H_4] = 2$, H_2, H_3, and H_4 are normal subgroups of Q_8. Now $ba^2b^{-1} = baab^{-1} = a^3bab^{-1} = a^3a^3bb^{-1} = a^2 \in H_1$. Since $Q_8 = \langle a, b \rangle$, H_1 is a normal subgroup of Q_8. Thus, every subgroup of Q_8 is a normal subgroup of G. It is also interesting to observe that all proper subgroups of Q_8 are cyclic.

Theorem 5.3.8 $D_4 \not\cong Q_8$.

Proof. We note from the above discussion that Q_8 contains six elements of order 4 while D_4 contains only two elements of order 4. Hence, $D_4 \not\cong Q_8$. ∎

The next theorem classifies all noncommutative groups of order 8.

Theorem 5.3.9 *There exist (up to isomorphism) only two noncommutative nonisomorphic groups of order 8.*

Proof. Let G be a noncommutative group of order 8. Since $|G|$ is even, there exists an element $u \in G$, $u \neq e$, such that $u^2 = e$. If $x^2 = e$ for all $x \in G$, then G is commutative, a contradiction. Thus, there exists $a \in G$ such that $a^2 \neq e$. Since $\circ(a)|8$, $\circ(a) = 4$ or 8. If $\circ(a) = 8$, then G is cyclic and hence commutative, a contradiction. Thus, $\circ(a) = 4$. Let $H = \{e, a, a^2, a^3\}$. Then H

is a subgroup of G of index 2 and so H is a normal subgroup of G. Let $b \in G$ be such that $b \notin H$. Then $G = H \cup Hb$ and $H \cap Hb = \phi$. This implies that

$$G = \{e, a, a^2, a^3, b, ab, a^2b, a^3b\} = \langle a, b \rangle .$$

Now $bab^{-1} \in H$. If $bab^{-1} = e$, then $a = e$, a contradiction. Thus, $bab^{-1} \neq e$. If $bab^{-1} = a$, then $ab = ba$ and hence G is commutative, a contradiction. If $bab^{-1} = a^2$, then $ba^2b^{-1} = (bab^{-1})^2 = a^4 = e$ and so $a^2 = e$, a contradiction. Therefore, $bab^{-1} = a^3$ and so $ba = a^3b$. Since $|G/H| = 2$ and $b \notin H$, $o(Hb) = 2$. Hence, $b^2 \in H$. If $b^2 = a$ or a^3, then $o(b) = 8$ and so G is commutative, a contradiction. Therefore, either $b^2 = e$ or $b^2 = a^2$. It now follows that if G is a noncommutative group of order 8, then either

$$G = \langle a, b \rangle \text{ such that } o(a) = 4, \ o(b) = 2 \text{ ,and } ba = a^3b$$

or

$$G = \langle a, b \rangle \text{ such that } o(a) = 4, \ b^2 = a^2, \text{ and } ba = a^3b.$$

In the first case, $G \simeq D_4$ and in the second case, $G \simeq Q_8$. ∎

5.3.1 Worked-Out Exercises

◇ **Exercise 1** Find $Z(D_4)$.

Solution: It is known that $Z(D_4)$ is a normal subgroup of D_4. Now D_4 has five normal subgroups: D_4, $\{e\}$, $H_1 = \{e, a^2\}$, $T_1 = \{e, a, a^2, a^3\}$, $T_2 = \{e, a^2, b, a^2b\}$, $T_3 = \{e, ab, a^2, a^3b\}$. Since $ab \neq ba$, D_4, T_1, and T_2 cannot be $Z(D_4)$. If $(ab)b = b(ab)$, then $a = (ba)b = a^3b^2 = a^3$ and so $a^2 = e$, a contradiction. Hence, $T_3 \neq Z(D_4)$. Now $a^2b = a^6b = a^3(a^3b) = a^3(ba) = (ba)a = ba^2$. Hence, $a^2 \in Z(D_4)$. Thus, $Z(D_4) = \{e, a^2\} = H_1$.

◇ **Exercise 2** Find $\text{Inn}(D_4)$.

Solution: By Corollary 5.2.18, $\text{Inn}(D_4) \simeq D_4/Z(D_4)$. Now $D_4/Z(D_4)$ is a group of order 4 and

$$D_4/Z(D_4) = \{eZ(D_4), \ aZ(D_4), \ bZ(D_4), \ abZ(D_4)\}.$$

Since $a^2 \in Z(D_4)$, $b^2 = e$, and $(ab)^2 = e$, we find that each nonidentity element of $D_4/Z(D_4)$ is of order 2. Hence, $D_4/Z(D_4) \simeq K_4$, the Klein 4-group.

5.3.2 Exercises

1. In D_4, find subgroups H and K such that K is a normal subgroup of H and H is a normal subgroup of D_4, but K is not a normal subgroup of D_4.

2. Show that Q_8 is the union of three subgroups each of index 2.

3. Find all homomorphic images of D_4.

4. Find all homomorphic images of Q_8.

5.4 Group Actions

As previously mentioned, the theory of groups first dealt with permutation groups. Later the notion of an abstract group was introduced in order to examine properties of permutation groups which did not refer to the set on which the permutations acted. However, one is primarily interested in permutation groups in geometry. Also, permutation groups are used in counting techniques that are important in finite group theory. An example of this can be seen in the proof of Lagrange's theorem. We extend the notion of a permutation on a set to a group action on a set. We use the notion of a group action on a set to determine, via counting techniques, important properties of finite groups.

Let G be a group and S a nonempty set. A **(left) action** of G on S is a function $\cdot : G \times S \rightarrow S$ (usually denoted by $\cdot(g, x) \rightarrow g \cdot x$) such that

(i) $(g_1 g_2) \cdot x = g_1 \cdot (g_2 \cdot x)$, and
(ii) $e \cdot x = x$, where e is the identity of G
for all $x \in S$, $g_1, g_2 \in G$.

Note: If no confusion arises, we write gx for $g \cdot x$.

If there is a left action of G on S, we say that G acts on S on the left and S is a **G-set.**

Example 5.4.1 *Let G be a permutation group on a set S. Define a left action of G on S by*

$$\sigma x = \sigma(x)$$

for all $\sigma \in G$, $x \in S$. Let $x \in S$. Now $ex = e(x) = x$, where e is the identity permutation on S. Let $\sigma_1, \sigma_2 \in G$. Then $(\sigma_1 \circ \sigma_2) \cdot x = (\sigma_1 \circ \sigma_2)(x) = \sigma_1(\sigma_2(x)) = \sigma_1 \cdot (\sigma_2(x)) = \sigma_1 \cdot (\sigma_2 \cdot x)$. Hence, S is a G-set.

Example 5.4.2 *Let G be a group and H be a normal subgroup of G. Define a left action of G on H by*

$$(g, h) \rightarrow ghg^{-1}$$

for all $g \in G$, $h \in H$. We denote this by $g \cdot h = ghg^{-1}$. Let $h \in H$. Now $e \cdot h = ehe^{-1} = ehe = h$. Let $g_1, g_2 \in G$. Then $(g_1 g_2) \cdot h = (g_1 g_2)h(g_1 g_2)^{-1} = (g_1 g_2)h(g_2^{-1} g_1^{-1}) = g_1(g_2 h g_2^{-1})g_1^{-1} = g_1(g_2 \cdot h)g_1^{-1} = g_1 \cdot (g_2 \cdot h)$. Hence, H is a G-set.

Theorem 5.4.3 *Let S be a G-set, where G is a group and S is a nonempty set. Define a relation $\sim$ on S by for all $a, b \in S$,*

$$a \sim b \text{ if and only if } ga = b \text{ for some } g \in G.$$

Then $\sim$ is an equivalence relation on S.

Proof.   Since for all $a \in S$, $ea = a$, $a \sim a$ for all $a \in S$. Thus, $\sim$ is reflexive. Let $a, b, c \in S$. Suppose $a \sim b$. Then $ga = b$ for some $g \in G$, which implies that $g^{-1}b = g^{-1}(ga) = (g^{-1}g)a = ea = a$. Hence, $b \sim a$ and so $\sim$ is symmetric. Now suppose $a \sim b$ and $b \sim c$. Then there exist $g_1, g_2 \in G$ such that $g_1 a = b$ and $g_2 b = c$. Thus, $(g_2 g_1)a = g_2(g_1 a) = g_2 b = c$ and so $a \sim c$. Hence, $\sim$ is transitive. Consequently, $\sim$ is an equivalence relation. ■

Definition 5.4.4 *Let S be a G-set, where G is a group and S is a nonempty set. The equivalence classes determined by the equivalence relation of Theorem 5.4.3 are called the **orbits** of G on S.*

For $a \in S$, the orbit containing a is denoted by $[a]$.

Lemma 5.4.5 *Let G be a group and S be a G-set. For all $a \in S$, the subset*

$$G_a = \{g \in G \mid ga = a\}$$

is a subgroup of G.

Proof.   Let $a \in S$. Since $ea = a$, $e \in G_a$ and so $G_a \neq \phi$. Let $g, h \in G_a$. Then $ga = a$ and $ha = a$. This implies that $(gh)a = g(ha) = ga = a$ and so $gh \in G_a$. Now $h^{-1}a = h^{-1}(ha) = (h^{-1}h)a = ea = a$. Thus, $h^{-1} \in G_a$. Hence, G_a is a subgroup of G. ■

The subgroup G_a of Lemma 5.4.5 is called the **stabilizer** of a or the **isotropy group** of a.

Lemma 5.4.6 *Let G be a group and S be a G-set. For all $a \in S$,*

$$[G : G_a] = |[a]|.$$

Proof.   Let $a \in S$. Let $\mathcal{L}$ be the set of all left cosets of G_a in G. Now

$$[a] = \{b \in S \mid a \sim b\} = \{b \in S \mid ga = b \text{ for some } g \in G\} = \{ga \mid g \in G\}.$$

We now show that there exists a one-one function from $\mathcal{L}$ onto $[a]$. Define

$$f : \mathcal{L} \rightarrow [a]$$

by

$$f(gG_a) = ga$$

for all $gG_a \in \mathcal{L}$. Let $g_1, g_2 \in G$. Then $g_1 G_a = g_2 G_a$ if and only if $g_2^{-1}g_1 \in G_a$ if and only if $g_2^{-1}(g_1 a) = (g_2^{-1}g_1)a = a$ if and only if $g_1 a = g_2 a$. Thus, f is a one-one function from $\mathcal{L}$ into $[a]$. Let $b \in [a]$. Then there exists $g \in G$ such that $ga = b$. Thus, $f(gG_a) = ga = b$. This implies that f is onto $[a]$. Consequently, $[G : G_a] = |\mathcal{L}| = |[a]|$. ∎

Theorem 5.4.7 *Let G be a group and S be a G-set. If S is finite, then*

$$|S| = \sum_{a \in A}[G : G_a],$$

where A is a subset of S containing exactly one element from each orbit $[a]$.

Proof. By Theorem 5.4.3, S can be partitioned as the union of orbits. Therefore,

$$S = \cup_{a \in A}[a].$$

Hence,

$$|S| = \sum_{a \in A} |[a]| = \sum_{a \in A}[G : G_a] \text{ by Lemma 5.4.6. } ∎$$

Theorem 5.4.8 *Let G be a group and S be a G-set. Then the left action of G on S induces a homomorphism from G onto $A(S)$, where $A(S)$ is the group of all permutations of S.*

Proof. Let $g \in G$. Define $\tau_g : S \to S$ by $\tau_g(a) = ga$ for all $a \in S$. Let $a, b \in S$. Then $\tau_g(a) = \tau_g(b)$ if and only if $ga = gb$ if and only if $a = b$. Therefore, τ_g is a one-one function. Now $b = g(g^{-1}b) = \tau_g(g^{-1}b)$ and $g^{-1}b \in S$. This shows that τ_g is onto S. Thus, $\tau_g \in A(S)$. Let $g_1, g_2 \in G$. Then $\tau_{g_1 g_2}(a) = (g_1 g_2)a = g_1(g_2 a) = \tau_{g_1}(g_2 a) = \tau_{g_1}(\tau_{g_2}(a)) = (\tau_{g_1} \circ \tau_{g_2})(a)$ for all $a \in S$. This implies that $\tau_{g_1 g_2} = \tau_{g_1} \circ \tau_{g_2}$. Define

$$\psi : G \to A(S)$$

by

$$\psi(g) = \tau_g$$

for all $g \in G$. Then ψ is a function. Now $\psi(g_1 g_2) = \tau_{g_1 g_2} = \tau_{g_1} \circ \tau_{g_2} = \psi(g_1) \circ \psi(g_2)$ for all $g_1, g_2 \in G$. This proves that ψ is a homomorphism. ∎

The following corollary, which is known as the extended Cayley's theorem, follows from the above theorem.

Corollary 5.4.9 *Let G be a group and H be a subgroup of G. Let $S = \{aH \mid a \in G\}$. Then there exists a homomorphism ψ from G into $A(S)$ (the group of all permutations on S) such that $\text{Ker } \psi \subseteq H$.*

Proof.   First we note that S is a G-set, where the left action of G on S is defined by $g(aH) = (ga)H$ for all $g \in G$. This left action induces the homomorphism ψ of Theorem 5.4.8. Now

$$\begin{aligned}
\text{Ker } \psi &= \{g \in G \mid \psi(g) = \tau_g = \text{ the identity mapping on } S\} \\
&= \{g \in G \mid \tau_g(aH) = aH \text{ for all } aH \in S\} \\
&= \{g \in G \mid g(aH) = aH \text{ for all } aH \in S\}.
\end{aligned}$$

Let $g \in \text{Ker } \psi$. Then $g(aH) = aH$ for all $aH \in S$. In particular, $gH = H$. Thus, $g \in H$. Hence, Ker $\psi \subseteq H$. ∎

Corollary 5.4.10 *Let G be a finite group and H be a proper subgroup of G of index n such that $|G|$ does not divide $n!$ Then G contains a nontrivial normal subgroup.*

Proof.   From Corollary 5.4.9, Ker $\psi \subseteq H$ and $G/\text{Ker } \psi$ is isomorphic to a subgroup of S_n, where ψ is as defined in Corollary 5.4.9. Therefore, $|G/\text{Ker } \psi|$ divides $n!$ But $|G|$ does not divide $n!$ Hence, $|\text{Ker } \psi| \neq 1$, proving that Ker ψ is a nontrivial normal subgroup of G.

Definition 5.4.11 *Let G be a group and S be a G-set. Let $a \in S$, $g \in G$. Then a is called **fixed** by g if $ga = a$. If $ga = a$ for all $g \in G$, then a is called fixed by G.*

Theorem 5.4.12 (Burnside) *Let S be a finite nonempty set and G be a finite group. If S is a G-set, then the number of orbits of G is*

$$\frac{1}{|G|} \sum_{g \in G} F(g),$$

where $F(g)$ is the number of elements of S fixed by g.

Proof.   Let $T = \{(g, a) \in G \times S \mid ga = a\}$. Since $F(g)$ is the number of elements $a \in S$ such that $(g, a) \in T$, it follows that $|T| = \sum_{g \in G} F(g)$. Also, $|G_a|$ is the number of elements $g \in G$ such that $(g, a) \in T$. Hence, $|T| = \sum_{a \in S} |G_a|$.

Let $S = [a_1] \cup [a_2] \cup \cdots \cup [a_k]$, where $\{[a_1], [a_2], \ldots, [a_k]\}$ is the set of all distinct orbits of G on S. Then

$$\sum_{g \in G} F(g) = \sum_{a \in [a_1]} |G_a| + \sum_{a \in [a_2]} |G_a| + \cdots + \sum_{a \in [a_k]} |G_a|.$$

Suppose a, b are in the same orbit. Then $[a] = [b]$ and $[G : G_a] = |[a]| = |[b]| = [G : G_b]$. This implies

$$\frac{|G|}{|G_a|} = \frac{|G|}{|G_b|}$$

and so $|G_a| = |G_b|$. Thus,

$$
\begin{aligned}
\sum_{g \in G} F(g) &= |[a_1]|\,|G_{a_1}| + |[a_2]|\,|G_{a_2}| + \cdots + |[a_k]|\,|G_{a_k}| \\
&= \frac{|G|}{|G_{a_1}|}|G_{a_1}| + \frac{|G|}{|G_{a_2}|}|G_{a_2}| + \cdots + \frac{|G|}{|G_{a_k}|}|G_{a_k}| \\
&= k\,|G|,
\end{aligned}
$$

where k is the number of distinct orbits. Consequently,

$$
k = \frac{1}{|G|} \sum_{g \in G} F(g). \quad \blacksquare
$$

5.4.1 Worked-Out Exercises

$\diamond$ **Exercise 1** Let S be a finite G-set, where G is a group of order p^n (p a prime). Let $S_0 = \{a \in S \mid ga = a \text{ for all } g \in G\}$. Show that

$$
|S| \equiv_p |S_0|.
$$

Solution: By Lemma 5.4.7,

$$
|S| = \sum_{a \in A} [G : G_a],
$$

where A is a subset of S containing exactly one element from each orbit $[a]$ of G. Now $a \in S_0$ if and only if $ga = a$ for all $g \in G$, i.e., if and only if $[a] = \{a\}$. Hence,

$$
|S| = |S_0| + \sum_{a \in A \setminus S_0} \frac{|G|}{|G_a|}.
$$

Since $|G_a| \neq |G|$ for all $a \in A \setminus S_0$, $\frac{|G|}{|G_a|}$ is some power of p for all $a \in A \setminus S_0$. Thus, $\frac{|G|}{|G_a|}$ is divisible by p, proving that $|S| \equiv_p |S_0|$.

$\diamond$ **Exercise 2** Let S be a finite G-set, where G is a group of order p^n (p a prime) such that p does not divide $|S|$. Show that there exists $a \in S$ such that a is fixed.

Solution: Let $S_0 = \{a \in S \mid ga = a \text{ for all } g \in G\}$. By Worked-Out Exercise 1, $|S| \equiv_p |S_0|$. Since p does not divide $|S|$, p does not divide $|S_0|$. Thus, $|S_0| \neq 0$. This shows that there exists $a \in S_0$. Thus, a is fixed by G.

$\diamond$ **Exercise 3** Let G be a finite group and H be a subgroup of G such that $|H| = p^k$, where p is a prime and k is a nonnegative integer.

(i) Show that
$$
[G : H] \equiv_p [N(H) : H],
$$
where $N(H) = \{g \in G \mid gHg^{-1} = H\}$.

(ii) If $p | [G : H]$, show that $N(H) \neq H$.

Solution: (i) Let $S = \{xH \mid x \in G\}$. Define a left action of H on S by $h(xH) = (hx)H$ for all $h \in H$, $xH \in S$. Then S is an H-set. Let $S_0 = \{xH \in S \mid h(xH) = xH \text{ for all } h \in H\}$. By Worked-Out Exercise 1, $|S| \equiv_p |S_0|$. Now $xH \in S_0$ if and only if $h(xH) = xH$ for all $h \in H$ if and only if $x^{-1}hx \in H$ for all $h \in H$ if and only if $x^{-1}Hx \subseteq H$. Now $|x^{-1}Hx| = |H|$. Hence, $xH \in S_0$ if and only if $x^{-1}Hx \subseteq H$ if and only if $x^{-1}Hx = H$ (since H is finite and $|x^{-1}Hx| = |H|$) if and only if $x \in N(H)$. This shows that S_0 is the set of all left cosets of H in $N(H)$. Thus, $|S_0| = [N(H) : H]$. Also, $|S| = [G : H]$. Hence, $[G : H] \equiv_p [N(H) : H]$.

(ii) By (i), $[G : H] \equiv_p [N(H) : H]$. Now p divides $[G : H]$. Thus, p divides $[N(H) : H]$. Since $[N(H) : H] \geq 1$, it follows that $N(H) \neq H$.

Exercise 4 Let G be a finite group. Let H be a subgroup of G of index p, where p is the smallest prime dividing the order of G. Show that H is a normal subgroup of G.

Solution: Let $S = \{aH \mid a \in G\}$. Since $[G : H] = p$, $|S| = p$. Thus, $|A(S)| = p!$, where $A(S)$ is the group of all permutations on S. Define a left action of G on S by $g(aH) = (ga)H$ for all $g \in G$, $aH \in S$. Now $e(aH) = aH$ and $(g_1g_2)aH = ((g_1g_2)a)H = g_1(g_2aH)$. Hence, S is a G-set. Now the left action induces a homomorphism $\psi : G \to A(S)$ defined by $\psi(g) = \tau_g$, where $\tau_g(aH) = (ga)H$ for all $g \in G$, $aH \in S$. Let $g \in \text{Ker } \psi$. Then $g(aH) = eH$ for all $aH \in S$, in particular, $gH = H$. Hence, $g \in H$. Thus, $\text{Ker } \psi \subseteq H$. Now $G/\text{Ker } \psi$ is isomorphic to a subgroup of $A(S)$. Therefore, $|G/\text{Ker } \psi|$ divides $|A(S)| = p!$. Let $|G/\text{Ker } \psi| = n$. Then $n = [G : H][H : \text{Ker } \psi] \geq p$. Let $n = p_1 p_2 \cdots p_k$, where p_i are prime integers, $i = 1, 2, \ldots, k$. Since p_i divides $|G|$ and p is the smallest prime dividing the order of G, $p_i \geq p$ for all $i = 1, 2, \ldots, k$. Since n divides $p!$, we have each p_i divides $p!$. Since each p_i is a prime and $p_i \geq p$, we must have $i = 1$ and $p_i = p$. Thus, $n = p$. This implies that $[H : \text{Ker } \psi] = 1$. Hence, $H = \text{Ker } \psi$ and so H is a normal subgroup of G.

$\diamond$ **Exercise 5** Let G be a group of order pn, p a prime, and $p > n$. If H is a subgroup of order p in G, prove that H is a normal subgroup of G.

Solution: Let $S = \{aH \mid a \in G\}$. Now $|S| = [G : H] = \frac{|G|}{|H|} = \frac{pn}{p} = n$. Define a left action of G on S by $g(aH) = (ga)H$ for all $g \in G$, $aH \in S$. Then S is a G-set. Now the left action induces a homomorphism $\psi : G \to A(S)$ defined by $\psi(g) = \tau_g$, where $\tau_g(aH) = (ga)H$ for all $g \in G$, $aH \in S$. As in Worked-Out Exercise 4, $\text{Ker } \psi \subseteq H$. Since $|H| = p$, either $\text{Ker } \psi = \{e\}$ or $\text{Ker } \psi = H$. If $\text{Ker } \psi = \{e\}$, then G is isomorphic to a subgroup of $A(S)$. This implies that $|G|$ divides $|A(S)|$, i.e., $pn|n!$. Therefore, $p|(n-1)!$ Since $p > n$, p does not divide $(n-1)!$ Thus, $\text{Ker } \psi = H$. Hence, H is a normal subgroup of G.

Exercise 6 Let G be a group. Show that G is isomorphic to a subgroup of $A(G)$. (This is Cayley's theorem. Here we want to prove this result by the group action method.)

Solution: G is a G-set, where the left action of G on G is defined by the group operation. This left action induces a homomorphism $\psi : G \rightarrow A(G)$ defined by $\psi(g) = \tau_g$, where $\tau_g(a) = ga$ for all $a, g \in G$. Now Ker $\psi = \{g \in G \mid \tau_g = $ identity permutation on $G\} = \{g \in G \mid ga = a$ for all $a \in G\} = \{e\}$. Hence, ψ is a monomorphism.

$\diamondsuit$ **Exercise 7** Let G be a group of order $2m$, where m is an odd integer. Show that G has a normal subgroup of order m.

Solution: By Cayley's theorem, G is isomorphic to a subgroup H of $A(G)$, where the isomorphism $\psi : G \rightarrow A(G)$ is given by $\psi(g) = \tau_g$, $\tau_g(a) = ga$ for all $a, g \in G$. Since G is of even order, there exists $g \in G$ such that $o(g) = 2$. Now $\tau_g(a) = ga$ and $\tau_g(ga) = g^2 a = a$. Hence, τ_g is the product of transpositions of the form $(a \ ga)$. Since $|G| = 2m$, the number of transpositions appearing in the factorization of τ_g is m. Thus, τ_g is an odd permutation. Therefore, H contains an odd permutation. Define

$$f : H \rightarrow \{1, -1\}$$

by for all $\sigma \in H$,

$$f(\sigma) = \begin{cases} 1 & \text{if } \sigma \text{ is an even permutation} \\ -1 & \text{if } \sigma \text{ is a odd permutation} \end{cases}$$

where $\{1, -1\}$ is a group under multiplication. Then f is an epimorphism of H onto $\{1, -1\}$. Hence,

$$H/\text{Ker } f \simeq \{-1, 1\}.$$

Thus,

$$2 = |\{-1, 1\}| = |H/\text{Ker } f| = \frac{|H|}{|\text{Ker } f|} = \frac{2m}{|\text{Ker } f|}.$$

Hence, $|\text{Ker } f| = m$. Consequently, H contains a normal subgroup of order m and so G contains a normal subgroup of order m.

5.4.2 Exercises

1. Show that $I_3 = \{1, 2, 3\}$ is a S_3-set, where the left action is defined by $\sigma a = \sigma(a)$ for all $\sigma \in S_3$, $a \in I_3$. Find all distinct orbits of S_3. Find G_1, G_2, and G_3.

2. Let H be a subgroup of order 11 and index 4 of a group G. Prove that H is a normal subgroup of G.

3. Let H be a subgroup of a group G of index n. If H does not contain any nontrivial normal subgroups of G, prove that H is isomorphic to a subgroup of S_n.

4. Let $G = GL(2, \mathbf{R})$ and $S = \mathbf{R}^2$. Show that S is a G-set under the left action defined by

$$\begin{bmatrix} a & b \\ c & d \end{bmatrix} (x, y) = (ax + by, \ cx + dy)$$

for all $\begin{bmatrix} a & b \\ c & d \end{bmatrix} \in G, \ (x, y) \in \mathbf{R}^2$.

5. Let G be a group of order 77 acting on a set S of 20 elements. Show that G must have a fixed point.

6. Let G be a group. The left action of G on the set G is defined by conjugation, i.e., $(g, x) \to gxg^{-1}$ for all $g, x \in G$. Show that the kernel of the homomorphism $\psi : G \to A(G)$ induced by this action is $Z(G)$.

7. Let G be a group of order 80 such that G has a subgroup of order 16. Show that G is not a simple group.

8. Show that a group of order 22 is not a simple group.

9. Show that there are no simple groups of orders 6, 10, 14, 26, 34, and 58.

10. Show that a group of order 8 cannot be a simple group.

11. Show that a simple group of order 63 cannot contain a subgroup of order 21.

12. Let G be a group of order 70 such that G has a subgroup of order 14. Show that G has a nontrivial normal subgroup.

Arthur Cayley(1821–1895) was born on August 16, 1821, in Cambridge, England. He was the second son. He entered Trinity College at the age of 17, as a pensioner. In 1842, he graduated as senior wrangler. Later he went to a law school and in 1849 he became a lawyer. As a lawyer, he made a comfortable living and in fourteen years, during which he practiced his law profession, he wrote approximately 300 mathematical papers.

In 1863, Cayley was elected to the new Sadlerian chair of pure mathematics at Cambridge, where he remained until his death. He died on January 26, 1895.

For most of his life, Cayley worked on mathematics, theoretical dynamics, and mathematical astronomy. In 1876, he published his only book, *Treatise* on *Elliptic* Functions. Cayley wrote 966 papers; there are thirteen volumes of his collected papers.

Cayley's mathematical style was terse. He usually wrote out his results and published them without delay. He, along with J. J. Sylvester, his lifelong friend, is considered to be the founder of invariant theory. He is also responsible for matrix theory. The square notation used for determinants is due to Cayley. He proved many important theorems of matrix theory, such as the Cayley-Hamilton theorem. He is one of the first mathematicians to consider geometry of more than three dimensions.

In 1854, Cayley published, "On the theory of groups depending on the symbolic equation $\theta^n = 1$." In this paper, he considered a group as a set of symbols, $1, \alpha, \beta, \ldots$, all of them different and such that the product of any two of them (no matter in what order), or the product of any one of them into itself, belongs to the set. This formulation of a group as a set of symbols and multiplications is different from the formulation considered by the earlier mathematicians. The paper is generally regarded as the earliest work on abstract group theory and Cayley is regarded as the founder of abstract group theory. He is best known for the theorem that every finite group is isomorphic to a suitable permutation group. In his article of 1854, he introduced a procedure for defining a finite group by listing its elements in the form of a multiplication table, known as a Cayley table. Cayley also proved a number of important theorems.

Chapter 6

Direct Product of Groups

6.1 External and Internal Direct Product

In Section 2.1, Exercise 25, we defined the direct product $G \times H$ of two groups G and H. In this section, we extend this concept to any finite family of groups and obtain their basic properties.

The notion of a direct product is used to factor a group into a product of smaller groups. This factorization gives structural properties of a group. In some cases, it allows for the complete characterization of a certain type of group. In Chapter 9, the concept of direct product is used to give a complete system of invariants for a finitely generated Abelian group, i.e., a finite set of positive integers which implies the isomorphism of any two finitely generated Abelian groups that have this set of integers.

Recall that $I_n = \{1, 2, \ldots, n\}$.

Let $\{G_i \mid i \in I_n\}$ be a family of groups. Let

$$G = G_1 \times G_2 \times \cdots \times G_n = \{(a_1, a_2, \ldots, a_n) \mid a_i \in G_i, i \in I_n\}.$$

Define $*$ on G as follows: for all $(a_1, a_2, \ldots, a_n), (b_1, b_2, \ldots, b_n) \in G$

$$(a_1, a_2, \ldots, a_n) * (b_1, b_2, \ldots, b_n) = (a_1 b_1, a_2 b_2, \ldots, a_n b_n).$$

In the following theorem, we show that $*$ is a binary operation on G and that the set G together with the binary operation $*$ is a group. We also obtain several important properties of G.

Theorem 6.1.1 *Let $\{G_i \mid i \in I_n\}$ be a family of groups and $G = G_1 \times G_2 \times \cdots \times G_n$. Let e_i be the identity of G_i for all $i \in I_n$. Then $(G, *)$, where $*$ is defined above, is a group with $e = (e_1, e_2, \ldots, e_n)$ the identity element, and for all $(a_1, a_2, \ldots, a_n) \in G$,*

$$(a_1, a_2, \ldots, a_n)^{-1} = (a_1^{-1}, a_2^{-1}, \ldots, a_n^{-1}).$$

Furthermore, let

$$H_i = \{(e_1, e_2, \ldots, e_{i-1}, a_i, e_{i+1}, \ldots, e_n) \mid a_i \in G_i\}$$

for all $i \in I_n$. Then the following assertions hold.

(i) H_i is a normal subgroup of G for all $i \in I_n$.

(ii) For all $a \in G$, a can be uniquely expressed as $a = h_1 h_2 \cdots h_n$, where $h_i \in H_i$, $i \in I_n$.

(iii) $H_i \cap (H_1 H_2 \cdots H_{i-1} H_{i+1} \cdots H_n) = \{e\}$ for all $i \in I_n$.

(iv) $G = H_1 H_2 \cdots H_n$.

Proof. First we note that $*$ is single-valued and if $(a_1, \ldots, a_n)$, $(b_1, \ldots, b_n)$ $\in G$, then $(a_1, \ldots, a_n) * (b_1, \ldots, b_n) = (a_1 b_1, \ldots, a_n b_n) \in G$ since $a_i b_i \in G_i$ for all i. Thus, $*$ is a binary operation on G. We ask the reader to verify that $*$ is associative. Now $e = (e_1, e_2, \ldots, e_n) \in G$ and for all $a = (a_1, a_2, \ldots, a_n) \in G$,

$$\begin{aligned}
ae &= (a_1, a_2, \ldots, a_n)(e_1, e_2, \ldots, e_n) \\
&= (a_1 e_1, a_2 e_2, \ldots, a_n e_n) \\
&= (a_1, a_2, \ldots, a_n) \\
&= a.
\end{aligned}$$

Similarly, $ea = a$. Hence, e is the identity of G. To show that every element of G has an inverse in G, let $(a_1, a_2, \ldots, a_n) \in G$. Then $(a_1^{-1}, a_2^{-1}, \ldots, a_n^{-1}) \in G$ since $a_i^{-1} \in G_i$ for all i and

$$\begin{aligned}
(a_1, a_2, \ldots, a_n)(a_1^{-1}, a_2^{-1}, \ldots, a_n^{-1}) &= (a_1 a_1^{-1}, a_2 a_2^{-1}, \ldots, a_n a_n^{-1}) \\
&= (e_1, e_2, \ldots, e_n) \\
&= e.
\end{aligned}$$

Similarly, $(a_1^{-1}, a_2^{-1}, \ldots, a_n^{-1})(a_1, a_2, \ldots, a_n) = e$. Thus, every element of G has an inverse. Consequently, $(G, *)$ is a group. We also note that by the uniqueness of the inverse of an element

$$(a_1, a_2, \ldots, a_n)^{-1} = (a_1^{-1}, a_2^{-1}, \ldots, a_n^{-1}).$$

(i) Let $i \in I_n$. Since $(e_1, e_2, \ldots, e_n) \in H_i$, $H_i \neq \phi$. Let $a = (e_1, \ldots, a_i, \ldots, e_n)$, $b = (e_1, \ldots, b_i, \ldots, e_n) \in H_i$. Then

$$\begin{aligned}
ab^{-1} &= (e_1, \ldots, a_i, \ldots, e_n)(e_1, \ldots, b_i, \ldots, e_n)^{-1} \\
&= (e_1, \ldots, a_i, \ldots, e_n)(e_1, \ldots, b_i^{-1}, \ldots, e_n) \\
&= (e_1, \ldots, a_i b_i^{-1}, \ldots, e_n) \in H_i.
\end{aligned}$$

Thus, H_i is a subgroup of G by Theorem 4.1.3. Let $g = (g_1, g_2, \ldots, g_n) \in G$. Then

$$\begin{aligned}
gag^{-1} &= (g_1, g_2, \ldots, g_n)(e_1, \ldots, a_i, \ldots, e_n)(g_1, g_2, \ldots, g_n)^{-1} \\
&= (g_1, g_2, \ldots, g_i a_i, \ldots, g_n)(g_1^{-1}, g_2^{-1}, \ldots, g_n^{-1}) \\
&= (e_1, \ldots, g_i a_i g_i^{-1}, \ldots, e_n) \in H_i \text{ since } g_i a_i g_i^{-1} \in G_i.
\end{aligned}$$

Hence, H_i is a normal subgroup of G.

(ii) Let $a = (a_1, a_2, \ldots, a_n) \in G$. Let $h_i = (e_1, \ldots, a_i, \ldots, e_n)$ for all $i \in I_n$. Then $a = h_1 h_2 \cdots h_n$. To show that the representation of a is unique, let $a = k_1 k_2 \cdots k_n$ be another representation of a, where $k_i \in H_i$ for all $i \in I_n$. Let $k_i = (e_1, \ldots, b_i, \ldots, e_n) \in H_i$ for all $i \in I_n$. Then

$$(a_1, a_2, \ldots, a_n) = h_1 h_2 \cdots h_n = a = k_1 k_2 \cdots k_n = (b_1, b_2, \ldots, b_n).$$

This implies that $a_i = b_i$ for all $i \in I_n$ and so $h_i = k_i$ for all $i \in I_n$. Hence, the representation of a is unique.

(iii) Suppose $a \in H_i \cap (H_1 \cdots H_{i-1} H_{i+1} \cdots H_n)$. Then $a \in H_i$ and

$$a \in H_1 \cdots H_{i-1} H_{i+1} \cdots H_n.$$

Since $a \in H_i$, $a = (e_1, \ldots, a_i, \ldots, e_n) \in H_i$ for some $a_i \in G_i$ and since

$$a \in H_1 \cdots H_{i-1} H_{i+1} \cdots H_n,$$

we have $a = h_1 h_2 \cdots h_{i-1} h_{i+1} \cdots h_n$, where $h_j = (e_1, \ldots, a_j, \ldots, e_n) \in H_j$ for some $a_j \in G_j$. Thus,

$$(e_1, \ldots, a_i, \ldots, e_n) = a = h_1 \cdots h_{i-1} h_{i+1} \cdots h_n = (a_1, \ldots, a_{i-1}, e_i, a_{i+1}, \ldots, a_n).$$

This implies that $a_i = e_i$ for all $i \in I_n$. Hence,

$$H_i \cap (H_1 H_2 \cdots H_{i-1} H_{i+1} \cdots H_n) = \{e\}.$$

(iv) The desired result follows from (ii). ∎

Definition 6.1.2 *The group G of Theorem 6.1.1 is called the **external direct product** of the groups G_i, $i = 1, 2, \ldots, n$.*

Theorem 6.1.1 motivates the following definition.

Definition 6.1.3 *Let G be a group and $\{N_i \mid i \in I_n\}$ be a family of normal subgroups of G. Then G is called the **internal direct product of** $N_1, N_2, \ldots, N_n$ if every $a \in G$ can be uniquely expressed as $a = a_1 a_2 \cdots a_n$, where $a_i \in N_i$ for all $i \in I_n$.*

Let $G = G_1 \times G_2 \times \cdots \times G_n$ be the external direct product of the groups G_i. Let H_i be defined as in Theorem 6.1.1. Then G is the internal direct product of $H_1, H_2, \ldots, H_n$ by Theorem 6.1.1(ii).

Theorem 6.1.4 *Let G be a group and $\{N_i \mid i \in I_n\}$ be a family of normal subgroups of G. Then G is an internal direct product of $\{N_i \mid i \in I_n\}$ if and only if $G = N_1 N_2 \cdots N_n$ and $N_i \cap (N_1 \cdots N_{i-1} N_{i+1} \cdots N_n) = \{e\}$ for all $i \in I_n$.*

Proof. Let G be an internal direct product of $\{N_i \mid i \in I_n\}$. Let $a \in G$. Then $a = a_1 a_2 \cdots a_n$ for some $a_i \in N_i$, $i \in I_n$. Thus, $a \in N_1 N_2 \cdots N_n$ and this implies that $G = N_1 N_2 \cdots N_n$. We now show that $N_i \cap (N_1 \cdots N_{i-1} N_{i+1} \cdots N_n) = \{e\}$ for all $i \in I_n$. Let $i \in I_n$ and $a \in N_i \cap (N_1 \cdots N_{i-1} N_{i+1} \cdots N_n)$. Then $a \in N_i$ and $a \in N_1 \cdots N_{i-1} N_{i+1} \cdots N_n$. This implies that we can write $a = a_1 a_2 \cdots a_{i-1} a_{i+1} \cdots a_n$ for some $a_j \in N_j$, $j \in I_n \backslash \{i\}$. Hence,

$$ee \cdots a \cdots e = a = a_1 a_2 \cdots a_{i-1} e a_{i+1} \cdots a_n$$

are two representations of a, where $a_j \in N_j$, $j \in I_n \backslash \{i\}$. Since the representation of a is unique, $a = e$. Hence, $N_i \cap (N_1 \cdots N_{i-1} N_{i+1} \cdots N_n) = \{e\}$.

Conversely, suppose $G = N_1 N_2 \cdots N_n$ and $N_i \cap (N_1 \cdots N_{i-1} N_{i+1} \cdots N_n) = \{e\}$ for all $i \in I_n$. Then $N_i \cap N_j = \{e\}$ for all $i \neq j$ and hence $uv = vu$ for all $u \in N_i$ and for all $v \in N_j$ by Exercise 12 (page 137). Let $a = a_1 a_2 \cdots a_n = b_1 b_2 \cdots b_n$ be two representations of a, where $a_i, b_i \in N_i$, $i \in I_n$. Then

$$
\begin{aligned}
e &= a^{-1} a \\
&= (a_1 a_2 \cdots a_n)^{-1} (b_1 b_2 \cdots b_n) \\
&= a_n^{-1} a_{n-1}^{-1} \cdots a_1^{-1} b_1 b_2 \cdots b_n \\
&= a_1^{-1} b_1 a_2^{-1} b_2 \cdots a_n^{-1} b_n
\end{aligned}
$$

since for all $i \neq j$ if $u \in N_i$ and $v \in N_j$, then $uv = vu$. This implies that

$$b_i^{-1} a_i = a_1^{-1} b_1 \cdots a_{i-1}^{-1} b_{i-1} a_{i+1}^{-1} b_{i+1} \cdots a_n^{-1} b_n \in N_i \cap N_1 N_2 \cdots N_{i-1} N_{i+1} \cdots N_n$$

for all $i \in I_n$. Since $N_i \cap N_1 N_2 \cdots N_{i-1} N_{i+1} \cdots N_n = \{e\}$, we must have $b_i^{-1} a_i = e$ or $a_i = b_i$ for all $i \in I_n$. Thus, a can be written uniquely as $a_1 a_2 \cdots a_n$, where $a_i \in N_i$, $i \in I_n$. Hence, G is an internal direct product of $\{N_i \mid i \in I_n\}$. ∎

In the following theorem, we show that if a group G is an internal direct product of a family of normal subgroups $\{N_i \mid i \in I_n\}$, then G can be viewed as an external direct product of the groups N_i's.

Theorem 6.1.5 *Let G be an internal direct product of a family of normal subgroups $\{N_i \mid i \in I_n\}$. Then*

$$G \simeq N_1 \times N_2 \times \cdots \times N_n.$$

Proof. Let $a \in G$. Then a can be expressed uniquely as $a = a_1 a_2 \cdots a_n$, where $a_i \in N_i$, $i \in I_n$. Define

$$f : G \to N_1 \times N_2 \times \cdots \times N_n$$

by

$$f(a) = (a_1, a_2, \ldots, a_n)$$

for all $a \in G$. From the definition of f, it follows that f is well defined and onto $N_1 \times N_2 \times \cdots \times N_n$. And from the uniqueness of the representation of a, it follows that f is one-one. We now show that f is a homomorphism. Let $a = a_1 a_2 \cdots a_n$ and $b = b_1 b_2 \cdots b_n$ be two elements of G, where $a_i, b_i \in N_i, i \in I_n$. Now $N_i \cap N_j = \{e\}$ for all $i \neq j$ and so $uv = vu$ for all $u \in N_i$, $v \in N_j$. This implies that

$$ab = a_1 a_2 \cdots a_n b_1 b_2 \cdots b_n = a_1 b_1 a_2 b_2 \cdots a_n b_n.$$

Thus,

$$\begin{aligned} f(ab) &= (a_1 b_1, a_2 b_2, \ldots, a_n b_n) \\ &= (a_1, a_2, \ldots, a_n)(b_1, b_2, \ldots, b_n) \\ &= f(a)f(b) \end{aligned}$$

and so f is a homomorphism. Consequently, $G \simeq N_1 \times N_2 \times \cdots \times N_n$. ∎

Considering Theorem 6.1.5, let us agree to write $G = N_1 \times N_2 \times \cdots \times N_n$ when G is an internal direct product of a family of normal subgroups $\{N_i \mid i \in I_n\}$.

6.1.1   Worked-Out Exercises

◇ **Exercise 1** Let G and G_1 be groups and $f : G \to G_1$ be a homomorphism. Let H be a normal subgroup of G. Suppose that $f|_H : H \to G_1$ is an isomorphism of H onto G_1. Prove that $G = H \times \mathrm{Ker}\ f$. Give an example to show that this result need not be true if H is not a normal subgroup.

Solution: Let $a \in G$. Then $f(a) \in G_1 = f(H)$. Thus, there exists $h \in H$ such that $f(a) = f(h)$. Now $f(a) = f(h)$ implies that $f(h^{-1}a) = e_1$ and hence $h^{-1}a \in \mathrm{Ker}\ f$. Therefore, there exists $b \in \mathrm{Ker}\ f$ such that $b = h^{-1}a$ or $a = hb$. Hence, $G = H\mathrm{Ker}\ f$. Suppose $a \in H \cap \mathrm{Ker}\ f$. Then $a \in H$ and $f(a) = e_1 = f(e)$. Since $f|_H$ is one-one, $f(a) = f(e)$ implies that $a = e$. Therefore, $H \cap \mathrm{Ker}\ f = \{e\}$. Thus, H and $\mathrm{Ker}\ f$ are normal subgroups of G such that $G = H\mathrm{Ker}\ f$ and $H \cap \mathrm{Ker}\ f = \{e\}$. Consequently, $G = H \times \mathrm{Ker}\ f$.

This result need not be true if H is not a normal subgroup of G. For let $G = S_3$ and $G_1 = \langle g' \rangle$ be such that $o(g') = 2$, i.e., G_1 is a cyclic group of order 2. Let $H = \langle (1\ 2) \rangle$. Define $f : G \to G_1$ by $f(e) = e$, $f(x) = e$ if x is an element of order 3, and $f(x) = g'$ if x is an element of order 2. Then $f|_H : H \to G_1$ is an isomorphism of H onto G_1. Now $\mathrm{Ker}\ f = \{e, (1\ 2\ 3), (1\ 3\ 2)\} = \langle (1\ 2\ 3) \rangle$. But $G \neq H \times \mathrm{Ker}\ f$ (see Exercise 14, page 188.)

Exercise 2 Let G be a group and H and K be subgroups of G such that $G = H \times K$. Let N be a normal subgroup of G such that $N \cap H = \{e\}$ and $N \cap K = \{e\}$. Prove that N is commutative.

Solution: Since $G = H \times K$, H and K are normal subgroups of G. Now for all $n \in N, h \in H, k \in K, nh = hn$, and $nk = kn$ by Exercise 12 (page 137). Let $a, b \in N$. Then there exist $h \in H, k \in K$ such that $b = hk$. Now $ab = a(hk) = (ah)k = (ha)k = h(ak) = h(ka) = (hk)a = ba$. Hence, N is commutative.

◇ **Exercise 3** Let G be a group and A and B be subgroups of G. If

(i) $G = AB$,

(ii) $ab = ba$ for all $a \in A$, $b \in B$, and

(iii) $A \cap B = \{e\}$,

prove that G is an internal direct product of A and B.

Solution: Let us first show that A and B are normal subgroups of G. For this, let $a \in A$, $g \in G$. There exist $c \in A$ and $b \in B$ such that $g = cb$ by (i). Now $gag^{-1} = (cb)a(cb)^{-1} = cbab^{-1}c^{-1} = cabb^{-1}c^{-1} = cac^{-1} \in A$. Hence, A is a normal subgroup of G. Similarly, B is a normal subgroup of G. Let $g \in G$. Then $g = ab$ for some $a \in A$, $b \in B$. Suppose $g = a_1 b_1$, where $a_1 \in A$, $b_1 \in B$. Then $ab = a_1 b_1$, which implies that $a_1^{-1}a = b_1 b^{-1} \in A \cap B = \{e\}$. Thus, $a = a_1$ and $b = b_1$. Therefore, we find that every element g of G can be expressed uniquely as $g = ab$, $a \in A$, $b \in B$. Consequently, G is an internal direct product of A, B.

◇ **Exercise 4** Let G be a cyclic group of order mn, where m, n are positive integers such that $\gcd(m, n) = 1$. Show that $G \simeq \mathbf{Z}_m \times \mathbf{Z}_n$.

Solution: Since m divides $|G|$ and G is cyclic, there exists a unique cyclic subgroup A of G of order m by Theorem 4.2.10. Similarly, there exists a unique cyclic subgroup B of G of order n. Now $|A \cap B|$ divides $|A| = m$ and $|A \cap B|$ divides $|B| = n$. Since $\gcd(m, n) = 1$, $|A \cap B| = 1$. Thus, by Theorem 4.3.15,

$$|AB| = \frac{|A| |B|}{|A \cap B|} = \frac{mn}{1} = mn = |G|.$$

Since $AB \subseteq G$, $|AB| = |G|$, and G is finite, we must have $G = AB$. Hence, $G = AB$, $A \cap B = \{e\}$, and A and B are normal subgroups of G. Thus, $G = A \times B \simeq \mathbf{Z}_m \times \mathbf{Z}_n$.

◇ **Exercise 5** Let A and B be two cyclic groups of order m and n, respectively. Show that $A \times B$ is a cyclic group if and only if $\gcd(m, n) = 1$.

Solution: Let $A = \langle a \rangle$ for some $a \in A$ and $B = \langle b \rangle$ for some $b \in B$. Suppose $\gcd(m, n) = 1$. Let $g = (a, b)$. Then $g^{mn} = (a, b)^{mn} = (a^{mn}, b^{mn}) = (e_A, e_B)$, where e_A denotes the identity of A and e_B denotes the identity of B. Suppose

$o(g) = t$. Then $(a, b)^t = (e_A, e_B)$. This implies that $a^t = e_A$ and $b^t = e_B$. Thus, $m|t$ and $n|t$. Since $\gcd(m, n) = 1$, $mn|t$. Hence, mn is the smallest positive integer such that $g^{mn} = e$. Thus, $o(g) = mn$. Now $|A \times B| = mn$ and $A \times B$ contains an element g of order mn. As a result, $A \times B$ is cyclic. Conversely, assume that $A \times B$ is cyclic and $\gcd(m, n) = d \neq 1$. Let $(a, b) \in A \times B$. Then $o(a)|m$ and $o(b)|n$. Now $\frac{mn}{d} = \frac{m}{d}n = m\frac{n}{d}$ is an integer and $\frac{mn}{d} < mn$. Also,

$$(a, b)^{\frac{mn}{d}} = (a^{m\frac{n}{d}}, b^{n\frac{m}{d}}) = (e_A, e_B).$$

Hence, $A \times B$ does not contain any element of order mn. This implies that $A \times B$ is not cyclic, a contradiction. Therefore, $\gcd(m, n) = 1$.

Exercise 6 Show that $|\text{Aut}(\mathbf{Z}_2 \times \mathbf{Z}_2)| = 6$.

Solution: First note that $\mathbf{Z}_2 \times \mathbf{Z}_2$ has four elements, $e = ([0], [0])$, $a = ([1], [0])$, $b = ([0], [1])$, $c = ([1], [1])$, and $o(a) = o(b) = o(c) = 2$. Let $f \in \text{Aut}(\mathbf{Z}_2 \times \mathbf{Z}_2)$. Then $o(f(x)) = o(x)$ for all $x \in \mathbf{Z}_2 \times \mathbf{Z}_2$. Hence, f maps $\{a, b, c\}$ onto $\{a, b, c\}$. Thus, f is a permutation of $\{a, b, c\}$. Since there are only six permutations of $\{a, b, c\}$, it follows that $|\text{Aut}(\mathbf{Z}_2 \times \mathbf{Z}_2)| \leq 6$. Now $a + b = c$, $a + c = b$, $b + c = a$, and $a + a = e = b + b = c + c$. Thus, any permutation of $\{a, b, c\}$ gives rise to an automorphism of $\mathbf{Z}_2 \times \mathbf{Z}_2$. For example, let $\alpha : a \to b$, $b \to c$, $c \to a$, and $e \to e$. Now $\alpha(a + b) = \alpha(c) = a$ and $\alpha(a) + \alpha(b) = b + c = a$. Therefore, $\alpha(a + b) = \alpha(a) + \alpha(b)$. Similarly, $\alpha(a + c) = \alpha(a) + \alpha(c)$, $\alpha(b + c) = \alpha(b) + \alpha(c)$, $\alpha(a + a) = \alpha(a) + \alpha(a)$, $\alpha(b + b) = \alpha(b) + \alpha(b)$, and $\alpha(c + c) = \alpha(c) + \alpha(c)$. Hence, α is an automorphism. Thus, $|\text{Aut}(\mathbf{Z}_2 \times \mathbf{Z}_2)| = 6$.

6.1.2 Exercises

1. Prove that the direct product of two groups A and B is commutative if and only if both groups A and B are commutative.

2. Let A, B, C, and D be four groups such that $A \simeq C$ and $B \simeq D$. Show that $A \times B \simeq C \times D$.

3. Let G be a group such that $G = H_1 \times H_2 \times \cdots \times H_n$, where H_i is a subgroup of G. Let K_i be a normal subgroup of G such that $K_i \subseteq H_i$, $1 \leq i \leq n$. Let $K = K_1 \times K_2 \times \cdots \times K_n$. Show that

$$\frac{G}{K} \simeq \frac{H_1}{K_1} \times \frac{H_2}{K_2} \times \cdots \times \frac{H_n}{K_n}.$$

4. Let G_i be a group, $1 \leq i \leq n$. Show that

$$Z(G_1 \times G_2 \times \cdots \times G_n) = Z(G_1) \times Z(G_2) \times \cdots \times Z(G_n).$$

5. Let G be a group and H and K be subgroups of G such that $G = H \times K$. Show that $G/K \simeq H$ and $G/H \simeq K$.

6. Let G be a finite cyclic group of order mn, where m and n are relatively prime. Let H and K be subgroups of G such that $|H| = m$ and $|K| = n$. Show that $G = H \times K$.

7. Prove that $\text{Aut}(\mathbf{Z}_2 \times \mathbf{Z}_2) \simeq S_3$.

8. Let G be a group and H and K be normal subgroups of G such that $G = HK$. Let $H \cap K = N$. Show that

$$G/N \simeq H/N \times K/N.$$

9. Prove that a finite Abelian group G is the internal direct product of subgroups H and K if and only if (i) $H \cap K = \{e\}$ and (ii) $|G| = |H||K|$.

10. Show that the Klein 4-group is isomorphic to the direct product of a cyclic group of order 2 with itself.

11. Show that a cyclic group of order 4 cannot be expressed as an internal direct product of two subgroups of order 2.

12. Show that a cyclic group of order 8 cannot be expressed as an internal direct product of two subgroups of order 4 and 2, respectively.

13. Can the cyclic group $\mathbf{Z}_{12}$ be expressed as an internal direct product of two proper subgroups?

14. Show that S_3 cannot be written as a direct product of proper subgroups.

15. Show that D_4 cannot be expressed as an internal direct product of two proper subgroups.

16. Consider the groups $\mathbf{Z}_2 \times S_3$, $\mathbf{Z}_2 \times \mathbf{Z}_6$, and $\mathbf{Z}_{12}$. Are any two of these groups isomorphic? Is any one noncommutative?

17. Show that the additive group $(\mathbf{Z}, +)$ cannot be expressed as an internal direct product of two nontrivial subgroups.

18. Show that the additive group $(\mathbf{Q}, +)$ cannot be expressed as an internal direct product of two nontrivial subgroups.

Heinrich Weber (1842–1913) was born on May 5, 1892, in Heidelberg, Germany. In 1860, he studied mathematics and physics at the University of Heidelberg. He received his Ph.D. in 1863. He was appointed as extraordinary professor at the University of Heidelberg in 1869 and also taught at Edgenössische Polytechnikum in Zurich, the University of Königsberg, the Technische Hochschule in Charlottenburg, and the universities of Marburg, Göttingen, and Strasbourg.

Weber was a friend of Richard Dedekind and they often collaborated. Together they edited the work of Riemann in 1876. Herman Minkowski and David Hilbert were among Weber's students.

Weber's main research interests were in analysis and its applications to mathematical physics and number theory. He was encouraged by von Neumann to investigate physical problems and by Richelot to study algebraic functions. Along the lines of Jacobi, he worked on the theory of differential equations. He proved Abel's theorem in its most general form. He also worked on physical problems concerning heat, static and current electricity, the motion of rigid bodies in liquids, and electrolytic displacement.

Weber's most profound and penetrating work is in algebra and number theory. He, jointly with Dedekind, did work of fundamental importance on algebraic functions.

In 1891, Weber gave the "modern" definition of an abstract finite group. One of his outstanding accomplishments was the proof of Kronecker's theorem, which states that absolute Abelian fields are cyclotomic.

Weber was an enthusiastic and inspiring teacher who took great interest in educational questions. He died on May 17, 1913.

Chapter 7

Sylow Theorems

In general, the converse of Lagrange's theorem does not hold (Exercise 19, page 138). In this chapter, we prove the Sylow theorems, which are very helpful in determining whether a given finite group has subgroups of specific orders. There are several known proofs of the Sylow theorems. In this text, we give two different proofs of the Sylow theorems, one based on the notion of group action (Section 5.4) and another based on the notion of conjugacy classes (Section 7.1). In Section 7.4, we will apply the Sylow theorems to determine certain simple groups.

7.1 Conjugacy Classes

In this section, we define an equivalence relation commonly known as a conjugacy relation on a group. This relation partitions the group into disjoint equivalence classes, which helps us to obtain a decomposition of the order of a finite group. This particular decomposition of the order of a finite group is known as the class equation. The class equation is very useful in determining the nature and structure of finite groups. The results obtained in this section will be used throughout this chapter.

Definition 7.1.1 *Let G be a group and a be an element of G. Then the **centralizer** or **normalizer** of a in G, denoted by $C(a)$, is the set of all elements of G which commute with a, i.e.,*

$$C(a) = \{b \in G \mid ba = ab\}.$$

We note that $C(a) = G$ if and only if a is in the center of G.

Let G be a group and $a \in G$. An element $b \in G$ is said to be a **conjugate** of a in G if there exists $c \in G$ such that $b = cac^{-1}$.

In the following theorem, we prove some basic properties of the centralizer of an element.

Theorem 7.1.2 *Let a be an element of a group G. Then*
(i) $C(a)$ *is a subgroup of G.*
(ii) *The relation ρ on G defined by*

$$\rho = \{(a, b) \in G \times G \mid b \text{ is a conjugate of } a\}$$

is an equivalence relation, known as **conjugacy**, *on G; the equivalence class* $[a]$ *of the relation ρ is called a* **conjugacy class** *of a in G. We denote the conjugacy class $[a]$ by $C_l(a)$.*
(iii) *The number of conjugates of a is equal to the index of $C(a)$ in G, i.e.,* $|C_l(a)| = [G : C(a)]$.

Proof. (i) Since $ea = a = ae$, $e \in C(a)$ and so $C(a) \neq \phi$. Let $b, c \in C(a)$. Then $ab = ba$ and $ac = ca$. Also, $ac = ca$ implies that $ac^{-1} = c^{-1}a$. Now $a(bc^{-1}) = (ab)c^{-1} = (ba)c^{-1} = b(ac^{-1}) = b(c^{-1}a) = (bc^{-1})a$. Therefore, $bc^{-1} \in C(a)$. Hence, $C(a)$ is a subgroup of G by Theorem 4.1.3.

(ii) Note that for all $a \in G$, $a = eae^{-1}$. Thus, for all $a \in G$, a is a conjugate of a. Hence, ρ is reflexive. For symmetry, let $(a, b) \in \rho$. Then there exists $c \in G$ such that $b = cac^{-1}$. This implies that $a = c^{-1}bc$ and so $(b, a) \in \rho$. Hence, ρ is symmetric. To show that ρ is transitive, let $(a, b), (b, c) \in \rho$. Then there exist $u, v \in G$ such that $b = uau^{-1}$ and $c = vbv^{-1}$. This implies that $c = (vu)a(vu)^{-1}$ and so $(a, c) \in \rho$. Thus, ρ is transitive. Consequently, ρ is an equivalence relation.

(iii) Let $a \in G$. Let $\mathcal{H}$ denote the set of all distinct left cosets of $C(a)$ in G. Then $|\mathcal{H}| = [G : C(a)]$. Now $bab^{-1} \in C_l(a)$ for all $b \in G$. Define $f : \mathcal{H} \rightarrow C_l(a)$ by $f(bC(a)) = bab^{-1}$. Let $b, c \in G$. Now $bC(a) = cC(a)$ if and only if $c^{-1}b \in C(a)$, which in turn is equivalent to $(c^{-1}b)a = a(c^{-1}b)$. Now $(c^{-1}b)a = a(c^{-1}b)$ if and only if $bab^{-1} = cac^{-1}$. Therefore, f is a one-one function. From the definition of f, it follows that f maps $\mathcal{H}$ onto $C_l(a)$. Hence, f is a one-one function of $\mathcal{H}$ onto $C_l(a)$. Consequently, $|C_l(a)| = |\mathcal{H}| = [G : C(a)]$. ∎

Corollary 7.1.3 *Let G be a finite group. Then*

$$|G| = \sum_a [G : C(a)],$$

where the summation is over a complete set of distinct conjugacy class representatives.

Proof. By Theorem 7.1.2(ii), $G = \cup_a C_l(a)$, where the union runs over a complete set of distinct conjugacy class representatives. The corollary follows since the distinct conjugacy classes are mutually disjoint and $|C_l(a)| = [G : C(a)]$ for all $a \in G$ by Theorem 7.1.2(iii). ∎

Corollary 7.1.4 *Let G be a finite group. Then*

$$|G| = |Z(G)| + \sum_{a \notin Z(G)} [G : C(a)], \tag{7.1}$$

where $Z(G)$ denotes the center of G and the summation runs over a complete set (possibly empty) of distinct conjugacy class representatives, which do not belong to $Z(G)$.

Proof. First observe that $a \in Z(G)$ if and only if $C(a) = G$ if and only if $[G : C(a)] = 1$. By Corollary 7.1.3

$$|G| = \sum_a [G : C(a)],$$

where the summation is over a complete set of distinct conjugacy class representatives. This implies that

$$|G| = \sum_{a \in Z(G)} [G : C(a)] + \sum_{a \notin Z(G)} [G : C(a)].$$

Since $a \in Z(G)$ if and only if $[G : C(a)] = 1$, it follows that $\sum_{a \in Z(G)}[G : C(a)] = |Z(G)|$. Hence,

$$|G| = |Z(G)| + \sum_{a \notin Z(G)} [G : C(a)],$$

where the summation runs over a complete set (possibly empty) of distinct conjugacy class representatives which do not belong to $Z(G)$. ∎

Eq. (7.1) in Corollary 7.1.4 is called the (**conjugacy**) **class equation.**

Example 7.1.5 *Consider S_3. By Worked-Out Exercise 1 (page 94), it follows that S_3 has three conjugacy classes, namely,*

$$\left\{ \begin{pmatrix} 1 & 2 & 3 \\ 1 & 2 & 3 \end{pmatrix} \right\},$$

$$\left\{ \begin{pmatrix} 1 & 2 & 3 \\ 2 & 3 & 1 \end{pmatrix}, \begin{pmatrix} 1 & 2 & 3 \\ 3 & 1 & 2 \end{pmatrix} \right\}, \text{ and}$$

$$\left\{ \begin{pmatrix} 1 & 2 & 3 \\ 2 & 1 & 3 \end{pmatrix}, \begin{pmatrix} 1 & 2 & 3 \\ 1 & 3 & 2 \end{pmatrix}, \begin{pmatrix} 1 & 2 & 3 \\ 3 & 2 & 1 \end{pmatrix} \right\}$$

The class equation reads

$$|S_3| = |Z(G)| + [S_3 : C(\begin{pmatrix} 1 & 2 & 3 \\ 2 & 3 & 1 \end{pmatrix})] + [S_3 : C(\begin{pmatrix} 1 & 2 & 3 \\ 2 & 1 & 3 \end{pmatrix})]$$

$$6 = 1 + 2 + 3.$$

Example 7.1.6 *Consider the group of symmetries of the square. The distinct conjugacy classes are* $\{r_{180}\}, \{r_{360}\}, \{r_{90}, r_{270}\}, \{d_1, d_2\}, \{h, v\}$. *The class equation reads*

$$\begin{aligned} |G| &= |Z(G)| &+& [G:C(r_{90})] &+& [G:C(d_1)] &+& [G:C(h)] \\ 8 &= (1+1) &+& 2 &+& 2 &+& 2. \end{aligned}$$

Until now our discussion focused on the conjugacy class of an element of a group. We now extend our discussion to the conjugate subgroup of a group. We will be mainly interested in determining the number of distinct conjugates of a subgroup induced by the elements of another subgroup. We begin with the following theorem.

Theorem 7.1.7 *Let H be a subgroup of a group G and $a \in G$. Then aHa^{-1} is a subgroup of G, called a **conjugate** of H. Furthermore, $H \simeq aHa^{-1}$.*

Proof. By Worked-Out Exercise 1(i) (page 106), aHa^{-1} is a subgroup of G. Now define $f : H \to aHa^{-1}$ by $f(h) = aha^{-1}$ for all $h \in H$. As in Worked-Out Exercise 1(ii) (page 106), f is a one-one function from H onto aHa^{-1}. To show that f is a homomorphism, let $h_1, h_2 \in H$. Then $f(h_1h_2) = a(h_1h_2)a^{-1} = (ah_1a^{-1})(ah_2a^{-1}) = f(h_1)f(h_2)$. Hence, $H \simeq aHa^{-1}$. ∎

Definition 7.1.8 *Let H be a subgroup of a group G and $a \in G$. If $aHa^{-1} = H$, then H is called **invariant** under a.*

Definition 7.1.9 *Let H and K be subgroups of a group G. Let $N_K(H)$ denote the set*

$$N_K(H) = \{k \in K \mid kHk^{-1} = H\}.$$

$N_K(H)$ *is called the **normalizer** of H in K.*

It follows that $N_K(H) = N_G(H) \cap K$.

Theorem 7.1.10 *Let H and K be subgroups of a group G. Then $N_K(H)$ is a subgroup of K.*

Proof. Since $e \in K$ and $eHe^{-1} = H$, $e \in N_K(H)$ and so $N_K(H) \neq \phi$. Let $k_1, k_2 \in N_K(H)$. Then $k_1Hk_1^{-1} = H = k_2Hk_2^{-1}$. Now $H = k_2Hk_2^{-1}$ implies that $H = k_2^{-1}Hk_2$. Thus,

$$\begin{aligned} H &= k_1Hk_1^{-1} \\ &= k_1(k_2^{-1}Hk_2)k_1^{-1} \\ &= (k_1k_2^{-1})H(k_1k_2^{-1})^{-1}. \end{aligned}$$

Hence, $k_1 k_2^{-1} \in N_K(H)$. Thus, $N_K(H)$ is a subgroup of G. ∎

When $K = G$, we write $N(H)$ for $N_G(H)$ and refer to the subgroup $N(H)$ simply as the normalizer of H. By Exercise 11 (page 137), $N(H)$ is the largest subgroup of G in which H is normal. Of course $N(H) = G$ when H is a normal subgroup of G or when G is commutative.

Example 7.1.11 *Consider the symmetric group S_3. In Example 4.3.2, the subgroup*

$$H' = \left\{ \begin{pmatrix} 1 & 2 & 3 \\ 1 & 2 & 3 \end{pmatrix}, \begin{pmatrix} 1 & 2 & 3 \\ 1 & 3 & 2 \end{pmatrix} \right\}$$

is not a normal subgroup of S_3. We note that $N(H') = H'$.

Theorem 7.1.12 *Let H and K be subgroups of a group G. The number of distinct conjugates of H induced by the elements of K is equal to $[K : N_K(H)]$, the index of $N_K(H)$ in K.*

Proof. Let $\mathcal{T}$ be the set of distinct conjugates of H induced by the elements of K, i.e., $\mathcal{T} = \{kHk^{-1} \mid k \in K\}$ and let $\mathcal{S}$ be the set of distinct left cosets of $N_K(H)$ in K, i.e., $\mathcal{S} = \{aN_K(H) \mid a \in K\}$. To show that the number of distinct conjugates of H induced by the elements of K is equal to $[K : N_K(H)]$, the index of $N_K(H)$ in K, we need to show that there exists a one-one function of $\mathcal{T}$ onto $\mathcal{S}$.

Define $f : \mathcal{T} \to \mathcal{S}$ by $f(aHa^{-1}) = aN_K(H)$ for all $aHa^{-1} \in \mathcal{T}$. Let k_1, $k_2 \in K$. Then $k_1 Hk_1^{-1} = k_2 Hk_2^{-1}$ if and only if $H = (k_1^{-1}k_2)H(k_1^{-1}k_2)^{-1}$. Now $H = (k_1^{-1}k_2)H(k_1^{-1}k_2)^{-1}$ if and only if $k_1^{-1}k_2 \in N_K(H)$ and the latter is true if and only if $k_1 N_K(H) = k_2 N_K(H)$. Thus, we have shown that f is a one-one function. From the definition of f, it is immediate that f is onto $\mathcal{S}$. Hence, the number of distinct conjugate subgroups of H by the elements of K is equal to the number of distinct cosets of $N_K(H)$ in K. ∎

Corollary 7.1.13 *Let H and K be finite subgroups of a group G. If H is invariant under n elements of K, then H has $|K|/n$ conjugates by elements of K.*

Proof. By hypothesis, $|N_K(H)| = n$. Hence, $|K| = [K : N_K(H)] \cdot |N_K(H)|$ by Lagrange's theorem. The corollary is now immediate by Theorem 7.1.12. ∎

7.1.1 Worked-Out Exercises

◇ **Exercise 1** Let G be a finite group and $a \in G$ be such that a has only two conjugates. Prove that $C(a)$ is a normal subgroup of G.

Solution: By Theorem 7.1.2, $[G : C(a)] = |C_l(a)|$. Now $|C_l(a)| = 2$. Hence, $[G : C(a)] = 2$, proving that $C(a)$ is a normal subgroup of G.

$\diamondsuit$ **Exercise 2** Let G be a finite group that has only two conjugate classes. Show that $|G| = 2$.

Solution: Let $|G| = n$. Let $a \in G$ and $a \neq e$. Then $G = C_l(e) \cup C_l(a)$. Since $|C_l(e)| = 1$, $|C_l(a)| = n - 1$. Hence, $n - 1 = |C_l(a)| = [G : C(a)]$ divides $|G| = n$. This is possible only if $n = 2$.

Exercise 3 Prove that there exists no finite nontrivial group every nonidentity element of which commutes with exactly half the elements of the group.

Solution: Let G be a group of order $n > 1$ such that every nonidentity element of G commutes with exactly half the elements of G. Let $a \in G$ and $a \neq e$. Then $|C(a)| = n/2$. Hence, $|C_l(a)| = [G : C(a)] = 2$. Now $|G| = |C_l(e)| + \sum_{a \neq e} |C_l(a)|$, where the summation runs over a complete set of distinct conjugacy class representatives. Since $|C_l(e)| = 1$ and $|C_l(a)| = 2$ for all $e \neq a \in G$, we find that $|G|$ is odd. But $|C(a)| = \frac{n}{2} = \frac{|G|}{2}$ shows that $|G|$ is even. This contradiction shows that there cannot exist any group of this type.

7.1.2   Exercises

1. Let G be a group and $a \in G$. Prove that $a \in Z(G)$ if and only if $C_l(a) = \{a\}$.

2. Let G be a finite group. Prove that if there exists an element $a \in G$ with exactly two conjugates, then G contains a nontrivial normal subgroup.

3. Prove that a subgroup H of a group G is a normal subgroup if and only if H is the union of conjugacy classes of G.

4. Let G be a group, H a subgroup of G, and $a \in G$. Prove that $N(aHa^{-1}) = aN(H)a^{-1}$.

5. Let H and K be subgroups of a group G. Prove that H is normal in K if and only if $H \subseteq K \subseteq N_G(H)$.

6. Let G be a group and H and K be subgroups of G. Prove that if H and K are conjugates, then $N_G(H)$ and $N_G(K)$ are conjugates.

7. Find the class equation for S_5.

7.2 Cauchy's Theorem and p-groups

In this section, we prove an important theorem which gives a partial converse of Lagrange's theorem. This interesting theorem is due to Cauchy. First we will prove this theorem for finite Abelian groups and then with the help of the class equation extend it to any finite group. The proof of Cauchy's theorem given in this book is intended to show the reader the use of the ideas of quotient groups and the class equation. With the help of Cauchy's theorem, we also prove that the converse of Lagrange's theorem holds for finite Abelian groups.

Lemma 7.2.1 *If G is a finite commutative group of order n such that n is divisible by a prime p, then G contains an element of order p (whence a subgroup of order p).*

Proof. The proof is by induction on the order of G. If $|G| = p$, a prime, then every element of G, other than the identity, has order p. Thus, in particular, the lemma is true when $|G| = 2$. Now make the induction hypothesis that the lemma is true for all groups of order r, where $2 \leq r < n$. Suppose G is a group of order n. Let $a \in G$ with $a \neq e$ and let m denote the order of a. Then either $p|m$ or $p \nmid m$. If $p|m$, then $m = pk$ for some positive integer k. In this case, $(a^k)^p = a^m = e$, from which it follows that $a^k \neq e$ and a^k is an element of order p. Now suppose $p \nmid m$. Since G is commutative, the cyclic subgroup $H = \langle a \rangle$ of G is of course a normal subgroup of G. Now $|G| = m \cdot [G : H]$. Since p does not divide m, we have $p|[G : H]$. Hence, p divides $|G/H|$. Since $|G/H| < n$, we have by the induction hypothesis that there exists $bH \in G/H$ such that $\circ(bH) = p$. Now $b^p H = (bH)^p = H$. Hence, $b^p \in H$. Thus, $(b^m)^p = (b^p)^m = e$, so that either $b^m = e$ or b^m has order p. But $b^m \neq e$ else $(bH)^m = H$ yielding $p|m$, a contradiction. Thus, b^m has order p and so b^m is the desired element of G. ∎

Theorem 7.2.2 (Cauchy) *Let G be a finite group of order n such that n is divisible by a prime p. Then G contains an element of order p and hence a subgroup of order p.*

Proof. The proof is by induction on n. If $n = 2$, then G is commutative and the result follows by Lemma 7.2.1. Make the induction hypothesis that the result is true for all groups of order m such that $2 \leq m < n$. Consider the class equation

$$|G| = |Z(G)| + \sum_{a \notin Z(G)} [G : C(a)]$$

for G. If $G = Z(G)$, then G is commutative and the result follows by Lemma 7.2.1. If $G \neq Z(G)$, then there exists $a \in G$ such that $a \notin Z(G)$. For such an element a, $G \neq C(a)$ and so $[G : C(a)] > 1$, whence by Lagrange's theorem

$$|G| = [G : C(a)] \cdot |C(a)| > |C(a)| \,.$$

If p divides $|C(a)|$, then by the induction hypothesis, $C(a)$ and thus G has an element of order p. If p does not divide $|C(a)|$ for all $a \notin Z(G)$, then p must divide $[G : C(a)]$ for all $a \notin Z(G)$. But in the class equation, p divides each term of the summation and also divides $|G|$. Thus, p divides $|Z(G)|$. Since $Z(G)$ is commutative, we have again by Lemma 7.2.1 that there exists $a \in Z(G)$ and hence $a \in G$ of order p. ∎

Next, we apply Cauchy's theorem to prove that the converse of Lagrange's theorem holds for finite commutative groups.

Theorem 7.2.3 *Let G be a finite commutative group of order n. If m is a positive integer such that $m|n$, then G has a subgroup of order m.*

Proof. If $m = 1$, then $\{e\}$ is the required subgroup of order m. If $n = 1$, then $m = n = 1$ and the result follows easily. We now assume that $m > 1$, $n > 1$ and prove the result by induction on n. If $n = 2$, then $m = 2 = n$ and G is the required subgroup of order m. Suppose the theorem is true for all finite commutative groups of order k such that $2 \leq k < n$. Let p be a prime integer such that $p|m$. Then there exists an integer m_1 such that $m = pm_1$. By Cauchy's theorem, G has a subgroup H of order p. Since G is commutative, H is normal and hence G/H is a group. Now

$$1 \leq |G/H| = \frac{|G|}{|H|} < |G|$$

and $|G/H| = \frac{n}{p}$. Now $n = mm_2$ for some positive integer m_2. Thus, $|G/H| = \frac{pm_1m_2}{p} = m_1m_2$ shows that m_1 divides $|G/H|$. Hence, from the induction hypothesis, G/H has a subgroup K/H such that $|K/H| = m_1$, where K is a subgroup of G. Now $|K| = |K/H||H| = m_1p = m$. Hence, K is a subgroup of G of order m. ∎

We now apply Cauchy's theorem to obtain some interesting properties of p-groups.

Definition 7.2.4 *Let p be a prime. A group G is said to be a **p-group** if the order of each element of G is a power of p. A subgroup H of a group G is called a **p-subgroup** if H is a p-group.*

Example 7.2.5 *The group of symmetries of a square and the Klein 4-group are p-groups, where $p = 2$. In fact, any group of order p^n (p a prime) is a p-group since the order of each element must divide the order of the group.*

The following theorem gives a necessary and sufficient condition for a finite group to be a p-group.

Theorem 7.2.6 *Let G be a nontrivial group. Then G is a finite p-group if and only if $|G| = p^k$ for some positive integer k.*

Proof. Suppose G is a finite p-group. If q divides $|G|$ for some prime $q \neq p$, then by Cauchy's theorem G has an element of order q, contradicting the fact that G is a p-group. Thus, p is the only prime divisor of $|G|$. Hence, $|G| = p^k$ for some positive integer k. Conversely, suppose $|G| = p^k$. Then by Lagrange's theorem, the order of each element of G is a power of p. ■

In the next theorem, we prove that the center of a p-group is nontrivial.

Theorem 7.2.7 *If G is a finite p-group with $|G| > 1$, then $Z(G)$, the center of G, has more than one element, i.e., if $|G| = p^k$ with $k \geq 1$, then $|Z(G)| > 1$.*

Proof. Consider the class equation

$$|G| = |Z(G)| + \sum_{a \notin Z(G)} [G : C(a)].$$

If $G = Z(G)$, then the theorem is immediate. Suppose $G \supset Z(G)$ and consider $a \in G$ such that $a \notin Z(G)$. Then $C(a)$ is a proper subgroup of G so that by Theorem 7.2.6 and by the fact that $C(a)$ is a subgroup of a p-group, $p|[G : C(a)]$ for all $a \notin Z(G)$. This implies that p divides $\sum_{a \notin Z(G)} [G : C(a)]$. Since p also divides $|G|$, p divides $|Z(G)|$. Hence, $|Z(G)| > 1$. ■

Corollary 7.2.8 *Let G be a group of order p^2, where p is a prime. Then G is commutative.*

Proof. By Theorem 7.2.7, $|Z(G)| > 1$. By Lagrange's theorem, $|Z(G)|$ divides p^2. Hence, $|Z(G)| = p$ or p^2. Suppose $|Z(G)| = p$. Then $Z(G) \neq G$ and so there exists $a \in G$ such that $a \notin Z(G)$. Now $C(a)$ is a subgroup of G and $a \in C(a)$. Hence, $Z(G) \subset C(a)$. This implies that $|C(a)| = p^2$ and so $G = C(a)$. However, this shows that $a \in Z(G)$, a contradiction. Therefore, $|Z(G)| = p^2$ and so $G = Z(G)$. Thus, G is commutative. ■

7.2.1 Worked-Out Exercises

◇ **Exercise 1** Show that every group of order pq, where p and q are primes, is not simple.

Solution: If $p = q$, then G is a group of order p^2. Hence, G is commutative. Also, Cauchy's theorem implies that G has a subgroup of order p, which must be normal. Therefore, G is not simple. Suppose now $p \neq q$. We may assume that $p > q$. By Exercise 8 (page 200), G has a normal subgroup of order p. Thus, G is not simple.

Exercise 2 Let H and K be subgroups of a commutative group G. Suppose $|H| = m$ and $|K| = n$. Let $d = \mathrm{lcm}(m, n)$. Show that G has a subgroup of order d.

Solution: Since G is commutative, HK is a subgroup of G, and since H and K are finite, HK is finite. Now H and K are subgroups of HK. Hence, $m | |HK|$ and $n | |HK|$. This implies that $d | |HK|$. Since HK is a finite commutative group and $d | |HK|$, HK has a subgroup of order d and so G has a subgroup of order d.

$\diamond$ **Exercise 3** Let G be a noncommutative group of order p^3, p a prime. Prove that $|Z(G)| = p$.

Solution: Write $Z = Z(G)$. Since $|G| = p^3$, $|Z| > 1$ by Theorem 7.2.7. Thus, $|Z| = p$, p^2 or p^3. If $|Z| = p^3$, then $G = Z$ and so G is commutative, which is a contradiction. If $|Z| = p^2$, then $|G/Z| = p$. Hence, G/Z is cyclic. But then G is commutative, again a contradiction. Thus, $|Z| = p$.

Exercise 4 Let G be a finite commutative group. Prove that the number of solutions of $x^n = e$ in G, where $n > 0$ and n divides $|G|$, is a multiple of n.

Solution: Let $H = \{x \mid x \in G, x^n = e\}$. Then H is a subgroup of G. Since n divides $|G|$ and G is commutative, there exists a subgroup K of G such that $|K| = n$. Let $a \in K$. Then $a^n = e$. Hence, $K \subseteq H$. By Lagrange's theorem, $|K|$ divides $|H|$. Thus, $|H| = nm$. Consequently, the number of solutions of $x^n = e$ is a multiple of n.

$\diamond$ **Exercise 5** Let G be a group of order p^n, p a prime, and $n \in \mathbf{Z}$, $n \geq 1$. Prove that any subgroup of G of order p^{n-1} is normal in G.

Solution: We will prove the result by induction on n. If $n = 1$, then G is a cyclic group of prime order and hence every subgroup of G is normal in G. Thus, the result is true if $n = 1$. Suppose the result is true for all groups of order p^m, where $1 \leq m < n$. Let H be a subgroup of order p^{n-1}. Consider $N(H)$. If $H \neq N(H)$, then $|N(H)| > p^{n-1}$. Thus, $|N(H)| = p^n$ and so $N(H) = G$. Hence, in this case H is normal in G. Suppose $H = N(H)$. Then $Z(G)$, the center of G, is a subset of H and $Z(G) \neq \{e\}$. By Cauchy's theorem and Theorem 7.2.7, there exists $a \in Z(G)$ such that $o(a) = p$. Let $K = \langle a \rangle$. Then K is a normal subgroup of G of order p. Now $|H/K| = p^{n-2}$ and $|G/K| = p^{n-1}$. Thus, by the induction hypothesis, H/K is a normal subgroup of G/K. Hence, H is a normal subgroup of G.

7.2.2   Exercises

1. Show that every group of order 14 contains only one normal subgroup of order 7.

2. How many elements of order 7 are there in a group of order 28?

3. Show that a group of order 15 is commutative.

4. Let G be a group of order p^n, where p is a prime and n is a positive integer. Show that G contains a subgroup of order p^i, $0 \le i \le n$.

5. Find all 2-subgroups and 3-subgroups of $(\mathbf{Z}_{12}, +_{12})$.

6. Find all 2-subgroups of A_4.

7. Show that every commutative group of order 36 contains an element of order 6.

8. Let G be a group of order pn, where p is a prime and $p > n$. Show that G contains a normal subgroup of order p.

9. Let G be a commutative group of order pq, where p and q are distinct primes. Show that G is cyclic. Is this result true when $p = q$?

10. For any prime p, prove that any group of order p^2 is either cyclic or a direct product of cyclic groups.

11. Show that every group of order 28 with a unique subgroup of order 4 is commutative.

12. Show that a group of order 81 contains a nontrivial normal subgroup with more than three elements.

13. Let G be a group of order 99. Prove the following.

    (i) G has a unique normal subgroup H of order 11.

    (ii) $H \subseteq Z(G)$.

    (iii) G has an element of order 33.

7.3   Sylow Theorems

M.L. Sylow did work of fundamental importance in determining the structure of finite groups. We can use his results to answer the problem now posed.

If G is a finite group of order n and if H is a subgroup of G, then we know by Lagrange's theorem that the order of H divides n. In this section, we give some answers to the question, "If m is a positive integer, which divides n, does G contain a subgroup of order m?"

It is interesting to note that Sylow's theorem was proved by Sylow for permutation groups. George Frobenius established the theorem in the general setting. He was influenced to do so by Cayley's theorem.

Theorem 7.3.1 (Sylow's First Theorem) *Let G be a finite group of order $p^r m$, where p is a prime, r and m are positive integers, and p and m are relatively prime. Then G has a subgroup of order p^k for all k, $0 \le k \le r$.*

Proof.   **First Proof of Sylow's First Theorem:** Let $|G| = n = p^r m$. We prove the result by induction on n. If $n = 1$, then $r = 0$ and $\{e\}$ is the required subgroup of order p^r. Suppose the result is true for all groups T of order less than $|G|$. If $r = 0$, then $\{e\}$ is the required subgroup of order p^r. We now assume that $r \ge 1$. First suppose p divides $|Z(G)|$, where $Z(G)$ is the center of G. Since p divides $|Z(G)|$, there exists $a \in Z(G)$ such that $o(a) = p$ by Cauchy's theorem. Let $H = \langle a \rangle$. Then H is a normal subgroup of G since $a \in Z(G)$. Now $|G/H| = p^{r-1} m$. Hence, by the induction hypothesis, G/H has subgroups K_i/H of order p^i for all $i = 0, 2, \ldots, r-1$. Then $\{e\}, H, K_1, \ldots, K_{r-1}$ are the subgroups of G of the required order.

Now suppose $p \nmid |Z(G)|$. Consider the class equation,

$$|G| = |Z(G)| + \sum_{a \notin Z(G)} [G : C(a)],$$

where the summation runs over a complete set (possibly empty) of distinct conjugacy class representatives which do not belong to $Z(G)$. From the hypothesis, p divides $|G|$. If $p|[G : C(a)]$ for all $a \notin Z(G)$, then from the class equation, it follows that p divides $|Z(G)|$, a contradiction to our assumption. Hence, there exists $a \notin Z(G)$ such that p does not divide $[G : C(a)]$. Now

$$|G| = [G : C(a)] \cdot |C(a)|.$$

This implies that p^r divides $|C(a)|$. Since $a \in C(a)$, $|C(a)| > 1$. Also, $C(a) \ne G$ since $a \notin Z(G)$. Hence, $|C(a)| < |G|$. Thus, by the induction hypothesis, $C(a)$ has a subgroup of order p^i for all i, $0 \le i \le r$. Hence, G has a subgroup of order p^i for all i, $0 \le i \le r$. ∎

Second Proof of Sylow's First Theorem: If $r = 0$, then $\{e\}$ is the required subgroup of order p^r. Suppose $r \geq 1$. Since $p|\,|G|$, G has a subgroup of order p by Cauchy's theorem. We now show that if G has a subgroup of order p^i, then G has a subgroup order p^{i+1}, where $1 \leq i < r$. Suppose G has a subgroup H of order p^i, $1 \leq i < r$. Then H is a proper subgroup of G. By Worked-Out Exercise 3 (page 176), $[N(H) : H] \equiv_p [G : H]$ and $H \neq N(H)$. Since $p|[G : H]$, it follows that $p|[N(H) : H]$, i.e., $p|\,|N(H)/H|$. Thus, $N(H)/H$ has a subgroup K/H of order p by Cauchy's theorem. Now $|K| = |K/H|\,|H| = pp^i = p^{i+1}$. Therefore, K is a subgroup of G of order p^{i+1}. The result now follows by induction. $\blacksquare$

The following corollary is immediate from Theorem 7.3.1

Corollary 7.3.2 *Let G be a finite group and p a prime. If p^n divides $|G|$, then G has a subgroup of order p^n.* $\blacksquare$

Definition 7.3.3 *Let G be a finite group and p a prime. A subgroup P of G is called a **Sylow p-subgroup** of G, if P is a p-subgroup and is not properly contained in any other p-subgroup of G, i.e., P is a maximal p-subgroup of G.*

Example 7.3.4 *The symmetric group S_3 has three Sylow 2-subgroups, namely*

$$H_1 = \left\{ \begin{pmatrix} 1 & 2 & 3 \\ 1 & 2 & 3 \end{pmatrix}, \begin{pmatrix} 1 & 2 & 3 \\ 2 & 1 & 3 \end{pmatrix} \right\},$$

$$H_2 = \left\{ \begin{pmatrix} 1 & 2 & 3 \\ 1 & 2 & 3 \end{pmatrix}, \begin{pmatrix} 1 & 2 & 3 \\ 3 & 2 & 1 \end{pmatrix} \right\},$$

and

$$H_3 = \left\{ \begin{pmatrix} 1 & 2 & 3 \\ 1 & 2 & 3 \end{pmatrix}, \begin{pmatrix} 1 & 2 & 3 \\ 1 & 3 & 2 \end{pmatrix} \right\}.$$

Thus, a Sylow p-subgroup of a given group need not be unique.

The following theorem shows the existence of Sylow p-subgroups in a finite group.

Theorem 7.3.5 *For each prime p, a finite group G has a Sylow p-subgroup.*

Proof. If $|G| = 1$ or p does not divide $|G|$, then $\{e\}$ is the required Sylow p-subgroup of G. If p divides $|G|$, then by Cauchy's theorem, there is at least one subgroup H of G of order p. Since G is finite, there are a finite number of subgroups of G, which contain H. Hence, one of these subgroups is a Sylow p-subgroup of G. $\blacksquare$

From Sylow's first theorem, every group of order $p^r m$ (p a prime, $\gcd(p, m) = 1$) contains a subgroup of order p^r. We now show that every subgroup of order p^r is a Sylow p-subgroup in G and every Sylow p-subgroup of G is of order p^r.

Theorem 7.3.6 *Let G be a finite group of order $p^r m$, where p is a prime, r and m are positive integers, and p and m are relatively prime.*

(i) Let H be a subgroup of G of order p^i, $1 \le i < r$. Then there exists a subgroup K of G such that $|K| = p^{i+1}$ and H is a normal subgroup of K.

(ii) Let H be a subgroup of G. Then H is a Sylow p-subgroup of G if and only if $|H| = p^r$.

Proof. (i) By Worked-Out Exercise 3 (page 176), $[N(H) : H] \equiv_p [G : H]$. Since $p|[G : H]$, $p|\,|N(H)/H|$. Thus, $N(H)/H$ has a subgroup K/H of order p by Cauchy's theorem. Now $|K| = |H|\,|K/H| = p^{i+1}$. Since H is normal in $N(H)$ and $K \subseteq N(H)$, H is normal in K. Hence, K is the desired subgroup of G.

(ii) Suppose H is a Sylow p-subgroup. Then H is a p-subgroup of G and so $|H| = p^k$ for some positive integer k. Suppose $k \ne r$. By (i), there exists a subgroup K of G such that $H \subset K$ and $|K| = p^{k+1}$. This implies that H is not a maximal p-subgroup of G, a contradiction. Thus, $k = r$. Conversely, suppose that $|H| = p^r$. Since $|G| = p^r m$ and p and m are relatively prime, it follows that H is a maximal p-subgroup of G. Hence, H is a Sylow p-subgroup of G. ∎

Theorem 7.3.7 *Let G be a finite group of order $p^r m$, where p is a prime, r and m are positive integers, and p and m are relatively prime, and P be a subgroup of G.*

(i) If P is a p-group, then any conjugate of P is a p-group.

(ii) If P is a Sylow p-subgroup, then any conjugate of P is a Sylow p-subgroup.

(iii) If P is the only Sylow p-subgroup of G, then P is a normal subgroup of G.

Proof. (i) Since $|P| = |aPa^{-1}|$ and aPa^{-1} is a subgroup of G, the desired result follows from Theorem 7.2.6.

(ii) Let P be a Sylow p-subgroup. Then $|P| = p^r$. This implies that $|aPa^{-1}| = p^r$ and so by Theorem 7.3.6(ii), aPa^{-1} is a Sylow p-subgroup.

(iii) Let $a \in G$. Then aPa^{-1} is a Sylow p-subgroup of G by (ii). Since P is the only Sylow p-subgroup of G, $aPa^{-1} = P$. Hence, P is a normal subgroup of G. ∎

Lemma 7.3.8 *Let H be a normal subgroup of a group G. If H and G/H are both p-groups, then G is a p-group.*

Proof. Let $a \in G$. Then $aH \in G/H$ and so aH has order some power of p, say, p^k. Thus, $(aH)^{p^k} = H$ and so $a^{p^k} \in H$. Now every element of H has order a power of p. Let us say a^{p^k} has order p^m. Thus, $(a^{p^k})^{p^m} = e$ or $a^{p^{m+k}} = e$. This implies that $\circ(a)$ has order some power of p. Since a was arbitrary in G, G is a p-group. ∎

Lemma 7.3.9 *Let G be a finite group. Let P be a Sylow p-subgroup of G and $a \in G$ be such that the order of a is a power of p. If $aPa^{-1} = P$, then $a \in P$.*

Proof. Since $aPa^{-1} = P$, $a \in N(P)$. Now $N(P) \supseteq P$, so if we show that no element of $N(P)\backslash P$ has order a power of p, then $a \in P$. Suppose there exists $b \in N(P)\backslash P$ such that the order of b is a power of p. Now P is a normal subgroup of $N(P)$ so that we may consider the quotient group $N(P)/P$ and the coset bP. The order of bP as an element of $N(P)/P$ divides the order of b. Hence, bP has order a power of p in $N(P)/P$. Thus, the cyclic subgroup $\langle bP \rangle$ of $N((P)/P$ has order a power of p and thus is a p-group. By Corollary 5.2.12, there is a subgroup K of $N(P)$ such that $K \supseteq P$ and $K/P = \langle bP \rangle$. Since $b \notin P$, $K \supset P$. By Lemma 7.3.8, K is a p-group since both P and $\langle bP \rangle$ are p-groups. However, this contradicts the fact that P is a maximal p-subgroup of G. Hence, no element of $N(P)\backslash P$ can have order a power of p. ∎

We now prove two more theorems due to Sylow.

Theorem 7.3.10 (Sylow's Second Theorem) *Let G be a finite group of order $p^r m$, where p is a prime, r and m are positive integers, and p and m are relatively prime. Then any two Sylow p-subgroups of G are conjugate, and therefore isomorphic.*

Proof. **First Proof of Sylow's Second Theorem:** By Theorem 7.3.5, G has a Sylow p-subgroup, say, P. Let S be the set of all conjugates of P. We show that S contains all Sylow p-subgroups. Let H be a Sylow p-subgroup of G such that $H \notin S$ and let $Q \in S$. Now Q is a Sylow p-subgroup of G and $|Q| = p^r$. Since $Q \neq H$, it follows that $Q \not\subseteq H$. Thus, there exists $h \in H$ such that $h \notin Q$. Now $\circ(h) = p^k$ for some positive integer k. By Lemma 7.3.9, $hQh^{-1} \neq Q$. Thus, the number of conjugates of Q induced by the elements of H is more than 1. Hence, by Theorem 7.1.12, $[H : N_H(Q)] > 1$. Now $p^r = |H| = [H : N_H(Q)]|N_H(Q)|$ and so $[H : N_H(Q)]$ is a positive multiple of p.

Let us now define a relation ρ on S by $\rho = \{(A, B) \in S \times S \mid A = hBh^{-1}$ for some $h \in H\}$. Then ρ is an equivalence relation on S and for all $A \in S$, the equivalence class, $[A]$, consists of all conjugates of A induced by the elements of H. Thus, as shown before, $\|[A]\|$ is a nonnegative multiple of p. Since S is a disjoint union of such equivalence classes, it follows that $|\mathcal{S}|$ is a positive multiple of p and so $p \mid |\mathcal{S}|$. By Theorem 7.1.12, $|\mathcal{S}| = [G : N(P)]$. Thus,

$$m = [G : P] = [G : N(P)][N(P) : P] = |\mathcal{S}|\,[N(P) : P].$$

From this, it follows that $p \mid m$, a contradiction. Hence, $\mathcal{S}$ is the set of all Sylow p-subgroups of G. ∎

Second Proof of Sylow's Second Theorem: Let H and K be Sylow p-subgroups of G and $\mathcal{S}$ be the set of all left cosets of H in G. Then $|\mathcal{S}| = [G : H]$. Let K act on $\mathcal{S}$ by for all $k \in K$, $aH \in \mathcal{S}$,

$$k(aH) = (ka)H.$$

Then $\mathcal{S}$ is a K-set. Let $\mathcal{S}_0 = \{aH \in \mathcal{S} \mid k(aH) = aH \text{ for all } k \in K\}$. By Worked-Out Exercise 1 (page 176),

$$|\mathcal{S}_0| \equiv_p |\mathcal{S}|.$$

Since H is a Sylow p-subgroup of G, $|\mathcal{S}| = [G : H]$ is not divisible by p. Thus, $|\mathcal{S}_0| \neq 0$. Let $aH \in \mathcal{S}_0$. Then $k(aH) = aH$ for all $k \in K$. From this, it follows that $a^{-1}kaH = H$ for all $k \in K$ and so $a^{-1}ka \in H$ for all $k \in K$. Therefore, $a^{-1}Ka \subseteq H$. Since $|a^{-1}Ka| = |K| = |H|$, $a^{-1}Ka = H$. Hence, H and K are conjugate. ∎

The following corollary is an immediate consequence of Sylow's second theorem.

Corollary 7.3.11 *Let G be a finite group and H be a Sylow p-subgroup of G. Then H is a unique Sylow p-subgroup of G if and only if H is a normal subgroup of G.* ∎

Theorem 7.3.12 (Sylow's Third Theorem) *Let G be a finite group of order $p^r m$, where p is a prime, r and m are positive integers, and p and m are relatively prime. Then the number n_p of Sylow p-subgroups of G is $1 + kp$ for some nonnegative integer k and $n_p \mid p^r m$.*

Proof.   First Proof of Sylow's Third Theorem: Let $\mathcal{S}$ be the set of all Sylow p-subgroups of G and $P \in \mathcal{S}$. Define a relation ρ on $\mathcal{S}$ by $\rho = \{(A, B) \in \mathcal{S} \times \mathcal{S} \mid A = aBa^{-1} \text{ for some } a \in P\}$. Then as in the first proof of Sylow's second theorem, ρ is an equivalence relation on $\mathcal{S}$ and for all $A \in \mathcal{S}$, $A \neq P$, the number of elements in the equivalence class, $[A]$, is a multiple of p. Now $[P] = \{A \in \mathcal{S} \mid A = xPx^{-1} \text{ for some } x \in P\} = \{P\}$. Thus, $|[P]| = 1$. Consequently, $|\mathcal{S}| = 1 + kp$ for some nonnegative integer k. Now by

Theorem 7.1.12, $n_p = |\mathcal{S}| = [G : N(P)]$. This implies that n_p divides $|G|$. $\blacksquare$

Second Proof of Sylow's Third Theorem: Let $\mathcal{S}$ be the set of all Sylow p-subgroups of G and $P \in \mathcal{S}$. Let P act on $\mathcal{S}$ by conjugation, i.e., for all $a \in P$, $Q \in \mathcal{S}$, $a \cdot Q = aQa^{-1}$. Let $\mathcal{S}_0 = \{Q \in \mathcal{S} \mid a \cdot Q = Q \text{ for all } a \in P\} = \{Q \in \mathcal{S} \mid aQa^{-1} = Q \text{ for all } a \in P\}$. By Worked-Out Exercise 1 (page 176),

$$|\mathcal{S}| \equiv_p |\mathcal{S}_0|.$$

Since $P \in \mathcal{S}_0$, $\mathcal{S}_0 \neq \phi$. Let $Q \in \mathcal{S}_0$. Then $Q = aQa^{-1}$ for all $a \in P$. Hence, $P \subseteq N(Q)$ and so P and Q are Sylow p-subgroups of $N(Q)$ since P and Q are Sylow p-subgroups of G. Thus, by Sylow's second theorem, $aQa^{-1} = P$ for some $a \in N(Q)$. But then $P = Q$. Thus, $\mathcal{S}_0 = \{P\}$ and so $|\mathcal{S}_0| = 1$. Hence, $|\mathcal{S}| \equiv_p 1$ and so $|\mathcal{S}| = 1 + kp$ for some integer k.

Let G act on $\mathcal{S}$ by conjugation. By Sylow's second theorem, any two Sylow p-subgroups are conjugate. Therefore, there is only one orbit of $\mathcal{S}$ under G. Let $P \in \mathcal{S}$. Then $G_P = \{g \in G \mid g \cdot P = P\} = \{g \in G \mid gPg^{-1} = P\} = N(P)$. Thus, by Lemma 5.4.6,

$$|\mathcal{S}| = \text{number of elements in the orbit of } P = [G : G_P].$$

But $[G : G_P]$ divides $|G|$. Consequently, the number of Sylow p-subgroups of G divides $|G|$. $\blacksquare$

7.3.1 Worked-Out Exercises

$\diamond$ **Exercise 1** Show that every group of order 45 has a normal subgroup of order 9.

Solution: Let G be a group of order $45 = 3^2 \cdot 5$ and n_3 denote the number of Sylow 3-subgroups of G. Then $n_3 = 3k + 1$ for some integer $k \geq 0$ and $n_3 | 45$. If $k = 0$, then $n_3 = 1$, which divides 45. But for any $k > 1$, n_3 does not divide 45. Hence, G contains a unique Sylow 3-subgroup H of order 9. Consequently, G has a normal subgroup of order 9.

$\diamond$ **Exercise 2** Let G be a finite group of order $p^m q$, where p and q are relatively prime, and P be a subgroup of order p^m, where p is a prime. Show that P is the only Sylow p-subgroup of order p^m lying in $N(P)$.

Solution: Clearly $|N(P)| = p^m r$ for some $r \leq q$ and p and r are relatively prime. Let P' be any other Sylow p-subgroup of G such that $P' \subseteq N(P)$. Then P and P' are Sylow p-subgroups of $N(P)$. Thus, there exists $x \in N(P)$ such that $P' = xPx^{-1}$. Since P is normal in $N(P)$, $P = xPx^{-1}$. Hence, $P' = P$.

◇ **Exercise 3** Let G be a finite group and p a prime such that p divides $|G|$.

(i) Let K be a normal subgroup of G. Show that for any Sylow p-subgroup P of G, $P \cap K$ is a Sylow p-subgroup of K. Conversely, if B is any Sylow p-subgroup of K, show that there exists a Sylow p-subgroup P of G such that $B = P \cap K$.

(ii) Let H be a normal subgroup of G. If P is a Sylow p-subgroup of G, show that PH/H is a Sylow p-subgroup of G/H. Conversely, show that any Sylow p-subgroup of G/H is of the form PH/H, where P is a Sylow p-subgroup of G.

Solution: (i) Let $|G| = p^m q$, where p and q are relatively prime. Let P be a Sylow p-subgroup of G. Then $|P| = p^m$. Since $|P \cap K|$ divides $|P|$, $|P \cap K| = p^i$ for some $i \leq m$. Hence, $P \cap K$ is a p-group. Let $|K| = p^s t$, where p and t are relatively prime and $s \geq i$. Suppose $s > i$. Now $|PK| = \frac{|P||K|}{|P \cap K|} = \frac{p^m p^s t}{p^i} = p^m p^j t = p^{m+j}$, $j = s - i \geq 1$, which is impossible since $|G| = p^m q$ and PK is a subgroup of G. Thus, $s = i$. Hence, $|P \cap K| = p^s$, i.e., $P \cap K$ is a Sylow p-subgroup of K. Conversely, let B be a Sylow p-subgroup of K. Let $|K| = p^s t$, where p and t are relatively prime. Then $|B| = p^s$. Now $P \cap K$ is a Sylow p-subgroup of K for any Sylow p-subgroup P of G. Then there exists $a \in K$ such that $B = a^{-1}(P \cap K)a = a^{-1}Pa \cap a^{-1}Ka = Q \cap K$, where $Q = a^{-1}Pa$. Clearly Q is a Sylow p-subgroup of G.

(ii) Let $|G| = p^m q$, where p and q are relatively prime. Let P be a Sylow p-subgroup of G. Then $|P| = p^m$. Let $|H| = p^s t$, where p and t are relatively prime. Now $P \cap H$ is a Sylow p-subgroup of H. Hence, $|P \cap H| = p^s$. Now $|PH/H| = \frac{|PH|}{|H|} = \frac{|P||H|}{|H||P \cap H|} = \frac{|P|}{|P \cap H|} = \frac{p^m}{p^s} = p^{m-s}$. Also, $|G/H| = \frac{|G|}{|H|} = \frac{p^m q}{p^s t} = p^{m-s} r$. Hence, PH/H is a Sylow p-subgroup of G/H. Conversely, let B/H be a Sylow p-subgroup of G/H. Now PH/H is a Sylow p-subgroup of G/H for any Sylow p-subgroup P of G. Therefore, there exists $aH \in G/H$ such that $B/H = a^{-1}H(PH/H)aH$. Now for all $b \in PH$, $a^{-1}HbHaH \in B/H$, and hence for all $b \in PH$, $a^{-1}ba \in B$. Thus, $a^{-1}(PH)a \subseteq B$. Let $Q = a^{-1}Pa$. Then Q is a Sylow p-subgroup of G. Also, $a^{-1}Ha = H$ since H is normal. Now $QH = (a^{-1}Pa)(a^{-1}Ha) = a^{-1}(PH)a \subseteq B$. Let $c \in B$. Then $cH \in B/H = a^{-1}H(PH/H)aH$. Therefore, $cH = a^{-1}HbHaH = a^{-1}baH$ for some $b \in PH$. Let $b = uv$ for some $u \in P$, $v \in H$. Then $a^{-1}ba = a^{-1}uva = (a^{-1}ua)(a^{-1}va) \in (a^{-1}Pa)H = QH$. Now $cH = a^{-1}baH$ implies $c^{-1}(a^{-1}ba) \in H \subseteq QH$. Hence, $c^{-1} \in QH$ or $c \in QH$. Thus, $B = QH$.

◇ **Exercise 4** Let H be a normal subgroup of a finite group G and p be a prime dividing the order of G. If $[G : H]$ and p are relatively prime, prove that H contains all Sylow p-subgroups of G. Show by an example that the result need not be true if H is not normal in G.

Solution: Let $|G| = p^k m$, where p and m are relatively prime. Let $|G/H| = [G : H] = q$. Then it follows that $q|m$. Thus, p^k divides $|H|$ since $|G| = q|H|$. Hence, $|H| = p^k r$, where p and r are relatively prime. Let P be a Sylow p-subgroup of H. Then $|P| = p^k$. Hence, P is a Sylow p-subgroup of G. If Q is any other Sylow p-subgroup of G, then there exists $x \in G$ such that $Q = x^{-1}Px$. Hence, $Q = x^{-1}Px \subseteq x^{-1}Hx = H$.

Consider $G = S_3$ and let $H = \{e, (1\ 2)\}$. Then H is a subgroup of G, which is not normal. Now $[G : H] = 3$, $p = 2$ divides $|G|$. But H does not contain all Sylow 2-subgroups of G. The Sylow 2-subgroups of G are $\{e, (1\ 2)\}, \{e, (1\ 3)\}$, and $\{e, (2\ 3)\}$.

$\diamond$ **Exercise 5** Show that a group of order 96 has a normal subgroup of order 16 or 32.

Solution: Let G be a group of order $96 = 2^5 \cdot 3$. Let n_2 denote the number of Sylow 2-subgroups of G. Now $n_2 = 2k+1$ for some integer $k \geq 0$ and n_2 divides 96. Then $n_2 = 1$ or 3. If $n_2 = 1$, then G contains a unique Sylow 2-subgroup of order 32. This subgroup of order 32 must be a normal subgroup by Theorem 7.3.7. Suppose $n_2 = 3$. Then G has three Sylow 2-subgroups A, B, and C, each of order 32. Let us now show that $|A \cap B| = 16$. Since $A \neq B$ and $|A \cap B|$ divides $|A|$, $|A \cap B| = 1, 2, 4, 8$, or 16. If $|A \cap B| \leq 8$, then $|AB| = \frac{|A||B|}{|A \cap B|}$ shows that $|AB| \geq \frac{32 \cdot 32}{8} = 128 > 96 = |G|$, a contradiction. Hence, $|A \cap B| = 16$. Since $[A : A \cap B] = 2$ and $[B : A \cap B] = 2$, $A \cap B$ is a normal subgroup of A and B. Thus, $A, B \subseteq N(A \cap B)$. Therefore, $AB \subseteq N(A \cap B)$. This implies that $|N(A \cap B)| \geq |AB| = \frac{|A||B|}{|A \cap B|} = \frac{32 \cdot 32}{16} = 64$. Since $N(A \cap B)$ is a subgroup of G, it follows that $|N(A \cap B)| = 96$. Thus, $N(A \cap B) = G$ and so $A \cap B$ is a normal subgroup of G of order 16.

$\diamond$ **Exercise 6** If a group G of order 52 contains a normal subgroup of order 4, show that G is a commutative group.

Solution: Suppose G contains a normal subgroup H of order 4. Then H is a commutative group. Now $|G| = 13 \cdot 4$. Let n_{13} denote the number of Sylow 13-subgroups of G. Then $n_{13} = 13k + 1$ for some integer $k \geq 0$ and n_{13} divides 52. Thus, $n_{13} = 1$ and so G contains a unique Sylow 13-subgroup, say, A. Then A is a normal subgroup of order 13 and $A \cap H = \{e\}$. Since $|AH| = \frac{|A||H|}{|A \cap H|} = 52$, we find that $G = A \times H$. Since A and H are both commutative, G is commutative.

Exercise 7 Let G be a finite group. Suppose that every Sylow subgroup of G is normal in G. Prove that G is the internal direct product of its Sylow subgroups.

Solution: Let $|G| = p_1^{n_1} p_2^{n_2} \cdots p_k^{n_k}$, where p_i are distinct primes. Since every Sylow p-subgroup of G is normal, there exists a unique Sylow p-subgroup for $p = p_i$ $(i = 1, 2, \ldots, k)$. Let $S(p_i)$ be the Sylow p_i-subgroup of G for all i. Then $S(p_i)$ is a normal subgroup of G and $S(p_i) \cap S(p_j) = \{e\}$ for all $i \neq j$. Hence, $a_i a_j = a_j a_i$ for all $a_i \in S(p_i)$ and $a_j \in S(p_j)$. Now consider, $S(p_i) \cap (S(p_1) \cdots S(p_{i-1}) S(p_{i+1}) \cdots S(p_k))$. Suppose

$$a \in S(p_i) \cap (S(p_1) \cdots S(p_{i-1}) S(p_{i+1}) \cdots S(p_k)).$$

Then $a \in S(p_i)$ and $a \in S(p_1) \cdots S(p_{i-1}) S(p_{i+1}) \cdots S(p_k)$. Hence,

$$a = a_1 \cdots a_{i-1} a_{i+1} \cdots a_k,$$

where $a_j \in S(p_j)$. Now

$$o(a) | p_1^{n_1} p_2^{n_2} \cdots p_{i-1}^{n_{i-1}} p_{i+1}^{n_{i+1}} \cdots p_k^{n_k}$$

and $o(a) | p_i^{n_i}$. Consequently, $o(a) = 1$, i.e., $a = e$. Thus,

$$S(p_i) \cap (S(p_1) \cdots S(p_{i-1}) S(p_{i+1}) \cdots S(p_k)) = \{e\}.$$

This implies that $|S(p_1) \cdots S(p_k)| = p_1^{n_1} p_2^{n_2} \cdots p_k^{n_k} = |G|$ and hence

$$G = S(p_1) \cdots S(p_k).$$

Thus, $G = S(p_1) \times S(p_2) \times \cdots \times S(p_k)$.

7.3.2 Exercises

1. Find the Sylow 3-subgroups of S_4.

2. Prove that if G is a group of order p^n, p a prime, then G contains a normal subgroup of order p^t for every nonnegative integer $t \leq n$.

3. Prove that a group G has only one proper subgroup if and only if G is a cyclic group of order p^2 for some prime p.

4. Prove that for any group G, $|G/Z(G)| \neq 91$.

5. Let G be a finite group and P be a Sylow p-subgroup of G. Let H be a subgroup of G such that $N_G(P) \subseteq H$. Prove that $N_G(H) = H$.

6. Let G be a finite group, P and H be subgroups of G such that P is a normal subgroup of H, and H is a normal subgroup of G. Show that if P is a Sylow p-subgroup of G, then P is a normal subgroup of G.

7. Let G be a group of order 143. Show that Sylow 11-subgroup of G is unique. Also, show that G is cyclic.

8. Let G be a finite group, H be a normal subgroup of G, and P be a Sylow p-subgroup of H. Show that $G = HN_G(P)$.

9. Let G be a finite commutative group. Show that G is the internal direct product of its Sylow subgroups.

10. Let G be a finite group and K be a normal subgroup of G. If K is a p-subgroup, prove that K is contained in every Sylow p-subgroup of G.

11. Let G be a finite group and suppose $|G| = p^k n$, where p is a prime and p and n are relatively prime. Prove that every p-subgroup of G is contained in some Sylow p-subgroup of G.

12. Let G be a group such that $|G| = p^m$, where p is a prime. Let H be a proper subgroup of G. Prove that there exists $a \in G$, $a \notin H$ such that $aHa^{-1} = H$.

7.4 Some Applications of the Sylow Theorems

We recall that a group $G \neq \{e\}$ is called simple if it has no nontrivial normal subgroups. If G is commutative, then it follows from Lagrange's theorem that G is simple if and only if G is of prime order. In Galois's mathematical legacy to us, he wrote in a letter to a friend on the eve of his death stating that the alternating group A_5 is the smallest noncommutative simple group. William Burnside conjectured in 1911 that no noncommutative simple group of odd order exists. The mathematicians John Thompson and Walter Feit proved in 1963 that Burnside's conjecture was true. John Thompson received the Fields Medal for his work on this and other problems.

In this section, we apply the Sylow theorems to determine some finite groups which are not simple.

Example 7.4.1 *Let G be a group of order 10. Now $10 = 5 \cdot 2$. Let n_5 denote the number of Sylow 5-subgroups of G. From Sylow Theorem 7.3.12, $n_5 = 5k+1$ for some integer $k \geq 0$ and n_5 divides $|G| = 10$. Thus, $n_5 = 1$ and so there exists only one Sylow 5-subgroup, say, H in G. Since H is a unique Sylow 5-subgroup, H is a normal subgroup of G by Corollary 7.3.11, proving that G is not simple. Thus, no group of order 10 is simple.*

Example 7.4.2 *Let G be a group of order 9. Then G is a p-group, where $p = 3$. From Theorem 7.2.7, we find that $Z(G) \neq \{e\}$. If $G = Z(G)$, then G is a commutative group. But commutative simple groups are precisely groups of prime order. Hence, in this case G is not simple. Suppose $Z(G) \neq G$. Then $Z(G)$ is a nontrivial normal subgroup of G. Thus, we find that a group of order 9 is not a simple group.*

In Example 7.4.2, we showed that a group of order $9 = 3^2$ is not simple. In the next theorem, we prove that, in general, if G is a p-group of order p^n, $n > 1$, then G is not simple.

Theorem 7.4.3 *Let p be a prime integer and $n > 1$ be any integer. Then no group of order p^n is simple.*

Proof. Let G be a group of order p^n. Consider the center $Z(G)$ of G. From Theorem 7.2.7, it follows that $Z(G) \neq \{e\}$. If $G = Z(G)$, then G is a commutative group. If G is simple, then $|G|$ is prime, which is a contradiction. Thus, in this case G is not simple. Suppose $Z(G) \neq G$. Then $Z(G)$ is a nontrivial normal subgroup of G, proving that G is not a simple group. ∎

Theorem 7.4.4 *Let p and q be two prime integers. Then no group of order pq is simple.*

Proof. Let G be a group of order pq. If $p = q$, then $|G| = p^2$ and so by Theorem 7.4.3, G is not simple. Suppose now $p \neq q$. Let $p > q$. Let n_p denote the number of Sylow p-subgroups of G. Then $n_p = pk + 1$ for some integer $k \geq 0$ and n_p divides pq. Since $\gcd(1 + kp, p) = 1$, n_p does not divide p. Hence, n_p divides q. Thus, $1 + kp \leq q$. But $p > q$. Therefore, $1 + kp \leq q$ holds only if $k = 0$. This implies that $n_p = 1$ and so G contains a unique Sylow p-subgroup of order p, which must be normal by Corollary 7.3.11. Hence, G is not simple. ∎

At this point let us recall the following result established in Worked-Out Exercise 5 (page 177).

In a group G of order pn, where p is a prime and $p > n$, if H is a subgroup of order p, then H is a normal subgroup. Now from Cauchy's theorem, any group of order pn, p prime, contains a subgroup of order p. Consequently, G contains a normal subgroup of order p.

Let G be a group of order $n \leq 60$. Applying the above result, we find that if $n = 6 \, (= 3 \cdot 2)$, $10 \, (= 5 \cdot 2)$, $14 \, (= 7 \cdot 2)$, $15 \, (= 5 \cdot 3)$, $20 \, (= 5 \cdot 4)$, $21 \, (= 7 \cdot 3)$, $22 \, (= 11 \cdot 2)$, $26 \, (= 13 \cdot 2)$, $28 \, (= 7 \cdot 4)$, $33 \, (= 11 \cdot 3)$, $34 \, (= 17 \cdot 2)$, $35 \, (= 7 \cdot 5)$, $38 \, (= 19 \cdot 2)$, $39 \, (= 13 \cdot 3)$, $42 \, (= 7 \cdot 6)$, $44 \, (= 11 \cdot 4)$, $46 \, (= 23 \cdot 2)$, $51 \, (= 17 \cdot 3)$, $52 \, (= 13 \cdot 4)$, $55 \, (= 11 \cdot 5)$, $57 \, (= 19 \cdot 3)$, or $58 \, (= 29 \cdot 2)$, then G is not simple.

In Worked-Out Exercise 7 (page 178), we have established that any group of order $2n$, where n is an odd integer, contains a normal subgroup of order n. Using this result, we find that no groups of order $6 \, (= 2 \cdot 3)$, $18 \, (= 2 \cdot 9)$, $50 \, (= 2 \cdot 25)$, $54 \, (= 2 \cdot 27)$, are simple.

Next, let us recall the following result established in Corollary 5.4.10. Let G be a finite group and H a proper subgroup of G of index n such that $|G|$ does not divide $n!$ Then G contains a nontrivial normal subgroup.

Now suppose G is a group of order $12 = 2^2 \cdot 3$. From Theorem 7.3.6, we find that G contains a Sylow 2-subgroup H of order 4. Thus, the index of H is 3. Now $|G| = 12$ does not divide 3! Therefore, G contains a nontrivial normal subgroup, proving that G is not simple. Proceeding this way with the help of the above result, we can show that no group of order 24 $(= 2^3 \cdot 3)$, 36 $(= 3^2 \cdot 4)$, 45 $(= 3^2 \cdot 5)$, or 48 $(= 2^4 \cdot 3)$ is a simple group.

Example 7.4.5 *In this example, we show that no group of order 40 is simple. Let G be a group of order $40 = 5 \cdot 8$. Let n_5 denote the number of Sylow 5-subgroups of G. By Sylow Theorem 7.3.12, $n_5 = 5k + 1$ for some integer $k \geq 0$ and n_5 divides 40. Hence, $n_5 = 1$. Thus, G has a unique Sylow 5-subgroup which must be normal by Corollary 7.3.11. Hence, G is not simple.*

Example 7.4.6 *In this example, we show that no group of order 56 is simple. Let G be a group of order $56 = 7 \cdot 2^3$. Let n_7 denote the number of Sylow 7-subgroups and n_2 denote the number of Sylow 2-subgroups of G. By Sylow's third theorem (Theorem 7.3.12), $n_7 = 7m + 1$ and $n_2 = 2k + 1$ for some integers m, $k \geq 0$. Now n_7 divides 56. Thus, $n_7 = 1$ or 8. If $n_7 = 1$, then G has a unique Sylow 7-subgroup which must be normal. Hence, G is not simple. Suppose $n_7 = 8$. Then G has eight Sylow 7-subgroups $A_1, A_2, \ldots, A_8$. Now $|A_i| = 7$, $i = 1, 2, \ldots, 8$. Also, $A_i \cap A_j = \{e\}$ for $i \neq j$ and for all $a \neq e$, $a \in A_i$, $o(a) = 7$. Thus, G contains 48 elements of order 7. Now $n_2 = 1$ or 7. If $n_2 = 1$, then G has a unique Sylow 2-subgroup which must be normal. Hence, G is not simple. Suppose $n_2 = 7$. Then G has seven Sylow 2-subgroups $B_1, B_2, \ldots, B_7$. Each B_i contains eight elements. Since $B_1 \neq B_2$, $|B_1 \cap B_2| \leq 4$. This implies that $B_1 \cup B_2$ contains at least 12 elements, none of which is of order 7. Hence, $|G| \geq 48 + 12 = 60$, a contradiction. Thus, we find that either $n_7 = 1$ or $n_2 = 1$, showing that G has either a normal subgroup of order 7 or a normal subgroup of order 8. Consequently, G is not simple.*

In Worked-Out Exercise 1 (page 216), we show that a group of order 30 is not simple. By Theorem 7.4.3, no group of order $4 = 2^2$, $8 = 2^3$, $9 = 3^2$, $16 = 2^4$, $25 = 5^2$, $27 = 3^3$, $32 = 2^5$, or $49 = 7^2$ is simple. We now summarize the above results.

Theorem 7.4.7 *Let n be an integer such that $1 \leq n < 60$ and n is not prime. Then no group of order n is simple.* ■

Let us now concentrate our discussion on $n = 60$. Since 60 is not prime, no commutative group of order 60 is simple. Now what is the answer if G is a noncommutative group of order 60? Recall that A_5 is a simple group of order 60. Hence, we find that there exists a noncommutative simple group of order 60. Next, let us ask the following question. Is A_5 the only (up to isomorphism) noncommutative simple group of order 60? To answer this question, we first prove the following result.

Lemma 7.4.8 *Let G be a simple group of order* 60. *Then G contains a subgroup of order* 12.

Proof. Suppose G has no subgroup of order 12. Now $|G| = 60 = 5 \cdot 3 \cdot 2^2$. Let n_5 denote the number of Sylow 5-subgroups and n_2 denote the number of Sylow 2-subgroups of G. By Sylow Theorem 7.3.12, $n_5 = 5m + 1$ for some integer $m \geq 0$ and n_5 divides 60. Thus, $n_5 = 1$ or 6. Since G is simple, $n_5 \neq 1$. Hence, $n_5 = 6$. Then G has six Sylow 5-subgroups $A_1, A_2, \ldots, A_6$. Now $|A_i| = 5$, $i = 1, 2, \ldots, 6$. Also, $A_i \cap A_j = \{e\}$ for $i \neq j$ and for all $e \neq a \in A_i$, $o(a) = 5$. Thus, G contains 24 elements of order 5. Now $n_2 = 1, 3, 5,$ or 15. Since G is simple, $n_2 \neq 1$. Suppose $n_2 = 15$. Let B_i, $i = 1, 2, \ldots, 15$, be the 15 Sylow 2-subgroups of G. If $B_i \cap B_j = \{e\}$ for $1 \leq i \neq j \leq 15$, then $\cup_{i=1}^{15} B_i$ contains 46 elements of order not equal to 5. Hence, $60 = |G| \geq 24 + 46 = 70$, a contradiction. Therefore, there exist i, j such that $B_i \cap B_j \neq \{e\}$. Then $|B_i \cap B_j| = 2$. This implies that $B_i \cap B_j$ is a normal subgroup of B_i and B_j. Thus, $B_i, B_j \subseteq N(B_i \cap B_j)$ and so $B_i B_j \subseteq N(B_i \cap B_j)$. Hence, $|N(B_i \cap B_j)| \geq |B_i B_j| = 8$. Since $N(B_i \cap B_j)$ is a subgroup of G and $|N(B_i \cap B_j)| \geq 8$, it follows that $|N(B_i \cap B_j)| = 12, 20, 30,$ or 60. Now $|N(B_i \cap B_j)| \neq 30$ for then $N(B_i \cap B_j)$ is normal in G. Also, from our assumption, $|N(B_i \cap B_j)| \neq 12$. If $|N(B_i \cap B_j)| = 20$, then from Corollary 5.4.10, G contains a nontrivial normal subgroup, which is a contradiction. Hence, $|N(B_i \cap B_j)| = 60$, proving that $B_i \cap B_j$ is a normal subgroup of G, which is also a contradiction. Suppose $n_2 = 3$ or 5. Let B be a Sylow 2-subgroup of G. Then $1 + 2k = n_2 = [G : N(B)]$. Thus, $N(B) \neq B$ and so $|N(B)| \neq 4$. But 4 divides $|N(B)|$ and $|N(B)|$ divides 60. Hence, $|N(B)| = 12, 20,$ or 60. Proceeding as above, we again get a contradiction. Consequently, G must contain a subgroup of order 12. ∎

Theorem 7.4.9 *Any simple group of order* 60 *is isomorphic to* A_5.

Proof. Let G be a simple group of order 60. By Lemma 7.4.8, G contains a subgroup H of order 12. Since $[G : H] = 5$, it follows that there exists a nontrivial homomorphism $f : G \to S_5$ such that Ker $f \subseteq H$ by Corollary 5.4.9. Since G is simple, Ker $f = \{e\}$. Hence, G is isomorphic to a subgroup, say, T, of S_5. We show that $T = A_5$. This will follow if we can show that T does not contain any odd permutation. Suppose T contains an odd permutation. Then the set of all even permutations is a normal subgroup of T of index 2. This implies that the group G, which is isomorphic to T, contains a nontrivial normal subgroup, a contradiction. Therefore, $T \subseteq A_5$. But $60 = |G| = |T| = |A_5|$. Consequently, $T = A_5$ and so $G \simeq A_5$. ∎

From Theorem 7.4.9, it follows that A_5 is the smallest noncommutative simple group.

The complete classification of simple groups was given in 1981. Hundreds of mathematicians contributed to this outstanding accomplishment. Two major contributors other than Thompson and Feit were M. Aschbacher and R.L. Griess. Certain troublesome groups appeared in the classification of simple groups. The largest of these **sporadic** groups was constructed by Griess. This group, known as the **monster**, has order approximately 8×10^{53}. Other names associated with the determination of simple groups are Emil Mathieu (1838–1890), F.N. Cole (1861–1927), G.A. Miller, Leonard Eugene Dickson (1874–1954), Jean Dieudonné, Claud Chevalley, Richard Brauer, F.A. Fowler, Daniel Gorenstein, and John H. Conway.

Let us now apply the Sylow theorems to classify some groups of small order.

Example 7.4.10 *Let G be a group of order $15 = 5 \cdot 3$. By Sylow's third theorem (Theorem 7.3.12), G has a Sylow 5-subgroup A and a Sylow 3-subgroup B. It is easy to check that A is a unique Sylow 5-subgroup and B is a unique Sylow 3-subgroup of G. Hence, A is a normal subgroup of order 5 and B is a normal subgroup of order 3. Now $A \cap B = \{e\}$. Thus, $|AB| = \frac{|A||B|}{|A \cap B|} = 15$. Hence, $G = AB$, $A \cap B = \{e\}$, and A and B are normal subgroups of G. Thus, $G = A \times B \simeq \mathbf{Z}_5 \times \mathbf{Z}_3 \simeq \mathbf{Z}_{15}$ since $\gcd(3,5) = 1$. Hence, G is a cyclic group.*

In the next theorem, we classify all groups of order pq, where p and q are distinct primes.

Theorem 7.4.11 *Let G be a group and p, q be primes with $p > q$. If $|G| = pq$, then G is either cyclic or generated by two elements a and b satisfying the following properties: $b^p = e$, $a^q = e$, and $a^{-1}ba = b^r$, where p does not divide $(r-1)$, but $p|(r^q - 1)$. The second possibility can occur only if $q|(p-1)$.*

Proof. By Cauchy's theorem, G contains an element b of order p. Set $P = \langle b \rangle$. Since P is a Sylow p-subgroup of G, it has $1 + mp$ conjugates for some nonnegative integer m. Now $1 + mp = [G : N(P)]$, which divides $|G| = pq$. Since $1 + mp$ and p are relatively prime, $(1 + mp)|q$. However, $q < p$ so that $m = 0$. Hence, P is a normal subgroup of G.

Now G contains an element a of order q. Set $S = \langle a \rangle$. Then S is a Sylow q-subgroup of G. Hence, $[G : N(S)] = 1 + kq$ for some nonnegative integer k. As above, $1 + kq$ divides p. Thus, either $k = 0$ or $q|(p-1)$. If $k = 0$, then S is a normal subgroup of G so that $G \simeq P \times S$. That is, $G \simeq \mathbf{Z}_p \times \mathbf{Z}_q \simeq \mathbf{Z}_{pq}$.

Suppose $q|(p-1)$. Then S is not a normal subgroup of G. However, since P is a normal subgroup of G, $a^{-1}ba = b^r$ for some integer r. We may assume $p \nmid (r-1)$ else we return to the commutative case. By induction on j, it follows that $a^{-j}ba^j = b^{r^j}$. In particular, if $j = q$, we have $b = b^{r^q}$ so that $p|(r^q - 1)$. ∎

Corollary 7.4.12 *Let G be a group of order pq, p and q be primes with $p > q$. If q does not divide $p - 1$, then G is cyclic.* ∎

In Chapter 5, we defined and studied D_4, the dihedral group of degree 4. Let us now define the dihedral group D_n of degree $n \geq 3$.

Definition 7.4.13 *A group G is called a **dihedral group** of degree $n \geq 3$ if G is generated by two elements a, b such that*
   (i) $o(a) = n$, $b^2 = e$, and
   (ii) $ba = a^{-1}b$.

We denote a dihedral group of degree $n \geq 3$ by D_n.

Example 7.4.14 *Consider the symmetric group S_n $(n \geq 3)$. The subgroup G generated by*

$$a = (1\ 2\ 3\ \cdots\ n),$$

$$b = \begin{pmatrix} 1 & 2 & 3 & \cdots & i & \cdots & n-1 & n \\ 1 & n & n-1 & \cdots & n+2-i & \cdots & 3 & 2 \end{pmatrix}$$

is an example of a dihedral group of degree n.

We leave the proof of the following theorem as an exercise.

Theorem 7.4.15 *Let G be a dihedral group of degree $n \geq 3$. Then G has $2n$ elements.* ∎

Theorem 7.4.16 *Let G be a group and p be an odd prime. If $|G| = 2p$, then G is either cyclic or dihedral.*

Proof. By Cauchy's theorem, G contains an element a of order p and an element b of order 2. Let $H = \langle a \rangle$. Then H is a normal subgroup of G since $[G : H] = 2$. Now $bab = bab^{-1} \in H$. Hence, there exists $a^i \in H$ such that $bab = a^i$, where $0 \leq i < p$. Now $a^{i^2} = (a^i)^i = (bab)^i = (bab^{-1})^i = ba^ib$. Again from $bab = a^i$, we find that $a = ba^ib$. Hence, $a = a^{i^2}$. This implies that $a^{i^2-1} = e$. Since $o(a) = p$, it follows that $p|(i^2 - 1)$. Therefore, $p|(i-1)$ or $p|(i+1)$ since p is prime. Suppose $p|(i-1)$. Then $i - 1 = 0$, i.e., $i = 1$. Thus, $bab = a$, which implies $ba = ab$. So in this case, we find that G contains an element of order $2p$ and so G is a cyclic group. If $p|(i+1)$, then $bab = a^{-1}$. Hence, G is generated by a, b such that $o(a) = p$, $o(b) = 2$, and $ba = a^{-1}b$. In this case, G is the dihedral group D_p. ∎

Let us now classify groups of order $n \leq 10$.

Let G be a group of order $n \leq 10$. If $n = 1$, then $G = \{e\}$ and thus is cyclic. If $n = 2, 3, 5$, or 7, then G is of prime order and hence cyclic. For $n = 4$, we know that G is isomorphic to either $\mathbf{Z}_4$ or $\mathbf{Z}_2 \times \mathbf{Z}_2$. If $n = 6$, then G is isomorphic to either $\mathbf{Z}_6$ or $S_3 \simeq D_3$. For $n = 8$, if G is noncommutative, then

G is isomorphic to either D_4 or Q_8. If G is commutative, then in Chapter 9 we will show that G is isomorphic to either $\mathbf{Z}_8$, $\mathbf{Z}_4 \times \mathbf{Z}_2$, or $\mathbf{Z}_2 \times \mathbf{Z}_2 \times \mathbf{Z}_2$.

Let us now consider the case $n = 9$. Then G has order 3^2. Since 3 is prime, G is commutative. Let $e \neq a \in G$. Then $o(a) = 3$ or 9. If $o(a) = 9$, then $G \simeq \mathbf{Z}_9$. Suppose G has no elements of order 9. Then $o(a) = 3$. Let $H = \{e, a, a^2\}$. Then H is a subgroup of G and $|H| = 3$. Let $b \in G$ be such that $b \notin H$. Let $K = \{e, b, b^2\}$. Now H and K are normal subgroups of G, $H \cap K = \{e\}$, and $G = HK$. Hence, $G = H \times K \simeq \mathbf{Z}_3 \times \mathbf{Z}_3$. Thus, either $G \simeq \mathbf{Z}_9$ or $G \simeq \mathbf{Z}_3 \times \mathbf{Z}_3$.

Suppose now $n = 10$. Then from Theorem 7.4.16, it follows that either $G \simeq \mathbf{Z}_{10}$ or $G \simeq D_5$. Hence, there are (up to isomorphism) two distinct groups of order 10.

We summarize the above discussion in the following table:

Order of the group	Number of Groups	Groups
1	1	$\{e\} = \mathbf{Z}_0$
2	1	$\mathbf{Z}_2$
3	1	$\mathbf{Z}_3$
4	2	$\mathbf{Z}_4, \mathbf{Z}_2 \times \mathbf{Z}_2$
5	1	$\mathbf{Z}_5$
6	2	$\mathbf{Z}_6, S_3$
7	1	$\mathbf{Z}_7$
8	5	$\mathbf{Z}_8, \mathbf{Z}_4 \times \mathbf{Z}_2, \mathbf{Z}_2 \times \mathbf{Z}_2 \times \mathbf{Z}_2, D_4, Q_8$
9	2	$\mathbf{Z}_9, \mathbf{Z}_3 \times \mathbf{Z}_3$
10	2	$\mathbf{Z}_{10}, D_5$

In the Worked-Out Exercises below, we illustrate several techniques that can be effectively used to find the Sylow subgroups of a group.

7.4.1 Worked-Out Exercises

◇ **Exercise 1** Let G be a group of order 30. Show that G is not simple.

Solution: Since $|G| = 30 = 2 \cdot 3 \cdot 5$, G has a Sylow 2-subgroup, a Sylow 3-subgroup, and a Sylow 5-subgroup. Consider Sylow 5-subgroups. The number of Sylow 5-subgroups is $1 + 5k$, where $1 + 5k | 6$. Thus, $k = 0$ or 1. If $k = 0$, then G has only one Sylow 5-subgroup, and hence this unique Sylow 5-subgroup must be normal in G. Therefore, in this case, G is not simple. Suppose $k = 1$. Then G has six distinct Sylow 5-subgroups, say, $H_1, H_2, \ldots, H_6$. Now for $i \neq j$, $|H_i \cap H_j| = 1$ since $H_i \cap H_j$ is a subgroup of H_i. Thus, the six Sylow 5-subgroups contain 24 distinct elements of order 5. Now consider Sylow 3-subgroups. The number of Sylow 3-subgroups is $1 + 3k_1$, where $1 + 3k_1 | 10$. Thus, $k_1 = 0$ or 3. If $k_1 = 0$, then G has a unique Sylow 3-subgroup, which must be normal in G, and hence, in this case, G is not simple. Suppose $k_1 = 3$. Then G has 10 distinct Sylow 3-subgroups. As in the case of Sylow 5-subgroups, we conclude that if

$k_1 = 3$, then G has 20 distinct elements of order 3. Thus, $|G| \geq 24 + 20 = 44$, a contradiction since G has only 30 elements. Hence, if $k = 1$, then $k_1 = 0$. Thus, G either has a Sylow 5-subgroup normal in G or a Sylow 3-subgroup normal in G.

◇ **Exercise 2** Let G be a group of order 36. Prove that G is not simple.

   Solution: (We have already established that a group of order 36 is not simple on page 212. Our objective here is to show some different techniques that can be used in other cases.) Since $|G| = 36 = 2^2 \cdot 3^2$, G has a Sylow 3-subgroup of order 9. The number of Sylow 3-subgroups is $1 + 3k$, where $(1 + 3k)|4$. Thus, $k = 0$ or 1. If $k = 0$, then G has only one Sylow 3-subgroup which must be normal in G. Suppose $k = 1$. Then G has four distinct Sylow 3-subgroups, say, H_1, H_2, H_3, H_4. Consider H_1 and H_2. Now $H_1 \cap H_2$ is a subgroup of H_1 (and also of H_2). Since $|H_1| = 9$ and the order of $H_1 \cap H_2$ divides the order of H_1, $|H_1 \cap H_2| = 1, 3$, or 9. If $|H_1 \cap H_2| = 9$, then $H_1 = H_2$, which is a contradiction. Suppose $|H_1 \cap H_2| = 1$. Then $|H_1 H_2| = \frac{|H_1||H_2|}{|H_1 \cap H_2|} = \frac{9 \cdot 9}{1} = 81$, i.e., $H_1 H_2$ has 81 elements, which is a contradiction since G has only 36 elements. Hence, $|H_1 \cap H_2| = 3$. By Worked-Out Exercise 5 (page 199), $H_1 \cap H_2$ is a normal subgroup of H_1 and H_2. Therefore, $H_1, H_2 \subseteq N(H_1 \cap H_2)$. As before, $H_1 H_2$ has 27 elements since $H_1 \cap H_2$ has three elements. Thus, $N(H_1 \cap H_2)$ has at least 27 elements. Since $N(H_1 \cap H_2)$ is a subgroup of G, the order of $N(H_1 \cap H_2)$ divides the order of G. Therefore, $|N(H_1 \cap H_2)| = 36$ and so $N(H_1 \cap H_2) = G$. Hence, $H_1 \cap H_2$ is a normal subgroup of G and so G is not simple.

◇ **Exercise 3** Let G be a group of order $231 = 3 \cdot 7 \cdot 11$.

   (i) Show that a Sylow 11-subgroup of G is normal in G.

   (ii) Show that a Sylow 7-subgroup of G is normal in G.

   (iii) Show that G has a cyclic subgroup of order 77.

   (iv) Let H be a Sylow 11-subgroup of G, K be a Sylow 7-subgroup of G, and L be a Sylow 3-subgroup of G. Show that $G = HKL$.

   (v) Show that $H \subseteq Z(G)$.

   Solution: By Theorem 7.3.5, G has a Sylow 11-subgroup, a Sylow 7-subgroup, and a Sylow 3-subgroup.
   (i) The number of Sylow 11-subgroups is $1 + 11k$, where $(1 + 11k)|3 \cdot 7$. Hence, $k = 0$ and so the number of Sylow 11-subgroups is 1. Let H be the Sylow 11-subgroup of G. Since H is a unique Sylow 11-subgroup of G, H is normal in G.

(ii) The number of Sylow 7-subgroups is $1+7k$, where $(1+7k)|3\cdot 11$. Hence, $k = 0$ and so the number of Sylow 7-subgroups is 1. Let K be the Sylow 7-subgroup of G. Since K is a unique Sylow 7-subgroup of G, K is normal in G.

(iii) Since H and K are normal subgroups of G, HK is a normal subgroup of G. Now $H \cap K = \{e\}$. Thus, $|HK| = 7 \cdot 11 = 77$. Since H and K are subgroups of order 11 and 7, respectively, H and K are cyclic groups. Note that $\gcd(7, 11) = 1$. Hence, HK is a cyclic group of order 77.

(iv) Let L be a Sylow 3-subgroup of G. Then $L \cap (HK) = \{e\}$ since nonidentity elements in L are of order 3 and nonidentity elements in HK are of order $7, 11$, or 77. Now

$$|HKL| = \frac{|HK| \cdot |L|}{|L \cap (HK)|} = \frac{77 \cdot 3}{1} = 231 = |G|.$$

Hence, $G = HKL$.

(v) Since H and K are normal subgroups of G and $H \cap K = \{e\}$, $hk = kh$ for all $h \in H$, $k \in K$. Now $|G/K| = 3 \cdot 11$. Thus, G/K is a cyclic group and hence G/K is commutative. Let $a \in L$ and $b \in H$ be nonidentity elements. Then $a, b \notin K$. Since G/K is commutative, $(aK)(bK) = (bK)(aK)$ or $(ab)K = (ba)K$. Hence, $(ab)^{-1}(ba) \in K$ and so $b^{-1}a^{-1}ba \in K$. Since H is a normal subgroup of G and $b \in H$, $b^{-1}a^{-1}ba \in H$. This implies that $b^{-1}a^{-1}ba \in H \cap K = \{e\}$. Hence, $b^{-1}a^{-1}ba = e$ and so $ba = ab$. Let $x \in G$ and $h \in H$. Now $G = HKL$ and so $x = abc$ for some $a \in H$, $b \in K$, and $c \in L$. Now $xh = (abc)h = ab(ch) = ab(hc) = a(bh)c = a(hb)c = (ah)bc = (ha)bc = hx$. Therefore, $h \in Z(G)$. Hence, $H \subseteq Z(G)$.

◇ **Exercise 4** Let G be a group of order 255. Show that G is cyclic.

Solution: Now $|G| = 255 = 3 \cdot 5 \cdot 17$. Let H be a Sylow 17-subgroup of G. The number of Sylow 17-subgroups is $1 + 17m$, where $1 + 17m|15$. Hence, $m = 0$ and so G has a unique Sylow 17-subgroup. Thus, H is a normal subgroup of G. Let K be a Sylow 5-subgroup of G and L be a Sylow 3-subgroup of G. The number of Sylow 5-subgroups is $1 + 5k$, where $1 + 5k|51$. Hence, $k = 0$ or 10. The number of Sylow 3-subgroups is $1 + 3l$, where $1 + 3l|85$. Therefore, $l = 0$ or 28. Suppose $k = 10$ and $l = 28$. Then G has 51 Sylow 5-subgroups and 85 Sylow 3-subgroups. Hence, in this case G would have $51 \cdot 4 = 204$ elements of order 5 and $85 \cdot 2 = 170$ elements of order 3. This is absurd since G has only 255 elements. Thus, either $k = 0$ or $l = 0$.

Case 1. $k = 0$. Then K is the unique Sylow 5-subgroup of G and so K is normal in G. Now $H \cap K = \{e\}$. Hence, $xy = yx$ for all $x \in H$ and $y \in K$. Now $|G/K| = 3 \cdot 17$. Since 3 does not divide $(17 - 1)$, G/K is cyclic and hence commutative. Let $a \in H$ and $b \in L$. Since G/K is commutative, $aba^{-1}b^{-1} \in K$. Since H is normal and $a \in H$, $aba^{-1}b^{-1} \in H$. Hence, $aba^{-1}b^{-1} \in H \cap K = \{e\}$.

Thus, $ab = ba$ for all $a \in H$ and $b \in L$. Clearly $G = HKL$. Since H is cyclic, H is commutative. Therefore, we have (i) H is commutative, (ii) $xy = yx$ for all $x \in H$ and $y \in K$, and (iii) $ab = ba$ for all $a \in H$ and $b \in L$. This implies that $H \subseteq Z(G)$. Hence, $|Z(G)| = 17, 51, 85$, or 255 and so $|G/Z(G)| = 15, 5, 3$, or 1. In either case, $G/Z(G)$ is cyclic and hence G is commutative. Thus, G has a unique Sylow 3-subgroup. Since $G = HKL$ and H, K, and L are normal subgroups of G, G is a direct product of cyclic groups such that the order of any two factors is relatively prime and hence G is cyclic.

Case 2. $l = 0$. This case is similar to Case 1.

$\Diamond$ **Exercise 5** Let G be a group of order 455. Show that G is cyclic.

Solution: Now $|G| = 455 = 5 \cdot 7 \cdot 13$. Let H be a Sylow 13-subgroup of G. The number of Sylow 13-subgroups is $1 + 13k$, where $1 + 13k | 35$. Hence, $k = 0$ and so G has a unique Sylow 13-subgroup. Thus, H is a normal subgroup of G. Hence, $N(H) = G$. Now $|\text{Aut}(H)| = 12$. Since $N(H)/C(H) \simeq$ to a subgroup of $\text{Aut}(H)$, $|N(H)/C(H)|$ divides 12. Also, $|N(H)/C(H)|$ divides 455. Hence, $|N(H)/C(H)| = 1$ and so $G = N(H) = C(H)$. Thus, $H \subseteq Z(G)$. This implies that $|Z(G)| = 13, 65, 91$, or 455. Hence, $|G/Z(G)| = 35, 7, 5$, or 1. In either case, $G/Z(G)$ is cyclic and hence G is commutative. It now follows that G has a unique Sylow 5-subgroup, say, K, and a unique Sylow 7-subgroup, say, L. Clearly $G = H \times K \times L$. Since H, K, and L are cyclic groups of prime order and their orders are relatively prime to each other, G is cyclic.

7.4.2 Exercises

1. Show that every group of order $20, 28, 36, 48$, or 56 contains a nontrivial normal subgroup.

2. Show that no group of order 125 is simple.

3. Show that no group of order 65 is simple.

4. Show that a group of order 130 contains a nontrivial normal subgroup.

5. Show that no group of order 75 is simple.

6. Show that a group of order 96, 150, or 200 is not simple.

7. Let G be a group of order 35. Show that G is cyclic.

8. Let G be a group of order 133. Show that G is cyclic.

9. Let G be a group of order $5 \cdot 7 \cdot 19$.

    (i) Show that G has a unique subgroup of order 5.

    (ii) Show that G is cyclic.

10. Let G be a group of order 100. Suppose that G has a unique Sylow 2-subgroup. Show that G is commutative.

11. Let G be a group of order 70.

    (i) Show that G has a unique Sylow 7-subgroup.

    (ii) Show that G has a unique Sylow 5-subgroup.

    (iii) Show that G has a cyclic subgroup of order 35.

12. Let G be a group of order 385. Show that a Sylow 7-subgroup of G is in the center of G.

13. Let G be a group of order $5 \cdot 11 \cdot 19$. Show that a Sylow 19-subgroup of G is in the center of G and a Sylow 11-subgroup of G is a normal subgroup of G.

14. Let G be a group of order $3 \cdot 11 \cdot 19$. Show that a Sylow 11-subgroup of G is in the center of G and a Sylow 19-subgroup of G is a normal subgroup of G.

15. Let G be a simple group of order 168.

    (i) Show that G has eight Sylow 7-subgroups.

    (ii) Let H be a Sylow 7-subgroup. Show that $|N_G(H)| = 21$.

    (iii) Show that G has no subgroup of order 14.

16. Show that there exists (up to isomorphism) only one group of order 77.

17. Let G be a group of order 123. Show that for every positive divisor n of 123, there exists a unique subgroup of order n in G.

18. Determine up to isomorphism all groups of order 70.

19. Let G be a group of order $p^n m$, p prime, $p > m$, $n \geq 1$. Show that G is not simple.

20. Let G be a group of order $p^2 q$, p and q are distinct primes. Show that G is not simple.

21. Classify all groups of order 14.

22. Prove that D_n is a noncommutative group of order $2n$.

23. Find $Z(D_n)$.

24. Find the conjugacy classes in D_{2n} and D_{2n+1}.

25. Let G be a group of order p^2q^2, where p and q are prime integers such that $p > q$. Prove the following.

    (i) The number of Sylow p-groups cannot be q.

    (ii) If the number of Sylow p-subgroups is q^2, then $p = 3$ and $q = 2$.

26. Show that no group of order p^2q^2, where p and q are prime integers, is simple.

27. Show that $\mathbf{Z}_8, \mathbf{Z}_4 \times \mathbf{Z}_2, \mathbf{Z}_2 \times \mathbf{Z}_2 \times \mathbf{Z}_2$, and D_4 are nonisomorphic groups of order 8. Prove that Q_8 is not isomorphic to the above groups.

28. Show that $\mathbf{Z}_{12}, \mathbf{Z}_2 \times \mathbf{Z}_6, \mathbf{Z}_2 \times S_3$, and A_4 are nonisomorphic groups.

29. Write the proof if the statement is true; otherwise, give a counterexample.

    (i) If a prime p divides the order of a group G, then G contains a normal subgroup of order p.

    (ii) Let G and H be groups of order 39 and 21, respectively. These two groups are not isomorphic, but their Sylow 3-subgroups are isomorphic.

    (iii) There exists only one (up to isomorphism) group of order 65.

    (iv) Every group of order 76 contains a unique element of order 19.

Peter Ludvig Mejdell Sylow (1832 –1918) was born on December 12, 1832, in Christiania (now Oslo), Norway. In 1850, he graduated from the Christiania Cathedral School. In 1853, he won a mathematics prize contest. In 1861, he traveled to Berlin and Paris after being awarded a traveling grant. He, jointly with Sophus Lie, prepared a new edition of Abel's work from 1873 to 1881. In 1902, he and Elling Holst published Abel's correspondence.

Sylow is best known for his work in finite group theory. In 1845, Cauchy proved that every finite group has a subgroup of any prime order dividing the order of the group. In 1872, Sylow published a ten-page paper extending Cauchy's result. The theorems proved in that paper are known as Sylow's theorems, which we discussed in Chapter 7. These theorems are fundamental for structural results in finite group theory. Sylow died on September 7, 1918.

Chapter 8

Solvable and Nilpotent Groups

8.1   Solvable Groups

The purpose of this chapter is to present the Jordan-Hölder theorem and the notion of solvable groups. The results chosen here lay groundwork for the determination of the solvability "by radicals" of a polynomial equation $f(x) = 0$. In this regard, we show that the symmetric group S_n on n symbols is not solvable for $n \geq 5$.

Definition 8.1.1 *Let G be a group and*

$$G = H_0 \supseteq H_1 \supseteq H_2 \supseteq \cdots \supseteq H_n = \{e\}$$

*be a chain of subgroups of G. The chain is called a **subnormal series (chain)** if each H_i is normal in H_{i-1}. The chain is called a **normal series (chain)** if each H_i is normal in G. The chain is called a **composition series** if each H_i is a maximal normal subgroup of H_{i-1}, i.e., $H_i \neq H_{i-1}$, and if $H_i \subset H \subseteq H_{i-1}$ and H is normal in H_{i-1}, then $H = H_{i-1}$, $i = 1, 2, \ldots, n$. The number of proper inclusions $\supset$ in the chain is called the **length** of the chain. The groups H_{i-1}/H_i are called the **factors** of the chain.*

In Definition 8.1.1, if $H_{i-1} = H_i$, then the group H_{i-1}/H_i consists of a single element and is called a **trivial factor** of the chain. Hence the length of the chain is the number of nontrivial factors H_{i-1}/H_i of the chain.

Every group G has a normal chain, namely, $G \supseteq \{e\}$, since $\{e\}$ is a normal subgroup of G. Furthermore, it can be shown by induction on $|G|$ that every finite group G has a composition series. The reader is asked to verify this in the exercises.

We see in a composition series $G = H_0 \supseteq H_1 \supseteq H_2 \supseteq \cdots \supseteq H_n = \{e\}$ for a group G that the factors H_{i-1}/H_i are simple groups. In some sense, the examination of G has been reduced to its composition factors.

Example 8.1.2 *Consider the symmetric group* S_4. *Set*

$$H_1 = \left\{ e, \begin{pmatrix} 1 & 2 & 3 & 4 \\ 2 & 1 & 4 & 3 \end{pmatrix}, \begin{pmatrix} 1 & 2 & 3 & 4 \\ 3 & 4 & 1 & 2 \end{pmatrix}, \begin{pmatrix} 1 & 2 & 3 & 4 \\ 4 & 3 & 2 & 1 \end{pmatrix} \right\}$$

and

$$H_2 = \left\{ e, \begin{pmatrix} 1 & 2 & 3 & 4 \\ 2 & 1 & 4 & 3 \end{pmatrix} \right\}.$$

Now $S_4 = H_0 \supset H_1 \supset H_2 \supset H_3 = \{e\}$ *is a subnormal chain which is not a normal chain since* H_2 *is not normal in* S_4 *even though* H_2 *is normal in* H_1.

Example 8.1.3 *Consider the group* $(\mathbf{Z}_{12}, +_{12})$. *Since* $\mathbf{Z}_{12}$ *is commutative, all subgroups are normal. Hence, the following chains are normal:*

$$\mathbf{Z}_{12} \supset \langle [6] \rangle \supset \langle [0] \rangle \,,$$
$$\mathbf{Z}_{12} \supset \langle [3] \rangle \supset \langle [6] \rangle \supset \langle [0] \rangle \,,$$
$$\mathbf{Z}_{12} \supset \langle [2] \rangle \supset \langle [4] \rangle \supset \langle [0] \rangle \,,$$
$$\mathbf{Z}_{12} \supset \langle [2] \rangle \supset \langle [6] \rangle \supset \langle [0] \rangle \,.$$

All chains except $\mathbf{Z}_{12} \supset \langle [6] \rangle \supset \langle [0] \rangle$ *are composition series.*

Definition 8.1.4 *Let* G *be a group and*

$$G = H_0 \supseteq H_1 \supseteq H_2 \supseteq \cdots \supseteq H_{n-1} \supseteq H_n = \{e\} \tag{8.1}$$

be a subnormal series in G. *A **one-step refinement** of this series is any series of the form*

$$G = H_0 \supseteq H_1 \supseteq \cdots \supseteq H_{i-1} \supseteq H \supseteq H_i \supseteq \cdots \supseteq H_{n-1} \supseteq H_n = \{e\},$$

where H *is a normal subgroup of* H_{i-1} *and* H_i *is a normal subgroup of* H, $i = 1, 2, \ldots, n$. *A **refinement** of (8.1) is a subnormal series which is obtained from (8.1) by a finite sequence of one-step refinements. A refinement*

$$G = K_0 \supseteq K_1 \supseteq K_2 \supseteq \cdots \supseteq K_{m-1} \supseteq K_m = \{e\} \tag{8.2}$$

*of (8.1) is called a **proper refinement** if there exists a subgroup* K_j *in (8.2) which is different from each* H_i *of (8.1).*

Thus, a chain of subgroups

$$G = K_0 \supseteq K_1 \supseteq K_2 \supseteq \cdots \supseteq K_{m-1} \supseteq K_m = \{e\}$$

of G is called a **refinement** of a chain of subgroups

$$G = H_0 \supseteq H_1 \supseteq H_2 \supseteq \cdots \supseteq H_{n-1} \supseteq H_n = \{e\}$$

of G if

$$\{H_0, H_1, H_2, \ldots, H_n\} \subseteq \{K_0, K_1, K_2, \ldots, K_m\}$$

and is called a **proper refinement** if

$$\{H_0, H_1, H_2, \ldots, H_n\} \subset \{K_0, K_1, K_2, \ldots, K_m\}.$$

Example 8.1.5 *(i) Consider the subnormal series*

$$\mathbf{Z} \supset 6\mathbf{Z} \supset 12\mathbf{Z} \supset 48\mathbf{Z} \supset \{0\}. \tag{8.3}$$

The subnormal series

$$\mathbf{Z} \supset 2\mathbf{Z} \supset 6\mathbf{Z} \supset 12\mathbf{Z} \supset 48\mathbf{Z} \supset \{0\} \tag{8.4}$$

is a one-step refinement of (8.3). Again the subnormal series

$$\mathbf{Z} \supset 2\mathbf{Z} \supset 6\mathbf{Z} \supset 12\mathbf{Z} \supset 24\mathbf{Z} \supset 48\mathbf{Z} \supset \{0\} \tag{8.5}$$

is a one-step refinement of (8.4). From the definition, it follows that both (8.4) and (8.5) are proper refinements of (8.3).

(ii) In Example 8.1.3, $\mathbf{Z}_{12} \supset \langle [3] \rangle \supset \langle [6] \rangle \supset \langle [0] \rangle$ is a refinement of

$$\mathbf{Z}_{12} \supset \langle [6] \rangle \supset \langle [0] \rangle$$

while $\mathbf{Z}_{12} \supset \langle [2] \rangle \supset \langle [4] \rangle \supset \langle [0] \rangle$ is not.

Example 8.1.6 *Consider the group $(\mathbf{Z}, +)$. Then $\mathbf{Z}$ does not have a composition series since every subgroup of $\mathbf{Z}$ is cyclic and every subgroup $\langle n \rangle$ of $\mathbf{Z}$ contains an infinite chain, namely,*

$$\langle n \rangle \supset \langle 2n \rangle \supset \langle 4n \rangle \supset \langle 8n \rangle \supset \cdots.$$

Theorem 8.1.7 *A subnormal series in a group G is a composition series if and only if it has no proper refinement.*

Proof.   Let

$$G = H_0 \supseteq H_1 \supseteq H_2 \supseteq \cdots \supseteq H_{n-1} \supseteq H_n = \{e\} \tag{8.6}$$

be a composition series. Let

$$G = H_0 \supseteq H_1 \supseteq \cdots \supseteq H_{i-1} \supseteq H \supseteq H_i \supseteq \cdots \supseteq H_{n-1} \supseteq H_n = \{e\} \tag{8.7}$$

be a one-step refinement of (8.6). Since (8.6) is a composition series, H_i is a maximal normal subgroup of H_{i-1}. Thus, either $H = H_{i-1}$ or $H = H_i$. Hence, it follows that (8.6) has no proper refinement.

Conversely, suppose that

$$G = H_0 \supseteq H_1 \supseteq H_2 \supseteq \cdots \supseteq H_{n-1} \supseteq H_n = \{e\} \tag{8.8}$$

is a subnormal series, which has no proper refinement. Suppose (8.8) is not a composition series. Then there exists a subgroup H_i in (8.8) such that H_i is not a maximal normal subgroup in H_{i-1}. Thus, there exists a subgroup H such that $H_{i-1} \neq H \neq H_i$, H is a normal subgroup of H_{i-1}, and H_i is a normal subgroup of H. This produces a proper refinement of (8.8), a contradiction. Hence, (8.8) is a composition series. ∎

Definition 8.1.8 *Two subnormal chains for a group* G

$$G = H_0 \supseteq H_1 \supseteq H_2 \supseteq \cdots \supseteq H_{n-1} \supseteq H_n = \{e\} \tag{8.9}$$

$$G = K_0 \supseteq K_1 \supseteq K_2 \supseteq \ldots \supseteq K_{m-1} \supseteq K_m = \{e\} \tag{8.10}$$

are called ***equivalent*** *if there is a one-one correspondence between the nontrivial factors of* (8.9) *and* (8.10) *such that corresponding factors are isomorphic.*

If the subnormal chains (8.9) and (8.10) are equivalent, then the length of (8.9) equals the length of (8.10).

Example 8.1.9 *Consider the subnormal series*

$$\mathbf{Z} \supset 4\mathbf{Z} \supset 12\mathbf{Z} \supset 24\mathbf{Z} \supset 120\mathbf{Z} \supset \{0\} \tag{8.11}$$

$$\mathbf{Z} \supset 2\mathbf{Z} \supset 8\mathbf{Z} \supset 24\mathbf{Z} \supset 120\mathbf{Z} \supset \{0\}. \tag{8.12}$$

The factors of (8.11) *are*

$$\mathbf{Z}/4\mathbf{Z} \simeq \mathbf{Z}_4, \ 4\mathbf{Z}/12\mathbf{Z} \simeq \mathbf{Z}_3, \ 12\mathbf{Z}/24\mathbf{Z} \simeq \mathbf{Z}_2,$$
$$24\mathbf{Z}/120\mathbf{Z} \simeq \mathbf{Z}_5, \ and \ 120\mathbf{Z}/\{0\} \simeq \mathbf{Z}$$

and the factors of (8.12) *are*

$$\mathbf{Z}/2\mathbf{Z} \simeq \mathbf{Z}_2, \ 2\mathbf{Z}/8\mathbf{Z} \simeq \mathbf{Z}_4, \ 8\mathbf{Z}/24\mathbf{Z} \simeq \mathbf{Z}_3,$$
$$24\mathbf{Z}/120\mathbf{Z} \simeq \mathbf{Z}_5, \ and \ 120\mathbf{Z}/\{0\} \simeq \mathbf{Z}.$$

Hence, there exists a one-one correspondence between the factors of (8.11) *and* (8.12). *Consequently,* (8.11) *and* (8.12) *are equivalent.*

Theorem 8.1.10 (Zassenhaus Lemma) *Let H', H, K', and K be subgroups of a group G such that H' is a normal subgroup of H and K' is a normal subgroup of K. Then $H'(H \cap K')$ is a normal subgroup of $H'(H \cap K)$ and $K'(H' \cap K)$ is a normal subgroup of $K'(H \cap K)$. Furthermore,*

$$\frac{H'(H \cap K)}{H'(H \cap K')} \simeq \frac{K'(H \cap K)}{K'(H' \cap K)}.$$

Proof.   From the hypothesis, it follows that $H \cap K'$ and $H' \cap K$ are normal subgroups of $H \cap K$. Thus, $(H \cap K')(H' \cap K)$ is a normal subgroup of $H \cap K$. Set $J = (H \cap K')(H' \cap K)$.

Define the function $f : H'(H \cap K) \to (H \cap K)/J$ as follows: If $a \in H'(H \cap K)$, then $a = h'b$, where $h' \in H'$ and $b \in H \cap K$. Set $f(a) = Jb$. Let a_1, $a_2 \in H'(H \cap K)$. Then $a_1 = h'_1 b_1$, $a_2 = h'_2 b_2$ for some $h'_1, h'_2 \in H'$ and $b_1, b_2 \in H \cap K$. Suppose $a_1 = a_2$. Then $h'_1 b_1 = h'_2 b_2$. Thus, $h'^{-1}_2 h'_1 = b_2 b^{-1}_1 \in H' \cap (H \cap K) \subseteq H' \cap K \subseteq J$. Hence, $Jb_1 = Jb_2$ and so $f(a_1) = f(a_2)$. Thus, f is well defined. Since H' is a normal subgroup of H, $b_1 h'_2 b^{-1}_1 \in H'$. Now $a_1 a_2 = h'_1 b_1 h'_2 b_2 = h'_1 b_1 h'_2 b^{-1}_1 b_1 b_2 = h' b_1 b_2$, where $h' = h'_1 b_1 h'_2 b^{-1}_1 \in H'$. Hence, $f(a_1 a_2) = Jb_1 b_2 = Jb_1 Jb_2 = f(a_1)f(a_2)$. Therefore, f is a homomorphism. From the definition of f, it follows that f maps $H'(H \cap K)$ onto $(H \cap K)/J$. Also, it is easy to verify that $\text{Ker } f = H'(H \cap K')$. Hence, by Theorem 5.2.2,

$$\frac{H'(H \cap K)}{H'(H \cap K')} \simeq \frac{(H \cap K)}{J}.$$

By symmetry,

$$\frac{K'(H \cap K)}{K'(H' \cap K)} \simeq \frac{(H \cap K)}{J}.$$

Finally, the desired isomorphism follows from these two isomorphisms. ■

Theorem 8.1.11 (Schreier) *Any two subnormal series*

$$G = H_0 \supseteq H_1 \supseteq H_2 \supseteq \cdots \supseteq H_{n-1} \supseteq H_n = \{e\} \qquad (8.13)$$

$$G = K_0 \supseteq K_1 \supseteq K_2 \supseteq \cdots \supseteq K_{m-1} \supseteq K_m = \{e\} \qquad (8.14)$$

of a group G have refinements which are equivalent.

Proof.   Between each H_i and H_{i+1}, insert the group $H_{i+1}(H_i \cap K_j)$, $j = 0, 1, 2, \ldots, m$. From the normality assertions of the Zassenhaus lemma, this refinement of (8.13) is a subnormal chain with mn (not necessarily strict) inclusions. Between each K_j and K_{j+1} insert the group $K_{j+1}(K_j \cap H_i)$, $i = 0$,

$1, 2, \ldots, n$. This refinement of (8.14) is also a subnormal chain with mn inclusions. The final refinements are

$$\cdots \supseteq H_{i+1}(H_i \cap K_j) \supseteq H_{i+1}(H_i \cap K_{j+1}) \supseteq \cdots$$

and

$$\cdots \supseteq K_{j+1}(K_j \cap H_i) \supseteq K_{j+1}(K_j \cap H_{i+1}) \supseteq \cdots.$$

From the Zassenhaus lemma,

$$H_{i+1}(H_i \cap K_j)/H_{i+1}(H_i \cap K_{j+1}) \simeq K_{j+1}(K_j \cap H_i)/K_{j+1}(K_j \cap H_{i+1}).$$

Hence, we have the desired result. ∎

Theorem 8.1.12 (Jordan-Hölder) *Any two composition series of a group are equivalent.*

Proof. Since composition series are subnormal series, any two composition series of G have equivalent refinements. Now a composition series has no proper refinements. Thus, a composition series is equivalent to every refinement of itself. Hence, any two composition series of a group are equivalent. ∎

By the Jordan-Hölder theorem, we find that if a group G has a composition series of length n, then the length of any composition series of G must be n. This n is called the **composition length** of the group G.

We now show that the fundamental theorem of arithmetic can be established from the Jordan-Hölder theorem. Let n be a positive integer greater than 1 and consider the group $(\mathbf{Z}_n, +_n)$. Since $\mathbf{Z}_n$ is finite, $\mathbf{Z}_n$ has a composition series. Let

$$\mathbf{Z}_n = H_0 \supseteq H_1 \supseteq H_2 \supseteq \cdots \supseteq H_{k-1} \supseteq H_k = \{[0]\}$$

be a composition series. The factors H_{i-1}/H_i are simple Abelian groups. Hence each factor is of prime order. Let $|H_{i-1}/H_i| = p_i$. Now

$$n = |\mathbf{Z}_n| = |H_0/H_1| \cdot |H_1/H_2| \cdots |H_{k-1}/H_k| = p_1 p_2 \cdots p_k.$$

This proves that every integer $n > 1$ can be expressed as a product of prime integers. The uniqueness of this factorization follows from the equivalence of the composition series.

Example 8.1.13 *Consider the group* $(\mathbf{Z}_{30}, +_{30})$. *Then* $\mathbf{Z}_{30}$ *has the following two composition series.*

$$\mathbf{Z}_{30} \supset \langle[5]\rangle \supset \langle[10]\rangle \supset \langle[0]\rangle$$

and

$$\mathbf{Z}_{30} \supset \langle [2] \rangle \supset \langle [6] \rangle \supset \langle [0] \rangle .$$

Now

$$\mathbf{Z}_{30}/ \langle [5] \rangle \not\simeq \mathbf{Z}_{30}/ \langle [2] \rangle ,$$

but we have the following isomorphisms:

$$\mathbf{Z}_{30}/ \langle [5] \rangle \simeq \langle [6] \rangle / \langle [0] \rangle$$
$$\mathbf{Z}_{30}/ \langle [2] \rangle \simeq \langle [5] \rangle / \langle [10] \rangle$$
$$\langle [2] \rangle / \langle [6] \rangle \simeq \langle [10] \rangle / \langle [0] \rangle .$$

Definition 8.1.14 *A group G is called **solvable** if it has a subnormal series*

$$G = H_0 \supseteq H_1 \supseteq H_2 \supseteq \cdots \supseteq H_{n-1} \supseteq H_n = \{e\}$$

*such that H_i/H_{i+1} is commutative, $i = 0, 1, \ldots, n-1$. Such a subnormal series is called a **solvable series** for G.*

Every commutative group is solvable since $G = H_0 \supseteq H_1 = \{e\}$ satisfies the above definition.

Example 8.1.15 *Consider the symmetric group S_3. Then*

$$S_3 \supset \left\{ e, \begin{pmatrix} 1 & 2 & 3 \\ 2 & 3 & 1 \end{pmatrix}, \begin{pmatrix} 1 & 2 & 3 \\ 3 & 1 & 2 \end{pmatrix} \right\} \supset \{e\}$$

is a solvable series for S_3. Hence, S_3 is solvable.

Example 8.1.16 *Consider the symmetric group S_4. Then*

$$S_4 \supset A_4 \supset \left\{ e, \begin{pmatrix} 1 & 2 & 3 & 4 \\ 2 & 1 & 4 & 3 \end{pmatrix}, \begin{pmatrix} 1 & 2 & 3 & 4 \\ 3 & 4 & 1 & 2 \end{pmatrix}, \begin{pmatrix} 1 & 2 & 3 & 4 \\ 4 & 3 & 2 & 1 \end{pmatrix} \right\} \supset$$
$$\left\{ e, \begin{pmatrix} 1 & 2 & 3 & 4 \\ 2 & 1 & 4 & 3 \end{pmatrix} \right\} \supset \{e\}$$

is a solvable series for S_4. Thus, S_4 and A_4 are solvable.

Since the symmetric groups S_1 and S_2 are commutative, they are solvable. Thus, S_n is solvable for $n \leq 4$. In Theorem 8.1.27 below, we show that S_n is not solvable for $n > 4$. The order of the alternating group A_3 is 3. Hence, A_3 is commutative and thus solvable. By Example 8.1.16, A_4 is solvable. Thus, A_n is solvable for $n \leq 4$.

In the next few theorems, we show how the solvability of a group is associated with the solvability of a normal subgroup and the quotient group created by the normal subgroup.

Theorem 8.1.17 *If G is a solvable group, then every subgroup of G is solvable and every homomorphic image of G is solvable.*

Proof.   Let

$$G = H_0 \supseteq H_1 \supseteq H_2 \supseteq \cdots \supseteq H_{n-1} \supseteq H_n = \{e\}$$

be a solvable series of G. Let K be any subgroup of G. Set $K_i = K \cap H_i$, $i = 0, 1, \ldots, n$. We shall show that the chain

$$K = K_0 \supseteq K_1 \supseteq K_2 \supseteq \cdots \supseteq K_{n-1} \supseteq K_n = \{e\}$$

is a solvable series for K. It follows that $H_{i+1} \cap K$ is a normal subgroup of the group $H_i \cap K$. That is, K_{i+1} is a normal subgroup of K_i. Now

$$K_{i+1} = K \cap H_{i+1} = K \cap H_i \cap H_{i+1} = K_i \cap H_{i+1}.$$

Thus, $K_i/K_{i+1} = K_i/(K_i \cap H_{i+1})$. Hence, by the second isomorphism theorem (Theorem 5.2.6), we have the isomorphism

$$K_i/K_{i+1} \simeq (K_i H_{i+1})/H_{i+1}.$$

The quotient group $(K_i H_{i+1})/H_{i+1}$ is commutative since it is a subgroup of the commutative group H_i/H_{i+1}. Thus, K_i/K_{i+1} is commutative and so K is solvable.

Let f be a homomorphism of G onto a group $\overline{G}$. Set $\overline{H}_i = f(H_i)$, $i = 0, 1, \ldots, n$. Since f is an epimorphism, $f(H_{i+1})$ is a normal subgroup of $f(H_i)$. Also, $H_i \supseteq H_{i+1}$ implies that $f(H_i) \supseteq f(H_{i+1})$. Hence,

$$\overline{G} = \overline{H}_0 \supseteq \overline{H}_1 \supseteq \overline{H}_2 \supseteq \cdots \supseteq \overline{H}_{n-1} \supseteq \overline{H}_n = \{e\} \tag{8.15}$$

is a subnormal series of $\overline{G}$. We now show that $f(H_i)/f(H_{i+1}) = \overline{H}_i/\overline{H}_{i+1}$ is commutative. Define $g : H_i \to \overline{H}_i/\overline{H}_{i+1}$ by $g(h_i) = f(h_i)\overline{H}_{i+1}$. Since f is an epimorphism, it follows that g is an epimorphism of H_i onto $\overline{H}_i/\overline{H}_{i+1}$. Note that for any $h_{i+1} \in \overline{H}_{i+1} \subseteq \overline{H}_i$, $g(h_{i+1}) = f(h_{i+1})\overline{H}_{i+1} = f(h_{i+1})f(H_{i+1}) = f(H_{i+1})$. Hence, $H_{i+1} \subseteq \operatorname{Ker} g$. Thus, g induces an epimorphism of H_i/H_{i+1} onto $\overline{H}_i/\overline{H}_{i+1}$. Since H_i/H_{i+1} is commutative, it follows that $\overline{H}_i/\overline{H}_{i+1}$ is commutative. Consequently, the subnormal series (8.15) is a solvable series, proving that $\overline{G}$ is a solvable group. ∎

The following corollary is immediate from Theorem 8.1.17.

Corollary 8.1.18 *If G is solvable and H is a normal subgroup of G, then H and G/H are solvable.* ∎

Theorem 8.1.19 *Let H be a normal subgroup of a group G. If both H and G/H are solvable, then G is solvable.*

Proof. Let

$$G/H = \overline{K}_0 \supseteq \overline{K}_1 \supseteq \overline{K}_2 \supseteq \cdots \supseteq \overline{K}_{m-1} \supseteq \overline{K}_m = \{eH\} = \{H\}$$

be a solvable series for G/H. By Corollary 5.2.12, there are subgroups K_i of G, $i = 0, 1, \ldots, m$, such that K_{i+1} is a normal subgroup of K_i, $\overline{K}_i = K_i/H$, $i = 0, 1, \ldots, m-1$, $G = K_0$, and $H = K_m$. Also, $K_i/K_{i+1} \simeq \overline{K}_i/\overline{K}_{i+1}$ by the third isomorphism theorem (Corollary 5.2.9). Since H is solvable, H has a solvable series, say,

$$H = H_0 \supseteq H_1 \supseteq H_2 \supseteq \cdots \supseteq H_{n-1} \supseteq H_n = \{e\}.$$

Thus,

$$G = K_0 \supseteq K_1 \supseteq \cdots \supseteq K_{m-1} \supseteq H \supseteq H_1 \supseteq \cdots \supseteq H_{n-1} \supseteq H_n = \{e\}$$

is a solvable series for G. That is, G is solvable. ∎

Theorem 8.1.20 *Let $G \neq \{e\}$ be a finite solvable group. Then the factor groups of any composition series of G are cyclic groups of prime order.*

Proof. The proof is by induction on $|G|$. If $|G|$ is a prime, then the theorem is valid since $G \supset \{e\}$ is the only composition series for G. Hence, the theorem is valid for $|G| = 2$. Suppose the theorem is true for all groups of order $< |G|$, where $|G| > 2$. If $|G|$ is not a prime, then G has a nontrivial normal subgroup H. (If G does not have a nontrivial normal subgroup, then $G \supset \{e\}$ is a composition series for G so that $G \simeq G/\langle e \rangle$ is commutative. Thus, G has no proper subgroups. Hence, $|G|$ is a prime, a contradiction.) By the induction hypothesis and Corollary 8.1.18, G/H and H have the composition series

$$G/H = \overline{K}_0 \supset \overline{K}_1 \supset \overline{K}_2 \supset \cdots \supset \overline{K}_{m-1} \supset \overline{K}_m = \{e\}$$

and

$$H = H_0 \supset H_1 \supset H_2 \supset \cdots \supset H_{n-1} \supset H_n = \{e\},$$

respectively, such that each $\overline{K}_i/\overline{K}_{i+1}$ and each H_i/H_{i+1} are cyclic groups of prime order. If we choose subgroups K_i of G corresponding to K_i' as in Theorem 8.1.19, then it follows by similar arguments that

$$G = K_0 \supset K_1 \supset \cdots \supset K_{m-1} \supset H \supset H_1 \supset \cdots \supset H_{n-1} \supset H_n = \{e\}$$

is a composition series of G satisfying the conditions of the theorem. Thus, by the Jordan-Hölder theorem, every composition series of G satisfies the conditions of the theorem. ∎

We now proceed to establish the unsolvability of S_n, $n \geq 5$. We first introduce the notion of the commutator subgroup of a group and obtain its basic properties. We also give a necessary and sufficient condition for the solvability of a group in terms of the commutator subgroup. We then apply these results to show that S_n, $n \geq 5$, is not solvable.

Definition 8.1.21 *Let G be a group and a, $b \in G$. The **commutator** of a and b is the element $aba^{-1}b^{-1}$. Set $A = \{aba^{-1}b^{-1} \mid a, b \in G\}$ and let G' be the subgroup of G generated by A. G' is called the **derived** or **commutator subgroup** of G.*

If G is commutative, then $A = \{e\}$ and so $G' = \{e\}$. Conversely, if $A = G' = \{e\}$, then $aba^{-1}b^{-1} = e$ for all a, $b \in G$. Therefore, $ab = ba$ for all $a, b \in G$, i.e., G is commutative. Thus, G is commutative if and only if $G' = \{e\}$.

Theorem 8.1.22 *The derived subgroup G' of a group G is a normal subgroup of G and G/G' is commutative.*

Proof. Let $a, b, g \in G$. Now

$$g(aba^{-1}b^{-1})g^{-1} = (gag^{-1})(gbg^{-1})(ga^{-1}g^{-1})(gb^{-1}g^{-1}) = cdc^{-1}d^{-1},$$

where $c = gag^{-1}$ and $d = gbg^{-1}$. This implies that for any commutator $aba^{-1}b^{-1}$ and for any $g \in G$, $g(aba^{-1}b^{-1})g^{-1}$ is a commutator. From this, it follows that $gG'g^{-1} \subseteq G'$ for all $g \in G$. Hence, G' is a normal subgroup of G. Next, we show that G/G' is commutative. Let $a, b \in G$. Then $(ba)^{-1}ab = a^{-1}b^{-1}ab \in G'$ and so $abG' = baG'$, i.e., $aG'bG' = bG'aG'$. Hence, G/G' is commutative. ∎

Theorem 8.1.23 *Let G' be the derived subgroup of a group G and H be a subgroup of G. Then $H \supseteq G'$ if and only if H is a normal subgroup of G and G/H is commutative.*

Proof. Suppose $H \supseteq G'$. Let $h \in H$ and $a \in G$. Then $aha^{-1}h^{-1} \in G' \subseteq H$. Thus, $aha^{-1} = (aha^{-1}h^{-1})h \in H$. Hence, H is a normal subgroup of G. Let us now show that G/H is commutative. To do this, let us consider two arbitrary elements aH, bH in G/H. Then $(aH)(bH)(aH)^{-1}(bH)^{-1} = aHbHa^{-1}Hb^{-1}H = aba^{-1}b^{-1}H$. Since $aba^{-1}b^{-1} \in G' \subseteq H$, it follows that $(aH)(bH)(aH)^{-1}(bH)^{-1} = H$. Therefore, $aHbH = bHaH$, proving that G/H is commutative. Conversely, suppose H is normal in G and G/H is commutative. Let a, $b \in G$. Then $(aH)(bH) = (bH)(aH)$. This implies that $a^{-1}b^{-1}ab \in H$. Hence, $G' \subseteq H$. ∎

Definition 8.1.24 *Let G' be the commutator subgroup of a group G. Set $G^{(1)} = G'$ and define inductively*

$$G^{(k+1)} = G^{(k)\prime},$$

*the commutator subgroup of $G^{(k)}$, $k > 0$. For any positive integer k, $G^{(k)}$ is called the **kth commutator subgroup** of G.*

The following theorem gives a necessary and sufficient condition for a group to be solvable in terms of a commutator subgroup.

Theorem 8.1.25 *Let G be a group. Then G is solvable if and only if there is a positive integer m such that $G^{(m)} = \{e\}$.*

Proof.   Suppose $G^{(m)} = \{e\}$. Then by Theorem 8.1.22, the chain

$$G \supseteq G^{(1)} \supseteq \cdots \supseteq G^{(m-1)} \supseteq G^{(m)} = \{e\}$$

is a solvable series. Thus, G is solvable. Conversely, suppose G is solvable. Then G has a solvable series, say,

$$G = H_0 \supseteq H_1 \supseteq H_2 \supseteq \cdots \supseteq H_{n-1} \supseteq H_n = \{e\}.$$

Since H_{i+1} is normal in H_i and H_i/H_{i+1} is commutative, we have by Theorem 8.1.23 that the commutator subgroup H_i' of H_i is contained in H_{i+1}. Thus,

$$H_1 \supseteq H_0' = G^{(1)}, \ \ H_2 \supseteq H_1' \supseteq G^{(2)}, \ldots, \{e\} = H_n \supseteq H_{n-1}' \supseteq G^{(n)}.$$

Hence, $G^{(n)} = \{e\}$. ∎

Lemma 8.1.26 *Let S_n be the symmetric group on n symbols. If $n \geq 5$, then $S_n^{(k)}$ contains every 3-cycle of S_n for $k = 1, 2, \ldots$.*

Proof.   Let $\pi = (a\ b\ c)$ be any 3-cycle in S_n. Since $n \geq 5$, there exist symbols d, f such that a, b, c, d, f are distinct. Set $\alpha = (a\ b\ d)$ and $\beta = (a\ c\ f)$. Let H be any subgroup of S_n with the property that H contains every 3-cycle of S_n. Then $\pi, \alpha, \beta \in H$. Hence,

$$(a\ b\ c) = (a\ b\ d) \circ (a\ c\ f) \circ (a\ d\ b) \circ (a\ f\ c) = \alpha\beta\alpha^{-1}\beta^{-1} \in H',$$

where H' is the derived subgroup of H. From this, it follows that $S_n^{(1)}$ contains every 3-cycle of S_n. We can employ induction to obtain the desired result. ∎

In the next theorem, we show that S_n is not solvable for $n \geq 5$.

Theorem 8.1.27 *The symmetric group S_n on n symbols is not solvable for $n \geq 5$.*

Proof. Since $S_n^{(k)}$ contains every 3-cycle of S_n for $k = 1, 2, \ldots$, there does not exist a positive integer m such that $S_n^{(m)} = \{e\}$. Thus, by Theorem 8.1.25, S_n is not solvable. ∎

8.1.1 Worked-Out Exercises

◇ **Exercise 1** Let G be a group of order pqr, where p, q, r are primes and $p > q > r$. Show that G is solvable.

Solution: The number of Sylow p-subgroups is $1 + kp$, where $1 + kp$ divides qr. Suppose $k \neq 0$. Since $p > q > r$, $1 + kp = qr$. The number of Sylow q-subgroups is $1 + k'q$, where $1 + k'q$ divides pr. Suppose $k' \neq 0$. Since $q > r$, either $1 + k'q = p$ or pr. In either case, $1 + k'q \geq p$. The number of Sylow r-subgroups is $1 + k''r$, where $1 + k''r$ divides pq. Suppose $k'' \neq 0$. Then either $1 + k''r = q$ or p or pq. Hence, in either case, $1 + k''r \geq q$. Thus, G has $qr(p-1)$ elements of order p, at least $p(q-1)$ elements of order q, and at least $q(r-1)$ elements of order r. Since G has pqr elements, $pqr \geq qr(p-1) + p(q-1) + q(r-1) + 1$. This implies that $0 \geq pq - p - q + 1$ or $0 \geq (p-1)(q-1)$. Therefore, $(p-1)(q-1) = 0$, which implies that either $p = 1$ or $q = 1$, a contradiction. Thus, either $k = 0$ or $k' = 0$ or $k'' = 0$. Suppose $k = 0$. Then G has a unique Sylow p-subgroup, say, H. Now H is a normal subgroup of G and G/H is of order qr. By Exercise 11 (page 238), we find that G/H is solvable. Since H is of order p, H is solvable. Hence, by Theorem 8.1.19, G is solvable. Similarly, if either $k' = 0$ or $k'' = 0$, then G is solvable.

◇ **Exercise 2** Let $H \neq \{e\}$ be a subgroup of a solvable group G. Prove that $H' \neq H$.

Solution: Suppose $H' = H$. Then $H^{(2)} = (H')' = H' = H \neq \{e\}$. Now by induction, we can show that $H^{(n)} = H \neq \{e\}$ for any positive integer n. On the other hand, H is a subgroup of a solvable group and so H is solvable. This implies that there exists a positive integer n such that $H^{(n)} = \{e\}$, a contradiction. Hence, $H' \neq H$.

Exercise 3 Let G be the group of all $n \times n$ invertible matrices over $\mathbf{R}$, $n \geq 3$. Show that G is not solvable.

Solution: Let E_{ij} be the $n \times n$ matrix whose (i, j) entry is 1 and all other entries are zero. Then

$$E_{ij}E_{rs} = \begin{cases} E_{is} \text{ if } j = r \\ 0 \text{ if } j \neq r. \end{cases}$$

Now for the identity matrix I and for $i \neq j$, $I + E_{ij} \in G$ and $(I + E_{ij})^{-1} = I - E_{ij}$. Let T be the subgroup generated by $\{I + E_{ij} \mid i \neq j\}$. Since $n \geq 3$, we can find an integer k such that $1 \leq i \neq k \neq j \leq n$. Now

$$
\begin{aligned}
(I + E_{ik})(I + E_{kj})(I + E_{ik})^{-1}(I + E_{kj})^{-1} &= (I + E_{ik})(I + E_{kj}) \\
&\quad (I - E_{ik})(I - E_{kj}) \\
&= (I + E_{kj} + E_{ik} + E_{ij}) \\
&\quad (I - E_{kj} - E_{ik} + E_{ij}) \\
&= (I + E_{ij}).
\end{aligned}
$$

Therefore, $(I + E_{ij}) \in T'$, proving that $T \subseteq T'$. As a result $T = T'$. Thus, T is not solvable and so G is not solvable.

◇ **Exercise 4** Let $GL(2, \mathbf{R})$ be the group of Example 2.1.10. Prove that the derived subgroup of $GL(2, \mathbf{R})$ is the subgroup

$$
SL(2, \mathbf{R}) = \left\{ \begin{bmatrix} a & b \\ c & d \end{bmatrix} \in GL(2, \mathbf{R}) \mid ad - bc = 1 \right\}.
$$

Solution: Let $\mathbf{R}^*$ be the multiplicative group of nonzero real numbers. Define $f : GL(2, \mathbf{R}) \to \mathbf{R}^*$ by

$$
f\left(\begin{bmatrix} a & b \\ c & d \end{bmatrix} \right) = ad - bc
$$

for all $\begin{bmatrix} a & b \\ c & d \end{bmatrix} \in GL(2, \mathbf{R})$. Now f is an epimorphism with $\mathrm{Ker}\ f = SL(2, \mathbf{R})$. Hence, $SL(2, \mathbf{R})$ is a normal subgroup of $GL(2, \mathbf{R})$ and

$$
GL(2, \mathbf{R})/SL(2, \mathbf{R}) \simeq \mathbf{R}^*.
$$

This implies that $GL(2, \mathbf{R})/SL(2, \mathbf{R})$ is a commutative group and $(GL(2, \mathbf{R}))'$ $\subseteq SL(2, \mathbf{R})$. Let us now show that $SL(2, \mathbf{R}) \subseteq (GL(2, \mathbf{R}))'$. For this, let $\begin{bmatrix} a & b \\ c & d \end{bmatrix}$ $\in SL(2, \mathbf{R})$. Then $ad - bc = 1$. If $c \neq 0$, then

$$
\begin{bmatrix} a & b \\ c & d \end{bmatrix} = \begin{bmatrix} 1 & \frac{a-1}{c} \\ 0 & 1 \end{bmatrix} \begin{bmatrix} 1 & 0 \\ c & 1 \end{bmatrix} \begin{bmatrix} 1 & \frac{d-1}{c} \\ 0 & 1 \end{bmatrix}.
$$

Now for any $r \in \mathbf{R}$,

$$
\begin{aligned}
\begin{bmatrix} 1 & r \\ 0 & 1 \end{bmatrix} &= \begin{bmatrix} 1 & -r \\ 0 & 1 \end{bmatrix} \begin{bmatrix} 1 & 0 \\ 0 & \frac{1}{2} \end{bmatrix} \begin{bmatrix} 1 & r \\ 0 & 1 \end{bmatrix} \begin{bmatrix} 1 & 0 \\ 0 & 2 \end{bmatrix} \\
&= \begin{bmatrix} 1 & -r \\ 0 & 1 \end{bmatrix} \begin{bmatrix} 1 & 0 \\ 0 & \frac{1}{2} \end{bmatrix} \begin{bmatrix} 1 & -r \\ 0 & 1 \end{bmatrix}^{-1} \begin{bmatrix} 1 & 0 \\ 0 & \frac{1}{2} \end{bmatrix}^{-1} \in (GL(2, \mathbf{R}))'
\end{aligned}
$$

and

$$\begin{bmatrix} 1 & 0 \\ r & 1 \end{bmatrix} = \begin{bmatrix} 1 & 0 \\ 0 & 2 \end{bmatrix} \begin{bmatrix} 1 & 0 \\ r & 1 \end{bmatrix} \begin{bmatrix} 1 & 0 \\ 0 & \frac{1}{2} \end{bmatrix} \begin{bmatrix} 1 & 0 \\ -r & 1 \end{bmatrix} \in (GL(2, \mathbf{R}))'.$$

Hence, $\begin{bmatrix} a & b \\ c & d \end{bmatrix} \in (GL(2, \mathbf{R}))'$.

Suppose $c = 0$. Then $ad = 1$. Thus, $a \neq 0$ and

$$\begin{bmatrix} a & b \\ 0 & d \end{bmatrix} = \begin{bmatrix} a & 0 \\ 0 & \frac{1}{a} \end{bmatrix} \begin{bmatrix} 1 & \frac{b}{a} \\ 0 & 1 \end{bmatrix},$$

$$\begin{bmatrix} a & 0 \\ 0 & \frac{1}{a} \end{bmatrix} = \begin{bmatrix} 0 & 1 \\ 1 & 0 \end{bmatrix} \begin{bmatrix} a & 2a \\ 2a^2 & a^2 \end{bmatrix} \begin{bmatrix} 0 & 1 \\ 1 & 0 \end{bmatrix} \begin{bmatrix} -\frac{1}{3a} & \frac{2}{3a^2} \\ \frac{2}{3a} & -\frac{1}{3a^2} \end{bmatrix} \in (GL(2, \mathbf{R}))'.$$

Also, from above, $\begin{bmatrix} 1 & \frac{b}{a} \\ 0 & 1 \end{bmatrix} \in (GL(2, \mathbf{R}))'$. As a result, $\begin{bmatrix} a & b \\ c & d \end{bmatrix} \in (GL(2, \mathbf{R}))'$.

Consequently, $SL(2, \mathbf{R}) = (GL(2, \mathbf{R}))'$.

Exercise 5 Prove that in a group G, any refinement of a solvable series is a solvable series.

Solution: Let

$$G = H_0 \supseteq H_1 \supseteq H_2 \supseteq \cdots \supseteq H_{n-1} \supseteq H_n = \{e\} \tag{8.16}$$

be a solvable series in G and let

$$G = H_0 \supseteq \cdots \supseteq H_{i-1} \supseteq H \supseteq H_i \supseteq \cdots \supseteq H_{n-1} \supseteq H_n = \{e\} \tag{8.17}$$

be a one-step refinement of (8.16). From (8.16) H_{i-1}/H_i is commutative. Now the group H/H_i is a subgroup of H_{i-1}/H_i. Hence, H/H_i is commutative. Again

$$(H_{i-1}/H_i)/(H/H_i) \simeq H_{i-1}/H$$

implies that H_{i-1}/H is commutative. Thus, (8.17) is a solvable series. Hence, any one-step refinement of (8.16) is a solvable series. By induction, any refinement of (8.16) is a solvable series.

◇ **Exercise 6** Find all composition series of the group $\mathbf{Z}/\langle 42 \rangle$. Verify that they are equivalent.

Solution: Now the subgroups of $\mathbf{Z}/\langle 42 \rangle$ are $\mathbf{Z}/\langle 42 \rangle$, $2\mathbf{Z}/\langle 42 \rangle$, $3\mathbf{Z}/\langle 42 \rangle$, $6\mathbf{Z}/\langle 42 \rangle$, $7\mathbf{Z}/\langle 42 \rangle$, $14\mathbf{Z}/\langle 42 \rangle$, $21\mathbf{Z}/\langle 42 \rangle$, and $\{\langle 42 \rangle\}$. Hence, the composition series are

$$\mathbf{Z}/\langle 42 \rangle \supset 2\mathbf{Z}/\langle 42 \rangle \supset 6\mathbf{Z}/\langle 42 \rangle \supset \{\langle 42 \rangle\}$$

$$\mathbf{Z}/\langle 42\rangle \supset 2\mathbf{Z}/\langle 42\rangle \supset 14\mathbf{Z}/\langle 42\rangle \supset \{\langle 42\rangle\}$$
$$\mathbf{Z}/\langle 42\rangle \supset 3\mathbf{Z}/\langle 42\rangle \supset 6\mathbf{Z}/\langle 42\rangle \supset \{\langle 42\rangle\}$$
$$\mathbf{Z}/\langle 42\rangle \supset 3\mathbf{Z}/\langle 42\rangle \supset 21\mathbf{Z}/\langle 42\rangle \supset \{\langle 42\rangle\}$$
$$\mathbf{Z}/\langle 42\rangle \supset 7\mathbf{Z}/\langle 42\rangle \supset 14\mathbf{Z}/\langle 42\rangle \supset \{\langle 42\rangle\}$$
$$\mathbf{Z}/\langle 42\rangle \supset 7\mathbf{Z}/\langle 42\rangle \supset 21\mathbf{Z}/\langle 42\rangle \supset \{\langle 42\rangle\}.$$

Each of the above six composition series has three factors. These factors are nothing but the groups $\mathbf{Z}_2$, $\mathbf{Z}_3$, and $\mathbf{Z}_7$. Hence, all these composition series are equivalent.

8.1.2 Exercises

1. Let G be the group of symmetries of the square. Prove that the following series are composition series for G :

$$G \supset \{r_{180}, r_{360}, h, v\} \supset \{r_{360}, h\} \supset \{r_{360}\}$$

and

$$G \supset \{r_{180}, r_{360}, d_1, d_2\} \supset \{r_{360}, d_1\} \supset \{r_{360}\}.$$

Establish the equivalence of these composition series. Verify that $\{r_{360}, d_1\}$ is normal in $\{r_{180}, r_{360}, d_1, d_2\}$, but not normal in G.

2. Find all composition series of the group $\mathbf{Z}/\langle 66\rangle$. Verify that they are equivalent.

3. Find all composition series of $\mathbf{Z}_{20}$.

4. Write all composition series of S_3, S_4, A_4, D_4, and $\mathbf{Z}_2 \times \mathbf{Z}_2$.

5. Prove that every finite group has a composition series.

6. Let G be a commutative group. Show that G has a composition series if and only if G is finite.

7. Let G be a group. Show that $G' = \{a_1 a_2 \cdots a_n a_1^{-1} a_2^{-1} \cdots a_n^{-1} \mid a_i \in G, n \geq 2\}$.

8. Show that a group G is commutative if and only if $G' = \{e\}$.

9. Let H be a subgroup of G. Show that $H' \subseteq G'$.

10. Let N be a normal subgroup of a group G such that $N \cap G' = \{e\}$. Show that

   (i) $N \subseteq Z(G)$,

   (ii) $Z(G/N) = Z(G)/N$.

11. Let G be a group of order pq (p, q primes). Show that G is solvable.

12. Let G be a group of order p^2q (p, q primes). Show that G is solvable.

13. Let G be a group of order p^2q^2 (p, q primes). Show that G is solvable.

14. Write a solvable series of $S_3 \times S_3$.

15. Let G be a simple and solvable group. Show that G is commutative.

16. Prove that a finite direct product of solvable groups is solvable. Hence, show that $S_3 \times \mathbf{Z}$ is an infinite noncommutative solvable group.

17. Let H be a normal subgroup of a group G. Prove that G has a composition series if and only if both H and G/H have composition series. Also, show that G has a composition series containing H.

18. Prove that a finite group G is solvable if and only if $H' \neq H$ for any subgroup $H \neq \{e\}$ of G.

19. Let G be a solvable group with a composition series. Show that G is finite.

20. Prove that a group G is solvable if and only if $G/Z(G)$ is solvable.

21. Let A and B be subgroups of a group G. If A and B are solvable and A is normal in G, prove that AB is a solvable subgroup of G.

22. For the following statement, write the proof if the statement is true; otherwise, give a counterexample.

    (i) If $G \neq \{e\}$ is a solvable group, then $Z(G) \neq \{e\}$.

    (ii) Let G be a solvable group of order m. Then for every positive divisor n of m, G has a subgroup of order n.

    (iii) Every group of order 15 is solvable.

    (iv) Every solvable group has a composition series.

    (v) Every solvable series is a composition series.

    (vi) Every composition series is a solvable series.

    (vii) If two groups have equivalent composition series, then the groups are isomorphic.

8.2 Nilpotent Groups

In this section, we study another class of groups called nilpotent groups. We show that the converse of Lagrange's theorem also holds for such groups.

Definition 8.2.1 *A chain $G_0 \subseteq G_1 \subseteq G_2 \subseteq \cdots \subseteq G_n$ of normal subgroups of a group G is called a **central series** if $G_{i+1}/G_i \subseteq Z(G/G_i)$ for all $i = 0, 1, \ldots, n-1$.*

Definition 8.2.2 *A group G is called **nilpotent** if G has a central series*

$$G_0 \subseteq G_1 \subseteq G_2 \subseteq \cdots \subseteq G_n$$

such that $G_0 = \{e\}$ and $G_n = G$.

From the definition of a nilpotent group and from the commutative property of $Z(G/G_i)$, it follows that every nilpotent group is solvable and also that every commutative group is nilpotent.

Example 8.2.3 *The symmetric group S_3 has only two normal series,*

$$\{e\} \subseteq S_3$$

and

$$\{e\} \subseteq \left\{ e, \begin{pmatrix} 1 & 2 & 3 \\ 2 & 3 & 1 \end{pmatrix}, \begin{pmatrix} 1 & 2 & 3 \\ 3 & 1 & 2 \end{pmatrix} \right\} \subseteq S_3.$$

For the first series, $S_3/\{e\} \simeq S_3 \nsubseteq Z(S_3/\{e\}) = \{e\}$. For the second series, let

$$H = \left\{ e, \begin{pmatrix} 1 & 2 & 3 \\ 2 & 3 & 1 \end{pmatrix}, \begin{pmatrix} 1 & 2 & 3 \\ 3 & 1 & 2 \end{pmatrix} \right\}.$$

Now $H/\{e\} \nsubseteq Z(S_3/\{e\})$. Hence, S_3 is not a nilpotent group. However, S_3 is solvable.

Finite p-groups are the most important examples of nilpotent groups.

Theorem 8.2.4 *Every finite p-group is nilpotent.*

Proof. Let G be a finite p-group. If $|G| = 1$, then G is nilpotent. Suppose $|G| > 1$. Then $Z_1 = Z(G) \neq \{e\}$ by Theorem 7.2.7. If $G \neq Z_1$, then $|G/Z_1| > 1$ and hence by Theorem 7.2.7, $|Z(G/Z_1)| > 1$. Now there exists a normal subgroup Z_2 of G such that $Z_1 \subset Z_2$ and $Z_2/Z_1 = Z(G/Z_1)$. Thus, we have $\{e\} \subset Z_1 \subset Z_2$. If $G \neq Z_2$, we repeat the above process and obtain a normal subgroup Z_3 of G such that $Z_3/Z_2 = Z(G/Z_2)$ and $\{e\} \subset Z_1 \subset Z_2 \subset Z_3$.

Since G is finite, this process must terminate after a finite number of steps. We obtain the normal series

$$\{e\} \subset Z_1 \subset Z_2 \subset \cdots \subset Z_n = G$$

such that $Z_{i+1}/Z_i = Z(G/Z_i)$. Hence, G is nilpotent. ∎

For a group G, let us define $Z_i(G)$ as follows:

$$Z_0(G) = \{e\}, \ Z_1(G) = Z(G).$$

Now $Z_1(G)$ is a normal subgroup of G and $Z(G/Z_1(G))$ is a normal subgroup of $G/Z_1(G)$. Hence, there exists a unique normal subgroup $Z_2(G)$ of G such that $Z_1(G) \subseteq Z_2(G)$ and $Z_2(G)/Z_1(G) = Z(G/Z_1(G))$. Suppose $Z_i(G)$, $i \geq 1$, has been defined, i.e., $Z_i(G)$ is the normal subgroup of G such that

$$Z_{i-1}(G) \subseteq Z_i(G) \text{ and } Z_i(G)/Z_{i-1}(G) = Z(G/Z_{i-1}(G)).$$

There exists a unique normal subgroup $Z_{i+1}(G)$ of G such that

$$Z_i(G) \subseteq Z_{i+1}(G) \text{ and } Z_{i+1}(G)/Z_i(G) = Z(G/Z_i(G)).$$

Thus, we have the chain of normal subgroups

$$\{e\} = Z_0(G) \subseteq Z_1(G) \subseteq Z_2(G) \subseteq \cdots \subseteq Z_n(G) \subseteq \cdots$$

and $Z_{i+1}(G)/Z_i(G) = Z(G/Z_i(G))$, $i \geq 0$. This chain of normal subgroups is called the **ascending central series** of G.

Theorem 8.2.5 *Let G be a group such that $Z_n(G) = G$ for some nonnegative integer n. Then G is nilpotent.*

Proof.   We have the normal series

$$\{e\} = Z_0(G) \subseteq Z_1(G) \subseteq Z_2(G) \subseteq \cdots \subseteq Z_n(G) = G$$

such that $Z_{i+1}(G)/Z_i(G) = Z(G/Z_i(G))$, $i = 0, 1, \ldots, n-1$. Hence, G is nilpotent. ∎

Let G be a group and $a, b \in G$. We denote by $[a, b]$ the commutator $aba^{-1}b^{-1}$. Let A and B be subgroups of G. We denote the subgroup generated by elements $[a, b]$, for all $a \in A$, $b \in B$, by $[A, B]$.

Lemma 8.2.6 *Let A and B be subgroups of a group G and A be normal in G. Then $[B, G] \subseteq A$ if and only if $AB/A \subseteq Z(G/A)$.*

Proof. Suppose $[B, G] \subseteq A$. Then for all $b \in B$, $g \in G$, $bgb^{-1}g^{-1} \in A$. This implies that $AbAg = AgAb$. Thus, $Ab \in Z(G/A)$. Let $a \in A$ and $b \in B$. Then $Aab = AaAb = AAb = Ab \in Z(G/A)$. Hence, $AB/A \subseteq Z(G/A)$. Conversely, suppose $AB/A \subseteq Z(G/A)$. Let $b \in B$, $g \in G$. Then $Abgb^{-1}g^{-1} = AbAgAb^{-1}Ag^{-1} = A$ since $Ab \in Z(G/A)$. This implies that $[b, g] \in A$. Thus, $[B, G] \subseteq A$. ∎

Theorem 8.2.7 *Let G be a nilpotent group. Then there exists a nonnegative integer n such that $G = Z_n(G)$.*

Proof. Since G is nilpotent, there exists a normal series

$$\{e\} = G_0 \subseteq G_1 \subseteq G_2 \subseteq \cdots \subseteq G_n = G$$

such that $G_i/G_{i-1} \subseteq Z(G/G_{i-1})$, $i = 1, 2, \ldots, n$, for some n. We now prove by induction on i that $G_i \subseteq Z_i(G)$ for all $i = 0, 1, \ldots, n$. If $i = 0$, then $G_0 = \{e\} = Z_0(G)$. Suppose that $G_i \subseteq Z_i(G)$ for some $i \geq 0$. Since $G_iG_{i+1}/G_i = G_{i+1}/G_i \subseteq Z(G/G_i)$, we have by Lemma 8.2.6 that $[G_{i+1}, G] \subseteq G_i \subseteq Z_i(G)$. Thus, by Lemma 8.2.6,

$$Z_i(G)G_{i+1}/Z_i(G) \subseteq Z(G/Z_i(G)) = Z_{i+1}(G)/Z_i(G).$$

This implies that $G_{i+1} \subseteq Z_i(G)G_{i+1} \subseteq Z_{i+1}(G)$. Hence, by induction, $G_i \subseteq Z_i(G)$ for all $i = 0, 1, \ldots, n$. Since $G_n = G$, $Z_n(G) = G$. ∎

Let G be a group. Define the subgroups $G^{[i]}$ of G inductively as follows: $G^{[1]} = G$, $G^{[2]} = [G^{[1]}, G]$, $\ldots$, $G^{[i]} = [G^{[i-1]}, G]$, $i \geq 1$. It can be easily seen that

$$G = G^{[1]} \supseteq G^{[2]} \supseteq G^{[3]} \supseteq \cdots$$

is a central series. This series is called the **descending central series** of G.

Theorem 8.2.8 *A group G is nilpotent if and only if there exists a nonnegative integer n such that $G^{[n+1]} = \{e\}$.*

Proof. If $G^{[n+1]} = \{e\}$ for some nonnegative integer n, then G has a central series

$$\{e\} = G^{[n+1]} \subseteq G^{[n]} \subseteq \cdots \subseteq G^{[1]} = G.$$

Hence, G is nilpotent. Conversely, suppose that G is nilpotent. Then there exists a central series

$$\{e\} = G_0 \subseteq G_1 \subseteq G_2 \subseteq \cdots \subseteq G_n = G$$

of G. We now show that $G^{[i]} \subseteq G_{n-i+1}$ for all $i = 1, 2, \ldots, n+1$. Clearly, $G^{[1]} = G = G_n$. Suppose $G^{[i]} \subseteq G_{n-i+1}$ for some i, $1 \leq i < n+1$. Now $G_{i+1}/G_i \subseteq$

$Z(G/G_i)$, $i = 0, 1, \ldots, n - 1$. Therefore, by Lemma 8.2.6, $[G_{i+1}, G] \subseteq G_i$, $i = 0, 1, \ldots, n - 1$. This implies that $G^{[i+1]} = [G^{[i]}, G] \subseteq [G_{n-i+1}, G] \subseteq G_{n-i}$. Thus, by induction, $G^{[i]} \subseteq G_{n-i+1}$ for all $i = 1, 2, \ldots, n + 1$. Consequently, $G^{[n+1]} \subseteq G_0 = \{e\}$. ∎

Theorem 8.2.9 *Let G be a nilpotent group. Then every subgroup of G is nilpotent.*

Proof. Let H be a subgroup of G. There exists a positive integer n such that $G^{[n+1]} = \{e\}$. Now $H^{[1]} = H \subseteq G = G^{[1]}$. Suppose $H^{[i]} \subseteq G^{[i]}$ for $1 \leq i < n + 1$. Then $H^{[i+1]} = [H^{[i]}, H] \subseteq [G^{[i]}, G] = G^{[i+1]}$. Therefore, by induction, $H^{[i]} \subseteq G^{[i]}$ for all $i = 1, 2, \ldots, n + 1$. Hence, $H^{[n+1]} \subseteq G^{[n+1]} = \{e\}$, proving that H is nilpotent. ∎

Lemma 8.2.10 *Let G, H, and K be groups such that $G = H \times K$. Then $Z_i(G) = Z_i(H) \times Z_i(K)$ for all $i = 1, 2, \ldots$.*

Proof. For $i = 1$, $Z_1(G) = Z(G) = Z(H \times K) = Z(H) \times Z(K) = Z_1(H) \times Z_1(K)$. Thus, the lemma is true for $i = 1$. Suppose $Z_i(G) = Z_i(H) \times Z_i(K)$ for some $i \geq 1$. Now $Z_{i+1}(G)$ is the unique normal subgroup of G such that $Z_i(G) \subseteq Z_{i+1}(G)$ and $Z_{i+1}(G)/Z_i(G) = Z(G/Z_i(G))$. Consider the isomorphism $\psi : H/Z_i(H) \times K/Z_i(K) \to (H \times K)/Z_i(H \times K)$. Now

$$
\begin{aligned}
Z(G/Z_i(G)) &= Z((H \times K)/Z_i(H \times K)) \\
&= Z((H \times K)/Z_i(H) \times Z_i(K)) \quad \text{(by the} \\
&\qquad\qquad\qquad\qquad\qquad\qquad \text{induction hypothesis)} \\
&= Z(\psi((H/Z_i(H)) \times (K/Z_i(K)))) \\
&= \psi(Z((H/Z_i(H)) \times (K/Z_i(K)))) \\
&= \psi(Z(H/Z_i(H)) \times Z(K/Z_i(K))) \\
&= \psi((Z_{i+1}(H)/Z_i(H)) \times (Z_{i+1}(K)/Z_i(K))) \\
&= (Z_{i+1}(H) \times Z_{i+1}(K))/(Z_i(H) \times Z_i(K)) \\
&= (Z_{i+1}(H) \times Z_{i+1}(K))/Z_i(H \times K) \\
&= (Z_{i+1}(H) \times Z_{i+1}(K))/Z_i(G).
\end{aligned}
$$

Hence, $Z_{i+1}(G) = Z_{i+1}(H) \times Z_{i+1}(K)$. ∎

Lemma 8.2.11 *The direct product of two nilpotent groups is a nilpotent group.*

Proof. Let H and K be two nilpotent groups. Then there exists a positive integer n such that $Z_n(H) = H$ and $Z_n(K) = K$. Hence, $Z_n(H \times K) = Z_n(H) \times Z_n(K) = H \times K$ by Lemma 8.2.10. Thus, $H \times K$ is nilpotent. ∎

Theorem 8.2.12 *Let G_i, $i = 1, 2, \ldots, n$, be a nilpotent group. Then $G_1 \times G_2 \times \cdots \times G_n$ is nilpotent.*

Proof. The desired result follows by Lemma 8.2.11 and induction. ∎

The following theorem gives several equivalent conditions of a finite group to be a nilpotent group. In particular, the following theorem describes all finite nilpotent groups in terms of p-groups. It is an analogue of the primary decomposition theorem for finite Abelian groups.

Theorem 8.2.13 *Let G be a finite group. Then the following conditions are equivalent.*

(i) G is nilpotent.
(ii) If H is a proper subgroup of G, then $H \subset N_G(H)$.
(iii) Every maximal subgroup of G is a normal subgroup of G.
(iv) Every Sylow subgroup of G is a normal subgroup of G.
(v) G is isomorphic to a direct product of p-groups.

Proof. (i)⇒(ii) Since G is nilpotent, G has a central series

$$\{e\} = G_0 \subseteq G_1 \subseteq G_2 \subseteq \cdots \subseteq G_n = G.$$

Now $G_0 \subseteq H \subset G = G_n$. Hence, we can find an integer $m \geq 0$ such that $G_m \subseteq H$, but $G_{m+1} \not\subseteq H$. Thus, there exists $a \in G_{m+1}$ such that $a \notin H$. Now $aG_m \in Z(G/G_m)$. Therefore, for all $h \in H$, $(aG_m)(hG_m) = (hG_m)(aG_m)$. This implies $h^{-1}a^{-1}ha = (ah)^{-1}ha \in G_m \subseteq H$. Hence, $a^{-1}ha \in H$, and so $a^{-1}Ha \subseteq H$. Similarly, $aHa^{-1} \subseteq H$. Thus, $H = a^{-1}(aHa^{-1})a \subseteq a^{-1}Ha \subseteq H$ and so $a^{-1}Ha = H$. Hence $a \in N(H)$. Consequently, $H \neq N(H)$.

(ii)⇒(iii) Let H be a maximal subgroup of G. Then $H \subset N(H) \subseteq G$. Since H is maximal, $N(H) = G$. Thus, H is normal.

(iii)⇒(iv) Let P be a Sylow p-subgroup of G such that P is not normal. Since G is finite, there exists a maximal subgroup H of G such that $N(P) \subseteq H$. By (iii), H is a normal subgroup of G. Let $a \in G$. Then $aPa^{-1} \subseteq aN(P)a^{-1} \subseteq aHa^{-1} = H$. Hence, P and aPa^{-1} are Sylow p-subgroups of H. Thus, there exists $h \in H$ such that $h(aPa^{-1})h^{-1} = P$. Therefore, $ha \in N(P) \subseteq H$. This implies that $a = h^{-1}(ha) \in H$. Hence, $G = H$, a contradiction. Thus, P is a normal subgroup of G.

(iv)⇒(v) By Worked-Out Exercise 7 (page 208), G is a direct product of its Sylow p-subgroups. Since every Sylow p-subgroup is a p-group, G is a direct product of p-groups.

(v)⇒(i) The result here follows by Theorems 8.2.4 and 8.2.12. ∎

We conclude this section by showing that the converse of Lagrange's theorem holds for finite nilpotent groups.

Theorem 8.2.14 *Let G be a nilpotent group of order m. If $n > 0$ and $n|m$, then G contains a subgroup of order n.*

Proof.   If $m = 1$, then the result is trivially true. Suppose $m > 1$. There exist distinct prime integers $p_1, p_2, \ldots, p_k$ such that $m = p_1^{r_1} \, p_2^{r_2} \cdots p_k^{r_k}$, where r_i are positive integers. Let H_i be the Sylow p-subgroup for $p = p_i$ ($i = 1, 2, \ldots, k$) in G. Thus, by Theorem 8.2.13 and Worked-Out Exercise 7 (page 208), $G = H_1 \times H_2 \times \cdots \times H_k$. Since $n|m$, there exist integers $t_1, t_2, \ldots, t_k$ such that $n = p_1^{t_1} \, p_2^{t_2} \cdots p_k^{t_k}$. Now $|H_i| = p_i^{r_i}$ and so by Theorem 7.3.1, H_i contains a subgroup A_i of order $p_i^{t_i}$ for $i = 1, 2, \ldots, k$. Thus, $B = A_1 \times A_2 \times \cdots \times A_k$ is a subgroup of G of order n. ∎

8.2.1   Worked-Out Exercises

◇ **Exercise 1** Find a central series $G_0 \subseteq G_1 \subseteq \cdots \subseteq G_n$ in D_4 such that $G_0 = \{e\}$ and $G_n = D_4$.

Solution: $D_4 = \langle a, b \rangle$ such that $\circ(a) = 4$, $\circ(b) = 2$, and $ba = a^3 b$. Now

$$\{e\} = G_0 \subseteq G_1 = \{e, a^2\} \subseteq G_2 = \{e, a, a^2, a^3\} \subseteq G_n = D_4$$

is a normal series in D_4. Since $|D_4/G_1| = 4$ and $|D_4/G_2| = 2$, it follows that D_4/G_1 and D_4/G_2 are commutative groups. Thus, $G_2/G_1 \subseteq D_4/G_1 = Z(D_4/G_1)$ and $D_4/G_2 \subseteq Z(D_4/G_2) = D_4/G_2$. Since $Z(D_4) = \{e, a^2\} = G_1$, it follows that $G_1/G_0 \subseteq Z(D_4/G_0)$. Hence, $\{e\} \subseteq \{e, a^2\} \subseteq \{e, a, a^2, a^3\} \subseteq D_4$ is a central series.

◇ **Exercise 2** Give an example of a group G such that G is not nilpotent, but G contains a normal subgroup H such that H and G/H are nilpotent.

Solution: The symmetric group S_3 is not nilpotent. Now A_3 is a normal subgroup. Since $|A_3| = 3$, A_3 is commutative and hence nilpotent. Also, $|S_3/A_3| = 2$. Thus, S_3/A_3 is commutative and so is nilpotent.

8.2.2   Exercises

1. Prove that a homomorphic image of a nilpotent group is nilpotent.

2. Prove that a group of order 65 is nilpotent.

3. Show that D_n is nilpotent if and only if $n = 2^m$ for some positive integer m.

4. Find ascending central series for S_3 and $S_3 \times \mathbf{Z}_2$.

5. Is $S_3 \times S_3$ nilpotent?

Camille Jordan (1838–1921) was born on January 5, 1838 in Lyons, France, into a well-to-do family. At the age of seventeen he entered the École Polytechnique to become an engineer. During his time as an engineer, he had ample opportunity to carry out his mathematical research and to write most of his 120 papers. He retired as an engineer in 1885. From 1873 until 1912 he taught at the École Polytechnique and the Collége de France.

Jordan was a universal mathematician. He published papers in all branches of mathematics of his time. In analysis, he originated the concept of a bounded function. In topology, he showed that a plane can be decomposed into two regions by a simple closed curve.

Primarily, Jordan was an algebraist. He became famous at the age of 30 and for the next 40 years he was considered the master of group theory. He was the first to develop the theory of finite groups and its applications in the direction of Galois. He originated the concept of composition series and proved the first half of the famous Jordan-Hölder theorem. He studied solvable groups in a very general sense. In 1870, he collected all his results on permutation groups for the previous ten years in *Traité des substitutions*. His *Traité des substitutions* became a bible in all areas of group theory. Jordan's deepest results in algebra were his finiteness theorems. He was joined by Felix Klein and Sophus Lie in the study of groups of movements in three-dimensional space.

His *Course d'analyse*, published in the early 1880s, had a great influence on mathematics and set the standard for rigor. In this book, he showed how multiple integrals can be evaluated by successive integrations.

In his study of solvable groups, he made extensive use of concepts such as normal subgroup, homomorphic images of a group, and quotient groups. He was the first one to use the term "simple group."

He died on January 22, 1921.

Otto Ludwig Hölder (1859–1937) was born on December 22, 1859, in Stuttgart, Germany. His father was a professor of French. He received his early education in Stuttgart. On a colleague's suggestion, his father sent him to Berlin in 1877. At that time, Weierstrass, Kronecker and Kummer were teaching there.

In his dissertation, presented in 1882, Hölder developed the continuity condition for volume density that bears his name. He gave the first complete general proof of Weierstrass's theorem and also examined the convergence of the Fourier series of a function, which was not assumed to be either continuous or bounded.

After receiving his doctorate, Hölder attended Kronecker's and Klein's seminar and became interested in group theory. He completed the proof of the so-called Jordan-Hölder theorem on composition series by showing the uniqueness of the factor group, which is now a fundamental concept in group theory. He also studied simple groups. Other than the known simple groups of order 60 and 168, he showed that there is no other simple group of composite order less that 200. He also investigated the structure of groups of orders p^3, pq^2, pqr, p^4, and n, where p, q, r are primes and n is a square free integer. He also worked on geometry and number theory. Hölder died on August 29, 1937.

Chapter 9

Finitely Generated Abelian Groups

The second source in the evolution of group theory, namely, number theory, led to the specialized theory of Abelian groups.

In this chapter, we determine the structural properties of finite Abelian groups and finitely generated Abelian groups. In Section 4.2, it was shown that every cyclic group is Abelian. In Section 5.1, it was proved that any two finite cyclic groups of the same order are isomorphic and thus for any positive integer n, $\mathbf{Z}_n$ is the only cyclic group of order n (up to isomorphism). That an infinite cyclic group is isomorphic to $\mathbf{Z}$ was shown in Section 5.1. Hence, all cyclic groups have been determined. In this chapter, it is proved that any finitely generated (and hence any finite) Abelian group can be expressed as a direct sum of cyclic groups. Thus, the structural properties of a finitely generated (finite) Abelian group can be determined from those of cyclic groups.

In this chapter, we use additive notation for the group operation. 0 will denote the identity element and $-a$ will denote the inverse of an element a. The direct product (internal or external) $G \times H$ of groups (subgroups) will be written as $G \oplus H$ and called the direct sum of G and H.

Let G be an Abelian group. By Theorem 6.1.4, G is the direct sum of subgroups $G_1, G_2, \ldots, G_n$ if and only if

(i) $G = G_1 + G_2 + \cdots + G_n$ (i.e., for all $g \in G$, $g = g_1 + g_2 + \cdots + g_n$ for some $g_i \in G_i$, $i = 1, 2, \ldots, n$) and

(ii) $G_i \cap (G_1 + \cdots + G_{i-1} + G_{i+1} + \cdots + G_n) = \{0\}$ for all $i = 1, 2, \ldots, n$.

If G is a direct sum of subgroups $G_1, G_2, \ldots, G_n$, then we write

$$G = G_1 \oplus G_2 \oplus \cdots \oplus G_n.$$

If $G = G_1 \oplus G_2 \oplus \cdots \oplus G_n$ and $G_i \simeq H_i$, where H_i is a group, $i = 1, 2, \ldots, n$, then

$$G \simeq H_1 \oplus H_2 \oplus \cdots \oplus H_n.$$

9.1 Finite Abelian Groups

Given a positive integer n, the cyclic groups of order n have been completely determined. We can determine the subgroups, homomorphic images, and generators of such groups. Now every cyclic group is Abelian, but not conversely. Given any positive integer n, what can we say about an Abelian group of order n? How many different Abelian groups of a given order are there? What can we say about the subgroups of such groups? In this section, we attempt to answer such questions. The main theorem of this section is that every finite Abelian group is a finite direct sum of a finite number of cyclic p-groups. We will use this theorem to answer some of the above questions. We begin with the following definition.

Let G be an Abelian group and A be a subgroup of G. Then A is called a **direct summand** of G if there exists a subgroup B of G such that

$$G = A \oplus B.$$

We leave the proof of the following theorem as an exercise.

Theorem 9.1.1 *Let G be an Abelian group. Let $r \in \mathbf{Z}$ and p be a prime.*
(i) Let $G[r] = \{g \in G \mid rg = 0\}$. Then $G[r]$ is a subgroup of G.
(ii) Let $rG = \{rg \mid g \in G\}$. Then rG is a subgroup of G.
(iii) Let $G(p) = \{g \in G \mid g$ is of order p^s for some $s \geq 0\}$. Then $G(p)$ is a subgroup of G.
(iv) $G/G[r] \simeq rG$. ∎

Definition 9.1.2 *The subgroup $G(p)$ of Theorem 9.1.1 is called a p-**primary component** of G.*

Let G be a finite Abelian group of order p^l for some $l \in \mathbf{N}$. Since the order of each element of G divides the order of G, the order of each element is p^r for some r, $0 \leq r \leq l$. Therefore, there exists $a \in G$ such that $o(a) \geq o(b)$ for all $b \in G$. Hence, the corresponding cyclic subgroup $\langle a \rangle$ is of maximal order in G. In the next theorem, we show that $\langle a \rangle$ is a direct summand of G.

Theorem 9.1.3 *Let G be a finite Abelian group of order p^l for some $l \in \mathbf{N}$, p a prime. Let $a \in G$ be such that $o(a) = p^k$ is the largest in G. Then $\langle a \rangle$ is a direct summand of G, i.e., there exists a subgroup B of G such that $G = \langle a \rangle \oplus B$.*

Proof. Let $0 \neq x \in G$. Since $|G| = p^l$, $o(x) = p^t$ for some positive integer t. Also, $o(a) \geq o(x)$ and so $t \leq k$. Therefore, $p^k x = 0$ for all $x \in G$. Let

$$\mathcal{C} = \{B \mid B \text{ is a subgroup of } G \text{ and } \langle a \rangle \cap B = \{0\}\}.$$

Since $\{0\} \in \mathcal{C}$, $\mathcal{C} \neq \phi$. Also, $\mathcal{C}$ contains only a finite number of subgroups. Hence, $\mathcal{C}$ has a maximal element, say, B. We show that $G = \langle a \rangle \oplus B$. Suppose there exists $g \in G$ such that $g \notin \langle a \rangle \oplus B$. Since $p^k g = 0 \in \langle a \rangle \oplus B$, there exists a positive integer s such that $p^s g \in \langle a \rangle \oplus B$. Let n be the smallest positive integer such that $p^n g \in \langle a \rangle \oplus B$, i.e., $p^n g \in \langle a \rangle \oplus B$, but $p^{n-1} g \notin \langle a \rangle \oplus B$. Write $d = p^{n-1} g$. Then $d \notin \langle a \rangle \oplus B$ and $pd \in \langle a \rangle \oplus B$. Now $pd = ta + b$ for some $t \in \mathbf{Z}$ and $b \in B$. Therefore, $0 = p^{k-1} pd = p^{k-1} ta + p^{k-1} b$. Thus, $p^{k-1} ta = -p^{k-1} b \in \langle a \rangle \cap B$ and so $p^{k-1} ta = 0$. Then $o(a) = p^k$ must divide $p^{k-1} t$ and so $p | t$. Let $t = pr$ and $a' = ra \in \langle a \rangle$. Then $pd = pa' + b$ or $p(d - a') = b \in B$. Write $x = d - a'$. Then $x = d - a' = d - ra \notin B$ and this shows that $\langle a \rangle \cap \langle B, x \rangle \neq \{0\}$. Hence, there exist $m, s \in \mathbf{Z}$ and $b_1 \in B$ such that $0 \neq ma = b_1 + sx$. If $\gcd(p, s) \neq 1$, then $s = pq$ for some $q \in \mathbf{Z}$. Since $px \in B$, $ma = b_1 + q(px) \in B$, which contradicts the fact that $\langle a \rangle \cap B = \{0\}$. Therefore, $\gcd(p, s) = 1$, which implies that there exist $u, v \in \mathbf{Z}$ such that $1 = us + vp$. Thus, $x = u(sx) + v(px) = u(ma - b_1) + v(px) = uma + (-ub_1 + v(px)) \in \langle a \rangle \oplus B$, i.e., $d - a' = x \in \langle a \rangle \oplus B$. But then $d = d - a' + a' \in \langle a \rangle \oplus B$, which is a contradiction since $d \notin \langle a \rangle \oplus B$. Hence, $G = \langle a \rangle \oplus B$. ∎

Example 9.1.4 *Let G be a noncyclic group of order p^2. Since $|G| = p^2$, G is Abelian. By Cauchy's theorem, there exists $a \in G$ such that $o(a) = p$. Since G is not cyclic, G does not contain any element of order p^2. Therefore, $o(a)$ is the largest in G. Thus, there exists a subgroup B of G such that*

$$G = \langle a \rangle \oplus B.$$

Since $|G| = |\langle a \rangle| \cdot |B|$, it follows that $|B| = p$. This shows that B is a cyclic group of order p and $\langle a \rangle \simeq \mathbf{Z}_p \simeq B$. Hence,

$$G \simeq \mathbf{Z}_p \oplus \mathbf{Z}_p.$$

In the next theorem, we prove that any nontrivial Abelian p-group can be expressed uniquely as a direct sum of nontrivial cyclic p-groups.

Theorem 9.1.5 *Let G be a finite Abelian p-group, p a prime. Then G is a direct sum of cyclic p-groups. Furthermore, if $G = G_1 \oplus G_2 \oplus \cdots \oplus G_r = H_1 \oplus H_2 \oplus \cdots \oplus H_s$, where G_i and H_j are cyclic p-groups, $|G_1| \geq |G_2| \geq \cdots \geq |G_r| > 1$, and $|H_1| \geq |H_2| \geq \cdots \geq |H_s| > 1$, then $r = s$ and $G_i \simeq H_i$, $1 \leq i \leq r$.*

Proof. Let $|G| = p^n$. We prove the result by induction on n. If $n = 1$, then G is a cyclic group of order p and so in this case the result is trivially true. Suppose the result is true for all p-groups of order less than the order of G. Let $a \in G$ be such that $o(a)$ is the largest in G. Then by Theorem 9.1.3, there exists a subgroup B of G such that $G = \langle a \rangle \oplus B$. Now B is a p-group and

$|B| < |G|$. Therefore, by the induction hypothesis, B is a direct sum of cyclic p-groups and therefore G is a direct sum of cyclic p-groups. We now prove the uniqueness part.

We first note that $G[p]$ and $G_i[p]$ are subgroups of G and G_i, respectively. Let $a \in G[p]$. Then $a = a_1 + a_2 + \cdots + a_r$ for some $a_i \in G_i$, $1 \leq i \leq r$. Now $pa_1 + pa_2 + \cdots + pa_r = pa = 0$. Hence, $pa_i = 0$ for all $1 \leq i \leq r$. Thus, $a_i \in G_i[p]$ for all $1 \leq i \leq r$. Therefore, $G[p] = G_1[p] \oplus G_2[p] \oplus \cdots \oplus G_r[p]$. Since $G_i[p]$ is a cyclic group such that every nonidentity element is of order p, $|G_i[p]| = p$ for all $1 \leq i \leq r$. Thus,

$$|G[p]| = |G_1[p]| \, |G_2[p]| \cdots |G_r[p]| = p^r.$$

By a similar argument, $|G[p]| = p^s$ since $G = H_1 \oplus H_2 \oplus \cdots \oplus H_s$. Thus, $p^r = p^s$ and so $r = s$. Now since cyclic groups of the same order are isomorphic, in order to show that $G_i \simeq H_i$, $1 \leq i \leq r$, it suffices to show that $|G_i| = |H_i|$, $1 \leq i \leq r$. We prove this by induction on n. If $n = 1$, then the result is trivially true. Suppose that the result is true for all p-groups of order less than p^n, where $n > 1$. By Theorem 9.1.1(iv), $G_i/G_i[p] \simeq pG_i$. Since G_i is cyclic, $G_i[p]$ is cyclic. Also, since every nonidentity element of $G_i[p]$ is of order p, $|G_i[p]| = p$. Thus, $|pG_i| = \frac{|G_i|}{p} < |G_i|$. This implies that $pG_i = \{0\}$ if and only if $|G_i| = p$. Now if $pG_i = \{0\}$, then $pG_l = \{0\}$ for all $i \leq l \leq r$. Thus, $pG = pG_1 \oplus \cdots \oplus pG_m$, where $m \leq r$, $pG_i \neq \{0\}$, $1 \leq i \leq m$, and $pG_l = \{0\}$, $m + 1 \leq l \leq r$. Similarly, $pG = pH_1 \oplus \cdots \oplus pH_t$, where $t \leq r$, $pH_l \neq \{0\}$, $1 \leq l \leq t$ and $pH_l = \{0\}$ for all $t + 1 \leq l \leq r$. Since $|pG| < |G|$, $m = t$ and $|pG_i| = |pH_i|$ for all $1 \leq i \leq m$, by the induction hypothesis, and therefore $|G_i| = |H_i|$ for all $1 \leq i \leq m$. Also, $|G_i| = p = |H_i|$ for all $m + 1 \leq i \leq r$. Consequently, $|G_i| = |H_i|$ for all $1 \leq i \leq r$. ∎

Example 9.1.6 *Let G be an Abelian group of order 8. Since $8 = 2^3$, G is a 2-group. There exists $a \in G$ such that $o(a)$ is the largest in G. By Cauchy's theorem, G has an element of order 2. Thus, $o(a) \geq 2$ and so $o(a) = 2$, 4 or 8. If $o(a) = 8$, then $G \simeq \mathbf{Z}_8$. If $o(a) = 4$, then $G \simeq \mathbf{Z}_4 \oplus \mathbf{Z}_2$. Now suppose that $o(a) = 2$. By Theorem 9.1.3, there exists a subgroup B of G such that*

$$G = \langle a \rangle \oplus B.$$

Then $|B| = 4 = 2^2$, proving that B is a 2-group. Since $o(a)$ is the largest in G, B has no element of order 4. Thus, $B \simeq \mathbf{Z}_2 \oplus \mathbf{Z}_2$. Hence,

$$G \simeq \mathbf{Z}_2 \oplus \mathbf{Z}_2 \oplus \mathbf{Z}_2.$$

Now $\mathbf{Z}_8$ has an element of order 8, $\mathbf{Z}_4 \oplus \mathbf{Z}_2$ has no element of order 8, but has an element of order 4 and $\mathbf{Z}_2 \oplus \mathbf{Z}_2 \oplus \mathbf{Z}_2$ has no element of order 4 or 8. Thus, $\mathbf{Z}_8$, $\mathbf{Z}_4 \oplus \mathbf{Z}_2$ and $\mathbf{Z}_2 \oplus \mathbf{Z}_2 \oplus \mathbf{Z}_2$ are nonisomorphic groups. Hence, there are exactly three (up to isomorphism) Abelian groups of order 8.

The next theorem is called **the fundamental theorem of finite Abelian groups**.

Theorem 9.1.7 *Let G be a finite Abelian group. Then G is a direct sum of cyclic p-groups. Furthermore, any two decompositions of G as a direct sum of nontrivial cyclic p-groups are the same except for the order in which the summands are arranged.*

Proof.   If $|G| = 1$, then the result follows easily. We now assume that $|G| > 1$. Let $|G| = p_1^{n_1} p_2^{n_2} \cdots p_l^{n_l}$, where the p_i's are distinct primes and the n_i's are positive integers. By Theorem 7.3.5, G has a Sylow p_i-subgroup, say, G_i for all $i = 1, 2, \ldots, l$. Since G is Abelian, G_i is a normal subgroup of G and hence G_i is unique for all $i = 1, 2, \ldots, l$. From Worked-Out Exercise 7 (page 208), it follows that G is the internal direct sum of G_i, $i = 1, 2, \ldots, l$. However, since we are using additive notation, we give details of the proof for the sake of completeness.

Now $|G_i| = p_i^{n_i}$ for all $i = 1, 2, \ldots, l$. Hence, $G_i \cap G_j = \{0\}$ for all $i \neq j$. We now show that

$$G_i \cap (G_1 + \cdots + G_{i-1} + G_{i+1} + \cdots + G_l) = \{0\}$$

for all $i = 1, 2, \ldots, l$. Suppose $a \in G_i \cap (G_1 + \cdots + G_{i-1} + G_{i+1} + \cdots + G_l)$. Then $a \in G_i$ and $a \in G_1 + \cdots + G_{i-1} + G_{i+1} + \cdots + G_l$. Hence,

$$a = a_1 + \cdots + a_{i-1} + a_{i+1} + \cdots + a_l,$$

where the $a_j \in G_j$. Now for all $j \neq i$, $\circ(a_j) = p^{r_j}$ for some r_j, $0 \leq r_j \leq n_j$. Let

$$r = p_1^{r_1} \cdots p_{i-1}^{r_{i-1}} p_{i+1}^{r_{i+1}} \cdots p_l^{r_l}.$$

Then $ra = 0$. Thus, $\circ(a)$ divides r. Since $a \in G_i$, $\circ(a)$ divides $p_i^{n_i}$. But r and $p_i^{n_i}$ are relatively prime. Therefore, $\circ(a) = 1$. This implies that $a = 0$. Hence,

$$G_i \cap (G_1 + \cdots + G_{i-1} + G_{i+1} + \cdots + G_l) = \{0\}.$$

From this, it follows that

$$|G_1 + \cdots + G_l| = |G_1| \cdots |G_l| = p_1^{n_1} p_2^{n_2} \cdots p_l^{n_l} = |G|.$$

Thus,

$$G = G_1 \oplus G_2 \oplus \cdots \oplus G_l.$$

Now each G_i is an Abelian p-group. Hence, by Theorem 9.1.5, G_i is a direct sum of cyclic p-groups, whence G is a direct sum of cyclic p-groups.

We now prove the uniqueness of the direct summands. We prove the result by induction on l, the number of distinct primes in the factorization of $|G|$. If

$l = 1$, then G is a p-group and the result is true by Theorem 9.1.5. Suppose the result is true for all nonzero finite Abelian groups H such that the number of distinct primes in the factorization of $|H|$ is less than l.

Let

$$G = G_1 \oplus G_2 \oplus \cdots \oplus G_r = H_1 \oplus H_2 \oplus \cdots \oplus H_t$$

be two decompositions of G as a direct sum of nontrivial cyclic p-groups. Since for groups $A \oplus B \simeq B \oplus A$, we may assume by rearranging if necessary that the summands $G_1, G_2, \ldots, G_m$ and $H_1, H_2, \ldots, H_s$ ($m \leq r$, $s \leq t$) are the cyclic p-groups for the prime p_1, the groups $G_{m+1}, \ldots, G_r$ and $H_{s+1}, \ldots, H_t$ are cyclic p-groups for the primes p different from p_1, $|G_1| \geq |G_2| \geq \cdots \geq |G_m|$, and $|H_1| \geq |H_2| \geq \cdots \geq |H_s|$. Let $A = G_1 \oplus G_2 \oplus \cdots \oplus G_m$, $B = H_1 \oplus H_2 \oplus \cdots \oplus H_s$, $C = G_{m+1} \oplus \cdots \oplus G_r$, and $D = H_{s+1} \oplus \cdots \oplus H_t$. Then

$$G = A \oplus C = B \oplus D.$$

We now show that $A = B$. First note that the order of a nonzero element of A and the order of a nonzero element of C are relatively prime. Similarly, the order of a nonzero element of B and the order of a nonzero element of D are relatively prime. Let $a \in A$, $a \neq 0$. Then $a \in G = B \oplus D$. Thus, $a = b + d$ for some $b \in B$ and $d \in D$. If $a - b \neq 0$, then the order of $a - b$ is some positive multiple of p_1 whereas the order of d is different from any positive multiple of p_1. Therefore, we have a contradiction and so $a - b = 0$ or $a = b \in B$. This implies that $A \subseteq B$. Similarly, $B \subseteq A$ and so $A = B$. A similar argument shows that $C = D$. Now $A = B$ is a p-group and hence by Theorem 9.1.5, $m = s$ and $G_i \simeq H_i$, $i = 1, 2, \ldots, m$. Now $C = D$ is an Abelian group of order $p_2^{n_2} \cdots p_l^{n_l}$. Hence, by the induction hypothesis, it follows that the two decompositions $G_{m+1} \oplus \cdots \oplus G_r$ and $H_{s+1} \oplus \cdots \oplus H_t$ of the group C are the same except for the order in which the summands are arranged. Consequently, the above two decompositions of G are also the same except for the order in which the summands are arranged. ∎

From Theorem 9.1.7, it follows that for any finite Abelian group $G \neq \{0\}$ there is a list of positive integers $p_1^{n_1}, p_2^{n_2}, \ldots, p_k^{n_k}$, which are unique except for their order, where $p_1, p_2, \ldots, p_k$ are primes (not necessarily distinct) and $n_1, n_2, \ldots, n_k$ are positive integers such that

$$G \simeq \mathbf{Z}_{p_1^{n_1}} \oplus \mathbf{Z}_{p_2^{n_2}} \oplus \cdots \oplus \mathbf{Z}_{p_k^{n_k}}.$$

The numbers $p_1^{n_1}, p_2^{n_2}, \ldots, p_k^{n_k}$ are called the **elementary divisors** of G.

Example 9.1.8 *Let G be the group $\mathbf{Z}_4 \oplus \mathbf{Z}_6 \oplus \mathbf{Z}_9$. Now*

$$G \simeq \mathbf{Z}_{2^2} \oplus \mathbf{Z}_3 \oplus \mathbf{Z}_2 \oplus \mathbf{Z}_{3^2} \simeq \mathbf{Z}_2 \oplus \mathbf{Z}_{2^2} \oplus \mathbf{Z}_3 \oplus \mathbf{Z}_{3^2}.$$

Hence, the elementary divisors of G are $2, 2^2, 3, 3^2$.

In Section 7.2, we proved by using Cauchy's theorem that the converse of Lagrange's theorem holds for finite Abelian groups. Next, we prove the same result by using the results developed in this chapter.

Corollary 9.1.9 *If G is a finite Abelian group of order n and m is a positive divisor of n, then G has a subgroup of order m.*

Proof.   If $n = 1$, then $m = 1$ and $\{e\}$ is the subgroup of order m. Suppose $n > 1$. By Theorem 9.1.7, there exist prime integers $p_1, p_2, \ldots, p_k$ and positive integers $n_1, n_2, \ldots, n_k$ such that $G \simeq \mathbf{Z}_{p_1^{n_1}} \oplus \mathbf{Z}_{p_2^{n_2}} \oplus \cdots \oplus \mathbf{Z}_{p_k^{n_k}}$. This implies that $n = p_1^{n_1} p_2^{n_2} \cdots p_k^{n_k}$. Since $m | n$, there exist integers $0 \le m_i \le n_i$, $i = 1, 2, \ldots, k$ such that $m = p_1^{m_1} p_2^{m_2} \cdots p_k^{m_k}$. Since $p_i^{m_i} | p_i^{n_i}$ for all i, by Theorem 4.2.10, the cyclic group $\mathbf{Z}_{p_i^{n_i}}$ has a unique subgroup G_i of order $p_i^{m_i}$ for all i. Thus, $G_1 + G_2 + \cdots + G_k = G_1 \oplus G_2 \oplus \cdots \oplus G_k$ is a subgroup of $\mathbf{Z}_{p_1^{n_1}} \oplus \mathbf{Z}_{p_2^{n_2}} \oplus \cdots \oplus \mathbf{Z}_{p_k^{n_k}}$ of order $p_1^{m_1} p_2^{m_2} \cdots p_k^{m_k} = m$. From this, it follows that G has a subgroup of order m. ∎

Let G be a finite Abelian group of order $n = p_1^{n_1} p_2^{n_2} \cdots p_k^{n_k}$, where the p_i's are distinct primes and the n_i's are nonnegative integers. Consider the subgroup G_i (as defined in the proof of Theorem 9.1.7). Now $|G_i| = p_i^{n_i}$. From this, it follows that $G_i \subseteq G(p_i)$. Thus, $|G(p_i)| \ge p_i^{n_i}$. Since $G(p_i)$ is a p_i-group, $|G(p_i)| = p_i^{t}$ for some integer t. Hence, $t \ge n_i$. Suppose $t > n_i$. By Lagrange's theorem, $|G(p_i)|$ divides $|G|$. This implies that $p_i^{t} | p_1^{n_1} p_2^{n_2} \cdots p_k^{n_k}$, which in turn implies that $p_i^{t-n_i} | p_1^{n_1} \cdots p_{i-1}^{n_{i-1}} p_{i+1}^{n_{i+1}} \cdots p_k^{n_k}$, a contradiction, since the p_i's are distinct primes. Hence, $t = n_i$ and so $G_i = G(p_i)$. From this, we conclude that G is a direct sum of its p-primary components.

Consider the cyclic group $\mathbf{Z}_n$. There exist distinct primes $p_1, p_2, \ldots, p_k$ and positive integers $n_1, n_2, \ldots, n_k$ such that $n = p_1^{n_1} p_2^{n_2} \cdots p_k^{n_k}$. For $p = p_i$, the p-primary component of $\mathbf{Z}_n$ is $\mathbf{Z}_{p_i^{n_i}}$. Hence, it follows that

$$\mathbf{Z}_n \simeq \mathbf{Z}_{p_1^{n_1}} \oplus \mathbf{Z}_{p_2^{n_2}} \oplus \cdots \oplus \mathbf{Z}_{p_k^{n_k}}.$$

Example 9.1.10 *(i) Let $G = \mathbf{Z}_{12}$. Now $12 = 2^2 \cdot 3$ and so by the previous paragraph, $G \simeq \mathbf{Z}_{2^2} \oplus \mathbf{Z}_3 = \mathbf{Z}_4 \oplus \mathbf{Z}_3$. Now $G(2) \simeq \mathbf{Z}_4$ and $G(3) \simeq \mathbf{Z}_3$. Hence, the primary components are $\mathbf{Z}_4$ and $\mathbf{Z}_3$.*

(ii) Let $G = \mathbf{Z}_{12} \oplus \mathbf{Z}_{18} \oplus \mathbf{Z}_{60}$. Now $12 = 2^2 \cdot 3$, $18 = 2 \cdot 3^2$, and $60 = 2^2 \cdot 3 \cdot 5$. Thus,

$$
\begin{aligned}
G &= \mathbf{Z}_{12} \oplus \mathbf{Z}_{18} \oplus \mathbf{Z}_{60} \\
  &\simeq (\mathbf{Z}_4 \oplus \mathbf{Z}_3) \oplus (\mathbf{Z}_2 \oplus \mathbf{Z}_9) \oplus (\mathbf{Z}_4 \oplus \mathbf{Z}_3 \oplus \mathbf{Z}_5) \\
  &\simeq (\mathbf{Z}_4 \oplus \mathbf{Z}_4 \oplus \mathbf{Z}_2) \oplus (\mathbf{Z}_9 \oplus \mathbf{Z}_3 \oplus \mathbf{Z}_3) \oplus \mathbf{Z}_5.
\end{aligned}
$$

This implies that $G(2) \simeq \mathbf{Z}_4 \oplus \mathbf{Z}_4 \oplus \mathbf{Z}_2$, $G(3) \simeq \mathbf{Z}_9 \oplus \mathbf{Z}_3 \oplus \mathbf{Z}_3$, and $G(5) \simeq \mathbf{Z}_5$. Hence, the primary components are $\mathbf{Z}_4 \oplus \mathbf{Z}_4 \oplus \mathbf{Z}_2$, $\mathbf{Z}_9 \oplus \mathbf{Z}_3 \oplus \mathbf{Z}_3$, and $\mathbf{Z}_5$.

Definition 9.1.11 *Let G be a finite Abelian p-group of order p^n $(n > 0)$. If $G = G_1 \oplus G_2 \oplus \cdots \oplus G_k$, where each G_i is a cyclic group of order p^{n_i} with $n_1 \geq n_2 \geq \cdots \geq n_k > 0$, then the integers $n_1, n_2, \ldots, n_k$ are called the* **invariants** *of G and the k-tuple $(n_1, n_2, \ldots, n_k)$ is called the* **type** *of G.*

We know that any two cyclic groups of the same order are isomorphic. However, this result does not hold for Abelian groups. For example, $\mathbf{Z}_8$ and $\mathbf{Z}_4 \oplus \mathbf{Z}_4$ are nonisomorphic Abelian groups of order $8 = 2^3$. In the next theorem, we obtain a necessary and sufficient condition for two finite Abelian p-groups of the same order to be isomorphic.

Theorem 9.1.12 *Two Abelian p-groups of order p^n $(n > 0)$ are isomorphic if and only if they have the same invariants.*

Proof. Let G and H be two Abelian p-groups of order p^n $(n > 0)$. Suppose G and H have the same invariants $n_1, n_2, \ldots, n_k$, where $n_1 \geq n_2 \geq \cdots \geq n_k > 0$. Then $G = G_1 \oplus G_2 \oplus \cdots \oplus G_k$, where each G_i is a cyclic group of order p^{n_i}, $1 \leq i \leq k$, and $H = H_1 \oplus H_2 \oplus \cdots \oplus H_k$, where each H_i is a cyclic group of order p^{n_i}, $1 \leq i \leq k$. Since cyclic groups of the same order are isomorphic, $G_i \simeq H_i$, $1 \leq i \leq k$. Hence, $G \simeq H$. Conversely, suppose $G \simeq H$. Let $G = G_1 \oplus G_2 \oplus \cdots \oplus G_k$, where each G_i is a cyclic group of order p^{n_i}, $1 \leq i \leq k$, $n_1 \geq n_2 \geq \cdots \geq n_k > 0$, and $H = H_1 \oplus H_2 \oplus \cdots \oplus H_t$, where each H_j is a cyclic group of order p^{r_j}, $1 \leq j \leq t$, $r_1 \geq r_2 \geq \cdots \geq r_t > 0$. Let $f : G \rightarrow H$ be an isomorphism of groups. Then $f^{-1}(H_i)$ is a cyclic subgroup of G of order p^{r_i} and also $G = f^{-1}(H_1) \oplus f^{-1}(H_2) \oplus \cdots \oplus f^{-1}(H_t)$. Hence, by Theorem 9.1.5, it follows that $t = k$ and $p^{r_i} = |f^{-1}(H_i)| = p^{n_i}$, $1 \leq i \leq k$. ∎

Example 9.1.13 $\mathbf{Z}_4 \oplus \mathbf{Z}_2$ *and* $\mathbf{Z}_2 \oplus \mathbf{Z}_2 \oplus \mathbf{Z}_2$ *are 2-groups of order 2^3. Now the invariants of $\mathbf{Z}_4 \oplus \mathbf{Z}_2$ are 2, 1 and the invariants of $\mathbf{Z}_2 \oplus \mathbf{Z}_2 \oplus \mathbf{Z}_2$ are 1, 1, 1. Hence, $\mathbf{Z}_4 \oplus \mathbf{Z}_2$ and $\mathbf{Z}_2 \oplus \mathbf{Z}_2 \oplus \mathbf{Z}_2$ are nonisomorphic groups.*

Let n be a positive integer. A **partition** of n is an s-tuple $(n_1, n_2, \ldots, n_s)$ of positive integers such that $n = n_1 + n_2 + \cdots + n_s$ and $n_1 \geq n_2 \geq \cdots \geq n_s$.

We find that any finite Abelian p-group G of order p^n $(n > 0)$ can be decomposed uniquely as $G = G_1 \oplus G_2 \oplus \cdots \oplus G_k$, where each G_i is a cyclic group of order p^{n_i}, $1 \leq i \leq k$, and $n_1 \geq n_2 \geq \cdots \geq n_k > 0$. It is also true that $n = n_1 + n_2 + \cdots + n_k$. Therefore, $n_1, n_2, \ldots, n_k$ determine a partition of n. Next, let $n = n_1 + n_2 + \cdots + n_k$, where each n_i is a positive integer and $n_1 \geq n_2 \geq \cdots \geq n_k$. Then $G = \mathbf{Z}_{p^{n_1}} \oplus \mathbf{Z}_{p^{n_2}} \oplus \cdots \oplus \mathbf{Z}_{p^{n_k}}$ is an Abelian p-group of order $p^{n_1 + n_2 + \cdots + n_k} = p^n$ such that the invariants of G are $n_1, n_2, \ldots, n_k$. It now follows that the number of nonisomorphic Abelian p-groups of order p^n $(n > 0)$ is equal to the number of partitions of n.

Example 9.1.14 *Let $p = 2$ and $n = 4$. In this example, we want to describe all Abelian groups of order 2^4. Now $1 + 1 + 1 + 1$, $2 + 1 + 1$, $3 + 1$, $2 + 2$, and 4 are all the partitions of 4. Thus, there are five nonisomorphic Abelian groups of order 2^4. They are*

$$\mathbf{Z}_{16}$$
$$\mathbf{Z}_8 \oplus \mathbf{Z}_2$$
$$\mathbf{Z}_4 \oplus \mathbf{Z}_4$$
$$\mathbf{Z}_4 \oplus \mathbf{Z}_2 \oplus \mathbf{Z}_2$$
$$\mathbf{Z}_2 \oplus \mathbf{Z}_2 \oplus \mathbf{Z}_2 \oplus \mathbf{Z}_2.$$

9.1.1   Worked-Out Exercises

◇ **Exercise 1** Describe all Abelian groups of order 2^5.

Solution: $5 = 1 + 1 + 1 + 1 + 1 = 2 + 1 + 1 + 1 = 3 + 1 + 1 = 4 + 1 = 3 + 2 = 2 + 2 + 1$. Thus, there are seven partitions of 5 and so there exist seven nonisomorphic 2-groups of order 2^5. They are

$$\mathbf{Z}_{32}$$
$$\mathbf{Z}_{16} \oplus \mathbf{Z}_2$$
$$\mathbf{Z}_8 \oplus \mathbf{Z}_4$$
$$\mathbf{Z}_8 \oplus \mathbf{Z}_2 \oplus \mathbf{Z}_2$$
$$\mathbf{Z}_4 \oplus \mathbf{Z}_4 \oplus \mathbf{Z}_2$$
$$\mathbf{Z}_4 \oplus \mathbf{Z}_2 \oplus \mathbf{Z}_2 \oplus \mathbf{Z}_2$$
$$\mathbf{Z}_2 \oplus \mathbf{Z}_2 \oplus \mathbf{Z}_2 \oplus \mathbf{Z}_2 \oplus \mathbf{Z}_2.$$

◇ **Exercise 2** Find all Abelian groups of order 20.

Solution: Let G be an Abelian group of order 20. Now $20 = 2^2 \cdot 5$. By Theorem 7.3.5, G has a Sylow 5-subgroup, say, $G(5)$ and a Sylow 2-subgroup, say, $G(2)$. Since G is Abelian, $G(2)$ and $G(5)$ are normal subgroups of G and hence are unique. Now $G(2) \cap G(5) = \{0\}$. This implies that $|G(2) + G(5)| = |G(2)| \cdot |G(5)| = 4 \cdot 5 = 20$. Thus, $G = G(2) + G(5)$. Hence, $G = G(2) \oplus G(5)$. Now $G(5) \simeq \mathbf{Z}_5$. Since $|G(2)| = 4 = 2^2$, either $G(2) \simeq \mathbf{Z}_4$ or $G(2) \simeq \mathbf{Z}_2 \oplus \mathbf{Z}_2$. Therefore, either $G \simeq \mathbf{Z}_5 \oplus \mathbf{Z}_4$ or $G \simeq \mathbf{Z}_5 \oplus \mathbf{Z}_2 \oplus \mathbf{Z}_2$. Thus, there are two Abelian groups of order 20 (up to isomorphism).

◇ **Exercise 3** Find all Abelian groups of order 63, which contain an element of order 21.

Solution: Let G be an Abelian group of order $63 = 3^2 \cdot 7$. Then $G = G(3) \oplus G(7)$, where $G(3)$ is a 3-group of order 3^2 and $G(7)$ is a 7-group of order 7. Now $2 = 1 + 1$ shows that either $G(3) \simeq \mathbf{Z}_{3^2}$ or $G(3) \simeq \mathbf{Z}_3 \oplus \mathbf{Z}_3$. Hence, $\mathbf{Z}_9 \oplus \mathbf{Z}_7$ and $\mathbf{Z}_3 \oplus \mathbf{Z}_3 \oplus \mathbf{Z}_7$ are the only two nonisomorphic Abelian groups of order 63. Now in $\mathbf{Z}_9 \oplus \mathbf{Z}_7$, $([3], [1])$ is an element of order 21 and in $\mathbf{Z}_3 \oplus \mathbf{Z}_3 \oplus \mathbf{Z}_7$, $([0], [1], [1])$ is an element of order 21.

◇ **Exercise 4** Find all Abelian groups of order 360.

Solution: Let G be an Abelian group of order $360 = 2^3 \cdot 3^2 \cdot 5$. Now G has a unique Sylow 2-subgroup, say, $G(2)$, a unique Sylow 3-subgroup, say, $G(3)$, and a unique Sylow 5-subgroup, say, $G(5)$. Thus, $G = G(2) \oplus G(3) \oplus G(5)$ and $|G(2)| = 2^3$, $|G(3)| = 3^2$, and $|G(5)| = 5$. Now $3 = 1 + 1 + 1 = 2 + 1$ and so there are three partitions of 3. This implies that there are three nonisomorphic Abelian groups of order 2^3. Hence,

$$G(2) \simeq \mathbf{Z}_8 \text{ or } G(2) \simeq \mathbf{Z}_4 \oplus \mathbf{Z}_2 \text{ or } G(2) \simeq \mathbf{Z}_2 \oplus \mathbf{Z}_2 \oplus \mathbf{Z}_2.$$

Similarly, since $2 = 1 + 1$, there are two partitions of 2. Therefore,

$$\text{either } G(3) \simeq \mathbf{Z}_9 \text{ or } G(3) \simeq \mathbf{Z}_3 \oplus \mathbf{Z}_3.$$

Since $|G(5)| = 5$,

$$G(5) \simeq \mathbf{Z}_5.$$

Hence, G is isomorphic to one of the following groups

$$\mathbf{Z}_8 \oplus \mathbf{Z}_9 \oplus \mathbf{Z}_5$$
$$\mathbf{Z}_4 \oplus \mathbf{Z}_2 \oplus \mathbf{Z}_9 \oplus \mathbf{Z}_5$$
$$\mathbf{Z}_2 \oplus \mathbf{Z}_2 \oplus \mathbf{Z}_2 \oplus \mathbf{Z}_9 \oplus \mathbf{Z}_5$$
$$\mathbf{Z}_8 \oplus \mathbf{Z}_3 \oplus \mathbf{Z}_3 \oplus \mathbf{Z}_5$$
$$\mathbf{Z}_4 \oplus \mathbf{Z}_2 \oplus \mathbf{Z}_3 \oplus \mathbf{Z}_3 \oplus \mathbf{Z}_5$$
$$\mathbf{Z}_2 \oplus \mathbf{Z}_2 \oplus \mathbf{Z}_2 \oplus \mathbf{Z}_3 \oplus \mathbf{Z}_3 \oplus \mathbf{Z}_5.$$

None of these groups is isomorphic to each other. Consequently, there are six Abelian groups of order 360 (up to isomorphism).

◇ **Exercise 5** Find the elementary divisors of the group $\mathbf{Z}_{20} \oplus \mathbf{Z}_8 \oplus \mathbf{Z}_{50}$.

Solution: Let $G = \mathbf{Z}_{20} \oplus \mathbf{Z}_8 \oplus \mathbf{Z}_{50}$. Then

$$\begin{aligned} G &= \mathbf{Z}_{20} \oplus \mathbf{Z}_8 \oplus \mathbf{Z}_{50} \\ &\simeq (\mathbf{Z}_5 \oplus \mathbf{Z}_4) \oplus \mathbf{Z}_8 \oplus (\mathbf{Z}_{25} \oplus \mathbf{Z}_2) \\ &\simeq \mathbf{Z}_5 \oplus \mathbf{Z}_{2^2} \oplus \mathbf{Z}_{2^3} \oplus \mathbf{Z}_{5^2} \oplus \mathbf{Z}_2 \\ &\simeq \mathbf{Z}_2 \oplus \mathbf{Z}_{2^2} \oplus \mathbf{Z}_{2^3} \oplus \mathbf{Z}_5 \oplus \mathbf{Z}_{5^2}. \end{aligned}$$

Hence, the elementary divisors are $2, 2^2, 2^3, 5, 5^2$.

◇ **Exercise 6** Let G and H be finite Abelian groups.

(i) Let $f : G \to H$ be a homomorphism. Show that $f(G(p)) \subseteq H(p)$ for all primes p.

(ii) Prove that $G \simeq H$ if and only if $G(p) \simeq H(p)$ for all primes p.

Solution: (i) Let $a \in G(p)$. Then $p^k a = 0$ for some $k \geq 0$. Thus, $0 = f(p^k a) = p^k f(a)$. Hence, $f(a) \in H(p)$. Thus, $f(G(p)) \subseteq H(p)$.

(ii) Suppose $G \simeq H$ and let $f : G \rightarrow H$ be the isomorphism of G onto H. Let p be a prime and $\alpha = f|_{G(p)}$, i.e., α is the restriction of f to $G(p)$. By (i), $\alpha : G(p) \rightarrow H(p)$. Clearly α is a monomorphism. Let $h \in H(p)$. There exists $a \in G$ such that $f(a) = h$. Also, $p^k h = 0$ for some $k \geq 0$. This implies that $f(p^k a) = p^k f(a) = p^k h = 0$, which in turn implies that $p^k a = 0$ since f is one-one. Hence, $a \in G(p)$ and so $h = f(a) = \alpha(a)$. Thus, α is an isomorphism of $G(p)$ onto $H(p)$, proving that $G(p) \simeq H(p)$. Conversely, suppose that $G(p) \simeq H(p)$ for all primes p. Let $G = G(p_1) \oplus G(p_2) \oplus \cdots \oplus G(p_k)$ and $H = H(p_1) \oplus H(p_2) \oplus \cdots \oplus H(p_k)$. Then $G(p_i) \simeq H(p_i)$ for all i. Let $f_i : G(p_i) \rightarrow H(p_i)$ be an isomorphism of $G(p_i)$ onto $H(p_i)$. Define $f : G \rightarrow H$ by $f(g_1 + g_2 + \cdots + g_k) = f_1(g_1) + f_2(g_2) + \cdots + f_k(g_k)$. Then f is an isomorphism of G onto H. Hence, $G \simeq H$.

9.1.2 Exercises

1. Let G be an Abelian group of order pq, where p and q are distinct primes. Show that $G \simeq \mathbf{Z}_p \oplus \mathbf{Z}_q$.

2. Find all Abelian groups of orders 9, 16, 27, and 32.

3. Find all Abelian groups of orders 15 and 21.

4. Find all Abelian groups of orders 60, 80, 240, and 540.

5. Prove that if G is an Abelian group of order $3 \cdot 7 \cdot 11$, then G is cyclic.

6. Find the elementary divisors of the following groups.

   (i) $\mathbf{Z}_{12} \oplus \mathbf{Z}_{144} \oplus \mathbf{Z}_8$.

   (ii) $\mathbf{Z}_{10} \oplus \mathbf{Z}_{30} \oplus \mathbf{Z}_{120}$.

7. Let A, B, and C be finite Abelian groups such that $A \oplus B \simeq A \oplus C$. Prove that $B \simeq C$.

8. Let G be an Abelian group such that $G = G_1 \oplus G_2$, where G_1 and G_2 are subgroups of G. Suppose that $G = H_1 \oplus H_2$, where H_i is a subgroup of G_i, $i = 1, 2$. Prove that $H_i = G_i$, $i = 1, 2$.

9. Determine all Abelian groups of order p^4, where p is a prime.

10. Find all Abelian groups of order $p^3 q^2$, where p and q are distinct primes.

11. Find all Abelian groups of order 72 which contain exactly three subgroups of order 2.

12. Prove that an Abelian group of order 8 is cyclic if and only if it has only one subgroup of order 2.

13. Prove that a finite Abelian group is cyclic if and only if all of its Sylow subgroups are cyclic.

14. Prove that a finite Abelian group of order n is cyclic if n is not divisible by p^2 for any prime p.

15. Find the number of elements of order 3 in a finite Abelian group of order 120.

16. Show that every Abelian group of order 28 has an element of order 14.

17. Find all Abelian groups of order 81 that have an element of order 27.

18. Which of the following statements are true? Justify your answer.

    (i) There is only one (up to isomorphism) Abelian group of order 35.

    (ii) The groups $\mathbf{Z}_5 \oplus \mathbf{Z}_3 \oplus \mathbf{Z}_5 \oplus \mathbf{Z}_3$ and $\mathbf{Z}_5 \oplus \mathbf{Z}_5 \oplus \mathbf{Z}_9$ are isomorphic.

    (iii) The number of nonisomorphic Abelian groups of order 3^4 is the same as the number of nonisomorphic Abelian groups of order 7^4.

9.2   Finitely Generated Abelian Groups

A finite direct sum of cyclic groups need not be a cyclic group. For example, $\mathbf{Z}_2 \oplus \mathbf{Z}_6 \oplus \mathbf{Z}$ is not a cyclic group. This group has elements of finite as well as of infinite orders. However, it is an Abelian group. Now

$$([1], [0], 0), ([0], [1], 0), ([0], [0], 1) \in \mathbf{Z}_2 \oplus \mathbf{Z}_6 \oplus \mathbf{Z}$$

and any element of this group can be expressed as

$$n_1([1], [0], 0) + n_2([0], [1], 0) + n_3([0], [0], 1)$$

for some integers n_1, n_2, n_3. A group of this kind is called a finitely generated Abelian group and is the subject of this section. Since a finite Abelian group has only finitely many elements, a finite Abelian group is obviously a finitely generated Abelian group. The main objective of this section is to give a complete description (up to isomorphism) of all possible types of finitely generated Abelian groups.

Definition 9.2.1 *A group G is called **finitely generated** if there exists a finite nonempty set $X \subseteq G$ such that $G = \langle X \rangle$. In this case, we call X a **generating set** for G.*

Let G be a finitely generated Abelian group generated by X, where $X = \{a_1, a_2, \ldots, a_k\}$. Then $G = \{n_1 a_1 + n_2 a_2 + \cdots + n_k a_k \mid n_i \in \mathbf{Z}, 1 \leq i \leq k\}$.

Definition 9.2.2 *Let G be an Abelian group. Let $X = \{a_1, a_2, \ldots, a_k\}$ be a finite nonempty subset of G. X is called a **basis for G** if $G = \langle X \rangle$ and for all $n_i \in \mathbf{Z}, 1 \leq i \leq k, n_1 a_1 + n_2 a_2 + \cdots + n_k a_k = 0$ implies that $n_i = 0, 1 \leq i \leq k$ (i.e., X is **linearly independent**).*

An Abelian group G is called a **finitely generated free Abelian group** if G has a finite basis.

Theorem 9.2.3 *Let G be an Abelian group. Then the following conditions are equivalent.*

(i) G has a finite basis.

(ii) G is the finite (internal) direct sum of a family of infinite cyclic subgroups.

(iii) G is isomorphic to a finite direct sum of finite copies of $\mathbf{Z}$.

Proof. (i)$\Rightarrow$(ii): Let $X = \{a_1, a_2, \ldots, a_k\}$ be a basis of G. Let $na_i = 0$ for some $n \in \mathbf{Z}$. Then $0a_1 + \cdots + na_i + \cdots + 0a_k = 0$. Hence, $n = 0$. This implies that a_i is of infinite order and $\langle a_i \rangle$ is an infinite cyclic group, $1 \leq i \leq k$. It is easy to verify that $G = \langle a_1 \rangle \oplus \cdots \oplus \langle a_k \rangle$.

(ii)$\Rightarrow$(iii): Let $G = G_1 \oplus \cdots \oplus G_k$, where G_i is an infinite cyclic subgroup of G, $1 \leq i \leq k$. Then $G_i \simeq \mathbf{Z}, 1 \leq i \leq k$. Hence, $G \simeq \mathbf{Z} \oplus \cdots \oplus \mathbf{Z}$.

(iii)$\Rightarrow$(i): Suppose $G \simeq \mathbf{Z} \oplus \cdots \oplus \mathbf{Z}$ is a finite direct sum of k copies of $\mathbf{Z}$. Let $\mathbf{Z}^{(k)}$ denote $\mathbf{Z} \oplus \cdots \oplus \mathbf{Z}$ and $f : G \longrightarrow \mathbf{Z}^{(k)}$ be an isomorphism. Let $u_i = (0, \ldots, 0, 1, 0, \ldots, 0) \in \mathbf{Z}^{(k)}$, with the ith component 1, $1 \leq i \leq k$. Then since f is onto $\mathbf{Z}^{(k)}$, there exists $a_i \in G$ such that $f(a_i) = u_i, 1 \leq i \leq k$. Now it is easy to verify that $X = \{a_1, a_2, \ldots, a_k\}$ is a basis of G. ∎

From the above theorem, it follows that in a finitely generated free Abelian group every nonzero element is of infinite order and that a finite Abelian group, though finitely generated, cannot be a finitely generated free Abelian group. Also, from the above theorem, we can draw an interesting conclusion that for every positive integer n, there exists a finitely generated free Abelian group with a basis consisting of n elements.

Consider the finitely generated free Abelian group $\mathbf{Z} \oplus \mathbf{Z}$. Now $\{(1,0), (0,1)\}$ and $\{(-1,0), (0,-1)\}$ are two different bases of $\mathbf{Z} \oplus \mathbf{Z}$. Thus, a finitely generated free Abelian group may have more than one basis. However, the number of elements in each basis is the same as proved in the next theorem.

Theorem 9.2.4 *Let F be a finitely generated free Abelian group. Then any two bases of F have the same number of elements.*

Proof.   Let $X = \{a_1, a_2, \ldots, a_k\}$ and $Y = \{b_1, b_2, \ldots, b_r\}$ be two bases of F. Then $F \simeq \mathbf{Z} \oplus \cdots \oplus \mathbf{Z}$ is a finite direct sum of k copies of $\mathbf{Z}$. Now $2F$ is a subgroup of F and $2F \simeq 2\mathbf{Z} \oplus \cdots \oplus 2\mathbf{Z}$. Hence,

$$\frac{F}{2F} \simeq \underbrace{\frac{\mathbf{Z}}{2\mathbf{Z}} \oplus \frac{\mathbf{Z}}{2\mathbf{Z}} \oplus \cdots \oplus \frac{\mathbf{Z}}{2\mathbf{Z}}}_{k \text{ summands}}.$$

This implies that $|F/2F| = 2^k$. Similarly, since Y is a basis of F, $|F/2F| = 2^r$. Thus, $2^k = 2^r$ and so $k = r$. $\blacksquare$

Let F be a finitely generated free Abelian group. The number of elements in a basis of F, which is unique by Theorem 9.2.4, is called the **rank** of F.

Theorem 9.2.5 *Every finitely generated Abelian group is a homomorphic image of a finitely generated free Abelian group.*

Proof.   Let G be a finitely generated Abelian group generated by $X = \{a_1, a_2, \ldots, a_k\}$. Let F be a finitely generated free Abelian group of rank k and let $\{x_1, x_2, \ldots, x_k\}$ be a basis for F. Define

$$f : F \to G$$

by

$$f(n_1 x_1 + n_2 x_2 + \cdots + n_k x_k) = n_1 a_1 + n_2 a_2 + \cdots + n_k a_k$$

for all $n_i \in \mathbf{Z}$, $1 \leq i \leq k$. Let $n_i, m_i \in \mathbf{Z}$, $1 \leq i \leq k$ be such that $n_1 x_1 + n_2 x_2 + \cdots + n_k x_k = m_1 x_1 + m_2 x_2 + \cdots + m_k x_k$. Then $(n_1 - m_1)x_1 + (n_2 - m_2)x_2 + \cdots + (n_k - m_k)x_k = 0$. Hence, $n_i - m_i = 0$ for all $1 \leq i \leq k$ and so $n_i = m_i$ for all $1 \leq i \leq k$. Thus, f is well defined. Also, f is an epimorphism and hence G is a homomorphic image of F. $\blacksquare$

Lemma 9.2.6 *Let $F = \langle x \rangle$, $x \in F$, be a free Abelian group. Then for all $m \in \mathbf{Z}$, $m \geq 0$,*

$$F/\langle mx \rangle \simeq \mathbf{Z}_m.$$

Proof.   Define

$$f : F \longrightarrow \mathbf{Z}_m$$

by

$$f(nx) = [n]$$

for all $n \in \mathbf{Z}$. Let $n_1 x = n_2 x$. Then $(n_1 - n_2)x = 0$ and so $n_1 = n_2$. Hence, $[n_1] = [n_2]$. Therefore, f is well defined. It is easy to verify that f is an epimorphism. Now $nx \in \text{Ker } f$ if and only if $f(nx) = [0]$ if and only if $[n] = [0]$ if and only if $m|n$ if and only if $n = ms$ for some $s \in \mathbf{Z}$ if and only if $nx = msx$ for some $s \in \mathbf{Z}$ if and only if $nx \in \langle mx \rangle$. This implies that Ker $f = \langle mx \rangle$. Thus, $F/\langle mx \rangle \simeq \mathbf{Z}_m$. ∎

The proof of the following lemma is straightforward and we leave it as an exercise.

Lemma 9.2.7 *Let F be a free Abelian group of rank k. Let $\{x_1, x_2, \ldots, x_k\}$ be a basis of F and $n \in \mathbf{Z}$. Then for all $i \neq j$, $1 \leq i, j \leq k$, $\{x_1, x_2, \ldots, x_{j-1}, x_j + nx_i, x_{j+1}, \ldots, x_k\}$ is also a basis of F.* ∎

Consider the group $\mathbf{Z}$. Now $\mathbf{Z}$ is a free Abelian group of rank 1 and $\{1\}$ is a basis of $\mathbf{Z}$. Every nonzero subgroup of $\mathbf{Z}$ is finitely generated and is generated by n for some positive integer n. Hence, every nonzero subgroup of $\mathbf{Z}$ is also free. We extend this result to any finitely generated free Abelian group in the next theorem.

Theorem 9.2.8 *Let F be a free Abelian group of rank k and H be a nonzero subgroup of F. Then there exists a basis $\{x_1, x_2, \ldots, x_k\}$ of F, an integer r $(1 \leq r \leq k)$, and positive integers $m_1, m_2, \ldots, m_r$ such that $m_{i-1}|m_i$, $2 \leq i \leq r$ such that $\{m_1 x_1, m_2 x_2, \ldots, m_r x_r\}$ is a basis of H.*

Proof. The proof is by induction on k. If $k = 1$, then $F = \langle x_1 \rangle$ and since a subgroup of a cyclic group is cyclic, H is cyclic. Clearly $H = \langle m_1 x_1 \rangle$ for some $m_1 > 0$. Suppose now that the theorem is true for all free Abelian groups of rank $< k$. Let

$$S = \{m \in \mathbf{Z} \mid m > 0 \text{ and there exists a basis } \{y_1, \ldots, y_k\} \text{ of } F$$
$$\text{such that } my_1 + n_2 y_2 + \cdots + n_k y_k \in H, \text{ for some } n_2, \ldots, n_k \in \mathbf{Z}\}.$$

Since $H \neq \{0\}$, $S \neq \phi$. Thus, S contains a smallest positive integer, say, m_1. This implies that there exists a basis $\{y_1, y_2, \ldots, y_k\}$ of F such that $m_1 y_1 + n_2 y_2 + \cdots + n_k y_k \in H$ for some $n_2, \ldots, n_k \in \mathbf{Z}$. Also, for any basis $\{z_1, z_2, \ldots, z_k\}$ of F, if $s_1 z_1 + s_2 z_2 + \cdots + s_k z_k \in H$ for some $s_1, \ldots, s_k \in \mathbf{Z}$, $s_1 > 0$, then $m_1 \leq s_1$. Let $h = m_1 y_1 + n_2 y_2 + \cdots + n_k y_k \in H$. Now by the division algorithm, there exist $q_i, r_i \in \mathbf{Z}$ such that

$$n_i = q_i m_1 + r_i, \ 0 \leq r_i < m_1, \ i = 2, 3, \ldots, k.$$

From this, it follows that $h = m_1(y_1 + q_2 y_2 + \cdots + q_k y_k) + r_2 y_2 + \cdots + r_k y_k$. Since $\{y_1 + q_2 y_2 + \cdots + q_k y_k, y_2, \ldots, y_k\}$ is a basis of F, we find that $r_i = 0$, $2 \leq i \leq k$, by the choice of m_1. Hence, $m_1 x_1 = h \in H$, where $x_1 = y_1 + q_2 y_2 + \cdots + q_k y_k$. Let $K = \langle y_2, \ldots, y_k \rangle$. Then K is a free Abelian group of rank $k - 1$ and $F = \langle x_1 \rangle \oplus K$. We now claim that $H = \langle m_1 x_1 \rangle \oplus (H \cap K)$. Let $a \in H$. Then

$a = t_1 x_1 + t_2 y_2 + \cdots + t_k y_k$ for some $t_i \in \mathbf{Z}$, $1 \leq i \leq k$. By the division algorithm, there exist $q_1, r_1 \in \mathbf{Z}$ such that $t_1 = q_1 m_1 + r_1$, $0 \leq r_1 < m_1$. This implies that $r_1 x_1 + t_2 y_2 + \cdots + t_k y_k = a - q_1 m_1 x_1 \in H$ and so $r_1 = 0$ by the minimality of m_1. Thus, $t_2 y_2 + \cdots + t_k y_k \in H$. Therefore,

$$a = q_1 (m_1 x_1) + t_2 y_2 + \cdots + t_k y_k \in \langle m_1 x_1 \rangle + (H \cap K).$$

It now follows from $F = \langle x_1 \rangle \oplus K$ that

$$H = \langle m_1 x_1 \rangle \oplus (H \cap K).$$

If $H \cap K = \{0\}$, then $H = \langle m_1 x_1 \rangle$ and the theorem is true. Suppose that $H \cap K \neq \{0\}$. Then $H \cap K$ is a nonzero subgroup of the finitely generated free Abelian group K. Hence, by the induction hypothesis, there exists a basis $\{x_2, \ldots, x_k\}$ of K and positive integers $r, m_2, \ldots, m_k$ such that $\{m_2 x_2, \ldots, m_r x_r\}$ is a basis of $H \cap K$ and $m_{i-1} | m_i$, $3 \leq i \leq r$. Clearly $\{x_1, \ldots, x_k\}$ is a basis of F and $\{m_1 x_1, m_2 x_2, \ldots, m_r x_r\}$ is a basis of H. It only remains to be shown that $m_1 | m_2$. By the division algorithm, there exist $q, r \in \mathbf{Z}$ such that $m_2 = q m_1 + r$, $0 \leq r < m_1$. Now $\{x_2, x_1 + q x_2, x_3, \ldots, x_k\}$ is a basis of F and $r x_2 + m_1 (x_1 + q x_2) = m_1 x_1 + m_2 x_2 + 0 m_3 x_3 + \cdots + 0 m_k x_k \in H$. Thus, by the minimality of m_1, $r = 0$, proving that $m_1 | m_2$. ∎

The next theorem is called **the fundamental theorem of finitely generated Abelian groups**.

Theorem 9.2.9 *Let G be a finitely generated nonzero Abelian group. Then G is isomorphic to a finite direct sum of cyclic groups, where the finite summands (if any) are of orders $m_1, m_2, \ldots, m_r$, $m_1 > 1$, and m_i divides m_{i+1}, $1 \leq i < r - 1$.*

Proof. Let G be generated by k elements. By Theorem 9.2.5, G is a homomorphic image of a free Abelian group F of rank k. Let $f : F \rightarrow G$ be a homomorphism of F onto G. Then $F/\text{Ker } f \simeq G$. If $\text{Ker } f = \{0\}$, then

$$G \simeq F \simeq \underbrace{\mathbf{Z} \oplus \cdots \oplus \mathbf{Z}}_{k \text{ copies}}.$$

Suppose now that $\text{Ker } f \neq \{0\}$. By Theorem 9.2.8, there exists a basis $\{x_1, x_2, \ldots, x_k\}$ of F, an integer r $(1 \leq r \leq k)$, and positive integers $m_1, m_2, \ldots, m_r$ such that $m_{i-1} | m_i$, $2 \leq i \leq r$, and $\{m_1 x_1, \ldots, m_r x_r\}$ is a basis of $\text{Ker } f$. Now $F = \langle x_1 \rangle \oplus \cdots \oplus \langle x_k \rangle$ and $\text{Ker } f = \langle m_1 x_1 \rangle \oplus \cdots \oplus \langle m_r x_r \rangle$. Hence,

$$
\begin{aligned}
G &\simeq F/\text{Ker } f &\simeq& \frac{\langle x_1 \rangle}{\langle m_1 x_1 \rangle} \oplus \cdots \oplus \frac{\langle x_r \rangle}{\langle m_r x_r \rangle} \oplus \langle x_{r+1} \rangle \oplus \cdots \oplus \langle x_k \rangle \\
& &\simeq& \mathbf{Z}_{m_1} \oplus \cdots \oplus \mathbf{Z}_{m_r} \oplus \mathbf{Z} \oplus \cdots \oplus \mathbf{Z}. \ \blacksquare
\end{aligned}
$$

Recall that a group G is torsion free if and only if every nonidentity element of G is of infinite order.

Theorem 9.2.10 *A finitely generated Abelian group $G \neq \{0\}$ is torsion free if and only if G is a finitely generated free Abelian group.*

Proof. Suppose that G is a finitely generated free Abelian group. Then there exists a positive integer r such that

$$G \simeq \underbrace{\mathbf{Z} \oplus \mathbf{Z} \oplus \cdots \oplus \mathbf{Z}}_{r \text{ copies}}.$$

Now every nonidentity element of $\mathbf{Z} \oplus \mathbf{Z} \oplus \cdots \oplus \mathbf{Z}$ is of infinite order. Hence, G is torsion free. Conversely, suppose that G is a finitely generated torsion free Abelian group. Then by Theorem 9.2.9,

$$G \simeq \mathbf{Z}_{m_1} \oplus \cdots \oplus \mathbf{Z}_{m_r} \oplus \underbrace{\mathbf{Z} \oplus \cdots \oplus \mathbf{Z}}_{s \text{ copies}}$$

for some positive integers $m_1, m_2, \ldots, m_r$ and a nonnegative integer s. If $r \neq 0$, then $\mathbf{Z}_{m_1} \oplus \cdots \oplus \mathbf{Z}_{m_r} \oplus \underbrace{\mathbf{Z} \oplus \cdots \oplus \mathbf{Z}}_{s \text{ copies}}$ and so G contains a nonzero element of finite order, which contradicts the hypothesis. Hence, $r = 0$. Thus, $G \simeq \underbrace{\mathbf{Z} \oplus \cdots \oplus \mathbf{Z}}_{s \text{ copies}}$, proving that G is a finitely generated free Abelian group. ∎

Theorem 9.2.11 *Let G be an Abelian group. Let*

$$T(G) = \{a \in G \mid \circ(a) \text{ is finite}\}.$$

Then $T(G)$ is a subgroup of G. Suppose G is finitely generated. If $G/T(G) \neq \{0\}$, then $G/T(G)$ is a finitely generated free Abelian group.

Proof. Clearly $T(G)$ is a subgroup of G. It is also a simple exercise to show that $G/T(G)$ is finitely generated. Suppose $G/T(G) \neq \{0\}$. Let $a + T(G) \in G/T(G)$. Now $n(a + T(G)) = 0 + T(G)$ if and only if $na \in T(G)$ if and only if $m(na) = 0$ for some positive integer m if and only if $a \in T(G)$ if and only if $a + T(G) = 0 + T(G)$. Hence, $G/T(G)$ is torsion free. By Theorem 9.2.10, $G/T(G)$ is a finitely generated free Abelian group. ∎

Definition 9.2.12 *Let G be an Abelian group. The subgroup $T(G)$ in Theorem 9.2.11 is called the **torsion subgroup** of G.*

Theorem 9.2.13 *Let G be a finitely generated nonzero Abelian group. Let*

$$G \simeq \mathbf{Z}_{m_1} \oplus \cdots \oplus \mathbf{Z}_{m_k} \oplus \underbrace{\mathbf{Z} \oplus \cdots \oplus \mathbf{Z}}_{r \text{ copies}}$$

and

$$G \simeq \mathbf{Z}_{n_1} \oplus \cdots \oplus \mathbf{Z}_{n_q} \oplus \underbrace{\mathbf{Z} \oplus \cdots \oplus \mathbf{Z}}_{s \ copies},$$

where $m_1 > 1$, m_i divides m_{i+1}, $1 \leq i < k - 1$, $n_1 > 1$, and n_i divides n_{i+1}, $1 \leq i < q - 1$. Then $k = q$, $r = s$, and $m_i = n_i$, $1 \leq i \leq k$.

Proof. Let

$$G_1 = \mathbf{Z}_{m_1} \oplus \cdots \oplus \mathbf{Z}_{m_k} \oplus \underbrace{\mathbf{Z} \oplus \cdots \oplus \mathbf{Z}}_{r \ copies}$$

and

$$G_2 = \mathbf{Z}_{n_1} \oplus \cdots \oplus \mathbf{Z}_{n_q} \oplus \underbrace{\mathbf{Z} \oplus \cdots \oplus \mathbf{Z}}_{s \ copies}.$$

We first show that the torsion subgroup $T(G_1)$ is isomorphic to $\mathbf{Z}_{m_1} \oplus \cdots \oplus \mathbf{Z}_{m_k}$ and the torsion subgroup $T(G_2)$ of G_2 is isomorphic to $\mathbf{Z}_{n_1} \oplus \cdots \oplus \mathbf{Z}_{n_q}$. Let $a \in G_1$. There exists $x_i \in \mathbf{Z}_{m_i}$, $i = 1, 2, \ldots, k$ and $y_j \in \mathbf{Z}$, $j = 1, 2, \ldots, r$, such that a can be written uniquely as

$$a = (x_1, x_2, \ldots, x_k, y_1, y_2, \ldots, y_r).$$

Let $m' \in \mathbf{N}$. Then $m'a = (m'x_1, m'x_2, \ldots, m'x_k, m'y_1, m'y_2, \ldots, m'y_r)$. Since $y_j \in \mathbf{Z}$, we find that $m'y_j = 0$ if and only if $y_j = 0$. Again for $m = m_1 m_2 \cdots m_k$, $mx_i = 0$, $i = 1, 2, \ldots, k$. Thus, it follows that $o(a)$ is finite if and only if $y_1 = y_2 = \cdots = y_r = 0$ and so $T(G_1)$ is the set of all elements $a = (x_1, x_2, \ldots, x_k, 0, 0, \ldots, 0) \in G_1$. Consequently, $T(G_1) \simeq \mathbf{Z}_{m_1} \oplus \cdots \oplus \mathbf{Z}_{m_k}$. Similarly, $T(G_2) \simeq \mathbf{Z}_{n_1} \oplus \cdots \oplus \mathbf{Z}_{n_q}$.

Next, let us show that $k = q$ and $m_i = n_i$, $i = 1, 2, \ldots, k$. Since for groups A and B, $A \oplus B \simeq B \oplus A$, we find that $\mathbf{Z}_{m_1} \oplus \cdots \oplus \mathbf{Z}_{m_k} \simeq \mathbf{Z}_{m_k} \oplus \cdots \oplus \mathbf{Z}_{m_1}$. For convenience, let us write $t_1 = m_k, \ldots, t_k = m_1$, and set $G_3 = \mathbf{Z}_{t_1} \oplus \cdots \oplus \mathbf{Z}_{t_k}$, where t_i are positive integers, $t_k > 1$ and $t_{i+1}|t_i$, $i = 1, 2, \ldots, k - 1$. Similarly, $\mathbf{Z}_{n_1} \oplus \cdots \oplus \mathbf{Z}_{n_q} \simeq \mathbf{Z}_{n_q} \oplus \cdots \oplus \mathbf{Z}_{n_1}$. For convenience, let us write $r_1 = n_q, \ldots, r_q = n_1$ and set $G_4 = \mathbf{Z}_{r_1} \oplus \cdots \oplus \mathbf{Z}_{r_q}$. For $x \in G_3$ there exist $x_i \in \mathbf{Z}_{t_i}$ such that $x = (x_1, \ldots, x_k)$. Now $|\mathbf{Z}_{t_i}| = t_i$ and $t_i|t_1$, $i = 1, 2, \ldots, k$. Also, note that $t_i x_i = 0$, $i = 1, 2, \ldots, k$. Hence, $t_1 a = 0$ for all $a \in G_3$. Again in $\mathbf{Z}_{t_1}$, there exists an element x_1 of order t_1, which implies that $a = (x_1, 0, \ldots, 0)$ is an element of order t_1 in G_3. Similarly, we can show that G_4 contains an element b such that $o(b) = r_1$ and $r_1 y = 0$ for all $y \in G_4$. Since $G_3 \simeq G_4$, there exists an isomorphism, say, $f : G_3 \to G_4$. Now $o(f(a)) = t_1$ and also $r_1 f(a) = 0$. Thus, $t_1 \leq r_1$. A similar argument shows that $r_1 \leq t_1$ and so $r_1 = t_1$. Suppose now that $r_2 = t_2, \ldots, r_{i-1} = t_{i-1}$, but $r_i \neq t_i$, where $1 \leq i \leq \min(k, q)$. Let $t_i < r_i$ and let $K = \{t_i x \mid x \in G_3\}$. It can be shown that K is a subgroup of G_3 and if $\mathbf{Z}_{t_i} = \langle a_i \rangle$, $i = 1, 2, \ldots, k$, then

$$K = \langle t_i a_1 \rangle \oplus \langle t_i a_2 \rangle \oplus \cdots \oplus \langle t_i a_k \rangle$$

and hence

$$
\begin{aligned}
|K| &= \frac{o(a_1)}{\gcd(o(a_1),t_i)} \frac{o(a_2)}{\gcd(o(a_2),t_i)} \cdots \frac{o(a_k)}{\gcd(o(a_k),t_i)} \\
&= \frac{t_1}{\gcd(t_1,t_i)} \frac{t_2}{\gcd(t_2,t_i)} \cdots \frac{t_k}{\gcd(t_k,t_i)} \\
&= \frac{t_1}{t_i} \frac{t_2}{t_i} \cdots \frac{t_{i-1}}{t_i} \frac{t_i}{t_i} \frac{t_{i+1}}{t_{i+1}} \cdots \frac{t_k}{t_k} \quad (\text{since } t_{i+1}|t_i, i = 1,2,\ldots,k-1) \\
&= \frac{t_1}{t_i} \frac{t_2}{t_i} \cdots \frac{t_{i-1}}{t_i}.
\end{aligned} \tag{9.1}
$$

Now $f(K) = \{t_i f(x) \mid x \in G_3\}$. If $\mathbf{Z}_{r_j} = \langle b_j \rangle$, $j = 1,2,\ldots,q$, then

$$
f(K) = \langle t_i b_1 \rangle \oplus \langle t_i b_2 \rangle \oplus \cdots \oplus \langle t_i b_q \rangle .
$$

Hence,

$$
\begin{aligned}
|f(K)| &= \frac{o(b_1)}{\gcd(o(b_1),t_i)} \frac{o(b_2)}{\gcd(o(b_2),t_i)} \cdots \frac{o(b_q)}{\gcd(o(b_q),t_i)} \\
&= \frac{r_1}{\gcd(r_1,t_i)} \frac{r_2}{\gcd(r_2,t_i)} \cdots \frac{r_q}{\gcd(r_q,t_i)} \\
&= \frac{t_1}{\gcd(t_1,t_i)} \cdots \frac{t_{i-1}}{\gcd(t_{i-1},t_i)} \frac{r_i}{\gcd(r_i,t_i)} \cdots \frac{r_q}{\gcd(r_q,t_i)} \\
&= \frac{t_1}{t_i} \frac{t_2}{t_i} \cdots \frac{t_{i-1}}{t_i} \frac{r_i}{\gcd(r_i,t_i)} \cdots \frac{r_q}{\gcd(r_q,t_i)}.
\end{aligned} \tag{9.2}
$$

Since $|K| = |f(K)|$, it follows from Eqs. (9.1) and (9.2) that

$$
\frac{r_i}{\gcd(r_i,t_i)} \cdots \frac{r_q}{\gcd(r_q,t_i)} = 1. \tag{9.3}
$$

Since $t_i < r_i$, $\gcd(r_i,t_i) < r_i$ and hence $\frac{r_i}{\gcd(r_i,t_i)} > 1$. Thus, we find that the left-hand side of Eq. (9.3) is greater than 1, whereas the right-hand side of Eq. (9.3) is 1. This is a contradiction. This contradiction implies that $t_i \nmid r_i$. Similarly, $r_i \nmid t_i$. Hence, $t_i = r_i$. But $G_3 \simeq G_4$ implies that $|G_3| = |G_4|$ and so $t_1 t_2 \cdots t_k = r_1 r_2 \cdots r_q$. Note that t_i and r_i are positive integers greater than 1. If $k < q$, then $t_i = r_i$, $i = 1,2,\ldots,k$ and hence $1 = r_{k+1} \cdots r_q$, which is not true. So $k \nless q$. Similarly, $q \nless k$. Consequently, $k = q$ and $t_i = r_i$, $i = 1,2,\ldots,k$.

Finally, let us show that $r = s$. From the assumption and from the above proof, it follows that

$$
G \simeq H \oplus F \simeq H \oplus F',
$$

where H is a finite direct sum of finite cyclic groups and F and F' are finitely generated free Abelian groups of rank r and s, respectively. The restriction of the isomorphism $G \simeq H \oplus F$ maps $T(G)$ onto H. Hence, $G/T(G) \simeq F$, which shows that $G/T(G)$ is a finitely generated free Abelian group of rank r. Similarly, $G/T(G) \simeq F'$ implies that $G/T(G)$ is a finitely generated free Abelian group of rank s. Thus, $r = s$. ∎

Corollary 9.2.14 *Let G be a nonzero finite Abelian group. Then there exists a unique list of positive integers (not necessarily distinct) $m_1, m_2, \ldots, m_k$ such that $m_1 > 1$, $m_i|m_{i+1}$, $i = 1,2,\ldots,k-1$, and $G \simeq \mathbf{Z}_{m_1} \oplus \cdots \oplus \mathbf{Z}_{m_k}$.*

Theorems 9.2.9 and 9.2.13 give a complete system of invariants for finitely generated Abelian groups. That is, the number r of Theorem 9.2.13 together with the integers $m_1, m_2, \ldots, m_k$ are invariants for finitely generated Abelian groups in the sense that any two finitely generated Abelian groups with these numbers must be isomorphic.

Let G be a finitely generated Abelian group. Then the unique number r of Theorem 9.2.13 is called the **betti number** of G and the integers $m_1, m_2, \ldots, m_k$, which are uniquely determined for the group G, are called the **torsion coefficients** of G.

9.2.1 Worked-Out Exercises

$\diamond$ **Exercise 1** Show that $(\mathbf{Q}, +)$ is not finitely generated.

Solution: Suppose $(\mathbf{Q}, +)$ is finitely generated. Then there exists a finite set
$$\left\{ \frac{a_1}{b_1}, \frac{a_2}{b_2}, \ldots, \frac{a_n}{b_n} \right\}$$
of rational numbers such that $\mathbf{Q} = \left\langle \frac{a_1}{b_1}, \frac{a_2}{b_2}, \ldots, \frac{a_n}{b_n} \right\rangle$. Now we can find a prime p such that p does not divide $b_1, b_2, \ldots, b_n$. Let $x \in \mathbf{Q}$. There exist integers $r_1, r_2, \ldots, r_n$ such that
$$x = r_1 \frac{a_1}{b_1} + r_2 \frac{a_2}{b_2} + \cdots + r_n \frac{a_n}{b_n} = \frac{c}{b_1 b_2 \cdots b_n}$$
for some integer c. Since p does not divide $b_1, b_2, \ldots, b_n$, we find that p does not divide $b_1 b_2 \cdots b_n$. Hence, p does not divide the denominator of any rational number (expressed in lowest terms) of $\left\langle \frac{a_1}{b_1}, \frac{a_2}{b_2}, \ldots, \frac{a_n}{b_n} \right\rangle$. This implies that $\frac{1}{p} \notin \left\langle \frac{a_1}{b_1}, \frac{a_2}{b_2}, \ldots, \frac{a_n}{b_n} \right\rangle$, a contradiction. Thus, $(\mathbf{Q}, +)$ is not finitely generated.

$\diamond$ **Exercise 2** Let G be a nonzero finitely generated Abelian group such that every nonzero element of G is of order p, where p is a prime. Show that $|G| = p^k$ for some positive integer p.

Solution: By Theorem 9.2.9, $G \simeq \mathbf{Z}_{m_1} \oplus \cdots \oplus \mathbf{Z}_{m_k} \oplus \underbrace{\mathbf{Z} \oplus \cdots \oplus \mathbf{Z}}_{r \text{ copies}}$. If $r \neq 0$, then G contains elements of infinite order. Hence, $r = 0$ and so $G \simeq \mathbf{Z}_{m_1} \oplus \cdots \oplus \mathbf{Z}_{m_k}$. Since each nonzero element of G is of order p, we find that $m_1 = \cdots = m_k = p$. Thus, $G \simeq \mathbf{Z}_p \oplus \cdots \oplus \mathbf{Z}_p$ and so $|G| = p^k$.

$\diamond$ **Exercise 3** Show that the torsion subgroup, $T(G)$, of $G = \mathbf{Z}_4 \oplus \mathbf{Z} \oplus \mathbf{Z}_5 \oplus \mathbf{Z}_3$ is a cyclic group. Find $|T(G)|$.

Solution: Recall that $\mathbf{Z}_{mn} \simeq \mathbf{Z}_m \oplus \mathbf{Z}_n$ if and only if $\gcd(m, n) = 1$. Now

$$
\begin{aligned}
G &= \mathbf{Z}_4 \oplus \mathbf{Z} \oplus \mathbf{Z}_5 \oplus \mathbf{Z}_3 \\
&\simeq \mathbf{Z}_4 \oplus \mathbf{Z}_5 \oplus \mathbf{Z}_3 \oplus \mathbf{Z} \\
&\simeq \mathbf{Z}_{20} \oplus \mathbf{Z}_3 \oplus \mathbf{Z} \\
&\simeq \mathbf{Z}_{60} \oplus \mathbf{Z}.
\end{aligned}
$$

Hence, $T(G) \simeq \mathbf{Z}_{60}$ and so $T(G)$ is a cyclic group. Also, $|T(G)| = |\mathbf{Z}_{60}| = 60$.

◇ **Exercise 4** Show that there are integers $d_1, \ldots, d_k$ such that $d_1 > 1, d_i | d_{i+1}$, $i = 1, 2, \ldots, k-1$, and $\mathbf{Z}_{2^2} \oplus \mathbf{Z}_{3^4} \oplus \mathbf{Z}_3 \oplus \mathbf{Z}_{5^2} \oplus \mathbf{Z}_2 \simeq \mathbf{Z}_{d_1} \oplus \mathbf{Z}_{d_2} \oplus \cdots \oplus \mathbf{Z}_{d_k}$.

Solution: Let $G = \mathbf{Z}_{2^2} \oplus \mathbf{Z}_{3^4} \oplus \mathbf{Z}_3 \oplus \mathbf{Z}_{5^2} \oplus \mathbf{Z}_2$. Then

$$
G \simeq \mathbf{Z}_2 \oplus \mathbf{Z}_{2^2} \oplus \mathbf{Z}_3 \oplus \mathbf{Z}_{3^4} \oplus \mathbf{Z}_{5^2}.
$$

Thus, the elementary divisors of G are $2, 2^2, 3, 3^4$, and 5^2. We form the following table:

$$
\begin{array}{ccc}
2 & 3 & \\
2^2 & 3^4 & 5^2
\end{array}
$$

From this table, we arrange the summands in the following way:

$$
\begin{aligned}
G &\simeq (\mathbf{Z}_2 \oplus \mathbf{Z}_3) \oplus (\mathbf{Z}_{2^2} \oplus \mathbf{Z}_{3^4} \oplus \mathbf{Z}_{5^2}) \\
&\simeq \mathbf{Z}_6 \oplus \mathbf{Z}_{8100}.
\end{aligned}
$$

Hence, $d_1 = 6$ and $d_2 = 8100$.

9.2.2   Exercises

1. Show that the group $\mathbf{Z} \oplus \mathbf{Z}_6$ is finitely generated, but has no basis.

2. Let G be a finitely generated nonzero Abelian group in which every non-identity element is of order 2. Show that $|G| = 2^k$ for some positive integer k.

3. Show that the torsion subgroup, $T(G)$, of $G = \mathbf{Z}_4 \oplus \mathbf{Z}_7 \oplus \mathbf{Z} \oplus \mathbf{Z} \oplus \mathbf{Z}_9$ is a cyclic group. Find $|T(G)|$.

4. Find the torsion coefficients and the betti number of the group $\mathbf{Z}_{20} \oplus \mathbf{Z} \oplus \mathbf{Z} \oplus \mathbf{Z}_{15} \oplus \mathbf{Z}_6$.

5. Find the elementary divisors of the group $G = \mathbf{Z}_{22} \oplus \mathbf{Z}_{15} \oplus \mathbf{Z}_{48}$ and find the positive integers $d_1, d_2, \ldots, d_k$ such that $d_1 > 1, d_i | d_{i+1}, i = 1, 2, \ldots, k-1$, and $G \simeq \mathbf{Z}_{d_1} \oplus \mathbf{Z}_{d_2} \oplus \cdots \oplus \mathbf{Z}_{d_k}$.

6. Find all Abelian groups of order 540. Express them as a direct sum of Abelian groups of the form $\mathbf{Z}_{d_1} \oplus \mathbf{Z}_{d_2} \oplus \cdots \oplus \mathbf{Z}_{d_k}$ such that $d_1, d_2, \ldots, d_k$ are positive integers and $d_1 > 1, d_i | d_{i+1}, i = 1, 2, \ldots, k-1$.

7. Are the following pairs of groups isomorphic?

    (i) $\mathbf{Z}_{20} \oplus \mathbf{Z}_{75} \oplus \mathbf{Z}_{90}$ and $\mathbf{Z}_{120} \oplus \mathbf{Z}_{25} \oplus \mathbf{Z}_{45}$.

    (ii) $\mathbf{Z}_{15} \oplus \mathbf{Z}_{12} \oplus \mathbf{Z}_{30} \oplus \mathbf{Z} \oplus \mathbf{Z}$ and $\mathbf{Z}_{108} \oplus \mathbf{Z}_{50} \oplus \mathbf{Z} \oplus \mathbf{Z}$.

8. Show that the group $\mathbf{Z}_{200} \oplus \mathbf{Z}_{30} \oplus \mathbf{Z}_{36}$ is isomorphic to $\mathbf{Z}_{120} \oplus \mathbf{Z}_{18} \oplus \mathbf{Z}_{100}$.

9. Let G be a finitely generated Abelian group generated by n elements. Let H be a subgroup of G. Prove that H is also finitely generated and H may be generated by m elements, where $m \leq n$.

10. Let H be a subgroup of an Abelian group G. If H is finitely generated and G/H is finitely generated, prove that G is finitely generated.

11. Prove that every homomorphic image of a finitely generated Abelian group is finitely generated.

12. Prove that two finitely generated free Abelian groups are isomorphic if and only if they have the same rank.

13. Prove or disprove:

    (i) In a finitely generated free Abelian group G of rank n, any linearly independent subset of n elements is a basis of G.

    (ii) In a finitely generated free Abelian group G of rank n, any linearly independent subset of m elements, $m \leq n$, can be extended to a basis of G.

    (iii) Every finite Abelian group is a finitely generated free Abelian group.

Chapter 10

Introduction to Rings

In the previous chapters, we investigated mathematical systems with one binary operation. There are many mathematical systems, called rings, with two binary operations. The notion of a ring is an outgrowth of such mathematical systems as the integers, rational numbers, real numbers, and complex numbers.

Although David Hilbert coined the term "ring," it was E. Noether who, under the influence of Hilbert, set down the axioms for rings. In 1914, Fraenkel gave the first definition of a ring. However, it is no longer commonly used.

As we shall see, a ring is a particular combination of a group and a semigroup. Hence, our previous work will prove helpful in our examination of rings. However, it is not enough to examine a set with two independent binary operations. In order to obtain the full power of the axiomatic approach, we need a dependency between the two operations—in particular, the distributive laws.

10.1   Elementary Properties

This section parallels Chapter 2. First we give a definition of a ring, followed by examples and elementary properties. We introduce several notations and definitions which will be used throughout the text.

The two binary operations that we consider on a nonempty set are usually denoted by $+$ (addition) and $\cdot$ (multiplication).

A ring is a mathematical system $(R, +, \cdot)$ such that $(R, +)$ is a commutative group, $(R, \cdot)$ is a semigroup, and the distributive laws hold, i.e., for all a, b, $c \in R$,

$$a \cdot (b + c) = (a \cdot b) + (a \cdot c),$$

$$(b + c) \cdot a = (b \cdot a) + (c \cdot a).$$

We denote the identity of $(R, +)$ by the symbol 0. The additive inverse of an element $a \in R$ is denoted by $-a$.

We now give a complete definition of a ring.

Definition 10.1.1 *A **ring** is an ordered triple $(R, +, \cdot)$ such that R is a nonempty set and $+$ and $\cdot$ are two binary operations on R satisfying the following axioms.*

   (R1) $(a + b) + c = a + (b + c)$ for all $a, b, c \in R$.
   (R2) $a + b = b + a$ for all $a, b \in R$.
   (R3) There exists an element 0 in R such that $a + 0 = a$ for all $a \in R$.
   (R4) For all $a \in R$, there exists an element $-a \in R$ such that

$$a + (-a) = 0.$$

   (R5) $(a \cdot b) \cdot c = a \cdot (b \cdot c)$ for all $a, b, c \in R$.
   (R6) $a \cdot (b + c) = (a \cdot b) + (a \cdot c)$ for all $a, b, c \in R$.
   (R7) $(b + c) \cdot a = (b \cdot a) + (c \cdot a)$ for all $a, b, c \in R$.

We call 0, the **zero element** of the ring $(R, +, \cdot)$.

During the development of the theory of rings, we will use the following conventions.

1. Multiplication is assumed to be performed before addition.

2. We write ab for $a \cdot b$.

3. We write $a - b$ for $a + (-b)$.

4. We refer to a ring $(R, +, \cdot)$ as a ring R.

Accordingly, $ab + c$ stands for $(a \cdot b) + c$, $ab + ac$ stands for $(a \cdot b) + (a \cdot c)$, $ab - ac$ stands for $(a \cdot b) + (-(a \cdot c))$, where $a, b, c \in R$.

Example 10.1.2 *Consider $\mathbf{Z}$, the set of integers, together with the usual addition, $+$, and multiplication, $\cdot$. By Example 2.1.3, $(\mathbf{Z}, +)$ is a group. Now multiplication of two integers is an integer and associativity holds for $\cdot$. Finally, we know that the distributive laws hold for the integers. Thus, $(\mathbf{Z}, +, \cdot)$ is a ring.*

The ring of Example 10.1.2 is called the **ring of integers**. This ring plays an important role in the study of ring theory. One of the basic problems in ring theory is to determine rings, which satisfy the same type of properties as the ring of integers.

Definition 10.1.3 *A ring R is called **commutative** if $ab = ba$ for all $a, b \in R$. A ring R which is not commutative is called a **noncommutative** ring.*

From the above definition, it follows that a ring R is commutative if and only if the semigroup $(R, \cdot)$ is commutative. The ring of integers is a commutative ring.

For a ring R, the set $C(R) = \{a \in R \mid ab = ba \text{ for all } b \in R\}$ is called the **center** of R. It follows that R is commutative if and only if $R = C(R)$.

Example 10.1.4 *Let $M_2(\mathbf{Z})$ denote the set of all 2×2 matrices over the ring of integers. Let $+$ and $\cdot$ denote the usual matrix addition and multiplication, respectively. Since addition (multiplication) of 2×2 matrices over $\mathbf{Z}$ is a 2×2 matrix over $\mathbf{Z}$, it follows that $+$ and $\cdot$ are binary operations on $M_2(\mathbf{Z})$. It is now easy to show that $(M_2(\mathbf{Z}), +, \cdot)$ is a ring. Now $\begin{bmatrix} 1 & 2 \\ 3 & 4 \end{bmatrix}, \begin{bmatrix} 5 & 1 \\ 7 & 8 \end{bmatrix} \in M_2(\mathbf{Z})$ and*

$$\begin{bmatrix} 1 & 2 \\ 3 & 4 \end{bmatrix}\begin{bmatrix} 5 & 6 \\ 7 & 8 \end{bmatrix} = \begin{bmatrix} 19 & 22 \\ 43 & 50 \end{bmatrix} \neq \begin{bmatrix} 23 & 34 \\ 31 & 46 \end{bmatrix} = \begin{bmatrix} 5 & 6 \\ 7 & 8 \end{bmatrix}\begin{bmatrix} 1 & 2 \\ 3 & 4 \end{bmatrix}.$$

Therefore, $M_2(\mathbf{Z})$ is not a commutative ring.

In a ring R, an element $e \in R$ is called an **identity** element if $ea = a = ae$ for all $a \in R$. An identity element of a ring R (if it exists) is an identity element of the semigroup $(R, \cdot)$. Therefore, a ring cannot contain more than one identity element (Theorem 1.6.11). The identity element of a ring (if it exists) is denoted by 1.

Definition 10.1.5 *A ring R is called a **ring with identity** if it has an identity.*

Example 10.1.6 *The ring $\mathbf{Z}$ of integers is a ring with identity. The integer 1 is the identity element of $\mathbf{Z}$.*

Example 10.1.7 *The ring $M_2(\mathbf{Z})$ of Example 10.1.4 is a ring with identity. The identity element of $M_2(\mathbf{Z})$ is $\begin{bmatrix} 1 & 0 \\ 0 & 1 \end{bmatrix}$.*

Example 10.1.8 *Let R denote the set of all functions $f : \mathbf{R} \to \mathbf{R}$. Define $+$, $\cdot$ on R by for all $f, g \in R$ and for all $a \in \mathbf{R}$,*

$$(f + g)(a) = f(a) + g(a),$$

$$(f \cdot g)(a) = f(a)g(a).$$

From the definition of $+$ and $\cdot$, it follows that $+$ and $\cdot$ are binary operations on R. Let $f, g, h \in R$. Then for all $a \in \mathbf{R}$, we have by using the associativity of $\mathbf{R}$ that $((f + g) + h)(a) = (f + g)(a) + h(a) = (f(a) + g(a)) + h(a) = f(a)+$

$(g(a)+h(a)) = f(a)+(g+h)(a) = (f+(g+h))(a)$. Thus, $(f+g)+h = f+(g+h)$. This shows that $+$ is associative. In a similar way, we can show that the other properties of a ring hold for R by using the fact that they hold for $\mathbf{R}$. Thus, $(R,+,\cdot)$ is a ring. We note that the function $i_0 : \mathbf{R} \to \mathbf{R}$, where $i_0(a) = 0$ for all $a \in \mathbf{R}$, is the additive identity of R and the element $i_1 \in R$, where $i_1(a) = 1$ for all $a \in \mathbf{R}$, is the identity of R. Also, for all $f, g \in R$ and for all $a \in \mathbf{R}$, $(f \cdot g)(a) = f(a)g(a) = g(a)f(a) = (g \cdot f)(a)$. Thus, for all $f, g \in R$, $f \cdot g = g \cdot f$. Consequently, $(R,+,\cdot)$ is a commutative ring with identity.

The addition and multiplication on R in Example 10.1.8 are the same as those encountered by the student in calculus.

Example 10.1.9 Let $(G, *)$ be a commutative group and $\text{Hom}(G, G)$ be the set of all homomorphisms of G into itself. Now the composition of two homomorphisms of G is again a homomorphism of G and so $\circ$ is a binary operation on $\text{Hom}(G, G)$. Also, $\circ$ is associative by Theorem 1.5.13 and $i_G \in \text{Hom}(G, G)$ is the identity. Thus, $(\text{Hom}(G, G), \circ)$ is a semigroup with identity. We now define a suitable $+$ on $\text{Hom}(G, G)$ so that $(\text{Hom}(G, G), +, \circ)$ becomes a ring with identity. Define $+$ on $\text{Hom}(G, G)$ by for all $f, g \in \text{Hom}(G, G)$,

$$(f+g)(a) = f(a) * g(a) \text{ for all } a \in G.$$

Let $f, g \in \text{Hom}(G, G)$. From the definition of $+$, it follows that $f + g$ is a mapping from G into G. Let $a, b \in G$. Then

$$\begin{aligned}(f+g)(ab) &= f(ab) * g(ab) \\ &= (f(a) * f(b)) * (g(a) * g(b)) \\ &= f(a) * g(a) * f(b) * g(b) \\ &= (f+g)(a) * (f+g)(b).\end{aligned}$$

This shows that $f + g$ is a homomorphism from G into G. We omit the routine verification that $+$ is associative. The identity of $(\text{Hom}(G, G), +)$ is the homomorphism that maps every element of G onto the identity of G. For any $f \in \text{Hom}(G, G)$, the mapping $-f$ defined by $(-f)(a) = f(a)^{-1}$ for all $a \in G$ is the additive inverse of f. Thus, $(\text{Hom}(G, G), +)$ is a group. We now show that the left distributive law holds. For any $a \in G$ and any elements $f, g, h \in \text{Hom}(G, G)$, $[f \circ (g+h)](a) = f((g+h)(a)) = f(g(a) * h(a)) = f(g(a)) * f(h(a)) = (f \circ g)(a) * (f \circ h)(a) = (f \circ g + f \circ h)(a)$. Hence, $f \circ (g+h) = (f \circ g) + (f \circ h)$. The right distributive law holds similarly. Consequently, $(\text{Hom}(G, G), +, \circ)$ is a ring.

We now prove some elementary properties of rings.

Theorem 10.1.10 Let R be a ring and $a, b, c \in R$. Then

(i) $a0 = 0a = 0$,

(ii) $a(-b) = (-a)b = -(ab)$,

(iii) $(-a)(-b) = ab$,

(iv) $a(b-c) = ab - ac$ and $(b-c)a = ba - ca$.

Proof. (i) Observe that $a0 + a0 = a(0+0) = a0$. Thus, $(a0+a0) + (-(a0)) = a0 + (-(a0))$ and so $a0 + (a0 + (-(a0))) = 0$. Hence, $a0 + 0 = 0$ or $a0 = 0$. Similarly, $0a = 0$.

(ii) $ab + a(-b) = a(b + (-b)) = a0 = 0 = a0 = a(-b + b) = a(-b) + ab$. Since the additive inverse of an element is unique, $a(-b) = -(ab)$. Similarly, $(-a)\, b = -(ab)$.

(iii) Using (ii), we have $(-a)(-b) = -(a(-b)) = -(-ab) = ab$.

(iv) Since $b - c = b + (-c)$, $a(b-c) = a(b + (-c)) = ab + a(-c) = ab + (-(ac))$ (by (ii)) $= ab - ac$. Similarly, $(b-c)a = ba - ca$. ∎

Corollary 10.1.11 *Let R be a ring with 1. Then $R \neq \{0\}$ if and only if the elements 0 and 1 are distinct.*

Proof. Suppose $R \neq \{0\}$. Let $a \in R$ be such that $a \neq 0$. Suppose $1 = 0$. Then $a = a1 = a0 = 0$, a contradiction. Thus, $1 \neq 0$. The converse follows since R has at least two distinct elements 0 and 1. ∎

Convention: From now on, we assume that the identity element 1 (if it exists) is different from the zero element of the ring.

From this convention, it follows that if R is a ring with 1, then R has at least two elements.

Let R be a ring with 1. An element $u \in R$ is called a **unit** (or an **invertible element**) if there exists $v \in R$ such that $uv = 1 = vu$. We note the following properties of invertible elements.

Theorem 10.1.12 *Let R be a ring with 1 and T be the set of all units of R. Then*

(i) $T \neq \phi$,

(ii) $0 \notin T$, and

(iii) $ab \in T$ for all $a, b \in T$.

Proof. (i) Since $1 \cdot 1 = 1 = 1 \cdot 1$, $1 \in T$. Hence, $T \neq \phi$.

(ii) Suppose that $0 \in T$. Then there exists $v \in R$ such that $0v = 1 = v0$. However, $0v = 0$ and so $0 = 1$, which is a contradiction. Thus, $0 \notin T$.

(iii) Let $a, b \in T$. There exist $c, d \in R$ such that $ac = 1 = ca$ and $bd = 1 = db$. Now $(ab)(dc) = a(bd)c = a1c = ac = 1$ and $(dc)(ab) = d(ca)b = d1b = db = 1$. Hence, $(ab)(dc) = 1 = (dc)(ab)$. Thus, ab is a unit and so $ab \in T$. ∎

Definition 10.1.13 *(i) A ring R with 1 is called a **division ring (skew-field)** if every nonzero element of R is a unit.*

*(ii) A commutative division ring R is called a **field**.*

Note that a ring R is a division ring (or skew-field) if and only if $(R\backslash\{0\}, \cdot)$ is a group. Therefore, if R is a division ring, then for all $a \in R$, $a \neq 0$, there exists a unique element denoted by $a^{-1} \in R$ such that $aa^{-1} = 1 = a^{-1}a$. We call a^{-1} the multiplicative inverse of a. Similarly, a ring R is a field if and only if $(R\backslash\{0\}, \cdot)$ is a commutative group.

Example 10.1.14 *(i) The ring $\mathbf{Z}$ of integers is not a field. In $\mathbf{Z}$, the only invertible elements are 1 and -1.*

*(ii) From Example 2.1.3, $(\mathbf{Q}, +, \cdot)$ is a field, where $+$ and $\cdot$ are the usual addition and multiplication, respectively. $\mathbf{Q}$ is called the **field of rational numbers**.*

*(iii) From Example 2.1.3, $(\mathbf{R}, +, \cdot)$ is a field, where $+$ and $\cdot$ are the usual addition and multiplication, respectively. $\mathbf{R}$ is called the **field of real numbers**.*

*(iv) From Example 2.1.3, $(\mathbf{C}, +, \cdot)$ is a field, where $+$ and $\cdot$ are the usual addition and multiplication, respectively. $\mathbf{C}$ is called the **field of complex numbers**.*

The following example is due to William Rowan Hamilton. Due to physical considerations, Hamilton constructed a consistent algebra in which the commutative law of multiplication fails to hold. At the time, such a construction seemed inconceivable. His work and H.G. Grossman's work on hypercomplex number systems began the liberation of algebra. Their work encouraged other mathematicians to create algebras, which broke with tradition, e.g., algebras in which $ab = 0$ with $a \neq 0$, $b \neq 0$ and algebras with $a^n = 0$, where $a \neq 0$ and n is a positive integer.

Example 10.1.15 *Let $Q_{\mathbf{R}} = \{(a_1, a_2, a_3, a_4) \mid a_i \in \mathbf{R}, i = 1, 2, 3, 4\}$. Define $+$ and $\cdot$ on $Q_{\mathbf{R}}$ as follows:*

$$(a_1, a_2, a_3, a_4) + (b_1, b_2, b_3, b_4) = (a_1 + b_1, a_2 + b_2, a_3 + b_3, a_4 + b_4)$$

$$(a_1, a_2, a_3, a_4) \cdot (b_1, b_2, b_3, b_4) = (a_1b_1 - a_2b_2 - a_3b_3 - a_4b_4, \ a_1b_2 + a_2b_1$$
$$+a_3b_4 - a_4b_3, \ a_1b_3 + a_3b_1 + a_4b_2 - a_2b_4, \ a_1b_4 + a_2b_3 - a_3b_2 + a_4b_1).$$

From the definition of $+$ and $\cdot$, it follows that $+$ and $\cdot$ are binary operations on $Q_{\mathbf{R}}$. Now $+$ is associative and commutative since addition is associative and commutative in $\mathbf{R}$. We also note that $(0, 0, 0, 0) \in Q_{\mathbf{R}}$ is the additive identity and if $(a_1, a_2, a_3, a_4) \in Q_{\mathbf{R}}$, then $(-a_1, -a_2, -a_3, -a_4) \in Q_{\mathbf{R}}$

and $-(a_1, a_2, a_3, a_4) = (-a_1, -a_2, -a_3, -a_4)$. *Hence,* $(Q_{\mathbf{R}}, +)$ *is a commutative group. Similarly,* $\cdot$ *is associative and* $(1, 0, 0, 0) \in Q_{\mathbf{R}}$ *is the multiplicative identity. Let* $(a_1, a_2, a_3, a_4) \in Q_{\mathbf{R}}$ *be a nonzero element. Then* $N = a_1^2 + a_2^2 + a_3^2 + a_4^2 \neq 0$ *and* $N \in \mathbf{R}$. *Thus,* $(a_1/N, -a_2/N, -a_3/N, -a_4/N)$ $\in Q_{\mathbf{R}}$. *We ask the reader to verify that* $(a_1/N, -a_2/N, -a_3/N, -a_4/N)$ *is the multiplicative inverse of* (a_1, a_2, a_3, a_4). *Thus,* $Q_{\mathbf{R}}$ *is a division ring and is called the ring of* **real quaternions**. *However,* $Q_{\mathbf{R}}$ *is not commutative since* $(0, 1, 0, 0)(0, 0, 1, 0) = (0, 0, 0, 1) \neq (0, 0, 0, -1) = (0, 0, 1, 0)(0, 1, 0, 0)$. *Therefore,* $Q_{\mathbf{R}}$ *is not a field.*

A nonzero element a in a ring R is called a **zero divisor** if there exists $b \in R$ such that $b \neq 0$ and either $ab = 0$ or $ba = 0$. We do not call 0 a zero divisor. An element cannot be a unit and zero divisor at the same time (Worked-Out Exercise 1, page 279). Thus, a field has no zero divisors.

Definition 10.1.16 *Let R be a commutative ring with 1. Then R is called an* **integral domain** *if R has no zero divisors.*

The ring of integers $\mathbf{Z}$ is an integral domain. The ring $M_2(\mathbf{Z})$ is not an integral domain since it is noncommutative. Also, $M_2(\mathbf{Z})$ has zero divisors. For example, $\begin{bmatrix} 1 & 0 \\ 0 & 0 \end{bmatrix}, \begin{bmatrix} 0 & 1 \\ 0 & 0 \end{bmatrix} \in M_2(\mathbf{Z})$ and $\begin{bmatrix} 0 & 1 \\ 0 & 0 \end{bmatrix}\begin{bmatrix} 1 & 0 \\ 0 & 0 \end{bmatrix} = \begin{bmatrix} 0 & 0 \\ 0 & 0 \end{bmatrix}$. We also note that every field F is an integral domain since every nonzero element of F is a unit.

Example 10.1.17 $\mathbf{Z}[\sqrt{3}] = \{a + b\sqrt{3} \mid a, b \in \mathbf{Z}\}$ *is an integral domain, where the operations $+$ and $\cdot$ are the usual operations of addition and multiplication. $0 + 0\sqrt{3}$ is the additive identity of $\mathbf{Z}[\sqrt{3}]$ and $1 + 0\sqrt{3}$ is the multiplicative identity of $\mathbf{Z}[\sqrt{3}]$. Suppose $\sqrt{3}$ is a unit in $\mathbf{Z}[\sqrt{3}]$. Then $(\sqrt{3})^{-1} = a + b\sqrt{3}$ for some $a, b \in \mathbf{Z}$. If $a = 0$, then $(\sqrt{3})^{-1} = b\sqrt{3}$ or $1 = 3b$, which is a contradiction since this equation has no solution in $\mathbf{Z}$. Therefore, $a \neq 0$ and so $1 = a\sqrt{3} + 3b$ or $\sqrt{3} = \frac{1-3b}{a} \in \mathbf{Q}$, a contradiction. Hence, $\sqrt{3}$ is not a unit, proving that $\mathbf{Z}[\sqrt{3}]$ is not a field.*

By arguments similar to the ones used in Example 10.1.17, we can show that the following sets are integral domains under the usual addition and multiplication.

$$\begin{aligned}
\mathbf{Z}[\sqrt{n}] &= \{a + b\sqrt{n} \mid a, b \in \mathbf{Z}\} \\
\mathbf{Z}[i\sqrt{n}] &= \{a + bi\sqrt{n} \mid a, b \in \mathbf{Z}\} \\
\mathbf{Z}[i] &= \{a + bi \mid a, b \in \mathbf{Z}\} \\
\mathbf{Q}[\sqrt{n}] &= \{a + b\sqrt{n} \mid a, b \in \mathbf{Q}\} \\
\mathbf{Q}[i\sqrt{n}] &= \{a + bi\sqrt{n} \mid a, b \in \mathbf{Q}\} \\
\mathbf{Q}[i] &= \{a + bi \mid a, b \in \mathbf{Q}\},
\end{aligned}$$

where n is a fixed positive integer and $i^2 = -1$. In fact, it can be shown that $\mathbf{Q}[\sqrt{n}], \mathbf{Q}[i\sqrt{n}]$, and $\mathbf{Q}[i]$ are fields.

Example 10.1.18 *The ring of even integers* $\mathbf{E}$ *is a commutative ring, without identity, and without zero divisors.*

The ring appearing in the following example is sometimes useful in the construction of counterexamples.

Example 10.1.19 *Let* $(R, +)$ *be a commutative group. Define multiplication on* R *by* $ab = 0$ *for all* $a, b \in R$, *where* 0 *denotes the identity element of the group* $(R, +)$. *Then* $(R, +, \cdot)$ *is a ring called the **zero** ring. If* R *contains more than one element, then* R *is a commutative ring without* 1 *and every nonzero element of* R *is a zero divisor.*

The following theorem establishes a relation between zero divisors and the cancellation property of a ring.

Theorem 10.1.20 *Let* R *be a ring. If* R *has no zero divisors, then the cancellation laws hold, i.e., for all* $a, b, c \in R$, $a \neq 0$, $ab = ac$ *implies* $b = c$ *(**left cancellation law**) and* $ba = ca$ *implies* $b = c$ *(**right cancellation law**). If either cancellation law holds, then* R *has no zero divisors.*

Proof. Suppose R has no zero divisors. Let a, b, $c \in R$ be such that $ab = ac$ and $a \neq 0$. Then $ab - ac = 0$ or $a(b - c) = 0$. Since R has no zero divisors and $a \neq 0$, we have $b - c = 0$ or $b = c$. Hence, the left cancellation law holds. Similarly, the right cancellation law holds. Conversely, suppose one of the cancellation laws hold, say, the left, i.e., if a, b, $c \in R$, $a \neq 0$, then $ab = ac$ implies $b = c$. Let a be a nonzero element of R and $b \in R$. Suppose $ab = 0$. Then $ab = a0$, from which $b = 0$ by canceling a. Suppose $ba = 0$ and $b \neq 0$. Then $ba = b0$ and by canceling b, we obtain $a = 0$, a contradiction. Therefore, $b = 0$. Hence, R has no zero divisors. Similarly, the right cancellation law implies that R has no zero divisors. ∎

Definition 10.1.21 *A ring* R *is called a **finite** ring if* R *has only a finite number of elements; otherwise* R *is called an **infinite ring**.*

The rings $\mathbf{Z}$ and $M_2(\mathbf{Z})$ are infinite.

Example 10.1.22 *Consider* $\mathbf{Z}_n$ *together with the binary operations* $+_n$ *and* $\cdot_n$ *as defined in Examples 2.1.5 and 2.1.6. By Example 2.1.5,* $(\mathbf{Z}_n, +_n)$ *is a commutative group and by Example 2.1.6,* $\cdot_n$ *is associative and commutative, and* $[1]$ *is the multiplicative identity of* $(\mathbf{Z}_n, +_n, \cdot_n)$. *Now for all* $[a], [b], [c] \in \mathbf{Z}_n$, $[a] \cdot_n ([b] +_n [c]) = [a] \cdot_n [b + c] = [a(b + c)] = [ab + ac] = [ab] +_n [ac] =$

$[a] \cdot_n [b] +_n [a] \cdot_n [c]$. Similarly, $([b] +_n [c]) \cdot_n [a] = [b] \cdot_n [a] +_n [c] \cdot_n [a]$. Hence, both distributive laws hold. Thus, $(\mathbf{Z}_n, +_n, \cdot_n)$ is a commutative ring with 1, called the **ring of integers mod** n. From Example 2.1.6, not every nonzero element of $\mathbf{Z}_n$ has an inverse. For example, suppose n is not prime, say, $n = 6$. Then $[4]$ has no multiplicative inverse in $\mathbf{Z}_6$. Also, $\mathbf{Z}_6$ has zero divisors. We have $[3] \neq [0] \neq [2]$. Since $[3] \cdot_6 [2] = [6] = [0]$, it follows that $[3]$ and $[2]$ are zero divisors. Thus, $\mathbf{Z}_6$ is not an integral domain and thus not a field. We can also conclude that $[2]$ and $[3]$ do not have multiplicative inverses since they are zero divisors.

The above example shows that for every positive integer n, there exists a commutative ring R with 1 such that the number of elements in R is n.

In the following result, we assume that the ring R is commutative. This assumption can be removed and the conclusion that R is a field remains valid. However, we have not developed the appropriate results to remove this assumption. We will prove the theorem in its most general form in Chapter 24.

Theorem 10.1.23 *A finite commutative ring R with more than one element and without zero divisors is a field.*

Proof. We must show that R has an identity and that every nonzero element of R is a unit. Let $a_1, a_2, \ldots, a_n$ be the distinct elements of R. Let $a \in R$, $a \neq 0$. Now $aa_i \in R$ for all i and so $\{aa_1, aa_2, \ldots, aa_n\} \subseteq R$. If $aa_i = aa_j$, then by Theorem 10.1.20, $a_i = a_j$. Therefore, the elements $aa_1, aa_2, \ldots, aa_n$ must be distinct and so $R = \{aa_1, aa_2, \ldots, aa_n\}$. This implies that one of the products must be equal to a, say, $aa_i = a$. Since R is commutative, we also have $a_i a = aa_i = a$. Let b be any element of R. Then there exists $a_j \in R$ such that $b = aa_j$. Thus,

$$
\begin{aligned}
ba_i &= a_i b && \text{(since } R \text{ is commutative)} \\
&= a_i(aa_j) && \text{(substituting for } b) \\
&= (a_i a)a_j \\
&= aa_j \\
&= b.
\end{aligned}
$$

This implies that a_i is the identity of R. We denote the identity of R by 1. Now $1 \in R = \{aa_1, aa_2, \ldots, aa_n\}$ and so one of the products, say, aa_j, must equal 1. By commutativity, $a_j a = aa_j = 1$. Hence, every nonzero element is a unit. Consequently, R is a field. ∎

The following corollary is immediate from above theorem.

Corollary 10.1.24 *Every finite integral domain is a field.* ∎

In Example 2.1.6, we showed that a nonzero element $[a]$ of $\mathbf{Z}_n$ has an inverse if and only if $\gcd(a, n) = 1$. Thus, the following corollary is an immediate consequence of this fact. We leave the details as an exercise.

Corollary 10.1.25 *Let n be a positive integer. Then $\mathbf{Z}_n$ is a field if and only if n is prime.* ∎

Let R be a ring and $a \in R$. Then for any integer n, define na as follows:

$$
\begin{aligned}
0a &= 0 \\
na &= a + (n-1)a \quad \text{if } n > 0 \\
na &= (-n)(-a) \qquad \text{if } n < 0.
\end{aligned}
$$

We emphasize that na is not a multiplication of elements of R since R may not contain $\mathbf{Z}$. We have the following properties holding for any $a, b \in R$ and any $m, n \in \mathbf{Z}$:

$$
\begin{aligned}
(m+n)a &= ma + na, \\
m(a+b) &= ma + mb, \\
(mn)a &= m(na), \\
m(ab) &= (ma)b = a(mb), \\
(ma)(nb) &= mn(ab).
\end{aligned}
$$

The proofs of the above properties can be obtained by induction and the defining conditions of a ring.

Definition 10.1.26 *If there exists a positive integer n such that for all $a \in R$, $na = 0$, then the smallest such positive integer is called the characteristic of R. If no such positive integer exists, then R is said to be of **characteristic zero**.*

Example 10.1.27 *The rings $\mathbf{Z}, \mathbf{Q}, \mathbf{R}, \mathbf{C}$ have characteristic 0. The ring $\mathbf{Z}_n$ ($n = 1, 2, 3, \ldots$) has characteristic n. Note that in $\mathbf{Z}_6$, $3[2] = [6] = [0]$ and $2[3] = [6] = [0]$. However, 6 is the smallest positive integer such that $6[a] = [0]$ for all $[a] \in \mathbf{Z}_6$. In particular, $[1]$ has additive order 6.*

Example 10.1.28 *Let X be a nonempty set and $\mathcal{P}(X)$ the power set of X. Then $(\mathcal{P}(X), \Delta, \cap)$ is a commutative ring with 1, where Δ is the operation "symmetric difference." In this example, Δ acts as $+$ and $\cap$ acts as $\cdot$. Now for all $A \in \mathcal{P}(X)$, $2A = A\Delta A = (A \backslash A) \cup (A \backslash A) = \phi$. Thus, $\mathcal{P}(X)$ has characteristic 2.*

Theorem 10.1.29 *Let R be a ring with 1. Then R has characteristic $n > 0$ if and only if n is the least positive integer such that $n1 = 0$.*

Proof. Suppose R has characteristic $n > 0$. Then $na = 0$ for all $a \in R$ and so, in particular, $n1 = 0$. If $m1 = 0$ for $0 < m < n$, then $ma = m(1a) = (m1)a = 0a = 0$ for all $a \in R$. However, this contradicts the minimality of n. Hence, n is the smallest positive integer such that $n1 = 0$. Conversely, suppose n is the smallest positive integer such that $n1 = 0$. Then for all $a \in R$, $na = n(1a) = (n1)a = 0a = 0$. By the minimality of n for 1, n must be the characteristic of R. ∎

Theorem 10.1.30 *The characteristic of an integral domain R is either zero or a prime.*

Proof. If there does not exist a positive integer n such that $na = 0$ for all $a \in R$, then R is of characteristic zero. Suppose there exists a positive integer n such that $na = 0$ for all $a \in R$. Let m be the smallest positive integer such that $ma = 0$ for all $a \in R$. Then $m1 = 0$. If m is not prime, then there exist integers m_1, m_2 such that $0 < m_1, m_2 < m$ and $m = m_1 m_2$. Hence, $0 = (m_1 m_2)1 = (m_1 1)(m_2 1)$. Since R has no zero divisors, either $m_1 1 = 0$ or $m_2 1 = 0$. This contradicts the minimality of m. Thus, m is a prime. ∎

10.1.1 Worked-Out Exercises

◇ **Exercise 1** Let R be a ring. An element $a \in R$ is called **idempotent** if $a^2 = a$ and **nilpotent** if $a^n = 0$ for some positive integer n.

(i) Let $a \in R$ be a nonzero idempotent. Show that a is not nilpotent.

(ii) Let R be with 1. Let $a \in R$ be such that a has an inverse. Show that a cannot be a zero divisor.

(iii) Let R be with 1 and suppose R has no zero divisors. Show that the only idempotents in R are 0 and 1.

Solution: (i) From the hypothesis, $a^2 = a$. By induction, $a^n = a$ for all positive integers n. Suppose a is nilpotent. Then $a^m = 0$ for some positive integer m and so $a = a^m = 0$, which is a contradiction and so a is not nilpotent.

(ii) There exists $b \in R$ such that $ab = 1 = ba$. Suppose that a is a zero divisor. Then there exists $c \in R$, $c \neq 0$, such that $ac = 0$. Thus, $0 = b0 = b(ac) = (ba)c = c$, which is a contradiction. Hence, a is not a zero divisor.

(iii) Clearly 0 and 1 are idempotent elements. Let $e \in R$ be an idempotent. Then $e^2 = e$ and so $e(e - 1) = 0$. Since R has no zero divisors, either $e = 0$ or $e - 1 = 0$, i.e., either $e = 0$ or $e = 1$. Therefore, the only idempotents of R are 0 and 1.

◇ **Exercise 2** Determine positive integers n such that $\mathbf{Z}_n$ has no nonzero nilpotent elements.

Solution: We claim that n is a square free integer, i.e., $n = p_1 p_2 \cdots p_k$, where the p_i's are distinct primes.

Suppose that $n = p_1 p_2 \cdots p_k$, p_i's are distinct primes. Let $[a] \in \mathbf{Z}_n$ be nilpotent. Then $[a]^m = [0]$ for some integer m. Hence, n divides a^m and so $p_1 p_2 \cdots p_k$ divides a^m. Then $p_i | a^m$ for all $i = 1, 2, \ldots, k$. Since the p_i's are prime, $p_i | a$ for all $i = 1, 2, \ldots, k$. Since $p_1, p_2, \ldots, p_k$ are distinct primes, we must have $p_1 p_2 \cdots p_k | a$, i.e., $n | a$ and so $[a] = [0]$. This implies that $\mathbf{Z}_n$ has no nonzero nilpotent elements. Conversely, suppose that $\mathbf{Z}_n$ has no nonzero nilpotent elements. Let $n = p_1^{m_1} p_2^{m_2} \cdots p_k^{m_k}$, where the p_i's are distinct primes and $m_i \geq 1$. Let $m = \max\{m_1, m_2, \ldots, m_k\}$. Now $[p_1 p_2 \cdots p_k]^m = [p_1^m p_2^m \cdots p_k^m] = [0]$ since $n | (p_1^m p_2^m \cdots p_k^m)$. Also, since $\mathbf{Z}_n$ has no nonzero nilpotent elements, $[p_1 p_2 \cdots p_k] = [0]$. Hence, $n | (p_1 \cdots p_k)$ and so $(p_1^{m_1} p_2^{m_2} \cdots p_k^{m_k}) | (p_1 \cdots p_k)$. Thus, $m_i \leq 1$ for all $i = 1, 2, \ldots, k$. Hence, $m_i = 1$ for all $i = 1, 2, \ldots, k$ and so n is a square free integer.

$\diamond$ **Exercise 3** Show that the number of idempotent elements in $\mathbf{Z}_{mn}$, where $m > 1$, $n > 1$, and m and n are relatively prime, is at least 4.

Solution: Clearly, $[0]$ and $[1]$ are idempotent elements. Since m and n are relatively prime, there exist integers a and b such that $am + bn = 1$. We now show that n does not divide a and m does not divide b. Suppose that $n | a$. Then $a = nr$ for some integer r. Thus, $n(rm + b) = nrm + nb = am + nb = 1$. This implies that $n = 1$, which is a contradiction. Therefore, n does not divide a and similarly m does not divide b. Now $m^2 a = m(1 - nb)$. This implies that $[m^2 a] = [m]$. Hence, $[ma]^2 = [ma]$. If $[ma] = [0]$, then $mn | ma$ and so $n | a$, which is a contradiction. Consequently, $[ma] \neq [0]$. If $[ma] = [1]$, then $mn | (ma - 1)$. Hence, $ma + mnt = 1$ for some integer t. Thus, $m(a + nt) = 1$. This implies $m = 1$, which is a contradiction. Hence, $[ma] \neq [1]$. Thus, $[ma]$ is an idempotent such that $[ma] \neq [0]$ and $[ma] \neq [1]$. Similarly, $[nb]$ is an idempotent such that $[nb] \neq [0]$ and $[nb] \neq [1]$. Clearly $[ma] \neq [nb]$. Thus, we find that $[0]$, $[1]$, $[ma]$, and $[nb]$ are idempotent elements of $\mathbf{Z}_{mn}$.

$\diamond$ **Exercise 4** Determine the positive integers n such that $\mathbf{Z}_n$ has no idempotent elements other than $[0]$ and $[1]$.

Solution: We show that $n = p^r$ for some prime p and some integer $r > 0$.

First assume that $n = p^r$ for some prime p and some positive integer r and $[x] \in \mathbf{Z}_n$ be an idempotent. Then $[x]^2 = [x]$. Thus, $p^r | (x^2 - x)$ or $p^r | x(x - 1)$. Since x and $x - 1$ are relatively prime, $p^r | x$ or $p^r | (x - 1)$. If $p^r | x$, then $[x] = [0]$ and if $p^r | (x - 1)$, then $[x] = [1]$. Thus, $[0]$ and $[1]$ are the only two idempotent elements. Conversely, suppose that $[0]$ and $[1]$ are the only two idempotent elements. Let $n = p_1^{m_1} p_2^{m_2} \cdots p_k^{m_k}$, where the p_i's are distinct primes, $m_i \geq 1$, and $k > 1$. Let $t = p_1^{m_1}$ and $s = p_2^{m_2} \cdots p_k^{m_k}$. Then t and s are relatively prime and $n = ts$. By Worked-Out Exercise 3, $\mathbf{Z}_n = \mathbf{Z}_{ts}$ must have at least four idempotents, which is a contradiction. Therefore, $k = 1$. Thus, $n = p^r$ for some prime p and some positive integer r.

Exercise 5 Let R be a ring. Show that the following conditions are equivalent.

(i) R has no nonzero nilpotent elements.

(ii) For all $a \in R$, if $a^2 = 0$, then $a = 0$.

Solution: (i)$\Rightarrow$(ii) Let $a \in R$ and $a^2 = 0$. If $a \neq 0$, then a is a nonzero nilpotent element of R, a contradiction. Thus, $a = 0$.

(ii)$\Rightarrow$(i) Let $a \in R$ be such that $a^n = 0$ for some positive integer n. Suppose $a \neq 0$. Let n be the smallest positive integer such that $a^n = 0$. Suppose n is even, say, $n = 2m$ for some positive integer m. Then $(a^m)^2 = a^{2m} = 0$ and so $a^m = 0$, contradicting the minimality of n. Suppose n is odd. If $n = 1$, then $a = 0$, a contradiction. Therefore, $n > 1$. Suppose $n = 2m + 1$. Then $m + 1 < n$. Thus, $a^{2m+2} = a^{2m+1}a = a^n a = 0$. This implies that $a^{m+1} = 0$, which is a contradiction of the minimality of n. Hence, R has no nonzero nilpotent elements.

$\diamond$ **Exercise 6** An element e of a ring R is called a **left (right) identity**, if $ea = a$ $(ae = a)$ for all $a \in R$. Show that if a ring R has a unique left identity e, then e is also the right identity of R and hence the identity of R.

Solution: Let e be the unique left identity of R. Then $ex = x$ for all $x \in R$. Let $x \in R$. Now $(xe - x + e)x = xex - xx + ex = xx - xx + x = x$. This implies that $xe - x + e$ is a left identity. Since e is the unique left identity, $xe - x + e = e$ and so $xe = x$. Thus, e is a right identity.

Exercise 7 Let R be a commutative ring with 1 and $a, b \in R$. Suppose that a is invertible and b is nilpotent. Show that $a + b$ is invertible. Also, show that if R is not commutative, then the result may not be true.

Solution: There exists $c \in R$ such that $ac = 1 = ca$ and there exists a positive integer n such that $b^n = 0$. Let $d = c - c^2 b + c^3 b^2 + \cdots + (-1)^{n+1} c^n b^{n-1}$. Now $(a+b)d = ac - ac^2 b + ac^3 b^2 + \cdots + (-1)^{n+1} ac^n b^{n-1} + bc - bc^2 b + bc^3 b^2 + \cdots + (-1)^{n+1} bc^n b^{n-1} = 1 - cb + c^2 b^2 + \cdots + (-1)^{n+1} c^{n-1} b^{n-1} + bc - c^2 b^2 + c^3 b^3 + \cdots + (-1)^{n+1} c^n b^n = 1$. Similarly, $d(a + b) = 1$. Hence, $a + b$ is invertible.

Consider the ring $M_2(\mathbf{Z})$. Let $a = \begin{bmatrix} 0 & -1 \\ -1 & 0 \end{bmatrix}$ and $b = \begin{bmatrix} 0 & 1 \\ 0 & 0 \end{bmatrix}$. Then a is invertible and b is nilpotent. Now $a + b = \begin{bmatrix} 0 & 0 \\ -1 & 0 \end{bmatrix}$. Clearly $a + b$ is a nonzero nilpotent element. Hence, $a + b$ is not invertible.

10.1.2 Exercises

1. In the rings $\mathbf{Z}_8$ and $\mathbf{Z}_6$, find the following elements:

    (i) the units, (ii) the nilpotent elements, and (iii) the zero divisors.

2. Let R be the set of all 2×2 matrices over the field of complex numbers of the form $\begin{bmatrix} z_1 & z_2 \\ -\bar{z}_2 & \bar{z}_1 \end{bmatrix}$, where $\bar{z}$ denotes the complex conjugate of the complex number z. Show that $(R, +, \cdot)$ is a division ring, where $+$ and $\cdot$ are the usual matrix addition and matrix multiplication, respectively. Is R a field?

3. Let R be a ring with 1. Prove that

    (i) $(-1)a = -a = a(-1)$ and $(-1)(-1) = 1$,

    (ii) if a is a unit in R, then $-a$ is a unit in R and $(-a)^{-1} = -(a^{-1})$.

4. Prove that a ring R is commutative if and only if $(a+b)^2 = a^2 + 2ab + b^2$ for all $a, b \in R$.

5. Prove that a ring R is commutative if and only if $a^2 - b^2 = (a+b)(a-b)$ for all $a, b \in R$.

6. Let R be a ring. If $a^3 = a$ for all $a \in R$, prove that R is commutative.

7. Let R be a commutative ring and $a, b \in R$. Prove that for all $n \in \mathbf{N}$,

$$(a+b)^n = a^n + \binom{n}{1}a^{n-1}b + \cdots + \binom{n}{r}a^{n-r}b^r + \cdots + \binom{n}{n-1}ab^{n-1} + b^n.$$

8. If a and b are elements of a ring and m and n are integers, prove that

    (i) $(na)(mb) = (nm)(ab)$,

    (ii) $n(ab) = (na)b = a(nb)$,

    (iii) $n(-a) = (-n)a$.

9. If R is an integral domain of prime characteristic p, prove that $(a+b)^p = a^p + b^p$ for all $a, b \in R$.

10. Let R be a ring with 1 and without zero divisors. Prove that for all $a, b \in R$, $ab = 1$ implies $ba = 1$.

11. Let R be a ring with 1. If a is a nilpotent element of R, prove that $1 - a$ and $1 + a$ are units.

12. Let R be a division ring and $a, b \in R$. Show that if $ab = 0$, then either $a = 0$ or $b = 0$.

13. Let $a \in R$ be an idempotent element. Show that $(1 - a)ba$ is nilpotent for all $b \in R$.

14. Find all idempotent elements of the ring $M_2(\mathbf{R})$.

15. Let R be a ring with 1. Let $0 \neq a \in R$. If there exist two distinct elements b and c in R such that $ab = ac = 1$, show that there are infinitely many elements x in R such that $ax = 1$. (*American Mathematical Monthly* 70(1961) 315).

16. Let R be an integral domain and $a, b \in R$. Let $m, n \in \mathbf{Z}$ be such that m and n are relatively prime. Prove that $a^m = b^m$ and $a^n = b^n$ imply that $a = b$.

17. Let R and R' be rings. Define $+$ and $\cdot$ on $R \times R'$ by for all $(a, b), (c, d) \in R \times R'$

$$(a, b) + (c, d) = (a + c, b + d) \text{ and } (a, b) \cdot (c, d) = (a \cdot c, b \cdot d).$$

(i) Prove that $(R \times R', +, \cdot)$ is a ring. This ring is called the **direct sum** of R and R' and is denoted by $R \oplus R'$.

(ii) If R and R' are commutative with identity, prove that $R \oplus R'$ is commutative with identity.

18. Extend the notion of direct sum in Exercise 17 to any finite number of rings.

19. Prove that the characteristic of a finite ring R divides $|R|$.

20. Let R be a ring with 1. Prove that the characteristic of the matrix ring $M_2(R)$ is the same as that of R.

21. If p is a prime integer, prove that $(p - 1)! \equiv_p -1$.

22. In the following exercises, write the proof if the statement is true; otherwise, give a counterexample.

(i) In a ring R, if a and b are idempotent elements, then $a + b$ is an idempotent element.

(ii) In a ring R, if a and b are nilpotent elements, then $a + b$ is a nilpotent element.

(iii) Every finite ring with 1 is an integral domain.

(iv) There exists a field with seven elements.

(v) The characteristic of an infinite ring is always 0.

(vi) An element of a ring R which is idempotent, but not a zero divisor, is the identity element of R.

(vii) If a and b are two zero divisors, then $a + b$ is also a zero divisor in a ring R.

(viii) In a finite field F, $a^2 + b^2 = 0$ implies $a = 0$ and $b = 0$ for all $a, b \in F$.

(ix) In a field F, $(a + b)^{-1} = a^{-1} + b^{-1}$ for all nonzero elements a, b such that $a + b \neq 0$.

(x) There exists a field with six elements.

10.2 Some Important Rings

In this section, we introduce two important rings and study some of their basic properties.

10.2.1 Boolean Rings

We recall that in Worked-Out Exercise 1 (page 279), an element x of a ring R is called an **idempotent** element if $x^2 = x$. The zero element and identity element of a ring are idempotent elements. In the ring $\mathbf{Z}$, the only idempotent elements are 0 and 1. There exist rings, which contain idempotent elements different from 0 and 1. For example, in $M_2(\mathbf{Z})$, $\begin{bmatrix} 1 & 0 \\ 2 & 0 \end{bmatrix}$ is an idempotent element.

Definition 10.2.1 *A ring R with 1 is called a **Boolean ring** if every element of R is an idempotent.*

Example 10.2.2 *(i) $\mathbf{Z}_2$ is a Boolean ring.*
(ii) The ring $\mathcal{P}(X)$ of Example 10.1.28 is a Boolean ring since for all $A \in \mathcal{P}(X)$, $A \cap A = A$.

Theorem 10.2.3 *Let R be a Boolean ring. Then the characteristic of R is 2 and R is commutative.*

Proof. First we show that R is of characteristic 2. Let $x \in R$. Now $x + x = (x+x)^2 = (x+x)(x+x) = x(x+x)+x(x+x) = x^2+x^2+x^2+x^2 = x+x+x+x$. This implies that $2x = 4x$ and so $0 = 2x$. Hence, $2 \cdot 1 = 0$ since x was arbitrary. It follows that the characteristic of R is 2 by Theorem 10.1.29. To show R is commutative, let $x, y \in R$. Then $x + y = (x + y)^2 = (x + y)(x + y) = x^2 + xy + yx + y^2 = x + xy + yx + y$. This implies that $0 = xy + yx$. Hence, $xy = xy + 0 = xy + xy + yx$ or $xy = 2xy + yx = yx$ since $2xy = 0$. Thus, R is commutative. ■

10.2.2 Regular Rings

An element x of a ring R is called a **regular element** if there exists $y \in R$ such that $x = xyx$.

Definition 10.2.4 *A ring R is called a **regular ring** if every element of R is regular.*

In the ring $\mathbf{Z}$, the only regular elements are $0, 1$, and -1. Thus, $\mathbf{Z}$ is not a regular ring.

Example 10.2.5 *Let R be a division ring and $x \in R$. If $x = 0$, then $x = xxx$. Suppose $x \neq 0$. Then $xx^{-1} = 1$ and so $x = xx^{-1}x$. Thus, R is a regular ring.*

From the definition of a Boolean ring, it follows that every Boolean ring is a regular ring. The field $\mathbf{R}$ is a regular ring, but not a Boolean ring.

Example 10.2.6 *Consider $\mathbf{R}$, the field of real numbers and*

$$\mathbf{R} \times \mathbf{R} = \{(x, y) \mid x, y \in \mathbf{R}\}.$$

Define $+$ and $\cdot$ on $\mathbf{R} \times \mathbf{R}$ by

$$
\begin{aligned}
(x, y) + (z, w) &= (x + z, y + w) \\
(x, y) \cdot (z, w) &= (xz, yw)
\end{aligned}
$$

for all $x, y, z, w \in \mathbf{R}$. Then $\mathbf{R} \times \mathbf{R}$ is a commutative ring with identity. Now $(1, 0), (0, 1) \in \mathbf{R} \times \mathbf{R}$ and $(1, 0)(0, 1) = (0, 0)$. This shows that $\mathbf{R} \times \mathbf{R}$ contains zero divisors and so $\mathbf{R} \times \mathbf{R}$ is not a field. We claim that $\mathbf{R} \times \mathbf{R}$ is regular. Let $(x, y) \in \mathbf{R} \times \mathbf{R}$. If $x = 0 = y$, then $(x, y)(x, y)(x, y) = (x, y)$. If $x \neq 0$ and $y \neq 0$, then $(x, y)(x^{-1}, y^{-1})(x, y) = (x, y)$. If $x = 0$, but $y \neq 0$, then $(x, y)(x, y^{-1})(x, y) = (x, y)$. Similarly, if $x \neq 0$ and $y = 0$, then $(x, y)(x^{-1}, y)(x, y) = (x, y)$. Thus, in any case, (x, y) is a regular element. Hence, $\mathbf{R} \times \mathbf{R}$ is a regular ring.

Example 10.2.7 *Let $M_2(\mathbf{R})$ be the set of all 2×2 matrices over $\mathbf{R}$. Now $M_2(\mathbf{R})$ is a noncommutative ring with 1, where $+$ and $\cdot$ are the usual matrix addition and multiplication, respectively. We show that $M_2(\mathbf{R})$ is a regular ring. Let $A = \begin{bmatrix} x & y \\ z & w \end{bmatrix} \in M_2(\mathbf{R})$.*

Case 1: $xw - zy \neq 0$. *Then* $B = \begin{bmatrix} \frac{w}{xw-zy} & \frac{-y}{xw-zy} \\ \frac{-z}{xw-zy} & \frac{x}{xw-zy} \end{bmatrix} \in M_2(\mathbf{R})$ *and* $A = ABA$.

Case 2: $xw - zy = 0$.

Subcase 2a: *x, y, z, w are all zero. In this case,* $A = \begin{bmatrix} 0 & 0 \\ 0 & 0 \end{bmatrix}$ *and so for*

any $B \in M_2(\mathbf{R})$, $ABA = A$.

Subcase 2b: *x, y, z, w are not all zero. Suppose $x \neq 0$ and let* $B = \begin{bmatrix} \frac{1}{x} & 0 \\ 0 & 0 \end{bmatrix}$. *Then*

$$
\begin{aligned}
ABA &= \begin{bmatrix} x & y \\ z & w \end{bmatrix} \begin{bmatrix} \frac{1}{x} & 0 \\ 0 & 0 \end{bmatrix} \begin{bmatrix} x & y \\ z & w \end{bmatrix} \\
&= \begin{bmatrix} 1 & 0 \\ \frac{z}{x} & 0 \end{bmatrix} \begin{bmatrix} x & y \\ z & w \end{bmatrix} \\
&= \begin{bmatrix} x & y \\ z & \frac{zy}{x} \end{bmatrix} = \begin{bmatrix} x & y \\ z & w \end{bmatrix}
\end{aligned}
$$

since $xw - zy = 0$ and $x \neq 0$ implies $w = \frac{zy}{x}$. If $y \neq 0$, then let $B = \begin{bmatrix} 0 & 0 \\ \frac{1}{y} & 0 \end{bmatrix}$.

Then

$$
\begin{aligned}
ABA &= \begin{bmatrix} x & y \\ z & w \end{bmatrix} \begin{bmatrix} 0 & 0 \\ \frac{1}{y} & 0 \end{bmatrix} \begin{bmatrix} x & y \\ z & w \end{bmatrix} \\
&= \begin{bmatrix} 1 & 0 \\ \frac{w}{y} & 0 \end{bmatrix} \begin{bmatrix} x & y \\ z & w \end{bmatrix} \\
&= \begin{bmatrix} x & y \\ \frac{wx}{y} & w \end{bmatrix} = \begin{bmatrix} x & y \\ z & w \end{bmatrix}.
\end{aligned}
$$

Similarly, if $z \neq 0$ or $w \neq 0$, then we can find B such that $ABA = A$. Thus, $M_2(\mathbf{R})$ is a regular ring.

Since $M_2(\mathbf{R})$ is not a division ring, it follows that a regular ring need not be a division ring. However, a division ring is a regular ring as shown in Example 10.2.5. In the next theorem, we show that a regular ring under a suitable condition becomes a division ring.

Theorem 10.2.8 *Let R be a regular ring with more than one element. Suppose for all $x \in R$, there exists a unique $y \in R$ such that $x = xyx$. Then*
(i) *R has no zero divisors,*
(ii) *if $x \neq 0$ and $x = xyx$, then $y = yxy$ for all $x, y \in R$,*
(iii) *R has an identity,*
(iv) *R is a division ring.*

Proof. (i) Let x be a nonzero element of R and $xz = 0$ for some $z \in R$. Now by hypothesis, there exists a unique $y \in R$ such that $xyx = x$. Thus, $x(y - z)x = xyx - xzx = xyx$. Hence, by the uniqueness of y, $y - z = y$ and so $z = 0$. This proves that R has no zero divisors.

(ii) Let $x \neq 0$ and $xyx = x$. Then $x(y - yxy) = xy - xyxy = xy - xy = 0$. Since R has no zero divisors and $x \neq 0$, $y - yxy = 0$ and so $yxy = y$.

(iii) Let $0 \neq x \in R$. Then there exists a unique $y \in R$ such that $xyx = x$. Let $e = yx$. If $e = 0$, then $x = xyx = 0$, which is a contradiction. Therefore, $e \neq 0$. Also, $e^2 = yxyx = y(xyx) = yx = e$. Let $z \in R$. Then $(ze - z)e = ze^2 - ze = ze - ze = 0$. Thus, by (i), either $ze - z = 0$ or $ze = z$. Similarly, $e(ez - z) = 0$ implies that $ez = z$. Hence, e is the identity of R.

(iv) By (iii), R contains an identity element e. To show R is a division ring, it remains to be shown that every nonzero element of R has an inverse in R. Let x be a nonzero element in R. Then there exists a unique $y \in R$ such that $xyx = x$. Thus, $xyx = xe$, i.e., $x(yx - e) = 0$. Since R has no zero divisors and $x \neq 0$, $yx - e = 0$ and so $yx = e$. Similarly, $xyx = ex$ implies $xy = e$. Therefore, $xy = e = yx$. Hence, R is a division ring.

10.2.3 Exercises

1. Prove that a Boolean ring R is a field if and only if R contains only 0 and 1.

2. Prove that a ring R with 1 is a Boolean ring if and only if for all $a, b \in R$, $(a + b)ab = 0$.

3. Let R be a Boolean ring with more than two elements. Find all zero divisors of R.

4. Let $T = \{f \mid f : \mathbf{R} \to \mathbf{Z}_2\}$. Define $+$ and $\cdot$ on T by for all $f, g \in T$, $(f + g)(x) = f(x) + g(x)$ and $(fg)(x) = f(x)g(x)$ for all $x \in \mathbf{R}$. Show that $(T, +, \cdot)$ is a Boolean ring.

5. Prove that a nonzero element of a regular ring with 1 is either a unit or a zero divisor.

6. Prove that the center of a regular ring is regular.

7. Let R be a ring in which each element is idempotent. Let $\overline{R} = R \times \mathbf{Z}_2$. Define $+$ and $\cdot$ on $\overline{R}$ by $(a, [n]) + (b, [m]) = (a + b, [n + m])$ and $(a, [n]) \cdot (b, [m]) = (na + mb + ab, [nm])$ for all $(a, [n]), (b, [m]) \in \overline{R}$. Show that $+$ and $\cdot$ are well defined on $\overline{R}$ and $\overline{R}$ is a Boolean ring.

8. Let R be a regular ring with 1.

   (i) Prove that for any $a \in R$, there exists an idempotent $e \in R$ such that $Ra = Re$.

   (ii) Prove that for any two idempotents $e, f \in R$, there exists an idempotent $g \in R$ such that $Re + Rf = Rg$.

William Rowan Hamilton (1805–1865) was born on August 4, 1805, in Dublin, Ireland. He was the fourth of nine children. His early education from the age of three was provided by his uncle. By the age of five, he was proficient in Latin, Greek, and Hebrew.

Hamilton started reading Newton's *Principia* when he was about 15 and became interested in astronomy. In 1822, he discovered an error in Laplace's *Mécanique céleste*, which was conveyed to John Brinkley through a friend. Brinkley later helped Hamilton in getting appointed as his successor at Dunsink Observatory.

On April 23, 1827, while still an undergraduate at Trinity College, Hamilton presented his paper, "Theory of Systems of Rays," to the Royal Irish Academy. This paper is responsible for creating the field of mathematical optics. Hamilton introduced the characteristic function, his first discovery. On June 10, 1827, he was appointed astronomer royal at Dunsink Observatory and professor of astronomy at Trinity College, even though he did not have a degree.

Hamilton's major contributions were in the algebra of quaternions, optics, and dynamics. He gave few examples to illustrate his concepts and so his papers were hard to read. He spent most of his life on the study of quaternions.

Hamilton was interested in three-dimensional complex numbers, which he called "triplets." He had little success in this area, as he was able to add, but could not find a suitable multiplication rule. He then considered the so-called quaternions. While he was walking along the Royal Canal on October 16, 1843, the discovery of the quaternions flashed in his mind. He immediately scratched the multiplication formula for the quaternions on the stone of a bridge over the canal. Hamilton discovered that he could give up the commutative law of multiplication and still have a meaningful algebraic system. The geometric significance of the quaternions was realized when Hamilton and Cayley independently showed that the quaternion operators rotated vectors about a given axis. In 1837, Hamilton corrected Abel's proof of the impossibility of solving the general quintic equations.

Hamilton's name is associated with concepts such as Hamiltonian functions, Hamiltonian-Jacobi differential equations, Hamiltonian path in graph theory, and the Cayley-Hamilton theorem in linear algebra. He coined the terms "vector," "scalar," and "tensor." Hamilton died on September 2, 1865.

Chapter 11

Subrings, Ideals, and Homomorphisms

The most important substructure of a ring is a particular subset called an "ideal." The term ideal was coined by Dedekind in honor of Kummer's work on ideal numbers. This notion of Kummer and Dedekind was used to obtain unique factorization properties. Kummer introduced the idea of an ideal number in his work on Fermat's last theorem. Noether followed with some important results on the theory of ideals. Some of her ideas were inspired by the work not only of Dedekind, but also of Kronecker and Lasker.

11.1 Subrings and Subfields

In this section, we introduce the idea of a subring of a ring. This concept is analogous to the concept of a subgroup of a group.

Definition 11.1.1 *Let $(R, +, \cdot)$ be a ring. Let R' be a subset of R. Then $(R', +, \cdot)$ is called a **subring** of $(R, +, \cdot)$ if $(R', +)$ is a subgroup of $(R, +)$ and for all $x, y \in R'$, $x \cdot y \in R'$.*

Let $(R', +, \cdot)$ be a subring of the ring $(R, +, \cdot)$. Since $R' \subseteq R$ and since the associativity for $\cdot$ and the distributive laws are inherited, $(R', +, \cdot)$ is itself a ring. We will usually suppress the operations $+$ and $\cdot$ and call R' a subring of R. When R' and R are fields, R' is called a **subfield** of R.

The following theorem gives a necessary and sufficient condition for a subset to be a subring. With these conditions it is easy to verify whether a nonempty subset of a ring is a subring or not.

Theorem 11.1.2 *Let R be a ring. A nonempty subset R' of R is a subring of R if and only if $x - y \in R'$ and $xy \in R'$ for all $x, y \in R'$.*

Proof.   First suppose that R' is a subring of R. Then R' is a ring and so for all $x, y \in R$, $x - y$, $xy \in R'$. Conversely, suppose $x - y \in R'$ and $xy \in R'$ for all $x, y \in R'$. Since $x - y \in R'$ for all $x, y \in R'$, $(R', +)$ is a subgroup of $(R, +)$ by Theorem 4.1.3. By the hypothesis, $xy \in R'$ for all $x, y \in R'$. Hence, R' is a subring of R. ∎

Example 11.1.3 *(i) The ring **E** of even integers is a subring of* **Z**. **E** *is without* 1.

(ii) Consider the subset $\mathbf{E}_8 = \{[0], [2], [4], [6]\}$ *of* $\mathbf{Z}_8$. *Then* $\mathbf{E}_8$ *is a subring of* $\mathbf{Z}_8$. *Hence,* $\mathbf{E}_8$ *is commutative. However,* $\mathbf{E}_8$ *has no identity and* $\mathbf{E}_8$ *does have zero divisors, namely,* $[2], [4]$, *and* $[6]$.

Example 11.1.4 *Let* $Q_{\mathbf{Z}} = \{(a_1, a_2, a_3, a_4) \mid a_i \in \mathbf{Z}, \ i = 1, 2, 3, 4\}$. *Define* $+$ *and* $\cdot$ *on* $Q_{\mathbf{Z}}$ *as in Example 10.1.15. Since the difference and product of integers is an integer, we have*

$$(a_1, a_2, a_3, a_4) - (b_1, b_2, b_3, b_4) \in Q_{\mathbf{Z}}$$

and

$$(a_1, a_2, a_3, a_4) \cdot (b_1, b_2, b_3, b_4) \in Q_{\mathbf{Z}}$$

for all $(a_1, a_2, a_3, a_4), (b_1, b_2, b_3, b_4) \in Q_{\mathbf{Z}}$. *Hence,* $Q_{\mathbf{Z}}$ *is a subring of* $Q_{\mathbf{R}}$. *We note that* $Q_{\mathbf{Z}}$ *is noncommutative, has an identity, and is without zero divisors. Now* $(0, 2, 0, 0) \in Q_{\mathbf{Z}}$ *and* $(0, 2, 0, 0)^{-1} = (0, -\frac{1}{2}, 0, 0) \notin Q_{\mathbf{Z}}$. *Thus,* $Q_{\mathbf{Z}}$ *is not a division ring.*

Example 11.1.5 *Set* $Q_{\mathbf{E}} = \{(a_1, a_2, a_3, a_4) \mid a_i \in \mathbf{E}, \ i = 1, 2, 3, 4\}$. *Define* $+$ *and* $\cdot$ *on* $Q_{\mathbf{E}}$ *as in Example 10.1.15. Since the difference and product of even integers is an even integer, we find that* $Q_{\mathbf{E}}$ *is a subring of* $Q_{\mathbf{Z}}$. *In fact,* $Q_{\mathbf{E}}$ *is a noncommutative ring without identity and without zero divisors.*

Example 11.1.6 *Consider the ring* $M_2(\mathbf{Z})$ *of Example 10.1.4. Let* $M_2(\mathbf{E})$ *denote the set of all* 2×2 *matrices with entries from* **E**. *Since the sum, difference, and product of even integers is an even integer, it follows that* $M_2(\mathbf{E})$ *is a subring of* $M_2(\mathbf{Z})$. *Also,* $M_2(\mathbf{E})$ *is a noncommutative ring without identity and with zero divisors.*

Following along the lines of Theorem 11.1.2, we can prove the next theorem. We leave its proof as an exercise.

Theorem 11.1.7 *Let* F *be a field. A nonempty subset* S *of* F *is a subfield of* F *if and only if*
(i) S contains more than one element,
(ii) $x - y, xy \in S$ for all $x, y \in S$, and
(iii) $x^{-1} \in S$ for all $x \in S$, $x \neq 0$. ∎

Example 11.1.8 $\mathbf{Q}$ *and* $\mathbf{Q}[\sqrt{2}] = \{a + b\sqrt{2} \mid a, b \in \mathbf{Q}\}$ *are subfields of* $\mathbf{R}$ *(see Worked-Out Exercise 4 below).*

Theorem 11.1.9 *Let* R *be a ring (field) and* $\{R_i \mid i \in \Lambda\}$ *be a nonempty family of subrings (subfields) of* R. *Then* $\cap_{i \in \Lambda} R_i$ *is a subring (subfield) of* R.

Proof.   Since $0 \in R_i$ for all $i \in \Lambda$, $0 \in \cap_{i \in \Lambda} R_i$ and so $\cap_{i \in \Lambda} R_i \neq \phi$. Let $x, y \in \cap_{i \in \Lambda} R_i$. Then $x, y \in R_i$ for all $i \in \Lambda$. Since each R_i is a subring, $x - y, xy \in R_i$ for all $i \in \Lambda$. Hence, $x - y, xy \in \cap_{i \in \Lambda} R_i$. Thus, $\cap_{i \in \Lambda} R_i$ is a subring of R.

Similarly, if each R_i is a subfield of the field R, then $\cap_{i \in \Lambda} R_i$ is a subfield of R. ∎

It is interesting to note that the intersection of all subfields of $\mathbf{R}$ is $\mathbf{Q}$.

11.1.1   Worked-Out Exercises

◇ **Exercise 1** Let X be an infinite set. Then $(\mathcal{P}(X), \Delta, \cap)$ is a ring with 1. Let

$$R = \{A \in \mathcal{P}(X) \mid A \text{ is finite}\}.$$

Prove the following assertions.

(i) R is a subring of $\mathcal{P}(X)$.

(ii) R is without identity.

(iii) For all $A \in R$, $A \neq \phi$, A is a zero divisor in R.

(iv) For all $A \in \mathcal{P}(X)$, $A \neq X$, $A \neq \phi$, A is a zero divisor in $\mathcal{P}(X)$.

Solution: (i) Since ϕ is finite, $\phi \in R$ and so R is nonempty. Let $A, B \in R$. Then A and B are finite and so $A \cap B$ is finite. Now $A \Delta B = (A \cup B) \backslash (A \cap B)$ and so $A \Delta B$ is finite. Therefore, $A \Delta B$, $A \cap B \in R$. Thus, R is closed under the operations Δ and $\cap$. Now it is easy to verify that $(R, \Delta, \cap)$ is a subring.

(ii) Suppose R has an identity, say, E. Then E is finite. Since X is infinite, there exists $a \in X$ such that $a \notin E$. Now $\{a\} \in R$. Thus, $\{a\} = E \cap \{a\} = \phi$, which is a contradiction. Hence, R has no identity.

(iii) Let $A \in R$ and $A \neq \phi$. Since A is finite and X is infinite, there exists $x \in X$ such that $x \notin A$. Now $\{x\} \in R$. Since $A \cap \{x\} = \phi$, A is a zero divisor.

(iv) Let $A \in \mathcal{P}(X)$ be such that $A \neq X$ and $A \neq \phi$. Then there exists $x \in X$ such that $x \notin A$. Hence, $A \cap \{x\} = \phi$ and so A is a zero divisor.

Exercise 2 Let R be a ring such that $a^2 + a$ is in the center of R for all $a \in R$. Show that R is commutative.

Solution: Let $x, y \in R$. Then $(x+y)^2 + x + y \in C(R)$, i.e., $x^2 + xy + yx + y^2 + x + y \in C(R)$. Since $x^2 + x$, $y^2 + y \in C(R)$ and $C(R)$ is a subring (Exercise 14, page 294), $xy + yx \in C(R)$. Therefore, $x(xy + yx) = (xy + yx)x$ and so $x^2y + xyx = xyx + yx^2$. Thus, $x^2y = yx^2$. Now $x^2 + x \in C(R)$ and so $y(x^2 + x) = (x^2 + x)y$. Hence, $yx^2 + yx = x^2y + xy$ and so $xy = yx$, proving that R is commutative.

$\diamond$ **Exercise 3** Find all subrings of the ring $\mathbf{Z}$ of integers. Find those subrings which do not contain the identity element.

Solution: Let n be a nonnegative integer and $T_n = n\mathbf{Z} = \{nt \mid t \in \mathbf{Z}\}$. Since $0 \in T_n$, $T_n \neq \phi$. Let $a = nt$, $b = ns$ be two elements in T_n. Then $a - b = nt - ns = n(t - s) \in T_n$ and $ab = (nt)(ns) = n(t(ns)) \in T_n$. Hence, T_n is a subring of $\mathbf{Z}$. We now show that if A is any subring of $\mathbf{Z}$, then $A = T_n$ for some nonnegative integer n.

Let A be a subring of $\mathbf{Z}$. If $A = \{0\}$, then $A = 0\mathbf{Z}$. Suppose $A \neq \{0\}$. Then there exists $m \in A$ such that $m \neq 0$. Now $-m \in A$ and so A contains a positive integer. By the well-ordering principle, A contains a smallest positive integer. Let n be the smallest positive integer in A. Then $n\mathbf{Z} \subseteq A$. Let $m \in A$. By the division algorithm, there exist integers q and r such that $m = nq + r$, $0 \leq r < n$. Since $n \in A$, $nq \in A$. Hence, $r = m - nq \in A$. The minimality of n implies that $r = 0$ and so $m = nq \in n\mathbf{Z}$. Thus, $A = n\mathbf{Z}$. If $n \neq 1$, then $n\mathbf{Z}$ does not contain identity.

$\diamond$ **Exercise 4** Show that $\mathbf{Q}[\sqrt{2}] = \{a + b\sqrt{2} \in \mathbf{R} \mid a, b \in \mathbf{Q}\}$ is a subfield of the field $\mathbf{R}$.

Solution: Since $0 = 0 + 0\sqrt{2} \in \mathbf{Q}[\sqrt{2}]$, $\mathbf{Q}[\sqrt{2}] \neq \phi$. Let $a + b\sqrt{2}$, $c + d\sqrt{2} \in \mathbf{Q}[\sqrt{2}]$. Then

$$(a + b\sqrt{2}) - (c + d\sqrt{2}) = (a - c) + (b - d)\sqrt{2} \in \mathbf{Q}[\sqrt{2}]$$

and

$$(a + b\sqrt{2})(c + d\sqrt{2}) = (ac + 2bd) + (ad + bc)\sqrt{2} \in \mathbf{Q}[\sqrt{2}].$$

Now $0 + 0\sqrt{2}$ and $1 + 0\sqrt{2}$ are distinct elements of $\mathbf{Q}[\sqrt{2}]$. Therefore, $\mathbf{Q}[\sqrt{2}]$ contains more than one element. Let $a + b\sqrt{2}$ be a nonzero element of $\mathbf{Q}[\sqrt{2}]$. Then a and b cannot both be zero simultaneously. We now show that $a - b\sqrt{2} \neq 0$. Suppose $a - b\sqrt{2} = 0$. Then $a = b\sqrt{2}$. If $b = 0$, then $a = 0$. Therefore, both a and b are zero, a contradiction. If $b \neq 0$, then $\sqrt{2} = \frac{a}{b} \in \mathbf{Q}$, a contradiction. Hence, $a - b\sqrt{2} \neq 0$. Similarly, $a + b\sqrt{2} \neq 0$. Thus, $a^2 - 2b^2 = (a + b\sqrt{2})(a - b\sqrt{2}) \neq 0$. Now

$$\frac{1}{a + b\sqrt{2}} = \frac{(a - b\sqrt{2})}{a^2 - 2b^2} = \frac{a}{a^2 - 2b^2} - \frac{b}{a^2 - 2b^2}\sqrt{2} \in \mathbf{Q}[\sqrt{2}].$$

Since $(a + b\sqrt{2})(\frac{1}{a+b\sqrt{2}}) = 1$, $(a + b\sqrt{2})^{-1}$ exists in $\mathbf{Q}[\sqrt{2}]$. Thus, we find that $\mathbf{Q}[\sqrt{2}]$ is a subfield of $\mathbf{R}$ by Theorem 11.1.7.

11.1.2 Exercises

1. Prove the following the statements.

   (i) $T_1 = \left\{ \begin{bmatrix} a & b \\ 0 & c \end{bmatrix} \mid a, b, c \in \mathbf{Z} \right\}$ is a subring of $M_2(\mathbf{Z})$.

   (ii) $T_2 = \left\{ \begin{bmatrix} a & b \\ -b & a \end{bmatrix} \mid a, b \in \mathbf{Z} \right\}$ is a subring of $M_2(\mathbf{Z})$.

   (iii) $T_3 = \left\{ \begin{bmatrix} a & 0 \\ 0 & a \end{bmatrix} \mid a \in \mathbf{Z} \right\}$ is a subring of $M_2(\mathbf{Z})$.

   (iv) $T_4 = \left\{ \begin{bmatrix} a & b \\ 0 & a \end{bmatrix} \mid a, b \in \mathbf{Z} \right\}$ is a subring of T_1.

2. In the ring $\mathbf{Z}$ of integers, find which of the following subsets of $\mathbf{Z}$ are subrings.

   (i) The set of integers of the form $4k + 2$, $k \in \mathbf{Z}$.

   (ii) The set of integers of the form $4k + 1$, $k \in \mathbf{Z}$.

   (iii) The set of integers of the form $4k$, $k \in \mathbf{Z}$.

3. Show that $T = \{[0], [5]\}$ is a subring of the ring $\mathbf{Z}_{10}$.

4. Let R be a ring with 1. Show that the subset $T = \{n1 \mid n \in \mathbf{Z}\}$ is a subring of R.

5. Let R be a ring and n be a positive integer. Show that the subset $T = \{a \in R \mid na = 0\}$ is a subring of R.

6. Show that $T = \left\{ \begin{bmatrix} a & b\sqrt{3} \\ -b\sqrt{3} & a \end{bmatrix} \mid a, b \in \mathbf{R} \right\}$ is a subring of $M_2(\mathbf{R})$.

7. Show that $\mathbf{Q}[\sqrt{3}]$ and $\mathbf{Q}[\sqrt{5}]$ are subfields of the field $\mathbf{R}$, but $\mathbf{Z}[\sqrt{2}] = \{a + b\sqrt{2} \mid a, b \in \mathbf{Z}\}$ is not a subfield of $\mathbf{R}$.

8. Show that $\mathbf{Q}(i) = \{a + bi \mid a, b \in \mathbf{Q}\}$ is a subfield of $\mathbf{C}$, where $i^2 = -1$.

9. Show that $F = \left\{ \begin{bmatrix} a & -b \\ b & a \end{bmatrix} \mid a, b \in \mathbf{Z}_5 \right\}$ is a subring of $M_2(\mathbf{Z}_5)$. Is F a field?

10. Let ω be a root of $x^2 + x + 1 = 0$. Prove that $T = \{a + b\omega \mid a, b \in \mathbf{Q}\}$ is a subfield of the field of complex numbers.

11. Let F be a field of characteristic $p > 0$. Show that $T = \{a \in F \mid a^p = a\}$ is a subfield of F.

12. Prove that $T = \left\{ \begin{bmatrix} x+y & y \\ -y & x \end{bmatrix} \mid x, y \in \mathbf{Z} \right\}$ is a subring of $M_2(\mathbf{Z})$. Also, show that every nonzero element of T is a unit in $M_2(\mathbf{R})$.

13. Let R be a commutative ring. Show that the set

$$T = \{r \in R \mid r^n = 0 \text{ for some integer } n\}$$

is a subring of R.

14. Prove that $C(R)$ is a subring of R and that $C(R)$ is commutative.

15. Let e be an idempotent of a ring R. Prove that the set

$$eRe = \{ere \mid r \in R\}$$

is a subring of R with e as the identity element.

16. Find the center of the ring $M_2(\mathbf{R})$.

17. Prove that the characteristic of a subfield is the same as the characteristic of the field.

18. Find all subrings with identity of the ring $\mathbf{Z}_{16}$.

19. Find all subfields of the field $\mathbf{Z}_p$, p a prime integer.

20. Let R be a ring without any nonzero nilpotent elements. Show that $(ara - ra)^2 = 0$ for all $r \in R$ and for all idempotent elements $a \in R$. Hence, show that $C(R)$ contains all idempotent elements.

21. Let $C = \{ f : \mathbf{R} \to \mathbf{R} \mid f \text{ is continuous on } \mathbf{R}\}$. Define $+$ and $\cdot$ on C by

$$\begin{aligned} (f+g)(x) &= f(x) + g(x), \\ (f \cdot g)(x) &= f(x)g(x) \end{aligned}$$

for all $f, g \in C$ and for all $x \in \mathbf{R}$.

(i) Show that C is a ring.

(ii) Let $D = \{f \in C \mid f \text{ is differentiable on } \mathbf{R}\}$. Show that D is a subring of C.

22. Let R be a ring and $f : R \to [0,1]$ be such that

$$\begin{aligned} f(a-b) &\geq \min\{f(a), f(b)\}, \\ f(ab) &\geq \min\{f(a), f(b)\} \end{aligned}$$

for all $a, b \in R$. Prove that for all $t \in \mathcal{I}(f)$, $R_t = \{x \in R \mid f(x) \geq t\}$ is a subring of R.

23. In the following exercises, write the proof if the statement is true; otherwise, give a counterexample.

   (i) The union of two subrings of a ring is a subring.

   (ii) The identity element of a subring is always the identity element of the ring.

   (iii) $\mathbf{Q}$ is the only subfield of the field $\mathbf{R}$.

   (iv) $\mathbf{Q}[\sqrt{3}] = \{a + b\sqrt{3} \mid a, b \in \mathbf{Q}\}$ is the intersection of all subfields of $\mathbf{R}$ which contain $\sqrt{3}$.

   (v) The set $\mathbf{Z}$ of integers is a subring of the field of real numbers.

   (vi) Every additive subgroup of $\mathbf{Z}$ is a subring of $\mathbf{Z}$.

11.2   Ideals and Quotient Rings

In this section, we introduce the notions of ideals and quotient rings. These concepts are analogous to normal subgroups and quotient groups.

The very famous problem called "Fermat's last theorem" led to the invention of ideals. Fermat (1601–1665) jotted many of his results in the margin of *Diophantus' Arithmetica*. For this particular "theorem," Fermat wrote that he discovered a remarkable theorem whose proof was too long to put in the margin. The theorem is stated as follows: If n is an integer greater than 2, then there exist no positive integers x, y, z such that $x^n + y^n = z^n$. However, no one was able to prove this result until recently; in 1994, Andrew Wiles found a proof after many years of work.

In 1843, Kummer (1810–1893) thought that he had found a proof of Fermat's last theorem. However, Kummer had incorrectly assumed uniqueness of the factorization of complex numbers of the form $x + \lambda y$, where $\lambda^p = 1$ for p an odd prime. Dirichlet (1805–1859) had made an incorrect assumption about factorization of numbers. Kummer continued his efforts to solve Fermat's last theorem. He was partially successful by introducing the concept of "ideal number." Dedekind (1831–1916) used Kummer's ideas to invent the notion of an ideal. Kronecker (1823–1891) also played an important part in the development of ring theory.

Definition 11.2.1 *Let R be a ring. A nonempty subset I of R is called a **left (right) ideal** of R if for all $a, b \in I$ and for all $r \in R$, $a - b \in I$, $ra \in I$ $(a - b \in I$, $ar \in I)$.*

A nonempty subset I of a ring R is called a **(two-sided) ideal** of R if I is both a left and a right ideal of R.

From the definition of a left (right) ideal, it follows that if I is a left (right) ideal of R, then I is a subring of R. Also, if R is a commutative ring, then every left ideal is also a right ideal and every right ideal is a left ideal. Thus, for commutative rings every left or right ideal is an ideal.

By Theorem 11.1.2, it is clear that a nonempty subset I of a ring R is an ideal if and only if $(I, +)$ is a subgroup of $(R, +)$ and for all $a \in I$ and for all $r \in R$, ar and $ra \in I$.

Example 11.2.2 *Let R be a ring. The subsets $\{0\}$ and R of R are (left, right) ideals. These ideals are called **trivial** ideals. All other (left, right) ideals are called **nontrivial**.*

An ideal I of a ring R is called a **proper** ideal if $I \neq R$.

Example 11.2.3 *Let $n \in \mathbf{Z}$ and $I = \{nk \mid k \in \mathbf{Z}\}$. As in Worked-Out Exercise 3 (page 292), I is a subring. Also, for all $r \in \mathbf{Z}$, $(nk)r = n(kr) \in I$ and $r(nk) = n(rk) \in I$. Hence, I is an ideal of $\mathbf{Z}$.*

Next, we give an example of a ring in which there exists a left ideal which is not a right ideal, a right ideal which is not a left ideal, and a subring which is not a left (right) ideal.

Example 11.2.4 *Consider the ring $M_2(\mathbf{Z})$. Let*

$$I_1 = \left\{ \begin{bmatrix} a & 0 \\ b & 0 \end{bmatrix} \mid a, b \in \mathbf{Z} \right\},$$

$$I_2 = \left\{ \begin{bmatrix} 0 & a \\ 0 & b \end{bmatrix} \mid a, b \in \mathbf{Z} \right\},$$

$$I_3 = \left\{ \begin{bmatrix} a & c \\ b & d \end{bmatrix} \mid a, b, c \text{ and } d \text{ are even integers} \right\},$$

and

$$I_4 = \left\{ \begin{bmatrix} a & 0 \\ 0 & 0 \end{bmatrix} \mid a \in \mathbf{Z} \right\}.$$

Since $\begin{bmatrix} 0 & 0 \\ 0 & 0 \end{bmatrix} \in I_1$, $I_1 \neq \phi$. *Let* $\begin{bmatrix} a & 0 \\ b & 0 \end{bmatrix}$, $\begin{bmatrix} c & 0 \\ d & 0 \end{bmatrix} \in I_1$ *and* $\begin{bmatrix} x & y \\ z & w \end{bmatrix} \in$ $M_2(\mathbf{Z})$. *Then*

$$\begin{bmatrix} a & 0 \\ b & 0 \end{bmatrix} - \begin{bmatrix} c & 0 \\ d & 0 \end{bmatrix} = \begin{bmatrix} a - c & 0 \\ b - d & 0 \end{bmatrix} \in I_1$$

and

$$\begin{bmatrix} x & y \\ z & w \end{bmatrix} \begin{bmatrix} a & 0 \\ b & 0 \end{bmatrix} = \begin{bmatrix} xa + yb & 0 \\ za + wb & 0 \end{bmatrix} \in I_1,$$

proving that I_1 is a left ideal of $M_2(\mathbf{Z})$. Now $\begin{bmatrix} 1 & 0 \\ 1 & 0 \end{bmatrix} \in I_1$ *and* $\begin{bmatrix} 0 & 1 \\ 0 & 0 \end{bmatrix} \in$

$M_2(\mathbf{Z})$, *but*

$$\begin{bmatrix} 1 & 0 \\ 1 & 0 \end{bmatrix} \begin{bmatrix} 0 & 1 \\ 0 & 0 \end{bmatrix} = \begin{bmatrix} 0 & 1 \\ 0 & 1 \end{bmatrix} \notin I_1.$$

Hence, I_1 is not a right ideal of $M_2(\mathbf{Z})$. Similarly, I_2 is a right ideal of $M_2(\mathbf{Z})$, but not a left ideal, I_3 is an ideal of $M_2(\mathbf{Z})$, and I_4 is a subring, but not an ideal of $M_2(\mathbf{Z})$.

We remind the reader to notice the similarity of the next few results with corresponding results in linear algebra and group theory.

Theorem 11.2.5 *Let R be a ring and $\{I_\alpha \mid \alpha \in \Lambda\}$ be a nonempty collection of left (right) ideals of R. Then $\cap_{\alpha \in \Lambda} I_\alpha$ is a left (right) ideal of R.*

Proof.   Suppose $\{I_\alpha \mid \alpha \in \Lambda\}$ is nonempty a collection of left ideals of R. Since $0 \in I_\alpha$ for all α, $0 \in \cap_\alpha I_\alpha$ and so $\cap_\alpha I_\alpha \neq \phi$. Let $a, b \in \cap_\alpha I_\alpha$. Then $a, b \in I_\alpha$ for all α. Since each I_α is a left ideal, $a - b \in I_\alpha$ for all α. Hence, $a - b \in \cap_\alpha I_\alpha$. Let $r \in R$. Since each I_α is a left ideal of R, $ra \in I_\alpha$ for all α and so $ra \in \cap_\alpha I_\alpha$. Thus, $\cap_\alpha I_\alpha$ is a left ideal of R. Similarly, if $\{I_\alpha \mid \alpha \in \Lambda\}$ is a nonempty collection of right ideals of R, then $\cap_\alpha I_\alpha$ is a right ideal of R. ∎

Let $a_1, a_2, \ldots, a_n \in R$. Then by the notation $\sum_{i=1}^n a_i$, we mean the sum $a_1 + a_2 + \cdots + a_n$.

Definition 11.2.6 *Let S be a nonempty subset of a ring R. Define $\langle S \rangle_l$ to be the intersection of all left ideals of R which contain S. Then the left ideal $\langle S \rangle_l$ is called the **left ideal generated** by S. Similarly, we can define $\langle S \rangle_r$, the **right ideal generated** by S, and $\langle S \rangle$, the **ideal generated** by S.*

Note that $\langle S \rangle_l$ is the smallest left ideal of R which contains S.

Theorem 11.2.7 *Let R be a ring and S be a nonempty subset of R. Then*
(i)

$$\langle S \rangle_l = \{\sum_{i=1}^k r_i s_i + \sum_{j=1}^l n_j s'_j \mid r_i \in R, n_j \in \mathbf{Z}, s_i, s_j \in S,$$
$$1 \leq i \leq k, 1 \leq j \leq l, k, l \in \mathbf{N}\}.$$

(ii)

$$\langle S \rangle_r = \{\sum_{i=1}^k s_i r_i + \sum_{j=1}^l n_j s'_j \mid r_i \in R, n_j \in \mathbf{Z}, s_i, s_j \in S,$$
$$1 \leq i \leq k, 1 \leq j \leq l, k, l \in \mathbf{N}\}.$$

Proof.   (i) Let

$$A = \{\sum_{i=1}^{k} r_i s_i + \sum_{j=1}^{l} n_j s'_j \mid r_i \in R, n_j \in \mathbf{Z}, s_i, s_j \in S,$$
$$1 \le i \le k, 1 \le j \le l, k, l \in \mathbf{N}\}.$$

Since $\langle S \rangle_l$ is the intersection of all left ideals of R which contain S, we have $\langle S \rangle_l \supseteq S$. Also, since $\langle S \rangle_l$ is closed under addition and closed from the left under multiplication by elements of R, we have $A \subseteq \langle S \rangle_l$. We now show that A is a left ideal of R such that $A \supseteq S$. Then $A \supseteq \langle S \rangle_l$ since $\langle S \rangle_l$ is the smallest left ideal of R containing S. Let $s \in S$. Then $s = 0 \cdot s + 1s \in A$ and so $S \subseteq A$. Let $\sum_{i=1}^{k} r_i s_i + \sum_{j=1}^{l} n_j s'_j$ and $\sum_{i=1}^{t} \overline{r_i}\, \overline{s_i} + \sum_{j=1}^{m} \overline{n_j s'_j} \in A$. Then $(\sum_{i=1}^{k} r_i s_i + \sum_{j=1}^{l} n_j s'_j) - (\sum_{i=1}^{t} \overline{r_i}\, \overline{s_i} + \sum_{j=1}^{m} \overline{n_j s'_j}) = (\sum_{i=1}^{k} r_i s_i + \sum_{i=1}^{t} (-\overline{r_i})\, \overline{s_i}) + (\sum_{j=1}^{l} n_j s'_j + \sum_{j=1}^{m} (-\overline{n_j})\overline{s'_j}) \in A$. Let $r \in R$. Then $r(\sum_{i=1}^{k} r_i s_i + \sum_{j=1}^{l} n_j s'_j) = \sum_{i=1}^{k} (r r_i) s_i + \sum_{j=1}^{l} (n_j r) s'_j \in A$. Hence, A is a left ideal of R.

(ii) The proof is similar to (i). ∎

Corollary 11.2.8 *Let R be a ring and S be a nonempty subset of R. If R is with 1, then*

(i)

$$\langle S \rangle_l = \{\sum_{i=1}^{k} r_i s_i \mid r_i \in R, s_i \in S, 1 \le i \le k, \ n \in \mathbf{N}\}.$$

(ii)

$$\langle S \rangle_r = \{\sum_{i=1}^{k} s_i r_i \mid r_i \in R, s_i \in S, 1 \le i \le k, \ n \in \mathbf{N}\}.$$

Proof.   (i) Clearly $\langle S \rangle_l \supseteq \{\sum_{i=1}^{k} r_i s_i \mid r_i \in R, \ s_i \in S\}$. Let $\sum_{i=1}^{t} r_i s_i + \sum_{j=1}^{l} n_j s'_j \in \langle S \rangle_l$. Since R has an identity 1, $n_j s'_j = (n_j 1) s'_j$ and $n_j 1 \in R$. Thus, $\sum_{i=1}^{t} r_i s_i + \sum_{j=1}^{l} n_j s'_j = \sum_{i=1}^{t} r_i s_i + \sum_{j=1}^{l} (n_j 1) s'_j \in \{\sum_{i=1}^{t} r_i s_i \mid r_i \in R, s_i \in S, 1 \le i \le k, \ n \in \mathbf{N}\}$. Hence, $\langle S \rangle_l \subseteq \{\sum_{i=1}^{k} r_i s_i \mid r_i \in R, s_i \in S, 1 \le i \le k, n \in \mathbf{N}\}$.

(ii) The proof is similar to (i). ∎

If $S = \{a_1, a_2, \ldots, a_n\}$, then the left ideal $\langle S \rangle_l$ generated by S is denoted by $\langle a_1, a_2, \ldots, a_n \rangle_l$. In this case, we call $\langle S \rangle_l$ a **finitely generated left ideal**. Similar terminology is used for $\langle S \rangle_r$ and $\langle S \rangle$. If $S = \{a\}$, then $\langle a \rangle_l$ is called the **principal left ideal** generated by a, $\langle a \rangle_r$ is called the **principal right ideal** generated by a, and $\langle a \rangle$ is called the **principal ideal** generated by a.

Corollary 11.2.9 *Let R be a ring and $a \in R$.*

(i) Then

$$\langle a \rangle_l = \{ra + na \mid r \in R, n \in \mathbf{Z}\}.$$

(ii) If R is with 1, then

$$\langle a \rangle_l = \{ra \mid r \in R\}.$$

Proof.   (i) This assertion follows from the equality

$$\sum_{i=1}^{k} r_i a + \sum_{j=1}^{m} n_j a = (\sum_{i=1}^{k} r_i) a + (\sum_{j=1}^{m} n_j) a.$$

(ii) This follows from (i) and Corollary 11.2.8. ∎

Similarly, we can prove that $\langle a \rangle_r = \{ar + na \mid r \in R, n \in \mathbf{Z}\}$ and $\langle a \rangle = \{ra + as + na + \sum_{i=1}^{k} r_i a s_i \mid r, s, r_i, s_i \in R, n \in \mathbf{Z}, 1 \le i \le k, k \in \mathbf{N}\}$.

Consider the subsets $Ra = \{ra \mid r \in R\}$ and $aR = \{ar \mid r \in R\}$ of R. If R is without identity, then Ra (aR) is still a left (right) ideal of R (Exercise 4, page 306). It is not necessarily the case that $a \in Ra$ $(a \in aR)$ as illustrated by the next example.

Example 11.2.10 *Consider the ring* $\mathbf{E}$ *of even integers.* $\mathbf{E}$ *does not have an identity.* $\langle 2 \rangle = \{r2 + n2 \mid r \in \mathbf{E}, n \in \mathbf{Z}\} = \{0, \pm 2, \pm 4, \ldots\}$ *and* $2 \in \langle 2 \rangle$. *However,* $\{r2 \mid r \in \mathbf{E}\} = \{0, \pm 4, \pm 8, \ldots\}$, *which does not contain* 2.

In the next theorem, we obtain a necessary and sufficient condition for a ring with 1 to be a division ring.

Theorem 11.2.11 *Let R be a ring with* 1. *Then R is a division ring if and only if R has no nontrivial left ideals.*

Proof.   Suppose R is a division ring. Let I be a left ideal of R such that $I \supset \{0\}$. Then there exists $a \in I$ such that $a \neq 0$ and since I is a left ideal, $1 = a^{-1}a \in I$. Hence, for all $r \in R$, $r = r1 \in I$, whence $R = I$.

Conversely, suppose R has no nontrivial left ideals. Let $a \in R$ and $a \neq 0$. Then $\langle a \rangle_l = R$ and so $1 \in \langle a \rangle_l$. Now $\langle a \rangle_l = \{ra \mid r \in R\}$, whence there exists $r \in R$ such that $1 = ra$. This implies that $r \neq 0$. Proceeding as in the case of the nonzero element a, we find that $tr = 1$ for some $t \in R$. Therefore, $t = t1 = t(ra) = (tr)a = 1a = a$. Thus, $ra = 1 = ar$ and so a is a unit. Consequently, every nonzero element of R is a unit. Hence, R is a division ring. ∎

Following along the lines of the above theorem, we can prove that a ring R with 1 is a division ring if and only if R has no nontrivial right ideals.

The following corollary is immediate from the above theorem.

Corollary 11.2.12 *Let R be a commutative ring with 1. Then R is a field if and only if R has no nontrivial ideals.* ∎

Definition 11.2.13 *A ring R is called a* **simple** *ring if $R^2 \neq \{0\}$ and $\{0\}$ and R are the only ideals of R.*

Example 11.2.14 *Every division ring is a simple ring.*

Example 11.2.15 *In this example, we show that $M_2(\mathbf{R})$ is a simple ring. Let A be a nonzero ideal of $M_2(\mathbf{R})$. Then there exists a nonzero element $\begin{bmatrix} a & b \\ c & d \end{bmatrix} \in A$. Now at least one of a, b, c, d is nonzero. Since A is an ideal and $\begin{bmatrix} 0 & 0 \\ 1 & 0 \end{bmatrix}, \begin{bmatrix} 0 & 1 \\ 0 & 0 \end{bmatrix} \in M_2(\mathbf{R})$, we have*

$$\begin{bmatrix} a & b \\ c & d \end{bmatrix} \begin{bmatrix} 0 & 0 \\ 1 & 0 \end{bmatrix} = \begin{bmatrix} b & 0 \\ d & 0 \end{bmatrix} \in A,$$

$$\begin{bmatrix} 0 & 1 \\ 0 & 0 \end{bmatrix} \begin{bmatrix} a & b \\ c & d \end{bmatrix} = \begin{bmatrix} c & d \\ 0 & 0 \end{bmatrix} \in A,$$

and

$$\begin{bmatrix} 0 & 1 \\ 0 & 0 \end{bmatrix} \begin{bmatrix} a & b \\ c & d \end{bmatrix} \begin{bmatrix} 0 & 0 \\ 1 & 0 \end{bmatrix} = \begin{bmatrix} d & 0 \\ 0 & 0 \end{bmatrix} \in A.$$

Therefore, we find that A contains a matrix $\begin{bmatrix} a & b \\ c & d \end{bmatrix}$ such that $a \neq 0$. Now $a^{-1} \in \mathbf{R}$ and

$$\begin{bmatrix} 1 & 0 \\ 0 & 0 \end{bmatrix} \begin{bmatrix} a & b \\ c & d \end{bmatrix} \begin{bmatrix} a^{-1} & 0 \\ 0 & 0 \end{bmatrix} = \begin{bmatrix} 1 & 0 \\ 0 & 0 \end{bmatrix} \begin{bmatrix} 1 & 0 \\ ca^{-1} & 0 \end{bmatrix} = \begin{bmatrix} 1 & 0 \\ 0 & 0 \end{bmatrix} \in A.$$

Thus,

$$\begin{bmatrix} 1 & 0 \\ 0 & 0 \end{bmatrix} \begin{bmatrix} 0 & 1 \\ 0 & 0 \end{bmatrix} = \begin{bmatrix} 0 & 1 \\ 0 & 0 \end{bmatrix} \in A.$$

Finally,

$$\begin{bmatrix} 0 & 0 \\ 1 & 0 \end{bmatrix} \begin{bmatrix} 0 & 1 \\ 0 & 0 \end{bmatrix} = \begin{bmatrix} 0 & 0 \\ 0 & 1 \end{bmatrix} \in A.$$

Hence,

$$\begin{bmatrix} 1 & 0 \\ 0 & 1 \end{bmatrix} = \begin{bmatrix} 1 & 0 \\ 0 & 0 \end{bmatrix} + \begin{bmatrix} 0 & 0 \\ 0 & 1 \end{bmatrix} \in A.$$

This implies that $A = M_2(\mathbf{R})$.

The above example shows that there are simple rings, which are not division rings.

For $a \in R$, aRa denotes the set $\{ara \mid r \in R\}$.

We now consider the sum and product of left (right) ideals.

Let A and B be two nonempty subsets of a ring R. Define the **sum** and **product** of A and B as follows:

$$A + B = \{a + b \mid a \in A,\ b \in B\}$$

$$AB = \{a_1 b_1 + a_2 b_2 + \cdots + a_n b_n \mid a_i \in A,\ b_i \in B,\ i = 1, 2, \ldots, n,\ n \in \mathbf{N}\}.$$

Thus, AB denotes the set of all finite sums of the form $\sum a_i b_i$, $a_i \in A$, $b_i \in B$.

Let $n \in \mathbf{N}$. Inductively, we define

$$\begin{aligned} A^1 &= A, \\ A^n &= AA^{n-1} \quad \text{if } n > 1. \end{aligned}$$

We now list some interesting properties of these two operations.

Theorem 11.2.16 *Let A, B, and C be left (right) ideals of a ring R. Then the following assertions hold.*

(i) $A + B = B + A$ is a left (right) ideal of R.

(ii) $A + A = A$.

(iii) $(A + B) + C = A + (B + C)$.

(iv) AB is a left (right) ideal of R.

(v) $(AB)C = A(BC)$.

(vi) If A, B and C are ideals, then $A(B + C) = AB + AC$, $(B + C)A = BA + CA$.

(vii) If A is a right ideal and B is a left ideal, then $AB \subseteq A \cap B$.

(viii) R is a regular ring if and only if for any right ideal A and for any left ideal B, $AB = A \cap B$.

(ix) The set $I(R)$ of all ideals of R forms a modular lattice with respect to set inclusion as a partial ordering.

Proof. We only prove (viii) and (ix) and leave the other properties as exercises.

(viii) Suppose R is a regular ring. Let $a \in A \cap B$. There exists $b \in R$ such that $a = aba$. Since B is a left ideal and $a \in B$, $ba \in B$. Thus, $a = a(ba) \in AB$, whence $A \cap B \subseteq AB$. By (vii), $AB \subseteq A \cap B$. Consequently, $AB = A \cap B$. Conversely, assume that $AB = A \cap B$ for any right ideal A and left ideal B of R. Let $a \in R$ and consider $\langle a \rangle_r$, the right ideal generated by a. Since $\langle a \rangle_r$ is a right ideal, $\langle a \rangle_r R \subseteq \langle a \rangle_r$. Also, by our assumption $\langle a \rangle_r \cap R = \langle a \rangle_r R$. Hence,

$$a \in \langle a \rangle_r \cap R = \langle a \rangle_r R.$$

Therefore, $a = \sum_{i=1}^{n} a_i b_i$ for some $a_i \in \langle a \rangle_r$, $b_i \in R$, $i = 1, 2, \ldots, n$. From the statements following Corollary 11.2.9, $a_i = at_i + n_i a$ for some $t_i \in R$, $n_i \in \mathbf{Z}$, $i = 1, 2, \ldots, n$. Thus, $a = \sum_{i=1}^{n} a_i b_i = \sum_{i=1}^{n}(at_i + n_i a)b_i = a(\sum_{i=1}^{n}(t_i b_i + n_i b_i)) \in aR$. This implies that $\langle a \rangle_r = aR$. Since $aR \subseteq \langle a \rangle_r$, $\langle a \rangle_r = aR$. Similarly, $\langle a \rangle_l = Ra$. It now follows that $a \in aR \cap Ra = (aR)(Ra) \subseteq aRa$. Hence, there exists $b \in R$ such that $a = aba$, i.e., a is regular. Consequently, R is regular.

(ix) By using arguments similar to the proof of Theorem 4.1.16, we can show that $(I(R), \subseteq)$ is a poset. To show $(I(R), \subseteq)$ is a lattice, let $A, B \in I(R)$. Now $A \cap B, A + B \in I(R)$. Also, $A, B \subseteq A + B$. Let $C \in I(R)$ be such that $A, B \subseteq C$. Since C is an ideal, $A + B \subseteq C$. Hence, $A + B = A \vee B$, the lub of $\{A, B\}$. Similarly, $A \cap B = A \wedge B$, the glb of $\{A, B\}$. Thus, $I(R)$ is a lattice. To show $(I(R), \subseteq)$ is a modular lattice, let A, B, C be three elements in $I(R)$ such that $A \subseteq C$. Note that $A \vee (B \wedge C) = A + (B \cap C)$ and $(A \vee B) \wedge C = (A + B) \cap C$. Now $A + (B \cap C) \subseteq (A + B) \cap C$ and so $A \vee (B \wedge C) \subseteq (A \vee B) \wedge C$. Let $x \in (A + B) \cap C$. Then $x \in C$ and $x \in A + B$. Thus, $x = a + b$ for some $a \in A \subseteq C$ and $b \in B$. This implies that $b = x - a \in C$ and so $b \in B \cap C$, which shows that $x \in A + (B \cap C)$. Hence, $(A + B) \cap C \subseteq A + (B \cap C)$, i.e., $(A \vee B) \wedge C \subseteq A \vee (B \wedge C)$. Thus, $A \vee (B \wedge C) = (A \vee B) \wedge C$. Consequently, $I(R)$ is a modular lattice. ■

We now give the analogue of quotient groups for rings. Let R be a ring and I an ideal of R. Then $(I, +)$ is a normal subgroup of $(R, +)$ since the latter group is commutative. Hence, if R/I denotes the set of all cosets $r + I = \{r + a \mid a \in I\}$ for all $r \in R$, then $(R/I, +)$ is a commutative group, where

$$(r + I) + (r' + I) = (r + r') + I$$

for all $r + I, r' + I \in R/I$. Now define multiplication on R/I by $(r + I) \cdot (r' + I) = rr' + I$ for all $r + I, r' + I \in R/I$. Then $(R/I, +, \cdot)$ forms a ring. We leave the details as an exercise.

Definition 11.2.17 *If R is a ring and I is an ideal of R, then the ring $(R/I, +, \cdot)$ is called the **quotient ring** of R by I.*

Theorem 11.2.18 *Let $n \in \mathbf{Z}$ be a fixed positive integer. Then the following conditions are equivalent.*
  (i) n is prime.
  (ii) $\mathbf{Z}/\langle n \rangle$ is an integral domain.
  (iii) $\mathbf{Z}/\langle n \rangle$ is a field.

Proof. (i) $\Rightarrow$ (ii): Let $a + \langle n \rangle, b + \langle n \rangle \in \mathbf{Z}/\langle n \rangle$. Suppose

$$(a + \langle n \rangle)(b + \langle n \rangle) = 0 + \langle n \rangle.$$

Then $ab + \langle n \rangle = 0 + \langle n \rangle$ and so $ab \in \langle n \rangle$. Thus, there exists $r \in \mathbf{Z}$ such that $ab = rn$. This implies that $n|ab$. Since n is prime, either $n|a$ or $n|b$, i.e., either $a \in \langle n \rangle$ or $b \in \langle n \rangle$ and hence either $a + \langle n \rangle = 0 + \langle n \rangle$ or $b + \langle n \rangle = 0 + \langle n \rangle$. This implies that $\mathbf{Z}/\langle n \rangle$ has no zero divisors, proving that $\mathbf{Z}/\langle n \rangle$ is an integral domain.

(ii)$\Rightarrow$(iii): Since $\mathbf{Z}/\langle n \rangle$ is a finite integral domain, the result follows from Theorem 10.1.23.

(iii)$\Rightarrow$(i): Suppose n is not prime. Then $n = n_1 n_2$ for some $1 < n_1 < n$ and $1 < n_2 < n$. Now $n_1 + \langle n \rangle$ and $n_2 + \langle n \rangle$ are nonzero elements of $\mathbf{Z}/\langle n \rangle$ and

$$(n_1 + \langle n \rangle)(n_2 + \langle n \rangle) = n_1 n_2 + \langle n \rangle = n + \langle n \rangle = 0 + \langle n \rangle.$$

Since $\mathbf{Z}/\langle n \rangle$ is a field, $\mathbf{Z}/\langle n \rangle$ has no zero divisors. Thus, either $n_1 + \langle n \rangle = 0 + \langle n \rangle$ or $n_2 + \langle n \rangle = 0 + \langle n \rangle$, a contradiction. Therefore, n is prime. ∎

Definition 11.2.19 *Let I be an ideal of a ring R.*

*(i) I is called a **nil** ideal if each element of I is a nilpotent element.*

*(ii) I is called a **nilpotent** ideal if $I^n = \{0\}$ for some positive integer n.*

Example 11.2.20 *In the ring $\mathbf{Z}_8$, the ideal $I = \{[0], [4]\}$ is a nil ideal and also a nilpotent ideal. $I^2 = \{\sum_{i=1}^{k}[a_i][b_i] \mid [a_i], [b_i] \in I, k \in \mathbf{N}\} = \{0\}$ since $16|a_i b_i$.*

From the definition, it follows that every nilpotent ideal is a nil ideal. The following example shows that the converse is not true. In this example, we construct a ring R from the rings $\mathbf{Z}_{p^n}$, $n = 1, 2, \ldots$, i.e., from the rings $\mathbf{Z}_p, \mathbf{Z}_{p^2}, \mathbf{Z}_{p^3}, \ldots$, where p is a fixed prime.

Example 11.2.21 *Let p be a fixed prime. Let R be the collection of all sequences $\{a_n\}$ such that $a_n \in \mathbf{Z}_{p^n}$ $(n \geq 1)$ and there exists a positive integer m (dependent on $\{a_n\}$) such that $a_n = [0]$ for all $n \geq m$. Define addition and multiplication on R by*

$$\begin{aligned} \{a_n\} + \{b_n\} &= \{a_n + b_n\}, \\ \{a_n\}\{b_n\} &= \{a_n b_n\} \end{aligned}$$

for all $\{a_n\}, \{b_n\} \in R$. We ask the reader to verify that R is a commutative ring under these two operations, where the zero element is the sequence $\{a_n\}$ such that $a_n = [0]$ for all n and the additive inverse of the sequence $\{a_n\}$ is the sequence $\{-a_n\}$. Now in $\mathbf{Z}_{p^n}$, $[p]$ is a nilpotent element since $[p]^n = [p^n] = [0]$. Thus, for any $[r] \in \mathbf{Z}_{p^n}$, $[p][r] = [pr]$ is a nilpotent element. Therefore, we find that each element of $[p]\mathbf{Z}_{p^n}$ is a nilpotent element.

Let

$$I = \{\{[p]a_1, [p]a_2, \ldots, [p]a_n, [0], [0], \ldots\} \in R \mid n \in \mathbf{N}, \ a_i \in \mathbf{Z}_{p^i}, i = 1, \ldots, n\}.$$

Then I is an ideal of R. Also, every element of I is nilpotent. Let us now show that I is not nilpotent. Suppose I is nilpotent. Then there exists a positive integer m such that $I^m = \{0\}$. Now the sequence $\{a_n\}$ such that $a_n = [p]$ for $n = 1, 2, \ldots, m + 1$ and $a_n = 0$ for all $n \geq m + 2$ is an element of I. Then $\{a_n\}^m = \{[0], [0], \ldots, [0], [p^m], [0], [0], \ldots\}$, where the $(m + 1)$th term of this sequence is $[p^m]$ and all other terms are 0. Since $[p^m]$ is not zero in $\mathbf{Z}_{p^{m+1}}$, we find that $\{a_n\}^m \neq 0$ and $\{a_n\}^m \in I^m = \{0\}$, a contradiction. This implies that I is not nilpotent.

Theorem 11.2.22 *Let R be a commutative ring with 1 and I denote the set of all nilpotent elements of R. Then*

    (i) I is a nil ideal of R,

    (ii) the quotient ring R/I has no nonzero nilpotent elements.

Proof. (i) Since $0 \in I$, $I \neq \phi$. Let $a, b \in I$. There exist positive integers m and n such that $a^n = 0$ and $b^m = 0$. Since R is commutative, we can write

$$(a - b)^{n+m} = a^{n+m} + \cdots + (-1)^r \binom{n + m}{r} a^{n+m-r} b^r + \cdots + (-1)^{n+m} b^{n+m}.$$

The general term of the above expression is $(-1)^r \binom{n+m}{r} a^{n+m-r} b^r$, where $0 \leq r \leq m + n$. If $r \leq m$, then $n + m - r \geq n$ and hence $a^{n+m-r} = a^n a^{m-r} = 0$. Again, if $r > m$, then $b^r = b^{m+(r-m)} = b^m b^{r-m} = 0$. Therefore, we find that $(-1)^r \binom{n+m}{r} a^{n+m-r} b^r = 0$, $r = 0, 1, 2, \ldots, n+m$. This implies that $(a-b)^{n+m} = 0$, i.e., $a-b$ is nilpotent and so $a-b \in I$. Let $r \in R$. Then $(ra)^n = r^n a^n = r^n 0 = 0$. Since R is commutative, $(ar)^n = (ra)^n = 0$. Thus, $ar, ra \in I$. Consequently, I is an ideal of R. Since every element of I is nilpotent, I is nil.

(ii) Let $a + I$ be a nilpotent element of R/I. Then $(a + I)^n = I$ for some positive integer n. But $a^n + I = (a + I)^n$. Thus, $a^n + I = I$, which implies that $a^n \in I$. Since every element of I is nilpotent, there exists a positive integer m such that $(a^n)^m = 0$, i.e., $a^{nm} = 0$, which shows that a is nilpotent and so $a \in I$. This implies $a+I = I$. Hence, R/I has no nonzero nilpotent elements. ∎

Theorem 11.2.23 *Let A and B be two nil ideals of a commutative ring R with 1. Then $A + B$ is a nil ideal.*

Proof. By Theorem 11.2.16, we know that $A + B$ is an ideal of R. Let I be the set of all nilpotent elements of R. Then $A \subseteq I$, $B \subseteq I$ and by Theorem 11.2.22, I is an ideal. Hence, $A + B \subseteq I$. Since I is nil, $A + B$ is nil. ∎

11.2.1 Worked-Out Exercises

◇ **Exercise 1** Find all ideals of **Z**.

Solution: From Worked-Out Exercise 3 (page 292), we know that the subrings of $\mathbf{Z}$ are the subsets $n\mathbf{Z}$, $n = 0, 1, 2, \ldots$. Let us now show that these subrings are precisely the ideals of $\mathbf{Z}$. If I is an ideal of $\mathbf{Z}$, then I is a subring of $\mathbf{Z}$ and so $I = n\mathbf{Z}$ for some nonnegative integer n. Now, let $I = n\mathbf{Z}$ (n is a nonnegative integer). Then I is a subring. If $r \in \mathbf{Z}$, then $rI = r(n\mathbf{Z}) = n(r\mathbf{Z}) \subseteq n\mathbf{Z} = I$. Similarly, $Ir \subseteq I$. Hence, I is an ideal of $\mathbf{Z}$.

Exercise 2 Let R be a ring such that R has no zero divisors. Show that if every subring of R is an ideal of R, then R is commutative.

Solution: Let $0 \neq a \in R$. Then $C(a) = \{x \in R \mid xa = ax\}$ is a subring of R and hence an ideal of R. Thus, $ra \in C(a)$ for all $r \in R$. Let $r \in R$. Now $ara = ra^2$ implies that $(ar - ra)a = 0$. Since R has no zero divisors and $a \neq 0$, $ar - ra = 0$ and so $ar = ra$. Hence, a is in the center of R. Since a is arbitrary, R is commutative.

$\diamondsuit$ **Exercise 3** Give an example of a ring R and ideals A_i, $i \in I$, such that $A_i \cap A_j = \{0\}$ if $i \neq j$, but $A_i \cap (\sum_{j \neq i} A_j) \neq \{0\}$.

Solution: Let $R = \{0, a, b, c\}$. Define $+$ and $\cdot$ on R by

$$2a = 2b = 2c = 0, \quad xy = 0, \text{ for all } x, y \in R \text{ and}$$
$$a + b = b + a = c, \; a + c = c + a = b, \text{ and } b + c = c + b = a.$$

Then $(R, +, \cdot)$ is a ring. Let $A_1 = \{0, a\}$, $A_2 = \{0, b\}$, and $A_3 = \{0, c\}$. Then $A_1 + A_2 = A_1 + A_3 = A_2 + A_3 = R$ and $A_1 \cap A_2 = A_1 \cap A_3 = A_2 \cap A_3 = \{0\}$.

$\diamondsuit$ **Exercise 4** Give an example of a ring R and ideals A and B such that $AB \subset A \cap B$.

Solution: Let R be the ring of Worked-Out Exercise 3. Let $A = B = \{0, a\}$. Then $AB = \{0\} \subset \{0, a\} = A \cap B$.

$\diamondsuit$ **Exercise 5** Characterize all commutative rings R such that R has only two ideals R and $\{0\}$.

Solution: Let R be a commutative ring such that the only ideals of R are R and $\{0\}$. Now R^2 is an ideal of R. Thus, $R^2 = \{0\}$ or $R^2 = R$.
   Case 1. $R^2 = \{0\}$. Then $ab = 0$ for all $a, b \in R$. In this case, every subgroup of $(R, +)$ is an ideal. Hence, $(R, +)$ has no proper subgroups and so $(R, +)$ is a cyclic group of prime order by Exercise 21 (page 138).
   Case 2. $R^2 = R$. Let $0 \neq a \in R$. Then aR is an ideal of R. Hence, either $aR = \{0\}$ or $aR = R$. Suppose $aR = \{0\}$. Let $T = \langle a \rangle$. Then T is an ideal of R and $a \in T$. Thus, $T = R$. Now $aR = \{0\}$ implies that $TR = \{0\}$ and

hence $R^2 = \{0\}$, which is a contradiction. Therefore, $aR = R$. Thus, for all $0 \neq a \in R$, $aR = R$. We now show that R has no zero divisors. Let a, b be two nonzero elements of R such that $ab = 0$. Let $T = \{c \in R \mid ac = 0\}$. It is easy to see that T is a nonzero ideal of R. Hence, by the hypothesis, $T = R$. This implies that $R = aR = aT = \{0\}$, a contradiction to the fact that $R = R^2 \neq \{0\}$. Consequently, R has no zero divisors. Next, for $0 \neq a \in R$, $aR = R$ and so we find that $ae = a$ for some $e \in R$. Since $a \neq 0$, we must have $e \neq 0$. Also, since R has no zero divisors, $a(e^2 - e) = 0$ implies that $e^2 = e$. Now for any $b \in R$, $eb = e^2 b$ implies that $e(b - eb) = 0$ and hence $b = eb = be$. This shows that e is the identity element of R. Also, $aR = R$ implies that $e = ab$ for some $b \in R$. Hence, a^{-1} exists in R. Consequently, R is a field.

So from the above two cases we conclude that either R is the zero ring with a prime number of elements or R is a field.

11.2.2   Exercises

1. Let $T_2(\mathbf{Z}) = \left\{ \begin{bmatrix} a & b \\ 0 & c \end{bmatrix} \mid a, b, c \in \mathbf{Z} \right\}$ be the ring of all upper triangular matrices over $\mathbf{Z}$.

   (i) Prove that $I = \left\{ \begin{bmatrix} 0 & b \\ 0 & c \end{bmatrix} \mid b, c \in \mathbf{Z} \right\}$ is an ideal of $T_2(\mathbf{Z})$. Find the quotient ring $T_2(\mathbf{Z})/I$.

   (ii) Prove that $I = \left\{ \begin{bmatrix} 0 & a \\ 0 & 0 \end{bmatrix} \mid a \in \mathbf{Z} \right\}$ is an ideal of $T_2(\mathbf{Z})$. Find the quotient ring $T_2(\mathbf{Z})/I$.

2. In the ring $\mathbf{Z}_{24}$, show that $I = \{[0], [8], [16]\}$ is an ideal. Find all elements of the quotient ring $\mathbf{Z}_{24}/I$.

3. Show that the set $I = \{a + bi\sqrt{5} \mid a, b \in \mathbf{Z}$ and $a - b$ is even$\}$ is an ideal of the ring $\mathbf{Z}[i\sqrt{5}]$.

4. Let R be a ring and $a \in R$. Show that aR is a right ideal of R and Ra is a left ideal of R.

5. Let R be a ring. Let A be a left ideal of R and B be a right ideal of R. Show that AB is an ideal of R and $BA \subseteq A \cap B$.

6. Let R be a ring such that $R^2 \neq \{0\}$. Prove that R is a division ring if and only if R has no nontrivial left ideals.

7. Let R be a ring with 1. Prove that R has no nontrivial left ideals if and only if R has no nontrivial right ideals.

8. Let I_1, I_2 be ideals of a ring R. Prove that $I_1 \cup I_2$ is an ideal of R if and only if either $I_1 \subseteq I_2$ or $I_2 \subseteq I_1$.

9. Let I and J be ideals of a ring R. Prove that $I + J$ is an ideal of R and that $I + J = \langle I \cup J \rangle$, the ideal of R generated by $I \cup J$.

10. Let I be an ideal of a commutative ring R and $a \in R$. Prove that $\langle I \cup \{a\} \rangle = \{i + ra + na \mid i \in I, \, r \in R, \, n \in \mathbf{Z}\}$.

11. Let m and n be positive integers in $\mathbf{Z}$. Prove that

    (i) $\langle m, n \rangle = \langle m \rangle + \langle n \rangle = \langle d \rangle$, where d is the greatest common divisor of m and n;

    (ii) $\langle m \rangle \cap \langle n \rangle = \langle q \rangle$, where q is the least common multiple of m and n.

12. Find all ideals of the Cartesian product $F_1 \times F_2$ of two fields F_1 and F_2.

13. Consider the Cartesian product ring $R_1 \times R_2$ of the rings R_1 and R_2.

    (i) If I_1 is an ideal of R_1 and I_2 is an ideal of R_2, prove that $I_1 \times I_2$ is an ideal of $R_1 \times R_2$.

    (ii) Suppose R_1 and R_2 are with 1 and I is an ideal of $R_1 \times R_2$. Does there exist ideals I_1 of R_1 and I_2 of R_2 such that $I = I_1 \times I_2$?

14. Let R be an ideal of a ring R. Prove that the quotient ring R/I is a commutative ring if and only if $ab - ba \in I$ for all $a, b \in R$.

15. Let $T = \{\frac{a}{b} \mid \frac{a}{b} \in \mathbf{Q}$, a and b are relatively prime and 5 does not divide $b\}$. Show that T is a ring under the usual addition and multiplication. Also, prove that $I = \{\frac{a}{b} \in T \mid 5$ divides $a\}$ is an ideal of T and the quotient ring T/I is a field.

16. Let I be an ideal of a ring R. Prove that if R is a commutative ring with identity, then R/I is a commutative ring with identity. If R has no zero divisors, is the same necessarily true for R/I?

17. Let I be an ideal of a commutative ring R. Define the **annihilator** of I to be the set

$$\operatorname{ann} I = \{r \in R \mid ra = 0 \text{ for all } a \in I\}.$$

    Prove that $\operatorname{ann} I$ is an ideal of R.

18. In the ring $\mathbf{Z}_{20}$, prove that $I = \{[n] \mid n$ is even$\}$ is an ideal. Find $\operatorname{ann} I$.

19. In the ring $\mathbf{Z}[i]$, show that $I = \{a + bi \mid a, b \in \mathbf{Z}$ and a, b are even$\}$ is an ideal. Find $\operatorname{ann} I$.

20. In a commutative regular ring R with 1, prove that every principal ideal I is generated by an idempotent and for every principal ideal I, there exists a principal ideal J such that $R = I + J$ and $I \cap J = \{0\}$.

21. Prove that every ideal of a regular ring is regular.

22. Prove that a ring R is regular if and only if every principal left ideal of R is generated by an idempotent.

23. Prove that in a commutative regular ring with 1 every finitely generated ideal is a principal ideal.

24. In a ring R, prove that $\{0\}$ is the only nilpotent ideal if and only if for all ideals A and B of R, $AB = \{0\}$ implies $A \cap B = \{0\}$.

25. Let R be a ring and $f : R \rightarrow [0, 1]$ be such that

$$
\begin{aligned}
f(a - b) &\geq \min\{f(a), f(b)\}, \\
f(rb) &\geq f(b)
\end{aligned}
$$

for all $a, b, r \in R$. Prove the following:

(i) $f(0) \geq f(a)$ for all $a \in R$;

(ii) $f(a) = f(-a)$ for all $a \in R$;

(iii) for all $t \in \mathcal{I}(f)$, $R_t = \{x \in R \mid f(x) \geq t\}$ is a left ideal of R;

(iv) $R_0 = \{a \in R \mid f(a) = f(0)\}$ is a left ideal of R.

26. Let R be a ring. A relation ρ on R is called a congruence relation on the ring R if ρ is an equivalence relation on R and for all $a, b, c \in R$, $a\rho b$ implies that $ac\rho bc$, $ca\rho cb$, and $(a + c)\rho(b + c)$. Let I be an ideal of R and ρ be the relation on R defined by $a\rho b$ if and only if $a - b \in I$. Show that ρ is a congruence relation on R.

27. In each of the following exercises, write the proof if the statement is true; otherwise, give a counterexample.

(i) If $\{I_i \mid i \in \mathbf{N}\}$ is a collection of ideals of R, then $\cup_{i \in \mathbf{N}} I_i$ is an ideal of R.

(ii) $\mathbf{Z}$ is a subring of $\mathbf{R}$, but not an ideal of $\mathbf{R}$.

(iii) If I is a nontrivial ideal of an integral domain R, then the quotient ring R/I is an integral domain.

11.3 Homomorphisms and Isomorphisms

In this section, we introduce the ideas of homomorphisms and isomorphisms of rings. These concepts are the analogs of homomorphisms and isomorphisms for groups.

Definition 11.3.1 *Let $(R, +, \cdot)$ and $(R', +', \cdot')$ be rings and f a function from R into R'. Then f is called a **homomorphism** of R into R' if*

$$f(a + b) = f(a) +' f(b),$$

$$f(a \cdot b) = f(a) \cdot' f(b)$$

for all $a, b \in R$.

A homomorphism f of a ring R into a ring R' is called
(i) a **monomorphism** if f is one-one,
(ii) an **epimorphism** if f is onto R', and
(iii) an **isomorphism** if f is one-one and maps R onto R'.

If f is an isomorphism of a ring R onto a ring R', then f^{-1} is an isomorphism of R' onto R.

An isomorphism of a ring R onto R is called an **automorphism**.

Definition 11.3.2 *Two rings R and R' are said to be **isomorphic** if there exists an isomorphism of R onto R'.*

We write $R \simeq R'$ when R and R' are isomorphic.

When speaking of two rings R and R', from now on we usually use the operations $+$ and $\cdot$ for both rings. Let $f : R \to R'$ be a homomorphism of rings. Since f preserves $+$, f is a also a homomorphism of the groups $(R, +)$ and $(R', +)$. Hence, we can immediately apply Theorem 5.1.2 to conclude that f maps 0 to $0'$, i.e., $f(0) = 0'$, and for all $a \in R$, $-f(a) = f(-a)$. We list some properties of homomorphisms in the following theorem. The proofs are similar to the proof of Theorem 5.1.2 and so we leave them as an exercise for the reader.

Theorem 11.3.3 *Let f be a homomorphism of a ring R into a ring R'. Then the following assertions hold.*
  (i) $f(0) = 0'$, where $0'$ is the zero of R'.
  (ii) $f(-a) = -f(a)$ for all $a \in R$.
  (iii) $f(R) = \{f(a) \mid a \in R\}$ is a subring of R'.
  (iv) If R is commutative, then $f(R)$ is commutative.
  Suppose R has an identity and $f(R) = R'$. Then

(v) R' has an identity, namely, $f(1)$.

(vi) If $a \in R$ is a unit, then $f(a)$ is a unit in R' and

$$f(a)^{-1} = f(a^{-1}). \blacksquare$$

We point out that in (v) of Theorem 11.3.3, if f is not onto, then R' may or may not have an identity. Even if R' has an identity, the identity of R need not map onto the identity of R'. We illustrate this point later in Example 11.3.7.

Definition 11.3.4 *Let f be a homomorphism of a ring R into a ring R'. Then the **kernel** of f, written Ker f, is defined to be the set*

$$Ker\ f = \{a \in R \mid f(a) = 0'\}.$$

From Theorem 11.3.3, we know that $0 \in$ Ker f.

Example 11.3.5 *The identity map of a ring R is a homomorphism (in fact, an isomorphism). Its kernel is $\{0\}$. Let R and R' be rings and $f : R \to R'$ be defined by $f(a) = 0'$ for all $a \in R$. Then f is a homomorphism of R into R' and Ker $f = R$.*

Example 11.3.6 *Let f be the mapping from $\mathbf{Z}$ onto $\mathbf{Z}_n$ defined by $f(a) = [a]$ for all $a \in \mathbf{Z}$. From Example 5.1.4, $f(a+b) = f(a) +_n f(b)$ for all $a, b \in \mathbf{Z}$. Also, $f(a \cdot b) = [ab] = [a] \cdot_n [b] = f(a) \cdot_n f(b)$ for all $a, b \in \mathbf{Z}$. Thus, f is a homomorphism of $\mathbf{Z}$ onto $\mathbf{Z}_n$. As in Example 5.1.4, Ker $f = \{qn \mid q \in \mathbf{Z}\}$.*

In the following example, we show that if f is a homomorphism from a ring R with 1 into a ring R' with 1 and f is not onto, then the identity of R need not map onto the identity of R'.

Example 11.3.7 *Consider the direct sum $\mathbf{Z} \oplus \mathbf{Z}$ of $\mathbf{Z}$ with itself (see Exercise 17, page 283). Define $f : \mathbf{Z} \to \mathbf{Z} \oplus \mathbf{Z}$ by $f(a) = (a, 0)$ for all $a \in \mathbf{Z}$. From the definition of f, f is well defined. Now for all $a, b \in \mathbf{Z}$, $f(a+b) = (a+b, 0) = (a, 0) + (b, 0) = f(a) + f(b)$ and $f(ab) = (ab, 0) = (a, 0)(b, 0) = f(a)f(b)$. Thus, f is a homomorphism. Also, Ker $f = \{0\}$. Now $f(1) = (1, 0)$, but $(1, 1)$ is the identity of $\mathbf{Z} \oplus \mathbf{Z}$. Therefore, the identity of $\mathbf{Z}$ does not map onto the identity of $\mathbf{Z} \oplus \mathbf{Z}$.*

Consider the rings $\mathbf{Z}$ and $\mathbf{Q}$. Suppose $\mathbf{Z} \simeq \mathbf{Q}$. Then the groups $(\mathbf{Z}, +)$ and $(\mathbf{Q}, +)$ are isomorphic. However, this is not possible since $(\mathbf{Z}, +)$ is a cyclic group and $(\mathbf{Q}, +)$ is not a cyclic group. In the following example, we give another argument to show that $\mathbf{Z}$ is not isomorphic to $\mathbf{Q}$.

Example 11.3.8 *Suppose* $\mathbf{Z} \simeq \mathbf{Q}$. *Let* $f : \mathbf{Z} \to \mathbf{Q}$ *be an isomorphism. Then* $f(1) = 1$ *and* $f(0) = 0$. *Let* n *be a positive integer. Then* $f(n) = f(\underbrace{1 + \cdots + 1}_{n \ times})$
$= f(1) + f(1) + \cdots + f(1) = nf(1) = n1 = n$. *Now suppose that* n *is a negative integer. Let* $n = -m$, *where* m *is positive. Then* $f(n) = f(-m) = f(-1 - 1 - \cdots - 1) = -f(1) - f(1) - \cdots - f(1) = m(-f(1)) = -mf(1) = -m1 = -m = n$. *Hence,* $f(n) = n$ *for all* $n \in \mathbf{Z}$. *Let* $0 \neq \frac{a}{b} \in \mathbf{Q} \backslash \mathbf{Z}$. *Since* f *is onto* $\mathbf{Q}$, *there exists* $n \in \mathbf{Z}$ *such that* $\frac{a}{b} = f(n) = n$, *which is a contradiction. Hence,* $\mathbf{Q}$ *is not isomorphic to* $\mathbf{Z}$.

In the following example, we consider two rings which look similar, but which are not isomorphic.

Example 11.3.9 *In this example, we show that the ring* $\mathbf{Z}[\sqrt{3}] = \{a + b\sqrt{3} \mid a, b \in \mathbf{Z}\}$ *and the ring* $\mathbf{Z}[\sqrt{5}] = \{a + b\sqrt{5} \mid a, b \in \mathbf{Z}\}$ *are not isomorphic. Suppose there exists an isomorphism* $f : \mathbf{Z}[\sqrt{3}] \to \mathbf{Z}[\sqrt{5}]$. *Now* $3 = (0 + \sqrt{3})^2$. *Thus,* $f(3) = f((\sqrt{3})^2) = (f(\sqrt{3}))^2$. *Since* f *is an isomorphism, we have* $f(1) = 1$. *This implies that* $f(3) = 3$. *Hence,* $3 = (f(\sqrt{3}))^2$. *Since* $f(\sqrt{3}) \in \mathbf{Z}[\sqrt{5}]$, $f(\sqrt{3}) = a + b\sqrt{5}$ *for some* $a + b\sqrt{5} \in \mathbf{Z}[\sqrt{5}]$. *Therefore,* $3 = (a + b\sqrt{5})^2$ *and so* $3 = a^2 + 5b^2 + 2ab\sqrt{5}$. *If* $ab = 0$, *then* $3 = a^2 + 5b^2$. *But there do not exist integers* a *and* b *such that* $ab = 0$ *and* $3 = a^2 + 5b^2$. *If* $ab \neq 0$, *then* $\sqrt{5} = \frac{3 - a^2 - 5b^2}{2ab} \in \mathbf{Q}$, *which is a contradiction. Hence,* $\mathbf{Z}[\sqrt{3}]$ *and* $\mathbf{Z}[\sqrt{5}]$ *are not isomorphic.*

The next example shows that the ring $\mathbf{Z}_n$ and the ring $\mathbf{Z}/\langle n \rangle$ are isomorphic.

Example 11.3.10 *Consider the ideal* $\langle n \rangle$ *generated by a fixed positive integer* $n \in \mathbf{Z}$. *By Corollary 11.2.9,* $\langle n \rangle = \{qn \mid q \in \mathbf{Z}\}$. *The cosets of* $\langle n \rangle$ *in* $\mathbf{Z}$ *are* $a + \langle n \rangle = \{a + qn \mid q \in \mathbf{Z}\}$. *Now*

$$\mathbf{Z}/\langle n \rangle = \{a + \langle n \rangle \mid a \in \mathbf{Z}\}.$$

Define $f : \mathbf{Z}_n \to \mathbf{Z}/\langle n \rangle$ *by* $f([a]) = a + \langle n \rangle$ *for all* $[a] \in \mathbf{Z}_n$. *We recall that* f *is an isomorphism of* $(\mathbf{Z}_n, +_n)$ *onto* $(\mathbf{Z}/\langle n \rangle, +)$ *(Example 5.1.15). Now* $f([a] \cdot_n [b]) = f([ab]) = ab + \langle n \rangle = (a + \langle n \rangle)(b + \langle n \rangle) = f([a])f([b])$. *Thus,* f *is a ring isomorphism of* $\mathbf{Z}_n$ *onto* $\mathbf{Z}/\langle n \rangle$.

Theorem 11.3.11 *Let* f *be a homomorphism of a ring* R *into a ring* R'. *Then* $Ker \ f$ *is an ideal of* R.

Proof. Since $0 \in Ker \ f$, $Ker \ f \neq \phi$. Let $a, b \in Ker \ f$. Then $f(a - b) = f(a) - f(b) = 0' - 0' = 0'$ and so $a - b \in Ker \ f$. Let $r \in R$. Then $f(ra) = f(r) \cdot f(a) = f(r) \cdot 0' = 0'$ and so $ra \in R$. Similarly, $ar \in Ker \ f$. Hence, $Ker \ f$

is an ideal of R. ■

In the remainder of the section, we consider isomorphism theorems which are parallel to those for groups (Section 5.2).

Theorem 11.3.12 *Let R be a ring and I be an ideal of R. Define the mapping $g : R \to R/I$ by $g(a) = a + I$ for all $a \in R$. Then g is a homomorphism, called the **natural homomorphism**, of R onto R/I. Furthermore, Ker $g = I$.*

Proof. Now for all $a, b \in R$, $g(a + b) = (a + b) + I = (a + I) + (b + I) = g(a) + g(b)$ and $g(ab) = ab + I = (a + I)(b + I) = g(a)g(b)$. That Ker $g = I$ follows from Theorem 5.1.12 in group theory. ■

Theorem 11.3.13 *Let f be a homomorphism of a ring R onto a ring R' and I be an ideal of R contained in Ker f. Let g be the natural homomorphism of R onto R/I. Then there exists a unique homomorphism h of R/I onto R' such that $f = h \circ g$. Furthermore, h is one-one if and only if $I = $ Ker f.*

Proof. Once again, we use the work already done for groups. Define $h : R/I \to R'$ by $h(a + I) = f(a)$ for all $a \in R$. We have the desired results by Theorem 5.2.1, once we verify that h preserves multiplication. Now $h((a + I)(b + I)) = h(ab + I) = f(ab) = f(a)f(b) = h(a + I)h(b + I)$. ■

The proof of the following theorem is similar to that of the first isomorphism theorem for groups. We omit the proof. This theorem is also known as **the fundamental theorem of homomorphisms** for rings.

Theorem 11.3.14 (First Isomorphism Theorem) *Let f be a homomorphism of a ring R into a ring R'. Then $f(R)$ is an ideal of R' and*

$$R/\text{Ker } f \simeq f(R). ■$$

We state the following theorem without proof. Its proof is a direct translation of the proof of the corresponding theorem for groups.

Theorem 11.3.15 (Correspondence Theorem) *Let f be a homomorphism of a ring R onto a ring R'. Then f induces a one-one inclusion preserving correspondence between the ideals of R containing Ker f and the ideals of R' in such a way that if I is an ideal of R containing Ker f, then $f(I)$ is the corresponding ideal of R', and if I' is an ideal of R', then $f^{-1}(I')$ is the corresponding ideal of R. ■*

An example similar to Example 5.2.13 can be developed to illustrate Theorem 11.3.15

The next two isomorphism theorems for rings correspond to Theorems 5.2.8 and 5.2.6, respectively.

Theorem 11.3.16 *Let f be a homomorphism of a ring R onto a ring R', I be an ideal of R such that $I \supseteq \operatorname{Ker} f$, g, and g' be the natural homomorphisms of R onto R/I and R' onto $R'/f(I)$, respectively. Then there exists a unique isomorphism h of R/I onto $R'/f(I)$ such that $g' \circ f = h \circ g$.* ∎

Corollary 11.3.17 *Let I_1, I_2 be ideals of a ring R such that $I_1 \subseteq I_2$. Then*

$$(R/I_1)/(I_2/I_1) \simeq R/I_2.$$ ∎

Theorem 11.3.18 *If I and J are ideals of the ring R, then $I/(I \cap J) \simeq (I + J)/J$.* ∎

11.3.1 Worked-Out Exercises

◇ **Exercise 1** Show that the function $f : \mathbf{Z}_6 \to \mathbf{Z}_{10}$ defined by $f([a]) = 5[a]$ for all $[a] \in \mathbf{Z}_6$ is a ring homomorphism of $\mathbf{Z}_6$ into $\mathbf{Z}_{10}$.

Solution: We first show that f is well defined. Let $[a] = [b]$ in $\mathbf{Z}_6$. Then $a - b$ is divisible by 6. Thus, $a = 6k + b$ for some $k \in \mathbf{Z}$. Now $5a = 30k + 5b$ shows that $5[a] = [5a] = [30k + 5b] = [30k] +_{10} [5b] = [0] +_{10} 5[b] = 5[b]$ in $\mathbf{Z}_{10}$. Therefore, $f([a]) = f([b])$. Thus, we find that f is well defined. Let $[a], [b] \in \mathbf{Z}_6$. Then $f([a] +_6 [b]) = f([a+b]) = 5[a+b] = 5([a] +_{10} [b]) = 5[a] +_{10} 5[b] = f(a) +_{10} f(b)$ and $f([a] \cdot_6 [b]) = f([ab]) = 5[ab] = 25[ab]$ (since $\mathbf{Z}_{10}$ is of characteristic 10) $= (5[a]) \cdot_{10} (5[b]) = f(a) \cdot_{10} f(b)$. Hence, f is a homomorphism.

◇ **Exercise 2** Let $\mathbf{R}$ be the field of real numbers. Let α be an automorphism of $\mathbf{R}$. Show that $\alpha(x) = x$ for all $x \in \mathbf{R}$.

Solution: Since α is an automorphism of $\mathbf{R}$, $\alpha(0) = 0$, and $\alpha(1) = 1$. Let $n \in \mathbf{N}$. Then $\alpha(n) = \alpha(1 + 1 + \cdots + 1) = \alpha(1) + \alpha(1) + \cdots + \alpha(1) = 1 + 1 + \cdots + 1 = n$. Now let $m \in \mathbf{Z}$ and $m < 0$. Let $n = -m > 0$. Then $\alpha(m) = \alpha(-n) = -\alpha(n) = -n = m$. This shows that $\alpha(x) = x$ for all $x \in \mathbf{Z}$. Let $\frac{p}{q} \in \mathbf{Q}$. Then $\alpha(\frac{p}{q}) = \alpha(pq^{-1}) = \alpha(p)\alpha(q^{-1}) = p\alpha(q)^{-1} = pq^{-1} = \frac{p}{q}$. This shows that $\alpha(x) = x$ for all $x \in \mathbf{Q}$. Let $x \in \mathbf{R}$ be such that $x \geq 0$. Then $x = y^2$ for some $y \in \mathbf{R}$. Thus, $\alpha(x) = \alpha(y^2) = \alpha(yy) = \alpha(y)\alpha(y) = \alpha(y)^2 \geq 0$. Now let $a, b \in \mathbf{R}$ be such that $a \geq b$. Then $a - b \geq 0$. Hence, $\alpha(a - b) \geq 0$ and so $\alpha(a) - \alpha(b) \geq 0$, i.e., $\alpha(a) \geq \alpha(b)$. Therefore, α is order preserving. We now show that α is continuous. Let $\epsilon \in \mathbf{R}$ and $\epsilon > 0$. Since α is onto $\mathbf{R}$, there exists $\delta > 0$ such that $\alpha(\delta) = \epsilon$. Now let $x, y \in \mathbf{R}$ be such that $|x - y| < \delta$. Thus,

$$-\delta < x - y < \delta.$$

Since α is order preserving,

$$\alpha(-\delta) < \alpha(x - y) < \alpha(\delta).$$

Therefore,

$$-\epsilon < \alpha(x - y) < \epsilon$$

and so

$$-\epsilon < \alpha(x) - \alpha(y) < \epsilon.$$

This implies that

$$|\alpha(x) - \alpha(y)| < \epsilon.$$

Hence, α is continuous. Now let $x \in \mathbf{R}$. Since $\mathbf{Q}$ is dense in $\mathbf{R}$, there exists a sequence $\{a_n\}$ of rational numbers such that

$$\lim_{n \to \infty} a_n = x.$$

Since α is continuous,

$$\alpha(x) = \alpha(\lim_{n \to \infty} a_n) = \lim_{n \to \infty} \alpha(a_n) = \lim_{n \to \infty} a_n = x,$$

proving the result.

◇ **Exercise 3** Let R be a ring with 1. If the characteristic of R is 0, show that R contains a subring isomorphic to $\mathbf{Z}$.

Solution: Let $T = \{n1 \mid n \in \mathbf{Z}\}$. Since $0 = 01 \in T$, $T \neq \phi$. Let $a = n1$ and $b = m1$ be two elements of T. Then $a - b = n1 - m1 = (n - m)1$ and $ab = (n1)(m1) = (nm)1$. Hence, $a - b, ab \in T$. Thus, T is a subring of R. Suppose n, m are two integers such that $n1 = m1$. If $n > m$, then $(n - m)1 = 0$. This contradicts the assumption that R is of characteristic 0. Similarly, $m > n$ also leads to a contradiction. Hence, $n = m$. Thus, we find that for each $a \in T$, there exists a unique integer n such that $a = n1$. Hence, the mapping $f : \mathbf{Z} \to T$ defined by $f(n) = n1$ is an isomorphism.

Exercise 4 Let p be a prime integer. Show that there are only two nonisomorphic rings of p elements.

Solution: It is known that $(\mathbf{Z}_p, +_p)$ is the only group of order p (up to isomorphism). Define $\odot_1$ and $\odot_2$ on $\mathbf{Z}_p$ by $[a] \odot_1 [b] = [0]$ and $[a] \odot_2 [b] = [ab]$ for all $[a], [b] \in \mathbf{Z}_p$. Now $\odot_1$ and $\odot_2$ are well defined and $(\mathbf{Z}_p, +_p, \odot_1)$ and $(\mathbf{Z}_p, +_p, \odot_2)$ are rings. Let R be a ring with p elements. Then $(R, +) \simeq (\mathbf{Z}_p, +_p)$. If $R \not\simeq (\mathbf{Z}_p, +_p, \odot_1)$, then the multiplication of R is not $\odot_1$. Let $[a]$ be a generator of $(\mathbf{Z}_p, +_p)$. Now $[a]^2 = n[a]$ for some nonzero integer n. There exists an integer m such that $mn \equiv_p 1$. Let $[b] = m[a]$. Then $[b]^2 = m^2[a]^2 =$

$m^2 n[a] = m[a] = [b]$. Let g be an isomorphism from $(\mathbf{Z}_p, +_p)$ onto $(R, +)$. Define $f : \mathbf{Z}_p \to R$ by $f([u]) = ug([b])$ for all $[u] \in \mathbf{Z}_p$. Then $f([u] +_p [v]) = f([u+v]) = (u+v)g([b]) = ug([b]) + vg([b]) = f([u]) + f([v])$ and $f([u] \odot_2 [v]) = f([uv]) = (uv)g([b]) = uvg([b]^2) = uvg([b])g([b]) = ug([b])vg([b]) = f([u])f([v])$. Hence, f is a ring homomorphism. Let $c \in R$. Then there exists $[u] \in \mathbf{Z}_p$ such that $g([u]) = c$. Now $[u] = t[a]$ for some $t \in \mathbf{Z}$. Thus, $f([tn]) = tng([b]) = tn \, g(m[a]) = tg(mn[a]) = tg([a]) = g(t[a]) = g([u]) = c$. Hence, f is onto R. Since $|\mathbf{Z}_p| = |R|$, it follows that f is one-one. Thus, f is an isomorphism.

11.3.2 Exercises

1. Let R denote the set of all 2×2 matrices of the form $\begin{bmatrix} a & b \\ -b & a \end{bmatrix}$, where a and b are real numbers. Prove that R is a ring and the function $a + bi \to \begin{bmatrix} a & b \\ -b & a \end{bmatrix}$ is an isomorphism of $\mathbf{C}$ onto R.

2. Define the binary operations $\oplus$ and $\odot$ on $\mathbf{Z}$ by $a \oplus b = a + b - 1$ and $a \odot b = a + b - ab$ for all $a, b \in \mathbf{Z}$. Show that $(\mathbf{Z}, \oplus, \odot)$ is a ring isomorphic to the ring $(\mathbf{Z}, +, \cdot)$.

3. (i) Show that the rings $\mathbf{R}$ and $\mathbf{Q}$ are not isomorphic.

   (ii) Show that the rings $\mathbf{R}$ and $\mathbf{C}$ are not isomorphic.

   (iii) Are the rings $\mathbf{Z}_6$ and $\mathbf{Z}_3 \times \mathbf{Z}_2$ isomorphic?

4. Let $T_2(\mathbf{Z}) = \left\{ \begin{pmatrix} a & b \\ 0 & c \end{pmatrix} \mid a, b, c \in \mathbf{Z} \right\}$ be the ring of all upper triangular matrices over $\mathbf{Z}$. Define $f : T_2(\mathbf{Z}) \to \mathbf{Z}$ by for all $\begin{pmatrix} a & b \\ 0 & c \end{pmatrix} \in T_2(\mathbf{Z})$,

$$f\left(\begin{pmatrix} a & b \\ 0 & c \end{pmatrix} \right) = a.$$

   (i) Show that f is a homomorphism.

   (ii) Is f an epimorphism?

   (iii) Is f an isomorphism?

   (iv) Find Ker f.

5. Does there exist an epimorphism from the ring $\mathbf{Z}_{24}$ onto the ring $\mathbf{Z}_7$?

6. Show that there does not exist a monomorphism from the ring $\mathbf{Z}_6$ into the ring $\mathbf{Z}_{11}$.

7. Show that the ring $2\mathbf{Z}$ is not isomorphic to the ring $3\mathbf{Z}$.

8. Let R be a Boolean ring. If $\{0\}$ and R are the only ideals of R, prove that $R \simeq \mathbf{Z}_2$.

9. Show that the ring $\mathbf{Z}$ is not isomorphic to any proper subring of $\mathbf{Z}$.

10. Is the ring $\mathbf{Q}[\sqrt{2}]$ isomorphic to the ring $\mathbf{Q}[\sqrt{3}]$?

11. Let $f : R \to S$ be a nontrivial homomorphism from a field R onto a ring S. Prove that S is a field.

12. Let R be a ring with 1. If R is of characteristic $n > 0$, show that R contains a subring isomorphic to the ring $\mathbf{Z}_n$.

13. Show that there exist only two homomorphisms from $\mathbf{R}$ into $\mathbf{R}$.

14. Prove that every ring R is isomorphic to a subring of $M_n(R)$, the ring of $n \times n$ matrices over R.

15. Let f be a homomorphism of a ring R onto a ring R'. Prove that

    (i) if I is an ideal of R, then $f(I)$ is an ideal of R';

    (ii) if I' is an ideal of R', then $f^{-1}(I')$ is an ideal of R and $f^{-1}(I') \supseteq \mathrm{Ker}\, f$;

    (iii) if R is commutative and I and J are two ideals of R, then $f(I+J) = f(I) + f(J)$ and $f(IJ) = f(I)f(J)$.

16. In each of the following exercises, write the proof if the statement is true; otherwise, give a counterexample.

    (i) There exist only two homomorphisms from the ring of integers into itself.

    (ii) The mapping $f : \mathbf{Z} \to \mathbf{Z}$ defined by $f(n) = 3n$ is a group homomorphism, but not a ring homomorphism.

    (iii) The only isomorphism of a ring R onto itself is the identity mapping of R.

    (iv) Let R be a ring with 1. Let $f : R \to S$ be a ring homomorphism. Then $f(1)$ is the identity element of S.

    (v) A nonzero homomorphism from a field into a ring with more than one element is a monomorphism.

    (vi) Every nontrivial homomorphic image of an integral domain is an integral domain.

Richard Dedekind (1831–1916) was born on October 6, 1831, in Brunswick, Germany, the birthplace of Gauss. He was the youngest of four children.

In 1848, Dedekind went to Collegium Carolinum, an institution attended by Gauss, where he became a master in analytic geometry, algebraic analysis, differential and integral calculus, and higher mechanics. In 1849–1850, he gave private lessons in mathematics. He matriculated, in 1850, at the University of Göttingen.

After four semesters, in 1852 Dedekind completed his Ph.D. work under Gauss. His thesis was on the elements of the theory of Eulerian integrals. Later he determined that his knowledge in some areas of mathematics was lacking for advanced study at Göttingen. He then spent the next two years, following his graduation, filling the gaps in his education.

Dedekind started his teaching career in 1854. In 1855, Dirichlet succeeded Gauss in Göttingen. Dedekind attended his lectures on various areas of mathematics, including the theory of numbers, and became a close friend of Dirichlet. In 1855–1856, he also attended Riemann's lectures on Abelian and elliptic functions. Thus, along with being an instructor, he was also a student.

Dedekind was the first university teacher to lecture on Galois theory. He introduced the concept of a field, replaced the concept of a permutation group by the abstract group concept, and, in 1858, introduced a purely arithmetic definition of continuity.

Dedekind is most remembered for his concept of "Dedekind cut," which he introduced in 1872. He was criticized on this theory by mathematicians such as Kronecker, Weiestrass, and Russell.

Dedekind edited the works of Gauss, Dirichlet, and Riemann. In 1871, he supplemented Dirichlet's lectures, introducing the notion of an "ideal," a term he coined. Later he developed the theory of ideals. He is also credited for such fundamental concepts as ring and unit. His treatises on number fields stimulated further development of ideal theory. Dedekind also extended Kummer's work on unique factorization domains. His work on abstract algebra influenced Emmy Noether's work on algebra.

Dedekind died on February 12, 1916.

Chapter 12

Ring Embeddings

12.1 Embedding of Rings

Sometimes it is worthwhile to study the properties of a ring by considering it as a subring of some ring with more ring properties than itself. A ring without identity lacks important arithmetic properties, in particular, a fundamental theorem of arithmetic. As another example, in the ring $\mathbf{E}$ of even integers, we cannot say that 2 divides 2 since $1 \notin \mathbf{E}$. Now $\mathbf{E}$ is a subring of $\mathbf{Z}$ and $1 \in \mathbf{Z}$. In $\mathbf{Z}$, it is true that 2 divides 2. The main aim of this section is to embed a ring into a suitable ring with additional properties. The main feature of this section is that any integral domain can be embedded in a field. The proof of this result yields a rigorous construction of the rational numbers from the integers.

Definition 12.1.1 *A ring R is said to be **embedded** in a ring S if there exists a monomorphism of R into S.*

From the above definition, it follows that a ring R can be embedded in a ring S if there exists a subring T of S such that $R \simeq T$.

In the next theorem, we show that any ring R can be embedded in a ring with identity.

Theorem 12.1.2 *Any ring R can be embedded in a ring S with 1 such that R is an ideal of S. If R is commutative, then S is commutative.*

Proof.   Set $S = R \times \mathbf{Z}$. Define addition and multiplication as follows:

$$
\begin{aligned}
(a, m) + (b, n) &= (a + b, m + n), \\
(a, m) \cdot (b, n) &= (ab + na + mb, mn)
\end{aligned}
$$

for all $a, b \in R$ and $m, n \in \mathbf{Z}$. (Here na means a adds to itself n times if n is positive, $-a$ adds to itself $|n|$ times if n is negative, and $0a = 0$.) Then S forms a ring under these definitions of addition and multiplication, a fact we ask the reader to prove in the exercises. We do note that $(0,0)$ is the additive identity and that $(0,1)$ is the multiplicative identity of S.

Consider the subset $R \times \{0\}$ of S. Since $(0,0) \in R \times \{0\}$, $R \times \{0\} \neq \phi$. Also, for all $(a,0),(b,0) \in R \times \{0\}$, $(a,0) - (b,0) = (a-b,0) \in R \times \{0\}$, and $(a,0) \cdot (b,0) = (ab,0) \in R \times \{0\}$. Thus, $R \times \{0\}$ is a subring of S. Now for all $(a,0) \in R \times \{0\}$ and $(c,n) \in S$, $(a,0) \cdot (c,n) = (ac+na,0) \in R \times \{0\}$ and $(c,n) \cdot (a,0) = (ca+na,0) \in R \times \{0\}$. This proves that $R \times \{0\}$ is an ideal of S.

Now define $f : R \to R \times \{0\}$ by $f(a) = (a,0)$ for all $a \in R$. Then f is an isomorphism of R onto $R \times \{0\}$ and so $R \simeq R \times \{0\}$. Therefore, R can be embedded in S. By identifying $a \in R$ with $(a,0) \in R \times \{0\}$, we can regard R to be an ideal of S. To show that the commutativity of R implies that of S, let $(a,m),(b,n) \in S$ and R be commutative. Then $(a,m) \cdot (b,n) = (ab+na+mb,mn) = (ba+mb+na,nm)$ (since R is commutative, $ab = ba$) $= (b,n) \cdot (a,m)$. Thus, S is commutative. ∎

Our main objective in this section is to embed a ring in a field. By Theorem 12.1.2, every ring can be embedded in a ring with identity. If S were a field, then S is commutative and has no zero divisors. This in turn implies that R is commutative and has no zero divisors. Thus, if we were to embed a ring R in a field S, then R must have at least these two properties, i.e., R must be commutative and have no zero divisors. In the next theorem, we embed a commutative ring with no zero divisors into an integral domain and then we will embed an integral domain in a field.

Theorem 12.1.3 *Let R be a commutative ring with no zero divisors. Then R can be embedded in an integral domain.*

Proof. Let S be the ring as defined in Theorem 12.1.2. Let A be the annihilator of R in S. Then A is an ideal of S by Exercise 17 (page 307). If $R \cap A = \{0\}$, then the natural homomorphism of R onto the quotient ring S/A must map R one-one into S/A, i.e., R can be embedded in S/A. We now show that $R \cap A = \{0\}$ and that S/A is an integral domain. Let $a \in R \cap A$. Then $ar = 0$ for all $r \in R$. Since R has no zero divisors, $a = 0$. Therefore, $R \cap A = \{0\}$. Let $b + A$, $c + A \in S/A$. If $(b+A)(c+A) = 0+A$, then $bc \in A$. Thus, $(bc)r = 0$ for all $r \in R$. Suppose $c + A \neq 0 + A$, i.e., $c \notin A$. Then there exists $r \in R$ such that $cr \neq 0$. Since R is an ideal of S, $cr \in R$, and for all $s \in R$, $bs \in R$. Now $(cr)(bs) = (bcr)s = 0s = 0$. Also, R has no zero divisors and $cr \neq 0$. Therefore, we must have $bs = 0$. This implies that $b \in A$ and so $b + A = 0 + A$. Hence, S/A is an integral domain. ∎

Suppose we are given the ring of integers $\mathbf{Z}$ and we are asked to construct the rational numbers from $\mathbf{Z}$. We can think of any integer as $n/1$, i.e., n divided by 1. However, we must somehow pick up the fractions which cannot be reduced to having a 1 for a denominator. One idea that suggests itself is to consider the Cartesian product $\mathbf{Z} \times \mathbf{Z}$ and consider the first component of the elements of $\mathbf{Z} \times \mathbf{Z}$ as the numerator and the second component as the denominator. However, the ordered pairs $(3,2)$ and $(6,4)$ are distinct. A common technique used in mathematics suggests putting these elements in the same equivalence class so that they become "equal." This is precisely what we shall do. Let's also remember not to have 0 in the denominator.

Theorem 12.1.4 *Any integral domain R can be embedded in a field.*

Proof. Let $S = R \times (R\backslash\{0\})$. Define the relation $\sim$ on S by for all $(a,b), (c,d) \in S$, $(a,b) \sim (c,d)$ if and only if $ad = bc$. Then $\sim$ is an equivalence relation. The reflexive and symmetric properties are immediate. Suppose that $(a,b) \sim (c, d)$ and $(c,d) \sim (e, f)$. Then $ad = bc$ and $cf = de$. This implies that $adf = bcf$ and $bcf = bde$ and so $adf = bde$. Canceling d, we obtain $af = be$, i.e., $(a,b) \sim (e, f)$. Hence, $\sim$ is transitive. Now $\sim$ partitions S into equivalence classes. Denote the equivalence class $\{(c,d) \in S \mid (c,d) \sim (a,b)\}$ by a/b. Set

$$F = \{a/b \mid (a,b) \in S\}.$$

Define $+$ and $\cdot$ on F as follows:

$$\begin{aligned} a/b + c/d &= (ad + bc)/bd, \\ a/b \cdot c/d &= ac/bd \end{aligned}$$

for all $a/b, c/d \in F$. We show that $+$ is well defined. Let $a/b, c/d, a'/b', c'/d' \in F$. Suppose $a/b = a'/b'$ and $c/d = c'/d'$. Then $ab' = ba'$ and $cd' = dc'$. Therefore, $ab'dd' = ba'dd'$ and $cd'bb' = dc'bb'$. Hence,

$$ab'dd' + cd'bb' = ba'dd' + dc'bb',$$

and so

$$(ad + bc)b'd' = bd(a'd' + b'c').$$

Thus,

$$(ad + bc, bd) \sim (a'd' + b'c', b'd')$$

and so

$$(ad + bc)/bd = (a'd' + b'c')/b'd'.$$

A similar proof shows that $\cdot$ is well defined.

The reader is asked to verify the associative, commutative, and distributive laws for F. The additive identity of F is $0/b$ and the multiplicative identity of F is b/b, where $b \neq 0$. For $a/b \in F$, the additive inverse is

$$(-a)/b = a/(-b)$$

and the multiplicative inverse is b/a (when $a \neq 0$). Thus, F is a field.

We now show that R can be embedded in F. Let

$$R' = \{a/1 \mid a \in R\} \subseteq F.$$

Then R' is a subring of F. Define $f : R \to R'$ by $f(a) = a/1$ for all $a \in R$. Then $a = b$ if and only if $a \cdot 1 = 1 \cdot b$ if and only if $a/1 = b/1$ if and only if $f(a) = f(b)$. Hence, f is a one-one function. Now

$$f(a+b) = (a+b)/1 = (a \cdot 1 + 1 \cdot b)/1 \cdot 1 = a/1 + b/1 = f(a) + f(b)$$

and

$$f(ab) = ab/1 = ab/1 \cdot 1 = a/1 \cdot b/1 = f(a) \cdot f(b).$$

From the definition of f, f is onto R'. Thus, f is an isomorphism of R onto $R' \subseteq F.$ ∎

The above theorem gives another instance of the power of the concept of an equivalence relation. We have once again used the notion of an ordered pair in a fundamental manner.

Definition 12.1.5 *Let R be an integral domain. A field F is called a **quotient field** of R or a **field of quotients** of R if there exists a subring R_1 of F such that*

(i) $R \simeq R_1$ and
(ii) for all $x \in F$, there exists $a, b \in R_1$ with $b \neq 0$ such that $x = ab^{-1}$.

Let us now show that for the given integral domain R, the field constructed in Theorem 12.1.4 is a quotient field of R. Let $x \in F$. Then $x = a/b$, where $(a, b) \in S$. Now $(a, 1) \in S$ and $(b, 1) \in S$. Thus, $a/1$, $b/1 \in R'$ and $a/b = a/1 \cdot 1/b = (a/1) \cdot (b/1)^{-1}$. Hence, F is a quotient field of R. We call F the **quotient field** or the **field of quotients** or R.

Theorem 12.1.6 *Let R be an integral domain and F its field of quotients. Let R' be an integral domain contained in a field K' and set*

$$F' = \{a'(b')^{-1} \mid a', b' \in R', b' \neq 0\}.$$

Then F' is the smallest subfield of K' which contains R' and any isomorphism of R onto R' has a unique extension to an isomorphism of F onto F'.

Proof.   By Exercise 2 (page 323), F' is the smallest subfield of K' which contains R'. Let f be an isomorphism of R onto R'. Let $a/b \in F$. If $f(a) = a'$ and $f(b) = b'$, define $g : F \to F'$ by

$$g(a/b) = a'(b')^{-1} = f(a)f(b)^{-1}.$$

Identifying the ring R with the set $\{a/1 \mid a \in R\}$, it is clear that $f = g|_R$. Now $a/b = c/d$ if and only if $ad = bc$ if and only if $f(ad) = f(bc)$ if and only if $f(a)f(d) = f(b)f(c)$ if and only if $f(a)f(b)^{-1} = f(c)f(d)^{-1}$ if and only if $g(a/b) = g(c/d)$. Therefore, g is a one-one function. From the definition of g, it follows that g is onto F'. Now

$$
\begin{aligned}
g(a/b + c/d) &= g((ad + bc)/bd) \\
&= f(ad + bc)(f(bd))^{-1} \\
&= [f(a)f(d) + f(b)f(c)][f(b)^{-1}f(d)^{-1}] \\
&= f(a)f(b)^{-1} + f(c)f(d)^{-1} \\
&= g(a/b) + g(c/d)
\end{aligned}
$$

and

$$
\begin{aligned}
g(a/b \cdot c/d) &= g(ac/bd) \\
&= f(ac)(f(bd))^{-1} \\
&= [f(a)f(c)][f(b)^{-1}f(d)^{-1}] \\
&= f(a)f(b)^{-1}f(c)f(d)^{-1} \\
&= g(a/b)g(c/d)
\end{aligned}
$$

for all $a/b, c/d \in F$. Thus, g is an isomorphism of F onto F'.

Let g' be any other isomorphism of F onto F' such that $f = g'|_R$. Then

$$
\begin{aligned}
g'(a/b) &= g'(a/1 \cdot (b/1)^{-1}) \\
&= g'(a/1)g'((b/1)^{-1}) \\
&= g'(a/1)g'(b/1)^{-1} \\
&= f(a)f(b)^{-1} \\
&= g(a/b)
\end{aligned}
$$

for all $a/b \in F$ and so $g' = g$. Thus, there is a unique extension of f. ∎

We can conclude from this result that the field of quotients F of an integral domain R is "the" smallest field containing R in the sense that there does not exist a field K such that $R \subset K \subset F$.

The field F' in Theorem 12.1.6 is called the **quotient field** of R' in K. In view of Theorem 12.1.6 and the comments preceding it, we do not differentiate between the notation a/b and ab^{-1} for the elements of F.

12.1.1   Worked-Out Exercises

$\Diamond$ **Exercise 1** Let $D = \{\frac{a}{b} \in \mathbf{Q} \mid 5 \text{ does not divide } b\}$. Show that D is a subring of $\mathbf{Q}$ with 1. Find the quotient field of D.

Solution: Let $a/b, c/d \in D$. Since 5 does not divide b and 5 does not divide d, 5 does not divide bd. Thus, $(ad - bc)/bd \in D$ and $ac/bd \in D$. Hence, D is a subring of $\mathbf{Q}$. Also, $1 = 1/1 \in D$. Since $\mathbf{Z} \subseteq D \subseteq \mathbf{Q}$ and $\mathbf{Q}$ is the quotient field of $\mathbf{Z}$, $\mathbf{Q}$ is the quotient field of D.

Exercise 2 Let S be a ring and f a one-one function of S onto a set T. Show that suitable addition and multiplication can be defined on T so that T becomes a ring isomorphic to S under f.

Solution: Define binary operations $+$ and $\cdot$ on T as follows: Let $t_1, t_2 \in T$. Since f maps S onto T, there exist $s_1, s_2 \in S$ such that $f(s_1) = t_1$ and $f(s_2) = t_2$. Define

$$\begin{aligned} t_1 + t_2 &= f(s_1 + s_2) \text{ and} \\ t_1 \cdot t_2 &= f(s_1 s_2). \end{aligned}$$

First we show that both these binary operations are well defined. Let $t_1, t_2, t_3, t_4 \in T$ be such that $t_1 = t_3$ and $t_2 = t_4$. Since f maps S onto T, there exist $s_1, s_2, s_3, s_4 \in S$ such that $f(s_1) = t_1$, $f(s_2) = t_2$, $f(s_3) = t_3$, and $f(s_4) = t_4$. Therefore, $f(s_1) = f(s_3)$ and $f(s_2) = f(s_4)$. Since f is one-one, $s_1 = s_3$ and $s_2 = s_4$. Hence, $t_1 + t_2 = f(s_1 + s_2) = f(s_3 + s_4) = t_3 + t_4$ and $t_1 \cdot t_2 = f(s_1 s_2) = f(s_3 s_4) = t_3 \cdot t_4$. Thus, $+$ and $\cdot$ are well defined. It is now a routine verification to show that $(T, +, \cdot)$ is a ring. We verify some of the properties and leave others as an exercise. First we show that $+$ is associative. Now $t_2 + t_3 = f(s_2 + s_3)$ and $t_1 + t_2 = f(s_1 + s_2)$. Thus, $t_1 + (t_2 + t_3) = f(s_1 + (s_2 + s_3)) = f((s_1 + s_2) + s_3)$ (since $+$ is associative for S) $= (t_1 + t_2) + t_3$. Hence, $+$ is associative for T. Also, $f(0) + t_1 = f(0 + s_1) = f(s_1) = f(s_1 + 0) = t_1 + f(0)$. This implies that $f(0)$ is the additive identity. Similarly, we can verify the other properties of a ring. It is immediate that f is a homomorphism and since f is one-one and f maps S onto T, S is isomorphic to T.

12.1.2   Exercises

1. Prove the associative, commutative, and distributive laws in Theorem 12.1.4.

2. Let R be an integral domain, which is a subring of a field F. Let $F' = \{ab^{-1} \mid a, b \in R, b \neq 0\}$. Show that F' is a subfield of F. Furthermore, show that F' is the smallest subfield of F which contains R.

3. Let R and R' be integral domains contained in fields. Set $F = \{ab^{-1} \mid a, b \in R, b \neq 0\}$ and $F' = \{a'b'^{-1} \mid a', b' \in R', b' \neq 0'\}$. Suppose f is an isomorphism of R onto R'. Prove that f has a unique extension to an isomorphism of F onto F'.

4. Prove that any field R is equal to its field of quotients F in the sense that $f(R) = F$, where f is the isomorphism defined in Theorem 12.1.4.

5. Prove that isomorphic integral domains have isomorphic fields of quotients.

6. Find the field of quotients of the integral domains $\mathbf{Z}[i]$ and $\mathbf{Z}[\sqrt{2}]$.

7. Let R be a ring of characteristic $n > 0$ and

$$R \times \mathbf{Z}_n = \{(r, [m]) \mid r \in R \text{ and } [m] \in \mathbf{Z}_n\}.$$

Define $+$ and $\cdot$ on $R \times \mathbf{Z}_n$ by

$$\begin{aligned} (a, [m]) + (b, [t]) &= (a+b, [m+t]), \\ (a, [m]) \cdot (b, [t]) &= (ab, [mt]) \end{aligned}$$

for all $a, b \in R$, $[m], [t] \in \mathbf{Z}_n$. Prove that

(i) the above two operations are well defined,

(ii) $(R \times \mathbf{Z}_n, +, \cdot)$ is a ring with 1,

(iii) $(R \times \mathbf{Z}_n, +, \cdot)$ is of characteristic n,

(iv) there exists a monomorphism from R into $(R \times \mathbf{Z}_n, +, \cdot)$.

8. Let S and R' be disjoint rings with the property that S contains a subring S' such that there is an isomorphism f' of S' onto R'. Prove that there is a ring R containing R' and an isomorphism f of S onto R such that $f' = f|_{S'}$.

David Hilbert (1862–1943) was born on January 23, 1862, in Königsberg, Germany. Hilbert's inclination toward mathematics is believed to be due to his mother. He attended the University of Königsberg from 1880 to 1884, and received his Ph.D. in 1885.

Heinrich Weber, Richard Dedekind's collaborator on the theory of algebraic functions, was a professor at the University of Königsberg while Hilbert was a student. In 1883, after Weber left, Lindeman was appointed as his successor. Lindeman's influence caused Hilbert to become interested in the theory of invariants.

Hilbert proved the famous Hilbert basis theorem—that is, if every ideal in a ring R has a finite basis, then so does every ideal in the polynomial ring $R[x]$. Hilbert's results connected the theory of invariants to the fields of algebraic functions and algebraic varieties. He also proved the Hilbert irreducibility theorem.

Hilbert also worked on algebraic number theory. This work centers on the reciprocity law, developed from Gauss's law of quadratic residues.

In 1893, Hilbert, along with Minkowski, was assigned to prepare a report on number theory. Minkowsky soon withdrew from this project. Hilbert summarized the known results in *Zahlbericht.* For half a century, it was a bible for anyone interested in learning algebraic number theory. In 1899, Hilbert published *Grundlagen der geometrie*, which went into its ninth edition in 1962. After 63 years, the book was still being read, although it was slowly modernized.

In 1900, while addressing the International Congress of Mathematicians on mathematical problems, Hilbert introduced 23 problems. These have since stimulated mathematical investigations.

Dirichlet's principle, which was used in boundary value problems, had been discredited by Weierstrass's criticism. Hilbert salvaged Dirichlet's principle by proving it in 1904.

Hilbert worked on algebraic forms, algebraic number theory, foundations of geometry, analysis, and theoretical physics. Many of his students became famous mathematicians, including Herman Weyl. Hilbert died on February 14, 1943.

Chapter 13

Direct Sum of Rings

In this chapter, we construct some new rings from a given family $\{R_i \mid i \in I\}$ of rings. For this purpose, we introduce the complete direct sum, the direct sum, and the subdirect sum of this family. The results developed in this chapter also help us to obtain structure results of rings.

13.1 Complete Direct Sum and Direct Sum

Let $\{R_i \mid i \in I\}$ be a family of rings indexed by a nonempty set I. The Cartesian product $\Pi\{R_i \mid i \in I\}$ of the sets R_i is the set of all functions $f : I \longrightarrow \cup\{R_i \mid i \in I\}$ such that $f(i) \in R_i$ for all $i \in I$. Let $f, g \in \Pi\{R_i \mid i \in I\}$. Define $f + g$, fg by

$$
\begin{aligned}
(f+g)(i) &= f(i) + g(i) \\
(fg)(i) &= f(i)g(i)
\end{aligned}
$$

for all $i \in I$. Then $f + g$, $fg \in \Pi\{R_i \mid i \in I\}$. It can be easily verified that $\Pi\{R_i \mid i \in I\}$ together with the above two operations is a ring. This ring is called the **complete direct sum** of the family of rings $\{R_i \mid i \in I\}$ and is denoted by $\Pi_{i \in I} R_i$. The zero element of $\Pi_{i \in I} R_i$ is the function $0 : I \longrightarrow \cup\{R_i \mid i \in I\}$ defined by $0(i) = 0_i$, the zero element of R_i, for all $i \in I$. The additive inverse of $f \in \Pi_{i \in I} R_i$ is the function $-f : I \longrightarrow \cup\{R_i \mid i \in I\}$ defined by $(-f)(i) = -f(i) \in R_i$ for all $i \in I$. Let $f \in \Pi_{i \in I} R_i$ and let $f(i) = a_i \in R_i$ for all $i \in I$. Usually f is identified with the image set $\{a_i \mid i \in I\}$. Using this notation, the above two operations can be defined by

$$
\begin{aligned}
\{a_i \mid i \in I\} + \{b_i \mid i \in I\} &= \{a_i + b_i \mid i \in I\} \\
\{a_i \mid i \in I\} \cdot \{b_i \mid i \in I\} &= \{a_i b_i \mid i \in I\}
\end{aligned}
$$

for all $a_i, b_i \in R_i$ for all $i \in I$.

Suppose now that I is a finite set, say, $I = \{1, 2, \ldots, n\}$. In this case, the complete direct sum is denoted by $\oplus_{i \in I} R_i = R_1 \oplus R_2 \oplus \cdots \oplus R_n$ and an element $\{a_i \mid i \in I\}$ is usually written as an n-tuple $(a_1, a_2, \ldots, a_n)$.

Definition 13.1.1 *The **direct sum** of a family of rings $\{R_i \mid i \in I\}$, denoted by $\oplus_{i \in I} R_i$, is the set*

$$\oplus_{i \in I} R_i \;=\; \{\{a_i \mid i \in I\} \in \Pi_{i \in I} R_i \mid a_i \neq 0 \text{ for at most finitely many } i \in I\}.$$

Theorem 13.1.2 *Let $\{R_i \mid i \in I\}$ be a family of rings. Then*
    (i) $\oplus_{i \in I} R_i$ is a subring of the complete direct sum of rings $\Pi_{i \in I} R_i$;
    (ii) for all $k \in I$, the function $i_k : R_k \to \oplus_{i \in I} R_i$ defined by

$$i_k(a) = \{\{a_i \mid i \in I\} \mid a_i = 0 \text{ for all } i \neq k \text{ and } a_k = a\}$$

for all $a \in R_k$, is a monomorphism of rings;
    (iii) for all $k \in I$, $i_k(R_k)$ is an ideal of $\oplus_{i \in I} R_i$.

Proof.   (i) Let $\{a_i \mid i \in I\}$ and $\{b_i \mid i \in I\}$ be two elements of $\oplus_{i \in I} R_i$. Since $a_i \neq 0$ for at most finitely many $i \in I$ and $b_i \neq 0$ for at most finitely many $i \in I$, it follows that $a_i - b_i \neq 0$ for at most finitely many $i \in I$ and $a_i b_i \neq 0$ for at most finitely many $i \in I$. Hence, $\{a_i \mid i \in I\} - \{b_i \mid i \in I\} \in \oplus_{i \in I} R_i$ and $\{a_i \mid i \in I\}\{b_i \mid i \in I\} \in \oplus_{i \in I} R_i$. Thus, $\oplus_{i \in I} R_i$ is a subring.
    (ii) Let $a, b \in R_k$. Then $i_k(a + b) = \{\{a_i \mid i \in I\} \mid a_i = 0 \text{ for all } i \neq k \text{ and } a_k = a + b\} = \{\{a_i' \mid i \in I\} \mid a_k' = 0 \text{ for all } i \neq k \text{ and } a_k' = a\} + \{\{b_i' \mid i \in I\} \mid b_i' = 0 \text{ for all } i \neq k \text{ and } b_k' = b\} = i_k(a) + i_k(b)$. Similarly, $i_k(ab) = i_k(a) i_k(b)$. Thus, i_k is a homomorphism. By the definition of i_k, we find that i_k is one-one. Hence, i_k is a monomorphism.
    (iii) Since i_k is a monomorphism, $i_k(R_k)$ is a subring of $\oplus_{i \in I} R_i$. Let $\{b_i \mid i \in I\} \in \oplus_{i \in I} R_i$ and $\{a_i \mid i \in I\} \in i_k(R_k)$. Since $a_i = 0$ for all $i \neq k$, $b_i a_i = 0$ for all $i \neq k$. Also, for $i = k$, $b_k, a_k \in R_k$. Therefore, $b_k a_k \in R_k$. Thus, $\{b_i \mid i \in I\}\{a_i \mid i \in I\} \in i_k(R_k)$, proving that $i_k(R_k)$ is a left ideal. Similarly, $\{a_i \mid i \in I\}\{b_i \mid i \in I\} \in i_k(R_k)$. Hence, $i_k(R_k)$ is an ideal. $\blacksquare$

    By Theorem 13.1.2, we find that R_k is isomorphic to the subring $i_k(R_k)$ of $\oplus_{i \in I} R_i$. Identifying R_k with $i_k(R_k)$, we can say that $\oplus_{i \in I} R_i$ contains R_k as an ideal.
    Let $I = \{1, 2, \ldots, n\}$ and $\{R_i \mid i \in I\}$ be a finite family of rings. From the definition of direct sum, it follows that the complete direct sum and the direct sum of this family is the same. Hence, by Theorem 13.1.2, we can say that the direct sum, $R_1 \oplus R_2 \oplus \cdots \oplus R_n$, contains each of $R_1, R_2, \ldots, R_n$ as an ideal.
    We now investigate the conditions under which a ring R is isomorphic to a direct sum of a family of ideals (considering each ideal as a ring) of R.

Definition 13.1.3 *Let I be a finite nonempty set, say, $\{1, 2, \ldots, n\}$, and $\{A_i \mid i \in I\}$ be a family of ideals of a ring R. Then the sum of this finite family, denoted by $\sum_{i \in I} A_i$, is the set*

$$\sum_{i \in I} A_i = \{a_1 + a_2 + \cdots + a_n \mid a_i \in A_i, \; i = 1, 2, \ldots, n\}.$$

If I is empty, then let us take $\sum_{i \in I} A_i = \{0\}$.

If $I = \{1, 2, \ldots, n\}$, then we also use the notation $A_1 + A_2 + \cdots + A_n$ to denote the sum $\sum_{i \in I} A_i$.

We leave the proof of the following theorem as an exercise.

Theorem 13.1.4 *Let $\{A_i \mid i \in I\}$ be a finite family of ideals of a ring R. Then*
(i) $\sum_{i \in I} A_i$ is an ideal of R,
(ii) $A_i \subseteq \sum_{j \in I} A_j$ for all $i \in I$,
(iii) if A is an ideal of R such that $A_i \subseteq A$ for all $i \in I$, then $\sum_{i \in I} A_i \subseteq A$. ■

Definition 13.1.5 *Let $\{A_i \mid i \in I\}$ be a family of ideals of a ring R, where I is finite or infinite. Then the sum of this family, denoted by $\sum_{i \in I} A_i$, is the set*

$$\sum_{i \in I} A_i = \{a \in R \mid a \in \sum_{i \in I_0} A_i \text{ for some finite subset } I_0 \text{ of } I\}.$$

Theorem 13.1.6 *Let $\{A_i \mid i \in I\}$ be a family of ideals of a ring R. Then $\sum_{i \in I} A_i$ is an ideal of R.*

Proof. Since $0 \in \sum_{i \in I} A_i$, $\sum_{i \in I} A_i \neq \phi$. Let $a, b \in \sum_{i \in I} A_i$ and $r \in R$. Then $a \in \sum_{i \in I_1} A_i$ and $b \in \sum_{i \in I_2} A_i$ for some finite subsets I_1 and I_2 of I. Let $I_3 = I_1 \cup I_2$. Then $a, b \in \sum_{i \in I_3} A_i$. By Theorem 13.1.4, $\sum_{i \in I_3} A_i$ is an ideal of R. Hence, $a - b, ar, ra \in \sum_{i \in I_3} A_i$. Thus, $a - b, ar, ra \in \sum_{i \in I} A_i$ and so $\sum_{i \in I} A_i$ is an ideal of R. ■

Definition 13.1.7 *Let $\{A_i \mid i \in I\}$ be a finite family of ideals of a ring R. A sum $\sum_{i \in I} A_i$ of $\{A_i \mid i \in I\}$ is called a **direct sum** if for all $k \in I$,*

$$A_k \cap \sum_{i \in I, \; i \neq k} A_i = \{0\}.$$

Lemma 13.1.8 *Let $\{A_i \mid i \in I\}$ be a finite family of ideals of a ring R. If $\sum_{i \in I} A_i$ is a direct sum, then for all $a \in A_k$, $b \in A_l$, $k \neq l$, $ab = 0$.*

Proof. Let $a \in A_k$, $b \in A_l$, and $k \neq l$. Since A_k and A_l are ideals, $ab \in A_k$ and $ab \in A_l$. Since $A_l \subseteq \sum_{i \in I, \; i \neq k} A_i$, $ab \in \sum_{i \in I, \; i \neq k} A_i$. Therefore, $ab \in A_k \cap \sum_{i \in I, \; i \neq k} A_i$. Since $\sum_{i \in I} A_i$ is a direct sum, $A_k \cap \sum_{i \in I, \; i \neq k} A_i = \{0\}$. Hence, $ab = 0$. ■

Theorem 13.1.9 *Let $\{A_i \mid i \in I\}$ be a family of ideals of a ring R, $I = \{1, 2, \ldots, n\}$. Then the following conditions are equivalent.*
(i) $\sum_{i \in I} A_i$ is a direct sum.
(ii) $a_1 + a_2 + \cdots + a_n = 0$, $a_i \in A_i$, $i \in I$, implies that $a_i = 0$ for all $i \in I$.
(iii) Each element $a \in \sum_{i \in I} A_i$ is uniquely expressible in the form

$$a = a_1 + a_2 + \cdots + a_n,$$

where $a_i \in A_i$, $i \in I$.

Proof.   (i)$\Rightarrow$(ii) Let $a_1 + a_2 + \cdots + a_n = 0$, $a_i \in A_i$, $i \in I$. Let $k \in I$. Now

$$-a_k = a_1 + a_2 + \cdots + a_{k-1} + a_{k+1} + \cdots + a_n \in A_k \cap \sum_{i \in I, \ i \neq k} A_i = \{0\}.$$

Hence, $a_k = 0$.

(ii)$\Rightarrow$(iii) Let $a = a_1 + a_2 + \cdots + a_n = b_1 + b_2 + \cdots + b_n$, where $a_i, b_i \in A_i$ for all $i \in I$. Then $(a_1 - b_1) + (a_2 - b_2) + \cdots + (a_n - b_n) = 0$. Hence, by (ii), $a_i - b_i = 0$ for all $i \in I$, i.e., $a_i = b_i$ for all $i \in I$.

(iii)$\Rightarrow$(i) Let $a \in A_k \cap \sum_{i \in I, \ i \neq k} A_i$. Then there exist $a_i \in A_i$, $i = 1, 2, \ldots, n$, such that

$$a = a_k = a_1 + a_2 + \cdots + a_{k-1} + a_{k+1} + \cdots + a_n.$$

This implies

$$a_1 + a_2 + \cdots + a_{k-1} + (-a_k) + a_{k+1} + \cdots + a_n = 0.$$

Also, $0 + 0 + \cdots + 0 = 0$. Therefore, by (iii), $a_i = 0$ for all $i \in I$ since 0 is uniquely expressible as a sum of elements of A_i. Thus, $A_k \cap \sum_{i \in I, \ i \neq k} A_i = \{0\}$ and so $\sum_{i \in I} A_i$ is a direct sum. ∎

Definition 13.1.10 *A ring R is said to be an **internal direct sum** of a finite family of ideals $\{A_1, A_2, \ldots, A_n\}$ if*
(i) $R = A_1 + A_2 + \cdots + A_n$ and
(ii) $A_1 + A_2 + \cdots + A_n$ is a direct sum.

Theorem 13.1.11 *Let R be a ring and $\{A_i \mid i \in I\}$ be a finite family of ideals of R. If R is an internal direct sum of $\{A_i \mid i \in I\}$, then*

$$R \simeq \oplus_{i \in I} A_i.$$

Proof.   Let $I = \{1, 2, \ldots, n\}$. Suppose R is an internal direct sum of ideals $A_1, A_2, \ldots, A_n$. Let $a \in R$. Then a is uniquely expressible in the form $a = a_1 + a_2 + \cdots + a_n$, where $a_i \in A_i$, $i \in I$. Now $(a_1, a_2, \ldots, a_n) \in \oplus_{i \in I} A_i$. Define $f : R \to \oplus_{i \in I} A_i$ by

$$f(a) = (a_1, a_2, \ldots, a_n).$$

Let $a, b \in R$. Then there exist $a_i, b_i \in A_i$, $i \in I$ such that $a = a_1 + a_2 + \cdots + a_n$ and $b = b_1 + b_2 + \cdots + b_n$. Now $a = b$ if and only if $a_1 + a_2 + \cdots + a_n = b_1 + b_2 + \cdots + b_n$ if and only if $a_i = b_i$ for all $i \in I$ if and only if $(a_1, a_2, \ldots, a_n) = (b_1, b_2, \ldots, b_n)$ if and only if $f(a) = f(b)$. This shows that f is a one-one function. Let $(a_1, a_2, \ldots, a_n) \in \oplus_{i \in I} A_i$. Then $a = a_1 + a_2 + \cdots + a_n \in \sum_{i \in I} A_i = R$ and $f(a) = (a_1, a_2, \ldots, a_n)$. Hence, f is onto $\oplus_{i \in I} A_i$. Finally, we show that f is a homomorphism. Since $a + b = (a_1 + b_1) + (a_2 + b_2) + \cdots + (a_n + b_n)$, we have $f(a + b) = ((a_1 + b_1), (a_2 + b_2), \ldots, (a_n + b_n)) = (a_1, a_2, \ldots, a_n) + (b_1, b_2, \ldots,$

$b_n) = f(a) + f(b)$. By Lemma 13.1.8, for all $i, j \in I$, $i \neq j$, $a_i b_j = 0$. From this, it follows that $ab = a_1 b_1 + a_2 b_2 + \cdots + a_n b_n$. Thus, $f(ab) = (a_1 b_1, a_2 b_2, \ldots, a_n b_n) = (a_1, a_2, \ldots, a_n)(b_1, b_2, \ldots, b_n) = f(a) f(b)$. Hence, f is an isomorphism of R onto $\oplus_{i \in I} A_i$, proving that $R \simeq \oplus_{i \in I} A_i$. $\blacksquare$

If R is an internal direct sum of ideals $A_1, A_2, \ldots, A_n$, then we identify R with $\oplus_{i \in I} A_i$ and we usually write

$$R = A_1 \oplus A_2 \oplus \cdots \oplus A_n.$$

Let us now characterize the direct sum of ideals of a ring R with 1 with the help of idempotent elements.

Theorem 13.1.12 *Let R be a ring with 1 and $\{A_1, A_2, \ldots, A_n\}$ be a finite family of ideals of R. Then $R = A_1 \oplus A_2 \oplus \cdots \oplus A_n$ if and only if there exist idempotents $e_i \in A_i$, $i = 1, 2, \ldots, n$, such that*
 (i) $1 = e_1 + e_2 + \cdots + e_n$,
 (ii) $Re_i = A_i$ for all $i = 1, 2, \ldots, n$, and
 (iii) $e_i e_j = e_j e_i = 0$ for $i \neq j$.

Proof.   Let $R = A_1 \oplus A_2 \oplus \cdots \oplus A_n$. Now $1 \in R$. Thus, there exist $e_i \in A_i$, $i = 1, 2, \ldots, n$, such that $1 = e_1 + e_2 + \cdots + e_n$. Then $e_i = e_1 e_i + e_2 e_i + \cdots + e_i^2 + \cdots + e_n e_i$. By Lemma 13.1.8, $e_j e_i = 0$ for all $j \neq i$. Hence, $e_i = e_i^2$, i.e., e_i is an idempotent for all $i = 1, 2, \ldots, n$. Since $e_i \in A_i$ and A_i is an ideal, $Re_i \subseteq A_i$. Let $a \in A_i$. Then

$$a = a1 = ae_1 + ae_2 + \cdots + ae_n = ae_i \in Re_i$$

since by Lemma 13.1.8, $ae_j = 0$ for all $j \neq i$. Thus, $A_i \subseteq Re_i$. Therefore, we find that $Re_i = A_i$.

Conversely, assume that there exist idempotents $e_i \in A_i$, $i = 1, 2, \ldots, n$, satisfying the given conditions. Let $a \in R$. Then $a = a1 = a(e_1 + e_2 + \cdots + e_n) = ae_1 + ae_2 + \cdots + ae_n \in Re_1 + Re_2 + \cdots + Re_n \subseteq A_1 + A_2 + \cdots + A_n$. Hence, $R = A_1 + A_2 + \cdots + A_n$. Let us now show that this sum is direct. Let $a \in A_i \cap (A_1 + A_2 + \cdots + A_{i-1} + A_{i+1} + \cdots + A_n)$. Then there exist $a_1, a_2, \ldots, a_n \in R$ such that $a_i e_i = a = a_1 e_1 + \cdots + a_{i-1} e_{i-1} + a_{i+1} e_{i+1} + \cdots + a_n e_n$. Thus, $a = a_i e_i$ implies that $ae_i = a_i e_i^2 = a_i e_i = a$ and $a = a_1 e_1 + \cdots + a_{i-1} e_{i-1} + a_{i+1} e_{i+1} + \cdots + a_n e_n$ implies that $ae_i = a_1 e_1 e_i + \cdots + a_{i-1} e_{i-1} e_i + a_{i+1} e_{i+1} e_i + \cdots + a_n e_n e_i = a0 + \cdots + a0 = 0$ (since by (iii), $e_i e_j = 0$ for $i \neq j$). Hence, $a = 0$, proving that $R = A_1 \oplus A_2 \oplus \cdots \oplus A_n$. $\blacksquare$

Let us now consider another type of subring of the complete direct sum $\Pi_{i \in I} R_i$ of a family of rings $\{R_i \mid i \in I\}$. For this, let us note that the mapping

$\pi_k : \Pi_{i \in I} R_i \longrightarrow R_k$ defined by

$$\pi_k(\{a_i \mid i \in I\}) = a_k$$

is an epimorphism of the ring $\Pi_{i \in I} R_i$ onto the ring R_k. π_k is called the kth **canonical projection.**

Definition 13.1.13 *A subring T of $\Pi_{i \in I} R_i$ is called a* **subdirect sum** *of the family of rings $\{R_i \mid i \in I\}$ if $\pi_i|_T$ (the restriction of π_i to T) is an epimorphism of T onto R_i. We denote T by $\oplus^s_{i \in I} R_i$.*

Theorem 13.1.14 *A ring S is isomorphic to a subdirect sum of a family $\{R_i \mid i \in I\}$ of rings if and only if S contains a family of ideals $\{A_i \mid i \in I\}$ such that $\cap_{i \in I} A_i = \{0\}$.*

Proof.   Suppose S is isomorphic to a subdirect sum of a family $\{R_i \mid i \in I\}$ of rings. Then there exists a subring T of $\Pi_{i \in I} R_i$ such that $S \simeq T$ and $T = \oplus^s_{i \in I} R_i$. Let α be the isomorphism of S onto T. Then $\pi_i \alpha : S \longrightarrow R_i$ is an epimorphism. Let $A_i = \text{Ker } \pi_i \alpha$. Then A_i is an ideal of S. Let $a \in \cap_{i \in I} A_i$. Then $(\pi_i \alpha)(a) = 0$ for all $i \in I$. Thus, $\pi_i(\alpha(a)) = 0$, i.e., the ith component of $\alpha(a)$ is 0 for all $i \in I$. Hence, $\alpha(a) = 0$. Since α is one-one, $a = 0$. This proves that $\cap_{i \in I} A_i = \{0\}$.

Conversely, suppose S contains a family of ideals $\{A_i \mid i \in I\}$ such that $\cap_{i \in I} A_i = \{0\}$. Consider the family $\{S/A_i \mid i \in I\}$ of quotient rings. Let $R = \Pi_{i \in I} S/A_i$. Define $\beta : S \longrightarrow R$ by

$$\beta(a) = \{a + A_i \mid i \in I\}$$

for all $a \in S$. Then β is a homomorphism. Let $a \in S$. Now $a \in \text{Ker } \beta$ if and only if $\beta(a) = 0$ if and only if $a + A_i = 0$ for all $i \in I$ if and only if $a \in A_i$ for all $i \in I$ if and only if $a \in \cap_{i \in I} A_i$ if and only if $a = 0$. Therefore, $\text{Ker } \beta = \{0\}$. Thus, β is a monomorphism. Let $\beta(S) = T$. Then T is a subring of R and also $\pi_i|_T$ is an epimorphism. ∎

13.1.1   Worked-Out Exercises

◇ **Exercise 1** An idempotent e of a ring R is called a **central idempotent** if $e \in C(R)$.

Let R be a ring with 1 and e be a central idempotent in R. Show that

(i) $1 - e$ is a central idempotent in R;

(ii) eR and $(1 - e)R$ are ideals of R;

(iii) $R = eR \oplus (1 - e)R$.

Solution: (i) $(1-e)(1-e) = 1 - e - e + e^2 = 1 - e - e + e = 1 - e$. Also, for all $a \in R$, $a(1-e) = a - ae = a - ea = (1-e)a$. Hence, $1-e$ is a central idempotent.

(ii) Now eR is a right ideal of R. Let $a \in R$. Then $a(eR) = (ae)R = (ea)R$ (since $e \in C(R)$) $= e(aR) \subseteq eR$. Hence, eR is also a left ideal. Thus, eR is an ideal of R. Similarly, $(1-e)R$ is an ideal of R.

(iii) Let $a \in R$. Then $a = ea + a - ea = ea + (1-e)a \in eR + (1-e)R$. Hence, $R = eR + (1-e)R$. Suppose $b \in eR \cap (1-e)R$. Then there exist $c, d \in R$ such that $b = ec = (1-e)d$. Hence, $eb = e^2c = ec = b$ and $eb = e(1-e)d = (e-e^2)d = (e-e)d = 0$. Thus, $b = 0$. As a result, $R = eR \oplus (1-e)R$.

$\Diamond$ **Exercise 2** Let A and B be two ideals of a ring R such that $R = A \oplus B$. Show that $R/A \simeq B$ and $R/B \simeq A$.

Solution: Let $x \in R$. Then x can be uniquely expressed as $x = a+b$, where $a \in A$ and $b \in B$. Define $f : R \to B$ by $f(x) = b$. Clearly f is well defined. Let $b \in B$. Then $b = 0+b \in A+B$. Hence, $f(b) = b$, which shows that f is onto B. Let $x, y \in R$. Then there exist $a_1, a_2 \in A$ and $b_1, b_2 \in B$ such that $x = a_1 + b_1$ and $y = a_2 + b_2$. Now $x + y = a_1 + b_1 + a_2 + b_2 = (a_1 + a_2) + (b_1 + b_2) \in A + B$ and $xy = (a_1 + b_1)(a_2 + b_2) = a_1a_2 + a_1b_2 + b_1a_2 + b_1b_2$. Since $a_1b_2, b_1a_2 \in A \cap B$ and $A \cap B = \{0\}$, $a_1b_2 = 0$ and $b_1a_2 = 0$. Therefore, $xy = a_1a_2 + b_1b_2 \in A + B$. Hence, $f(x + y) = b_1 + b_2 = f(x) + f(y)$ and $f(xy) = b_1b_2 = f(x)f(y)$. Thus, f is an epimorphism. Therefore, by the first isomorphism theorem (Theorem 11.3.14), $R/\mathrm{Ker} f \simeq B$. Let $x \in \mathrm{Ker}\ f$. Then $f(x) = 0$. Since $x \in \mathrm{Ker}\ f \subseteq R$, there exist $a \in A$ and $b \in B$ such that $x = a+b$. Now $f(x) = b$ and this implies that $b = 0$. Therefore, $x = a \in A$ and so $\mathrm{Ker}\ f \subseteq A$. On the other hand, let $a \in A$. Then $a = a + 0 \in A + B$. Therefore, $f(a) = 0$ and so $a \in \mathrm{Ker}\ f$. Thus, $A \subseteq \mathrm{Ker}\ f$. Hence, $A = \mathrm{Ker}\ f$ and so $R/A \simeq B$. Similarly, $R/B \simeq A$.

Exercise 3 Let $R = R_1 \oplus R_2 \oplus \cdots \oplus R_n$ be the direct of sum of rings $R_1, R_2, \ldots, R_n$ and $1 \in R$. Show that an element $a = (a_1, a_2, \ldots, a_n) \in R$ is a unit if and only if a_i is a unit in R_i for all $i = 1, 2, \ldots, n$.

Solution: Since $1 \in R = R_1 \oplus R_2 \oplus \cdots \oplus R_n$, $1 = (e_1, e_2, \ldots, e_n)$, where e_i is the identity of R_i for all $i = 1, 2, \ldots, n$. Suppose $a = (a_1, a_2, \ldots, a_n) \in R$ is a unit. Then there exists $b = (b_1, b_2, \ldots, b_n) \in R$ such that $ab = 1 = ba$. Thus, $(a_1, a_2, \ldots, a_n)(b_1, b_2, \ldots, b_n) = (e_1, e_2, \ldots, e_n) = (b_1, b_2, \ldots, b_n)(a_1, a_2, \ldots, a_n)$. From this, it follows that $a_ib_i = e_i = b_ia_i$ for all $i = 1, 2, \ldots, n$. Hence, a_i is a unit in R_i for all $i = 1, 2, \ldots, n$. Conversely, assume that a_i is a unit in R_i for all $i = 1, 2, \ldots, n$. Thus, there exists $b_i \in R_i$ such that $a_ib_i = e_i = b_ia_i$ for all $i = 1, 2, \ldots, n$. Let $b = (b_1, b_2, \ldots, b_n)$. Then $ab = 1 = ba$, proving that a is a unit.

◇ **Exercise 4** Let R be a direct of sum of rings $R_1, R_2, \ldots, R_n$ with identity. Let A be an ideal of R. Show that there exist ideals A_i in R_i, $i = 1, 2, \ldots, n$, such that $A = A_1 \oplus A_2 \oplus \cdots \oplus A_n$.

Solution: For all k, $1 \le k \le n$, define $\alpha_k : \oplus R_i \to R_k$ by

$$\alpha_k((a_1, a_2, \ldots, a_n)) = a_k$$

for all $(a_1, a_2, \ldots, a_n) \in \oplus R_i$. It can be easily verified that α_k is an epimorphism. Let $\alpha_k(A) = A_k$. Then A_k is an ideal of R_k. We now show that $A = A_1 \oplus A_2 \oplus \cdots \oplus A_n$. Let $a = (a_1, a_2, \ldots, a_n) \in A$. Now $\alpha_k(a) = a_k \in A_k$. Therefore, $a \in A_1 \oplus A_2 \oplus \cdots \oplus A_n$ and so $A \subseteq A_1 \oplus A_2 \oplus \cdots \oplus A_n$. Suppose now that $b = (b_1, b_2, \ldots, b_n) \in A_1 \oplus A_2 \oplus \cdots \oplus A_n$. Then $b_k \in A_k = \alpha_k(A)$. Therefore, there exists an element $a = (a_1, a_2, \ldots, a_{k-1}, b_k, a_{k+1}, \ldots, a_n) \in A$. Now $(0, 0, \ldots, 0, b_k, 0, \ldots, 0) = (0, 0, \ldots, 1, \ldots, 0)(a_1, a_2, \ldots, a_{k-1}, b_k, a_{k+1}, \ldots, a_n) \in A$ for all $k = 1, 2, \ldots, n$. Hence, $(b_1, b_2, \ldots, b_n) = (b_1, 0, \ldots, 0) + (0, b_2, \ldots, 0) + \cdots + (0, 0, \ldots, b_n) \in A$ showing that $A_1 \oplus A_2 \oplus \cdots \oplus A_n \subseteq A$. Thus, $A = A_1 \oplus A_2 \oplus \cdots \oplus A_n$.

◇ **Exercise 5** Let R be a ring with 1. Suppose that A and B are ideals of R such that $R = A + B$. Show that

$$R/(A \cap B) \simeq R/A \oplus R/B.$$

(This result is known as the **Chinese remainder theorem for rings**.)

Solution: Define $f : R \to R/A \oplus R/B$ by

$$f(x) = (x + A, x + B)$$

for all $x \in R$. Let $x, y \in R$. Then

$$
\begin{aligned}
f(x + y) &= ((x + y) + A, (x + y) + B) \\
&= ((x + A) + (y + A), (x + B) + (y + B)) \\
&= (x + A, x + B) + (y + A, y + B) \\
&= f(x) + f(y).
\end{aligned}
$$

Similarly, $f(xy) = f(x)f(y)$. Hence, f is a homomorphism. Now $R = A + B$ implies that $1 = a + b$ for some $a \in A$ and $b \in B$. Thus, $a + B = (1 - b) + B = (1 + B) + (-b + B) = 1 + B$ since $-b \in B$. Similarly, $b + A = 1 + A$. Let $(x + A, y + B) \in R/A \oplus R/B$. Now $xb + ya \in R$. Therefore,

$$
\begin{aligned}
f(xb + ya) &= ((xb + ya) + A, (xb + ya) + B) \\
&= ((xb + A) + (ya + A), (xb + B) + (ya + B)) \\
&= ((xb + A) + (0 + A), (0 + B) + (ya + B)) \text{ (since } a \in A,\ b \in B) \\
&= ((xb + A), (ya + B)) \\
&= ((x + A)(b + A), (y + B)(a + B)) \\
&= ((x + A)(1 + A), (y + B)(1 + B)) \\
&= (x + A,\ y + B).
\end{aligned}
$$

Hence, f is an epimorphism. By the first isomorphism theorem (Theorem 11.3.14),

$$R/\mathrm{Ker}\ f \simeq R/A \oplus R/B.$$

We now show that $\mathrm{Ker}\ f = A \cap B$.

$$
\begin{aligned}
\mathrm{Ker}\ f &= \{x \in R \mid f(x) = 0\} \\
&= \{x \in R \mid (x + A, x + B) = (A, B)\} \\
&= \{x \in R \mid x + A = A \text{ and } x + B = B\} \\
&= \{x \in R \mid x \in A \text{ and } x \in B\} \\
&= \{x \in R \mid x \in A \cap B\} \\
&= A \cap B.
\end{aligned}
$$

Consequently, $R/(A \cap B) \simeq R/A \oplus R/B$.

13.1.2   Exercises

1. Let $R = R_1 \oplus R_2 \oplus \cdots \oplus R_n$ be a direct sum of rings. If A_i is an ideal of R_i, $(1 \leq i \leq n)$, prove that $A = A_1 \oplus A_2 \oplus \cdots \oplus A_n$ is an ideal of R.

2. Let R be a direct of sum of rings $R_1, R_2, \ldots, R_n$ with 1. Let A be an ideal of R. Show that there exist ideals A_i of R_i, $i = 1, 2, \ldots, n$, such that $A = A_1 \oplus A_2 \oplus \cdots \oplus A_n$ and

$$R/A \simeq R_1/A_1 \oplus R_2/A_2 \oplus \cdots \oplus R_n/A_n.$$

3. Show that the ring $\mathbf{Z}$ cannot be expressed as a direct sum of a finite family of proper ideals of $\mathbf{Z}$.

4. If m and n are two positive integers such that $\gcd(m, n) = 1$, prove that $\mathbf{Z}_{mn} \simeq \mathbf{Z}_m \oplus \mathbf{Z}_n$.

Chapter 14

Polynomial Rings

The study of polynomials dates back to 1650 B.C., when Egyptians were solving certain linear polynomial equations. In 600 B.C., Hindus had learned how to solve quadratic equations. However, polynomials, as we know them today, i.e., polynomials written in our notation, did not exist until approximately 1700 A.D.

About 400 A.D., the use of symbolic algebra began to appear in India and Arabia. Some mark the use of symbols in algebra as the first level of abstraction in mathematics.

14.1 Polynomial Rings

An important class of rings is the so-called class of polynomial rings. We are all familiar with polynomials. We may be used to thinking of a polynomial as an expression of the form $a_0 + a_1 x + \cdots + a_n x^n$, where x is a symbol and the a_i are possibly real numbers, or as a function $f(x) = a_0 + a_1 x + \cdots + a_n x^n$. However, does one really know what a polynomial is? What really is the symbol x? Why are two polynomials $a_0 + a_1 x + \cdots + a_n x^n$ and $b_0 + b_1 x + \cdots + b_m x^m$ equal if and only if $n = m$ and $a_i = b_i$, $i = 1, 2, \ldots, n$? In this section, we answer these questions and give some basic properties of polynomials.

Definition 14.1.1 *For any ring R, let $R[x]$ denote the set of all infinite sequences $(a_0, a_1, a_2, \ldots)$, where $a_i \in R$, $i = 0, 1, 2, \ldots$, and where there is a nonnegative integer n (dependent on $(a_0, a_1, a_2, \ldots)$) such that for all integers $k \geq n$, $a_k = 0$. The elements of $R[x]$ are called* **polynomials** *over R.*

We now define addition and multiplication on $R[x]$ as follows:

$$
\begin{aligned}
(a_0, a_1, a_2, \ldots) + (b_0, b_1, b_2, \ldots) &= (a_0 + b_0, a_1 + b_1, a_2 + b_2, \ldots) \\
(a_0, a_1, a_2, \ldots) \cdot (b_0, b_1, b_2, \ldots) &= (c_0, c_1, c_2, \ldots),
\end{aligned}
$$

where

$$c_j = \sum_{i=0}^{j} a_i b_{j-i} \text{ for } j = 0, 1, 2, \ldots$$

We leave it to the reader to verify that $(R[x], +, \cdot)$ is a ring. We do note that $(0, 0, \ldots)$ is the additive identity of $R[x]$ and that the additive inverse of $(a_0, a_1, \ldots)$ is $(-a_0, -a_1, \ldots)$. The ring $R[x]$ is called a **ring of polynomials** or a **polynomial ring** over R. It is clear that $R[x]$ is commutative when R is commutative. Also, if R has an identity 1, then $R[x]$ has an identity, namely, $(1, 0, 0, 0, \ldots)$.

The mapping $a \to (a, 0, 0, \ldots)$ is a monomorphism of R into $R[x]$. Thus, R is embedded in $R[x]$. Therefore, we can consider R as a subring of $R[x]$ and we no longer distinguish between a and $(a, 0, 0, \ldots)$.

We now convert our notation of polynomials into a notation which is more familiar to the reader.

Let

$$a = ax^0 \text{ denote } (a, 0, 0, \ldots)$$
$$ax = ax^1 \text{denote } (0, a, 0, \ldots)$$
$$ax^2 \text{ denote } (0, 0, a, \ldots)$$
$$\vdots$$

Then

$(a_0, a_1, a_2, \ldots, a_n, 0, \ldots) = (a_0, 0, 0, \ldots) + (0, a_1, 0, 0, \ldots) + \cdots + (0, \ldots, 0, a_n, 0, \ldots) = a_0 + a_1 x + a_2 x^2 + \cdots + a_n x^n.$

The symbol x is called an **indeterminate** over R and the elements $a_0, a_1, \ldots, a_n$ of R are called the **coefficients** of $a_0 + a_1 x + a_2 x^2 + \cdots + a_n x^n$.

The reason two polynomials $a_0 + a_1 x + \cdots + a_n x^n$ and $b_0 + b_1 x + \cdots + b_m x^m$ are equal if and only if $n = m$ and $a_i = b_i$, $i = 1, 2, \ldots, n$, is that the two sequences $(a_0, a_1, \ldots)$ and $(b_0, b_1, \ldots)$ are equal if and only if $a_i = b_i$, $i = 1, 2, \ldots$. (One must recall that an infinite sequence of elements of R is a function from the set of nonnegative integers into R. Consequently, the concept of an ordered pair is again being used to give a rigorous definition of a mathematical concept.)

If R has an identity 1, then we can consider x an element of $R[x]$. We do this by identifying $1x$ with x, i.e., $(0, 1, 0, \ldots)$ is called x.

The reader can check that the definitions of addition and multiplication of two polynomials are the familiar ones. Thus, when R has an identity, $ax = (a, 0, 0, \ldots)(0, 1, 0, \ldots) = (0, a, 0, \ldots) = (0, 1, 0, \ldots)(a, 0, 0, \ldots) = xa.$

Theorem 14.1.2 *(i) If R is a commutative ring with 1, then $R[x]$ is a commutative ring with 1.*

(ii) If R is an integral domain, then $R[x]$ is also an integral domain.

Proof.   (i) Let $f(x) = a_0 + a_1 x + \cdots + a_n x^n$ and $g(x) = b_0 + b_1 x + \cdots + b_m x^m$ be two elements in $R[x]$. Let $f(x)g(x) = c_0 + c_1 x + \cdots + c_t x^t$ and $g(x)f(x) = d_0 + d_1 x + \cdots + d_s x^s$. Now $c_j = \sum_{i=0}^{j} a_i b_{j-i}$ and $d_j = \sum_{i=0}^{j} b_i a_{j-i}$. Since R is commutative, $c_j = a_0 b_j + a_1 b_{j-1} + \cdots + a_j b_0 = b_0 a_j + b_1 a_{j-1} + \cdots + b_j a_0 = d_j$ for all $j = 0, 1, 2, \ldots$. Thus, $R[x]$ is a commutative ring. Since $1 \in R$, $1 \in R[x]$ and $1f(x) = f(x)1 = f(x)$ for all $f(x) \in R[x]$. Hence, $R[x]$ is a commutative ring with 1.

(ii) Let R be an integral domain. Then by (i), $R[x]$ is a commutative ring with 1. Let $f(x) = a_0 + a_1 x + \cdots + a_n x^n$ and $g(x) = b_0 + b_1 x + \cdots + b_m x^m$ be two nonzero polynomials in $R[x]$. Then there exist a_i and b_j such that $a_i \neq 0$, $b_j \neq 0$, $a_{i+t} = 0$, and $b_{j+t} = 0$ for all $t \geq 1$. Consider the polynomial $f(x)g(x) = c_0 + c_1 x + \cdots + c_{n+m} x^{n+m}$. Now $c_{i+j} = a_0 b_{i+j} + a_1 b_{i+j-1} + \cdots + a_i b_j + \cdots + a_{i+j} b_0 = a_i b_j \neq 0$ since R is an integral domain. This implies that $f(x)g(x) \neq 0$. Thus, $R[x]$ is an integral domain. ∎

Definition 14.1.3 *Let R be a ring. If $f(x) = a_0 + a_1 x + \cdots + a_n x^n$, $a_n \neq 0$, is a polynomial in $R[x]$, then n is called the **degree** of $f(x)$, written $\deg f(x)$, and a_n is called the **leading coefficient** of $f(x)$. If R has an identity and $a_n = 1$, then $f(x)$ is called a **monic polynomial**.*

The polynomials of degree 0 in $R[x]$ are exactly those elements from $R \backslash \{0\}$. $0 \in R[x]$ has no degree. We call the elements of R **scalar** or **constant polynomials**.

Theorem 14.1.4 *Let $R[x]$ be a polynomial ring and $f(x)$, $g(x)$ be two nonzero polynomials in $R[x]$.*

(i) If $f(x)g(x) \neq 0$, then $\deg f(x)g(x) \leq \deg f(x) + \deg g(x)$.
(ii) If $f(x) + g(x) \neq 0$, then

$$\deg(f(x) + g(x)) \leq \max\{\deg f(x), \deg g(x)\}.$$

Proof.   (i) If $f(x) = a_0 + a_1 x + \cdots + a_n x^n$ and $g(x) = b_0 + b_1 x + \cdots + b_m x^m$, then $f(x)g(x) = a_0 b_0 + (a_0 b_1 + a_1 b_0)x + \cdots + a_n b_m x^{n+m}$. If $f(x)g(x) \neq 0$, then at least one of the coefficients of $f(x)g(x)$ is nonzero. Suppose $a_n b_m \neq 0$, then $\deg(f(x)g(x)) = n + m = \deg f(x) + \deg g(x)$. If $a_n b_m = 0$ (which can hold if R has zero divisors), then $\deg(f(x)g(x)) < \deg f(x) + \deg g(x)$.

(ii) If $\deg f(x) > \deg g(x)$, then $\deg(f(x)+g(x)) = \max\{\deg f(x), \deg g(x)\}$. If $\deg f(x) = \deg g(x)$, then it is possible that $f(x) + g(x) = 0$ or $\deg(f(x) + g(x)) < \max\{\deg f(x), \deg g(x)\}$. We leave the details as an exercise. ∎

From the proof of Theorem 14.1.4(i), it is immediate that if R is an integral domain, then equality holds in (i).

Example 14.1.5 *Consider the polynomial ring* $\mathbf{Z}_6[x]$. *Let* $f(x) = [1] + [2]x^2$ *and* $g(x) = [1]+[3]x$. *Then* $f(x)g(x) = [1]+[3]x+[2]x^2$. *Hence,* $\deg (f(x)g(x)) = 2 < 3 = \deg f(x) + \deg g(x)$. *Let* $h(x) = [5] + [4]x^2$. *Then* $f(x) + h(x) = [6] + [6]x^2 = [0]$ *and so* $\deg(f(x) + h(x))$ *is not defined.*

Theorem 14.1.6 (Division Algorithm) *Let R be a commutative ring with 1 and $f(x)$, $g(x)$ be polynomials in $R[x]$ with the leading coefficient of $g(x)$ a unit in R. Then there exist unique polynomials $q(x)$, $r(x) \in R[x]$ such that*

$$f(x) = q(x)g(x) + r(x),$$

where either $r(x) = 0$ or $\deg r(x) < \deg g(x)$.

Proof. If $f(x) = 0$ or $\deg f(x) < \deg g(x)$, then we take $q(x) = 0$ and $r(x) = f(x)$. We now assume that $\deg f(x) \geq \deg g(x)$ and prove the result by induction on $\deg f(x) = n$. If $\deg f(x) = \deg g(x) = 0$, then we have $q(x) = f(x)g(x)^{-1}$ and $r(x) = 0$. Make the induction hypothesis that the theorem is true for all polynomials of degree less than n. Let $f(x) = a_0 + a_1 x + \cdots + a_n x^n$ have degree n and $g(x) = b_0 + b_1 x + \cdots + b_m x^m$ have degree m, where $n \geq m$. The polynomial

$$f_1(x) = f(x) - (a_n b_m^{-1})x^{n-m}g(x) \tag{14.1}$$

has degree less than n since the coefficient of x^n is $a_n - (a_n b_m^{-1})b_m = 0$. Hence, by the induction hypothesis, there exist polynomials $q_1(x)$, $r_1(x) \in R[x]$ such that

$$f_1(x) = q_1(x)g(x) + r_1(x), \tag{14.2}$$

where $r_1(x) = 0$ or $\deg r_1(x) < \deg g(x)$. Substituting the representation of $f_1(x)$ in Eq. (14.2) into Eq. (14.1) and solving for $f(x)$, we obtain

$$f(x) = (q_1(x) + a_n b_m^{-1} x^{n-m})g(x) + r_1(x) = q(x)g(x) + r(x),$$

where $q(x) = q_1(x) + a_n b_m^{-1} x^{n-m}$ and $r(x) = r_1(x)$, the desired representation when $f(x)$ has degree n.

The uniqueness of $q(x)$ and $r(x)$ remains to be shown. Suppose there are polynomials $q'(x)$ and $r'(x) \in R[x]$ such that

$$f(x) = q(x)g(x) + r(x) = q'(x)g(x) + r'(x),$$

where $r(x) = 0$ or $\deg r(x) < \deg g(x)$, $r'(x) = 0$ or $\deg r'(x) < \deg g(x)$. Then

$$r(x) - r'(x) = (q'(x) - q(x))g(x).$$

Suppose $r(x) - r'(x) \neq 0$. Since the leading coefficient of $g(x)$ is a unit,

$$\deg((q'(x) - q(x))g(x)) = \deg(q'(x) - q(x)) + \deg g(x) \geq \deg g(x).$$

This implies that

$$\deg(r(x) - r'(x)) \geq \deg g(x),$$

which is impossible since $\deg r(x)$, $\deg r'(x) < \deg g(x)$. Thus,

$$r(x) - r'(x) = 0 \text{ or } r(x) = r'(x).$$

Therefore,

$$0 = (q'(x) - q(x))g(x). \tag{14.3}$$

Since b_m is a unit, $\deg(((q'(x) - q(x))g(x)) \geq 0$ unless $q'(x) - q(x) = 0$. Thus, from Eq. (14.3), we see that $q'(x) - q(x) = 0$ must be the case. ∎

The polynomials $q(x)$ and $r(x)$ in Theorem 14.1.6 are called the **quotient** and **remainder**, respectively, on division of $f(x)$ by $g(x)$.

Definition 14.1.7 *Let R be a commutative ring with 1 and $f(x) = a_0 + a_1 x + \cdots + a_n x^n \in R[x]$. For all $r \in R$, define*

$$f(r) = a_0 + a_1 r + \cdots + a_n r^n.$$

*When $f(r) = 0$, we call r a **root** or **zero** of $f(x)$.*

In Definition 14.1.7, we think of substituting r for x in $f(x)$. The student is used to doing this freely. However, certain difficulties arise when R is not commutative. For instance, let $f(x) = a - x$, $g(x) = b - x$. Set $h(x) = f(x)g(x)$. Then $h(x) = (a - x)(b - x) = ab - (a + b)x + x^2$. For $c \in R$, $h(c) = ab - (a+b)c + c^2 = ab - ac - bc + c^2$ while $f(c)g(c) = (a - c)(b - c) = ab - cb - ac + c^2$. Hence, we cannot draw the conclusion that $h(c) = f(c)g(c)$. However, if R is commutative (with identity), then we can conclude that $h(c) = f(c)g(c)$. Clearly if $k(x) = f(x) + g(x)$, then $k(c) = f(c) + g(c)$.

Definition 14.1.8 *Let R be a commutative ring with 1 and $f(x)$, $g(x) \in R[x]$ be such that $g(x) \neq 0$. We say that $g(x)$ **divides** $f(x)$ or that $g(x)$ is a **factor** of $f(x)$, and write $g(x)|f(x)$ if there exists $q(x) \in R[x]$ such that $f(x) = q(x)g(x)$.*

Theorem 14.1.9 (Remainder Theorem) *Let R be a commutative ring with identity. For $f(x) \in R[x]$ and $a \in R$, there exists $q(x) \in R[x]$ such that*

$$f(x) = (x - a)q(x) + f(a).$$

Proof.   By applying the division algorithm with $x - a = g(x)$, there exist unique $q(x), r(x) \in R[x]$ such that $f(x) = (x - a)q(x) + r(x)$, where $r(x) = 0$ or $\deg r(x) < 1$. Hence, $r(x)$ is a constant polynomial, say, $r(x) = d$. By substituting a for x, we obtain $f(a) = (a - a)q(a) + d = d$, which yields the desired result. ∎

Corollary 14.1.10 (Factorization Theorem) *Let R be a commutative ring with identity. For $f(x) \in R[x]$ and $a \in R$, $x - a$ divides $f(x)$ if and only if a is a root of $f(x)$.*

Proof.   Suppose $(x - a)|f(x)$. Then there exists $q(x) \in R[x]$ such that $f(x) = (x - a)q(x)$. Hence, $f(a) = (a - a)q(a) = 0$ and so a is a root of $f(x)$. Conversely, suppose a is a root of $f(x)$. Then by the remainder theorem (Theorem 14.1.9) and the fact that $f(a) = 0$, we have $f(x) = (x - a)q(x)$. Consequently, $(x - a)|f(x)$. ∎

Theorem 14.1.11 *Let R be an integral domain and $f(x)$ be a nonzero polynomial in $R[x]$ of degree n. Then $f(x)$ has at most n roots in R.*

Proof.   If $\deg f(x) = 0$, then $f(x)$ is a constant polynomial, say, $f(x) = c \neq 0$. Clearly c has no roots in R. Assume that the theorem is true for all polynomials of degree less than n, where $n > 0$ (the induction hypothesis). Suppose $\deg f(x) = n$. If $f(x)$ has no roots in R, then the theorem is true. Suppose $r \in R$ is a root of $f(x)$. Then by Corollary 14.1.10, $f(x) = (x - r)q(x)$, where $\deg q(x) = n - 1$. If there exists any other root $r' \in R$ of $f(x)$, then $0 = f(r') = (r' - r)q(r')$. Since $r' \neq r$ and R is an integral domain, $q(r') = 0$ and so r' is a root of $q(x)$. Therefore, any other root of $f(x)$ is also a root of $q(x)$. Since $f(x) = (x - r)q(x)$, any root of $q(x)$ is also a root of $f(x)$. By the induction hypothesis and the fact that $\deg q(x) = n - 1$, there are at most $n-1$ of these other roots r'. Hence, in all, $f(x)$ has at most n roots in R. ∎

We now extend the definition of a polynomial ring from one indeterminate to several indeterminates.

Definition 14.1.12 *For any ring R, we define recursively*

$$R[x_1, x_2, \ldots, x_n] = R[x_1, x_2, \ldots, x_{n-1}][x_n],$$

*where x_1 is an indeterminate over R and x_n is an indeterminate over $R[x_1, x_2, \ldots, x_{n-1}]$. $R[x_1, x_2, \ldots, x_n]$ is called a **polynomial ring in n indeterminates**.*

Before describing the ring $R[x_1, x_2, \ldots, x_n]$, we introduce some notation. We write $\sum_{i_n, \ldots, i_1} r_{i_1 \ldots i_n} x_1^{i_1} \cdots x_n^{i_n}$ for $\sum_{i_n=0}^{k_n} \cdots \sum_{i_1=0}^{k_1} r_{i_1 \ldots i_n} x_1^{i_1} \cdots x_n^{i_n}$, where each $r_{i_1 \ldots i_n} \in R$ and $k_1, \ldots, k_n$ are nonnegative integers.

The ring

$$R[x_1, x_2, \ldots, x_n] = \{ \sum_{i_n, \ldots, i_1} r_{i_1 \ldots i_n} x_1^{i_1} \cdots x_n^{i_n} \mid r_{i_1 \ldots i_n} \in R \}.$$

We have for $n = 2$ that

$$R[x_1, x_2] = R[x_1][x_2] = \{ \sum_{i_2} s_{i_2} x_2^{i_2} \mid s_{i_2} \in R[x_1] \}.$$

Now each s_{i_2} has the form $\sum_{i_1} r_{i_1 i_2} x_1^{i_1}$.

Thus,

$$
\begin{aligned}
R[x_1, x_2] &= \{ \sum_{i_2} (\sum_{i_1} r_{i_1 i_2} x_1^{i_1}) x_2^{i_2} \mid r_{i_1 i_2} \in R \} \\
&= \{ \sum_{i_2} \sum_{i_1} r_{i_1 i_2} x_1^{i_1} x_2^{i_2} \mid r_{i_1 i_2} \in R \} \\
&= \{ \sum_{i_2, i_1} r_{i_1 i_2} x_1^{i_1} x_2^{i_2} \mid r_{i_1 i_2} \in R. \}.
\end{aligned}
$$

Definition 14.1.13 *Let R be a subring of the ring S. Let $c_1, c_2, \ldots, c_n$ be elements of S. Define $R[c_1] = \{ \sum_i r_i \, c_1^i \mid r_i \in R \}$ and*

$$R[c_1, c_2, \ldots, c_n] = R[c_1, c_2, \ldots, c_{n-1}][c_n].$$

*We say that $c_1, c_2, \ldots, c_n$ are **algebraically independent** over R if*

$$\sum_{i_n, \ldots, i_1} r_{i_1 \ldots i_n} c_1^{i_1} \ldots c_n^{i_n} = 0$$

can occur only when each $r_{i_1 \ldots i_n} = 0$, where $r_{i_1 \ldots i_n} \in R$.

$R[c_1, c_2, \ldots, c_n]$ is a subring of S and equals the set of all finite sums of the form

$$\sum_{i_n, \ldots, i_1} r_{i_1 \ldots i_n} c_1^{i_1} \ldots c_n^{i_n},$$

where $r_{i_1 \ldots i_n} \in R$.

Theorem 14.1.14 *Let R be a subring of a commutative ring S such that R and S have the same identity. Let $c \in S$. Then there exists a unique homomorphism α of $R[x]$ onto $R[c]$ such that $\alpha(x) = c$ and $\alpha(a) = a$ for all $a \in R$.*

Proof. Define $\alpha : R[x] \to R[c]$ by $\alpha(\sum a_i x^i) = \sum a_i c^i$ for all $\sum a_i x^i \in R[x]$. Now $a_0 + a_1 x + \cdots + a_n x^n = b_0 + b_1 x + \cdots + b_m x^m$ implies that $n = m$ and $a_i = b_i$ for $i = 1, 2, \ldots, n$. Thus, $a_0 + a_1 c + \cdots + a_n c^n = b_0 + b_1 c + \cdots + b_n c^n$ and so α is well defined. By Definition 14.1.13, α clearly maps $R[x]$ onto $R[c]$. Since for any two polynomials $f(x), g(x) \in R[x]$, $k(x) = f(x) + g(x)$ implies $k(c) = f(c) + g(c)$ and $h(x) = f(x)g(x)$ implies $h(c) = f(c)g(c)$, it follows that α preserves $+$ and $\cdot$. Therefore, α is a homomorphism of $R[x]$ onto $R[c]$. Clearly $\alpha(x) = c$ and $\alpha(a) = a$ for all $a \in R$. Let β be a homomorphism of $R[x]$ onto $R[c]$ such that $\beta(x) = c$ and $\beta(a) = a$ for all $a \in R$. Then $\beta(\sum a_i x^i) = \sum \beta(a_i)\beta(x)^i = \sum a_i c^i = \alpha(\sum a_i x^i)$. Thus, $\beta = \alpha$ so α is unique. ∎

We emphasize that α is well defined in Theorem 14.1.14 because x is algebraically independent over R. We illustrate this in the following example.

Example 14.1.15 *Define $\alpha : \mathbf{Q}[\sqrt{2}] \to \mathbf{Q}[x]$ by $\alpha(\sum a_i \sqrt{2}) = \sum a_i x^i$. Then α is not a function since $\alpha(2) = 2$ and $\alpha(2) = \alpha((\sqrt{2})^2) = x^2$, but $2 \neq x^2$.*

14.1.1 Worked-Out Exercises

$\diamond$ **Exercise 1** Let R be a ring with 1. Show that

$$R[x]/\langle x \rangle \simeq R.$$

Solution: Define $f : R[x] \to R$ by

$$f(a_0 + a_1 x + a_2 x^2 + \cdots + a_n x^n) = a_0$$

for all $a_0 + a_1 x + a_2 x^2 + \cdots + a_n x^n \in R[x]$. Suppose that $a_0 + a_1 x + a_2 x^2 + \cdots + a_n x^n = b_0 + b_1 x + b_2 x^2 + \cdots + b_m x^m$. Then $a_0 = b_0$ and so $f(a_0 + a_1 x + a_2 x^2 + \cdots + a_n x^n) = f(b_0 + b_1 x + b_2 x^2 + \cdots + b_m x^m)$. Thus, f is well defined. Clearly f is an epimorphism. Now $a_0 + a_1 x + a_2 x^2 + \cdots + a_n x^n \in \text{Ker } f$ if and only if $f(a_0 + a_1 x + a_2 x^2 + \cdots + a_n x^n) = 0$ if and only if $a_0 = 0$ if and only if $a_0 + a_1 x + a_2 x^2 + \cdots + a_n x^n \in \langle x \rangle$. Therefore, Ker $f = \langle x \rangle$. Thus,

$$R[x]/\langle x \rangle \simeq R.$$

Exercise 2 Let F be a field and $\alpha : F[x] \to F[x]$ be an automorphism such that $\alpha(a) = a$ for all $a \in F$. Show that $\alpha(x) = ax + b$ for some $a, b \in F$.

Solution: By the division algorithm, $\alpha(x) = g(x)x + b$ for some $g(x) \in F[x]$ and $b \in F$. Since α is onto $F[x]$, there exist $h(x), p(x) \in F[x]$ such that $g(x) = \alpha(h(x))$ and $x = \alpha(p(x))$. Therefore, $\alpha(x) = g(x)x + b = \alpha(h(x))\alpha(p(x)) + \alpha(b) = \alpha(h(x)p(x) + b)$. Thus, $x = h(x)p(x) + b$ since α is one-one. Now $\deg(x) = \deg(h(x)p(x) + b)$ implies that $\deg(h(x)p(x)) = 1$. Hence, either

$\deg h(x) = 1$ and $\deg p(x) = 0$ or $\deg h(x) = 0$ and $\deg p(x) = 1$. Suppose $\deg p(x) = 0$. Then $p(x) = c$ for some $c \in F$. This implies that $x = \alpha(p(x)) = \alpha(c) = c$, which is a contradiction. Therefore, $\deg h(x) = 0$ and $\deg p(x) = 1$. Let $h(x) = a$ for some $a \in F$. Thus, $\alpha(x) = \alpha(h(x))x + b = \alpha(a)x + b = ax + b$.

$\diamond$ **Exercise 3** Let R be a commutative ring with 1 and $f(x) = a_0 + a_1 x + a_2 x^2 + \cdots + a_n x^n \in R[x]$. If a_0 is a unit and $a_1, a_2, \ldots, a_n$ are nilpotent elements, prove that $f(x)$ is invertible.

Solution: We prove this result by induction on $n = \deg f(x)$. If $n = 0$, then $f(x) = a_0$. Hence, $f(x)$ is invertible. Assume that the result is true for all polynomials of the above form and degree $< n$. Suppose now $f(x) = a_0 + a_1 x + a_2 x^2 + \cdots + a_n x^n \in R[x]$ such that a_0 is a unit and $a_1, a_2, \ldots, a_n$ are nilpotent elements and $\deg f(x) = n$. Let $g(x) = a_0 + a_1 x + a_2 x^2 + \cdots + a_{n-1} x^{n-1}$. Note that $\deg g(x) < n$. Hence, by the induction hypothesis, $g(x)$ is invertible. Since a_n is nilpotent there exists a positive integer m such that $a_n^m = 0$. Then $(g(x) + a_n x^n)(g(x)^{-1} - a_n g(x)^{-2} x^n + a_n^2 g(x)^{-3} x^{2n} - \cdots + (-1)^{m-1} a_n^{m-1} g(x)^{-(m-1)} x^{(m-1)n}) = 1$. It now follows that $f(x)$ is invertible.

14.1.2 Exercises

1. If I is an ideal of a ring R, prove that $I[x]$ is an ideal of the polynomial ring $R[x]$.

2. Let R be an integral domain. Prove that R and $R[x]$ have the same characteristic.

3. Let R be a commutative ring with 1. Describe, $\langle x \rangle$, the ideal of $R[x]$ generated by x.

4. (i) Let $f(x) = x^4 + 3x^3 + 2x^2 + 2$ and $g(x) = x^2 + 2x + 1 \in Q[x]$. Find the unique polynomials $q(x)$, $r(x) \in Q[x]$ such that $f(x) = q(x)g(x) + r(x)$, where either $r(x) = 0$ or $0 \le \deg r(x) < \deg g(x)$.

   (ii) Let $f(x) = x^4 + [3]x^3 + [2]x^2 + [2]$ and $g(x) = x^2 + [2]x + [1] \in Z_5[x]$. Find $q(x)$, $r(x) \in Z_5[x]$ such that $f(x) = q(x)g(x) + r(x)$, where either $r(x) = 0$ or $0 \le \deg r(x) < \deg g(x)$.

5. Let $f(x) = x^5 + x^4 + x^3 + x + [3]$, $g(x) = x^4 + x^3 + [2]x^2 + [2]x \in Z_5[x]$. Find $q(x), r(x) \in Z_5[x]$ such that $f(x) = q(x)g(x) + r(x)$, where either $r(x) = 0$ or $0 \le \deg r(x) < \deg g(x)$.

6. Let $R = Z \oplus Z$. Show that the polynomial $(1,0)x$ in $R[x]$ has infinitely many roots in R.

7. Show that the polynomial ring $Z_4[x]$ over the ring Z_4 is infinite, but $Z_4[x]$ is of finite characteristic.

8. In the ring $\mathbf{Z}_8[x]$, show that $[1] + [2]x$ is a unit.

9. Let R be a commutative ring with 1 and $f(x) = a_0 + a_1x + \cdots + a_nx^n \in R[x]$. If $f(x)$ is a unit in $R[x]$, prove that a_0 is a unit in R and a_i is nilpotent for all $i = 1, 2, \ldots, n$.

10. Use the result of Exercise 9 to show that $1 + 5x$ is not a unit in $\mathbf{Z}[x]$.

11. Find all units of $\mathbf{Z}[x]$.

12. Find all units of $\mathbf{Z}_6[x]$.

13. Let R be an integral domain. Prove that the units of $R[x]$ are contained in R.

14. In $\mathbf{Z}_8[x]$, prove the following.

    (i) $[4]x^2 + [2]x + [4]$ is a zero divisor.

    (ii) $[2]x$ is nilpotent.

    (iii) $[4]x + [1]$ and $[4]x + [3]$ are units.

15. Let R be a subring of a commutative ring S such that R has an identity.

    (i) In the polynomial ring $R[x_1, x_2, \ldots, x_n]$, prove that $x_1, x_2, \ldots, x_n$ are algebraically independent over R.

    (ii) Prove that the mapping

    $$\alpha : R[x_1, x_2, \ldots, x_n] \to R[c_1, c_2, \ldots, c_n]$$

    defined by $\alpha\left(\sum_{i_n \ldots i_1} r_{i_1 \ldots i_n} x_1^{i_1} \cdots x_n^{i_n} \right) = \sum_{i_n \ldots i_1} r_{i_1 \ldots i_n} c_1^{i_1} \cdots c_n^{i_n}$ is a homomorphism of $R[x_1, \ldots, x_n]$ onto $R[c_1, \ldots, c_n]$, where $c_1, \ldots, c_n \in S$.

    (iii) Prove that the homomorphism α in (ii) is an isomorphism if and only if $c_1, c_2, \ldots, c_n$ are algebraically independent over R.

16. Let $f(x)$ be a polynomial of degree $n > 0$ in a polynomial ring $K[x]$ over a field K. Prove that any element of the quotient ring $K[x]/\langle f(x) \rangle$ is of the form $g(x) + \langle f(x) \rangle$, where $g(x)$ is a polynomial of degree at most $n - 1$.

17. For the following statements, write the proof if the statement is true; otherwise, give a counterexample.

    (i) If a polynomial ring $R[x]$ has zero divisors, so does R.

    (ii) If R is a field, then $R[x]$ is a field.

    (iii) In $\mathbf{Z}_7[x]$, $(x + [1])^7 = x^7 + [1]$.

Chapter 15

Euclidean Domains

We have seen that both rings $\mathbf{Z}$ and $F[x]$, F a field, have a Euclidean or division algorithm. Because of the significance of these rings and the power of this common property, the concept of a division algorithm is worth abstracting.

15.1   Euclidean Domains

Definition 15.1.1 *A **Euclidean domain** $(E, +, \cdot, v)$ is an integral domain $(E, +, \cdot)$ together with a function $v : E\backslash\{0\} \to \mathbf{Z}^{\#}$ such that*
   (i) for all $a, b \in E$ with $b \neq 0$, there exist $q, r \in E$ such that $a = qb + r$, where either $r = 0$ or $v(r) < v(b)$ and
   (ii) for all $a, b \in E\backslash\{0\}$, $v(a) \leq v(ab)$.
   *v is called a **Euclidean valuation**.*

The next two results show that the ring $\mathbf{Z}$ and the polynomial ring $F[x]$, F a field, are Euclidean domains.

Example 15.1.2 *The ring $\mathbf{Z}$ of integers can be considered a Euclidean domain with $v(a) = |a|$, $a \neq 0$.*

Theorem 15.1.3 *If F is a field, then the polynomial ring $F[x]$ is a Euclidean domain.*

Proof.   By Theorem 14.1.2(ii), $F[x]$ is an integral domain. Define

$$v : F[x]\backslash\{0\} \longrightarrow \mathbf{Z}^{\#}$$

by

$$v(f(x)) = \deg f(x)$$

for all $f(x) \in F[x]\backslash\{0\}$. Since $\deg f(x) \geq 0$, $v(f(x)) \in \mathbf{Z}^{\#}$ for all $f(x) \in F[x]\backslash\{0\}$. Let $f(x), g(x) \in F[x]$, $g(x) \neq 0$. By Theorem 14.1.6, there exist $q(x), r(x) \in F[x]$ such that

$$f(x) = q(x)g(x) + r(x), \text{ where either } r(x) = 0 \text{ or } \deg r(x) < \deg g(x).$$

Hence,

$$f(x) \quad = \quad q(x)g(x) + r(x), \text{ where either } r(x) = 0 \text{ or } v(r(x)) < v(g(x)).$$

Let $f(x) = a_0 + a_1 x + \cdots + a_n x^n$, $a_n \neq 0$ and $g(x) = b_0 + b_1 x + \cdots + b_m x^m$, $b_m \neq 0$. Then $f(x)g(x) = a_0 b_0 + (a_0 b_1 + a_1 b_0)x + \cdots + a_n b_m x^{n+m}$. Since F is a field and $a_n \neq 0$, $b_m \neq 0$, we find that $a_n b_m \neq 0$. This implies that $\deg(f(x)g(x)) = n + m$. Thus, $v(f(x)) = \deg(f(x)) = n \leq n + m = \deg(f(x)g(x)) = v(f(x)g(x))$. Hence, $F[x]$ is a Euclidean domain. ∎

Example 15.1.4 *Any field can be considered as a Euclidean domain with* $v(a) = 1$ *for all* $a \neq 0$. $(a = (ab^{-1})b + 0.)$

Definition 15.1.5 *The subset* $\mathbf{Z}[i] = \{a + bi \mid a, b \in \mathbf{Z}\}$ *of the complex numbers is called the set of **Gaussian integers**.*

In the next theorem, we show that $\mathbf{Z}[i]$ is a subring of $\mathbf{C}$ and determine the units of $\mathbf{Z}[i]$. Gauss was the first to study $\mathbf{Z}[i]$ and hence in his honor $\mathbf{Z}[i]$ is called the **ring of Gaussian integers**.

Theorem 15.1.6 *The set* $\mathbf{Z}[i]$ *of Gaussian integers is a subring of* $\mathbf{C}$. *The units of* $\mathbf{Z}[i]$ *are* ± 1 *and* $\pm i$.

Proof.   It is easily verified that $\mathbf{Z}[i]$ is a subring of $\mathbf{C}$. Since $\mathbf{C}$ is a field, $\mathbf{Z}[i]$ is of course an integral domain. Suppose $a + bi$ is a unit of $\mathbf{Z}[i]$. Then there exists $c + di \in \mathbf{Z}[i]$ such that $(a + bi)(c + di) = 1$. This implies that $1 = \bar{1} = \overline{(a + bi)(c + di)} = \overline{(a + bi)} \, \overline{(c + di)} = (a - bi)(c - di)$, where the bar denotes complex conjugate. Thus, $1 = (a^2 + b^2)(c^2 + d^2)$ and therefore $1 = a^2 + b^2$. Hence, $a = 0, b = \pm 1$, or $a = \pm 1, b = 0$, proving that the only units of $\mathbf{Z}[i]$ are $\pm 1, \pm i$. ∎

Theorem 15.1.7 *The ring* $\mathbf{Z}[i]$ *of Gaussian integers becomes a Euclidean domain when we let the function,*

$$N : \mathbf{Z}[i]\backslash\{0\} \to \mathbf{Z}^{\#}$$

defined by $N(a + bi) = (a + bi)(a - bi) = a^2 + b^2$ *for all* $a, b \in \mathbf{Z}$, *serve as the function* v.

Proof. Clearly $N(a+bi)$ is a positive integer for any nonzero element $a+bi \in \mathbf{Z}[i]$. Let $a+bi, c+di \in \mathbf{Z}[i]$. Now $N((a+bi)(c+di)) = N(ac-bd+(bc+ad)i) = (ac-bd)^2 + (bc+ad)^2 = (a^2+b^2)(c^2+d^2) = N(a+bi)N(c+di)$. From this, it follows that $N(a+bi) \leq N((a+bi)(c+di))$.

It remains to be shown that for $a+bi$ and $c+di \neq 0$ in $\mathbf{Z}[i]$, there exist $q_0 + q_1 i, r_0 + r_1 i \in \mathbf{Z}[i]$ such that

$$a + bi = (q_0 + q_1 i)(c + di) + (r_0 + r_1 i),$$

where $r_0 + r_1 i = 0$ or $N(r_0 + r_1 i) < N(c + di)$. We work backward in order to see how to choose $q_0 + q_1 i$. If such an element $q_0 + q_1 i$ exists, then in $\mathbf{C}$

$$
\begin{aligned}
r_0 + r_1 i &= (a + bi) - (c + di)(q_0 + q_1 i) \\
&= (c + di)[(a + bi)(c + di)^{-1} - (q_0 + q_1 i)].
\end{aligned}
$$

Let $(a + bi)(c + di)^{-1} = u + vi$, where u and v are rational numbers. Then

$$
\begin{aligned}
r_0 + r_1 i &= (c + di)[(u + vi) - (q_0 + q_1 i)] \\
&= (c + di)[(u - q_0) + (v - q_1)i] \\
&= [c(u - q_0) - d(v - q_1)] + [c(v - q_1) + d(u - q_0)]i.
\end{aligned}
$$

Now

$$
\begin{aligned}
N(r_0 + r_1 i) &= [c(u - q_0) - d(v - q_1)]^2 + [c(v - q_1) + d(u - q_0)]^2 \\
&= (c^2 + d^2)[(u - q_0)^2 + (v - q_1)^2].
\end{aligned}
$$

Hence, $N(r_0 + r_1 i) < N(c + di)$ if $(u - q_0)^2 + (v - q_1)^2 < 1$. We now find an element $q_0 + q_1 i \in \mathbf{Z}[i]$ so that the latter inequality holds. Take integers q_0 and q_1 such that $(u - q_0)^2 \leq \frac{1}{4}$ and $(v - q_1)^2 \leq \frac{1}{4}$. Then $(u - q_0)^2 + (v - q_1)^2 < 1$. Let

$$r_0 + r_1 i = (a + bi) - (c + di)(q_0 + q_1 i).$$

Then $a + bi = (c + di)(q_0 + q_1 i) + (r_0 + r_1 i)$, where $r_0 + r_1 i = 0$ or $N(r_0 + r_1 i) < N(c + di)$. ∎

We now consider the ideals of a Euclidean domain.

Recall that an ideal I of a ring R is called a principal ideal if $I = \langle a \rangle$ for some $a \in I$.

Definition 15.1.8 *Let R be a commutative ring with 1. If every ideal of R is a principal ideal, then R is called a **principal ideal ring**. An integral domain which is also a principal ideal ring is called a **principal ideal domain (PID)**.*

Theorem 15.1.9 *Every Euclidean domain is a principal ideal domain.*

Proof.   Let E be a Euclidean domain with Euclidean valuation v. We want to show that every ideal of E is a principal ideal. Let I be an ideal of E. Since E is a commutative ring with 1, it is enough to show that $I = Ea$ for some $a \in E$. If I is the zero ideal, then $I = E0$. Suppose now $I \neq \{0\}$. Then I contains some nonzero element. Let $P = \{v(x) \mid 0 \neq x \in I\}$. This is a nonempty subset of the nonnegative integers. By the well-ordering principle, we find that P contains a least element. Therefore, there exists an element $a \in I$, $a \neq 0$ such that $v(a) \geq 0$ and $v(a) \leq v(b)$ for all $b \in I$, $b \neq 0$. We now show that $I = Ea$. Since I is an ideal and $a \in I$, it follows that $Ea \subseteq I$. Let $b \in I$. Since E is a Euclidean domain, there exist $q, r \in E$ such that $b = aq + r$, where $r = 0$ or $v(r) < v(a)$. Now $r = b - qa \in I$. If $r \neq 0$, then $v(r) \in P$. This is a contradiction of the minimality of $v(a)$ since $v(r) < v(a)$. Therefore, $r = 0$ and so $b = qa \in Ea$. This proves that $I \subseteq Ea$. Hence, $I = Ea$. ∎

By Theorem 15.1.9, $\mathbf{Z}$, $F[x]$ (F a field), and $\mathbf{Z}[i]$ are principal ideal domains.

Theorem 15.1.10 *Let R be a commutative ring with 1. The following conditions are equivalent.*
  (i) R is a field.
  (ii) $R[x]$ is a Euclidean domain.
  (iii) $R[x]$ is a PID.

Proof.   (i)$\Rightarrow$(ii) Follows from Theorem 15.1.3.
  (ii)$\Rightarrow$(iii) Follows from Theorem 15.1.9.
  (iii)$\Rightarrow$(i) Let $a \in R$ and $a \neq 0$. Consider $I = \langle a, x \rangle$, the ideal of $R[x]$ generated by a and x. Since $R[x]$ is a PID, there exists $f(x) \in R[x]$ such that $I = \langle f(x) \rangle$. Now $a, x \in \langle f(x) \rangle$. Therefore, there exist $g(x)$ and $h(x)$ in $R[x]$ such that $f(x)g(x) = a$ and $f(x)h(x) = x$. Since $f(x)g(x) = a$, we must have $\deg f(x) = 0$ and so $f(x) \in R$. Let $f(x) = b$. Now $bh(x) = x$ implies that $bc = 1$ for some $c \in R$. Thus, b is a unit and so $I = \langle b \rangle = R[x]$. From this, we have $1 \in I$. Therefore, $1 = af_1(x) + xf_2(x)$ for some $f_1(x), f_2(x) \in R[x]$. This implies that $1 = da$ for some $d \in R$. Hence, a is a unit in R and so R is a field. ∎

Corollary 15.1.11 $\mathbf{Z}[x]$ *is not a PID.*

Proof.   Now $\mathbf{Z}$ is a commutative ring with 1. Since $\mathbf{Z}$ is not a field, $\mathbf{Z}[x]$ is not a PID by Theorem 15.1.10. ∎

We conclude this section with the following remark.

Remark 15.1.12 *Consider $\mathbf{Z}[\sqrt{-19}] = \{a + b\sqrt{-19} \mid a, b \in \mathbf{Z}$ and a and b are either both even or both odd\}. It is known that $\mathbf{Z}[\sqrt{-19}]$ is a principal ideal* *domain, but not a Euclidean domain. The proof of this result is beyond the scope of this book. However, the interested reader can find the proof in, J.C. Wilson, "A principal ideal ring that is not a Euclidean ring," Mathematics Magazine* 46(1973), 34 − 38.

15.1.1 Worked-Out Exercises

◇ **Exercise 1** Let $(E, +, \cdot, v)$ be a Euclidean domain.

(i) Show that $v(a) = v(-a)$ for all $a \in E \backslash \{0\}$.

(ii) Show that for all $a \in E \backslash \{0\}$, $v(a) \geq v(1)$, where equality holds if and only if a is a unit in E.

(iii) Let n be an integer such that $v(1) + n \geq 0$. Show that the function

$$v_n : E \backslash \{0\} \to \mathbf{Z}^\#$$

defined by $v_n(a) = v(a) + n$ for all $a \in E \backslash \{0\}$ is a Euclidean valuation.

Solution: (i) For all $a \in E \backslash \{0\}$, $v(a) = v((-1)(-a)) \geq v(-a) = v((-1)a) \geq v(a)$. Hence, $v(a) = v(-a)$ for all $a \in E \backslash \{0\}$.

(ii) Let $a \in E \backslash \{0\}$. Now $v(a) = v(1a) \geq v(1)$. Suppose a is a unit. Then there exists an element $c \in E$ such that $ac = 1$. Thus, $v(1) = v(ac) \geq v(a)$. This implies that $v(a) = v(1)$. Conversely, suppose that $v(a) = v(1)$. Since $a \neq 0$, there exist $q, r \in E$ such that $1 = qa + r$, where $r = 0$ or $v(r) < v(1)$. Now $v(r) < v(1)$ is impossible. Hence, $r = 0$, showing that $1 = qa$. Thus, a is a unit.

(iii) Let $a \in E \backslash \{0\}$. Then $v_n(a) = v(a) + n \geq v(1) + n \geq 0$. Hence, $v_n(a) \in \mathbf{Z}^\#$. Suppose $a, b \in E$ with $b \neq 0$. There exist $q, r \in E$ such that $a = qb + r$, where either $r = 0$ or $v(r) < v(b)$. Now $v(r) < v(b)$ implies that $v(r) + n < v(b) + n$. Thus, $v_n(r) < v_n(b)$. Also, for $a, b \in E \backslash \{0\}$, $v_n(ab) = v(ab) + n \geq v(a) + n = v_n(a)$. Therefore, v_n is a Euclidean valuation on E.

◇ **Exercise 2** Let n be a square free integer (an integer different from 0 and 1, which is not divisible by the square of any integer). Let $\mathbf{Z}[\sqrt{n}] = \{a + b\sqrt{n} \mid a, b \in \mathbf{Z}\}$. Show that $\mathbf{Z}[\sqrt{n}]$ is an integral domain. Define a function $N : \mathbf{Z}[\sqrt{n}] \to \mathbf{Z}^\#$ by

$$N(a + b\sqrt{n}) = (a + b\sqrt{n})(a - b\sqrt{n}) = a^2 - nb^2.$$

(i) Let $x \in \mathbf{Z}[\sqrt{n}]$. Prove that $N(x) = 0$ if and only if $x = 0$.

(ii) Prove that $N(xy) = N(x)N(y)$ for all $x, y \in \mathbf{Z}[\sqrt{n}]$.

(iii) Let $x \in \mathbf{Z}[\sqrt{n}]$. Prove that $N(x) = \pm 1$ if and only if x is a unit in $\mathbf{Z}[\sqrt{n}]$.

Solution: Let $x = a + b\sqrt{n}$ and $y = c + d\sqrt{n}$ be two elements in $\mathbf{Z}[\sqrt{n}]$. Now $x - y = (a-c) + (b-d)\sqrt{n} \in \mathbf{Z}[\sqrt{n}]$ and $xy = (ac+nbd) + (ad+bc)\sqrt{n} \in \mathbf{Z}[\sqrt{n}]$. We have $0 = 0 + 0\sqrt{n} \in \mathbf{Z}[\sqrt{n}]$ and $1 = 1 + 0\sqrt{n} \in \mathbf{Z}[\sqrt{n}]$. Now it is easy to verify that $\mathbf{Z}[\sqrt{n}]$ is an integral domain.

(i) Let $x = a + b\sqrt{n}$. Then $N(x) = a^2 - nb^2$. Suppose $N(x) = 0$. If $b = 0$, then $a = 0$. If $b \neq 0$, then $n = \frac{a^2}{b^2} = (\frac{a}{b})^2$, which is a contradiction to the assumption that n is a square free integer. Therefore, $a = 0$ and $b = 0$. Thus, $x = 0$. The converse is trivial.

(ii) Let $x = a + b\sqrt{n}$ and $y = c + d\sqrt{n}$. Now

$$
\begin{aligned}
N(xy) &= [(ac+nbd) + (ad+bc)\sqrt{n}][(ac+nbd) - (ad+bc)\sqrt{n}] \\
&= (ac+nbd)^2 - (ad+bc)^2 n \\
&= a^2c^2 + n^2b^2d^2 - a^2d^2n - b^2c^2n \\
&= (a^2 - nb^2)(c^2 - nd^2) \\
&= N(x)N(y).
\end{aligned}
$$

(iii) Let $x = a + b\sqrt{n}$. $N(x) = \pm 1$ if and only if $(a + b\sqrt{n})(a - b\sqrt{n}) = \pm 1$ if and only if $a + b\sqrt{n}$ divides 1, i.e., if and only if $a + b\sqrt{n}$ is a unit in $\mathbf{Z}[\sqrt{n}]$.

$\diamondsuit$ **Exercise 3** Show that $\mathbf{Z}[\sqrt{n}]$ is a Euclidean domain for $n = -1, -2, 2, 3$.

Solution: By Worked-Out Exercise 2 (page 349), $\mathbf{Z}[\sqrt{n}]$ is an integral domain. Define $v : \mathbf{Z}[\sqrt{n}] \backslash \{0\} \to \mathbf{Z}^{\#}$ by $v(a + b\sqrt{n}) = |N(a + b\sqrt{n})|$, where N is defined as in Worked-Out Exercise 2. Let $a + b\sqrt{n}, c + d\sqrt{n} \in \mathbf{Z}[\sqrt{n}] \backslash \{0\}$. Now

$$
\begin{aligned}
v((a + b\sqrt{n})(c + d\sqrt{n})) &= |N((a + b\sqrt{n})(c + d\sqrt{n}))| \\
&= |(a^2 - nb^2)(c^2 - nd^2)| \\
&= |(a^2 - nb^2)||(c^2 - nd^2)| \\
&\geq |(a^2 - nb^2)| \\
&= v((a + b\sqrt{n})).
\end{aligned}
$$

Let $a + b\sqrt{n}, c + d\sqrt{n} \in \mathbf{Z}[\sqrt{n}]$ with $c + d\sqrt{n} \neq 0$. We want to show that there exist $q_0 + q_1\sqrt{n}, r_0 + r_1\sqrt{n} \in \mathbf{Z}[\sqrt{n}]$ such that

$$
a + b\sqrt{n} = (c + d\sqrt{n})(q_0 + q_1\sqrt{n}) + (r_0 + r_1\sqrt{n}),
$$

where either $r_0 + r_1\sqrt{n} = 0$ or $|(r_0^2 - nr_1^2)| < |(c^2 - nd^2)|$. We work backward in order to see how to choose $q_0 + q_1\sqrt{n}$. If such an element $q_0 + q_1\sqrt{n}$ exists in $\mathbf{Z}[\sqrt{n}]$, then in $\mathbf{Q}[\sqrt{n}]$

$$
\begin{aligned}
r_0 + r_1\sqrt{n} &= (a + b\sqrt{n}) - (c + d\sqrt{n})(q_0 + q_1\sqrt{n}) \\
&= (c + d\sqrt{n})[(a + b\sqrt{n})(c + d\sqrt{n})^{-1} - (q_0 + q_1\sqrt{n})].
\end{aligned}
$$

Let $(a + b\sqrt{n})(c + d\sqrt{n})^{-1} = u + v\sqrt{n}$, where u and v are rational numbers. Then

$$
\begin{aligned}
r_0 + r_1\sqrt{n} &= (c + d\sqrt{n})[(u + v\sqrt{n}) - (q_0 + q_1\sqrt{n})] \\
&= (c + d\sqrt{n})[(u - q_0) + (v - q_1)\sqrt{n}] \\
&= [c(u - q_0) + d(v - q_1)n] + [c(v - q_1) + d(u - q_0)]\sqrt{n}.
\end{aligned}
$$

Now

$$v(r_0 + r_1\sqrt{n}) = \left| [c(u - q_0) + d(v - q_1)n]^2 - [c(v - q_1) + d(u - q_0)]^2 n \right|$$
$$= \left| (c^2 - nd^2)[(u - q_0)^2 - n(v - q_1)^2] \right|$$
$$< \left| (c^2 - nd^2) \right|$$

if $\left| (u - q_0)^2 - n(v - q_1)^2 \right| < 1$. We now find an element $q_0 + q_1\sqrt{n} \in \mathbf{Z}[\sqrt{n}]$ such that $\left| (u - q_0)^2 - n(v - q_1)^2 \right| < 1$. Take integers q_0 and q_1 such that $(u - q_0)^2 \leq \frac{1}{4}$ and $(v - q_1)^2 \leq \frac{1}{4}$. For $n = -1$ or -2,

$$\left| (u - q_0)^2 - n(v - q_1)^2 \right| \leq \frac{1}{4} + (-n)\frac{1}{4} < 1.$$

For $n = 2$ or 3,

$$-\frac{n}{4} \leq (u - q_0)^2 - n(v - q_1)^2 \leq \frac{1}{4}.$$

Then $\left| (u - q_0)^2 - n(v - q_1)^2 \right| < 1$ for $n = -1, -2, 2$ or 3. Hence, there exist $q_0 + q_1\sqrt{n}, r_0 + r_1\sqrt{n} \in \mathbf{Z}[\sqrt{n}]$ such that

$$a + b\sqrt{n} = (c + d\sqrt{n})(q_0 + q_1\sqrt{n}) + (r_0 + r_1\sqrt{n}),$$

where either $r_0 + r_1\sqrt{n} = 0$ or $\left| (r_0^2 - nr_1^2) \right| < \left| (c^2 - nd^2) \right|$.

◇ **Exercise 4** Let $\mathbf{Z}[i\sqrt{3}] = \{a + bi\sqrt{3} \mid a, b \in \mathbf{Z}\}$. Show that $\mathbf{Z}[i\sqrt{3}]$ is an integral domain. Define $v : \mathbf{Z}[i\sqrt{3}]\backslash\{0\} \to \mathbf{Z}^{\#}$ by $v(a + bi\sqrt{3}) = a^2 + 3b^2$. Show that v is not a Euclidean valuation on $\mathbf{Z}[i\sqrt{3}]$.

   Solution: Proceeding as in Worked-Out Exercise 2 (page 349), we can show that $\mathbf{Z}[i\sqrt{3}]$ is an integral domain. Suppose v is a Euclidean valuation. Now 2 and $1 + i\sqrt{3}$ are elements of $\mathbf{Z}[i\sqrt{3}]$. Suppose there exist $q_0 + q_1 i\sqrt{3}$, $r_0 + r_1 i\sqrt{3} \in \mathbf{Z}[i\sqrt{3}]$ such that

$$2 = (1 + i\sqrt{3})(q_0 + q_1 i\sqrt{3}) + (r_0 + r_1 i\sqrt{3}),$$

where either $r_0 + r_1 i\sqrt{3} = 0$ or $r_0^2 + 3r_1^2 < 4$. If $r_0 + r_1 i\sqrt{3} = 0$, then

$$2 = (1 + i\sqrt{3})(q_0 + q_1 i\sqrt{3}).$$

This implies that

$$4 = v(2) = v((1 + i\sqrt{3})(q_0 + q_1 i\sqrt{3})) = 4(q_0^2 + 3q_1^2).$$

Then $q_0^2 + 3q_1^2 = 1$, which shows that $q_0 = \pm 1$, $q_1 = 0$. As a result, $2 = 1 + i\sqrt{3}$ or $2 = -(1 + i\sqrt{3})$, a contradiction. Suppose now $r_0^2 + 3r_1^2 < 4$. Then $r_0^2 + 3r_1^2 = 1$, 2, or 3. Since r_0 and r_1 are integers, $r_0^2 + 3r_1^2 \neq 2$. Suppose $r_0^2 + 3r_1^2 = 1$. Then $r_0 = \pm 1$, $r_1 = 0$. Thus,

$$2 = (1 + i\sqrt{3})(q_0 + q_1 i\sqrt{3}) + (r_0 + r_1 i\sqrt{3}),$$

whence

$$2 = q_0 - 3q_1 + r_0$$

and

$$0 = q_1 + q_0 + r_1.$$

If $r_0 = 1$ and $r_1 = 0$, then $q_0 - 3q_1 = 1$ and $q_1 + q_0 = 0$. This implies that $-2q_1 = 1$, which is impossible. Similarly, for each remaining case we can show a contradiction. Also, from $r_0^2 + 3r_1^2 = 3$, we can show a contradiction. Hence, v is not a Euclidean valuation on $\mathbf{Z}[i\sqrt{3}]$.

15.1.2   Exercises

1. Show that the mapping $v : \mathbf{Z}\setminus\{0\} \to \mathbf{N}$ defined by $v(a) = |a|^n$ for some fixed positive integer n is a Euclidean valuation on $\mathbf{Z}$.

2. In $\mathbf{Z}[\sqrt{3}]$, for $9 + 5\sqrt{3}$ and $1 + 7\sqrt{3}$, find $q_0 + q_1\sqrt{3}$, $r_0 + r_1\sqrt{3} \in \mathbf{Z}[\sqrt{3}]$ such that

$$9 + 5\sqrt{3} = (q_0 + q_1\sqrt{3})(1 + 7\sqrt{3}) + r_0 + r_1\sqrt{3},$$

where either $r_0 + r_1\sqrt{3} = 0$ or $|r_0^2 - 3r_1^2| < 146$.

3. Consider the integral domain $\mathbf{Z}[i]$. Find $q_0 + q_1 i$, $r_0 + r_1 i \in \mathbf{Z}[i]$ such that

$$3 + 7i = (q_0 + q_1 i)(1 + 2i) + r_0 + r_1 i,$$

where either $r_0 + r_1 i = 0$ or $|r_0^2 + r_1^2| < 5$.

4. Let $a = 3 + 8i$, $b = -2 + 3i \in \mathbf{Z}[i]$. Find $c, d = x + yi$ in $\mathbf{Z}[i]$ such that $a = bc + d$, where either $d = 0$ or $x^2 + y^2 < 9$.

5. Let $f : R \to S$ be an epimorphism of rings. If R is a principal ideal ring, prove that S is also a principal ideal ring.

6. Prove that the ring $\mathbf{Z}_n$ is a principal ideal ring for all $n \in \mathbf{N}$.

7. Which of the following statements are true? Justify.

   (i) $(\mathbf{Z}, +, \cdot, v)$ is a Euclidean domain, where $v(n) = n^2$ for all $n > 0$.

   (ii) $(\mathbf{Q}, +, \cdot, v)$ is a Euclidean domain, where $v(\frac{p}{q}) = |\frac{p}{q}|$ for all $\frac{p}{q} \neq 0$.

   (iii) If a ring R is a PID, then every subring of R with identity is a PID.

15.2   Greatest Common Divisors

Definition 15.2.1 *Let R be a commutative ring and a, $b \in R$ be such that $a \neq 0$. If there exists $c \in R$ such that $b = ac$, then a is said to **divide** b or a is said to be a **divisor** of b and we write $a|b$.*

When we write $a|b$, we mean that $a \neq 0$ and a divides b. The notation $a \nmid b$ will mean that a does not divide b.

Let R be a commutative ring with 1. By Definition 15.2.1, the following results follow immediately. For all $a, b, c \in R$,

(i) $a|a$, $1|a$ and $a|0$,

(ii) a is a unit if and only if $a|1$,

(iii) if $a|b$ and $b|c$, then $a|c$.

Definition 15.2.2 *Let R be a commutative ring with 1. A nonzero element $a \in R$ is said to be an **associate** of a nonzero element $b \in R$ if $a = bu$ for some unit $u \in R$.*

Example 15.2.3 *(i) In $\mathbf{Z}, 1$ and -1 are the only units. For every $a \in \mathbf{Z}$, a and $-a$ are associates.*

(ii) In $\mathbf{Z}[i]$, $1, -1, i, -i$ are the only units. Thus, $1+i, -1-i, -1+i, 1-i$ are all associates of $1+i$.

Example 15.2.4 *In the polynomial ring $F[x]$ over a field F, the units form the set $F\backslash\{0\}$. A nonconstant polynomial $f(x)$ has $uf(x)$ for an associate, where u is a unit in F.*

Theorem 15.2.5 *Let R be a commutative ring with 1 and $a, b, c \in R$.*

(i) If a is an associate of b, then b is an associate of a.

(ii) If a is an associate of b and b is an associate of c, then a is an associate of c.

(iii) Suppose R is an integral domain. Then a is an associate of b if and only if $a|b$ and $b|a$.

(iv) Suppose R is an integral domain. Then a and b are associates of each other if and only if $\langle a \rangle = \langle b \rangle$.

Proof.   (i) This result follows from the fact that the inverse of a unit is also a unit.

(ii) This result follows from the fact that the product of two units is also a unit.

(iii) Suppose a is an associate of b. Then $a = bu$ for some unit $u \in R$. This implies that $b = au^{-1}$. Hence, $a|b$ and $b|a$. Conversely, suppose that $a|b$ and $b|a$. Then there exist $q_1, q_2 \in R$ such that $a = q_1 b$ and $b = q_2 a$. Thus, $b = q_2 q_1 b$

and so $1 = q_2q_1$ by cancellation. This implies that q_1 and q_2 are units and so a and b are associates.

(iv) The result here follows from (iii) and the fact that $\langle a \rangle = \{q_2a \mid q_2 \in R\}$ and $\langle b \rangle = \{q_1b \mid q_1 \in R\}$. ∎

We now introduce the notion of a greatest common divisor in a commutative ring.

Definition 15.2.6 *Let R be a commutative ring and $a_1, a_2, \ldots, a_n$ be elements in R, not all zero. A nonzero element $d \in R$ is called a **common divisor** of $a_1, a_2, \ldots, a_n$ if $d|a_i$ for all $i = 1, 2, \ldots, n$. A nonzero element $d \in R$ is called a **greatest common divisor (gcd)** of $a_1, a_2, \ldots, a_n$ if*
(i) d is a common divisor of $a_1, a_2, \ldots, a_n$ and
(ii) if $c \in R$ is a common divisor of $a_1, a_2, \ldots, a_n$, then $c|d$.

The greatest common divisor (gcd) of two elements need not be unique. In fact, the gcd of two elements may not even exist.

Example 15.2.7 *Consider the ring $\mathbf{Z}_{10}$. Then $[4] = [4][6]$ and $[6] = [4][4]$. This shows that $[4]$ and $[6]$ are common divisors of each other. Hence, $[4]$ and $[6]$ must be greatest common divisors of $[4]$ and $[6]$. Now $[4]$ and $[6]$ are associates since $[9]$ is a unit and $[6] = [9][4]$.*

Example 15.2.8 *In the ring $\mathbf{E}$ of even integers, 2 has no divisor. Hence, 2 and no other even integer can have a common divisor.*

Example 15.2.9 *In a field F, $a|b$ and $b|a$ for all $a, b \in F$ with $a \neq 0$ and $b \neq 0$. Thus, every nonzero element is a gcd of any pair of elements.*

The next result shows that in a principal ideal ring, every pair of elements not both zero has a gcd.

Theorem 15.2.10 *Let R be a principal ideal ring and $a, b \in R$ not both zero. Then a and b have a gcd d. For every gcd d of a and b, there exist $s, t \in R$ such that $d = sa + tb$.*

Proof.   The ideal $\langle a, b \rangle$ of R must be a principal ideal, whence there exists $d \in R$ such that $\langle a, b \rangle = \langle d \rangle$. Thus, there exist $u, v \in R$ such that $a = ud$ and $b = vd$. Therefore, d is a common divisor of a and b. Since $d \in \langle a, b \rangle$, there exist $s, t \in R$ such that $d = sa + tb$. Now suppose c is any common divisor of a and b. Then there exist $u', v' \in R$ such that $a = u'c$ and $b = v'c$. Thus, $d = (su' + tv')c$ and so $c|d$. Hence, d is a gcd of a and b. Let d' be any gcd of a and b. Then $d|d'$ and $d'|d$, whence $\langle d' \rangle = \langle d \rangle = \langle a, b \rangle$. Thus, there exist $s', t' \in R$ such that $d' = s'a + t'b$. ∎

Corollary 15.2.11 *Let R be a Euclidean domain and $a, b \in R$, not both zero. Then a and b have a gcd d. For every gcd d of a and b, there exist $s, t \in R$ such that $d = sa + tb$.*

Proof. Since every Euclidean domain is a principal ideal ring, the corollary follows by Theorem 15.2.10. ∎

Proceeding as in the proof of Theorem 15.2.10, we can prove a similar result for any finite set of elements $a_1, a_2, \ldots, a_n$ (not all zero) of a principal ideal ring.

Let R be an integral domain and $a_1, a_2, \ldots, a_n \in R$, not all zero. Suppose that a gcd of $a_1, a_2, \ldots, a_n$ exists. Let d and d' be two greatest common divisors of $a_1, a_2, \ldots, a_n$. Then $d \mid d'$ and $d' \mid d$. We ask the reader to verify in Exercise 6 (page 359) that d and d' are associates. If d is a gcd of $a_1, a_2, \ldots, a_n$, then any associate of d is also a gcd of $a_1, a_2, \ldots, a_n$. Considering this, we can say that the gcd of $a_1, a_2, \ldots, a_n$ is unique in the sense that if d and d' are greatest common divisors of $a_1, a_2, \ldots, a_n$, then d and d' are associates. Hence, from now on, the gcd of $a_1, a_2, \ldots, a_n$ is denoted by $\gcd(a_1, a_2, \ldots, a_n)$. This outcome motivates the definition of associates. We will further motivate this concept when we examine unique factorization in integral domains.

In a Euclidean domain $(E, +, \cdot, v)$, we have seen that the $\gcd(a, b)$ of two elements $a, b \in E$ (a, b not both zero) exists in E. Next we give an algorithm similar to the algorithm of finding the gcd of two integers given in Chapter 1.

Let $a, b \in E$ with $b \neq 0$.

Step 1: Find q_1 and r_1 in E such that $a = q_1 b + r_1$, where $r_1 = 0$ or $v(r_1) < v(b)$. If $r_1 = 0$, then $b \mid a$ and so $\gcd(a, b) = b$. If $r_1 \neq 0$, then $\gcd(a, b) = \gcd(b, r_1)$. Thus, we need to find $\gcd(b, r_1)$.

Step 2: Find q_2 and r_2 in E such that $b = q_1 r_1 + r_2$, where $r_2 = 0$ or $v(r_2) < v(r_1)$. If $r_2 = 0$, then $\gcd(a, b) = \gcd(b, r_1) = r_1$. If $r_2 \neq 0$, then proceed to find $\gcd(r_1, r_2)$. Since $v(b) > v(r_1) > v(r_2) > \cdots$ is a strictly descending chain of nonnegative integers, the above process must stop after a finite number of steps. Therefore, there exists a positive integer n such that in the nth step there exist elements q_n and r_n in E such that $r_{n-2} = q_n r_{n-1} + r_n$, where $r_n = 0$. Thus,

$$
\begin{aligned}
\gcd(a, b) &= \gcd(b, r_1) & (a = q_1 b + r_1, \ v(r_1) < v(b)) \\
&= \gcd(r_1, r_2) & (b = q_2 r_1 + r_2, \ v(r_2) < v(r_1)) \\
&= \gcd(r_2, r_3) & (r_1 = q_3 r_2 + r_3, \ v(r_3) < v(r_2)) \\
&\ \ \vdots \qquad \vdots & \vdots \\
&= \gcd(r_{n-2}, r_{n-1}) & (r_{n-3} = q_{n-1} r_{n-2} + r_{n-1}, \\
& & \qquad v(r_{n-1}) < v(r_{n-2})) \\
&= \gcd(r_{n-1}, r_n) & (r_{n-2} = q_n r_{n-1} + r_n, \ r_n = 0).
\end{aligned}
$$

Next we find x, y in E such that $\gcd(a, b) = ax + by$.

$$
\begin{aligned}
r_{n-1} &= r_{n-3} - q_{n-1} r_{n-2} \\
&= r_{n-3} - q_{n-1}(r_{n-4} - q_{n-2} r_{n-3}) \\
&= r_{n-3}(1 + (-q_{n-1})(-q_{n-2})) + r_{n-4}(-q_{n-1}) \\
&\;\;\vdots \qquad \vdots \\
&= by + ax.
\end{aligned}
$$

15.2.1 Worked-Out Exercises

◇ **Exercise 1** Let E be a Euclidean domain. Let $a, b, q, r \in E$ be such that $b \neq 0$, $a = qb + r$, and $r \neq 0$. Show that $\gcd(a, b) = \gcd(b, r)$.

Solution: Let $\gcd(a, b) = d$ and $\gcd(b, r) = d'$. Now $d|a$ and $d|b$. Thus, $r = a - qb$ implies that $d|r$. Hence, we find that d is a common divisor of b and r and so $d'|d$. Now $d'|b$ and $d'|r$ and so $a = qb + r$ implies that $d'|a$. Therefore, d' is a common divisor of a and b and so $d|d'$. By Theorem 15.2.5(iii), it follows that d and d' are associates and so $\gcd(a, b) = \gcd(b, r)$.

Exercise 2 Let a, b, and c be three nonzero elements of a PID R. Show that there exist $x, y \in R$ such that $ax + by = c$ if and only if $\gcd(a, b)|c$.

Solution: Let $\gcd(a, b) = d$. Suppose there exist $x, y \in R$ such that $ax + by = c$. Since $d|a$ and $d|b$, we find that $d|c$. Conversely, suppose that $\gcd(a, b)|c$. Then $c = dd'$ for some $d' \in R$. Now there exist $x', y' \in R$ such that $d = ax' + by'$. Then $ax'd' + by'd' = dd' = c$. Let $x = x'd'$ and $y = y'd'$. Then $ax + by = c$.

◇ **Exercise 3** In the domain $\mathbf{Z}[i\sqrt{5}]$, prove the following:

(i) $\gcd(2, 1 + i\sqrt{5}) = 1$,

(ii) $\gcd$ of $6(1 - i\sqrt{5})$ and $3(1 + i\sqrt{5})(1 - i\sqrt{5})$ does not exist.

Solution: (i) In $\mathbf{Z}[i\sqrt{5}]$, the units are 1 and -1. Let $a + ib\sqrt{5} = \gcd(2, 1 + i\sqrt{5})$. Then $(a + ib\sqrt{5})|2$. Thus, $2 = (a + ib\sqrt{5})(c + id\sqrt{5})$ for some $c + id\sqrt{5} \in \mathbf{Z}[i\sqrt{5}]$. This implies that

$$
4 = (a^2 + 5b^2)(c^2 + 5d^2).
$$

Hence,

$$
a^2 + 5b^2 = 2, \qquad c^2 + 5d^2 = 2 \tag{15.1}
$$

or

$$
a^2 + 5b^2 = 4, \qquad c^2 + 5d^2 = 1 \tag{15.2}
$$

or

$$
a^2 + 5b^2 = 1, \qquad c^2 + 5d^2 = 4. \tag{15.3}
$$

Now Eqs. (15.1) cannot hold for any $c, d \in \mathbf{Z}$. The only integral solutions of $a^2 + 5b^2 = 4$ are $a = \pm 2$ and $b = 0$ and the only integral solutions of $a^2 + 5b^2 = 1$ are $a = \pm 1$ and $b = 0$. Thus, from Eqs. (15.2) and Eqs. (15.3) we find that $\gcd(2, 1 + i\sqrt{5}) = 1$ or 2. If $\gcd(2, 1 + i\sqrt{5}) = 2$, then $2|(1 + i\sqrt{5})$. Hence, $1 + i\sqrt{5} = 2(p + iq\sqrt{5})$ for some $p + iq\sqrt{5} \in \mathbf{Z}[i\sqrt{5}]$. This implies that $2p = 1 = 2q$. But there do not exist integers p and q such that $2p = 1 = 2q$. Therefore, $\gcd(2, 1 + i\sqrt{5}) = 1$.

(ii) Suppose $\gcd(6(1 - i\sqrt{5}), 3(1 + i\sqrt{5})(1 - i\sqrt{5}))$ exists. Then $\gcd(6(1 - i\sqrt{5}), 3(1 + i\sqrt{5})(1 - i\sqrt{5})) = 3(1 - i\sqrt{5}) \gcd(2, 1 + i\sqrt{5}) = 3(1 - i\sqrt{5})$. Now $(1 + i\sqrt{5})(1 - i\sqrt{5}) = 6$. Hence, 6 is a common divisor of $6(1 - i\sqrt{5})$ and $3(1 + i\sqrt{5})(1 - i\sqrt{5})$. Consequently, $6|3(1 - i\sqrt{5})$. This implies that $2|(1 - i\sqrt{5})$, which is not true in $\mathbf{Z}[i\sqrt{5}]$. Therefore, $\gcd(6(1 - i\sqrt{5}), 3(1 + i\sqrt{5})(1 - i\sqrt{5}))$ does not exist.

◇ **Exercise 4** In $\mathbf{Z}[i]$, find $\gcd(9 - 5i, -9 + 13i)$.

Solution: By Theorem 15.1.7, $\mathbf{Z}[i]$ is a Euclidean domain, where the valuation is defined by $N(a + bi) = a^2 + b^2$. Now $N(9 - 5i) = 106$ and $N(-9 + 13i) = 250$.

Step 1: $\frac{-9+13i}{9-5i} = \frac{(-9+13i)(9+5i)}{106} = \frac{-81-45i+117i-65}{106} = \frac{-146+72i}{106} = \frac{-146}{106} + \frac{72i}{106} = (-1 - \frac{40}{106}) + (1 - \frac{34}{106})i = (-1 + i) - \frac{40+34i}{106}$.

Thus, $-9 + 13i = (-1 + i)(9 - 5i) - \frac{40+34i}{106}(9 - 5i) = (-1 + i)(9 - 5i) - \frac{360+306i-200i+170}{106} = (-1 + i)(9 - 5i) - \frac{530+106i}{106} = (-1 + i)(9 - 5i) + (-5 - i)$. Note that $N(-5 - i) < N(9 - 5i)$.

Step 2: $\frac{9-5i}{-5-i} = \frac{9-5i}{-5-i} \frac{-5+i}{-5+i} = \frac{-45+9i+25i+5}{26} = \frac{-40+34i}{26} = \frac{-20+17i}{13} = \frac{-20}{13} + \frac{17}{13}i = (-1 - \frac{7}{13}) + (1 + \frac{4}{13})i = (-1 + i) + \frac{-7+4i}{13}$.

Thus, $9 - 5i = (-1 + i)(-5 - i) + \frac{-7+4i}{13}(-5 - i) = (-1 + i)(-5 - i) + \frac{35+7i-20i+4}{13} = (-1 + i)(-5 - i) + \frac{39-13i}{13} = (-1 + i)(-5 - i) + (3 - i)$. Note that $N(3 - i) < N(-5 - i)$.

Step 3: $\frac{-5-i}{3-i} = \frac{-5-i}{3-i} \frac{3+i}{3+i} = \frac{-15-5i-3i+1}{10} = \frac{-14-8i}{10} = \frac{-7-4i}{5} = \frac{-7}{5} - \frac{4i}{5} = (-1 - \frac{2}{5}) - (1 - \frac{1}{5})i = (-1 - i) + \frac{-2+i}{5}$.

Thus, $-5 - i = (-1 - i)(3 - i) + \frac{-2+i}{5}(3 - i) = (-1 - i)(3 - i) + \frac{-6+2i+3i+1}{5} = (-1 - i)(3 - i) + \frac{-5+5i}{5} = (-1 - i)(3 - i) + (-1 + i)$. Note that $N(-1 + i) < N(3 - i)$.

Step 4: $\frac{3-i}{-1+i} = \frac{3-i}{(-1+i)} \frac{-1-i}{(-1-i)} = \frac{-3-3i+i-1}{2} = \frac{-4-2i}{2} = -2 + i$.

Thus, $3 - i = (-2 + i)(-1 + i) + 0$.

Hence, $\gcd(9 - 5i, -9 + 13i) = -1 + i$.

◇ **Exercise 5** In $\mathbf{Z}[x]$, find two polynomials $f(x)$ and $g(x)$ such that $\gcd(f(x), g(x)) = 1$, but there do not exist $f_1(x)$ and $g_1(x)$ in $\mathbf{Z}[x]$ such that $1 = f(x)f_1(x) + g(x)g_1(x)$.

Solution: $x+6$ and $x+4$ are elements of $\mathbf{Z}[x]$. The $\gcd(x+6,\ x+4) = 1$. Suppose there exist $f_1(x)$ and $g_1(x)$ in $\mathbf{Z}[x]$ such that

$$1 = (x+6)f_1(x) + (x+4)g_1(x). \tag{15.4}$$

The constant term of the right-hand side in Eq. (15.4) is an even integer, whereas in the left-hand side, the constant term is 1, a contradiction. Hence, there do not exist $f_1(x)$ and $g_1(x)$ in $\mathbf{Z}[x]$ such that $1 = (x+6)f_1(x) + (x+4)g_1(x)$.

Exercise 6 Let R be a commutative ring with 1 and S denote the set of all infinite sequences $\{a_n\}$ of elements from R. Define $+$ and $\cdot$ on S by

$$\{a_n\} + \{b_n\} = \{a_n + b_n\} \text{ and}$$

$$\{a_n\} \cdot \{b_n\} = \{c_n\},$$

where

$$c_n = a_0 b_n + a_1 b_{n-1} + \cdots + a_n b_0 \text{ for all } n = 0, 1, 2, \ldots.$$

Show that

(i) S is a commutative ring with 1;

(ii) an element $\{a_n\}$ is a unit if and only if a_0 is a unit in R;

(iii) if R is a field, then S is a PID.

Solution: (i) It is easy to verify that S is a commutative ring with 1. The sequence $\{1, 0, 0, \ldots\}$ is the identity element of S.

(ii) Let $\{a_n\} \in S$. Suppose $\{a_n\}$ is a unit. Then there exists a sequence $\{b_n\}$ such that $\{a_n\}\{b_n\} = 1$. Hence, $a_0 b_0 = 1$ and so a_0 is a unit. Conversely, suppose that a_0 is a unit. We now consider the sequence $\{b_n\}$, where $b_0 = a_0^{-1}$, $b_1 = -a_0^{-1}(a_1 a_0^{-1}), \ldots, b_k = -a_0^{-1}(a_1 b_{k-1} + \cdots + a_k b_0)$, $k \geq 2$. Now $a_0 b_0 = 1$, $a_0 b_1 + a_1 b_0 = a_0(-a_0^{-1}(a_1 a_0^{-1})) + a_1 a_0^{-1} = 0, \ldots, a_k b_0 + a_{k-1} b_1 + \cdots + a_0 b_k = a_k b_0 + a_{k-1} b_1 + \cdots + a_0(-a_0^{-1}(a_1 b_{k-1} + \cdots + a_k b_0)) = 0$. Therefore, $\{a_n\}\{b_n\} = 1$, proving that $\{a_n\}$ is a unit.

(iii) Suppose R is a field. Let I be an ideal of S. If $I = \{0\}$, then I is a principal ideal. Suppose $I \neq \{0\}$. Let $\{a_n\}$ be a nonzero element of I. We define the order of a nonzero sequence $\{a_n\}$ as the first nonnegative integer n such that $a_n \neq 0$, i.e., n is a nonnegative integer such that $a_n \neq 0$ and $a_i = 0$ for $i < n$. There exists a sequence $\{a_n\}$ such that order of $\{a_n\} \leq$ order of $\{b_n\}$ for all $\{b_n\} \in I$. Suppose order of $\{a_n\} = k$. Let $\{c_n\}$ be a sequence such that $c_i = a_{k+i}$ for all $i \geq 0$. Then $\{c_n\}^{-1}$ exists and $\{c_n\}^{-1}\{a_n\} = \{d_n\} \in I$. Also, $d_k = 1$ and $d_i = 0$ for all $i \neq k$. We now show that $I = \langle\{d_n\}\rangle$. Clearly $\langle\{d_n\}\rangle \subseteq I$. Suppose $\{u_n\} \in I$. Let the order of $\{u_n\}$ be m. Then $m \geq k$. Let $\{r_n\} \in S$ be such that $r_{m-k+i} = u_{m+i}$ for all $i \geq 0$ and $r_i = 0$ for all $i \leq m - k$. It is easy to verify that $\{u_n\} = \{r_n\}\{d_n\} \in \langle\{d_n\}\rangle$. Hence, $I = \langle\{d_n\}\rangle$.

15.2.2 Exercises

1. Find all associates of (i) $3 - 2i$ in $\mathbf{Z}[i]$, (ii) $1 + i\sqrt{5}$ in $\mathbf{Z}[i\sqrt{5}]$, (iii) $[6]$ in $\mathbf{Z}_{10}$, (iv) $[4]$ in $\mathbf{Z}_5$, and (v) $[2] + x$ in $\mathbf{Z}_3[x]$.

2. Find all the units of the integral domain $\mathbf{Z}[i\sqrt{3}]$.

3. Find all the associates of $2 + x - 3x^2$ in $\mathbf{Z}[x]$.

4. Show that $[4]$ and $[6]$ are associates in $\mathbf{Z}_{10}$.

5. Find all units of the polynomial ring $\mathbf{Z}_7[x]$. Find all associates of $x^2 + [2]$ in $\mathbf{Z}_7[x]$.

6. Let R be an integral domain and $a_1, a_2, \ldots, a_n$ ($n \geq 2$) be elements of R not all zero. If d_1 and d_2 are two greatest common divisors of $a_1, a_2, \ldots, a_n$, prove that d_1 and d_2 are associates.

7. Let $(E, +, \cdot, v)$ be a Euclidean domain. Let $a, b \in E$ be such that a and b are associates. Prove that $v(a) = v(b)$.

8. Let $(E, +, \cdot, v)$ be a Euclidean domain and $a, b \in E$. If $a|b$ and $v(a) = v(b)$, prove that a and b are associates.

9. Let $(E, +, \cdot, v)$ be a Euclidean domain and a and b be nonzero elements of E. Prove that $v(ab) > v(a)$ if and only if b is not a unit.

10. Let E be a Euclidean domain. Let a, a', b, b', d be nonzero elements of E such that $a = a'd$ and $b = b'd$. Prove that $\gcd(a', b') = 1$ if and only if $\gcd(a, b) = d$.

11. In a PID R, prove that the congruence $ax \equiv b(\bmod\ c)$, where a, b, c are nonzero elements of R has a solution in R if and only if $\gcd(a, c)|b$. (Here $ax \equiv b(\bmod\ c)$ means $ax - b = cr$ for some $r \in R$.)

12. Let R be an integral domain. Let a, b, and c be nonzero elements of R such that $\gcd(a, b)$ and $\gcd(ca, cb)$ exist. Prove that $\gcd(ca, cb) = c\gcd(a, b)$.

13. In $\mathbf{Z}[i]$, find $\gcd(2 - 7i, 2 + 11i)$. Also, find x and y in $\mathbf{Z}[i]$ such that $\gcd(2 - 7i, 2 + 11i) = x(2 - 7i) + y(2 + 11i)$.

14. Let R be an integral domain and $a_1, a_2, \ldots, a_n$ ($n \geq 2$) be nonzero elements of R. An element $d \in R$ is called a **least common multiple** (**lcm**) of $a_1, a_2, \ldots, a_n$ if

    (i) $a_i|d$, $i = 1, 2, \ldots n$ and

    (ii) if $c \in R$ is such that $a_i|c$, $i = 1, 2, \ldots n$, then $d|c$.

    Prove the following in R.

(i) If d_1 and d_2 are two least common multiples of $a_1, a_2, \ldots, a_n$, then d_1 and d_2 are associates.

(ii) If d is a least common multiple of $a_1, a_2, \ldots, a_n$, then rd is a least common multiple of $ra_1, ra_2, \ldots, ra_n$, for all $r \in R$, $r \neq 0$.

15. Let I be the set of all nonunits of $\mathbf{Z}[i]$. Is I an ideal of $\mathbf{Z}[i]$? Show that for any nontrivial ideal P of $\mathbf{Z}[i]$, the quotient ring $\mathbf{Z}[i]/P$ is a finite ring.

16. Show that $\mathbf{Z}[\sqrt{2}]$ has no unit between 1 and $1 + \sqrt{2}$.

17. In the domain $\mathbf{Z}[\sqrt{2}]$, prove that an element $a + b\sqrt{2} \neq \pm 1$ is a unit if and only if $a + b\sqrt{2} = (1 + \sqrt{2})^k$ or $a + b\sqrt{2} = -(1 + \sqrt{2})^k$ for some positive integer k.

18. An integral domain R is said to satisfy the **gcd property** if every finite nonempty subset of R has a gcd. Prove that every PID satisfies the gcd property.

19. Prove that the integral domain $\mathbf{Z}[\sqrt{2}]$ satisfies the gcd property, where the gcd property is defined in Exercise 18.

15.3   Prime and Irreducible Elements

In this section, we introduce the concepts of prime elements and irreducible elements in a commutative ring with 1. We show that in a PID and hence in a Euclidean domain these two concepts coincide.

Definition 15.3.1 *Let R be a commutative ring with 1.*

*(i) An element p of R is called **irreducible** if p is nonzero and a nonunit, and $p = ab$ with $a, b \in R$ implies that either a or b is a unit. An element p of R is called **reducible** if p is not irreducible.*

*(ii) An element p of R is called **prime** if p is nonzero and a nonunit, and if whenever $p|ab$, $a, b \in R$, then either p divides a or p divides b.*

*(iii) Two elements a and b of R are called **relatively prime** if their only common divisors are units.*

Remark 15.3.2 *Let $p \in \mathbf{Z}$. If p is an ordinary prime, then both p and $-p$ are irreducible and prime in the sense of Definition 15.3.1.*

From the definition of an irreducible element, it follows that the only divisors of an irreducible element p are the associates of p and the unit elements of R. The converse of this result does not always hold in a commutative ring with 1.

Example 15.3.3 *The ring* $\mathbf{Z}_6$ *is a commutative ring with* 1. *In this ring, the unit elements are* [1] *and* [5]. *Since* [3] = [3][3] *and* [3] *is not a unit it follows that* [3] *is not irreducible. But* [3] *is an associate of* [3]. *Also, in* $\mathbf{Z}_6$, *it can be verified that* [3] *is divisible only by associates and the units of* $\mathbf{Z}_6$. *Next, we show that* [3] *is a prime element in* $\mathbf{Z}_6$. *Let* $[a], [b] \in \mathbf{Z}_6$ *and* $[3]|[a][b]$. *Then there exists* $[c] \in \mathbf{Z}_6$ *such that* $[a][b] = [3][c]$, *i.e.,* $[ab] = [3c]$. *From this, it follows that* $6|(ab - 3c)$. *This implies that* $3|(ab - 3c)$. *Since* $3|3c$, *we must have* $3|ab$. *Since* 3 *is prime in* $\mathbf{Z}$, $3|a$ *or* $3|b$. *Thus, either* $[3]|[a]$ *or* $[3]|[b]$. *Hence,* [3] *is a prime element in* $\mathbf{Z}_6$.

Theorem 15.3.4 *Let* R *be an integral domain and* $p \in R$ *be such that* p *is nonzero and a nonunit. Then* p *is irreducible if and only if the only divisors of* p *are the associates of* p *and the unit elements of* R.

Proof. Suppose the only divisors of p are the associates of p and the unit elements of R. Let $p = ab$ for some $a, b \in R$. Suppose a is not a unit. Then a is an associate of p. Therefore, $a = pu$ for some unit $u \in R$. Now $p = pub$. Since R is an integral domain, it follows that $ub = 1$. Hence, b is a unit and so p is irreducible. We leave the converse as an exercise. ∎

We now consider several examples of prime elements and irreducible elements.

Example 15.3.5 *In* $\mathbf{Z}$, 1 *and* -1 *are the only units, and therefore* 2 *is divisible by* ± 1 *and* ± 2. *It follows that* 2 *is not divisible by any other integer. Therefore,* 2 *is an irreducible element. Suppose now* $2|ab$ *and* 2 *does not divide* a *for some* $a, b \in \mathbf{Z}$. *Since* 2 *does not divide* a, a *is an odd integer and so* $\gcd(2, a) = 1$. *Therefore, there exist* $c, d \in \mathbf{Z}$ *such that* $1 = 2c + ad$. *Thus,* $b = 2cb + abd$. *Since* $2|ab$ *and* $2|2bc$, *it follows that* $2|b$. *Hence,* 2 *is prime.*

Example 15.3.6 *The polynomial* $x^2 + 1$ *is irreducible in* $\mathbf{R}[x]$, *but is reducible in* $\mathbf{C}[x]$. *If* $x^2 + 1$ *were reducible in* $\mathbf{R}[x]$, *then there would exist real numbers* a, b, c, d *such that*

$$x^2 + 1 = (ax + b)(cx + d) = acx^2 + (ad + bc)x + bd.$$

Then $ac = 1 = bd$ *and* $ad + bc = 0$. *Thus,* $1 = (ac)(bd) = (ad)(bc) = (ad)(-ad)$. *Hence,* $1 = -(ad)^2$, *which is impossible in* $\mathbf{R}$. *However,* $x^2 + 1 = (x + i)(x - i)$ *in* $\mathbf{C}[x]$.

Example 15.3.7 *The polynomial* $x^2 - 2$ *is irreducible in* $\mathbf{Q}[x]$ *and reducible in* $\mathbf{R}[x]$. *If* $x^2 - 2$ *were reducible in* $\mathbf{Q}[x]$, *then there would exist* $a, b, c, d \in \mathbf{Q}$ *such that*

$$x^2 - 2 = (ax + b)(cx + d) = acx^2 + (ad + bc)x + bd.$$

Then $ac = 1, ad + bc = 0,$ *and* $bd = -2.$ *Thus,* $(ad)^2 = (ad)(ad) = -(ad)(bc) = (ac)(-bd) = 2.$ *This implies that* $\sqrt{2} = ad \in \mathbf{Q}.$ *This is a contradiction since* $\sqrt{2} \notin \mathbf{Q}.$ *Therefore,* $x^2 - 2$ *is irreducible in* $\mathbf{Q}[x].$ *However,* $x^2 - 2 = (x - \sqrt{2})(x + \sqrt{2})$ *in* $\mathbf{R}[x].$

Example 15.3.8 *The polynomial* $ax + b$ *is irreducible in* $F[x],$ *where* F *is a field and* $a \neq 0.$ *Suppose* $ax + b = f(x)g(x).$ *Then* $\deg(f(x)g(x)) = 1 = \deg f(x) + \deg g(x).$ *We may assume that* $\deg f(x) = 0$ *and* $\deg g(x) = 1.$ *Since* $\deg f(x) = 0,$ $f(x)$ *is a nonzero constant polynomial and thus a unit. Hence,* $ax + b$ *is irreducible.*

Example 15.3.9 *Consider the polynomial ring* $\mathbf{Z}[x, y].$ *Then* x *and* y *are irreducible.* $2x$ *is not prime since* $2x|2x,$ *but* $2x$ *does not divide* 2 *and* $2x$ *does not divide* $x.$ *Also,* $2x$ *is reducible.* x^2 *and* y^2 *are relatively prime, but neither is irreducible nor prime.*

Theorem 15.3.10 *Let* R *be an integral domain and* p *be a prime element in* $R.$ *Then* p *is irreducible.*

Proof. Suppose $p = bc$ for some $b, c \in R.$ To show p is irreducible, we must show that either b is a unit or c is a unit. Now $p = bc$ implies that $p|bc.$ Since p is prime, $p|b$ or $p|c.$ If $p|b,$ then $b = pq$ for some $q \in R.$ Thus, $p = bc = pqc$ and so $p(1 - qc) = 0.$ Since R is an integral domain and $p \neq 0,$ $p(1 - qc) = 0$ and so $1 - qc = 0.$ Thus, $qc = 1,$ which implies that c is a unit. Similarly, if $p|c,$ then b is a unit. Hence, p is irreducible. ∎

The following example shows that the converse of Theorem 15.3.10 is not true.

Example 15.3.11 *Consider the integral domain*

$$\mathbf{Z}[i\sqrt{5}] = \{a + bi\sqrt{5} \mid a, b \in \mathbf{Z}\}.$$

Let us show that $3 = 3 + 0i\sqrt{5} \in \mathbf{Z}[i\sqrt{5}]$ *is irreducible, but not prime. Suppose* $3 = (a + bi\sqrt{5})(c + di\sqrt{5})$ *in* $\mathbf{Z}[i\sqrt{5}].$ *Then* $3 = \overline{3} = (a - bi\sqrt{5})(c - di\sqrt{5}).$ *Hence,* $9 = (a^2 + 5b^2)(c^2 + 5d^2).$ *Since* a, b, c, d *are integers, the previous equality implies that*

$$a^2 + 5b^2 = 3 \text{ and } c^2 + 5d^2 = 3 \tag{15.5}$$

or

$$a^2 + 5b^2 = 1 \text{ and } c^2 + 5d^2 = 9 \tag{15.6}$$

or

$$a^2 + 5b^2 = 9 \text{ and } c^2 + 5d^2 = 1. \tag{15.7}$$

Clearly there do not exist integers a, b, c, d satisfying Eqs. (15.5). The first equation of Eqs. (15.6) implies that $b = 0$ and $a = \pm 1$. Thus, it follows that $a + bi\sqrt{5}$ is a unit. Similarly, the second equation of Eqs. (15.7) implies that $c + di\sqrt{5}$ is a unit. Hence, 3 is irreducible. Now $3|6$ and $6 = (1 + i\sqrt{5})(1 - i\sqrt{5})$. Suppose $3|(1 + i\sqrt{5})$. Then $1 + i\sqrt{5} = 3(a + bi\sqrt{5})$ for some $a, b \in \mathbf{Z}$. This implies that $3a = 1$, a contradiction, since the equation $3a = 1$ has no solution in $\mathbf{Z}$. Hence, 3 does not divide $(1 + i\sqrt{5})$. Similarly, 3 does not divide $(1 - i\sqrt{5})$. Thus, 3 is not prime.

The following theorem show that the converse of Theorem 15.3.10 holds in a principal ideal ring.

Theorem 15.3.12 *Let R be a principal ideal ring and $p \in R$. If p is irreducible, then p is prime.*

Proof. Suppose p divides ab, where $a, b \in R$. Then there exists $r \in R$ such that $pr = ab$. Now $\langle p, b \rangle = \langle d \rangle$ for some $d \in R$. Therefore, there exists $q \in R$ such that $p = dq$. Since p is irreducible, either d or q must be a unit. If d is a unit, then $\langle p, b \rangle = \langle d \rangle = R$. Hence, $1 = sp + tb$ for some $s, t \in R$. Therefore, $a = asp + atb = asp + tpr = (as + tr)p$. This implies that p divides a. If, on the other hand, q is a unit, then $d = pq^{-1} \in \langle p \rangle$. Thus, $\langle d \rangle \subseteq \langle p \rangle \subseteq \langle p, b \rangle = \langle d \rangle$ so that $\langle p \rangle = \langle p, b \rangle$. Hence, $b \in \langle p \rangle$ and so p divides b. ∎

Corollary 15.3.13 *Let R be a principal ideal domain and $p \in R$. Then p is irreducible if and only if p is prime.*

Proof. The result follows by Theorems 15.3.10 and 15.3.12. ∎

Corollary 15.3.14 *Let R be a Euclidean domain and $p \in R$. Then p is irreducible if and only if p is prime.*

Proof. Since every Euclidean domain is a principal domain, the result follows from Corollary 15.3.13. ∎

Theorem 15.3.15 *Let R be a principal ideal ring and $a, b \in R$. If a and b are relatively prime, then there exist $s, t \in R$ such that $1 = sa + tb$.*

Proof. Since the common divisors are units, 1 is a gcd of a and b. The desired result follows from Theorem 15.2.10. ∎

We conclude this section by proving the following theorem, which characterizes irreducible polynomials over a field.

Theorem 15.3.16 *Consider the polynomial ring $F[x]$ over the field F and $p(x) \in F[x]$. Then the following conditions are equivalent.*
(i) $p(x)$ is irreducible.
(ii) $F[x]/\langle p(x) \rangle$ is an integral domain.
(iii) $F[x]/\langle p(x) \rangle$ is a field.

Proof. (i)$\Rightarrow$(iii). Let $\overline{f(x)} \in F[x]/\langle p(x) \rangle$ be such that $\overline{f(x)} \neq \overline{0}$, where $\overline{f(x)}$ denotes the coset $f(x) + \langle p(x) \rangle$. Now $up(x)$ and u, where $u \in F\backslash\{0\}$, are the only elements of $F[x]$ which divide $p(x)$. Since $f(x) \notin \langle p(x) \rangle$, $f(x)$ and $p(x)$ are relatively prime and so there exist $s(x), t(x) \in F[x]$ such that $1 = s(x)f(x) + t(x)p(x)$. Thus

$$\overline{1} = \overline{s(x)f(x)} + \overline{t(x)p(x)} \text{ (in } F[x]/\langle p(x) \rangle)$$

and so $\overline{1} = \overline{s(x)} \ \overline{f(x)}$. Hence, $\overline{f(x)}$ has an inverse, namely, $\overline{s(x)}$, and so $F[x]/\langle p(x) \rangle$ is a field.

(iii)$\Rightarrow$(ii): Immediate.

(ii)$\Rightarrow$(i): If $p(x)$ is a unit, then $\langle p(x) \rangle = F[x]$ and so $F[x]/\langle p(x) \rangle = \{0\}$, a contradiction to the hypothesis that $F[x]/\langle p(x) \rangle$ is an integral domain. Therefore, $p(x)$ is not a unit. Suppose $p(x) = f(x)g(x)$. Then $\overline{0} = \overline{p(x)} = \overline{f(x)g(x)} = \overline{f(x)} \ \overline{g(x)}$. Therefore, $\overline{f(x)} = \overline{0}$ or $\overline{g(x)} = \overline{0}$. This implies that $f(x) \in \langle p(x) \rangle$ or $g(x) \in \langle p(x) \rangle$, say, $f(x) \in \langle p(x) \rangle$. Thus, $f(x) = q(x)p(x)$ for some $q(x) \in F[x]$. Hence, $p(x) = q(x)p(x)g(x)$ and so by a degree argument $q(x), g(x) \in F\backslash\{0\}$ are units. Thus, the only factorization of $p(x)$ is $u^{-1}(up(x))$, where u is a unit in $F[x]$. Consequently, $p(x)$ is irreducible. ∎

15.3.1 Worked-Out Exercises

◇ **Exercise 1** Show that [2] is a prime element in $\mathbf{Z}_{10}$, but [2] is not irreducible in $\mathbf{Z}_{10}$.

Solution: In $\mathbf{Z}_{10}$, [1], [3], [7], and [9] are the only units. Now $[2] = [2] \cdot [6]$. Since neither [2] nor [6] is a unit, [2] is reducible. Suppose $[2] | [a][b]$. Then $[2] | [ab]$. Therefore, $[ab] = [k][2]$ for some $[k] \in \mathbf{Z}_{10}$. This implies that $ab - 2k = 10r$ for some $r \in \mathbf{Z}$, i.e., $ab = 2k + 10r = 2(k + 5r)$. Therefore, $2|ab$. Since 2 is prime in $\mathbf{Z}$, $2|a$ or $2|b$. Hence, $[2] | [a]$ or $[2] | [b]$. Thus, [2] is prime. Note that $\mathbf{Z}_{10}$ is not an integral domain.

◇ **Exercise 2** Let R be an integral domain such that any two elements $a, b \in R$, not both zero, have a gcd d expressible in the form $d = ra + tb$, $r, t \in R$. Let $p \in R$. Show that p is prime if and only if p is irreducible.

Solution: Every prime element in an integral domain is irreducible by Theorem 15.3.10. Let us prove the converse. Suppose p is irreducible. Let

$p|ab$, $a, b \in R$. Now $\gcd(p, a)$ exists in R. Let $d = \gcd(p, a)$. Since $d|p$ and p is irreducible, it follows that either d is an associate of p or d is a unit. Suppose d is an associate of p. Then $p|d$. This implies that $p|a$, since $d|a$. Suppose d is a unit. Since 1 is an associate of d, $1 = \gcd(p, a)$. Thus, there exist $s, t \in R$ such that $1 = ps + at$. This implies that $b = psb + abt$. Now $p|psb$ and $p|abt$. Hence, $p|b$.

$\diamond$ **Exercise 3** Let n be a square free integer (an integer different from 0 and 1, which is not divisible by the square of any integer). Let $\mathbf{Z}[\sqrt{n}] = \{a + b\sqrt{n} \mid a, b \in \mathbf{Z}\}$. Define a function $N : \mathbf{Z}[\sqrt{n}] \to \mathbf{Z}$ by

$$N(a + b\sqrt{n}) = (a + b\sqrt{n})(a - b\sqrt{n}) = a^2 - nb^2.$$

Show that if $N(x)$ is a prime integer, then x is irreducible for all $x \in \mathbf{Z}[\sqrt{n}]$.

Solution: Suppose $N(x) = p$, where p is a prime integer. Suppose $x = (a + b\sqrt{n})(c + d\sqrt{n})$. Now $p = N(a + b\sqrt{n})N(c + d\sqrt{n}) = (a^2 - nb^2)(c^2 - nd^2)$ by Worked-Out Exercise 2 (page 349). Hence, either $(a^2 - nb^2) = \pm 1$ or $(c^2 - nd^2) = \pm 1$, i.e., either $a + b\sqrt{n}$ is a unit or $c + d\sqrt{n}$ is a unit. Thus, x is irreducible.

15.3.2 Exercises

1. Show that in the integral domain $\mathbf{Z}[i\sqrt{5}]$, $2 + i\sqrt{5}$ is an irreducible element, but not a prime element.

2. Show that $2 - i$, $1 + i$, and 11 are irreducible elements in $\mathbf{Z}[i]$.

3. In $\mathbf{Z}[i\sqrt{5}]$, show that 3 is not a prime element.

4. In $\mathbf{Z}_{12}$, show that $[3]$ is a prime element, but is not irreducible.

5. Is the polynomial $x^2 + [1]$ irreducible in $\mathbf{Z}_2[x]$?

6. Let T be the set of all sequences $\{a_n\}$ of elements of $\mathbf{Z}$. Prove the following.

   (i) T is an integral domain with respect to addition and multiplication defined by for all $\{a_n\}, \{b_n\} \in T$,

   $$\begin{aligned} \{a_n\} + \{b_n\} &= \{a_n + b_n\} \\ \{a_n\} \cdot \{b_n\} &= \{c_n\}, \qquad \text{where } c_n = \sum_{i=0}^{n} a_i b_{n-i}. \end{aligned}$$

   (ii) $T_0 = \{\{a_n\} \in T \mid a_i = 0 \text{ for all but a finite number of indices}\}$ is a subring with identity.

   (iii) The element $(1, 1, 0, \ldots)$ is a unit in T, but not in T_0.

   (iv) $(2, 3, 1, 0, 0, \ldots)$ is irreducible in T, but not in T_0.

7. Let R be an integral domain. Show that (i) every associate of an irreducible element in R is irreducible and (ii) every associate of a prime element in R is prime.

8. In $\mathbf{Z}[i]$, show that 3 is a prime element, but 5 is not a prime element.

9. What are the prime elements of $\mathbf{Z}_9$? Are they irreducible?

10. In $\mathbf{Z}[i]$, if $a + bi$ is an element such that $a^2 + b^2$ is a prime integer, then show that $a + bi$ is a prime element.

11. Let $a + bi\sqrt{3} \in \mathbf{Z}[i\sqrt{3}]$. If $a^2 + 3b^2$ is a prime integer, show that $a + bi\sqrt{3}$ is an irreducible element in $\mathbf{Z}[i\sqrt{3}]$.

12. In the following exercises, write the proof if the statement is true; otherwise, give a counterexample.

    (i) 13 is an irreducible element in $\mathbf{Z}[i]$.

    (ii) Every prime element of $\mathbf{Z}$ is also a prime element of $\mathbf{Z}[i]$.

    (iii) In $\mathbf{Z}_{18}$, every prime element is an irreducible element.

    (iv) In $\mathbf{Z}[i]$, $a + bi$ is a prime element if and only if $a - bi$ is a prime element.

    (v) In a PID R, if p and q are two prime elements such that $p|q$, then p and q are associates.

Chapter 16

Unique Factorization Domains

16.1 Unique Factorization Domains

In this section, we study those integral domains in which an analogue of the fundamental theorem of arithmetic holds.

Definition 16.1.1 *A nonzero nonunit element a of an integral domain D is said to have a **factorization** if a can be expressed as*

$$a = p_1 p_2 \cdots p_n,$$

*where $p_1, p_2, \ldots, p_n$ are irreducible elements of D. The expression $p_1 p_2 \cdots p_n$ is called a **factorization** of a.*

An integral domain D is called a **factorization domain (FD)** if every nonzero nonunit element has a factorization.

In Chapter 15, we saw that in an integral domain D every nonzero element $a \in D$ is always divisible by the associates of a and the units of D. These are called the **trivial factors** of a. All other factors (if any) of a are called **nontrivial**. For example, ± 2 and ± 3 are nontrivial factors of 6 in $\mathbf{Z}$. In the following lemma, we show that a nonzero nonunit element that has no factorization as a product of irreducible elements can be expressed as a product of any number of nontrivial factors.

Lemma 16.1.2 *Let D be an integral domain. Let a be a nonzero nonunit element of D such that a does not have a factorization. Then for every positive integer n, there exist nontrivial factors $a_1, a_2, \ldots, a_n \in D$ of a such that $a = a_1 a_2 \cdots a_n$.*

Proof.   By the hypothesis, a is not irreducible. Therefore, $a = a_1b_1$, where a_1, $b_1 \in D$ are nontrivial factors of a. At least one of a_1 or b_1 does not have a factorization; otherwise the factorization of a_1 and b_1 put together produces a factorization of a. Suppose a_1 does not have a factorization. Then a_1 is a nonzero nonunit element and a_1 is not irreducible. There exist nontrivial factors a_2, $b_2 \in D$ of a_1 such that $a_1 = a_2b_2$. Then $a = a_2b_2b_1$. Now at least one of a_2 or b_2 does not have a factorization. If a_2 does not have a factorization, we repeat the above process with a_2. Proceeding this way, we can find nontrivial factors $a_1, a_2, \ldots, a_n \in D$ of a such that $a = a_1a_2 \cdots a_n$. ∎

Theorem 16.1.3 *Let D be an integral domain with a function $N : D\backslash\{0\} \to$ $\mathbf{Z}^{\#}$ such that for all $a,b \in D\backslash\{0\}$, $N(ab) \geq N(b)$, where equality holds if and only if a is a unit. Then D is a FD.*

Proof.   Suppose D contains a nonzero nonunit element a such that a does not have a factorization. Now $N(a) \in \mathbf{Z}^{\#}$. Let $N(a) = n$. By Lemma 16.1.2, a can be expressed as a product of $n + 2$ nontrivial factors $a_1, a_2, \ldots, a_{n+2} \in D$. Then $a = a_1a_2 \cdots a_{n+2}$ and

$$
\begin{aligned}
n &= N(a) \\
&> N(a_2 \cdots a_{n+2}) \quad \text{(since } a_1 \text{ is not a unit)} \\
&> N(a_3 \cdots a_{n+2}) \\
&> N(a_4 \cdots a_{n+2}) \\
&\ \ \vdots \\
&> N(a_{n+1}a_{n+2}) \\
&> N(a_{n+2}).
\end{aligned}
$$

This shows that there exist at least $n + 1$ distinct nonnegative integers strictly less than n, a contradiction. Thus, D is a FD. ∎

Example 16.1.4 *Consider the integral domain $\mathbf{Z}[i]$. Define*

$$N : \mathbf{Z}[i]\backslash\{0\} \to \mathbf{Z}^{\#}$$

by $N(a+bi) = a^2+b^2$ for all $a+bi \in \mathbf{Z}[i]$. It is easy to verify that $a+bi$ is a unit if and only if $N(a + bi) = 1$. Let $a + bi$, $c + di$ be two nonzero elements of $\mathbf{Z}[i]$. Then $N((a+bi)(c+di)) = N((ac-bd)+(ad+bc)i) = (ac-bd)^2 + (ad+bc)^2 = (a^2 + b^2)(c^2 + d^2) \geq (c^2 + d^2) = N(c + di)$, where the equality holds if and only if $N(a + bi)$ is a unit. Hence, $\mathbf{Z}[i]$ is a FD.

Definition 16.1.5 *An integral domain D is said to satisfy the **ascending chain condition for principal ideals (ACCP)**, if for each sequence of principal ideals, $\langle a_1 \rangle$, $\langle a_2 \rangle$, $\langle a_3 \rangle$, $\ldots$ such that*

$$\langle a_1 \rangle \subseteq \langle a_2 \rangle \subseteq \langle a_3 \rangle \subseteq \cdots,$$

there exists a positive integer n (depending on the sequence) such that $\langle a_n \rangle = \langle a_t \rangle$ for all $t \geq n$.

Lemma 16.1.6 *Every principal ideal domain D satisfies the ACCP.*

Proof. Let $\langle a_1 \rangle \subseteq \langle a_2 \rangle \subseteq \langle a_3 \rangle \subseteq \cdots$ be a chain of principal ideals in D. It can be easily verified that $I = \cup_{i \in \mathbf{N}} \langle a_i \rangle$ is an ideal of D. Since D is a PID, there exists an element $a \in D$ such that $I = \langle a \rangle$. Hence, $a \in \langle a_n \rangle$ for some positive integer n. Then $I \subseteq \langle a_n \rangle \subseteq I$. Therefore, $I = \langle a_n \rangle$. For $t \geq n$, $\langle a_t \rangle \subseteq I = \langle a_n \rangle \subseteq \langle a_t \rangle$. Thus, $\langle a_n \rangle = \langle a_t \rangle$ for all $t \geq n$. ∎

Theorem 16.1.7 *An integral domain D with the ACCP is a FD.*

Proof. Suppose D is not a FD. Then there exists a nonzero nonunit element a such that a does not have a factorization. Thus, a is not irreducible and so $a = a_1 b_1$, where $a_1, b_1 \in D$ are nontrivial factors of a. At least one of a_1 or b_1 must not have a factorization, otherwise the factorization of a_1 and b_1 put together will produce a factorization of a. Suppose a_1 does not have a factorization. Now a and a_1 are not associates. Therefore, $\langle a \rangle \subset \langle a_1 \rangle$. Since a_1 does not have a factorization, we can express $a_1 = a_2 b_2$, where $a_2, b_2 \in D$ are nontrivial factors of a_1. At least one of a_2 or b_2 does not have a factorization. Suppose a_2 does not have a factorization. Then $\langle a \rangle \subset \langle a_1 \rangle \subset \langle a_2 \rangle$. We now repeat the above process with a_2. Thus, we find that there exists an infinite strictly ascending chain of principal ideals in D, a contradiction. Hence, D is a FD. ∎

Corollary 16.1.8 *Every PID is a FD.*

Proof. The proof is immediate by Lemma 16.1.6 and Theorem 16.1.7. ∎

Definition 16.1.9 *An integral domain D is called a **unique factorization domain (UFD)** if the following two conditions hold in D:*
  (i) every nonzero nonunit element of D can be expressed as

$$a = p_1 p_2 \cdots p_n,$$

where $p_1, p_2, \ldots, p_n$ are irreducible elements of D and
  (ii) if $a = p_1 p_2 \cdots p_n = q_1 q_2 \cdots q_m$ are two factorizations of a as a finite product of irreducible elements of D, then $n = m$ and there is a permutation σ of $\{1, 2, \ldots, n\}$ such that p_i and $q_{\sigma(i)}$ are associates for all $i = 1, 2, \ldots, n$.

From the above definition, it follows that an integral domain D is a UFD if and only if D is a FD and every nonzero nonunit element of D is uniquely expressible (apart from unit factors and order of the factors) as a finite product of irreducible elements.

Let us first prove the following interesting property of a UFD.

Theorem 16.1.10 *In a unique factorization domain, every irreducible element is prime.*

Proof. Let D be a UFD. Let p be an irreducible element of D and $p|ab$ in D, where $a, b \in D$. If $a = 0$, then p divides a, and if $b = 0$, then p divides b. If a is a unit, then p divides b, and if b is a unit, then p divides a. We now assume that a and b are nonzero and nonunits. Now $ab = pc$ for some $c \in D$. Let $d = pc = ab$. Since neither a nor b is a unit, it follows that d is not a unit. If c is a unit, then d is irreducible and so either a or b must be a unit, a contradiction. Therefore, c is not a unit. Since D is a UFD, there exist irreducible elements $c_1, c_2, \ldots, c_n, a_1, a_2, \ldots, a_m$, and $b_1, b_2, \ldots, b_r$ in D such that $c = c_1 c_2 \cdots c_n$, $a = a_1 a_2 \cdots a_m$, and $b = b_1 b_2 \cdots b_r$. Hence, $d = pc_1 c_2 \cdots c_n = a_1 a_2 \cdots a_m b_1 b_2 \cdots b_r$ are two expressions of d as a finite product of irreducible elements. Since D is UFD, p must be an associate of one of the irreducible elements $a_1, a_2, \ldots, a_m$, $b_1, b_2, \ldots, b_r$. If one of $a_1, a_2, \ldots, a_m$ is an associate of p, then $p|a$, and if one of $b_1, b_2, \ldots, b_r$ is an associate of p, then $p|b$. Hence, p is prime. ∎

Example 16.1.11 *Consider the integral domain* $\mathbf{Z}[i\sqrt{5}] = \{a + bi\sqrt{5} \mid a, b \in \mathbf{Z}\}$. *Define*

$$N : \mathbf{Z}[i\sqrt{5}]\backslash\{0\} \to \mathbf{Z}^{\#}$$

by

$$N(a + bi\sqrt{5}) = a^2 + 5b^2.$$

We can show that $a + bi\sqrt{5}$ is a unit if and only if $N(a + bi\sqrt{5}) = 1$. Let $a + bi\sqrt{5}$, $c + di\sqrt{5}$ be two nonzero elements of $\mathbf{Z}[i\sqrt{5}]$. Then $N((a + bi\sqrt{5})(c + di\sqrt{5})) = N((ac - 5bd) + i(ad + bc)\sqrt{5}) = (ac - 5bd)^2 + 5(ad + bc)^2 = (a^2 + 5b^2)(c^2 + 5d^2) \geq (c^2 + 5d^2) = N((c + di\sqrt{5}))$, where equality holds if and only if $N((a + bi\sqrt{5})) = 1$, i.e., if and only if $a + bi\sqrt{5}$ is a unit. Hence, $\mathbf{Z}[i\sqrt{5}]$ is a FD by Theorem 16.1.3. In Example 15.3.11, we showed that 3 is an irreducible element. Now $3|(2 + i\sqrt{5})(2 - i\sqrt{5})$. Suppose $3|(2 + i\sqrt{5})$. Then $2 + i\sqrt{5} = 3(m + ni\sqrt{5})$ for some $m + ni\sqrt{5} \in \mathbf{Z}[i\sqrt{5}]$. This implies $2 = 3m$ and $1 = 3n$, which is impossible for integers m and n. Therefore, $3 \nmid (2 + i\sqrt{5})$. Similarly, $3 \nmid (2 - i\sqrt{5})$. Thus, 3 is not prime in $\mathbf{Z}[i\sqrt{5}]$. Hence, $\mathbf{Z}[i\sqrt{5}]$ is not a UFD by Theorem 16.1.10.

In this integral domain, we can also show that $2, 1 + i\sqrt{5}, 1 - i\sqrt{5}$ are irreducible elements and 2 is not an associate of any one of $1 + i\sqrt{5}$ and $1 - i\sqrt{5}$. Hence, $6 = 2 \cdot 3 = (1 + i\sqrt{5})(1 - i\sqrt{5})$ are two factorizations of 6, but there does not exist any correspondence between the irreducible factors such that the corresponding elements are associates.

Theorem 16.1.12 *A factorization domain D is a UFD if and only if every irreducible element of D is a prime element.*

Proof. Suppose the factorization domain D is a UFD. Then by Theorem 16.1.10, every irreducible element is a prime element.

Conversely, assume that every irreducible element is a prime element in the FD D. Suppose $a = p_1 p_2 \cdots p_n = q_1 q_2 \cdots q_m$ are two factorizations of a as a finite product of irreducible elements. Then $p_1 p_2 \cdots p_n = q_1(q_2 \cdots q_m)$ implies that $q_1 | (p_1 p_2 \cdots p_n)$. Since q_1 is also prime, at least one of $p_1, p_2, \ldots, p_n$ is divisible by q_1. Let $q_1 | p_1$. Now p_1 and q_1 are both irreducible. Hence, $p_1 = u_1 q_1$ for some unit u_1. Then $u_1 q_1 p_2 \cdots p_n = q_1 q_2 \cdots q_m$, from which it follows by the cancelation property that $u_1 p_2 \cdots p_n = q_2 \cdots q_m = q_2(q_3 \cdots q_m)$. Now $q_2 | (u_1 p_2 \cdots p_n)$. Since q_2 is prime, q_2 does not divide u_1. Hence, q_2 divides one of $p_2, \ldots, p_n$, say, $q_2 | p_2$. Then $p_2 = u_2 q_2$ for some unit u_2 and $u_1 u_2 q_2 p_3 \cdots p_n = q_2 \cdots q_m$. Canceling q_2 from this relation, we obtain $u_1 u_2 p_3 \cdots p_n = q_3 \cdots q_m$. If $n > m$, then proceeding this way we find that $u_1 u_2 \cdots u_m p_{m+1} \cdots p_n = 1$, which implies that each of $p_{m+1}, \ldots, p_n$ is a unit, a contradiction. If $n < m$, then we find that $u_1 u_2 \cdots u_n = q_{n+1} \cdots q_m$. This implies that each of $q_{n+1}, \ldots, q_m$ divides a unit, which is again a contradiction. Thus, $n = m$. Also, we have shown that the corresponding irreducible factors p_i, q_i, $i = 1, 2, \ldots, n$, in the factorizations $p_1 p_2 \cdots p_n$ and $q_1 q_2 \cdots q_n$ are associates. Hence, D is a UFD. ■

Theorem 16.1.13 *Every PID is a UFD.*

Proof. From Lemma 16.1.6, we find that every PID satisfies ACCP. Hence, by Theorem 16.1.7, every PID is a FD. Also, by Theorem 15.3.12, every irreducible element is prime in a PID. Thus, by Theorem 16.1.12, it follows that every PID is a UFD. ■

By Theorem 15.1.9, every Euclidean domain is a PID and hence by Theorem 16.1.13, every Euclidean domain is a UFD. This result is one of the important results in factorization theory. Let us prove this result independently. First we prove the following lemma.

Lemma 16.1.14 *Let E be a Euclidean domain and $a, b \in E$. If $a | b$, $b \neq 0$, and a is neither a unit nor an associate of b, then $v(a) < v(b)$.*

Proof. Since a is not an associate of b, it follows that $b \nmid a$. Hence, $a = bq + r$, where $r = 0$ or $v(r) < v(b)$. Now $b = ac$ for some $c \in E$. This implies that $r = a - bq = a - acq = a(1 - cq)$. If $1 - cq = 0$, then c is a unit and so b is an associate of a, a contradiction. Therefore, $1 - cq \neq 0$. Thus, $v(r) = v(a(1 - cq)) \geq v(a)$ and so $v(b) > v(a)$. ■

Theorem 16.1.15 *A Euclidean domain E is a unique factorization domain.*

Proof. Let v denote the Euclidean valuation of the Euclidean domain E. By induction on $v(a)$, we first show that every nonzero element a of E is either a unit or can be written as a finite product of irreducible elements. If $v(a) = v(1)$, then a is a unit. Assume that every nonzero element $b \in E$ is either a unit or expressible as a finite product of irreducible elements if $v(b) < v(a)$, where $v(a) > v(1)$ (the induction hypothesis). If a is irreducible, there is nothing to prove. Suppose that a is not irreducible. Then $a = bc$, where neither b nor c is a unit. Suppose b is an associate of a. Then $b = au$ for some unit $u \in E$. Thus, $a = bc = auc$ and so $1 = uc$, i.e., c is a unit, a contradiction. Therefore, b is not an associate of a. Similarly, c is not an associate of a. By Lemma 16.1.14, it now follows that $v(b) < v(a)$ and $v(c) < v(a)$. Thus, by our induction hypothesis, b and c are expressible as a finite product of irreducible elements of E. Hence, so is a.

The uniqueness of the factorization follows as in Theorem 16.1.12 ∎

From Theorem 15.1.9, we know that every Euclidean domain is a principal ideal domain. We noted in the remark on page 348 that the converse of this result is not true. In Theorem 16.1.13, we showed that every principal ideal domain is a unique factorization domain. The converse of this result is also not true. There is a class of rings for which the converse is true. Call a complex number an **algebraic integer** if it is a root of a monic polynomial $p(x)$ in $\mathbf{Z}[x]$. The set of all algebraic integers in a finite field extension (Chapter 24) of $\mathbf{Q}$ is such a ring. However, most of these rings are not unique factorization domains. For example, the ring $\mathbf{Z}[i\sqrt{5}]$ in Example 16.1.11 is a ring in which there is no unique factorization. Here $6 = (1 - i\sqrt{5})(1 + i\sqrt{5}) = 2 \cdot 3$ are two factorizations of 6 as a product of two irreducible elements. However, the ideal $\langle 6 \rangle$ has a unique (up to order) factorization as a product of prime ideals (defined in Chapter 17), $\langle 6 \rangle = \langle 3, 1 + i\sqrt{5} \rangle \langle 3, 1 - i\sqrt{5} \rangle \langle 2, 1 + i\sqrt{5} \rangle^2$. As a matter of fact, the entire class of rings in question has the property that every ideal has a unique factorization as a product of prime ideals.

16.1.1 Worked-Out Exercises

◇ **Exercise 1** Show that the integral domain $\mathbf{Z}[\sqrt{10}] = \{a + b\sqrt{10} \mid a, b \in \mathbf{Z}\}$ is a FD.

Solution: Define $N : \mathbf{Z}[\sqrt{10}]\backslash\{0\} \to \mathbf{Z}^{\#}$ by for all $a + b\sqrt{10} \in \mathbf{Z}[\sqrt{10}]$,

$$N(a + b\sqrt{10}) = \left| a^2 - 10b^2 \right|.$$

Now $N(a + b\sqrt{10}) = 1$ if and only if $|a^2 - 10b^2| = 1$ if and only if $(a + b\sqrt{10})(a - b\sqrt{10}) = \pm 1$ if and only if $a + b\sqrt{10}$ is a unit. Let $a + b\sqrt{10}$, $c + d\sqrt{10}$ be two nonzero elements of $\mathbf{Z}[\sqrt{10}]$. Then $N((a+b\sqrt{10})(c+d\sqrt{10})) =$

$|a^2 - 10b^2| |c^2 - 10d^2| \geq |c^2 - 10d^2| = N((c + d\sqrt{10}))$, where equality holds if and only if $N((a + b\sqrt{10})) = 1$, i.e., if and only if $a + b\sqrt{10}$ is a unit. Hence, $\mathbf{Z}[\sqrt{5}]$ is a FD by Theorem 16.1.3.

$\diamondsuit$ **Exercise 2** Show that in a UFD, every nonzero nonunit has only a finite number of nonassociated nontrivial factors.

Solution: Let D be a UFD. Suppose a is a nonzero nonunit element of D. Then a can be expressed uniquely as

$$a = p_1^{r_1} p_2^{r_2} \cdots p_k^{r_k},$$

where $p_1, p_2, \ldots, p_k$ are distinct primes and $r_1, r_2, \ldots, r_k$ are positive integers. Let $d = p_1^{t_1} p_2^{t_2} \cdots p_k^{t_k}$, where $0 \leq t_i \leq r_i$, $i = 1, 2, \ldots, k$. Then d is a divisor of a. Now suppose d is any divisor of a and d is a nonunit. Then d can be expressed uniquely as $d = q_1^{t_1} q_2^{t_2} \cdots q_m^{t_m}$, where $q_1, q_2, \ldots, q_m$ are distinct primes and $t_1, t_2, \ldots, t_m$ are positive integers. Since $d|a$, for all $i = 1, 2, \ldots, m$, $q_i^{t_i} | p_j^{r_j}$ for some j, $1 \leq j \leq k$. Then $q_i | p_j^{r_j}$ and so $q_i | p_j$. Therefore, q_i is an associate of p_j. Also, we find that $t_i \leq r_j$. Thus, d is an associate of $p_1^{l_1} p_2^{l_2} \cdots p_k^{l_k}$, $0 \leq l_i \leq r_i$, $i = 1, 2, \ldots, k$. Consequently, a has only a finite number of nonassociated nontrivial divisors.

$\diamondsuit$ **Exercise 3** Let $R = \{a_0 + a_1 x + \cdots + a_n x^n \in \mathbf{Q}[x] \mid a_0 \in \mathbf{Z}, n \in \mathbf{Z}^{\#}\}$. Show that R is not a UFD.

Solution: Clearly R is a subring of $\mathbf{Q}[x]$ and R contains 1. Hence, R is an integral domain. Now any unit of R is also a unit of $\mathbf{Q}[x]$. In $\mathbf{Q}[x]$, the units are the nonzero elements of $\mathbf{Q}$. Since $R \cap \mathbf{Q} = \mathbf{Z}$, it follows that 1 and -1 are the only units of R. For any nonnegative integer n, $\frac{1}{2^n} x \in R$ and $\frac{1}{2^n} x$ is not an associate of $\frac{1}{2^m} x$ when $n \neq m$. Now $x = 2^n (\frac{1}{2^n} x)$ shows that $\frac{1}{2^n} x$ is a divisor of x. Hence, x has infinite number of nontrivial divisors in R. If R is a UFD, then x cannot have an infinite number of nontrivial divisors. Thus, R is not a UFD.

$\diamondsuit$ **Exercise 4** In a UFD, show that the gcd of any two nonzero elements exists.

Solution: Let R be a UFD and a, b be nonzero elements of R. If one of a or b is a unit, then $\gcd(a, b) = 1$. Suppose a and b are nonunits. Then a can be expressed uniquely as

$$a = p_1^{t_1} p_2^{t_2} \cdots p_k^{t_k},$$

where $p_1, p_2, \ldots, p_k$ are irreducible elements such that p_i is not an associate of p_j when $i \neq j$ and $t_1, t_2, \ldots, t_k$ are positive integers. Similarly, b can be expressed uniquely (up to associates) as

$$b = q_1^{r_1} q_2^{r_2} \cdots q_n^{r_n},$$

where $q_1, q_2, \ldots, q_n$ are irreducible and $r_1, r_2, \ldots, r_n$ are positive integers. Now if q_1 is not an associate of any of $p_1, \ldots, p_k$, then we write $a = p_1^{t_1} \cdots p_k^{t_k} \cdot q_1^0$. Next if q_2 is not an associate of any of $p_1, p_2, \ldots, p_k$, then we write $a = p_1^{t_1} p_2^{t_2} \cdots p_k^{t_k} q_1^0 q_2^0$. But, if q_2 is an associate of one of $p_1, p_2, \ldots, p_k$, then skip q_2 and consider q_3. Continue the process for $q_3, \ldots, q_n$. We do the same thing for b. So we can write

$$
\begin{aligned}
a &= u_1^{n_1} u_2^{n_2} \cdots u_m^{n_m} \\
b &= u_1^{l_1} u_2^{l_2} \cdots u_m^{l_m},
\end{aligned}
$$

where $u_1, u_2, \ldots, u_m$ are irreducible elements such that u_i is not an associate of u_j when $i \neq j$ and $n_1, n_2, \ldots, n_m, l_1, l_2, \ldots, l_m$ are nonnegative integers. Let $d = u_1^{k_1} u_2^{k_2} \cdots u_m^{k_m}$, where $k_i = \min\{n_i, l_i\}$, $i = 1, 2, \ldots, m$. Then $d|a$ and $d|b$. Let $c|a$ and $c|b$, $c \in R$. Since any irreducible divisor of c is an associate of one of $u_1, u_2, \ldots, u_m$, it follows that c must be of the form

$$
c = u_1^{h_1} u_2^{h_2} \cdots u_m^{h_m},
$$

where $h_i \geq 0$, and $h_i \leq n_i$, $h_i \leq l_i$, $i = 1, 2, \ldots, m$. Thus, $h_i \leq k_i$, $i = 1, 2, \ldots, m$. Hence, $c|d$. Thus, $d = \gcd(a, b)$.

16.1.2 Exercises

1. Show that $\mathbf{Z}$ satisfies the ACCP.

2. If the integral domain R satisfies the ACCP, prove that the polynomial ring $R[x]$ satisfies the ACCP.

3. Prove that an integral domain D is a UFD if and only if D satisfies the ACCP and every irreducible element is prime in D.

4. Show that the integral domains $\mathbf{Z}[i\sqrt{6}]$, $\mathbf{Z}[i\sqrt{7}]$, and $\mathbf{Z}[i\sqrt{10}]$ are factorization domains, but not unique factorization domains.

5. Let a, b be two nonzero elements of a UFD D. If $\gcd(a, b) = 1$ and $a|c$, $b|c$, prove that $ab|c$ in D, where $c \in D$.

6. For the following statements, write the proof if the statement is true; otherwise, give a counterexample.

   (i) Any subring of a UFD with identity is also a UFD.

   (ii) 1 and -1 are the only units of a UFD.

16.2 Factorization of Polynomials over a UFD

In this section, we show that every polynomial of degree ≥ 1 over a UFD R can be uniquely expressed as a product of irreducible polynomials over R.

Definition 16.2.1 *Let* $f(x) = a_0 + a_1 x + \cdots + a_n x^n$ *be a nonzero polynomial in* $R[x]$. *Then the* $gcd\{a_0, a_1, \ldots, a_n\}$ *is called the* **content** *of* $f(x)$.

It is known that the gcd of $\{a_0, a_1, \ldots, a_n\}$ is not unique. If u and v are two gcd's of $\{a_0, a_1, \ldots, a_n\}$, then u and v are associates. Hence, if c_1 and c_2 are two contents of $f(x)$, then c_1 and c_2 are associates and any associate of c_1 is also a content of $f(x)$. If a and b are two elements of R such that a is an associate of b, then we write $a \sim b$.

The content of $f(x)$ is denoted by $\operatorname{cont} f(x)$.

Definition 16.2.2 *A nonzero polynomial* $f(x) \in R[x]$ *is called a* **primitive** *polynomial if* $\operatorname{cont} f(x)$ *is a unit.*

Lemma 16.2.3 *Let* R *be a UFD. Let* $f(x)$ *and* $g(x)$ *be two primitive polynomials in* $R[x]$. *Then* $f(x)g(x)$ *is also a primitive polynomial in* $R[x]$.

Proof.   Let $f(x) = a_0 + a_1 x + \cdots + a_n x^n$ and $g(x) = b_0 + b_1 x + \cdots + b_m x^m$. Let $c_f \sim \operatorname{cont} f(x)$ and $c_g \sim \operatorname{cont} g(x)$. Since $f(x)$ and $g(x)$ are primitive, c_f and c_g are unit elements in R. Suppose that $f(x)g(x)$ is not a primitive polynomial. Let $f(x)g(x) = c_0 + c_1 x + \cdots + c_{n+m} x^{n+m}$, where $c_0 = a_0 b_0$, $c_1 = a_0 b_1 + a_1 b_0, \ldots, c_i = \sum_{j=0}^{i} a_j b_{i-j}$, where $a_j = 0$ if $j > n$, and $b_{i-j} = 0$ if $i - j > m$. Now $\operatorname{cont} f(x)g(x)$ is not a unit. Let p be a prime element in R such that p divides $\operatorname{cont} f(x)g(x)$. Then p divides c_i for all $i = 0, 1, \ldots, n + m$. Since c_f and c_g are unit elements, p does not divide each of $a_0, a_1, \ldots, a_n$ and also p does not divide each of $b_0, b_1, \ldots, b_m$. Let t be the smallest nonnegative integer such that p does not divide a_t. Then p divides a_i, for $i = 0, 1, \ldots, t - 1$, and p does not divide a_t. Similarly, let r be the smallest nonnegative integer such that p does not divide b_r. Then p divides b_j, for $j = 0, 1, \ldots, r - 1$, and p does not divide b_r. Therefore, p does not divide $a_t b_r$. Now $c_{t+r} = a_0 b_{t+r} + a_1 b_{t+r-1} + \cdots + a_{t-1} b_{r+1} + a_t b_r + a_{t+1} b_{r-1} + \cdots + a_{t+r} b_0$, where $b_i = 0$ if $i > m$ and $a_i = 0$ if $i > n$. Now p divides a_i, for $i = 0, 1, \ldots, t - 1$, p divides b_j, for $j = 0, 1, \ldots, r - 1$, and p divides c_{t+r}. Hence, p divides $a_t b_r$, which is a contradiction. Thus, $\operatorname{cont} f(x)g(x)$ is a unit and so $f(x)g(x)$ is a primitive polynomial. ∎

Example 16.2.4 *In* $\mathbf{Z}[x]$, $6x^2 + 3x - 9 = 3(2x^2 + x - 3)$. *Hence,* $6x^2 + 3x - 9$ *is not a primitive polynomial. But* $2x^2 + x - 3$ *is a primitive polynomial.*

Theorem 16.2.5 *Let* R *be a UFD. Let* $f(x)$ *and* $g(x)$ *be two nonzero polynomials in* $R[x]$. *Then there exists a unit* $u \in R$ *such that*

$$cont(f(x)g(x)) = u \operatorname{cont} f(x) \operatorname{cont} g(x).$$

Proof. Let c_f denote $\mathrm{cont}\, f(x)$ and c_g denote $\mathrm{cont}\, g(x)$. Then $f(x) = c_f f_1(x)$ and $g(x) = c_g g_1(x)$, where $f_1(x)$ and $g_1(x)$ are primitive polynomials in $R[x]$. Now $\mathrm{cont}(f(x)g(x))$ and $\mathrm{cont}(c_f c_g f_1(x)g_1(x))$ are associates. Since $c_f c_g$ is a nonzero element of R, it follows that

$$\mathrm{cont}(c_f c_g f_1(x)g_1(x))$$

and

$$c_f c_g \mathrm{cont}(f_1(x)g_1(x))$$

are associates. By Lemma 16.2.3, $\mathrm{cont}(f_1(x)g_1(x))$ is a unit. Hence,

$$\mathrm{cont}(f(x)g(x)) = u c_f c_g$$

for some unit u. ∎

It is known that the polynomial ring $F[x]$ over a field F is a Euclidean domain, and hence a unique factorization domain. To take advantage of this result, let us extend an integral domain R to its quotient field $Q(R)$ and establish the relationship between elements of $Q(R)[x]$ and $R[x]$.

In the remainder of the section, we let $Q(R)$ denote the quotient field of R.

Lemma 16.2.6 *Let R be a UFD. If $f(x)$ is a nonzero polynomial in $Q(R)[x]$, then there exist nonzero elements $a, b \in R$ and a primitive polynomial $f_1(x)$ in $R[x]$ such that $f(x) = ab^{-1}f_1(x)$, where b^{-1} is the inverse of b in $Q(R)[x]$.*

Proof. Let $f(x) = c_0 + c_1 x + \cdots + c_n x^n \in Q(R)[x]$ be a nonzero polynomial. Then $c_i \in Q(R)$, $i = 0, 1, \ldots, n$. Therefore, there exist $a_i, b_i \in R$ such that $c_i = a_i b_i^{-1}$, $b_i \neq 0$, $i = 0, 1, \ldots, n$. Now $f(x) = a_0 b_0^{-1} + a_1 b_1^{-1} x + \cdots + a_n b_n^{-1} x^n$. Let $b = b_0 b_1 \cdots b_n$. Then

$$bf(x) = a_0 b_1 \cdots b_n + a_1 b_0 b_2 \cdots b_n x + \cdots + a_n b_0 b_1 \cdots b_{n-1} x^n \in R[x].$$

Clearly $bf(x)$ is nonzero. Let $a = \mathrm{cont}(bf(x))$. Then $bf(x) = a f_1(x)$, where $\mathrm{cont}\, f_1(x)$ is a unit and $f_1(x) \in R[x]$. Hence, $f(x) = b^{-1} a f_1(x)$, where $b, a \in R$ and $f_1(x)$ is a primitive polynomial in $R[x]$. ∎

Lemma 16.2.7 *Let R be a UFD. Let $f(x)$ be a nonzero polynomial in $R[x]$. If $f(x) = d_1 f_1(x) = d_2 f_2(x)$, where $f_1(x)$ and $f_2(x)$ are primitive polynomials in $R[x]$ and $d_1, d_2 \in Q(R)$, then $d_1 = u d_2$ for some unit $u \in R$.*

Proof. Since $d_1, d_2 \in Q(R)$, we can write $d_1 = ab^{-1}$ and $d_2 = cd^{-1}$ for some $a, b, c, d \in R$. Thus, $f(x) = ab^{-1}f_1(x) = cd^{-1}f_2(x)$. This implies that $adf_1(x) = cbf_2(x)$. Since $f_1(x)$ and $f_2(x)$ are primitive, $ad = ucb$ for some unit $u \in R$ by Theorem 16.2.5. Thus, $d_1 = ab^{-1} = ucd^{-1} = ud_2$. ∎

Lemma 16.2.8 *Let R be a UFD. Let $f(x)$ be a nonconstant primitive polynomial in $R[x]$. Then $f(x)$ is irreducible in $R[x]$ if and only if $f(x)$ is irreducible in $Q(R)[x]$.*

Proof. Suppose $f(x)$ is irreducible in $R[x]$ and $f(x)$ is not irreducible in $Q(R)[x]$. Then there exist $h(x), g(x) \in Q(R)[x]$ such that $f(x) = h(x)g(x)$, $\deg h(x) \geq 1$, and $\deg g(x) \geq 1$. By Lemma 16.2.6, there exist $a, b, c, d \in R$ with $b \neq 0$, $d \neq 0$, and primitive polynomials $h_1(x), g_1(x) \in R[x]$ such that $h(x) = ab^{-1}h_1(x)$ and $g(x) = cd^{-1}g_1(x)$. Hence, $f(x) = ab^{-1}cd^{-1}h_1(x)g_1(x)$. This implies that $bdf(x) = ach_1(x)g_1(x)$. Now $f(x)$ is primitive and so $\text{cont} f(x)$ is a unit. Thus, $\text{cont}(bdf(x)) = bdu$ for some unit u. Now

$$
\begin{aligned}
\text{cont}(ach_1(x)g_1(x)) &= vac \; \text{cont}(h_1(x)g_1(x)) \text{ for some unit } v \in R \\
&= v_1ac \; \text{cont}(h_1(x)) \; \text{cont}(g_1(x)) \text{ for some unit } v_1 \in R \\
&= v_1acv_2v_3 \text{ for some units } v_2, v_3 \in R.
\end{aligned}
$$

Hence, $bd = acw$ for some unit $w \in R$. Thus, $f(x) = wh_1(x)g_1(x)$ for some unit $w \in R$. This shows that $f(x)$ is not irreducible in $R[x]$, which is a contradiction. Therefore, $f(x)$ is irreducible in $Q(R)[x]$. Conversely, let $f(x)$ be irreducible in $Q(R)[x]$. Suppose $f(x)$ is reducible in $R[x]$. Now $f(x) = rg(x)$, where $r \in R$ and r is a not a unit is impossible since $f(x)$ is primitive. Thus, there exist polynomials $f_1(x), f_2(x)$ in $R[x]$ such that $\deg f_1(x) \geq 1$, $\deg f_2(x) \geq 1$, and $f(x) = f_1(x)f_2(x)$. Now $f_1(x)$ and $f_2(x)$ are also nonconstant polynomials in $Q(R)[x]$. Hence, $f(x)$ is not irreducible in $Q(R)[x]$, a contradiction. Consequently, $f(x)$ is irreducible in $R[x]$. ∎

Example 16.2.9 *Consider the polynomial $4x + 4$ in $\mathbf{Q}[x]$. Now $4x + 4 = 4(x + 1)$. 4 is a unit in $\mathbf{Q}[x]$ and $x + 1$ is irreducible in $\mathbf{Q}[x]$. Hence, $4x + 4$ is irreducible in $\mathbf{Q}[x]$. But 4 is not a unit in $\mathbf{Z}[x]$. Hence, $4x + 4$ is not irreducible in $\mathbf{Z}[x]$. Also, 3 is irreducible in $\mathbf{Z}[x]$, but 3 is not irreducible in $\mathbf{Q}[x]$.*

We are now in a position to prove our main result of this section. Before proving this theorem, let us recall the following assertions concerning the polynomial ring $R[x]$ so that we can enjoy the beauty and depth of this theorem.

(i) If R is a commutative ring with 1, then $R[x]$ is a commutative ring with 1.

(ii) If R is an integral domain, then $R[x]$ is an integral domain.

(iii) If R is a field, then $R[x]$ is not a field, but $R[x]$ is a Euclidean domain.

(iv) If R is a PID, then $R[x]$ may not be a PID.

Theorem 16.2.10 *Let R be a UFD. Then $R[x]$ is a UFD.*

Proof.   Let $f(x)$ be a polynomial of degree $n \geq 1$. Let $f(x) = c_f f_1(x)$, where c_f is a content of $f(x)$ and $f_1(x)$ is a primitive polynomial in $R[x]$. Now $Q(R)[x]$ is a UFD and $f_1(x) \in R[x] \subseteq Q(R)[x]$. Therefore, there exist irreducible polynomials $g_1(x), g_2(x), \ldots, g_r(x)$ in $Q(R)[x]$ such that $f_1(x) = g_1(x)g_2(x) \cdots g_r(x)$. By Lemma 16.2.7, $g_i(x) = a_i b_i^{-1} h_i(x)$, $a_i, b_i \in R$, $b_i \neq 0$, and $h_i(x)$ is a primitive polynomial in $R[x]$, $i = 1, 2, \ldots, r$. Also, by Lemma 16.2.8, $h_i(x)$ is irreducible in $R[x]$, $i = 1, 2, \ldots, r$. Hence,

$$f_1(x) = a_1 a_2 \cdots a_r b_1^{-1} b_2^{-1} \cdots b_r^{-1} h_1(x) \cdots h_r(x).$$

Let $a = a_1 a_2 \cdots a_r$ and $b = b_1 b_2 \cdots b_r$. Then

$$b f_1(x) = a h_1(x) \cdots h_r(x). \tag{16.1}$$

By Lemma 16.2.3, $h_1(x) \cdots h_r(x)$ is primitive. This implies that $a = ub$ for some unit $u \in R$ and so

$$f_1(x) = u h_1(x) \cdots h_r(x).$$

This shows that

$$f(x) = u c_f h_1(x) \cdots h_r(x). \tag{16.2}$$

Since an associate of an irreducible polynomial is also an irreducible polynomial, it follows that $u h_1(x)$ is irreducible. Thus, for any polynomial $f(x)$ of degree ≥ 1, there exist irreducible polynomials $g_1(x), \ldots, g_k(x)$ in $R[x]$ such that

$$f(x) = c_f g_1(x) \cdots g_k(x),$$

where $c_f = \text{cont} f(x)$. If c_f is not a unit, then there exist irreducible elements $a_1, a_2, \ldots, a_t \in R$ such that

$$f(x) = a_1 a_2 \cdots a_t g_1(x) \cdots g_k(x). \tag{16.3}$$

Suppose now that

$$f(x) = a_1 a_2 \cdots a_t g_1(x) \cdots g_k(x) = b_1 b_2 \cdots b_l h_1(x) \cdots h_q(x), \tag{16.4}$$

where a_i, b_j are irreducible elements in R, $i = 1, \ldots, t$, $j = 1, \ldots, l$ and

$$g_1(x), \ldots, g_k(x), \ h_1(x), \ldots, h_q(x)$$

are irreducible elements in $R[x]$. Now $a_1 a_2 \cdots a_t$ and $b_1 b_2 \cdots b_l$ are two factorizations as a product of irreducible elements in R of c_f. Therefore, by (16.4)

$$g_1(x) \cdots g_k(x) = d h_1(x) \cdots h_q(x), \tag{16.5}$$

where d is a unit in R. Now $g_1(x), \ldots, g_k(x), h_1(x), \ldots, h_q(x)$ are primitive and irreducible in $R[x]$. Hence, these polynomials are also irreducible in $Q(R)[x]$. Since $Q(R)[x]$ is a UFD, Eq. (16.5) implies that $k = q$ and there exists a one-one correspondence between $\{g_1(x), \ldots, g_k(x)\}$ and $\{h_1(x), \ldots, h_q(x)\}$ such that the corresponding factors are associates in $Q(R)[x]$ and hence by Lemma 16.2.7, they are also associates in $R[x]$. Thus, the factorization (16.4) of $f(x)$ in $R[x]$ is unique. Consequently, $R[x]$ is a UFD. ∎

Corollary 16.2.11 *Let R be a UFD. The polynomial ring $R[x_1, \ldots, x_n]$ is a UFD.* ∎

We see that the polynomial ring $F[x, y]$ is a unique factorization domain. However, $F[x, y]$ is not a Euclidean domain. This can be verified by showing that $F[x, y]$ is not a principal ideal ring. We ask the reader to show in the exercises that the ideal $\langle x, y \rangle$ in $F[x, y]$ is not a principal ideal.

As shown in Example 16.1.11, $\mathbf{Z}[i\sqrt{5}]$ is not a UFD. Thus, even though the polynomial ring $F[x]$ is a unique factorization domain, a ring of the form $F[c]$ need not be one. Thus, the homomorphic image of a unique factorization domain need not be a unique factorization domain.

16.2.1 Worked-Out Exercises

◇ **Exercise 1** Let $f(x)$ be a nonzero polynomial in $\mathbf{Z}[x]$. Show that $f(x)$ can be expressed as a product of two polynomials $g(x)$ and $h(x)$ of $\mathbf{Q}[x]$ with $\deg g(x) < \deg f(x)$ and $\deg h(x) < \deg f(x)$ if and only if there exist $g_1(x), h_1(x) \in \mathbf{Z}[x]$ such that $\deg g(x) = \deg g_1(x)$, $\deg h(x) = \deg h_1(x)$, and $f(x) = g_1(x)h_1(x)$.

Solution: Suppose there exist $g(x)$ and $h(x)$ in $\mathbf{Q}[x]$ with $\deg g(x) < \deg f(x)$, $\deg h(x) < \deg f(x)$, and $f(x) = g(x)h(x)$. There exist nonzero elements $a, b, c, d \in \mathbf{Z}$ and primitive polynomials $g_2(x)$, $h_2(x) \in \mathbf{Z}[x]$ such that $g(x) = ab^{-1}g_2(x)$ and $h(x) = cd^{-1}h_2(x)$ by Lemma 16.2.6. Hence, $f(x) = ab^{-1}cd^{-1}g_2(x)h_2(x)$. This implies that $bdf(x) = acg_2(x)h_2(x)$. Let d_1 be the content of $f(x)$. Then we can write $f(x) = d_1 f_1(x)$, where $f_1(x)$ is a primitive polynomial in $\mathbf{Z}[x]$. Hence, $bdd_1 f_1(x) = acg_2(x)h_2(x)$. Now $g_2(x)h_2(x)$ is also a primitive polynomial. Then $bdd_1 = uac$ for some unit $u \in \mathbf{Z}$. This implies $bdd_1 = ac$ or $bdd_1 = -ac$. Hence, $f_1(x) = g_2(x)h_2(x)$ or $f_1(x) = -g_2(x)h_2(x)$. Let $g_1(x) = d_1 g_2(x)$. Now $f(x) = d_1 f_1(x) = d_1 g_2(x)h_2(x) = g_1(x)h_1(x)$, where $h_1(x) = h_2(x)$ or $f(x) = d_1 f_1(x) = -d_1 g_2(x)h_2(x) = g_1(x)h_1(x)$, where $h_1(x) = -h_2(x)$. Also, from the construction, it follows that $\deg g_2(x) = \deg g_1(x) = \deg g(x) < \deg f(x)$ and $\deg h_2(x) = \deg h_1(x) = \deg h(x) < \deg f(x)$. The converse is trivial.

Exercise 2 Show that $\mathbf{Z}[x]$ is a UFD, but not a PID.

Solution: Since $\mathbf{Z}$ is a UFD, $\mathbf{Z}[x]$ is a UFD by Theorem 16.2.10. (By Corollary 15.1.11, $\mathbf{Z}[x]$ is not a PID. However, here we want to show that $\mathbf{Z}[x]$ is not a PID by showing the existence of ideals in $\mathbf{Z}[x]$, which are not principal.) Consider

$$I = \langle x \rangle + \langle n \rangle,$$

where $n \in \mathbf{Z}$, $n \notin \{0, 1, -1\}$. We claim that I is not a principal ideal. Suppose $I = \langle f(x) \rangle$, where $f(x) \in \mathbf{Z}[x]$. Then $\langle n \rangle \subseteq \langle f(x) \rangle$. Therefore, $n = f(x)g(x)$ for some $g(x) \in \mathbf{Z}$. Since $\deg n = 0$, $\deg f(x) = 0$ and hence $f(x) \in \mathbf{Z}$. Let $f(x) = a \in \mathbf{Z}$. Now $\langle x \rangle \subseteq \langle a \rangle$. Then $x = ah(x)$ for some $h(x) \in \mathbf{Z}[x]$. Again by a degree argument, $\deg h(x) = 1$. Let $h(x) = a_0 + a_1 x$, where $a_0, a_1 \in \mathbf{Z}$, $a_1 \neq 0$. Then $x = a(a_0 + a_1 x)$. Hence, $1 = aa_1 \in \langle a \rangle = I = \langle x \rangle + \langle n \rangle$. Thus, $1 = xs(x) + nt(x)$ for some $s(x), t(x) \in \mathbf{Z}[x]$. Let $t(x) = t_0 + t_1 x + \cdots + t_r x^r$. Then by comparing coefficients in $1 = xs(x) + nt(x)$, we get $1 = nt_0$. Hence, n divides 1, which is a contradiction. Therefore, I is not a principal ideal.

16.2.2   Exercises

1. Let $f(x) \in \mathbf{Z}[x]$ be irreducible. Prove that $f(x)$ is primitive.

2. Let $f(x)$ be a nonconstant primitive polynomial in $\mathbf{Z}[x]$. Prove that if $f(x)$ is not irreducible in $\mathbf{Q}[x]$, then $f(x)$ is not irreducible in $\mathbf{Z}[x]$.

3. Show that the polynomial ring $\mathbf{Q}[x, y]$ is a UFD, but not a PID.

4. Let R be a UFD. Let $f(x)$ be a primitive polynomial in $R[x]$. Show that any nonconstant divisor of $f(x)$ is also a primitive polynomial.

16.3   Irreducibility of Polynomials

In the previous section, we proved that any polynomial of degree ≥ 1 over a UFD can be expressed as a product of irreducible polynomials. Thus, irreducible polynomials play an important role in polynomial rings. But it is not always easy to determine if a polynomial is irreducible over a UFD. In this section, we establish some criteria for irreducibility of polynomials. We first note that any polynomial of degree 1 over a field F is always irreducible. If $f(x) = ax + b \in F[x]$ with $a \neq 0$, then $x = -a^{-1}b$ is a root of $f(x)$ in F. In this connection, let us point out that a linear polynomial over a UFD D may not be irreducible in $D[x]$. For example $2x + 4 = 2(x + 2)$ is not irreducible in $\mathbf{Z}[x]$. We now consider polynomials of degree 2 and 3. For these polynomials, we can apply the following test to check irreducibility. Let F denote a field.

Theorem 16.3.1 *Let $f(x) \in F[x]$ be a polynomial of degree 2 or 3. Then $f(x)$ is irreducible over F if and only if $f(x)$ has no roots in F.*

Proof.   Suppose that $\deg f(x) = 3$ and $f(x)$ is irreducible. If $f(x)$ has a root in F, say a, then $x - a$ divides $f(x)$ in $F[x]$ and so $f(x)$ is reducible over F. Conversely, suppose $f(x)$ has no roots in F. Assume that $f(x)$ is reducible. Then $f(x) = g(x)h(x)$ for some $g(x)$, $h(x) \in F[x]$, $\deg g(x) \geq 1$ and $\deg h(x) \geq 1$. Now $\deg(g(x)h(x)) = 3$. Therefore, either $\deg g(x) = 1$ and $\deg h(x) = 2$ or $\deg h(x) = 1$ and $\deg g(x) = 2$. To be specific, let $\deg g(x) = 1$ and $\deg h(x) = 2$. Then $g(x) = ax + b$ for some $a, b \in F$, $a \neq 0$. Now $-a^{-1}b \in F$ and $g(-a^{-1}b) = 0$. Thus, $-a^{-1}b$ is a root of $g(x)$ and hence $-a^{-1}b$ is a root of $f(x)$ in F. This is a contradiction to our assumption that $f(x)$ has no roots in F. Hence, $f(x)$ is irreducible over F. A similar argument can be used for the case when $\deg f(x) = 2$. $\blacksquare$

Example 16.3.2 *(i) Let $f(x) = x^2 + x + [1] \in \mathbf{Z}_2[x]$. Now*

$$f([0]) = [0]^2 + [0] + [1] \neq [0],$$

$$f([1]) = [1]^2 + [1] + [1] = [1] \neq [0].$$

Hence, $f(x)$ has no roots in $\mathbf{Z}_2$. Thus, by Theorem 16.3.1, $f(x)$ is irreducible over $\mathbf{Z}_2$.
    (ii) Let $g(x) = x^3 + [2]x + [1] \in \mathbf{Z}_3[x]$. Now

$$g([0]) = [0]^3 + [2][0] + [1] \neq [0],$$

$$g([1]) = [1]^3 + [2][1] + [1] = [4] = [1] \neq [0],$$

and

$$g([2]) = [2]^3 + [2][2] + [1] = [13] = [1] \neq [0].$$

Hence, $g(x)$ has no roots in $\mathbf{Z}_3$. Thus, by Theorem 16.3.1, $g(x)$ is irreducible over $\mathbf{Z}_3$.

Instead of considering polynomials over an arbitrary field, let us now consider polynomials over the field $\mathbf{Q}$ of all rational numbers. By Lemma 16.2.8, a nonconstant primitive polynomial $f(x) \in \mathbf{Z}[x]$ is irreducible in $\mathbf{Q}[x]$ if and only if $f(x)$ is irreducible in $\mathbf{Z}[x]$. It is not difficult to decide whether or not a polynomial is primitive. In order to decide whether or not $f(x)$ is irreducible, we sometimes consider the corresponding polynomial in $\mathbf{Z}_p[x]$ for some prime p.

Theorem 16.3.3 *Let $f(x) = a_0 + a_1x + \cdots + a_nx^n \in \mathbf{Z}[x]$ be of degree $n > 1$. If there exists a prime p such that $\overline{f}(x) = [a_0] + [a_1]x + \cdots + [a_n]x^n$ is irreducible in $\mathbf{Z}_p[x]$ and $\deg f(x) = \deg \overline{f}(x)$, then $f(x)$ is irreducible in $\mathbf{Q}[x]$.*

Proof.   Suppose $f(x)$ satisfies the given conditions of the theorem for some prime p. Suppose $f(x)$ is reducible in $\mathbf{Q}[x]$. Then there exist polynomials $g(x) = b_0 + b_1 x + \cdots + b_m x^m$ and $h(x) = c_0 + c_1 x + \cdots + c_k x^k$ in $\mathbf{Z}[x]$, $0 < m < n$, $0 < k < n$ such that $f(x) = g(x)h(x)$ by Worked-Out Exercise 1 (page 379). Thus, $[a_0] + [a_1]x + \cdots + [a_n]x^n = ([b_0] + [b_1]x + \cdots + [b_m]x^m)([c_0] + [c_1]x + \cdots + [c_k]x^k)$. Since $\deg \overline{f}(x) = \deg f(x) = n = k + m$, it follows that $[b_m][c_k] \neq 0$ in $\mathbf{Z}_p$. Hence, $[b_m] \neq [0]$ and $[c_k] \neq [0]$. Consequently, $\overline{g}(x)$ and $\overline{h}(x)$ are nonconstant polynomials in $\mathbf{Z}_p[x]$. Since the units of $\mathbf{Z}_p[x]$ are the nonzero elements of $\mathbf{Z}_p$, it follows that $\overline{g}(x)$ and $\overline{h}(x)$ are nonunits. Therefore, $\overline{f}(x)$ is not irreducible in $\mathbf{Z}_p[x]$, a contradiction. Hence, $f(x)$ is irreducible in $\mathbf{Q}[x]$. ∎

Example 16.3.4 *Consider the polynomial $f(x) = \frac{5}{7}x^3 - \frac{1}{2}x + 1$ in $\mathbf{Q}[x]$. Then $14f(x) = 10x^3 - 7x + 14$. Let $f_1(x) = 10x^3 - 7x + 14$. Now in $\mathbf{Z}_3[x]$, $\overline{f}_1(x) = [10]x^3 - [7]x + [14] = x^3 - x + [2]$. Since $\overline{f}_1([0]) = [2]$, $\overline{f}_1([1]) = [2]$, $\overline{f}_1([2]) = [2]^3 - [2] + [2] = [2]$, it follows that $\overline{f}_1(x)$ has no root in $\mathbf{Z}_3[x]$. As a result $14f(x)$ is irreducible in $\mathbf{Q}[x]$. But 14 is a unit in $\mathbf{Q}[x]$. Hence, $f(x)$ is irreducible in $\mathbf{Q}[x]$.*

Let $f(x) \in \mathbf{Q}[x]$ and $\deg f(x) \geq 2$. If $f(x)$ has a root in $\mathbf{Q}$, then $f(x)$ is reducible. The following theorem will help us to see whether a polynomial $f(x) \in \mathbf{Q}[x]$ has a root in $\mathbf{Q}$.

Theorem 16.3.5 *Let $f(x) = a_0 + a_1 x + \cdots + a_n x^n \in \mathbf{Z}[x]$ be of degree n and $a_0 \neq 0$. Let $\frac{u}{v} \in \mathbf{Q}$ be a root of $f(x)$, where u and v are relatively prime. Then*

$$u \mid a_0 \text{ and } v \mid a_n.$$

Proof.   Since $\frac{u}{v}$ is a root of $f(x)$,

$$0 = f(\frac{u}{v}) = a_0 + a_1 \frac{u}{v} + \cdots + a_n (\frac{u}{v})^n.$$

Thus,

$$0 = a_0 v^n + a_1 u v^{n-1} + \cdots + a_{n-1} u^{n-1} v + a_n u^n.$$

Hence,

$$v(a_0 v^{n-1} + a_1 u v^{n-2} + \cdots + a_{n-1} u^{n-1}) = -a_n u^n.$$

This implies that $v \mid a_n u^n$. Since u and v are relatively prime, $v \mid a_n$. Similarly, $u \mid a_0$. ∎

Example 16.3.6 *Let $f(x) = 2x^3 - 7x + 1$ and $\frac{u}{v} \in \mathbf{Q}$ be a root of $f(x)$ with $\gcd(u, v) = 1$. Then $u \mid 1$ and $v \mid 2$. Hence, $u = \pm 1$ and $v = \pm 1, \pm 2$. This implies that $\frac{u}{v} = \pm 1, \pm \frac{1}{2}$. Now $f(1) \neq 0$, $f(-1) \neq 0$, $f(\frac{1}{2}) = \frac{1}{4} - \frac{7}{2} + 1 \neq 0$, and $f(-\frac{1}{2}) = -\frac{1}{4} + \frac{7}{2} + 1 \neq 0$. So we find that $f(x)$ has no root in $\mathbf{Q}$. Thus, by Theorem 16.3.1, $f(x)$ is irreducible in $\mathbf{Q}[x]$. Since $f(x)$ is primitive, $f(x)$ is also irreducible in $\mathbf{Z}[x]$.*

Let us now give another criterion for irreducibility. This famous criterion is known as Eisenstein's irreducibility criterion.

Theorem 16.3.7 (Eisenstein's Irreducibility Criterion) *Let D be a UFD and $Q(D)$ be its quotient field. Let*

$$f(x) = a_0 + a_1 x + \cdots + a_n x^n$$

be a nonconstant polynomial in $D[x]$. Suppose that D contains a prime p such that

(i) $p|a_i$, $i = 0, 1, \ldots, n - 1$,
(ii) $p \nmid a_n$, and
(iii) $p^2 \nmid a_0$.
Then $f(x)$ is irreducible in $Q(D)[x]$.

Proof.   **Case 1.** $f(x)$ is a primitive polynomial in $D[x]$. Under this assumption, if we can show that $f(x)$ is irreducible in $D[x]$, then by Lemma 16.2.8, it will follow that $f(x)$ is irreducible in $Q(D)[x]$. Suppose that $f(x)$ is not irreducible in $D[x]$. Then there exist polynomials

$$\begin{aligned} g(x) &= b_0 + b_1 x + \cdots + b_t x^t \\ h(x) &= c_0 + c_1 x + \cdots + c_k x^k \end{aligned}$$

in $D[x]$ such that $f(x) = g(x)h(x)$ and $g(x)$ and $h(x)$ are nonunits in $D[x]$. Now $n = t + k$. If $t = 0$, then $g(x) = b_0$, a nonunit element of D. Thus, $f(x) = b_0 h(x)$ implies that $f(x)$ is not primitive. Therefore, $t \neq 0$. Similarly, $k \neq 0$. Hence, $0 < t < n$ and $0 < k < n$. Now from $f(x) = g(x)h(x)$, we find that $a_0 = b_0 c_0$. Since p is a prime such that $p|a_0$ and $p^2 \nmid a_0$, it follows that p divides one of b_0, c_0, but not both. Suppose $p|b_0$ and $p \nmid c_0$. Since $p \nmid a_n$ and $a_n = b_t c_k$, $p \nmid b_t$ and $p \nmid c_k$. Thus, $p|b_0$ and $p \nmid b_t$. Let m be the smallest positive integer such that $p \nmid b_m$. Then $p|b_i$ for $0 \leq i < m \leq t$. Now considering the coefficient of x^m in $f(x)$ and $g(x)h(x)$, it follows that

$$a_m = b_0 c_m + b_1 c_{m-1} + \cdots + b_{m-1} c_1 + b_m c_0.$$

Since $p|b_i$, $0 \leq i < m$, we find that $p|(a_m - b_m c_0)$. Since $m \leq t < n$, $p|a_m$. Hence, $p|b_m c_0$ and so $p|b_m$ or $p|c_0$ since p is prime. This is a contradiction. Therefore, $f(x)$ is irreducible in $D[x]$ and hence in $Q(D)[x]$.
   Case 2. $f(x)$ is not a primitive polynomial in $D[x]$. Let $d = \gcd\{a_0, a_1, \ldots, a_n\}$ in D. Then $f(x) = df_1(x)$, where $f_1(x)$ is a primitive polynomial in $D[x]$. Let $f_1(x) = d_0 + d_1 x + \cdots + d_n x^n$. Then $a_i = dd_i$, for all $i = 1, 2, \ldots, n$. Since p does not divide a_n, p does not divide d. Therefore, it now follows that $p|d_i$, $i = 0, 1, \ldots, n-1$, $p \nmid d_n$ and $p^2 \nmid d_0$. Thus, by Case 1, $f_1(x)$ is irreducible in $Q(D)[x]$. Now d is a unit in $Q(D)$. Hence, $f(x)$ is irreducible in $Q(D)[x]$. ∎

Corollary 16.3.8 *Let D be a UFD and $f(x) = a_0 + a_1x + \cdots + a_nx^n$ be a nonconstant primitive polynomial in $D[x]$. Suppose that D contains a prime p such that*

    (i) $p|a_i$, $i = 0, 1, \ldots, n-1$,
    (ii) $p \nmid a_n$, and
    (iii) $p^2 \nmid a_0$.
    Then $f(x)$ is irreducible in $D[x]$. ∎

Corollary 16.3.9 *Let $f(x) = a_0 + a_1x + \cdots + a_nx^n$ be a nonconstant polynomial in $\mathbf{Z}[x]$. If there exists a prime p such that*

    (i) $p|a_i$, $i = 0, 1, \ldots, n-1$,
    (ii) $p \nmid a_n$, and
    (iii) $p^2 \nmid a_0$,
    then $f(x)$ is irreducible in $\mathbf{Q}[x]$. ∎

Corollary 16.3.10 *The **cyclotomic polynomial***

$$\phi_p(x) = 1 + x + \cdots + x^{p-1} = \frac{x^p - 1}{x - 1}$$

is irreducible in $\mathbf{Z}[x]$, where p is a prime.

Proof.   Since the content of $\phi_p(x)$ is 1, we find that $\phi_p(x)$ is a primitive polynomial. Suppose $\phi_p(x)$ is not irreducible in $\mathbf{Z}[x]$. Then there exist nontrivial factors $h(x)$ and $g(x)$ of $\phi_p(x)$ such that $\phi_p(x) = h(x)g(x)$. This implies that $\phi_p(x + 1) = h(x + 1)g(x + 1)$ is a nontrivial factorization of $\phi_p(x + 1)$. However,

$$
\begin{aligned}
\phi_p(x + 1) &= \frac{(x+1)^p - 1}{(x+1) - 1} \\
&= \frac{x^p + px^{p-1} + \cdots + \binom{p}{i}x^i + \cdots + px}{x} \\
&= p + \cdots + \binom{p}{i}x^{i-1} + \cdots + px^{p-2} + x^{p-1}
\end{aligned}
$$

is clearly irreducible by Eisenstein's criterion. Hence, $\phi_p(x)$ is irreducible in $\mathbf{Z}[x]$. ∎

Gauss is said to have placed Eisenstein at the same mathematical level as Newton and Archimedes. However, Eisenstein's influence on mathematics is considered to be small in comparison to that of the giants of mathematics.

16.3.1   Worked-Out Exercises

◇ **Exercise 1** Show that $f(x) = x^3 + [2]x + [4]$ is irreducible in $\mathbf{Z}_5[x]$.

Solution: $f([0]) = [4]$, $f([1]) = [7] = [2]$, $f([2]) = [3] + [4] + [4] = [1]$, $f([3]) = [2] + [1] + [4] = [2]$, $f([4]) = [4] + [3] + [4] = [1]$. Hence, $f(x)$ has no roots in $\mathbf{Z}_5$. Thus, by Theorem 16.3.1, $f(x)$ is irreducible in $\mathbf{Z}_5[x]$.

◇ **Exercise 2** Let $f(x) = x^6 + x^3 + 1 \in \mathbf{Z}[x]$. Show that $f(x)$ is irreducible over $\mathbf{Q}$.

Solution: Now $f(x+1) = x^6 + 6x^5 + 15x^4 + 21x^3 + 18x^2 + 9x + 3$. Let $p = 3$. Then by Eisenstein's criterion, $f(x+1)$ is irreducible over $\mathbf{Q}$. Hence, $f(x)$ is irreducible over $\mathbf{Q}$.

◇ **Exercise 3** Show that $f(x) = x^4 - 5x^2 + x + 1$ is irreducible in $\mathbf{Z}[x]$.

Solution: Let us first show that $f(x)$ is irreducible in $\mathbf{Q}[x]$. If $f(x)$ has a linear factor, then $f(x)$ has a root in $\mathbf{Q}$. Let $\frac{a}{b}$ (a, b are relatively prime) be a root of $f(x)$ in $\mathbf{Q}$. Then $b|1$ and $a|1$ by Theorem 16.3.5. Hence, $\frac{a}{b} = 1$ or -1. But $f(1) = 1 - 5 + 1 + 1 = -2 \neq 0$ and $f(-1) = 1 - 5 - 1 + 1 = -4 \neq 0$. Therefore, $f(x)$ has no linear factors in $\mathbf{Q}[x]$. Let $f(x) = (x^2 + ax + b)(x^2 + cx + d)$ in $\mathbf{Z}[x]$. Equating coefficients of powers of x, we find that

$$c + a = 0, \ d + b + ac = -5, \ ad + bc = 1, \ bd = 1.$$

Now $bd = 1$ implies that either $b = d = 1$ or $b = d = -1$. Suppose $b = d = 1$. Then $a + c = 1$. But we also have $a + c = 0$, a contradiction. Suppose $b = d = -1$. Then $ad + bc = 1$ implies that $a + c = -1$. Thus, $a + c = -1$ and $a + c = 0$, a contradiction. Hence, we find that there are no integers a, b, c, d such that $f(x) = (x^2 + ax + b)(x^2 + cx + d)$. This also implies that $f(x)$ cannot be factored as a product of two quadratic polynomials in $\mathbf{Q}[x]$ (see Worked-Out Exercise 1, page 379). Thus, $f(x)$ is irreducible in $\mathbf{Q}[x]$. Hence, by Lemma 16.2.8, $f(x)$ is irreducible in $\mathbf{Z}[x]$.

◇ **Exercise 4** Show that $f(x) = x^5 + 15x^3 + 10x + 5$ is irreducible in $\mathbf{Z}[x]$.

Solution: The content of $f(x)$ is 1. Therefore, $f(x)$ is a primitive polynomial. Now 5 is a prime integer and $5|5$, $5|10$, $5|0$, $5|15$, $5 \nmid 1$, $5^2 \nmid 5$. Hence, by Corollary 16.3.8, $f(x)$ is irreducible in $\mathbf{Z}[x]$.

◇ **Exercise 5** Give an example of a primitive polynomial which has no root in $\mathbf{Q}$, but is reducible over $\mathbf{Z}$.

Solution: Let $f(x) = x^4 + 2x^2 + 1$. This is a primitive polynomial in $\mathbf{Z}[x]$. If possible, let $\frac{a}{b}$ be a root of $f(x)$, where $a \neq 0$, $b \neq 0$ and $\gcd(a, b) = 1$. Then $a|1$ and $b|1$ by Theorem 16.3.5. Hence, $\frac{a}{b} = \pm 1$. But $f(1) \neq 0$ and $f(-1) \neq 0$. Therefore, $f(x)$ has no root in $\mathbf{Q}$. Since $f(x) = (x^2 + 1)(x^2 + 1)$, $f(x)$ is reducible in $\mathbf{Z}[x]$.

Exercise 6 Show that $x^2 + x + [1]$ is the only irreducible polynomial of degree 2 over $\mathbf{Z}_2$.

Solution: Any polynomial of degree 2 over $\mathbf{Z}_2$ is of the form $ax^2 + bx + c$, where $a, b, c \in \mathbf{Z}_2 = \{[0], [1]\}$. Now $a \neq [0]$. Therefore, $a = [1]$. Then x^2, $x^2 + x$, $x^2 + [1]$, and $x^2 + x + [1]$ are the only polynomials of degree 2 over $\mathbf{Z}_2$. Now $x^2 = xx$, $x^2 + x = x(x + [1])$, and $x^2 + [1] = (x + [1])(x + [1])$ showing that x^2, $x^2 + x$, and $x^2 + [1]$ are reducible. Let $f(x) = x^2 + x + [1]$. Then $f([0]) = [1] \neq 0$ and $f([1]) = [3] = [1] \neq 0$. Therefore, $f(x)$ has no root in $\mathbf{Z}_2$. Thus, $x^2 + x + [1]$ is irreducible over $\mathbf{Z}_2$.

16.3.2   Exercises

1. Find all irreducible polynomials of degree ≤ 2 in $\mathbf{Z}_2[x]$. Is $x^3 + [1]$ irreducible in $\mathbf{Z}_2[x]$? If not, then express it as a product of irreducible polynomials in $\mathbf{Z}_2[x]$.

2. Show that the polynomial $x^5 + x^2 + [1]$ is irreducible in $\mathbf{Z}_2[x]$. Hence, prove that $x^5 - x^2 + 9$ is irreducible in $\mathbf{Z}[x]$.

3. Show that the polynomial $x^2 + [2]x + [6]$ is reducible in $\mathbf{Z}_2[x]$ even though $x^2 + 2x + 6$ is irreducible in $\mathbf{Z}[x]$.

4. Use Eisenstein's criterion to prove that the polynomials $x^2 + 2x + 6$ and $2x^4 + 6x^3 - 9x^2 + 15$ are irreducible over $\mathbf{Z}$.

5. For $f(x) \in D[x]$, D a UFD, prove that $f(x)$ is irreducible in $D[x]$ if and only if $f(x - c)$ is irreducible in $D[x]$ for any $c \in D$.

6. Show that the polynomials $x^3 - x^2 + 1$, $x^3 - x + 1$, and $x^3 + 2x^2 + 3$ are irreducible in $\mathbf{Z}[x]$.

7. Show that the polynomial $2x^3 - x^2 + 4x - 2$ is not irreducible in $\mathbf{Z}[x]$.

8. Show that the polynomial $x^2 + \frac{1}{3}x - \frac{2}{5}$ is irreducible in $\mathbf{Q}[x]$.

9. Prove that the polynomial $f(x) = 1 - x + x^2 - x^3 + \cdots + (-1)^{p-1}x^{p-1}$ is irreducible in $\mathbf{Z}[x]$ for any prime p.

10. Let D be a UFD and $f(x) = a_0 + a_1x + \cdots + a_nx^n \in D[x]$ be of degree n and $a_0 \neq 0$. Let $uv^{-1} \in Q(D)$ be a root of $f(x)$, where $u, v \in D$ and $\gcd(u, v) = 1$. Prove that $u | a_0$ and $v | a_n$ in D.

11. Show that for any positive integer $n > 1$, $f(x) = x^n + 2$ is irreducible in $\mathbf{Z}[x]$.

12. Find all irreducible polynomials of degree 2 over the field $\mathbf{Z}_3$.

13. If $f(x)$ is an irreducible polynomial over $\mathbf{R}$, prove that either $f(x)$ is linear or $f(x)$ is quadratic.

14. Show that there are only three irreducible monic quadratic polynomials over $\mathbf{Z}_3$.

15. (i) Show that there are only 10 irreducible monic quadratic polynomials over $\mathbf{Z}_5$.

    (ii) Let p be a prime. Find the number of irreducible monic quadratic polynomials over $\mathbf{Z}_p$.

Leopold Kronecker (1823–1891) was born on December 7, 1823, in Liegnitz, Germany, to a wealthy family. He was provided with private tutoring at home. He later entered Liegnitz Gymnasium, where E. E. Kummer was his mathematics teacher. Kummer recognized his talent and encouraged him to do independent research.

In 1841, he matriculated at the University of Berlin. There he attended Dirichlet's and Steiner's mathematics lectures. He was also attracted to astronomy and in 1843 attended the University of Bonn. He returned to Berlin in 1845, the year he received his Ph.D. His thesis was on complex units.

On Kummer's nomination, Kronecker became a full member of the Berlin Academy in 1861. He was very influential at the Academy and personally helped fifteen mathematicians, including Riemann, Sylvester, Dedekind, Hermite, and Fuchs, to get various memberships.

Kronecker's primary work is in algebraic number theory. He is believed to be one of the inventors of algebraic number theory along with Kummer and Dedekind. He was the first mathematician who clearly understood Galois's work. He also proved the fundamental theorem of finite Abelian groups.

Briefly Kronecker withdrew from academic life to manage the family business. However, he continued to do mathematics as a recreation. In 1855, he returned to the academic life in Berlin. In 1880, he became editor of the *Journal für die reine and angewandte Mathematik.*

Kronecker and Weierstrass were good friends. While Weierstrass and Cantor were creating modern analysis, Kronecker's remark that "God himself made the whole numbers—everything else is the work of men" deeply affected Cantor, who was very sensitive. His remarks in opposition to Cantor's work are believed to be a factor in Cantor's nervous breakdown.

Kronecker died on December 29, 1891.

Chapter 17

Maximal, Prime, and Primary Ideals

17.1   Maximal, Prime, and Primary Ideals

In this section, we introduce certain special ideals. These ideals are motivated in large part by certain arithmetic properties of the integers. Throughout the section, we assume that the ring R contains at least two elements.

Definition 17.1.1 *An ideal P of a ring R is called **prime** if for any two ideals A and B of R, $AB \subseteq P$ implies that either $A \subseteq P$ or $B \subseteq P$.*

The following theorem gives a useful characterization of a prime ideal with the help of elements of R. Let us first recall that if A is a left ideal and B is a right ideal of a ring R, then AB is an ideal of R. Let $a \in R$. Then Ra is a left ideal of R and aR is a right ideal of R. Thus, $R(aR)$ is an ideal of R. We denote $R(aR)$ by RaR. Also, for $a \in R$, $aRa = \{ara \mid r \in R\}$.

Theorem 17.1.2 *An ideal P of a ring R is a prime ideal if and only if for all $a, b \in R$, $aRb \subseteq P$ implies that either $a \in P$ or $b \in P$.*

Proof.   Suppose P is a prime ideal and $aRb \subseteq P$, where $a, b \in R$. Let $A = RaR$ and $B = RbR$. Then A and B are ideals of R. Also, $AB = (RaR)(RbR) \subseteq R(aRb)R \subseteq RPR \subseteq P$. Since P is a prime ideal, it follows that either $A \subseteq P$ or $B \subseteq P$. Suppose $A \subseteq P$. Now $\langle a \rangle^3 \subseteq RaR = A \subseteq P$. Since P is a prime ideal, $\langle a \rangle \subseteq P$ and so $a \in P$. Similarly, if $B \subseteq P$, then $b \in P$. Thus, either $a \in P$ or $b \in P$. Conversely, suppose that the ideal P satisfies the given condition of the theorem. Let A and B be two ideals of R such that $AB \subseteq P$. Suppose that $A \not\subseteq P$. Then there exists $a \in A$ such that $a \notin P$. Let $b \in B$. Now $aRb = (aR)b \subseteq AB \subseteq P$. This implies that $a \in P$ or $b \in P$. But $a \notin P$. Therefore, $b \in P$. Hence, $B \subseteq P$. ∎

Corollary 17.1.3 *Let R be a commutative ring. An ideal P of R is a prime ideal if and only if for all $a, b \in R$, $ab \in P$ implies that either $a \in P$ or $b \in P$.* ■

Example 17.1.4 *In the ring $\mathbf{Z}$ of integers, the ideal $P = \{3k \mid k \in \mathbf{Z}\}$ is a prime ideal. For, $ab \in P$ if and only if ab is divisible by 3 if and only if a is divisible by 3 or b is divisible by 3 (since 3 is prime) if and only if a is a multiple of 3 or b is a multiple of 3 if and only if $a \in P$ or $b \in P$. In $\mathbf{Z}$, the ideal $J = \{6k \mid k \in \mathbf{Z}\}$ is not a prime ideal since $3 \cdot 2 = 6 \in J$, but $3 \notin J$ and $2 \notin J$.*

Theorem 17.1.5 *Let R be a PID and P be a nonzero ideal of R. Then P is prime and $P \neq R$ if and only if P is generated by a prime element.*

Proof.   Let R be a PID and $P = \langle p \rangle$ be a nonzero proper prime ideal of R. Then $p \neq 0$. Since $P \neq R$, p is not a unit. Let $a, b \in R$ be such that $p|ab$. Then $ab = pc$ for some $c \in R$. Hence, $ab \in P$. Since P is a prime ideal, either $a \in P$ or $b \in P$. Therefore, either $p|a$ or $p|b$. Thus, p is a prime element. Conversely, suppose that $P = \langle p \rangle$ is a nonzero ideal of R such that p is a prime element. Since p is not a unit, $P \neq R$. Let a, b be two elements of R such that $ab \in P$. Then $p|ab$. Since p is a prime element, either $p|a$ or $p|b$. Therefore, either $a \in P$ or $b \in P$. Hence, P is a prime ideal of R. ■

As a consequence of Theorem 17.1.5 and Theorem 15.1.9, the prime ideals of $\mathbf{Z}$ are precisely those ideals generated by primes and the ideals $\{0\}$ and $\mathbf{Z}$. Also, by Theorem 15.3.16, the prime ideals in the polynomial ring $F[x]$ over a field F are those ideals generated by irreducible polynomials and the ideals $\{0\}$ and $F[x]$.

Definition 17.1.6 *Let R be a ring and M be an ideal of R. Then M is called a **maximal ideal** of R if $M \neq R$ and there does not exist any ideal I of R such that $M \subset I \subset R$.*

Let $\mathcal{I}(R)$ be the collection of all proper ideals of R. Since $\{0\} \in \mathcal{I}(R)$, $\mathcal{I}(R) \neq \phi$. Now $(\mathcal{I}(R), \leq)$ is a lattice, where $\leq$ is the set inclusion relation. Clearly a maximal element (if one exists) of this lattice is a maximal ideal of the ring R.

Theorem 17.1.7 *Let R be a commutative ring with 1. Then every maximal ideal of R is a prime ideal of R.*

Proof.   Let I be a maximal ideal of R and a and b be two elements of R such that $ab \in I$ and $a \notin I$. Now $\langle I, a \rangle = \{u + ra \mid u \in I, \, r \in R\}$ is the ideal generated by $I \cup \{a\}$. Since $a \notin I$, $I \subset \langle I, a \rangle$. Also, since I is a maximal ideal,

$\langle I, a \rangle = R$. Thus, there exist $u \in I$ and $r \in R$ such that $1 = u + ra$. This implies that $b = ub + rab \in I$. Hence, I is a prime ideal. ∎

The converse of the above theorem is not true, as shown by the following examples.

Example 17.1.8 *In the ring $\mathbf{Z}$ of integers, $\{0\}$ is a prime ideal, but not a maximal ideal.*

Example 17.1.9 *Let $R = \{(a, b) \mid a, b \in \mathbf{Z}\}$. Then $(R, +, \cdot)$ is a ring, where $+$ and $\cdot$ are defined by*

$$
\begin{aligned}
(a, b) + (c, d) &= (a + c, b + d), \\
(a, b) \cdot (c, d) &= (ac, bd)
\end{aligned}
$$

for all $a, b, c, d \in \mathbf{Z}$. Let $I = \{(a, 0) \mid a \in \mathbf{Z}\}$. Then I is a prime ideal of R, but not a maximal ideal since $I \subset \langle I, (0, 2) \rangle \subset R$.

Theorem 17.1.10 *Let R be a principal ideal domain. Then a nonzero ideal $P \ (\neq R)$ of R is prime if and only if it is maximal.*

Proof. Suppose $P \ (\neq R)$ is a nonzero prime ideal. By Theorem 17.1.5, $P = \langle p \rangle$ for some prime element $p \in R$. We now show that there is no ideal I of R such that $P \subset I \subset R$. Suppose I is an ideal of R such that $P \subset I$. Since $P \neq I$, there exists an element $a \in I$ such that $a \notin P$. Then a and p are relatively prime and so there exist $s, t \in R$ such that $1 = sa + tp$. Since $sa \in I$ and $tp \in P \subset I$, we must have $1 \in I$. This implies that $I = R$. Hence, P is maximal. ∎

We now give characterizations of prime ideals and maximal ideals in a commutative ring with identity by the quotient rings of the ideals.

Theorem 17.1.11 *Let R be a commutative ring with 1 and P be an ideal of R such that $P \neq R$. Then P is a prime ideal if and only if R/P is an integral domain.*

Proof. Let P be a prime ideal of R. Since R is a commutative ring with 1, the quotient ring R/P is also a commutative ring with 1. Now $P \neq R$ and so the identity element $1 + P$ of R/P is different from the zero element $0 + P$. Let us now show that R/P has no zero divisors. Let $a + P, \ b + P \in R/P$, and $(a + P)(b + P) = 0 + P$. Then $ab + P = 0 + P$, which implies that $ab \in P$. Since P is a prime ideal, either $a \in P$ or $b \in P$, i.e., either $a + P = 0 + P$ or $b + P = 0 + P$. Thus, R/P has no zero divisors. This implies that R/P is an integral domain. Conversely, suppose R/P is an integral domain. Let $ab \in P$. Then $0 + P = ab + P = (a+P)(b+P)$, whence $a + P = 0 + P$ or $b + P = 0 + P$. Thus, $a \in P$ or $b \in P$ and so P is a prime ideal. ∎

Theorem 17.1.12 *Let R be a commutative ring with 1 and M be an ideal of R. Then M is a maximal ideal if and only if R/M is a field.*

Proof. Suppose that M is a maximal ideal. Since R is a commutative ring with 1, R/M is a commutative ring with 1. For all $a \in R$, let $\bar{a}$ denote the coset $a + M$ in R/M. Let $\bar{a} \in R/M$ be such that $\bar{a} \neq \bar{0}$. Then $a \notin M$. Hence, the ideal $\langle M, a \rangle$ generated by $M \cup \{a\}$ properly contains M. Since M is a maximal ideal, we have $\langle M, a \rangle = R$. This implies that there exist $m \in M$ and $r \in R$ such that $m + ra = 1$. Thus, $\bar{m} + \overline{ra} = \bar{1}$ and so $\overline{ra} = \bar{1}$. Hence, $\bar{a}$ has an inverse. This shows that every nonzero element of R/M is a unit and so R/M is a field. Conversely, suppose R/M is a field. Since R/M is a field, $R \neq M$. Let I be an ideal of R such that $M \subset I \subseteq R$. There exists $a \in I$ such that $a \notin M$. Then $\bar{a} \neq \bar{0}$ and so there exists $\bar{r} \in R/M$ such that $\overline{ar} = \bar{1}$. Thus, $(a + M)(r + M) = 1 + M$, which implies $1 - ar \in M$. Hence, $1 = m + ar$ for some $m \in M$. Thus, $1 = m + ar \in M + I \subseteq I$. This implies that $I = R$. Therefore, M is maximal. ∎

As a consequence of Theorems 15.1.9 and 17.1.10, the maximal ideals of **Z** are precisely those ideals generated by primes. Also, by Theorem 15.3.16, the maximal ideals in the polynomial ring $F[x]$ over a field F are those ideals generated by irreducible polynomials.

Example 17.1.13 *Consider the polynomial ring $R[x, y]$ over an integral domain R. Then $R[x, y]/\langle x \rangle \simeq R[y]$ and $R[x, y]/\langle y \rangle \simeq R[x]$, which are integral domains. Thus, $\langle x \rangle$ and $\langle y \rangle$ are prime ideals. Since $R[x, y]/\langle x \rangle$ and $R[x, y]/\langle y \rangle$ are not fields, $\langle x \rangle$ and $\langle y \rangle$ are not maximal ideals.*

Example 17.1.14 *Consider $\mathbf{E}$, the ring of even integers. The ideal $\langle 4 \rangle$ is maximal, but not prime in $\mathbf{E}$ since $2 \cdot 2 \in \langle 4 \rangle$, but $2 \notin \langle 4 \rangle$. Note that $\mathbf{E}$ is commutative without identity.*

We now show the existence of maximal ideals in certain rings. In order to accomplish this, we require Zorn's lemma.

Theorem 17.1.15 *Let R be a commutative ring with 1. Then every proper ideal of R is contained in a maximal ideal of R.*

Proof. Let I be a proper ideal of R and set $\mathcal{A} = \{J \mid I \subseteq J, J \text{ is a proper}$ ideal of $R\}$. Since $I \in \mathcal{A}$, $\mathcal{A} \neq \phi$. Also, $\mathcal{A}$ is a partially ordered set, where the partial order $\leq$ is the usual set inclusion. We now show that any chain in $\mathcal{A}$ has an upper bound in $\mathcal{A}$. Let $\mathcal{C} = \{J_\alpha \mid \alpha \in K\}$ be a chain in $\mathcal{A}$. Since $I \subseteq J_\alpha$ for all α, $I \subseteq \cup_\alpha J_\alpha$. Let $a, b \in \cup_\alpha J_\alpha$. Then $a \in J_\alpha$ and $b \in J_\beta$ for some α, β. Since $\mathcal{C}$ is a chain, either $J_\alpha \subseteq J_\beta$ or $J_\beta \subseteq J_\alpha$, say, $J_\alpha \subseteq J_\beta$. Thus, $a, b \in J_\beta$. Since J_β is an ideal of R, $a - b \in J_\beta \subseteq \cup_\alpha J_\alpha$. Let $r \in R$. Then $ra \in J_\alpha \subseteq \cup_\alpha J_\alpha$, whence $\cup_\alpha J_\alpha$ is an ideal of R. Now $\cup_\alpha J_\alpha \neq R$ else $1 \in J_\alpha$ for some α, which is impossible since $J_\alpha \neq R$. Hence, $\cup_\alpha J_\alpha \in \mathcal{A}$, which is clearly an upper bound of $\mathcal{C}$ and so by Zorn's lemma, $\mathcal{A}$ has a maximal element, say, M. We now show that M is a maximal ideal. If there exists an ideal J of R such that $M \subset J \subset R$, then $J \in \mathcal{A}$ and so M is not maximal in $\mathcal{A}$, a contradiction. Thus, no such J exists and so M is a maximal ideal. $\blacksquare$

Corollary 17.1.16 *Let R be a commutative ring with 1 and $a \in R$. Then a is in a maximal ideal of R if and only if a is not a unit.*

Proof. Suppose a is not a unit. Then $\langle a \rangle \subset R$ else $1 = ra$ for some r. By Theorem 17.1.15, there exists a maximal ideal M such that $\langle a \rangle \subseteq M$. Now $a \in \langle a \rangle \subseteq M$. Conversely, suppose $a \in M$, where M is a maximal ideal. If a is a unit, then $1 = a^{-1}a \in M$ and so $M = R$, a contradiction. $\blacksquare$

Corollary 17.1.17 *Let R be a commutative ring with 1. Then R has a maximal ideal.*

Proof. In R, $\{0\}$ is a proper ideal. Hence, by Theorem 17.1.15, there exists a maximal ideal M of R such that $\{0\} \subseteq M$. $\blacksquare$

The fundamental theorem of arithmetic says that any integer n has a prime factorization $n = p_1^{e_1} \cdots p_s^{e_s}$, where $p_1, \ldots, p_s$ are primes and $e_1, \ldots, e_s$ are positive integers. The ideals $\langle p_i \rangle$ are prime ideals of $\mathbf{Z}$. The ideals $\langle p_i^{e_i} \rangle$ are also special ideals of $\mathbf{Z}$. Their study is motivated in part by the fundamental theorem of arithmetic.

Definition 17.1.18 *Let R be a commutative ring and Q be an ideal of R. Then Q is called a **primary ideal** if for all a, $b \in R$, $ab \in Q$ and $a \notin Q$ implies that there exists a positive integer n such that $b^n \in Q$.*

From the definition of primary ideal, it follows immediately that every prime ideal in a commutative ring is a primary ideal. Now in the ring $\mathbf{Z}$, for any prime integer p, the ideal $\langle p^n \rangle$ contains p^n but not p, where n is a positive integer and $n \geq 2$. Hence, $\langle p^n \rangle$ is not a prime ideal. The following example shows that $\langle p^n \rangle$ is a primary ideal.

Example 17.1.19 *Let p be a prime in $\mathbf{Z}$ and n be a positive integer. We show that $\langle p^n \rangle$ is a primary ideal. Let $ab \in \langle p^n \rangle$ and $a \notin \langle p^n \rangle$. Then there exists $r \in \mathbf{Z}$ such that $ab = rp^n$. Since p^n does not divide a, $p|b$ and so $b = qp$ for some $q \in \mathbf{Z}$. Thus, $b^n = q^n p^n$ and so $b^n \in \langle p^n \rangle$.*

Example 17.1.20 *Let $p(x)$ be irreducible in $F[x]$, F a field, and n be a positive integer. Then $\langle p(x)^n \rangle$ is a primary ideal by an argument entirely similar to the one used in Example 17.1.19.*

Definition 17.1.21 *Let R be a commutative ring and I be an ideal of R. Then the **radical** of I, denoted by $\sqrt{I}$, is defined to be the set*

$$\sqrt{I} = \{a \in R \mid a^n \in I \text{ for some positive integer } n\}.$$

Theorem 17.1.22 *Let Q be an ideal of a commutative ring R. Then*
(i) $\sqrt{Q}$ is an ideal of R and $\sqrt{Q} \supseteq Q$,
(ii) if Q is a primary ideal, then $\sqrt{Q}$ is a prime ideal.

Proof.   (i) Clearly $\sqrt{Q} \supseteq Q$. Let $a, b \in \sqrt{Q}$. Then there exist positive integers n, m such that $a^n, b^m \in Q$. Thus, $(a - b)^{n+m} \in Q$ and so $a - b \in \sqrt{Q}$. Let $r \in R$. Then $(ra)^n = r^n a^n \in Q$ and so $ra \in \sqrt{Q}$. Hence, $\sqrt{Q}$ is an ideal of R.

(ii) Let $a, b \in R$ be such that $ab \in \sqrt{Q}$ and $a \notin \sqrt{Q}$. There exists a positive integer n such that $a^n b^n = (ab)^n \in Q$. But $a^n \notin Q$. Since Q is primary, there exists a positive integer m such that $b^{nm} = (b^n)^m \in Q$. Therefore, $b \in \sqrt{Q}$ and so $\sqrt{Q}$ is prime. ∎

Definition 17.1.23 *Let Q be a primary ideal of a commutative ring R. Then the radical $P = \sqrt{Q}$ of Q is called the **associated prime ideal** of Q and Q is called a **primary ideal belonging to** (or **primary for**) the prime ideal P.*

Example 17.1.24 *Let i be a positive integer. In $\mathbf{Z}$, we show that $\langle p^i \rangle$ is primary for $\langle p \rangle$, where p is a prime. It suffices to show that $\langle p \rangle = \sqrt{\langle p^i \rangle}$. Let $a \in \sqrt{\langle p^i \rangle}$. Then there exists a positive integer n such that $a^n \in \langle p^i \rangle$. Therefore, $a^n = rp^i$ for some $r \in \mathbf{Z}$. This implies that $p|a$ and so $a \in \langle p \rangle$. Hence, $\sqrt{\langle p^i \rangle} \subseteq \langle p \rangle$. Let $a \in \langle p \rangle$. Then there exists $t \in \mathbf{Z}$ such that $a = tp$. This implies that $a^i = t^i p^i \in \langle p^i \rangle$ and so $a \in \sqrt{\langle p^i \rangle}$. Thus, $\langle p \rangle \subseteq \sqrt{\langle p^i \rangle}$.*

In $F[x]$ (F a field), a similar argument shows that $\langle p(x)^i \rangle$ is primary for $\langle p(x) \rangle$, where $p(x)$ is irreducible and $\langle p(x) \rangle = \sqrt{\langle p(x)^i \rangle}$.

Theorem 17.1.25 *Let Q and P be ideals of a commutative ring R. Then Q is primary and $P = \sqrt{Q}$ if and only if*
(i) $Q \subseteq P \subseteq \sqrt{Q}$ and
(ii) $ab \in Q$, $a \notin Q$ implies $b \in P$.

Proof. The necessity of (i) and (ii) is immediate. Suppose (i) and (ii) hold. Let $ab \in Q$, $a \notin Q$. Then $b \in P \subseteq \sqrt{Q}$ and so there exists a positive integer n such that $b^n \in Q$, whence Q is primary. We now show that $P = \sqrt{Q}$. Let $b \in \sqrt{Q}$. Then there exists a positive integer n such that $b^n \in Q \subseteq P$. Let n be the smallest positive integer such that $b^n \in Q$. If $n = 1$, then $b \in P$. So assume that $n \geq 2$. Then $bb^{n-1} \in Q$ and $b^{n-1} \notin Q$ implies that $b \in P$. Hence, $\sqrt{Q} \subseteq P$ and so $P = \sqrt{Q}$. ∎

We now show that every primary ideal I of a commutative ring R can be characterized with the help of some properties of the quotient ring R/I.

Theorem 17.1.26 *Let R be a commutative ring and I be an ideal of R. Then I is a primary ideal if and only if every zero divisor of R/I is nilpotent.*

Proof. First suppose that I is a primary ideal. Let $a + I$ be a zero divisor in R/I. Then there exists an element $b + I \in R/I$, $b + I \neq I$, such that $(a + I)(b + I) = I$. Now $ab \in I$ and $b \notin I$. Since I is a primary ideal, it follows that $a^n \in I$ for some positive integer n. Hence, $(a + I)^n = a^n + I = I$, showing that $a + I$ is nilpotent.

Conversely, suppose that every zero divisor of R/I is nilpotent. Let $a, b \in R$ be such that $ab \in I$ and $a \notin I$. Then $a + I \neq I$. Now $(a + I)(b + I) = ab + I = I$. If $b + I = I$, then $b \in I$. Suppose $b + I \neq I$. This implies that $b + I$ is a zero divisor and so is nilpotent. Therefore, there exists a positive integer n such that $b^n + I = (b + I)^n = I$. Thus, $b^n \in I$. Consequently, I is a primary ideal. ∎

Consider **Z**. For the prime factorization of an integer n, $n = p_1^{e_1} \cdots p_s^{e_s}$, we have

$$\langle n \rangle = \langle p_1^{e_1} \rangle \cdots \langle p_s^{e_s} \rangle = \langle p_1^{e_1} \rangle \cap \cdots \cap \langle p_s^{e_s} \rangle$$

and $\sqrt{\langle p_i^{e_i} \rangle} = \langle p_i \rangle$, $i = 1, 2, \ldots, s$. However, in the polynomial ring $\mathbf{Z}[x, y]$, it can be shown that the ideal $\langle x^2, xy, 2 \rangle$ is an intersection of primary ideals, but not a product of primary ideals. These concepts involving prime and primary ideals are used in the study of nonlinear equations. For example, consider the following nonlinear equations:

$$x^2 - y = 0$$
$$x^2 z = 0.$$

In the polynomial ring $\mathbf{R}[x, y]$, let $I = \langle x^2 - y, x^2 z \rangle$. It can be shown that $\langle x^2 - y, z \rangle$ and $\langle x^2, y \rangle$ are primary ideals and that $I = \langle x^2, y \rangle \cap \langle x^2 - y, z \rangle$. In fact, it can be shown in any polynomial ring $F[x_1, \ldots, x_n]$ over a field F that every ideal is a finite intersection of primary ideals. This latter result is a type of fundamental theorem of arithmetic for ideals. It can also be shown that

$\sqrt{\langle x^2 - y, z \rangle} = \langle x^2 - y, z \rangle$ and $\sqrt{\langle x^2, y \rangle} = \langle x, y \rangle$. The solution to the above system of equations is

$$\{(x, x^2, 0) \mid x \in \mathbf{R}\} \cup \{(0, 0, z) \mid z \in \mathbf{R}\}.$$

The ideal $\langle x^2 - y, z \rangle$ corresponds to $\{(x, x^2, 0) \mid x \in R\}$, while the ideal $\langle x, y \rangle$ corresponds to $\{(0, 0, z) \mid z \in \mathbf{R}\}$.

We conclude this section by mentioning the following differences between the ideals of $\mathbf{Z}$ and $\mathbf{Z}[x]$.

1. In the ring $\mathbf{Z}$, every ideal is a principal ideal, but in $\mathbf{Z}[x]$ there exist ideals (for example, $\langle x, 2 \rangle$), which are not principal.

2. In the ring $\mathbf{Z}$, a nontrivial ideal is a prime ideal if and only if it is a maximal ideal. In the ring $\mathbf{Z}[x]$, there are prime ideals (for example $\langle x \rangle$), which are not maximal.

3. In the ring $\mathbf{Z}$, a nontrivial ideal I is a primary ideal if and only if $I = \langle p^n \rangle$ for some prime p and for some positive integer n. Hence, in $\mathbf{Z}$, if I is a primary ideal, then I is expressible as some power of its associated prime ideal. In $\mathbf{Z}[x]$, this is not true, as $\langle x, 4 \rangle$ is a primary ideal with $\langle x, 2 \rangle$ as its associated prime ideal, but $\langle x, 4 \rangle \neq \langle x, 2 \rangle^n$ for any $n \geq 1$.

17.1.1 Worked-Out Exercises

$\diamond$ **Exercise 1** Let R be an integral domain. Prove that if every ideal of R is a prime ideal, then R is a field.

Solution: Let $0 \neq a \in R$. Then $a^2 R$ is an ideal of R and hence it is a prime ideal. Now $a^2 \in a^2 R$. Since $a^2 R$ is a prime ideal, $a \in a^2 R$. Thus, $a = a^2 b$ for some $b \in R$. Then $a(1 - ab) = 0$. Since R is an integral domain and $a \neq 0$, $1 - ab = 0$ and so $ab = 1$, proving that a is a unit. Hence, R is a field.

$\diamond$ **Exercise 2** Let R be a commutative ring with 1. Suppose that $\langle x \rangle$ is a prime ideal of $R[x]$. Show that R is an integral domain.

Solution: Since $\langle x \rangle$ is a prime ideal $R[x]/\langle x \rangle$ is an integral domain. Since $R[x]/\langle x \rangle \simeq R$, R is an integral domain.

Exercise 3 Let R be a commutative ring and I be an ideal of R. Let P be a prime ideal of I. Show that P is an ideal of R.

Solution: Let $a \in P \subseteq I$ and $r \in R$. Then $rar \in I$. Therefore, $a(rar) \in P$ and so $(ar)^2 \in P$. Since P is a prime ideal of I, $ar \in P$. Hence, P is an ideal of R.

Exercise 4 Show that a proper ideal I of a ring R is a maximal ideal if and only if for any ideal A of R either $A \subseteq I$ or $A + I = R$.

Solution: Suppose I is a maximal ideal of R and let A be any ideal of R. If $A \nsubseteq I$, then $A + I$ is an ideal of R such that $I \subset A + I$. Since I is maximal, it follows that $A + I = R$.

Conversely, assume that the proper ideal I satisfies the given condition. Let J be an ideal of R such that $I \subset J$. Now $J \nsubseteq I$. Therefore, $I + J = R$. But $I + J = J$. Thus, $J = R$. Hence, I is a maximal ideal of R.

◇ **Exercise 5** Let R be a PID which is not a field. Prove that any nontrivial ideal I of R is a maximal ideal if and only if it is generated by an irreducible element.

Solution: Since R is not a field, there exists an element $0 \neq a \in R$ such that a is not a unit. Then $\langle 0 \rangle \subset \langle a \rangle \subset R$. Therefore, $\langle 0 \rangle$ is not a maximal ideal. Let I be a maximal ideal of R. Then $I \neq \{0\}$ and $I = \langle p \rangle$ for some $p \in R$, where p is irreducible by Theorem 17.1.5 and Corollary 15.3.13. Conversely, let $I = \langle p \rangle$ and p be irreducible. Let $I \subset J \subseteq R$. Since R is a PID, $J = \langle a \rangle$ for some $a \in R$. Since $p \in \langle a \rangle$, a divides p. Thus, $p = ab$ for some $b \in R$. Since p is irreducible, either a is a unit or b is a unit. If b is a unit, then $a = pb^{-1} \in \langle p \rangle$. Thus, $J \subseteq I$, which is a contradiction. Hence, a is a unit and so $J = R$. Thus, I is a maximal ideal.

◇ **Exercise 6** Show that the ideal $\langle x \rangle$ in $\mathbf{Z}[x]$ is a prime ideal, but not a maximal ideal.

Solution: Let $f(x) = a_0 + a_1 x + \cdots + a_n x^n$ and $g(x) = b_0 + b_1 x + \cdots + b_m x^m$ be two elements in $\mathbf{Z}[x]$ such that $f(x)g(x) \in \langle x \rangle$. Then $a_0 b_0 = 0$. Thus, either $a_0 = 0$ or $b_0 = 0$. Hence, either $f(x) \in \langle x \rangle$ or $g(x) \in \langle x \rangle$, showing that $\langle x \rangle$ is a prime ideal. Now the ideal $\langle x, 2 \rangle$ of $\mathbf{Z}[x]$ is such that $\langle x \rangle \subset \langle x, 2 \rangle \subset \mathbf{Z}[x]$. Hence, $\langle x \rangle$ is not a maximal ideal.

◇ **Exercise 7** Let R be a commutative ring with 1. Let A and B be two distinct maximal ideals of R. Show that $AB = A \cap B$.

Solution: Since $AB \subseteq A$ and $AB \subseteq B$, $AB \subseteq A \cap B$. Since A and B are distinct maximal ideals, there exists $b \in B$ such that $b \notin A$. Then $\langle A, b \rangle = \{a + br \mid a \in A, r \in R\}$ is an ideal of R such that $A \subset \langle A, b \rangle$. Since A is maximal, $\langle A, b \rangle = R$. This implies that $1 = a + br$ for some $a \in A$ and $r \in R$. Let $x \in A \cap B$. Then $x = x1 = xa + xbr = xa + (xb)r \in AB$. Hence, $A \cap B \subseteq AB$. Thus, $AB = A \cap B$.

◇ **Exercise 8** Let $f(x) = x^5 + 12x^4 + 9x^2 + 6$. Show that the ideal $I = \langle f(x) \rangle$ is maximal in $\mathbf{Z}[x]$.

Solution: I will be a maximal ideal if we can prove that $f(x)$ is an irreducible polynomial in $\mathbf{Z}[x]$. The content of $f(x)$ is 1. Hence, $f(x)$ is a primitive polynomial in $\mathbf{Z}[x]$. Also, for the prime 3, we find that $3|6$, $3|9$, $3|12$, $3 \nmid 1$, $3^2 \nmid 6$. Hence, $f(x)$ is irreducible in $\mathbf{Z}[x]$, by Eisenstein's criterion.

◇ **Exercise 9** (i) Find all maximal ideals of the ring $\mathbf{Z}_6$.

(ii) Find all ideals and all maximal ideals of the ring $\mathbf{Z}_8$.

Solution: (i) The mapping $\beta : \mathbf{Z} \to \mathbf{Z}_6$ defined by $\beta(n) = [n]$ is a homomorphism of $\mathbf{Z}$ onto $\mathbf{Z}_6$ and Ker $\beta = 6\mathbf{Z}$. If I is any ideal of $\mathbf{Z}_6$, then there exists a unique ideal A of $\mathbf{Z}$ such that Ker $\beta \subseteq A$ and $\beta(A) = I$. Now $\mathbf{Z}$, $2\mathbf{Z}$, $3\mathbf{Z}$, and $6\mathbf{Z}$ are the only ideals of $\mathbf{Z}$ which contain $6\mathbf{Z}$. Also, $\beta(\mathbf{Z}) = \mathbf{Z}_6$, $\beta(2\mathbf{Z}) = \{[0], [2], [4]\}$, $\beta(3\mathbf{Z}) = \{[0], [3]\}$, and $\beta(6\mathbf{Z}) = \{[0]\}$. Hence, $\{[0], [2], [4]\}$ and $\{[0], [3]\}$ are the only maximal ideals of $\mathbf{Z}_6$ since $2\mathbf{Z}$ and $3\mathbf{Z}$ are maximal ideals of $\mathbf{Z}$.

(ii) The mapping $\beta : \mathbf{Z} \to \mathbf{Z}_8$ defined by $\beta(n) = [n]$ is an epimorphism of rings and Ker $\beta = 8\mathbf{Z}$. Now $\mathbf{Z}$, $2\mathbf{Z}$, $4\mathbf{Z}$, and $8\mathbf{Z}$ are the only ideals of $\mathbf{Z}$ which contain $8\mathbf{Z}$. Also, $\beta(\mathbf{Z}) = \mathbf{Z}_8$, $\beta(2\mathbf{Z}) = \{[0], [2], [4], [6]\}$, $\beta(4\mathbf{Z}) = \{[0], [4]\}$, and $\beta(8\mathbf{Z}) = \{[0]\}$. Hence, the ideals of $\mathbf{Z}_8$ are $\mathbf{Z}_8$, $\{[0], [2], [4], [6]\}$, $\{[0], [4]\}$, and $\{[0]\}$. Now $\{[0]\} \subset \{[0], [4]\} \subset \{[0], [2], [4], [6]\} \subset \mathbf{Z}_8$. This implies that $\mathbf{Z}_8$ has only one maximal ideal, which is $\{[0], [2], [4], [6]\}$.

◇ **Exercise 10** Show that $\langle x^2 \rangle$ is a primary ideal in $\mathbf{Z}[x]$ with $\langle x \rangle$ as its associated prime ideal.

Solution: Let $f(x) = a_0 + a_1 x + \cdots + a_n x^n$ and $g(x) = b_0 + b_1 x + \cdots + b_m x^m$ be two elements in $\mathbf{Z}[x]$ such that $f(x)g(x) \in \langle x^2 \rangle$ and $f(x) \notin \langle x^2 \rangle$. Then $f(x)g(x) = x^2 h(x)$ for some $h(x) \in \mathbf{Z}[x]$. Hence, $a_0 b_0 = 0$ and $a_0 b_1 + a_1 b_0 = 0$. Since $f(x) \notin \langle x^2 \rangle$, it follows that either $a_0 \neq 0$ or $a_1 \neq 0$. If $a_0 \neq 0$, then $b_0 = 0$ and $b_1 = 0$ and so $g(x) \in \langle x^2 \rangle$. If $a_0 = 0$, then $a_1 \neq 0$. Hence, $a_0 b_1 + a_1 b_0 = 0$ shows that $b_0 = 0$. So we find that $b_0^2 = 0$, $b_0 b_1 + b_1 b_0 = 0$ and thus $(g(x))^2 \in \langle x^2 \rangle$. Hence, $\langle x^2 \rangle$ is a primary ideal. Now $\langle x^2 \rangle \subseteq \langle x \rangle$ and $f(x) \in \sqrt{\langle x^2 \rangle}$ if and only if $(f(x))^n \in \langle x^2 \rangle$ for some positive integer n. This is true if and only if the constant term of $f(x)$ is zero, i.e., if and only if $f(x) \in \langle x \rangle$.

Exercise 11 Show that a commutative ring R with 1 is isomorphic to a subdirect sum of a family of fields if and only if the intersection of all maximal ideals of R is $\{0\}$.

Solution: Suppose R is isomorphic to a subdirect sum of a family of fields $\{F_i \mid i \in I\}$. Then there exists a subring T of $\Pi_{i \in I} F_i$ such that $T = \oplus_{i \in I}^s F_i$ and $R \simeq T$. Let $\alpha : R \to T$ be an isomorphism. Then $\pi_i \circ \alpha : R \to F_i$ is an epimorphism for all $i \in I$, where π_i is the ith canonical projection. Proceeding as in the proof of Theorem 13.1.14, we can show that

$$\cap_{i \in I} A_i = \{0\},$$

where $A_i =$ Ker $\pi_i \circ \alpha$ for all $i \in I$. Now $R/A_i \simeq F_i$. Since F_i is a field, A_i is a maximal ideal for all $i \in I$. If A is the intersection of all maximal ideals of R, then $A \subseteq \cap_{i \in I} A_i = \{0\}$. Hence, $A = \{0\}$. Conversely, suppose that $A = \{0\}$, where $A = \cap_{i \in J} \{M_i \mid M_i$ is a maximal ideal of $R\}$. By Theorem 13.1.14, R is monomorphic to the subdirect sum of a family of rings $\{R/M_i \mid i \in J\}$. Since each M_i is a maximal ideal, we find that R/M_i is a field.

17.1.2 Exercises

1. Find all maximal and prime ideals of $\mathbf{Z}_{10}$.

2. Prove that $I = \{(5n, m) \mid n, m \in \mathbf{Z}\}$ is a maximal ideal of $\mathbf{Z} \times \mathbf{Z}$.

3. Find all ideals and maximal ideals of $\mathbf{Z}_{p^k}$, where p is a prime and k is a positive integer.

4. Let $I = \{a_0 + a_1 x + \cdots + a_n x^n \in \mathbf{Z}[x] \mid 3$ divides $a_0\}$. Show that I is a prime ideal of $\mathbf{Z}[x]$. Is I a maximal ideal?

5. Let I be an ideal of a ring R. Prove that the following conditions are equivalent.

   (i) I is a prime ideal.

   (ii) If $a, b \in R \backslash I$, then there exists $c \in R$ such that $acb \in R \backslash I$.

6. Let R be a finite commutative ring with 1. Show that in R, every prime ideal $I \neq R$ is a maximal ideal.

7. Let R be a Boolean ring. Prove that a nonzero proper ideal I of R is a prime ideal if and only if it is a maximal ideal.

8. Let R be a ring with 1. Prove that a nonzero proper ideal I of R is a maximal ideal if and only if the quotient ring R/I is a simple ring.

9. Let I be an ideal of a ring R. If P is a prime ideal of the quotient ring R/I, prove that there exists a prime ideal J of R such that $I \subseteq J$ and $J/I = P$.

10. Let R be a commutative ring with 1. Prove that there exists an epimorphism from R onto some field.

11. Let I be an ideal of a ring R with 1. Prove that the quotient ring R/I is a division ring if and only if I is a maximal ideal.

12. For all $r \in \mathbf{R}$, show that $I_r = \{f(x) \in \mathbf{R}[x] \mid f(r) = 0\}$ is a maximal ideal of $\mathbf{R}[x]$ and $\mathbf{R}[x]/I_r \simeq \mathbf{R}$. Also, prove that $\cap_{r \in \mathbf{R}} I_r = \{0\}$.

13. Consider the polynomial ring $K[x]$ over a field K. Let $a \in K$. Define the mapping $\phi_a : K[x] \to K$ by $\phi_a(f(x)) = f(a)$ for all $f(x) \in K[x]$. Show that ϕ_a is an epimorphism and Ker ϕ_a is a maximal ideal of $K[x]$.

14. Let R be a PID.

    (i) Prove that every nonzero nonunit element is divisible by a prime element.

    (ii) If $\{I_n\}_{n \in \mathbf{N}}$ is a sequence of ideals of R such that $I_1 \subseteq I_2 \subseteq \cdots \subseteq I_n \subseteq \cdots$, prove that there exists a positive integer n such that $I_n = I_{n+1} = \cdots$.

    (iii) Prove that every nonzero nonunit can be expressed as a finite product of prime elements.

15. Let $\{I_\alpha\}$ be a collection of prime ideals in a commutative ring R such that $\{I_\alpha\}$ forms a chain. Prove that $\cap_\alpha I_\alpha$ and $\cup_\alpha I_\alpha$ are prime ideals of R.

16. If I_1 and I_2 are ideals of a commutative ring 1, prove that $\sqrt{I_1 \cap I_2} = \sqrt{I_1} \cap \sqrt{I_2}$.

17. Let R be a commutative ring with 1 and Q_i, $i = 1, 2, \ldots, n$, be ideals in R. Set $Q = \cap_{i=1}^n Q_i$. Prove that if $\sqrt{Q_i} = P$ for some ideal P of R, $i = 1, 2, \ldots, n$, then $\sqrt{Q} = P$. If $\sqrt{Q_i} = P$, $i = 1, 2, \ldots, n$, and each Q_i is primary, prove that Q is primary.

18. If I is an ideal of a commutative ring R with 1 such that $\sqrt{I}$ is a maximal ideal, prove that I is a primary ideal.

19. In the polynomial ring $\mathbf{Z}[x]$, prove the following.

    (i) $I = \{f(x) \in \mathbf{Z}[x] \mid$ the constant term of $f(x)$ is divisible by 4$\}$ is a primary ideal with $J = \langle x, 2 \rangle$ as its associated prime ideal.

    (ii) The ideal $\langle x, 6 \rangle$ is not a primary ideal.

20. Prove that every prime ideal is a primary ideal in a commutative ring.

21. Let M be an ideal of a commutative ring R. Prove that R/M is a field if and only if M is a maximal ideal and $x^2 \in M$ implies $x \in M$ for all $x \in R$.

22. Prove that in a PID every nontrivial ideal I can be expressed as a finite product of prime ideals $I = P_1 \cdots P_n$ such that $P_1, P_2, \ldots, P_n$ are determined uniquely up to order.

23. An ideal P of a ring R is called a **semiprime ideal** if for any ideal I of R, $I^2 \subseteq P$ implies that $I \subseteq P$.

    (i) Prove that an ideal P of R is a semiprime ideal if and only if the quotient ring R/P contains no nonzero nilpotent ideals.

    (ii) If R is a commutative ring with 1, prove that an ideal P of R is a semiprime ideal if and only if $\sqrt{P} = P$.

24. A commutative ring R with 1 is called a **local ring** if R has only one maximal ideal. Prove the following.

    (i) $\mathbf{Z}_8$ and $\mathbf{Z}_9$ are local rings.

    (ii) In a local ring, all nonunits form a maximal ideal.

    (iii) In a local ring R, for all $r, s \in R, r + s = 1$ implies either r is a unit or s is a unit.

25. Let p be a prime integer and $\mathbf{Q}_p = \{\frac{a}{b} \in \mathbf{Q} \mid p$ does not divide $b\}$. Show that $\mathbf{Q}_p$ is a local ring under the usual addition and multiplication of rational numbers.

26. Let R be a field and T be the set of all sequences $\{a_n\}$ of elements of R. Then $(T, +, \cdot)$ is a ring, where $+$ and $\cdot$ are defined as in Worked-Out Exercise 6 (page 358). Prove the following.

    (i) The set I of all nonunits of T is a maximal ideal of T.

    (ii) I is the only maximal ideal of T.

    (iii) T is a local ring.

27. Let $R = R_1 \oplus R_2 \oplus \cdots \oplus R_n$ be the direct sum of the finite family of rings $\{R_1, R_2, \ldots, R_n\}$, where each R_i contains an identity. Prove the following:

    (i) If M_i is a maximal ideal of R_i $(1 \leq i \leq n)$, then $R_1 \oplus R_2 \oplus \cdots \oplus R_{i-1} \oplus M_i \oplus R_{i+1} \oplus \cdots \oplus R_n$ is a maximal ideal of R.

    (ii) Every maximal ideal M of R is of the form

    $$R_1 \oplus R_2 \oplus \cdots \oplus R_{i-1} \oplus M_i \oplus R_{i+1} \oplus \cdots \oplus R_n,$$

    where M_i is a maximal ideal of R_i for some i $(1 \leq i \leq n)$.

28. Show that the ring $\mathbf{Z}$ is isomorphic to a subdirect sum of a family of fields.

29. An ideal I of a ring R is called a **minimal ideal** if $I \neq \{0\}$ and there does not exist any ideal J of R such that $\{0\} \neq J \subset I$. If I is a minimal ideal of a commutative ring R with 1, prove that either $I^2 = \{0\}$ or $I = eR$ for some idempotent $e \in R$.

30. In the following exercises, write the proof if the statement is true; otherwise, give a counterexample.

   (i) Let R be a commutative ring with 1 and P be a prime ideal of R such that $P \neq R$. If the quotient ring R/P contains a finite number of elements, then R/P is a field.

   (ii) In a PID, there exists a prime element.

   (iii) In a PID, every proper prime ideal is a maximal prime ideal.

   (iv) The intersection of two prime ideals of a ring R is a prime ideal of R.

   (v) If I is a prime ideal of a ring R, then $I[x]$ is also a prime ideal of $R[x]$.

   (vi) If I is a maximal ideal of a ring R, then $I[x]$ is also a maximal ideal of $R[x]$.

   (vii) A commutative ring with 1 and with only a finite number of maximal ideals is a field.

   (viii) In the ring $\mathbf{Z}$, the ideal $\langle 5 \rangle$ is a maximal ideal, but in the ring $\mathbf{Z}[i]$, the ideal $\langle 5 \rangle$ is not a maximal ideal.

17.2   Jacobson Semisimple Ring

In this section, we introduce an interesting class of commutative rings and give a simple characterization of this class.

Throughout the section, we assume that R is a commutative ring with 1.

Definition 17.2.1 *The **Jacobson radical** of a ring R, denoted by $radR$, is the set*

$$radR = \cap\{M \mid M \text{ is a maximal ideal of } R\}.$$

Since the ring R contains 1, maximal ideals in R exist and thus $radR$ is well defined.

The following theorem gives a characterization of $radR$ with the help of elements of $radR$.

Theorem 17.2.2 *Let $y \in R$. Then $y \in radR$ if and only if $1 - xy$ is a unit in R for all $x \in R$.*

Proof.   Suppose $y \in radR$ and there exists an element $x \in R$ such that $1 - xy$ is not a unit. Then from the Corollary 17.1.16, there exists a maximal ideal M of R such that $1 - xy \in M$. Since $y \in radR$, $y \in M$. Therefore, $xy \in M$. This implies that $1 = 1 - xy + xy \in M$, which is a contradiction. Hence, $1 - xy$ is a unit.

Conversely, assume that $1 - xy$ is a unit in R for all $x \in R$. Suppose $y \notin$ rad$R$. Then there exists a maximal ideal M of R such that $y \notin M$. Consider $\langle M, y \rangle = \{m + ry \mid m \in M, r \in R\}$, the ideal generated by $M \cup \{y\}$. Clearly $M \subset \langle M, y \rangle$. Since M is a maximal ideal, it follows that $\langle M, y \rangle = R$. Hence, $1 = m + ry$ for some $m \in M$ and $r \in R$. Thus, $1 - ry = m \in M$. By the hypothesis, $m = 1 - ry$ is a unit. Hence, $1 = mm^{-1} \in M$. This implies that $M = R$, which is a contradiction. Hence, $y \in$ rad$R$. ∎

Corollary 17.2.3 *0 is the only idempotent element in rad$R$.*

Proof.   Let $y \in$ rad$R$ and y be an idempotent. Now $1 - y = 1 - 1y$ is a unit. Hence, there exists $u \in R$ such that $(1 - y)u = 1$. Then $y = y(1 - y)u = (y - y^2)u = (y - y)u = 0u = 0$. ∎

Corollary 17.2.4 *rad$R$ contains all nil ideals of R.*

Proof.   Let I be a nil ideal of R. Now every element of I is nilpotent. Hence, for all $a \in I$, $r \in R$, ar is nilpotent. This implies that $1 + ar$ is a unit for all $r \in R$ by Exercise 11 (page 282). Hence, $a \in$ rad$R$. Thus, $I \subseteq$ rad$R$. ∎

Theorem 17.2.5 *rad$R$ is an ideal of R and $\mathrm{rad}(R/\mathrm{rad}R) = \{\bar{0}\}$, where $\bar{0} = 0 + \mathrm{rad}R$.*

Proof.   Since the intersection of a family of ideals of R is an ideal, rad$R$ is an ideal of R. Denote $I = $ rad$R$. Now $I \neq R$ and R/I is a commutative ring with identity. Let $a + I \in \mathrm{rad}(R/I)$ and $x + I \in R/I$. Then $1 + I - (x + I)(a + I)$ is a unit in R/I. Thus, there exists $u + I \in R/I$ such that $(1 + I - (x + I)(a + I))(u + I) = 1 + I$ or $(1 - xa + I)(u + I) = 1 + I$. Hence, $1 - (1 - xa)u \in I$ and so by Theorem 17.2.2, $1 - (1 - (1 - xa)u)$ is a unit in R, i.e., $(1 - xa)u$ is a unit in R. This implies that $1 - xa$ is a unit in R for all $x \in R$. Therefore, $a \in I$ and so $a + I = I$. Consequently, $\mathrm{rad}(R/\mathrm{rad}R) = \{\bar{0}\}$. ∎

Let us now consider those commutative rings with 1 for which rad$R = \{0\}$.

Definition 17.2.6 *R is called a **Jacobson semisimple ring (J-semisimple ring)** if rad$R = \{0\}$.*

Example 17.2.7 *(i) For any ring R, the quotient ring $R/\mathrm{rad}R$ is J-semisimple by Theorem 17.2.5.*

(ii) The ring $\mathbf{Z}$ of integers is J-semisimple. In $\mathbf{Z}$, the maximal ideals are of the form $\langle p \rangle$, where p is a prime. Let $n \in \mathrm{rad}\mathbf{Z}$. Then $p \mid n$ for all $p \in \mathbf{Z}$, p is prime. Since $\mathbf{Z}$ is a UFD, $n = 0$. Hence, rad$\mathbf{Z} = \{0\}$.

(iii) Every commutative regular ring R is J-semisimple. Let $a \in \text{rad} R$. Then there exists $b \in R$ such that $a = aba$. Now $ab \in \text{rad} R$ and $(ab)^2 = abab = ab$. By Corollary 17.2.3, $ab = 0$. Hence, $a = aba = 0$. Thus, $\text{rad} R = \{0\}$.

(iv) Every field is J-semisimple.

(v) Every polynomial ring $F[x]$ over a field F is J-semisimple.

Theorem 17.2.8 *A commutative ring R with 1 is J-semisimple if and only if it is isomorphic to a subdirect sum of a family of fields.*

Proof.   Suppose R is J-semisimple. Let $\{M_i \mid i \in I\}$ be the collection of all maximal ideals of R. Then $\text{rad} R = \cap_{i \in I} M_i = \{0\}$. Hence, by Theorem 13.1.14, R is isomorphic to a subdirect sum of a family $\{R/M_i \mid i \in I\}$ of rings. But R/M_i is a field for all $i \in I$. Hence, R is isomorphic to a subdirect sum of a family of fields. Conversely, suppose that R is isomorphic to a subdirect sum of a family of fields $\{F_i \mid i \in I\}$. Then there exists a family of ideals M_i such that each F_i is isomorphic to R/M_i and $\cap_{i \in I} M_i = \{0\}$. Since each M_i is a maximal ideal of R and $\cap_{i \in I} M_i = \{0\}$, it follows that $\text{rad} R \subseteq \cap_{i \in I} M_i = \{0\}$. Thus, R is J-semisimple. ∎

17.2.1   Worked-Out Exercises

◇ **Exercise 1** Find $\text{rad} \mathbf{Z}_{12}$. Is the ring $\mathbf{Z}_{12}$ a J-semisimple ring?

Solution: The mapping $\beta : \mathbf{Z} \rightarrow \mathbf{Z}_{12}$ defined by $\beta(n) = [n]$ is an epimorphism of rings and $\text{Ker } \beta = 12\mathbf{Z}$. Now $\text{Ker } \beta$ is contained in the ideals, $\mathbf{Z}$, $2\mathbf{Z}$, $3\mathbf{Z}$, $4\mathbf{Z}$, $6\mathbf{Z}$ and $12\mathbf{Z}$. Also, $\beta(\mathbf{Z}) = \mathbf{Z}_{12}$, $\beta(2\mathbf{Z}) = \{[0], [2], [4], [6], [8], [10]\}$, $\beta(3\mathbf{Z}) = \{[0], [3], [6], [9]\}$, $\beta(4\mathbf{Z}) = \{[0], [4], [8]\}$, $\beta(6\mathbf{Z}) = \{[0], [6]\}$, and $\beta(12\mathbf{Z}) = \{[0]\}$. Hence, $I = \{[0], [2], [4], [6], [8], [10]\}$ and $J = \{[0], [3], [6], [9]\}$ are the only maximal ideals of $\mathbf{Z}_{12}$. Now $\text{rad} \mathbf{Z}_{12} = I \cap J = \{[0], [6]\}$. Since $\text{rad} \mathbf{Z}_{12} \neq \{[0]\}$, $\mathbf{Z}_{12}$ is not J-semisimple.

◇ **Exercise 2** Is the ring $\mathbf{Z}_{15}$ a J-semisimple ring?

Solution: Proceeding as in Worked-Out Exercise 1, we can show that $I = \{[0], [3], [6], [9], [12]\}$, $J = \{[0], [5], [10]\}$ are the only maximal ideals of $\mathbf{Z}_{15}$. Now $\text{rad} \mathbf{Z}_{15} = I \cap J = \{[0]\}$. Hence, $\mathbf{Z}_{15}$ J-semisimple.

Exercise 3 Let R be a commutative ring with 1 and A be an ideal of R such that $A \subset \text{rad} R$. Show that $\text{rad}(R/A) = \text{rad} R/A$.

Solution: Let $b + A \in \text{rad}(R/A)$. Let $r \in R$. Then $(1 + A) - (b + A)(r + A)$ is a unit. Hence, there exists $d + A \in R/A$ such that $((1 - br) + A)(d + A) = 1 + A$. This implies that $1 - (d - dbr) \in A \subseteq \text{rad} R$. Hence, $1 - (1 - (d - dbr))$ is a unit in R, i.e., $d(1 - br)$ is a unit in R. Thus, $1 - br$ is a unit in R for all $r \in R$. Hence,

$b \in \mathrm{rad}R$, and so $b + A \in \mathrm{rad}R/A$. Thus, $\mathrm{rad}(R/A) \subseteq \mathrm{rad}R/A$. Now let $b + A \in \mathrm{rad}R/A$, where $b \in \mathrm{rad}R$. Then, $1 - bc$ is a unit for all $c \in R$. Let $c \in R$. Now there exists $d \in R$ such that $(1 - bc)d = 1$. Thus, $((1 - bc) + A)(d + A) = 1 + A$ in R/A, i.e., $(1 + A) - (b + A)(c + A)$ is a unit in R/A. Hence, $b + A \in \mathrm{rad}(R/A)$. Thus, $\mathrm{rad}R/A \subseteq \mathrm{rad}(R/A)$. Hence, $\mathrm{rad}(R/A) = \mathrm{rad}R/A$.

17.2.2   Exercises

1. Prove that the ring $\mathbf{Z}_n$, $n > 1$, is J-semisimple if and only if n is a square free integer.

2. Is the ring $\mathbf{Z}_{10}$ a J-semisimple ring?

3. Let R be a PID. If R has an infinite number of maximal ideals, prove that R is J-semisimple.

4. Let $R = R_1 \times R_2$ be the direct product of two commutative rings R_1 and R_2 with 1. Prove that $\mathrm{rad}R = \mathrm{rad}R_1 \times \mathrm{rad}R_2$.

5. Let $\mathcal{F}(\mathbf{R}) = \{f \mid f : \mathbf{R} \to \mathbf{R}\}$. $\mathcal{F}(\mathbf{R})$ is a commutative ring with 1, where $+$ and $\cdot$ are defined by

$$
\begin{aligned}
(f + g)(x) &= f(x) + g(x) \\
(f \cdot g)(x) &= f(x)g(x)
\end{aligned}
$$

for all $f, g \in \mathcal{F}(\mathbf{R})$ and for all $x \in \mathbf{R}$. Let $t \in \mathbf{R}$.

(i) Show that $I_t = \{f \in \mathcal{F}(\mathbf{R}) \mid f(t) = 0\}$ is a maximal ideal and $\cap_{t \in \mathbf{R}} I_t = \{0\}$.

(ii) Prove that $\mathcal{F}(\mathbf{R})$ is a J-semisimple ring.

6. Which of the following statements are true? Justify your answer.

(i) If F_1 and F_2 are two fields, then $F_1 \times F_2$ is a J-semisimple ring, but not a field.

(ii) If R_1 and R_2 are two J-semisimple rings, then $R_1 \times R_2$ is a J-semisimple ring.

(iii) A J-semisimple ring may contain a nonzero nil ideal.

(iv) In a commutative ring R with 1, for any two ideals A and B, $AB = \{0\}$ may not imply $A \cap B = \{0\}$, but in a J-semisimple ring this is always true.

Chapter 18

Noetherian and Artinian Rings

In Hilbert's work on invariant theory is the result that in certain polynomial rings, every ideal is finitely generated. Lasker, a student of Hilbert and a former world chess champion, showed that in certain polynomial rings, every ideal is a finite intersection of primary ideals. Noether generalized Lasker's result to commutative rings in which any strictly ascending chain of ideals is finite.

18.1   Noetherian and Artinian Rings

In the present section, we introduce two special classes of rings—Noetherian rings and Artinian rings. Noetherian rings satisfy an ascending chain condition of ideals, whereas Artinian rings satisfy a descending chain condition of ideals. We first define these two properties of ideals.

Definition 18.1.1 *A ring R is said to satisfy the **ascending chain condition** (**ACC**) for left (right) ideals if for each sequence of left (right) ideals $A_1, A_2, \ldots$ of R with $A_1 \subseteq A_2 \subseteq \cdots$, there exists a positive integer n (depending on the sequence) such that $A_n = A_{n+1} = \cdots$. R is said to satisfy the **descending chain condition** (**DCC**) for left (right) ideals if for each sequence of left (right) ideals $A_1, A_2, \ldots$ of R with $A_1 \supseteq A_2 \supseteq \cdots$, there exists a positive integer n (depending on the sequence) such that $A_n = A_{n+1} = \cdots$.*

Clearly the ACC on left ideals is equivalent to the statement that any sequence of left ideals $A_1, A_2, \ldots$ of R such that $A_1 \subset A_2 \subset \cdots$ must be finite. A similar equivalence holds for the DCC.

Let $\mathcal{I}$ be a nonempty set of left ideals of a ring R. Then $(\mathcal{I}, \leq)$ is a partially ordered set, where $\leq$ is defined by the set inclusion relation. This partially ordered set may have a maximal element, i.e., there may exist an element

$A \in \mathcal{I}$ such that A is not contained in any other element of $\mathcal{I}$. Also, this partially ordered set may contain a minimal element, i.e., an element of $\mathcal{I}$ that does not contain any other element of $\mathcal{I}$. Considering all these conditions, let us introduce the following conditions on a ring.

Definition 18.1.2 *A ring R is said to satisfy the **maximal condition** (MC) for left (right) ideals if in any nonempty set of left (right) ideals of R, there exist some left (right) ideal which is maximal in the set, i.e., not contained in any other ideal of the set. R is said to satisfy the **minimal condition** (mC) for left (right) ideals if in any nonempty set of left (right) ideals of R, there exist some left (right) ideal which is minimal in the set, i.e., does not contain any other left (right) ideal of the set.*

Example 18.1.3 *The ring $\mathbf{Z}$ of integers satisfies the maximal condition for ideals, but does not satisfy the minimal condition for ideals. Let $\mathcal{I}$ be any nonempty collection of ideals of $\mathbf{Z}$. Let $A_1 \in \mathcal{I}$. Then there exists a nonnegative integer n such that $A_1 = \langle n \rangle$. If A_1 is not maximal, then there exists an ideal $A_2 = \langle m \rangle$ such that $A_1 \subset A_2$. Then $m \neq n$ and m divides n. Again, if A_2 is not maximal, then there exists an ideal $A_3 = \langle r \rangle$ such that $A_2 \subset A_3$. Then $r \neq m$, $r \neq n$, and r divides m and n. If A_3 is not maximal, then we repeat this process. Since $\mathbf{Z}$ is a UFD, n has finitely many distinct divisors. Hence, the above process must terminate after finitely many steps. Thus, $\mathcal{I}$ contains a maximal element. Consider the set $\mathcal{J} = \{m\mathbf{Z} \mid m \text{ in a positive even integer}\}$ of ideals of $\mathbf{Z}$. For any $m\mathbf{Z} \in \mathcal{J}$, $2m\mathbf{Z} \in \mathcal{J}$, and $m\mathbf{Z} \supset 2m\mathbf{Z}$. Therefore, it follows that $\mathcal{J}$ has no minimal element.*

Theorem 18.1.4 *In any ring R, the following conditions are equivalent.*
  (i) R satisfies the ACC for left ideals.
  (ii) R satisfies the MC for left ideals.
  (iii) Every left ideal of R is finitely generated.

Proof.   (i)$\Rightarrow$(ii): Let $\mathcal{A}$ be any collection of left ideals of R. Let $A_1 \in \mathcal{A}$. Then A_1 is either maximal in $\mathcal{A}$ or there exists a left ideal $A_2 \in \mathcal{A}$ such that $A_1 \subset A_2$. If A_1 is maximal in $\mathcal{A}$, then we have proved the assertion. If A_1 is not maximal in $\mathcal{A}$, then either A_2 is maximal in $\mathcal{A}$ or there exists $A_3 \in \mathcal{A}$ such that $A_2 \subset A_3$. By the ACC, this process must terminate in a finite number of steps, say, n steps. Then A_n is maximal in $\mathcal{A}$.

(ii)$\Rightarrow$(iii): Let A be any left ideal of R. Let $a_1 \in A$. Then either $\langle a_1 \rangle_l = A$ or $\langle a_1 \rangle_l \subset A$. If $\langle a_1 \rangle_l = A$, then A is finitely generated. Suppose $\langle a_1 \rangle_l \subset A$. Let $a_2 \in A$ and $a_2 \notin \langle a_1 \rangle_l$. Then $\langle a_1 \rangle_l \subset \langle a_1, a_2 \rangle_l$ and either $\langle a_1, a_2 \rangle_l = A$ or $\langle a_1, a_2 \rangle_l \subset A$. If $\langle a_1, a_2 \rangle_l = A$, then A is finitely generated. If $\langle a_1, a_2 \rangle_l \subset A$, then we continue this process. If in a finite number of steps, say, n, we obtain

$\langle a_1, a_2, \ldots, a_n \rangle_l = A$, then A is finitely generated. If this is not the case, then there exist elements $a_1, a_2, \ldots$ in A such that

$$\langle a_1 \rangle_l \subset \langle a_1, a_2 \rangle_l \subset \cdots \subset \langle a_1, a_2, \ldots, a_n \rangle_l \subset \cdots.$$

In this case the set $\mathcal{A} = \{\langle a_1 \rangle_l, \langle a_1, a_{2l} \rangle, \ldots, \langle a_1, a_2, \ldots, a_n \rangle_l, \ldots\}$ is a collection of left ideals of R which does not have a maximal element. However, this contradicts our assumption that R satisfies (ii).

(iii)$\Rightarrow$(i): Let A_1, A_2, $\ldots$ be any sequence of left ideals of R such that

$$A_1 \subseteq A_2 \subseteq \cdots.$$

Then $A = \cup_{i=1}^{\infty} A_i$ is a left ideal of R and is finitely generated, say, $A = \langle a_1, a_2, \ldots, a_n \rangle_l$. Now $a_j \in A_{i_j}$ for some A_{i_j}, $j = 1, 2, \ldots, n$. Let k be the maximum of i_1, $\ldots$, i_n. Then a_1, a_2, $\ldots, a_n \in A_k$. This implies that $A = \langle a_1, a_2, \ldots, a_n \rangle_l \subseteq A_k \subseteq A$. Hence, we must have that $A = A_k$. Thus, for any positive integer i, $A_{k+1} \supseteq A_k = A \supseteq A_{k+i}$. Consequently,

$$A_1 \subseteq A_2 \subseteq \cdots \subseteq A_k = A_{k+1} = \cdots = A. \blacksquare$$

Corollary 18.1.5 *Any principal ideal ring satisfies the ACC.* $\blacksquare$

Corollary 18.1.5 provides us with many examples of rings satisfying the ACC. For instance, $\mathbf{Z}$ and the polynomial ring $F[x]$ over a field F satisfy the ACC since they are principal ideal rings.

Definition 18.1.6 *A ring which satisfies the ACC for left (right) ideals is called a **left(right) Noetherian ring**.*

A ring which is both left Noetherian and right Noetherian is called a **Noetherian ring**.

The following theorem follows from Theorem 18.1.4.

Theorem 18.1.7 *In any ring R, the following conditions are equivalent.*
   (i) R is a left Noetherian ring.
   (ii) R satisfies the MC for left ideals.
   (iii) Every left ideal of R is finitely generated. $\blacksquare$

Example 18.1.8 *(i) A principal ideal ring is a Noetherian ring.*
   (ii) A polynomial ring over a field is a Noetherian ring.

We now study the homomorphic images, quotients, and finite direct sums of Noetherian rings.

Theorem 18.1.9 *If R is a left Noetherian ring, then any homomorphic image of R is a left Noetherian ring.*

Proof. Let R be a left Noetherian ring and $f : R \to S$ be an epimorphism of rings. Let

$$J_1 \subseteq J_2 \subseteq \cdots$$

be any ascending chain of left ideals of S. Let $I_k = f^{-1}(J_k)$ for all $k \geq 1$. Then I_k is a left ideal of R for all k and $I_1 \subseteq I_2 \subseteq \cdots$. Since R is left Noetherian, there exists a positive integer n such that $I_n = I_{n+i}$ for all $i \geq 1$. Let $y \in J_{n+i}$, $i \geq 1$. Since f is onto, there exists $x \in R$ such that $f(x) = y$. Then $x \in I_{n+i} = I_n$ and so $y \in J_n$. Therefore, $J_n = J_{n+i}$ for all $i \geq 1$, proving that S is left Noetherian. ∎

Theorem 18.1.10 *Let I be an ideal of a ring R. If I and R/I are both left Noetherian rings, then R is left Noetherian.*

Proof. Let $A_1 \subseteq A_2 \subseteq \cdots$ be an ascending chain of left ideals in R. Let $\psi : R \to R/I$ be the natural homomorphism of R onto R/I. Then $\psi(A_1) \subseteq \psi(A_2) \subseteq \cdots$ is an ascending chain of left ideals in R/I. Since R/I is left Noetherian, there exists a positive integer n such that $\psi(A_n) = \psi(A_{n+i})$ for all $i \geq 1$. Also, $A_1 \cap I \subseteq A_2 \cap I \subseteq \cdots$ is an ascending chain of left ideals in I. Since I is left Noetherian, there exists a positive integer m such that $A_m \cap I = A_{m+i} \cap I$ for all $i \geq 1$. Let k be the larger of m and n. Then $\psi(A_k) = \psi(A_{k+i})$ and $A_k \cap I = A_{k+i} \cap I$ for all $i \geq 1$. Let $b \in A_{k+i}$. There exists $x \in A_k$ such that $\psi(b) = \psi(x)$, i.e., $b + I = x + I$. Therefore, $b - x \in I$ and also $b - x \in A_{k+i}$. This implies that $b - x \in A_{k+i} \cap I = A_k \cap I$. Hence, $b - x \in A_k$ and so $b \in A_k$. Thus, $A_k = A_{k+i}$ for all $i \geq 1$. Consequently, R is left Noetherian. ∎

Theorem 18.1.11 *A finite direct sum of left Noetherian rings is left Noetherian.*

Proof. Let $R = R_1 \oplus R_2 \oplus \cdots \oplus R_n$ be a finite direct sum of left Noetherian rings. We show the result for $n = 2$. The general case will follow by induction. Let $R = R_1 \oplus R_2$, where R_1 and R_2 are left Noetherian. Now $(R_1 \oplus R_2)/R_1 \simeq R_2$. Thus, $(R_1 \oplus R_2)/R_1$ is left Noetherian. Since $(R_1 \oplus R_2)/R_1$ and R_1 are left Noetherian, $R_1 \oplus R_2$ is left Noetherian by Theorem 18.1.10. ∎

Note: All the results which are established for left Noetherian rings can also be proved for right Noetherian rings by simply replacing left ideals with right ideals.

Since every ideal of **Z** is a principal ideal, **Z** is Noetherian. Every ideal of **Z** is generated by a single element, but the ideals of **Z**$[x]$ may not be principal ideals, i.e., may not be generated by a single element. Interestingly, the ideals of $\mathbf{Z}[x]$ are finitely generated. Thus, $\mathbf{Z}[x]$ is not a principal ideal ring, but nevertheless is a Noetherian ring. This result follows from the following theorem.

Theorem 18.1.12 *If R is a commutative Noetherian ring with 1, then the polynomial ring $R[x]$ is a Noetherian ring.*

Proof.   We show that every ideal A of $R[x]$ is finitely generated. For each integer $n \geq 0$, let I_n be the set of all $a \in R$ such that either $a = 0$ or a is the coefficient of x^n of a polynomial $f(x) \in A$ of degree n. Suppose $a, b \in I_n$ and $a \neq 0, b \neq 0$. Then there exist $f(x), g(x) \in A$ such that $\deg f(x) = \deg g(x) = n$ and a is the coefficient of x^n in $f(x)$ and b is the coefficient of x^n in $g(x)$. If $a - b = 0$, then $a - b \in I_n$. Assume $a - b \neq 0$. Now $a - b$ is the coefficient of x^n of $f(x) - g(x) \in A$. Therefore, $a - b \in I_n$. Also, for $r \in R$, if $ra \neq 0$, then ra is the coefficient of x^n of $rf(x) \in A$ and so $ra \in I_n$. Hence, I_n is an ideal of R for $n \geq 0$. We now show that $I_n \subseteq I_{n+1}$. Let a be a nonzero element of I_n. There exists $f(x) = a_0 + a_1 x + \cdots + a_{n-1} x^{n-1} + ax^n \in A$. Then $xf(x) = a_0 x + a_1 x^2 + \cdots + a_{n-1} x^n + ax^{n+1} \in A$. Therefore, $a \in I_{n+1}$ and so $I_n \subseteq I_{n+1}$. Thus, we obtain an ascending chain

$$I_0 \subseteq I_1 \subseteq I_2 \subseteq \cdots$$

of ideals of R. Since R is Noetherian, there exists an integer m such that $I_m = I_n$ for all $n \geq m$. Again, every ideal of a Noetherian ring is finitely generated. Hence, each of the ideals $I_0, I_1, \ldots, I_m$ is finitely generated. Let

$$I_k = \langle a_{k1}, a_{k2}, \ldots, a_{kt_k} \rangle$$

for $k = 0, 1, \ldots, m$, where a_{kj} is the leading coefficient of $f_{kj}(x) \in A$, a polynomial of degree k. Note that $a_{0j} = f_{0j}(x)$ for $k = 0$ and $j = 1, 2, \ldots, t_k$ are the polynomials of degree 0 in $R[x]$. Let

$$S = \{ f_{kj}(x) \mid 0 \leq k \leq m,\ 1 \leq j \leq t_k \}$$

and B be the ideal generated by S. Then

$$B = \langle f_{01}(x), \ldots, f_{0t_0}(x), \ldots, f_{m1}(x), \ldots, f_{mt_m}(x) \rangle \subseteq A$$

since each $f_{kj}(x) \in A$. Next we show that $A \subseteq B$. We prove this by induction on the degree of the polynomials in A. Let $f(x)$ be any polynomial of degree 0 in A. Then $f(x) \in I_0 \subseteq B$. Hence, any polynomial of degree 0 in A is also in B. Now assume that any polynomial of degree less than r in A is also in B. Consider a polynomial

$$f(x) = b_0 + b_1 x + \cdots + b_r x^r \in A$$

with $b_r \neq 0$. If $r \leq m$, then $b_r \in I_r$ and hence $b_r = c_1 a_{r1} + c_2 a_{r2} + \cdots + c_r a_{rt_r}$ for some $c_1, c_2, \ldots, c_r \in R$. Hence, the polynomial

$$h_r(x) = c_1 f_{r1}(x) + c_2 f_{r2}(x) + \cdots + c_r f_{rt_r}(x) \in B$$

and the coefficient of x^r of this polynomial is $b_r \neq 0$. Thus, $h_r(x)$ is of degree r with b_r as the coefficient of x^r. Therefore, $f(x) - h_r(x)$ is a polynomial of degree less than r and $f(x) - h_r(x) \in A$. Thus, by the induction hypothesis, $f(x) - h_r(x) \in B$. But $h_r(x) \in B$. Consequently, $f(x) \in B$. Hence, by induction, if $f(x) \in A$ with $\deg f(x) \leq m$, then $f(x) \in B$. If $m < r$, then $b_r \in I_r = I_m = \langle a_{m1}, a_{m2}, \ldots, a_{mt_m} \rangle$. Therefore, there exist $d_{m1}, d_{m2}, \ldots, d_{mt_m} \in R$ such that

$$b_r = d_{m1} a_{m1} + d_{m2} a_{m2} + \cdots + d_{mt_m} a_{mt_m}.$$

This implies that the polynomial

$$h_2(x) = f(x) - x^{r-m}(d_{m1} f_{m1}(x) + d_{m2} f_{m2}(x) + \cdots + d_{mt_m} f_{mt_m}(x)) \quad (18.1)$$

is a polynomial in A of degree less than r. Hence, by the induction hypothesis, $h_2(x) \in B$. But $d_{m1} f_{m1}(x) + d_{m2} f_{m2}(x) + \cdots + d_{mt_m} f_{mt_m}(x) \in B$. From Eq. (18.1), it follows that $f(x) \in B$. Therefore, by induction, if $f(x) \in A$ with $\deg f(x) > m$, then $f(x) \in B$. Consequently, we find that $A = B$, proving that A is finitely generated. ∎

Corollary 18.1.13 (Hilbert Basis Theorem) *Let R be a commutative ring with 1. If R is a Noetherian ring, then the polynomial ring $R[x_1, \ldots, x_n]$ is a Noetherian ring.* ∎

If F is a field, then F is clearly Noetherian since it has only two ideals. Thus, the polynomial ring $F[x_1, \ldots, x_n]$ is a Noetherian ring.

Thus, we find that the Hilbert basis theorem gives us a wide class of Noetherian rings.

We now introduce another class of rings called Artinian rings. First we note the following equivalence.

Theorem 18.1.14 *In any ring R, the following conditions are equivalent.*
   (i) R satisfies the DCC for left ideals.
   (ii) R satisfies the mC for left ideals.

Proof.   The proof is similar to the proof of Theorem 18.1.4. We leave the proof as an exercise. ∎

Example 18.1.15 $\mathbb{Z}$ *does not satisfy the DCC since* $\langle 2 \rangle \supset \langle 4 \rangle \supset \langle 8 \rangle \supset \cdots$ *is an infinite chain with* $\langle 2^n \rangle \supset \langle 2^{n+1} \rangle$, $n = 1, 2, \ldots$. *The polynomial ring* $F[x]$ *does not satisfy the DCC since* $\langle x \rangle \supset \langle x^2 \rangle \supset \langle x^3 \rangle \supset \cdots$ *is an infinite chain with* $\langle x^n \rangle \supset \langle x^{n+1} \rangle$, $n = 1, 2, \ldots$.

Example 18.1.16 *Let R be the ring of Example 10.1.8. For $r \in \mathbf{R}$, define*

$$I_r = \{f \in R \mid f(x) = 0 \text{ for all } -r \leq x \leq r\}.$$

Now I_r is an ideal of R and $I_r \subset I_t$ if $t < r$. Therefore,

$$I_1 \supset I_2 \supset I_3 \supset \cdots$$

is an infinite strictly descending chain of ideals and

$$I_1 \subset I_{\frac{1}{2}} \subset I_{\frac{1}{3}} \subset \cdots$$

is an infinite strictly ascending chain of ideals of R. Hence, R satisfies neither the ACC nor the DCC.

Definition 18.1.17 *A ring which satisfies the DCC for left (right) ideals is called a **left (right) Artinian ring.***

A ring which is both left Artinian and right Artinian is called an **Artinian ring**.

By Theorem 18.1.14, it follows that a ring R is left Artinian if and only if the mC holds for left ideals in R.

Example 18.1.18 *Let p be a fixed prime and*

$$\mathbf{Z}(p^\infty) = \left\{\frac{a}{p^n} \in \mathbf{Q} \mid 0 \leq a < p^n, \ n \in \mathbf{N}\right\}.$$

Then $(\mathbf{Z}(p^\infty), +, \cdot)$ is a commutative ring without identity, where $+$ (addition) is modulo 1 and $a \cdot b = 0$ for all $a, b \in \mathbf{Z}(p^\infty)$. From the definition of multiplication, it follows that every subgroup of $(\mathbf{Z}(p^\infty), +)$ is an ideal. Hence, the ideals of $\mathbf{Z}(p^\infty)$ are precisely the subgroups of $(\mathbf{Z}(p^\infty), +)$.

Let I be a nontrivial ideal of $\mathbf{Z}(p^\infty)$. Let k be the smallest positive integer such that $\frac{q}{p^k} \notin I$ for some integer q, $0 \leq q < p^k$. If $p|q$, then $\frac{a}{p^{k-1}} \notin I$ for some integer a, $0 \leq a < p^{k-1}$, contrary to the choice of k. Therefore, $\gcd(p, q) = 1$. Now

$$J = \left\{0, \ \frac{1}{p^{k-1}}, \ \frac{2}{p^{k-1}}, \ \cdots, \ \frac{p^{k-1}-1}{p^{k-1}}\right\}$$

is a subset of I. Let us show that $I = J$.

Consider the rational number $\frac{r}{p^n}$, where $\gcd(p, r) = 1$ and $n > k$. Suppose that $\frac{r}{p^n} \in I$. Since $\gcd(p, r) = 1$, there exist integers x and y such that $rx + py = 1$. Now

$$\frac{xr}{p^k} = \frac{(xp^{n-k})r}{p^n} \quad \text{and}$$

$$\frac{py}{p^k} = \frac{y}{p^{k-1}}$$

(both the numbers are reduced modulo 1) are in I. Hence,

$$\frac{1}{p^k} = \frac{xr + yp}{p^k} \in I.$$

This is contrary to the choice of k. Hence, $I = J = \{0, \frac{1}{p^{k-1}}, \frac{2}{p^{k-1}}, \cdots, \frac{p^{k-1}-1}{p^{k-1}}\}$. We denote this ideal by I_k. It is also clear that for any positive integer k, I_k is an ideal of $\mathbf{Z}(p^\infty)$. These ideals form the following strictly ascending chain

$$\{0\} \subset I_1 \subset I_2 \subset \cdots \subset I_k \subset \cdots$$

in $\mathbf{Z}(p^\infty)$, proving that $\mathbf{Z}(p^\infty)$ is not Noetherian. Since every proper ideal is finite, every descending chain of ideals must be finite. Therefore, $\mathbf{Z}(p^\infty)$ is Artinian.

It is known that any finite ring with more than one element and without zero divisors is a division ring. The following theorem generalizes this result.

Theorem 18.1.19 *Let R be a left Artinian ring with more than one element. If R does not contain zero divisors, then R is a division ring.*

Proof.   Let $0 \neq a \in R$. Now

$$\langle a \rangle_l \supseteq \langle a^2 \rangle_l \supseteq \langle a^3 \rangle_l \supseteq \cdots,$$

where $\langle a^n \rangle_l$ is the left ideal generated by a^n. Since R is left Artinian, there exists a positive integer n such that $\langle a^n \rangle_l = \langle a^{n+1} \rangle_l = \cdots$. Therefore, $a^n \in \langle a^{n+1} \rangle_l$. Thus, there exist $r \in R$ and $m \in \mathbf{Z}$ such that

$$a^n = ra^{n+1} + ma^{n+1}.$$

Now $a \neq 0$ and R has no zero divisors. Therefore, $a^n \neq 0$. This implies that $a = ra^2 + ma^2 = (ra + ma)a$. Let $e = ra + ma$. Then $bea = ba$ implies that $be = b$ for all $b \in R$ and $e^2 = e$. This also shows that $e \neq 0$. Now $eb = e^2 b$ implies $b = eb$. Hence, e is the identity element of R. Now $e = ra + ma = (r + me)a$. This implies that left inverse exists for each nonzero element of R. Hence, a^{-1} exists in R for all nonzero element $a \in R$. Consequently, R is a division ring. ∎

Corollary 18.1.20 *A commutative Artinian ring is a field if and only if it is an integral domain.* ∎

We now want to characterize J-semisimple rings which are either Noetherian or Artinian.

Theorem 18.1.21 *Let R be a commutative ring with 1. If R is an Artinian ring, then radR is a nilpotent ideal.*

Proof. Let $J = \text{rad}R$. Now J^n is an ideal of R for all positive integers n and

$$J \supseteq J^2 \supseteq J^3 \supseteq \cdots.$$

Since R is Artinian, there exists a positive integer n such that $J^n = J^{n+1} = \cdots$. Let $I = J^n$. Then $I^2 = I$. Suppose that $I \neq \{0\}$. Let

$$\mathcal{F} = \{T \mid T \text{ is an ideal of } R, \ T \subseteq I \text{ and } IT \neq \{0\}\}.$$

Now $I \in \mathcal{F}$ and so $\mathcal{F} \neq \phi$. The minimal property of R on ideals implies that $\mathcal{F}$ contains a minimal element, say, T_0. Then $T_0 \subseteq I$, $IT_0 \neq \{0\}$, and T_0 is minimal in $\mathcal{F}$. Now $IT_0 \neq \{0\}$ implies that $Ia \neq \{0\}$ for some nonzero $a \in T_0$. Now Ia is an ideal of R. Also, $I(Ia) = I^2a = Ia \neq \{0\}$ and $Ia \subseteq T_0 \subseteq I$. Thus, $Ia \in \mathcal{F}$. By the minimality of T_0, $T_0 = Ia$. Therefore, there exists $b \in I$ such that $a = ba$. Now $b \in I = J^n \subseteq J = \text{rad}R$. Thus, $1 - xb$ is a unit for all $x \in R$ and so $(1 - b)^{-1}$ exists in R. As a result, we deduce that $a = 0$ since $a(1 - b) = 0$. However, this is a contradiction. Hence, $J^n = \{0\}$ and so J is nilpotent. ∎

Corollary 18.1.22 *Let R be a commutative ring with 1. If R is an Artinian ring, then every nil ideal of R is nilpotent.*

Proof. Let I be a nil ideal of R. Then $I \subseteq \text{rad}R$ by Corollary 17.2.4. Since $\text{rad}R$ is nilpotent, there exists a positive integer n such that $(\text{rad}R)^n = \{0\}$. Then $I^n \subseteq (\text{rad}R)^n = \{0\}$. Hence, $I^n = \{0\}$ and so I is nilpotent. ∎

Theorem 18.1.23 *Let R be a commutative ring with 1. If R is J-semisimple Artinian, then R is a direct sum of a finite number of fields.*

Proof. Let $\mathcal{F}$ be the collection of all maximal ideals of R. Then $\mathcal{F} \neq \phi$. We now show that $\mathcal{F}$ has only a finite number of elements. Suppose that $|\mathcal{F}| = \infty$. Then $\mathcal{F}$ contains an infinite set $\{M_i \mid i \in \mathbf{N}\}$ of distinct maximal ideals of R. Also,

$$M_1 \supseteq M_1 M_2 \supseteq M_1 M_2 M_3 \supseteq \cdots.$$

Since R is Artinian, there exists a positive integer n such that

$$M_1 M_2 \cdots M_n = M_1 M_2 \cdots M_{n+i}$$

for all $i \geq 1$. Therefore, $M_1 M_2 \cdots M_n \subseteq M_{n+1}$. Since M_{n+1} is also a prime ideal, $M_i \subseteq M_{n+1}$ for some i, $1 \leq i \leq n$. This contradicts the assumption that $M_1, M_2, \ldots, M_n, M_{n+1}$ are all distinct maximal ideals. Therefore, R has a finite number of maximal ideals. Since R is J-semisimple,

$$\cap\{M \mid M \text{ is a maximal ideal of } R\} = \{0\}.$$

We can find maximal ideals $M_1, M_2, \ldots, M_n$ such that $M_1 \cap M_2 \cap \cdots \cap M_n = \{0\}$, but

$$I_i = M_1 \cap M_2 \cap \cdots \cap M_{i-1} \cap M_{i+1} \cap \cdots \cap M_n \neq \{0\}$$

for all i, $1 \leq i \leq n$. Thus, $M_i \cap I_i = \{0\}$ for all i, $1 \leq i \leq n$. Since M_i is maximal and $I_i \not\subseteq M$, $I_i + M_i = R$ for all i, $1 \leq i \leq n$. Hence, $R = I_i \oplus M_i$ for all i, $1 \leq i \leq n$. This implies that $R/M_i \simeq I_i$ for all i, $1 \leq i \leq n$. Since R/M_i is a field, I_i is a field for all i, $1 \leq i \leq n$. Let $x \in R$. Then $x = a_i + m_i$ for some $a_i \in I_i$ and $m_i \in M_i$, for all i, $1 \leq i \leq n$. Let $y = a_1 + a_2 + \cdots + a_n$. Then

$$
\begin{aligned}
x - y &= (x - a_i) - a_1 - a_2 - \cdots - a_{i-1} - a_{i+1} - \cdots - a_n \\
&= m_i - a_1 - a_2 - \cdots - a_{i-1} - a_{i+1} - \cdots - a_n \in M_i
\end{aligned}
$$

since $a_k \in M_i$ for $k \neq i$, $1 \leq k \leq n$. Therefore, $x - y \in \cap_{i=1}^{n} M_i = \{0\}$ and so $x = y \in \sum_{i=1}^{n} I_i$. This implies that $R = \sum_{i=1}^{n} I_i$. Now $I_k \cap \sum_{\substack{i=1 \\ i \neq k}}^{n} I_i \subseteq I_k \cap M_k = \{0\}$. Hence, $R = \oplus_{i=1}^{n} I_i$. ∎

Theorem 18.1.24 *Let R be a commutative ring with 1. If R is J-semisimple Artinian, then R is Noetherian.*

Proof. Let R be a J-semisimple commutative Artinian ring. Then R is isomorphic to a direct sum of a finite number of fields. Let $R \simeq F_1 \oplus F_2 \oplus \cdots \oplus F_n$, where F_i is a field, $1 \leq i \leq n$. Now each F_i contains only two ideals and hence F_i is Noetherian. Thus, R is a finite direct sum of Noetherian rings. Hence, by Theorem 18.1.11, R is Noetherian. ∎

Remark 18.1.25 *In this book, we proved Theorem 18.1.24 for J-semisimple commutative Artinian rings for the sake of simplicity. However, it is known, in general, that any Artinian ring with 1 is Noetherian.*

18.1.1 Worked-Out Exercises

◇ **Exercise 1** Show that a subring of an Artinian ring may not be Artinian.

Solution: The field $\mathbf{Q}$ of rational numbers has only two ideals and hence is Artinian. The ring $\mathbf{Z}$ is a subring of $\mathbf{Q}$, but $\mathbf{Z}$ is not Artinian.

◇ **Exercise 2** Show that the sum of all nilpotent ideals of a commutative Noetherian ring R is a nilpotent ideal.

Solution: If R is nilpotent, then the result is immediate. Suppose R is not nilpotent. Let $A = \sum_{i \in I} A_i$ be the sum of all nilpotent ideals of R. Then A is the ideal generated by $\cup_{i \in I} A_i$. Let $\mathcal{F} = \{A_i \mid i \in I\}$. Since R is Noetherian, $\mathcal{F}$ has a maximal element, say, B. Let us show that $A_i \subseteq B$ for all $i \in I$. Let

$A_i \in \mathcal{F}$. Now $A_i + B$ is nilpotent and $B \subseteq A_i + B$. Since B is a maximal element of $\mathcal{F}$ and $A_i + B \in \mathcal{F}$, it follows that $A_i + B = B$. Hence, $A_i \subseteq B$ for all $i \in I$. This implies that $A = \langle \cup_{i \in I} A_i \rangle \subseteq B$. But $B = A_k$ for some $k \in I$. Therefore, $A = B$. Thus, A is nilpotent.

Exercise 3 Let R be a commutative Noetherian ring with 1. Show that every ideal of R contains a finite product of prime ideals.

Solution: Let $\mathcal{F} = \{A \mid A$ is an ideal of R and A does not contain any finite product of prime ideals of $R\}$. We show that $\mathcal{F} = \phi$. Suppose $\mathcal{F} \neq \phi$. Then $\mathcal{F}$ has a maximal element, say, A_0. Now A_0 cannot be a prime ideal. Thus, there exist ideals B and C of R such that $BC \subseteq A_0$, but $B \not\subseteq A_0$ and $C \not\subseteq A_0$. Now $A_0 \subset A_0 + B$ and $A_0 \subset A_0 + C$. Hence, $A_0 + B$ and $A_0 + C$ are ideals of R such that $A_0 + B$, $A_0 + C \notin \mathcal{F}$. Then $A_0 + B$ and $A_0 + C$ contain a finite product of prime ideals. This implies that $(A_0 + B)(A_0 + C)$ contains a finite product of prime ideals. Now $(A_0 + B)(A_0 + C) = A_0 A_0 + A_0 C + A_0 B + BC \subseteq A_0$. This implies that A_0 contains a finite product of prime ideals, which is a contradiction. Thus, $\mathcal{F} = \phi$. Hence, every ideal of R contains a finite product of prime ideals.

Exercise 4 Let f be an epimorphism of a Noetherian ring R onto itself. Show that f is an isomorphism.

Solution: For each positive integer n, f^n is an epimorphism and

$$\text{Ker } f \subseteq \text{Ker } f^2 \subseteq \text{Ker } f^3 \subseteq \cdots$$

is an ascending chain of ideals in R. Since R is Noetherian, there exists a positive integer m such that $\text{Ker } f^m = \text{Ker } f^{m+i}$ for all $i \geq 1$. Thus, $\text{Ker } f^m = \text{Ker } f^{m+1}$. Let $x \in \text{Ker } f$. Since f^m is onto R, there exists an element $y \in R$ such that $f^m(y) = x$. Now $0 = f(x) = f^{m+1}(y)$. This implies that $y \in \text{Ker } f^{m+1} = \text{Ker} f^m$. Hence, $x = f^m(y) = 0$. Thus, f is one-one and so is an isomorphism.

$\diamond$ **Exercise 5** Let $R = \left\{ \begin{bmatrix} a & b \\ 0 & c \end{bmatrix} \mid a \in \mathbf{Z}, b, c \in \mathbf{Q} \right\}$.

(i) Show that R is a subring of $M_2(\mathbf{Q})$.

(ii) Show that R is not left Noetherian.

(iii) Let A be a nonzero right ideal of R such that every element of A is of the form $\begin{bmatrix} 0 & b \\ 0 & c \end{bmatrix}$. Show that A is finitely generated.

(iv) Let A be a nonzero right ideal of R such that every element of A is of the form $\begin{bmatrix} a & b \\ 0 & c \end{bmatrix}$ with $a \neq 0$. Show that A is finitely generated.

(v) Show that R is right Noetherian.

Solution: (i) It is a routine verification to show that R is a subring of $M_2(\mathbf{Q})$.

(ii) For any positive integer n, let

$$I_n = \left\{ \begin{bmatrix} 0 & \frac{m}{2^n} \\ 0 & 0 \end{bmatrix} \mid m \in \mathbf{Z} \right\}.$$

Then each I_n is a left ideal of R. Since $\frac{m}{2^n} = \frac{2m}{2^{n+1}}$, $I_n \subseteq I_{n+1}$. But $\begin{bmatrix} 0 & \frac{1}{2^{n+1}} \\ 0 & 0 \end{bmatrix} \notin I_n$. Therefore, $I_n \subset I_{n+1}$. Thus,

$$I_1 \subset I_2 \subset I_3 \subset \cdots$$

is an infinite strictly ascending chain of left ideals of R. Hence, R is not left Noetherian.

(iii) **Case 1:** Suppose $c = 0$ for all $\begin{bmatrix} 0 & b \\ 0 & c \end{bmatrix}$ In this case, the elements of A are of the form $\begin{bmatrix} 0 & b \\ 0 & 0 \end{bmatrix}$, where $b \in \mathbf{Q}$. Since $A \neq \{0\}$, there exists a nonzero rational number b such that $\begin{bmatrix} 0 & b \\ 0 & 0 \end{bmatrix} \in A$. Thus, $\begin{bmatrix} 0 & 1 \\ 0 & 0 \end{bmatrix} = \begin{bmatrix} 0 & b \\ 0 & 0 \end{bmatrix} \begin{bmatrix} 0 & 0 \\ 0 & \frac{1}{b} \end{bmatrix} \in A$. Hence, $A = \begin{bmatrix} 0 & 1 \\ 0 & 0 \end{bmatrix} R$.

Case 2: Suppose $b = 0$ for all $\begin{bmatrix} 0 & b \\ 0 & c \end{bmatrix} \in A$. In this case, proceeding as in Case 1, we can show that $A = \begin{bmatrix} 0 & 0 \\ 0 & 1 \end{bmatrix} R$.

Case 3: Suppose A contains an element $\begin{bmatrix} 0 & b \\ 0 & c \end{bmatrix}$ such that $b \neq 0$ and $c \neq 0$. If $A = \begin{bmatrix} 0 & b \\ 0 & c \end{bmatrix} R$, then A is generated by $\begin{bmatrix} 0 & b \\ 0 & c \end{bmatrix}$. Suppose $\begin{bmatrix} 0 & b \\ 0 & c \end{bmatrix} R \subset A$. Now A contains an element $\begin{bmatrix} 0 & u \\ 0 & v \end{bmatrix}$ such that $bv \neq cu$. Then $bv - cu \neq 0$ and

$$\begin{bmatrix} 0 & bv - cu \\ 0 & 0 \end{bmatrix} = \begin{bmatrix} 0 & b \\ 0 & c \end{bmatrix} \begin{bmatrix} 0 & u \\ 0 & v \end{bmatrix} - \begin{bmatrix} 0 & u \\ 0 & v \end{bmatrix} \begin{bmatrix} 0 & b \\ 0 & c \end{bmatrix} \in A.$$

Now $\begin{bmatrix} 0 & 1 \\ 0 & 0 \end{bmatrix} = \begin{bmatrix} 0 & bv - cu \\ 0 & 0 \end{bmatrix} \begin{bmatrix} 0 & 0 \\ 0 & \frac{1}{bv-cu} \end{bmatrix} \in A$ and

$$\begin{bmatrix} 0 & b \\ 0 & 0 \end{bmatrix} = \begin{bmatrix} 0 & 1 \\ 0 & 0 \end{bmatrix} \begin{bmatrix} 0 & 0 \\ 0 & b \end{bmatrix} \in A.$$

Thus, we find that $\begin{bmatrix} 0 & 0 \\ 0 & c \end{bmatrix} = \begin{bmatrix} 0 & b \\ 0 & c \end{bmatrix} - \begin{bmatrix} 0 & b \\ 0 & 0 \end{bmatrix} \in A$. Since $c \neq 0$, this

implies that $\begin{bmatrix} 0 & 0 \\ 0 & 1 \end{bmatrix} \in A$. Hence, A is generated by $\{ \begin{bmatrix} 0 & 1 \\ 0 & 0 \end{bmatrix}, \begin{bmatrix} 0 & 0 \\ 0 & 1 \end{bmatrix} \}$,

i.e., A is finitely generated.

(iv) Let n_0 be the smallest positive integer such that $\begin{bmatrix} n_0 & b \\ 0 & c \end{bmatrix} \in A$ for

some $b, c \in \mathbf{Q}$. We show that A is generated by either $\{ \begin{bmatrix} n_0 & 0 \\ 0 & 0 \end{bmatrix}, \begin{bmatrix} 0 & 1 \\ 0 & 0 \end{bmatrix},$

$\begin{bmatrix} 0 & 0 \\ 0 & 1 \end{bmatrix} \}$ or $\{ \begin{bmatrix} n_0 & 0 \\ 0 & 0 \end{bmatrix}, \begin{bmatrix} 0 & 1 \\ 0 & 0 \end{bmatrix} \}$.

Now $\begin{bmatrix} 0 & n_0 \\ 0 & 0 \end{bmatrix} = \begin{bmatrix} n_0 & b \\ 0 & c \end{bmatrix} \begin{bmatrix} 0 & 1 \\ 0 & 0 \end{bmatrix} \in A$. This implies that

$$\begin{bmatrix} 0 & 1 \\ 0 & 0 \end{bmatrix} = \begin{bmatrix} 0 & n_0 \\ 0 & 0 \end{bmatrix} \begin{bmatrix} 0 & 0 \\ 0 & \frac{1}{n_0} \end{bmatrix} \in A.$$

Thus, $\begin{bmatrix} 0 & b \\ 0 & 0 \end{bmatrix} = \begin{bmatrix} 0 & 1 \\ 0 & 0 \end{bmatrix} \begin{bmatrix} 0 & 0 \\ 0 & b \end{bmatrix} \in A$. Again

$$\begin{bmatrix} n_0 & 0 \\ 0 & 0 \end{bmatrix} = \begin{bmatrix} n_0 & b \\ 0 & c \end{bmatrix} \begin{bmatrix} 1 & 0 \\ 0 & 0 \end{bmatrix} \in A.$$

Hence,

$$\begin{bmatrix} 0 & 0 \\ 0 & c \end{bmatrix} = \begin{bmatrix} n_0 & b \\ 0 & c \end{bmatrix} - \begin{bmatrix} n_0 & 0 \\ 0 & 0 \end{bmatrix} - \begin{bmatrix} 0 & b \\ 0 & 0 \end{bmatrix} \in A.$$

If $c \neq 0$, then it follows that $\begin{bmatrix} 0 & 0 \\ 0 & 1 \end{bmatrix} \in A$ and A is generated by $\{ \begin{bmatrix} n_0 & 0 \\ 0 & 0 \end{bmatrix},$

$\begin{bmatrix} 0 & 1 \\ 0 & 0 \end{bmatrix}, \begin{bmatrix} 0 & 0 \\ 0 & 1 \end{bmatrix} \}$. If every element of A is of the form $\begin{bmatrix} a & b \\ 0 & 0 \end{bmatrix}$ with $a \neq 0$,

then A is generated by $\{ \begin{bmatrix} n_0 & 0 \\ 0 & 0 \end{bmatrix}, \begin{bmatrix} 0 & 1 \\ 0 & 0 \end{bmatrix} \}$, where n_0 is the smallest positive

integer such that $\begin{bmatrix} n_0 & b \\ 0 & 0 \end{bmatrix} \in A$ for some $b \in \mathbf{Q}$.

(v) From (iii) and (iv), it follows that every right ideal of R is finitely generated. Hence, R is right Noetherian.

18.1.2 Exercises

1. Give an example of a ring R with the following properties.

    (i) R is left Noetherian, but not right Noetherian.

    (ii) R is left Artinian, but not right Artinian.

    (iii) R is right Artinian, but not left Artinian.

    (iv) R is noncommutative and both Noetherian and Artinian.

2. Show that a subring of a Noetherian ring may not be Noetherian.

3. Give an example of a ring R in which every proper ideal is finitely generated, but R is not Noetherian.

4. In a right Artinian ring with 1, if $ab = 1$ for $a, b \in R$, prove that $ba = 1$.

5. Prove that every homomorphic image of a left Artinian ring is left Artinian.

6. Let R be a commutative Artinian ring with 1. Show that in R, every nonzero prime ideal is a maximal ideal and show that R has only a finite number of prime ideals.

7. Show that a Noetherian domain in which the sum of two principal ideals is a principal ideal is a PID.

8. Prove that every homomorphic image of an Artinian ring is Artinian.

9. Let I be an ideal of a ring R. If I and R/I are both Artinian rings, prove that R is Artinian.

10. Let R be a right Artinian ring and I be a nonnilpotent right ideal of R. Prove the following.

    (i) The collection $\mathcal{F}_1$ of all nonnilpotent right ideals of R which are contained in I, contains a minimal element I_0 such that $I_0^2 = I_0$.

    (ii) Let $\mathcal{F} = \{J \mid J \text{ is a right ideal of } R, \, JI_0 \neq 0, \, J \subseteq I_0\}$. Then $\mathcal{F}$ contains a minimal element I_1 and I_1 contains an element $u \neq 0$ such that $uI_0 = I_1$.

    (iii) I is not a nil right ideal.

    (iv) I contains a nonzero idempotent element.

11. Prove that a commutative ring R with 1 is Noetherian if and only if every prime ideal is finitely generated.

12. Let R be a ring with 1. Let $f : R \to [0, 1]$ be such that

$$\begin{aligned} f(a - b) &\geq \min\{f(a), f(b)\}, \\ f(ra) &\geq f(a) \end{aligned}$$

for all $a, b, r \in R$.

(i) Prove that R is left Artinian if and only if for every mapping $f : R \to [0, 1]$ that satisfies the above conditions, $|\mathcal{I}(f)| < \infty$.

(ii) Prove that R is left Noetherian if and only if for every mapping $f : R \to [0, 1]$ that satisfies the above conditions, $|\mathcal{I}(f)|$ is a well-ordered subset of $[0, 1]$.

13. Write the proof if the statement is true; otherwise give a counterexample.

(i) Every finite ring is both Noetherian and Artinian.

(ii) Every Noetherian domain is a field.

(iii) Let R be a commutative ring with 1. If R is J-semisimple and Artinian, then R is regular.

Amalie Emmy Noether (1882–1935) was born on March 23, 1882, in Erlangen, Germany, the oldest child. Her father, Max Noether, a noted mathematician himself, was a professor at the University of Erlangen. She studied mathematics and foreign languages at Erlangen from 1900 to 1902.

In 1903, Noether started her mathematics career at the University of Göttingen. Since at that time girls could not be admitted as regular students, she was a nonmatriculated auditor. In 1904, she was permitted to enroll at the University of Erlangen

and in 1907 she received her Ph.D, *summa cum laude*. Her thesis was on algebraic invariants.

In 1915, on Hilbert's invitation, she went to Göttingen. There she lectured on courses given under Hilbert's name. She applied her invariant theoretic knowledge on problems considered by Hilbert and Klein. Hilbert made several personal attempts to get her a regular position, but prejudice against women at that time thwarted his efforts. Finally, in 1922, she was appointed as an unofficial associate professor; later, she received a modest salary. She taught at Göttingen from 1922 to 1933. Due to the Nazi regime uprising, all Jewish professors were dismissed in April 1933. Through the efforts of Herman Weyl, she was able to get a visiting professor's position at Bryn Mawr College and left for the United States in October 1933. She lectured and did research at Bryn Mawr College and at the Institute of Advanced Study. Noether died of surgical complications on April 14, 1935.

Influenced by Hilbert's axiomatization of Euclidean geometry, Noether became interested in an abstract axiomatic approach to ring theory. Between 1922 and 1926, she published a series of papers focusing on "the general theory of ideals." In her paper "Abstract construction of ideal theory in the domain of algebraic number fields," published in 1926, she characterized rings in which every ideal is uniquely expressed as a product of prime ideals. This is analogous to Euclid's fundamental theorem of arithmetic. Two of the generalized structures she associated with ideals are the "group" and the "ring." She introduced the present-day definition of a ring in her paper, "Theory of ideals in a ring," published in 1921. She showed that the ascending chain condition is important to ideal theory. She introduced the concept of a primary ideal and proved that in a commutative ring satisfying the ascending chain condition, every ideal can be expressed as an intersection of primary ideals.

In 1932, while working on noncommutative rings in linear algebra with Richard Brauer and Helmut Hasse, she proved that every simple algebra over an ordinary algebraic number field is cyclic. From 1932 to 1934, she worked on noncommutative algebras by means of cross products.

Noether published 45 research papers.

Chapter 19

Modules and Vector Spaces

19.1   Modules and Vector Spaces

Our main interest here is to set down only the results of vector spaces which are needed for our study of fields in the next chapter. We do this in such a way that the reader will become acquainted with the notion of a module.

Definition 19.1.1 *Let R be a ring. A commutative group $(M, +)$ is called a* **left R-module** *or a* **left module** *over R with respect to a mapping $\cdot : R \times M \to M$ if for all $r, s \in R$ and $m, m' \in M$,*
    (i) $r \cdot (m + m') = r \cdot m + r \cdot m'$,
    (ii) $r \cdot (s \cdot m) = (rs) \cdot m$,
    (iii) $(r + s) \cdot m = r \cdot m + s \cdot m$.
    If R has an identity 1 and if $1 \cdot m = m$ for all $m \in M$, then M is called a **unitary** *or* **unital left R-module.**

A **right R-module** can be defined in a similar fashion.
In the above definition, we used the same notation for the addition in the ring R and the addition in the group M. We also used the same notation for the multiplication in R and the multiplication between the elements of R and M. It should be clear to the reader by now that there are actually four distinct operations involved. We write rm for $r \cdot m$.

Example 19.1.2 *In a ring R, every left ideal is a left R-module and every right ideal is a right R-module. In particular, R is a left and right R-module.*

Example 19.1.3 *Every commutative group M is a module over the ring of integers $\mathbf{Z}$. For $n \in \mathbf{Z}$ and $a \in M$, the element na is defined to be a added to itself n times if n is positive and $-a$ added to itself $|n|$ times if n is negative. $0a$ is defined to be the zero element of M. Under these definitions, M becomes a unitary left $\mathbf{Z}$-module.*

Let M be any commutative group and R be any ring. If we define $rm = 0$ for all $r \in R$, $m \in M$, then M forms a left R-module, called a trivial module.

Since all results that are true for left R-modules are also true for right R-modules, we prove results only for left R-modules. From now on, unless stated otherwise, by an R-module, we mean a left R-module.

Definition 19.1.4 *Let M be an R-module and N be a nonempty subset of M. Then N is called a **submodule** of M if N is a subgroup of M and for all $r \in R$, $a \in N$, we have $ra \in N$.*

It is clear that a submodule of an R-module is itself an R-module.

Using arguments similar to those used for subgroups and ideals, one can show that the intersection of any nonempty collection of submodules of an R-module is again a submodule.

Definition 19.1.5 *Let X be a subset of an R-module M. Then the submodule of M **generated** by X is defined to be the intersection of all submodules of M which contain X and is denoted by $\langle X \rangle$. X is called a **basis** of $\langle X \rangle$ if no proper subset of X generates $\langle X \rangle$. If $M = \langle X \rangle$ and X is a finite set, then M is said to be **finitely generated**. When $X = \{x\}$ and $M = \langle \{x\} \rangle$, then M is called a **cyclic** R-module and in this case we write $M = \langle x \rangle$.*

We ask the reader to prove that any finitely generated module has a finite basis.

The proof of the following theorem is similar to that of the corresponding theorem for ideals, Theorem 11.2.7. Hence, we omit its proof.

Theorem 19.1.6 *Let M be an R-module and X be a nonempty subset of M. Then*

$$\langle X \rangle = \{\textstyle\sum_{i=1}^{k} r_i x_i + \sum_{j=1}^{l} n_j x_j' \mid r_i \in R, n_j \in \mathbf{Z}, x_i, x_j' \in X,$$
$$1 \le i \le k, 1 \le j \le l, k, l \in \mathbf{N}\}.$$

If M is a unitary R-module, then

$$\langle X \rangle = \{\sum_{i=1}^{k} r_i x_i \mid r_i \in R,\ x_i \in X, 1 \le i \le k,\ k \in \mathbf{N}\}. \blacksquare$$

Example 19.1.7 *(i) $\mathbf{Q}$ is a $\mathbf{Q}$-module. If N is a submodule of $\mathbf{Q}$, then N is a left ideal of $\mathbf{Q}$. Since $\mathbf{Q}$ is a field, the only left ideals of $\mathbf{Q}$ are $\{0\}$ and $\mathbf{Q}$. Hence the submodules of $\mathbf{Q}$ are $\{0\}$ and $\mathbf{Q}$.*

(ii) We know that $\mathbf{Q} \oplus \mathbf{Q}$ is a commutative group. For all $x \in \mathbf{Q}$ and for all $(a, b) \in \mathbf{Q} \oplus \mathbf{Q}$, define $x(a, b) = (xa, xb)$. Then $\mathbf{Q} \oplus \mathbf{Q}$ is a $\mathbf{Q}$-module. We now determine all submodules of $\mathbf{Q} \oplus \mathbf{Q}$. Let M be a nonzero $\mathbf{Q}$-submodule of $\mathbf{Q} \oplus \mathbf{Q}$.

Case 1: Suppose for all $(a, b) \in M$, $b = 0$. Now there exists $(a, 0) \in M$ such that $a \neq 0$. Then $(1, 0) = \frac{1}{a}(a, 0) \in M$. Thus, $M = \mathbf{Q} \oplus \{0\}$.

Case 2: Suppose for all $(a, b) \in M$, $a = 0$. Now there exists $(0, b) \in M$ such that $b \neq 0$. Then $(0, 1) = \frac{1}{b}(0, b) \in M$. Thus, $M = \{0\} \oplus \mathbf{Q}$.

Case 3: Suppose there exists $(a, b) \in M$ such that $a \neq 0$, $b \neq 0$.

Case 3a: Suppose $M = \langle (a, b) \rangle$. Then M is a cyclic submodule of $\mathbf{Q} \oplus \mathbf{Q}$ generated by (a, b).

Case 3b: Suppose $M \neq \langle (a, b) \rangle$. Then $\langle (a, b) \rangle \subset M$. Thus, there exists $(a', b') \in M \setminus \langle (a, b) \rangle$. Then $a' \neq 0$ or $b' \neq 0$. Suppose that $a' = 0$. Then $(0, 1) = \frac{1}{b'}(0, b') \in M$. Therefore, $(a, 0) = (a, b) - (0, 1)b \in M$. Hence, $(1, 0) = \frac{1}{a}(a, 0) \in M$. Thus, $(1, 0), (0, 1) \in M$. This implies that $M = \mathbf{Q} \oplus \mathbf{Q}$. Similarly, if $b' = 0$, then $M = \mathbf{Q} \oplus \mathbf{Q}$.

Now suppose that $a' \neq 0$ and $b' \neq 0$. If $\frac{a}{a'} = \frac{b}{b'} = t$ (say), then $t(a', b') = (ta', tb') = (\frac{a}{a'}a', \frac{b}{b'}b') = (a, b) \in \langle (a, b) \rangle$, which is a contradiction. Therefore, $\frac{a}{a'} \neq \frac{b}{b'}$ and so $ab' - ba' \neq 0$. Let $(p, q) \in \mathbf{Q} \oplus \mathbf{Q}$. Choose $t = \frac{pb' - qa'}{ab' - ba'}$ and $s = \frac{qa - pb}{ab' - ba'}$. Then $(p, q) = t(a, b) + s(a', b') \in M$. Thus, $\mathbf{Q} \oplus \mathbf{Q} \subseteq M$. Hence, $M = \mathbf{Q} \oplus \mathbf{Q}$.

Consequently, if M is a $\mathbf{Q}$-submodule of $\mathbf{Q} \oplus \mathbf{Q}$, then M is of the following form:

(i) $M = \{0\}$, or

(ii) $M = \{0\} \oplus \mathbf{Q} = \langle (0, 1) \rangle$, or

(iii) $M = \mathbf{Q} \oplus \{0\} = \langle (1, 0) \rangle$, or

(iv) $M = \langle (a, b) \rangle$, $a \neq 0$, $b \neq 0$, $a, b \in \mathbf{Q}$, or

(v) $M = \mathbf{Q} \oplus \mathbf{Q}$.

This also proves that M is finitely generated.

Definition 19.1.8 *Let F be a field. A unitary (left) F-module M is called a* **(left) vector space** *over F. The elements of M are called* **vectors** *and the elements of F are called* **scalars**. *A submodule of M is called a* **subspace** *of M. If X is a subset of M such that $M = \langle X \rangle$, then X is said to* **span** *or* **generate** *M and M is called the* **span** *of X over F.*

Example 19.1.9 *Let F be any field and F^n denote the Cartesian product of F with itself n times. Then F^n becomes a vector space over F under the following definitions: For all $(a_1, a_2, \ldots, a_n)$, $(b_1, b_2, \ldots, b_n) \in F^n$ and $a \in F$*

$$(a_1, a_2, \ldots, a_n) + (b_1, b_2, \ldots, b_n) = (a_1 + b_1, +a_2 + b_2, \ldots, a_n + b_n),$$
$$a(a_1, a_2, \ldots, a_n) = (aa_1, aa_2, \ldots, aa_n).$$

The set

$$X = \{(1, 0, 0, \ldots, 0), (0, 1, 0, \ldots, 0), \ldots, (0, 0, 0, \ldots, 1)\}$$

spans F^n since for all $(a_1, a_2, \ldots, a_n) \in F^n$,

$$(a_1, a_2, \ldots, a_n) = a_1(1, 0, 0, \ldots, 0) + a_2(0, 1, 0, \ldots, 0) + \cdots + a_n(0, 0, 0, \ldots, 1).$$

When $n = 2$ or 3 and F is the field of real numbers, then the vector space F^n over F is the one usually encountered in elementary analytical geometry.

By Example 19.1.9, $\mathbf{R}^3$ is a vector space over $\mathbf{R}$.

Example 19.1.10 *Consider the vector space $\mathbf{R}^3$ over $\mathbf{R}$. Let*

$$U = \{(a, b, c) \in \mathbf{R}^3 \mid 2a + 3b + 5c = 0\}.$$

Then U is a subspace of $V_3(\mathbf{R})$. Let

$$U_1 = \{(a, b, c) \in \mathbf{R}^3 \mid 2a + 3b + 5c = 5\}.$$

Now $(0, 0, 1)$ and $(1, 1, 0) \in U_1$, but $(0, 0, 1) + (1, 1, 0) \notin U_1$. Hence, U_1 is not a subspace of $\mathbf{R}^3$.

Example 19.1.11 *Let V be a vector space over F. Then $\{0\}$ and V are subspaces of V. These are called **trivial subspaces** of V.*

Theorem 19.1.12 *Let V be a vector space over F and S be a nonempty subset of V. Then S is a subspace of V if and only if for all $a \in F$ and for all $x, y \in S$, $ax + y \in S$.*

Proof. Suppose S is a subspace of V. Then for all $a \in F$ and for all $x, y \in S$, $ax \in S$ and so $ax + y \in S$. Conversely, suppose for all $a \in F$ and for all $x, y \in S$, $ax + y \in S$. Since $S \neq \phi$, there exists $x \in S$. By Exercise 2 (page 431), $-x = (-1)x$. Therefore, $0 = -x + x = (-1)x + x \in S$. Hence, for all $x \in S$, $-x = (-1)x + 0 \in S$. Also, for all $x, y \in S$, $x + y = 1x + y \in S$. S inherits the associative and commutative laws. Thus, $(S, +)$ is a commutative group. Now for all $a \in F$ and for all $x \in S$, $ax = ax + 0 \in S$. Therefore, S is a vector space over F since the other properties are inherited. ∎

Theorem 19.1.13 *Let V be a vector space over F and $\{U_\alpha \mid \alpha \in I\}$ be any nonempty collection of subspaces of V. Then $\cap_{\alpha \in I} U_\alpha$ is a subspace of V.*

Proof. First note that $0 \in U_\alpha$ for all $\alpha \in I$ and so $0 \in \cap_{\alpha \in I} U_\alpha$. Therefore, $\cap_{\alpha \in I} U_\alpha \neq \phi$. Let $a \in F$ and $x, y \in \cap_{\alpha \in I} U_\alpha$. Then $x, y \in U_\alpha$ for all α. Since U_α is a subspace of V, $ax + y \in U_\alpha$ for all $\alpha \in I$ and so $ax + y \in \cap_{\alpha \in I} U_\alpha$. Thus, $\cap_{\alpha \in I} U_\alpha$ is a subspace of V by Theorem 19.1.12. ∎

Theorem 19.1.14 *Let V be a vector space over F and S be a nonempty subset of V. Then*

$$\langle S \rangle = \{\textstyle\sum a_i s_i \mid a_i \in F, \ s_i \in S\},$$

where $\sum a_i s_i$ is a finite sum.

Proof.   Let $U = \{\sum a_i s_i \mid a_i \in F, s_i \in S\}$. Let $a \in F$ and $\sum a_i s_i, \sum b_j s_j \in U$.
Then $a(\sum a_i s_i) + \sum b_j s_j = \sum (aa_i)s_i + \sum b_j s_j \in U$ and so U is a subspace of
V by Theorem 19.1.12. Since for all $s \in S$, $s = 1s \in U$, $U \supseteq S$. Thus, $U \supseteq \langle S \rangle$
since $\langle S \rangle$ is the smallest subspace of V containing S. Let $\sum a_i s_i \in U$. Then
since $s_i \in S \subseteq \langle S \rangle$, $a_i s_i \in \langle S \rangle$. Thus, $\sum a_i s_i \in \langle S \rangle$, whence $U \subseteq \langle S \rangle$. ∎

Definition 19.1.15 *Let V be a vector space over the field F. A subset X of V
is called **linearly independent** over F if for every finite number of distinct
elements $x_1, x_2, \ldots, x_n \in X$, $a_1 x_1 + a_2 x_2 + \cdots + a_n x_n = 0$ implies that $a_1 =
a_2 = \cdots a_n = 0$ for any finite set of scalars $\{a_1, a_2, \ldots, a_n\}$. Otherwise X is
called **linearly dependent** over F.*

The set X in Example 19.1.9 is linearly independent over F. $\{0\}$ is linearly
dependent over F.

Definition 19.1.16 *Let V be a vector space over F. A subset A of V is called
a **basis** for V over F if A spans V, i.e., $V = \langle A \rangle$, and A is linearly independent
over F.*

Consider the zero vector space, $\{0\}$, over the field F. We note that the
empty subset, ϕ, is linearly independent over F vacuously and that ϕ spans
$\{0\}$. Hence, ϕ is a basis for $\{0\}$.

Example 19.1.17 *The set*

$$X = \{(1, 0, 0, \ldots, 0), (0, 1, 0, \ldots, 0), \ldots, (0, 0, 0, \ldots, 1)\}$$

of Example 19.1.9 *is a basis for F^n. We showed there that X spans F^n over F.
Suppose*

$$(0, 0, \ldots, 0) = a_1(1, 0, 0, \ldots, 0) + a_2(0, 1, 0, \ldots, 0) + \cdots + a_n(0, 0, 0, \ldots, 1).$$

*Then $(0, 0, \ldots, 0) = (a_1, a_2, \ldots, a_n)$. Therefore, we must have $a_i = 0$ for $i =
1, 2, \ldots, n$. Thus, X is linearly independent.*

Theorem 19.1.18 *Let V be a vector space over F and S be a subset of V. If
$s \in \langle S \rangle$, then $\langle S \cup \{s\} \rangle = \langle S \rangle$.*

Proof.   Clearly $\langle S \rangle \subseteq \langle S \cup \{s\} \rangle$. If $S = \phi$, then $\langle S \rangle = \{0\}$ and so $s = 0$.
Hence, $\langle S \cup \{s\} \rangle = \langle \{0\} \rangle = \{0\} = \langle S \rangle$. Suppose $S \neq \phi$. Let $\sum a_i s_i + as \in
\langle S \cup \{s\} \rangle$, where $s_i \in S$. Then $\sum a_i s_i, as \in \langle S \rangle$ and so $\sum a_i s_i + as \in \langle S \rangle$.
Hence, $\langle S \cup \{s\} \rangle = \langle S \rangle$. ∎

Theorem 19.1.19 *Let V be a vector space over F and $A = \{x_1, x_2, \ldots, x_r\}$
be a subset of V which spans V. Let B be any linearly independent set of vectors
in V. Then B contains at most r vectors.*

Proof.   If B contains less than r vectors, the theorem is true. Suppose B contains at least r vectors, say, $y_1, y_2, \ldots, y_r \in B$. Then since A spans V,

$$y_1 = \sum_{i=1}^{r} a_{i1} x_i$$

and since $y_1 \neq 0$, not all $a_{i1} = 0$, say, $a_{11} \neq 0$. Thus,

$$x_1 = \sum_{i=2}^{r} (-a_{11}^{-1} a_{i1}) x_i + a_{11}^{-1} y_1.$$

This implies that $x_1 \in \langle \{y_1, x_2, \ldots, x_r\} \rangle$. Hence, $\langle \{y_1, x_2, \ldots, x_r\} \rangle = V$ by Theorem 19.1.18. Assume $\langle \{y_1, y_2, \ldots, y_k, x_{k+1}, \ldots, x_r\} \rangle = V$, the induction hypothesis. Then

$$y_{k+1} \in \langle \{y_1, y_2, \ldots, y_k, x_{k+1}, \ldots, x_r\} \rangle.$$

Thus,

$$y_{k+1} = \sum_{i=1}^{k} a_{i,k+1} y_i + \sum_{i=k+1}^{r} a_{i,k+1} x_i$$

and not all $a_{i,k+1} = 0$ for $i = k+1, \ldots, r$, say, $a_{k+1,k+1} \neq 0$. This implies that

$$x_{k+1} = \sum_{i=1}^{k} (-a_{k+1,k+1}^{-1} a_{i,k+1}) y_i + \sum_{i=k+2}^{r} (-a_{k+1,k+1}^{-1} a_{i,k+1}) x_i + a_{k+1,k+1}^{-1} y_{k+1}.$$

Thus, $x_{k+1} \in \langle \{y_1, y_2, \ldots, y_k, y_{k+1}, x_{k+2}, \ldots, x_r\} \rangle$. Hence,

$$V = \langle \{y_1, y_2, \ldots, y_k, y_{k+1}, x_{k+2}, \ldots, x_r\} \rangle$$

by Theorem 19.1.18. Thus, $\langle \{y_1, y_2, \ldots, y_r\} \rangle = V$ by induction. If there exists $y \in B$ such that $y \neq y_i$, $i = 1, 2, \ldots, r$, then $y = \sum_{i=1}^{r} a_i y_i$ and so $0 = \sum_{i=1}^{r} a_i y_i + (-1) y$ and since $-1 \neq 0$, $y_1, y_2, \ldots, y_r, y$ are not linearly independent, a contradiction. Therefore, y does not exist and so $B = \{y_1, y_2, \ldots, y_r\}$. ∎

Theorem 19.1.20 *Let V be a vector space over F, $A = \{x_1, \ldots, x_r\}$, and $B = \{y_1, \ldots, y_s\}$ be two bases for V. Then $r = s$.*

Proof.   Since A spans V and B is linearly independent, $s \leq r$ by Theorem 19.1.19. Similarly, $r \leq s$. ∎

Definition 19.1.21 *Let V be a vector space over F. If V is spanned by a finite set of vectors, then V is called **finite dimensional** over F.*

Lemma 19.1.22 *Let V be a vector space over F and A be a linearly independent subset of V. If $x \in V$ and $x \notin \langle A \rangle$, then $A \cup \{x\}$ is linearly independent.*

Proof. Let $x_1, \ldots, x_n \in A$. Suppose $0 = a_1x_1 + a_2x_2 + \cdots + a_rx_r + ax$. Suppose $a \neq 0$. Then

$$x = (-a)^{-1}a_1x_1 + \cdots + (-a)^{-1}a_rx_r \in \langle A \rangle,$$

a contradiction. Thus, $a = 0$. Hence, $0 = a_1x_1 + a_2x_2 + \cdots + a_rx_r$. Since $\{x_1, x_2, \ldots, x_r\}$ is linearly independent, $a_1 = 0, \ldots, a_r = 0$. Thus, $A \cup \{x\}$ is linearly independent. ∎

Theorem 19.1.23 *Let V be a finite dimensional vector space over F. Then V has a basis.*

Proof. If $V = \{0\}$, then ϕ is a basis for V. We now assume that $V \neq \{0\}$. Let $x_1 \in V$ be such that $x_1 \neq 0$. Then x_1 is linearly independent. If $\langle x_1 \rangle \neq V$, then there exists $x_2 \in V$ such that $x_2 \notin \langle x_1 \rangle$. By Lemma 19.1.22, x_1 and x_2 are linearly independent. Suppose $x_1, \ldots, x_k \in V$ are linearly independent and $\langle \{x_1, \ldots, x_k\} \rangle \neq V$. Then there exists $x_{k+1} \in V$ such that $x_{k+1} \notin \langle \{x_1, \ldots, x_k\} \rangle$. Therefore, $x_1, \ldots, x_k, x_{k+1}$ are linearly independent. Since V is finite dimensional, V is spanned by, say, r vectors. By Theorem 19.1.19, any linearly independent set of vectors in V cannot have more than r vectors. Hence, if we continue the above process of constructing x_i's, then there must exist a positive integer s such that $\{x_1, \ldots, x_s\}$ is linearly independent, $\langle \{x_1, \ldots, x_s\} \rangle = V$, and $s \leq r$. Thus, $\{x_1, \ldots, x_s\}$ is a basis of V. ∎

Theorem 19.1.23 gives us a method for constructing a basis for a finite dimensional vector space V of dimension n over F. We first take any nonzero vector x_1 of V. If $\langle x_1 \rangle = V$, then $\{x_1\}$ is a basis of V. If $\langle x_1 \rangle \subset V$, then we take any $x_2 \in V$, $x_2 \notin \langle x_1 \rangle$. Then by Lemma 19.1.22 $\{x_1, x_2\}$ is linearly independent over F. If $\langle \{x_1, x_2\} \rangle = V$, then $\{x_1, x_2\}$ is a basis for V over F. If $\langle \{x_1, x_2\} \rangle \subset V$, we can choose $x_3 \in V$, $x_3 \notin \langle \{x_1, x_2\} \rangle$ and so on. In a finite number of steps, precisely n steps, we must arrive at a basis for V over F.

Definition 19.1.24 *Let V be a finite dimensional vector space over F. The **dimension** V is the number of elements in a basis for V.*

From the statements following Definition 19.1.16, it follows that the zero vector space, $\{0\}$, is of dimension 0.

Theorem 19.1.25 *Let V be a vector space of dimension n over the field F. Then $X = \{x_1, x_2, \ldots, x_n\}$ is a basis of V if and only if every vector in V is a unique linear combination of $x_1, x_2, \ldots, x_n$ over F.*

Proof. Suppose X is a basis of V over F. Then by Theorem 19.1.14, every vector $v \in V$ is a linear combination of $x_1, x_2, \ldots, x_n$. Let

$$v = a_1 x_1 + \cdots + a_n x_n = b_1 x_1 + \cdots + b_n x_n$$

be any two linear combinations of $x_1, x_2, \ldots, x_n$. Then

$$0 = (a_1 - b_1)x_1 + \cdots + (a_n - b_n)x_n.$$

The linear independence of X over F implies that $a_1 - b_1 = 0, \ldots, a_n - b_n = 0$. That is, the representation of v as a linear combination of $x_1, x_2, \ldots, x_n$ is unique. Conversely, suppose every vector in V is a unique linear combination of $x_1, x_2, \ldots, x_n$ over F. Then clearly X generates V over F. Suppose $0 = a_1 x_1 + \cdots + a_n x_n$ for $a_i \in F$. Since also $0 = 0x_1 + \cdots + 0x_n$, we have $a_i = 0$, $i = 1, \ldots, n$. Thus, X is linearly independent over F. By definition, X is a basis of V over F. ■

We now show that every nonzero vector space, not necessarily finite dimensional, has a basis. For this we prove the following lemma.

Lemma 19.1.26 *Let V be a vector space over a field F and X be a nonempty subset of V. Then X is a basis for V if and only if X is a maximal linearly independent set over F.*

Proof. If X is a basis for V, then X is linearly independent over F and $\langle X \rangle = V$. Let $y \in V$, $y \notin X$. Then $V = \langle X \rangle \subseteq \langle X \cup \{y\} \rangle \subseteq V$ so that $V = \langle X \cup \{y\} \rangle$. Since the proper subset X of $X \cup \{y\}$ also generates V, $X \cup \{y\}$ cannot be linearly independent over F. Thus, X is a maximal linearly independent set over F. Conversely, let X be a maximal linearly independent set over F. It suffices to show that $V = \langle X \rangle$. If $V \supset \langle X \rangle$, then there exists $y \in V$, $y \notin \langle X \rangle$. By Lemma 19.1.22, $X \cup \{y\}$ is linearly independent over F, which contradicts the maximality of X. Thus, $V = \langle X \rangle$. ■

Theorem 19.1.27 *Let V be a vector space over the field F. Then V has a basis.*

Proof. If $V = \{0\}$, then ϕ is a basis for V. We now assume that $V \neq \{0\}$. Let x be a nonzero element of V. Then $\{x\}$ is a linearly independent subset of V. Let T be the set of all linearly independent subsets of V that contain $\{x\}$. Clearly $T \neq \phi$. T is a poset with respect to the set inclusion relation. By Zorn's lemma, we can show that T has a maximal element, say, X. Then X is a maximal linearly independent subset of V and by Lemma 19.1.26, it follows that X is a basis of V. ■

Finally, we state the following theorem without proof. The finite dimensional case was proved in Theorem 19.1.20.

Theorem 19.1.28 *Let V be a vector space over a field F. If A and B are two bases of V, then $|A| = |B|$.* ∎

From Theorem 19.1.27, we find that a vector space V over a field F has a basis B. If B is a basis for V over F, then $|B|$ is called the **dimension** of V over F.

19.1.1 Worked-Out Exercises

◇ **Exercise 1** Let V be a vector space of dimension n. Show that any set of n linearly independent vectors is a basis of V.

Solution: Let B be a set of n linearly independent vectors. Suppose $V \neq \langle B \rangle$. Let $y \in V$ be such that $y \notin \langle B \rangle$. Then $B \cup \{y\}$ is a set of $n+1$ linearly independent vectors by Lemma 19.1.22, a contradiction to Theorem 19.1.19. Hence, B is a basis of V.

◇ **Exercise 2** Let $u_1 = (0,1,1,0)$, $u_2 = (1,0,1,0)$, and $u_3 = (-1,-2,0,0)$ be three vectors in $\mathbf{R}^4$. Show that $\{u_1, u_2, u_3\}$ is a linearly independent set. Extend this set to a basis of $\mathbf{R}^4$.

Solution: Let $a_1, a_2, a_3 \in \mathbf{R}$ be such that

$$a_1 u_1 + a_2 u_2 + a_3 u_3 = 0.$$

Then $a_2 - a_3 = 0$, $a_1 - 2a_3 = 0$, and $a_1 + a_2 = 0$. From this, it follows that $a_1 = a_2 = a_3 = 0$. Hence, $\{u_1, u_2, u_3\}$ is a linearly independent set. Suppose

$$(0,0,0,1) \in \langle \{u_1, u_2, u_3\} \rangle.$$

Then there exists $b_1, b_2, b_3 \in \mathbf{R}$ such that

$$b_1 u_1 + b_2 u_2 + b_3 u_3 = (0,0,0,1).$$

Thus, $b_2 - b_3 = 0$, $b_1 - 2b_3 = 0$, $b_1 + b_2 = 0$, and $1 = 0$, a contradiction. Therefore, $e_4 = (0,0,0,1) \notin \langle \{u_1, u_2, u_3\} \rangle$. Hence, $\{u_1, u_2, u_3, e_4\}$ is a linearly independent set of vectors in $\mathbf{R}^4$. Since the dimension of $\mathbf{R}^4$ is 4, $\{u_1, u_2, u_3, e_4\}$ is a basis.

◇ **Exercise 3** Let V be a nonzero vector space of dimension n. Let X be a finite subset of V such that $V = \langle X \rangle$. Show that X contains a subset Y such that Y is a basis of V.

Solution: Let $X = \{x_1, x_2, \ldots, x_t\}$. Clearly $t \geq n$. Since $V \neq \{0\}$, X contains a nonzero element. Thus, X contains a linearly independent subset. If X is linearly independent, then X is a basis of V and $n = t$. Suppose X is not linearly independent. Then there exists x_i, say, x_t, such that $x_t \in \langle\{x_1, x_2, \ldots, x_{t-1}\}\rangle$. Then $V = \langle\{x_1, x_2, \ldots, x_{t-1}\}\rangle$. Let $s = t - n - 1$. By repeating the process finitely many times, we can show that there are s vectors $x_{i_1}, \ldots, x_{i_s} \in \{x_1, x_2, \ldots, x_{t-1}\}$ such that

$$x_{i_1}, \ldots, x_{i_s} \in \langle\{x_1, x_2, \ldots, x_{t-1}\}\backslash\{x_{i_1}, \ldots, x_{i_s}\}\rangle.$$

Let

$$Y = \{x_1, x_2, \ldots, x_{t-1}\}\backslash\{x_{i_1}, \ldots, x_{i_s}\}.$$

Then $Y \subseteq X$, $|Y| = n$, and $V = \langle Y \rangle$. If Y is not linearly independent, then there exists $y \in Y$ such that $y \in \langle Y\backslash\{y\}\rangle$. Then $V = \langle Y\backslash\{y\}\rangle$ and $|Y\backslash\{y\}| = n - 1$, a contradiction to the fact that the dimension of V is n.

Exercise 4 Let $T = \{(x, y, z) \in \mathbf{R}^3 \mid 2x + 3y + z = 0\}$. Show that T is a subspace of $V_3(\mathbf{R})$. Find a basis for T.

Solution: Since $(0, 0, 0) \in T$, $T \neq \phi$. Let (x_1, y_1, z_1), $(x_2, y_2, z_2) \in T$ and $r \in \mathbf{R}$. Then $2x_1 + 3y_1 + z_1 = 0$ and $2x_2 + 3y_2 + z_2 = 0$. Hence, $2(x_1 + x_2) + 3(y_1 + y_2) + (z_1 + z_2) = 0$ and $2rx_1 + 3ry_1 + rz_1 = r(2x_1 + 3y_1 + z_1) = 0$. Therefore, $(x_1, y_1, z_1) + (x_2, y_2, z_2) \in T$ and $r(x_1, y_1, z_1) \in T$. Thus, T is a subspace of $V_3(\mathbf{R})$. Now $2x_1 + 3y_1 + z_1 = 0$ implies that $(x_1, y_1, z_1) = (x_1, y_1, -2x_1 - 3y_1) = x_1(1, 0, -2) + y_1(0, 1, -3)$. Since $(1, 0, -2)$, $(0, 1, -3) \in T$ and (x_1, y_1, z_1) is an arbitrary element of T, $T = \langle\{(1, 0, -2), (0, 1, -3)\}\rangle$. It is easy to verify that $\{(1, 0, -2), (0, 1, -3)\}$ is a linearly independent set. Hence, $\{(1, 0, -2), (0, 1, -3)\}$ is a basis of T.

19.1.2 Exercises

1. For the vector space $\mathbf{R}^3$ over $\mathbf{R}$, determine whether or not the sets listed are bases of $\mathbf{R}^3$.

   (i) $\{(1, 1, 0), (1, 1, 1), (1, 0, 0)\}$.

   (ii) $\{(2, 0, 0), (0, 2, 0), (0, 0, 2)\}$.

   (iii) $\{(-1, 0, 0), (0, -1, 0), (0, 0, -1)\}$.

   (iv) $\{(1, 0, 0), (1, 1, 0), (1, 1, 1), (0, 1, 0)\}$.

2. Let M be an R-module, $m \in M$ and $r \in R$. Prove that $r0 = 0$, $0m = 0$, and $-(rm) = (-r)m = r(-m)$.

3. Show that the intersection of two submodules of an R-module M is a submodule.

4. Show that the **Z**-module **Q** has no finite set of generators.

5. Find all subspaces of the real vector space $\mathbf{R}^2$. Is is true that for any elements $u = (a, b)$ and $v = (c, d)$ of $\mathbf{R}^2$, there exists a nontrivial subspace W of $\mathbf{R}^2$ such that $u, v \in W$?

6. Let A, B, and C be submodules of an R-module M.

   (i) Prove that $A + B = \{a + b \mid a \in A, \, b \in B\}$ is a submodule of M.

   (ii) If $A \subseteq C$, prove that $A + (B \cap C) = (A + B) \cap C$.

7. Let M be an R-module and $a \in M$. Show that $T = \{ra + na \mid r \in R, \, n \in \mathbf{Z}\}$ is a submodule of M.

8. Let M be a unitary R-module. M is called a **simple** R-module if $M \neq \{0\}$ and the only submodules of M are M and $\{0\}$. Prove that M is simple if and only if M is generated by any nonzero element of M.

9. Let M be a unitary R-module. M is called **Noetherian** if for any sequence
$$A_1 \subseteq A_2 \subseteq \cdots \subseteq A_n \subseteq \cdots$$
of submodules of M, there exists a positive integer n such that $A_n = A_{n+1} = \ldots$. Prove that M is Noetherian if and only if every submodule of M is finitely generated.

10. Let M be a unitary R-module. M is said to satisfy the maximal condition on submodules if any nonempty collection of submodules of M has a maximal element. Prove that M is Noetherian if and only if M satisfies the maximal condition for submodules.

11. Let M be a unitary R-module. M is called **Artinian** if for any sequence
$$A_1 \supseteq A_2 \supseteq \cdots \supseteq A_n \supseteq \cdots$$
of submodules of M, there exists a positive integer n such that $A_n = A_{n+1} = \ldots$. Prove that M is Artinian if and only if any nonempty set of submodules of M has a minimal element.

12. Let N be a submodule of a unitary R-module M and $a \in M$. Let
$$a + N = \{a + b \mid b \in N\}.$$
Prove the following.

   (i) $a \in a + N$.

   (ii) For all $a, b \in M$, $a + N = b + N$ if and only if $a - b \in N$.

   (iii) For all $a, b \in M$, either $(a + N) \cap (b + N) = \phi$ or $a + N = b + N$.

13. Let N be a submodule of an R-module M. Let

$$M/N = \{a + N \mid a \in N\}.$$

Define the following operations on M/N

$$(a + N) + (b + N) = (a + b) + N$$
$$r(a + N) = ra + N$$

for all $a + N, b + N \in M/N$, $r \in R$. Prove that M/N is an R-module.

14. Let N be a submodule of an R-module M. Prove that M is Artinian (Noetherian) if and only if N and M/N are Artinian (Noetherian).

15. Let V be a finite dimensional vector space over F. If U and W are two subspaces of V, prove the following:

(i) $U + W = \{u + w \mid u \in U, w \in W\}$ is a subspace of V.

(ii) $\dim U + \dim W = \dim(U + W) - \dim(U \cap W)$.

16. Let N be a submodule of an R-module M. N is called a **direct summand** of M if there exists a submodule P of M such that $M = N + P$ and $N \cap P = \{0\}$. In a finite dimensional vector space V over F, show that every subspace is a direct summand of V.

17. Write the proof if the statement is true; otherwise give a counterexample.

(i) If $\{u, v, w\}$ is a linearly independent subset of a vector space V, then $\{u, u + v, u + v + w\}$ is also a linearly independent subset.

(ii) If W is a subspace of a finite dimensional vector space V such that $\dim W = \dim V$, then $W = V$.

(iii) Let V be a vector space over a field F. If $0 \neq v \in V$, then there exists a basis containing v.

(iv) If S and T are two basis of a vector space V, then $S \cup T$ is a basis of V.

Chapter 20

Rings of Matrices

In this chapter, we study some elementary properties of rings of matrices. Rings of matrices provide a rich source of examples for noncommutative ring theory. They are also useful for the understanding of noncommutative ring theory since they often appear in representation theorems.

20.1   Full Matrix Rings

Let R be a ring with 1. Let $M_n(R)$ be the ring of all $n \times n$ matrices with entries from R. Let E_{ij} be the element of $M_n(R)$ whose (i, j) entry is 1 and all other entries are 0, $1 \le i, j \le n$. Let $E_{ij}, E_{kl} \in M_n(R)$. Then the following can be easily verified.

$$E_{ij} E_{kl} = \begin{cases} E_{il} \text{ if } j = k \\ 0 \text{ if } j \ne k. \end{cases}$$

Let $(a_{ij}) \in M_n(R)$. Then $(a_{ij}) = \sum_{i,j=1}^{n} a_{ij} E_{ij}$.

The following describes ideals of $M_n(R)$.

Theorem 20.1.1 *Let M be an ideal in $M_n(R)$. Then there exists an ideal I of R such that $M = M_n(I)$, i.e., M is the set of all $n \times n$ matrices with entries from I.*

Proof.   Let $I = \{a \in R \mid a = a_{11} \text{ for some } (a_{ij}) \in M\}$. Since $0 \in M$, $0 \in I$. Thus, $I \ne \phi$. Clearly if $a, b \in I$, then $a - b \in I$. Let $a \in I$ and $r \in R$. Then $a = a_{11}$ for some $\sum_{i,j=1}^{n} a_{ij} E_{ij} \in M$. Since M is an ideal,

$$ar E_{11} = E_{11}(\sum_{i,j=1}^{n} a_{ij} E_{ij}) r E_{11} \in M.$$

Hence, $ar \in I$. Similarly, since

$$ra E_{11} = r E_{11} \left(\sum_{i,j=1}^{n} a_{ij} E_{ij} \right) E_{11} \in M,$$

$ra \in I$. Thus, I is an ideal of R. We now proceed to show that $M = M_n(I)$. Let $\sum_{i,j=1}^{n} a_{ij} E_{ij} \in M$. Let $1 \leq k, l \leq n$. Now

$$a_{kl} E_{11} = E_{1k} \left(\sum_{i,j=1}^{n} a_{ij} E_{ij} \right) E_{l1} \in M.$$

Therefore, $a_{kl} \in I$ for $1 \leq k, l \leq n$. This implies that $\sum_{i,j=1}^{n} a_{ij} E_{ij} \in M_n(I)$. Thus, $M \subseteq M_n(I)$. Conversely, let $\sum_{i,j=1}^{n} b_{ij} E_{ij} \in M_n(I)$. Let $1 \leq k, l \leq n$. Then $b_{kl} = c_{11}$ for some $\sum_{i,j=1}^{n} c_{ij} E_{ij} \in M$. Since M is an ideal,

$$b_{kl} E_{kl} = c_{11} E_{kl} = E_{k1} \left(\sum_{i,j=1}^{n} c_{ij} E_{ij} \right) E_{1l} \in M.$$

Therefore, $\sum_{i,j=1}^{n} b_{ij} E_{ij} \in M$. Thus, $M_n(I) \subseteq M$. Consequently, $M = M_n(I)$. ∎

Corollary 20.1.2 *Let R be a ring with 1. If R is simple, then $M_n(R)$ is simple.* ∎

Theorem 20.1.3 *Let R be a ring with 1. Let I be an ideal of R. Then*

$$M_n(R)/M_n(I) \simeq M_n(R/I).$$

Proof. Define $f : M_n(R) \to M_n(R/I)$ by

$$f((a_{ij})) = (a_{ij} + I)$$

for all $(a_{ij}) \in M_n(R)$. Then it can be easily verified that f is an epimorphism and Ker $f = M_n(I)$. Hence,

$$M_n(R)/M_n(I) \simeq M_n(R/I). \blacksquare$$

In the next theorem, we describe the center of $M_n(R)$ when R is a commutative ring with 1.

Theorem 20.1.4 *Let R be a commutative ring with 1. Then $C(M_n(R))$, the center of $M_n(R)$, is the set of all scalar matrices in $M_n(R)$.*

Proof. Let $\sum_{k=1}^{n} aE_{kk}$, $\sum_{i,j=1}^{n} b_{ij}E_{ij} \in M_n(R)$. Then

$$(\sum_{k=1}^{n} aE_{kk})(\sum_{i,j=1}^{n} b_{ij}E_{ij}) = \sum_{k,j=1}^{n} ab_{kj}E_{kj} = \sum_{r,s=1}^{n} ab_{rs}E_{rs}$$

and

$$
\begin{aligned}
(\sum_{i,j=1}^{n} b_{ij}E_{ij})(\sum_{k=1}^{n} aE_{kk}) &= \sum_{k,i=1}^{n} b_{ik}aE_{ik} \\
&= \sum_{r,s=1}^{n} b_{rs}aE_{rs} \\
&= \sum_{r,s=1}^{n} ab_{rs}E_{rs}
\end{aligned}
$$

since R is commutative. Therefore,

$$(\sum_{k=1}^{n} aE_{kk})(\sum_{i,j=1}^{n} b_{ij}E_{ij}) = (\sum_{i,j=1}^{n} b_{ij}E_{ij})(\sum_{k=1}^{n} aE_{kk}).$$

Thus, $\sum_{k=1}^{n} aE_{kk} \in C(M_n(R))$.

Now, let $g = \sum_{i,j=1}^{n} a_{ij}E_{ij} \in C(M_n(R))$. Let $E_{kk} \in M_n(R)$, $1 \leq k \leq n$. Then

$$E_{kk}(\sum_{i,j=1}^{n} a_{ij}E_{ij}) = (\sum_{i,j=1}^{n} a_{ij}E_{ij})E_{kk}$$

implies that $\sum_{j=1}^{n} a_{kj}E_{kj} = \sum_{i=1}^{n} a_{ik}E_{ik}$ and hence by comparing the corresponding entries, we get $a_{ik} = 0 = a_{kj}$ for all $i,j = 1,2,\ldots,n$, $i \neq k$, $j \neq k$, $1 \leq k \leq n$. Thus, all entries in g are zero except (possibly) the diagonal entries. Hence, $g = \sum_{k=1}^{n} a_{kk}E_{kk}$. Let $E_{rs} \in M_n(R)$ be such that $r \neq s$, $1 \leq r,s \leq n$. Then

$$(\sum_{k=1}^{n} a_{kk}E_{kk})E_{rs} = E_{rs}(\sum_{k=1}^{n} a_{kk}E_{kk})$$

implies that $a_{rr}E_{rs} = a_{ss}E_{rs}$. Therefore, $a_{rr} = a_{ss}$, $r \neq s$, $1 \leq r,s \leq n$. Consequently, g is a scalar matrix. ∎

20.1.1 Worked-Out Exercises

◇ **Exercise 1** Let R be a ring with 1. Let A and B be ideals of R. Show that $M_n(AB) = M_n(A)M_n(B)$.

Solution: Let $\sum_{i,j=1}^{n} c_{ij}E_{ij} \in M_n(AB)$. Then $c_{ij} \in AB$. Let

$$c_{ij} = \sum_{l=1}^{s} a_{i_l}b_{i_l} \in AB,$$

$a_{i_l} \in A$, $b_{i_l} \in B$, $1 \leq l \leq s$. Then

$$
\begin{aligned}
c_{ij}E_{ij} &= (\sum_{l=1}^{s} a_{i_l}b_{i_l})E_{ij} \\
&= \sum_{l=1}^{s}(a_{i_l}b_{i_l}E_{ij}) \\
&= \sum_{l=1}^{s}(a_{i_l}E_{i1}b_{i_l}E_{1j}) \in M_n(A)M_n(B).
\end{aligned}
$$

Thus, $\sum_{i,j=1}^{n} c_{ij} E_{ij} \in M_n(A) M_n(B)$ and so $M_n(AB) \subseteq M_n(A) M_n(B)$. Let $\sum_{i,j=1}^{n} a_{ij} E_{ij} \in M_n(A)$ and $\sum_{i,j=1}^{n} b_{ij} E_{ij} \in M_n(B)$. Let $c_{ij} = \sum_{k=1}^{n} a_{ik} b_{kj}$, $1 \le i, j \le n$. Then $c_{ij} = \sum_{k=1}^{n} a_{ik} b_{kj} \in AB$, $1 \le i, j \le n$. This implies that

$$(\sum_{i,j=1}^{n} a_{ij} E_{ij})(\sum_{i,j=1}^{n} b_{ij} E_{ij}) = \sum_{i,j=1}^{n} c_{ij} E_{ij} \in M_n(AB).$$

Thus, $M_n(A) M_n(B) \subseteq M_n(AB)$. Consequently, $M_n(AB) = M_n(A) M_n(B)$.

◇ **Exercise 2** A ring R is called a **prime ring** if $\{0\}$ is a prime ideal. Prove that a ring R with 1 is a prime ring if and only if $M_n(R)$ is a prime ring.

Solution: Suppose R is a prime ring and $M_n(R)$ is not a prime ring. There exist nonzero ideals P and Q of $M_n(R)$ such that $PQ =$ the zero ideal of $M_n(R)$. There exist nonzero ideals A and B of R such that $P = M_n(A)$ and $Q = M_n(B)$. Thus, $PQ = M_n(A) M_n(B)$ implies that $AB = \{0\}$. Since R is a prime ring, $A = \{0\}$ or $B = \{0\}$, which is a contradiction. Hence, $M_n(R)$ is a prime ring. Conversely, suppose that $M_n(R)$ is a prime ring and R is not a prime ring. Thus, there exist nonzero ideals A and B of R such that $AB = \{0\}$. Then $M_n(A)$, $M_n(B)$ are nonzero ideals of $M_n(R)$ such that $M_n(A) M_n(B) =$ the zero ideal of $M_n(R)$, a contradiction. Hence, R is a prime ring.

20.1.2   Exercises

1. If R is a field, find all ideals of $M_n(R)$.

2. If R is a Noetherian ring, prove that $M_n(R)$ is a Noetherian ring.

3. If R is a ring with 1, prove that $M_n(R)[x] \simeq M_n(R[x])$.

20.2   Rings of Triangular Matrices

Let A, B, and C be rings with identity such that $_C B_A$ is a unital bimodule, i.e., B is a unitary right A-module, B is a unitary left C-module and for all $a \in A$, for all $b \in B$, and for all $c \in C$, $c(ba) = (cb)a$.

Let

$$R = \left\{ \begin{bmatrix} a & 0 \\ b & c \end{bmatrix} \mid a \in A, \ b \in B, \ c \in C \right\} = \begin{bmatrix} A & 0 \\ B & C \end{bmatrix}.$$

Define $+$ and $\cdot$ as the usual matrix addition and multiplication, i.e., if $\begin{bmatrix} a & 0 \\ b & c \end{bmatrix}$,

$\begin{bmatrix} a' & 0 \\ b' & c' \end{bmatrix} \in R$, then

$$\begin{bmatrix} a & 0 \\ b & c \end{bmatrix} + \begin{bmatrix} a' & 0 \\ b' & c' \end{bmatrix} = \begin{bmatrix} a+a' & 0 \\ b+b' & c+c' \end{bmatrix}$$

$$\begin{bmatrix} a & 0 \\ b & c \end{bmatrix} \cdot \begin{bmatrix} a' & 0 \\ b' & c' \end{bmatrix} = \begin{bmatrix} aa' & 0 \\ ba'+cb' & cc' \end{bmatrix}.$$

Since $ba' + cb' \in B$, $+$ and $\cdot$ are well defined. It is easy to check that $(R, +, \cdot)$ is a ring.

Consider $A \oplus B$, the (external) direct sum of the rings A and B. Let $(a, b) \in A \oplus B$ and $a' \in A$. Define $(a, b)a' = (aa', ba')$. Then $A \oplus B$ is a unital right A-module.

We now proceed to describe all right ideals of R.

Let I be a right ideal of R. Let

$$M = \left\{ (a, b) \in A \oplus B \ \middle| \ \begin{bmatrix} a & 0 \\ b & c \end{bmatrix} \in I, \text{ for some } c \in C \right\}$$

and

$$K = \left\{ c \in C \ \middle| \ \begin{bmatrix} a & 0 \\ b & c \end{bmatrix} \in I, \text{ for some } a \in A, \ b \in B \right\}.$$

We now claim the following:

(i) M is a right A-submodule of $A \oplus B$;

(ii) K is a right ideal of C;

(iii) $\{0\} \oplus KB \subseteq M$.

(i): Clearly $M \neq \phi$ and $(M, +)$ is an Abelian group. Let $(a, b) \in M$ and $a' \in A$. Then $\begin{bmatrix} a & 0 \\ b & c \end{bmatrix} \in I$ for some $c \in C$. Now $\begin{bmatrix} a & 0 \\ b & c \end{bmatrix} \in I$ and $\begin{bmatrix} a' & 0 \\ 0 & 0 \end{bmatrix} \in R$. Since I is a right ideal of R,

$$\begin{bmatrix} aa' & 0 \\ ba' & 0 \end{bmatrix} = \begin{bmatrix} a & 0 \\ b & c \end{bmatrix} \begin{bmatrix} a' & 0 \\ 0 & 0 \end{bmatrix} \in I.$$

Thus, $(a, b)a' = (aa', ba') \in M$. It is now easy to verify that M is a right A-submodule of $A \oplus B$.

(ii): Clearly $K \neq \phi$, and $(K, +)$ is an Abelian group. Let $k \in K$ and $c \in C$. Then $\begin{bmatrix} a & 0 \\ b & k \end{bmatrix} \in I$ for some $a \in A$, $b \in B$. Since I is a right ideal of R,

$$\begin{bmatrix} 0 & 0 \\ 0 & kc \end{bmatrix} = \begin{bmatrix} a & 0 \\ b & k \end{bmatrix} \begin{bmatrix} 0 & 0 \\ 0 & c \end{bmatrix} \in I.$$

Thus, $kc \in K$. Hence, K is a right ideal of C.

(iii): Let $k \in K$, $y \in B$. Then $\begin{bmatrix} a & 0 \\ b & k \end{bmatrix} \in I$ for some $a \in A$, $b \in B$. Since I is a right ideal of R,

$$\begin{bmatrix} 0 & 0 \\ ky & 0 \end{bmatrix} = \begin{bmatrix} a & 0 \\ b & k \end{bmatrix} \begin{bmatrix} 0 & 0 \\ y & 0 \end{bmatrix} \in I.$$

Hence, $(0, ky) \in M$. Thus, $\{0\} \oplus KB \subseteq M$.

Conversely, let M and K be defined as in (i), (ii), and (iii). Let

$$I = \left\{ \begin{bmatrix} a & 0 \\ b & c \end{bmatrix} \in R \mid (a, b) \in M,\ c \in K \right\}.$$

Let $\begin{bmatrix} a & 0 \\ b & c \end{bmatrix} \in I$, $\begin{bmatrix} x & 0 \\ y & z \end{bmatrix} \in R$. Then

$$\begin{bmatrix} a & 0 \\ b & c \end{bmatrix} \begin{bmatrix} x & 0 \\ y & z \end{bmatrix} = \begin{bmatrix} ax & 0 \\ bx + cy & cz \end{bmatrix}.$$

Now $(a, b) \in M$, $x \in A$. Since M is a right A-submodule of $A \oplus B$, $(ax, bx) = (a, b)x \in M$. Now $c \in K$ and $y \in B$. Hence, $(0, cy) \in \{0\} \oplus KB \subseteq M$. Thus, $(ax, bx + cy) = (ax, bx) + (0, cy) \in M$. Since K is a right ideal of C, $cz \in K$.

Therefore, $\begin{bmatrix} a & 0 \\ b & c \end{bmatrix} \begin{bmatrix} x & 0 \\ y & z \end{bmatrix} = \begin{bmatrix} ax & 0 \\ bx + cy & cz \end{bmatrix} \in I$. Now it can be easily verified that I is a right ideal of R.

We summarize the above discussion in the following theorem.

Theorem 20.2.1 *Let R, A, B, and C be defined as above. Let M be a right A-submodule of $A \oplus B$, K be a right ideal of C, and $\{0\} \oplus KB \subseteq M$. Let*

$$I = \left\{ \begin{bmatrix} a & 0 \\ b & c \end{bmatrix} \in R \mid (a, b) \in M,\ c \in K \right\}.$$

Then I is a right ideal of R. Conversely, let I be a right ideal of R. Then there exists a right A-submodule M of $A \oplus B$ and a right ideal K of C such that $\{0\} \oplus KB \subseteq M$ and

$$I = \left\{ \begin{bmatrix} a & 0 \\ b & c \end{bmatrix} \in R \mid (a, b) \in M,\ c \in K \right\}.\ \blacksquare$$

The following theorem, which is dual to the above theorem, can be proved in a similar manner. We leave its proof as an exercise.

Theorem 20.2.2 *Let R, A, B, and C be defined as above. Let M be a left C-submodule of $B \oplus C$, K be a left ideal of A, and $BK \oplus \{0\} \subseteq M$. Let*

$$I = \left\{ \begin{bmatrix} a & 0 \\ b & c \end{bmatrix} \in R \mid (b, c) \in M, \ a \in K \right\}.$$

Then I is a left ideal of R. Conversely, let I be a left ideal of R. Then there exists a left A-submodule M of $B \oplus C$ and a left ideal K of A such that $BK \oplus \{0\} \subseteq M$ and

$$I = \left\{ \begin{bmatrix} a & 0 \\ b & c \end{bmatrix} \in R \mid (b, c) \in M, \ a \in K \right\}. \ \blacksquare$$

Let M and N be right R-modules and $f : M \to N$. Then f is called a R-**homomorphism** if (i) $f(a + b) = f(a) + f(b)$ and (ii) $f(ar) = f(a)r$ for all $a, b \in M$, $r \in R$. If f is a one-one function from M onto N and f is a R-homomorphism, then f is called an R-**isomorphism** or simply an **isomorphism** from M onto N. M and N are **isomorphic** as right R-modules, if there exists an R-isomorphism from M onto N. Similar conventions hold for left R-modules.

Theorem 20.2.3 *Let R, A, B, and C be defined as above. Let $I = \begin{bmatrix} 0 & 0 \\ B & 0 \end{bmatrix}$ and $J = \begin{bmatrix} A & 0 \\ 0 & 0 \end{bmatrix}$. Then the following assertions hold.*

(i) I is an ideal of R.

(ii) $R/I \simeq A \oplus C$, where $A \oplus C$ is the (external) direct sum of rings.

(iii) I is a Noetherian (Artinian) right R-module if and only if B is a Noetherian (Artinian) right A-module.

(iv) I is a Noetherian (Artinian) left R-module if and only if B is a Noetherian (Artinian) left C-module.

(v) J is a right ideal of R.

(vi) $R/J \simeq B \oplus C$ as a right R-module.

Proof. (i) Let $M = \{0\} \oplus B$ and $K = \{0\}$. Then by Theorems 20.2.1 and 20.2.2, I is an ideal of R.

(ii) Define $f : R \longrightarrow A \oplus C$ by

$$f\left(\begin{bmatrix} a & 0 \\ b & c \end{bmatrix} \right) = (a, c)$$

for all $\begin{bmatrix} a & 0 \\ b & c \end{bmatrix} \in R$. Clearly, f is an epimorphism. Now $\begin{bmatrix} a & 0 \\ b & c \end{bmatrix} \in \text{Ker } f$ if and only if $f\left(\begin{bmatrix} a & 0 \\ b & c \end{bmatrix} \right) = (0, 0)$ if and only if $(a, c) = (0, 0)$ if and only if

$a = 0$ and $c = 0$. Thus, Ker $f = I$. Hence,

$$R/I \simeq A \oplus C.$$

(iii) Suppose I is a Noetherian right R-module. Let $B_1 \subseteq B_2 \subseteq \cdots$ be an ascending sequence of right A-submodules of B. Let

$$I_i = \begin{bmatrix} 0 & 0 \\ B_i & 0 \end{bmatrix}, \quad i = 1, 2, \ldots.$$

Then I_i is a right ideal of R and $I_i \subseteq I_{i+1} \subseteq I$, $i = 1, 2, \ldots$. Thus, $I_1 \subseteq I_2 \subseteq \cdots$ is an ascending sequence of right R-submodules of I. Since I is a Noetherian right R-module, there exists a positive integer n such that $I_n = I_{n+k}$ for all $k \geq 0$. Therefore, $B_n = B_{n+k}$ for all $k \geq 0$. Hence, B is a Noetherian right A-module. Conversely, let B be a Noetherian right A-module. Let $I_1 \subseteq I_2 \subseteq \cdots$ be an ascending sequence of right R-submodules of I. Let

$$B_i = \left\{ b \in B \;\middle|\; \begin{bmatrix} 0 & 0 \\ b & 0 \end{bmatrix} \in I_i \right\}, \quad i = 1, 2, \ldots.$$

Clearly $(B_i, +)$ is an Abelian group. Let $b \in B_i$ and $a \in A$. Then $\begin{bmatrix} 0 & 0 \\ b & 0 \end{bmatrix} \in$ I_i and $\begin{bmatrix} a & 0 \\ 0 & 0 \end{bmatrix} \in R$. Since I_i is a right R-module, $\begin{bmatrix} 0 & 0 \\ ba & 0 \end{bmatrix} = \begin{bmatrix} 0 & 0 \\ b & 0 \end{bmatrix}$ $\begin{bmatrix} a & 0 \\ 0 & 0 \end{bmatrix} \in I_i$. Hence, $ba \in B_i$. Thus, B_i is a right A-submodule of B, $i =$ $1, 2, \ldots$. Clearly $B_i \subseteq B_{i+1}$, $i = 1, 2, \ldots$. Therefore, $B_1 \subseteq B_2 \subseteq \cdots$ is an ascending sequence of right A-submodules of B. Since B is a Noetherian right A-module, there exists a positive integer n such that $B_n = B_{n+k}$ for all $k \geq 0$. Thus, $I_n = I_{n+k}$ for all $k \geq 0$. Hence, I is a Noetherian right A-module.

(iv) The proof of this part is analogous to the proof of part (iii).

(v) Let $\begin{bmatrix} a & 0 \\ 0 & 0 \end{bmatrix}, \begin{bmatrix} b & 0 \\ 0 & 0 \end{bmatrix} \in J$ and $\begin{bmatrix} a & 0 \\ b & c \end{bmatrix} \in R$. Then

$$\begin{bmatrix} a & 0 \\ 0 & 0 \end{bmatrix} - \begin{bmatrix} b & 0 \\ 0 & 0 \end{bmatrix} = \begin{bmatrix} a-b & 0 \\ 0 & 0 \end{bmatrix} \in J$$

and

$$\begin{bmatrix} a & 0 \\ 0 & 0 \end{bmatrix} \begin{bmatrix} a' & 0 \\ b' & c' \end{bmatrix} = \begin{bmatrix} aa' & 0 \\ 0 & 0 \end{bmatrix} \in J.$$

Thus, J is a right ideal of R and hence a right R-submodule.

(vi) Clearly R/J is a right R-module. Let $(b, c) \in B \oplus C$ and $\begin{bmatrix} a' & 0 \\ b' & c' \end{bmatrix} \in R$. Define a binary operation $\cdot$

$$(b, c) \cdot \begin{bmatrix} a' & 0 \\ b' & c' \end{bmatrix} = (ba' + cb', cc').$$

Clearly $(ba' + cb', cc') \in B \oplus C$. It can now be easily checked that $B \oplus C$ is a right R-module under $\cdot$. Define

$$f : R \to B \oplus C$$

by

$$f\left(\begin{bmatrix} a & 0 \\ b & c \end{bmatrix}\right) = (b, c)$$

for all $\begin{bmatrix} a & 0 \\ b & c \end{bmatrix} \in R$. Let $\begin{bmatrix} a & 0 \\ b & c \end{bmatrix}, \begin{bmatrix} a' & 0 \\ b' & c' \end{bmatrix} \in R$. Then

$$\begin{aligned}
f\left(\begin{bmatrix} a & 0 \\ b & c \end{bmatrix} + \begin{bmatrix} a' & 0 \\ b' & c' \end{bmatrix}\right) &= f\left(\begin{bmatrix} a+a' & 0 \\ b+b' & c+c' \end{bmatrix}\right) \\
&= (b+b', c+c') \\
&= (b, c) + (b', c') \\
&= f\left(\begin{bmatrix} a & 0 \\ b & c \end{bmatrix}\right) + f\left(\begin{bmatrix} a' & 0 \\ b' & c' \end{bmatrix}\right).
\end{aligned}$$

Now

$$\begin{aligned}
f\left(\begin{bmatrix} a & 0 \\ b & c \end{bmatrix}\begin{bmatrix} a' & 0 \\ b' & c' \end{bmatrix}\right) &= f\left(\begin{bmatrix} aa' & 0 \\ ba'+cb' & cc' \end{bmatrix}\right) \\
&= (ba'+cb', cc') \\
&= (b, c) \cdot \begin{bmatrix} a' & 0 \\ b' & c' \end{bmatrix} \\
&= f\left(\begin{bmatrix} a & 0 \\ b & c \end{bmatrix}\right)\begin{bmatrix} a' & 0 \\ b' & c' \end{bmatrix}.
\end{aligned}$$

Hence, f is an R-homomorphism. Clearly f is onto $B \oplus C$ and $\text{Ker } f = J$. Consequently,

$$R/J \simeq B \oplus C. \blacksquare$$

20.2.1 Worked-Out Exercises

$\diamond$ **Exercise 1** Let $R = \left\{ \begin{bmatrix} a & 0 \\ b & n \end{bmatrix} \mid a, b \in \mathbf{Q},\ n \in \mathbf{Z} \right\} = \begin{bmatrix} \mathbf{Q} & 0 \\ \mathbf{Q} & \mathbf{Z} \end{bmatrix}$.

(i) Find all right ideals of R.

(ii) Show that R is right Noetherian, but not left Noetherian.

Solution: (i) Let J be a right ideal of R. Suppose $J \neq R$. By Theorem 20.2.1, there exists a right $\mathbf{Q}$-submodule M of $\mathbf{Q} \oplus \mathbf{Q}$, a right ideal K of $\mathbf{Z}$ such that $\{0\} \oplus K\mathbf{Q} \subseteq M$, and

$$I = \left\{ \begin{bmatrix} a & 0 \\ b & c \end{bmatrix} \in R \mid (a, b) \in M, \ c \in K \right\}.$$

Since $\mathbf{Z}$ is a PID, $K = \langle n \rangle$ for some $n \in \mathbf{Z}$. Now as in Example 19.1.7(ii), we can show that M is of the following form:

(1) $M = \{0\}$, or
(2) $M = \{0\} \oplus \mathbf{Q}$, or
(3) $M = \mathbf{Q} \oplus \{0\}$, or
(4) $M = \langle (a, b) \rangle$ for some $a, b \in \mathbf{Q}$, $a \neq 0$, $b \neq 0$, or
(5) $M = \mathbf{Q} \oplus \mathbf{Q}$.

Case 1. $M = \{0\}$. Since $\{0\} \oplus K\mathbf{Q} \subseteq M$, $K = \{0\}$. Therefore, in this case, $J = \{0\}$.

Case 2. $M = \{0\} \oplus \mathbf{Q} = \langle (0, 1) \rangle$. In this case,

$$J = \begin{bmatrix} 0 & 0 \\ 1 & n \end{bmatrix} R,$$

i.e., J is a principal right ideal of R.

Case 3. $M = \mathbf{Q} \oplus \{0\}$. Since $\{0\} \oplus K\mathbf{Q} \subseteq M$, $K = \{0\}$. Therefore, in this case,

$$J = \begin{bmatrix} 0 & 0 \\ 1 & 0 \end{bmatrix} R,$$

i.e., J is a principal right ideal of R.

Case 4. $M = \langle (a, b) \rangle$, for some $a, b \in \mathbf{Q}$, $a \neq 0$, $b \neq 0$. Since $\{0\} \oplus K\mathbf{Q} \subseteq M$, $(0, n) \in M$. Thus, $(0, n) = u(a, b)$ for some $u \in \mathbf{Q}$. Hence, $ua = 0$ and $n = ub$. Since $a \neq 0$, $ua = 0$ implies that $u = 0$ and hence $n = 0$. Thus, $K = \{0\}$. Therefore, in this case,

$$J = \begin{bmatrix} a & 0 \\ b & 0 \end{bmatrix} R,$$

i.e., J is a principal right ideal of R.

Case 5. $M = \mathbf{Q} \oplus \mathbf{Q}$.

Case 5a. $K \neq \{0\}$. Then $n \neq 0$. Let $\begin{bmatrix} a & 0 \\ b & c \end{bmatrix} \in J$. Then $c \in K$ and so $c = nm$ for some $m \in \mathbf{Z}$. Now

$$\begin{bmatrix} a & 0 \\ b & c \end{bmatrix} = \begin{bmatrix} 1 & 0 \\ 1 & n \end{bmatrix} \begin{bmatrix} a & 0 \\ \frac{b-a}{n} & m \end{bmatrix}.$$

Hence, $\begin{bmatrix} a & 0 \\ b & c \end{bmatrix} \in \begin{bmatrix} 1 & 0 \\ 1 & n \end{bmatrix} R$. Thus, $J \subseteq \begin{bmatrix} 1 & 0 \\ 1 & n \end{bmatrix} R$. Clearly $\begin{bmatrix} 1 & 0 \\ 1 & n \end{bmatrix} R$ $\subseteq J$. Therefore,

$$J = \begin{bmatrix} 1 & 0 \\ 1 & n \end{bmatrix} R.$$

Consequently, J is a principal right ideal of R.

Case 5b. $K = \{0\}$. Then $J = \begin{bmatrix} \mathbf{Q} & 0 \\ \mathbf{Q} & 0 \end{bmatrix}$. In this case, J is generated by

$\begin{bmatrix} 1 & 0 \\ 0 & 0 \end{bmatrix}, \begin{bmatrix} 0 & 0 \\ 1 & 0 \end{bmatrix}$, i.e., J is finitely generated.

(ii) Let $I = \begin{bmatrix} 0 & 0 \\ \mathbf{Q} & 0 \end{bmatrix}$. Then I is an ideal of R. Now $\mathbf{Q}$ is a left $\mathbf{Z}$-module.
Let $A_k = \langle \frac{1}{2^k} \rangle$, k is a positive integer. Then A_k is a left $\mathbf{Z}$-submodule of $\mathbf{Q}$.
Since $\frac{1}{2^{k+1}} \notin A_k$, $A_k \subset A_{k+1}$. Thus,

$$A_1 \subset A_2 \subset \cdots \subset A_k \subset A_{k+1} \subset \cdots$$

is a strictly ascending sequence of left $\mathbf{Z}$-submodules of $\mathbf{Q}$. Hence, $\mathbf{Q}$ is not a
Noetherian left $\mathbf{Z}$-module. Therefore, by Theorem 20.2.3, I is not a Noetherian
left R-module. Thus, R is not left Noetherian.
By (i), if J is a right ideal of R, then J is a finitely generated. Since every
right ideal of R is finitely generated, R is right Noetherian.

20.2.2   Exercises

1. Consider the ring $R = \begin{bmatrix} \mathbf{Z} & 0 \\ \mathbf{Q} & \mathbf{Q} \end{bmatrix}$.

   (i) Find all left ideals of R.

   (ii) Show that R is left Noetherian, but not right Noetherian.

2. Consider the ring $R = \begin{bmatrix} \mathbf{Q} & 0 \\ \mathbf{R} & \mathbf{R} \end{bmatrix}$.

   (i) Find all left ideals of R.

   (ii) Show that R is left Artinian, but not right Artinian.

Chapter 21

Field Extensions

In this chapter, we study a special type of ring called a field. Results about fields have applications in number theory and the theory of equations. The theory of equations deals with roots of polynomials. It is here that our main interest lies. This interest leads us to an introduction of Galois theory.

The importance of the concept of a field was first recognized by Abel and Galois in their research on the solution of equations by radicals. However, the formal definition of a field appeared more than 70 years later. The works of Dedekind and Kronecker seem to be responsible for the entrance of the concept of a field into mathematics. However, in 1910, in his paper, *Algebraic Theorie der Köperer*, Steinitz gave the first abstract definition of a field. His work freed the concept of a field from the context of complex numbers.

21.1   Algebraic Extensions

Let us recall that the characteristic of a field F is either 0 or a prime p. By Theorem 11.1.9, the intersection of any collection of subfields of a field F is again a subfield of F. Hence, a field contains a subfield which has no proper subfield, namely, the intersection of all its subfields.

Definition 21.1.1 *A field F is called a **prime field** if F has no proper subfield.*

Theorem 21.1.2 *Let F be a field.*

(i) If the characteristic of F is 0, then F contains a subfield K such that $K \simeq \mathbf{Q}$.

(ii) If the characteristic of F is $p > 0$, then F contains a subfield K such that $K \simeq \mathbf{Z}_p$.

Proof. Define $f : \mathbf{Z} \to F$ by

$$f(n) = n1$$

for all $n \in \mathbf{Z}$, where 1 denotes the identity of F. Then f is a homomorphism.

(i) Suppose the characteristic of F is 0. Then Ker $f = \{0\}$ and so f is one-one. Define $f^* : \mathbf{Q} \to F$ by

$$f^*\left(\frac{a}{b}\right) = f(a)f(b)^{-1}$$

for all $\frac{a}{b} \in \mathbf{Q}$. Let $\frac{a}{b}, \frac{c}{d} \in \mathbf{Q}$. Now $\frac{a}{b} = \frac{c}{d}$ if and only if $ad = bc$ if and only if $f(ad) = f(bc)$ if and only if $f(a)f(d) = f(c)f(b)$ if and only if $f(a)f(b)^{-1} = f(c)f(d)^{-1}$ if and only if $f^*(\frac{a}{b}) = f^*(\frac{c}{d})$. Hence, f^* is a one-one function. Now

$$
\begin{aligned}
f^*\left(\frac{a}{b} + \frac{c}{d}\right) &= f^*\left(\frac{ad+bc}{bd}\right) \\
&= f(ad + bc)f(bd)^{-1} \\
&= (f(a)f(d) + f(b)f(c))f(b)^{-1}f(d)^{-1} \\
&= f(a)f(b)^{-1} + f(c)f(d)^{-1} \\
&= f^*\left(\frac{a}{b}\right) + f^*\left(\frac{c}{d}\right).
\end{aligned}
$$

Also,

$$
\begin{aligned}
f^*\left(\frac{a}{b} \cdot \frac{c}{d}\right) &= f^*\left(\frac{ac}{bd}\right) \\
&= f(ac)f(bd)^{-1} \\
&= f(a)f(c)f(b)^{-1}f(d)^{-1} \\
&= f(a)f(b)^{-1}f(c)f(d)^{-1} \\
&= f^*\left(\frac{a}{b}\right)f^*\left(\frac{c}{d}\right).
\end{aligned}
$$

Thus, f^* is a homomorphism. Hence, $\mathbf{Q} \simeq \mathcal{I}(f^*)$, where $\mathcal{I}(f^*)$ is the image of f^*. Let $K = \mathcal{I}(f^*)$.

(ii) Suppose the characteristic of F is $p > 0$. Now

$$\mathbf{Z}/\text{Ker } f \simeq \mathcal{I}(f).$$

Since the characteristic of F is not zero, $\mathcal{I}(f) \neq \{0\}$. Therefore, $\mathcal{I}(f)$ is a nontrivial subring with 1 of the field F. Consequently, $\mathcal{I}(f)$ is an integral domain and so $\mathbf{Z}/\text{Ker } f$ is an integral domain. This implies Ker f is a prime ideal of $\mathbf{Z}$ and $\mathbf{Z} \neq \text{Ker } f$. There exists a prime q such that Ker $f = q\mathbf{Z}$. Now $q1 = 0$ implies that $p|q$ and so $q = p$. Hence, $\mathbf{Z}/\text{Ker } f \simeq \mathbf{Z}_p$. ∎

Let L be a subfield of $\mathbf{Q}$. Since $L \backslash \{0\}$ is a subgroup of $\mathbf{Q} \backslash \{0\}$ under multiplication, $1 \in L$. Hence, $\mathbf{Z} \subseteq L$ and so $\mathbf{Q} \subseteq L$. Thus, $\mathbf{Q}$ has no proper subfield. Similarly, $\mathbf{Z}_p$ has no proper subfield, where p is a prime.

Thus, the subfield K of the field F in Theorem 21.1.2 is the prime subfield of F.

The following theorem can be easily verified. We leave its proof as an exercise.

Theorem 21.1.3 *Let F be a field and K be a subfield of F. The following conditions are equivalent.*

(i) K is the prime subfield of F.

(ii) K is the intersection of all subfields of F. ∎

Let F be a field and K a subfield of F. The field F is called an **extension** of the field K. We express this by F/K and call F/K a **field extension** or an **extension field**.

Definition 21.1.4 *Let F/K be a field extension and C be a subset of F. Define $K(C)$ to be the intersection of all subfields of F which contain $K \cup C$. Then the subfield $K(C)$ of F is called the subfield of F **generated** by C over K. C is called a set of **generators** for $K(C)/K$.*

Let $K[C]$ be the smallest subring of F containing $K \cup C$. Since any subfield of F which contains $K \cup C$ must contain $K[C]$, we have that $K(C)$ equals the intersection of all subfields which contain $K[C]$. Now $K[C]$ is an integral domain since it is a subring (with identity) of a field. Thus, by Theorem 12.1.6,

$$K(C) = \{ab^{-1} \mid a, b \in K[C], \ b \neq 0\}.$$

That is, $K(C)$ is the set of all rational expressions of the elements of $K[C]$. Hence, $K(C)$ is a quotient field of $K[C]$.

Let F/K be a field extension and $c_1, c_2, \ldots, c_n \in F$. Considering Definition 21.1.4, it follows that $K(c_1, c_2, \ldots, c_n) = K(c_1, c_2, \ldots, c_{n-1})(c_n)$. Recall that $K(c_1) = \{ab^{-1} \mid a, b \in K[c_1], \ b \neq 0\}$.

Definition 21.1.5 *Let F/K be a field extension. An element $a \in F$ is said to be **algebraic** over K if there exist $k_0, k_1, \ldots, k_n \in K$, not all zero, such that $k_0 + k_1 a + \cdots + k_n a^n = 0$; otherwise a is called **transcendental** over K.*

Let F/K be a field extension and let $a \in F$. Then a is algebraic over K if and only if a is a root of a nonzero polynomial with coefficients from K.

Example 21.1.6 *The element $\sqrt{2}$ in $\mathbf{R}$ is algebraic over $\mathbf{Q}$ since $\sqrt{2}$ is a root of $x^2 - 2 \in \mathbf{Q}[x]$. The element $i \in \mathbf{C}$ is algebraic over $\mathbf{R}$ and $\mathbf{Q}$ since i is a root of $x^2 + 1 \in \mathbf{Q}[x]$.*

Example 21.1.7 *It can be shown that $\pi, e \in \mathbf{R}$ are transcendental over $\mathbf{Q}$. In the quotient field $F(x)$ of the polynomial ring $F[x]$, F a field, x is transcendental over F since $\sum_{i=0}^{n} a_i x^i = 0$ if and only if $a_i = 0$ for $i = 0, 1, \ldots, n$.*

Theorem 21.1.8 *Let F/K be a field extension and $c \in F$. Then c is algebraic over K if and only if c is a root of some unique irreducible monic polynomial $p(x)$ over K.*

Proof. Suppose c is algebraic over K. There exists a nonzero polynomial $f(x) \in K[x]$ such that c is a root of $f(x)$ and $f(x) \notin K$. By Theorem 16.1.15, there exist irreducible polynomials $f_1(x), f_2(x), \ldots, f_m(x) \in K[x]$ such that $f(x) = f_1(x)f_2(x) \cdots f_m(x)$. Thus,

$$0 = f(c) = f_1(c)f_2(c) \cdots f_m(c).$$

Since F has no zero divisors, we must have $f_i(c) = 0$ for some i. Thus, there exists an irreducible polynomial $h(x) = b_0 + b_1 x + \cdots + b_m x^m$, $b_m \neq 0$, such that $h(c) = 0$. Let $p(x) = b_m^{-1} h(x)$. Then $p(x)$ is an irreducible monic polynomial in $K[x]$ with c as a root.

Let $g(x)$ be any polynomial in $K[x]$, which has c as a root. Let $p(x)$ be a monic polynomial of smallest degree in $K[x]$, which has c as a root. There exist $q(x), r(x) \in K[x]$ such that $g(x) = q(x)p(x) + r(x)$, where either $r(x) = 0$ or $\deg r(x) < \deg p(x)$. Now

$$0 = g(c) = q(c)p(c) + r(c) = q(c) \cdot 0 + r(c).$$

Thus, $r(c) = 0$, whence $r(x) = 0$ else we contradict the minimality of the degree of $p(x)$. This implies that $p(x)|g(x)$ in $K[x]$. Let $s(x)$ be any irreducible polynomial in $K[x]$, which has c as a root (one such polynomial is $f_i(x)$ for some i, $1 \leq i \leq m$). Then $p(x)|s(x)$. Now $p(x)$ is not a constant polynomial in $K[x]$ since it has c as a root. Thus, since $s(x)$ is irreducible in $K[x]$, $p(x)$ must be irreducible in $K[x]$. Also, $p(x) = ks(x)$ for some $k \in K$. If we choose $s(x)$ monic, then $k = 1$ and so we have the desired uniqueness property of $p(x)$. The converse is immediate. ■

The proof of Theorem 21.1.8 yields the next result.

Corollary 21.1.9 *Let F/K be a field extension and $c \in F$ be such that c is algebraic over K. Then the unique monic irreducible polynomial $p(x)$ over K having c as a root satisfies the following properties:*

(i) There is no polynomial $g(x) \in K[x]$ having smaller degree than $p(x)$ and which has c as a root.

(ii) If c is a root of some $g(x) \in K[x]$, then $p(x)|g(x)$ in $K[x]$. ■

We call the polynomial $p(x)$, in Corollary 21.1.9, the **minimal polynomial** of c over K. The degree of $p(x)$ is called the **degree** of c over K

Example 21.1.10 *By Examples 21.1.6, 15.3.6, and 15.3.7, we have that $x^2 - 2$ is the minimal polynomial of $\sqrt{2}$ over $\mathbf{Q}$ and $x^2 + 1$ is the minimal polynomial of i over $\mathbf{R}$.*

Theorem 21.1.11 *Let F/K be a field extension and $c \in F$.*

(i) If c is transcendental over K, then $K(c) \simeq K(x)$, where $K(x)$ is the quotient field of the polynomial ring $K[x]$.

(ii) If c is algebraic over K, then $K[c] \simeq K[x]/\langle p(x) \rangle$, where $p(x)$ is the minimal polynomial of c over K.

Proof.   Define the mapping $\alpha : K[x] \longrightarrow K[c]$ by for all $f(x) \in K[x]$,

$$\alpha(f(x)) = f(c).$$

Then by Theorem 14.1.14, α is a homomorphism of $K[x]$ onto $K[c]$. Thus,

$$K[x]/\text{Ker } \alpha \simeq K[c].$$

(i) Now $f(x) \in \text{Ker } \alpha$ if and only if $f(c) = 0$, i.e., if and only if c is a root of $f(x)$. Hence, $\text{Ker } \alpha = \{0\}$ if and only if c is transcendental over K. Thus, c is transcendental over K implies α is an isomorphism of $K[x]$ onto $K[c]$ and so by Exercise 5 (page 324), α can be extended to an isomorphism of $K(x)$ onto $K(c)$. Consequently, if c is transcendental over K, then $K(x) \simeq K(c)$.

(ii) Suppose c is algebraic over K. Since $K[x]$ is a principal ideal domain, there exists $g(x) \in K[x]$ such that $\text{Ker } \alpha = \langle g(x) \rangle$. Now $\alpha(g(x)) = g(c) = 0$. Hence, c is a root of $g(x)$. Thus, $p(x)|g(x)$ and so there exists $q(x) \in K[x]$ such that $g(x) = q(x)p(x)$. This implies that $g(x) \in \langle p(x) \rangle$ and so

$$\text{Ker } \alpha = \langle g(x) \rangle \subseteq \langle p(x) \rangle .$$

Since $p(c) = 0$, $p(x) \in \text{Ker } \alpha$. Therefore, $\langle p(x) \rangle \subseteq \text{Ker } \alpha$. Consequently, $\text{Ker } \alpha = \langle p(x) \rangle$. $\blacksquare$

Corollary 21.1.12 *Let F/K be a field extension and $c \in F$. Then*

(i) $K[c] \subset K(c)$ if and only if c is transcendental over K,

(ii) $K[c] = K(c)$ if and only if c is algebraic over K.

Proof.   Since $K[c] \subseteq K(c)$ always holds, (i) and (ii) are equivalent statements. Hence, we show that (ii) holds. Suppose c is algebraic over K. Then by Theorem 21.1.11,

$$K[c] \simeq K[x]/\langle p(x) \rangle$$

and since $p(x)$ is irreducible, $K[x]/\langle p(x) \rangle$ is a field. Thus, $K[c] = K(c)$. Conversely, suppose $K[c] = K(c)$. If $c = 0$, then c is the root of the polynomial $x \in K[x]$. Suppose that $c \neq 0$. Then $c^{-1} \in K(c)$ and so $c^{-1} = k_0 + k_1 c + \cdots + k_n c^n$ for some $k_i \in K$. This implies that $0 = -1 + k_0 c + k_1 c^2 + \cdots + k_n c^{n+1}$ and so c is algebraic over K. $\blacksquare$

Let F/K be a field extension. Under the field operations of F, F can be considered as a vector space over K. The elements of F are thought of as "vectors" while those of K are thought of as "scalars." Recall that $(F, +)$ is a commutative group and that for all $k_1, k_2 \in K$ and $a_1, a_2 \in F$, $k_1(a_1 + a_2) = k_1a_1 + k_1a_2$, $(k_1 + k_2)a_1 = k_1a_1 + k_2a_1$ hold from the distributive laws and that $(k_1k_2)a_1 = k_1(k_2a_1)$ holds from the associative law of multiplication.

Definition 21.1.13 *Let F/K be a field extension. The dimension of the vector space F over K is called the* **degree** *or* **dimension** *of F/K and is denoted by $[F : K]$. If the dimension of F/K is finite, then F/K is called a* **finite extension**.

Theorem 21.1.14 *Let F/K be a field extension and $c \in F$ be algebraic over K. Let $p(x)$ be the minimal polynomial of c over K. If $\deg p(x) = n$, then $\{1, c, c^2, \ldots, c^{n-1}\}$ is a basis of $K(c)/K$.*

Proof.   By Corollary 21.1.12, $K[c] = K(c)$. Let $g(c) \in K[c]$ and $g(x)$ be the corresponding element in $K[x]$. There exist $q(x), r(x) \in K[x]$ such that $g(x) = q(x)p(x) + r(x)$, where either $r(x) = 0$ or $\deg r(x) < \deg p(x)$. Thus, $g(c) = q(c)p(c) + r(c) = r(c)$. Hence, $\{1, c, c^2, \ldots, c^{n-1}\}$ spans $K(c)/K$. Suppose $0 = \sum_{i=0}^{n-1} k_i c^i$, $k_i \in K$. If the k_i's are not all zero, then c is a root of a polynomial of degree $\leq n - 1 < n$, a contradiction. Thus, $k_i = 0$ for $i = 0, 1, \ldots, n - 1$ and so $\{1, c, c^2, \ldots, c^{n-1}\}$ is linearly independent over K. Hence, $\{1, c, c^2, \ldots, c^{n-1}\}$ is a basis of $K(c)/K$. ∎

Corollary 21.1.15 *Let F/K be a field extension. If $c \in F$ is algebraic and of degree n over K, then $[K(c) : K] = n$.* ∎

Example 21.1.16 *The field extension $\mathbf{Q}(\sqrt{2})/\mathbf{Q}$ is of degree 2 and $\{1, \sqrt{2}\}$ is a basis of $\mathbf{Q}(\sqrt{2})$ over $\mathbf{Q}$ since $p(x) = x^2 - 2$ is the minimal polynomial of $\sqrt{2}$ over $\mathbf{Q}$ by Example 21.1.10. Thus, $\mathbf{Q}(\sqrt{2}) = \{a + b\sqrt{2} \mid a, b \in \mathbf{Q}\}$.*

The student may recall from another mathematics course that $a + b\sqrt{2} = c + d\sqrt{2}$ if and only if $a = c$ and $b = d$, where $a, b, c, d \in \mathbf{Q}$. This becomes clear now since 1 and $\sqrt{2}$ are linearly independent over $\mathbf{Q}$ by Theorem 21.1.14.

Example 21.1.17 *By Theorem 21.1.14, the field extension $\mathbf{R}(i)/\mathbf{R}$ is of degree 2 and $\{1, i\}$ is a basis of $\mathbf{R}(i)$ over $\mathbf{R}$ since $p(x) = x^2 + 1$ is the minimal polynomial of i over $\mathbf{R}$. Thus, $\mathbf{R}(i) = \{a + bi \mid a, b \in \mathbf{R}\}$. Hence, we see that $\mathbf{R}(i)$ is $\mathbf{C}$, the field of complex numbers.*

Theorem 21.1.18 *Let F/K be a finite field extension. Then every element of F is algebraic over K.*

Proof. Let n be the dimension of F/K. Let $c \in F$ be such that $c \neq 0$, $c \neq 1$. (Clearly 0 and 1 are algebraic over K.) If the set $\{1, c, c^2, \ldots, c^n\}$ does not contain $n+1$ distinct elements, then $c^{j-i} = 1$ for some i, j ($0 \leq i < j \leq n$) and so c is a root of $x^{j-i} - 1$. Suppose $1, c, c^2, \ldots, c^n$ are distinct. Then they must be linearly dependent since they are more in number than the dimension of the vector space F over K. Hence, there exist $k_0, k_1, \ldots, k_n \in K$ not all zero such that $0 = \sum_{i=0}^{n} k_i c^i$. Thus, c is a root of the polynomial $\sum_{i=0}^{n} k_i x^i$ over K. ∎

The converse of Theorem 21.1.18 is not true, that is, it is not necessarily the case that if every element of F is algebraic over K, then F/K is a finite field extension. It can be shown that the set of all elements A of $\mathbf{R}$, which are algebraic over $\mathbf{Q}$ is a field such that $[A : \mathbf{Q}]$ is infinite (Theorem 21.1.22 and Example 21.1.25). A is called the **field of algebraic** numbers.

Theorem 21.1.19 *Let $K(c)/K$ be a field extension. Then $K(c)/K$ is finite if and only if c is algebraic over K.*

Proof. If $K(c)/K$ is finite, then c is algebraic over K by Theorem 21.1.18. If c is algebraic over K, then $K(c)/K$ is finite by Corollary 21.1.15. ∎

Let F/K be a field extension. A subfield L of F is called an **intermediate field** of F/K if $K \subseteq L \subseteq F$. Since $a - b \in L$ for all $a, b \in L$ and $ka \in L$ for all $k \in K$ and $a \in L$, it follows that L is a subspace of F over K. An intermediate field L of F/K is called **proper** if $L \neq F$.

Theorem 21.1.20 *Let F/K be a field extension and L be an intermediate field of F/K. Then*

$$[F : K] = [F : L][L : K].$$

Moreover, F/K is a finite extension if and only if F/L and L/K are finite extensions.

Proof. Let V be a basis of F/L and U be a basis of L/K. We show that

$$W = \{uv \mid u \in U, v \in V\}$$

is a basis of F/K. Let $c \in F$. Since V is a basis of F/L, there exist $v_1, v_2, \ldots, v_n \in V$ and $c_1, c_2, \ldots, c_n \in L$ such that

$$c = \sum_{j=1}^{n} c_j v_j. \tag{21.1}$$

Since U is a basis of L/K, there exist $u_1, u_2, \ldots, u_m \in U$ and $k_{1j}, k_{2j}, \ldots,$ $k_{mj} \in K$ such that

$$c_j = \sum_{i=1}^{m} k_{ij} u_i, \qquad j = 1, 2, \ldots, n. \tag{21.2}$$

Substituting Eq. (21.2) into Eq. (21.1), we obtain

$$c = \sum_{j=1}^{n} \sum_{i=1}^{m} k_{ij} u_i v_j.$$

Thus, W spans F over K. Suppose

$$0 = \sum_{j=1}^{n} \sum_{i=1}^{m} k_{ij} u_i v_j,$$

where $u_i \in U$, $v_j \in V$, and $k_{ij} \in K$ for all $i = 1, 2, \ldots, m$; $j = 1, 2, \ldots, n$. Then

$$0 = \sum_{j=1}^{n} (\sum_{i=1}^{m} k_{ij} u_i) v_j$$

and since V is linearly independent over L,

$$0 = \sum_{i=1}^{m} k_{ij} u_i, \ j = 1, 2, \ldots, n.$$

Thus, $k_{ij} = 0$ for $i = 1, 2, \ldots, m$; $j = 1, 2, \ldots, n$ since U is linearly independent over K. Hence, W is linearly independent over K, whence W is a basis of F over K. Let $u, u' \in U$ and $v, v' \in V$. If $v \neq v'$, then $uv \neq u'v'$ since v and v' are linearly independent over L. If $v = v'$, then $uv = u'v'$ if and only if $u = u'$. Consequently, for all $u, u' \in U$ and for all $v, v' \in V$ if either $u \neq u'$ or $v \neq v'$, then $uv \neq u'v'$. Hence, $[F : K] = |U \times V| = |U||V| = [F : L][L : K]$. Now if either U or V is infinite, then W is infinite. If U and V are finite sets, then W is a finite set. Hence, F/K is a finite extension if and only if F/L and L/K are finite extensions. ∎

Example 21.1.21 *Consider the field extension* $\mathbf{Q}(\sqrt{2}, \sqrt{3})/\mathbf{Q}$. *By Example 21.1.10,* $x^2 - 2$ *is the minimal polynomial of* $\sqrt{2}$ *over* $\mathbf{Q}$. *Also,* $x^2 - 3$ *is the minimal polynomial of* $\sqrt{3}$ *over* $\mathbf{Q}(\sqrt{2})$. *(That* $x^2 - 3$ *is irreducible over* $\mathbf{Q}(\sqrt{2})$ *follows by an argument that is similar to the one used in Worked-Out Exercise 1, page 454.) Thus,* $\{1, \sqrt{2}\}$ *is a basis of* $\mathbf{Q}(\sqrt{2})/\mathbf{Q}$ *and* $\{1, \sqrt{3}\}$ *is a basis of* $\mathbf{Q}(\sqrt{2}, \sqrt{3})/\mathbf{Q}(\sqrt{2})$. *By Theorem 21.1.20,* $\{1, \sqrt{2}, \sqrt{3}, \sqrt{6}\}$ *is a basis of* $\mathbf{Q}(\sqrt{2}, \sqrt{3})/\mathbf{Q}$. $[\mathbf{Q}(\sqrt{2}, \sqrt{3}) : \mathbf{Q}] = 4$, $[\mathbf{Q}(\sqrt{2}, \sqrt{3}) : \mathbf{Q}(\sqrt{2})] = 2$, *and* $[\mathbf{Q}(\sqrt{2}) : \mathbf{Q}] = 2$.

Theorem 21.1.22 *Let* F/K *be a field extension. If* L *is the set of all elements in* F, *which are algebraic over* K, *then* L *is an intermediate field of* F/K.

Proof. Any $k \in K$ is a root of the polynomial $x - k$ over K. Thus, $L \supseteq K$. Let a and b be elements of L, where a is of degree m over K and b is of degree n over K. Then $K(a)/K$ is of degree m and $K(a,b)/K(a)$ is of degree at most n. Hence, by Theorem 21.1.20, $K(a,b)/K$ is a finite extension. By Theorem 21.1.18, every element of $K(a,b)$ is algebraic over K. Since $a - b$ and ab^{-1} (for $b \neq 0$) are elements of $K(a,b)$, $a - b$ and ab^{-1}(for $b \neq 0$) are algebraic over K. Thus, $a - b$ and ab^{-1} (for $b \neq 0$) $\in L$ and so L is a field. ∎

Definition 21.1.23 *A field extension F/K is called **algebraic** if every element of F is algebraic over K; otherwise F/K is called **transcendental**.*

Theorem 21.1.24 *Let L be an intermediate field of the field extension F/K. Then F/K is an algebraic extension if and only if F/L and L/K are algebraic extensions.*

Proof. Suppose that F/K is algebraic. Let $a \in F$. Then a is a root of a nonzero polynomial $p(x) \in K[x]$. Since $K \subseteq L$, $p(x) \in L[x]$. Thus, a is algebraic over L and so F/L is algebraic. Every element of L is an element of F. Hence, L/K is algebraic. Conversely, suppose F/L and L/K are algebraic extensions. Let $c \in F$. Then c is a root of some nonzero polynomial $c_0 + c_1 x + \cdots + c_n x^n \in L[x]$. Thus, c is algebraic over $K(c_0, c_1, \ldots, c_n)$ whence $K(c_0, c_1, \ldots, c_n)(c)/K(c_0, c_1, \ldots, c_n)$ is a finite extension. Since $c_0, c_1, \ldots, c_n$ are algebraic over K, repeated application of Theorem 21.1.20 yields that $K(c_0, c_1, \ldots, c_n)(c)/K$ is a finite extension. Therefore, c is algebraic over K by Theorem 21.1.18. Hence, F/K is an algebraic extension. ∎

Example 21.1.25 *Let $F = \mathbf{Q}(\{\sqrt{p} \mid p \in \mathbf{Z}, p \text{ is a prime}\}) \subseteq \mathbf{R}$. We show that $F/\mathbf{Q}$ is algebraic and $[F : \mathbf{Q}] = \infty$. Now for any prime p, $\sqrt{p} \notin \mathbf{Q}$. Let $p_1, \ldots, p_n$ be any distinct primes. Suppose $p \neq p_i$, $i = 1, 2, \ldots, n$, and p is a prime. Assume that $\sqrt{p} \notin \mathbf{Q}(\sqrt{p_1}, \ldots, \sqrt{p_n})$, the induction hypothesis. (The case $n = 0$ is $\sqrt{p} \notin \mathbf{Q}$ and this case is described above.) We show that if $p_1, \ldots, p_{n+1}$ are distinct primes and $p \neq p_i$, $i = 1, 2, \ldots, n + 1$, then $\sqrt{p} \notin \mathbf{Q}(\sqrt{p_1}, \ldots, \sqrt{p_{n+1}})$. Suppose $\sqrt{p} \in \mathbf{Q}(\sqrt{p_1}, \ldots, \sqrt{p_{n+1}})$. Then there exist $a, b \in \mathbf{Q}(\sqrt{p_1}, \ldots, \sqrt{p_n})$ such that $\sqrt{p} = a + b\sqrt{p_{n+1}}$. If $a = 0$, then $p = b^2 p_{n+1}$, a contradiction since p and p_{n+1} are distinct primes. If $b = 0$, then $\sqrt{p} = a \in \mathbf{Q}(\sqrt{p_1}, \ldots, \sqrt{p_n})$, a contradiction to our induction hypothesis. Suppose $a \neq 0$ and $b \neq 0$. Then $p = a^2 + p_{n+1}b^2 + 2ab\sqrt{p_{n+1}}$. Hence, $\sqrt{p_{n+1}} = (p - a^2 - p_{n+1}b^2)/2ab \in \mathbf{Q}(\sqrt{p_1}, \ldots, \sqrt{p_n})$ and so $\sqrt{p} \in \mathbf{Q}(\sqrt{p_1}, \ldots, \sqrt{p_n})$, a contradiction of the hypothesis. Hence, $\sqrt{p} \notin \mathbf{Q}(\sqrt{p_1}, \ldots, \sqrt{p_{n+1}})$. Thus, by the induction hypothesis, we find that for any positive integer k, if $p_1, \ldots, p_k, p$ are distinct primes, then $\sqrt{p} \notin \mathbf{Q}(\sqrt{p_1}, \ldots, \sqrt{p_k})$. Hence,*

$$\mathbf{Q} \subset \mathbf{Q}(\sqrt{2}) \subset \mathbf{Q}(\sqrt{2}, \sqrt{3}) \subset \cdots$$

is an infinite strictly ascending chain of intermediate fields of $F/\mathbf{Q}$. Hence, $F/\mathbf{Q}$ must be of infinite dimension. Let $a \in F$. Then there exist primes $p_1, \ldots,$ p_n such that $a \in \mathbf{Q}(\sqrt{p_1}, \ldots, \sqrt{p_n})$. Since $\mathbf{Q}(\sqrt{p_1}, \ldots, \sqrt{p_n})/\mathbf{Q}$ is a finite field extension, a is algebraic over $\mathbf{Q}$ by Theorem 21.1.18. Hence, $F/\mathbf{Q}$ is algebraic. Note that from this example, it follows that $[\mathbf{R} : \mathbf{Q}] = \infty$.

The above example provides us with a field extension $F/\mathbf{Q}$ which shows that the converse of Theorem 21.1.18 is not true. Since the field of algebraic numbers A contains F, we have $[A : \mathbf{Q}] = \infty$.

Definition 21.1.26 *Let F/K and L/K be field extensions and $\sigma : F \to L$ be a homomorphism. Then σ is called a K-**homomorphism** if $\sigma(a) = a$ for all $a \in K$.*

Let F/K and L/K be field extensions and $\sigma : F \to L$ be a K-homomorphism. Since σ is a nonzero homomorphism, Ker $\sigma \neq F$. Therefore, Ker $\sigma = \{0\}$ since the only ideals of F are F and $\{0\}$. This implies that σ is one-one. Hence, σ is an isomorphism of F onto $\sigma(F)$. We simply call σ a K-**isomorphism** of F into L. If $L = F = \sigma(F)$ and σ is a K-isomorphism of F into L, then we call σ a K-**automorphism**.

Theorem 21.1.27 *Let F/K be an algebraic extension and $\sigma : F \to F$ be a K-homomorphism. Then σ is an automorphism.*

Proof.   As above σ is one-one. To show σ is an automorphism, it only remains to be shown that $\sigma(F) = F$, i.e., σ is onto F.
Let $a \in F$. Let $f(x) = a_0 + a_1x + \cdots + a_kx^k \in K[x]$ be the minimal polynomial of a over K. Let b be any root of $f(x)$ in F. Then $f(\sigma(b)) = a_0 + a_1\sigma(b) + \cdots + a_k\sigma(b)^k = \sigma(a_0 + a_1b + \cdots + a_kb^k) = 0$. Hence, $\sigma(b)$ is a root of $f(x)$. Let F' be the subfield of F generated by all roots of $f(x)$ over K that lie in F. Then F'/K is a finite extension. Since σ maps a root of $f(x)$ to a root of $f(x)$, σ maps F' into F'. Since $[F' : K] = [\sigma(F') : K]$, it now follows that $[F' : \sigma(F')] = 1$ by Theorem 21.1.20 and so $F' = \sigma(F')$. Hence, $a \in F' = \sigma(F') \subseteq \sigma(F)$. Thus, σ is onto F. $\blacksquare$

21.1.1   Worked-Out Exercises

$\diamond$ **Exercise 1** Show that the polynomial $x^2 - 7$ is irreducible in $\mathbf{Q}(\sqrt{3})[x]$.

Solution: Suppose $x^2 - 7 = (x - (a + b\sqrt{3}))(x - (c + d\sqrt{3}))$, where $a, b, c, d \in \mathbf{Q}$. Then $x^2 - 7 = x^2 - ((a+c) + (b+d)\sqrt{3})x + (ac + 3bd + ad\sqrt{3} + bc\sqrt{3})$. This implies that

$$(a + c) + (b + d)\sqrt{3} = 0$$
$$ac + 3bd + ad\sqrt{3} + bc\sqrt{3} = -7.$$

Since $\{1, \sqrt{3}\}$ is linearly independent over $\mathbf{Q}$, $a + c = 0$ and $b + d = 0$. Hence,

$$-a^2 - 3b^2 + (-2ab)\sqrt{3} = -7.$$

Thus, $-a^2 - 3b^2 = -7$ and $-2ab = 0$. Hence, $ab = 0$. Suppose $a = 0$. Then $3b^2 = 7$. Now $b = \frac{m}{n}$ for some integers m and n with $\gcd(m, n) = 1$. Therefore, $3m^2 = 7n^2$, which contradicts the fundamental theorem of arithmetic. Suppose $b = 0$. Then $a^2 = 7$, which again leads to a contradiction of the fundamental theorem of arithmetic. Thus, $x^2 - 7$ is irreducible in $\mathbf{Q}(\sqrt{3})[x]$.

$\Diamond$ **Exercise 2** Find $[\mathbf{Q}(\sqrt{3}, \sqrt{7}) : \mathbf{Q}(\sqrt{3})]$ and $[\mathbf{Q}(\sqrt{3}) : \mathbf{Q}]$. Also, find a basis for $\mathbf{Q}(\sqrt{3}, \sqrt{7})/\mathbf{Q}(\sqrt{3})$ and a basis for $\mathbf{Q}(\sqrt{3}, \sqrt{7})/\mathbf{Q}$.

Solution: By Worked-Out Exercise 1 (page 454), $x^2 - 7$ is irreducible over $\mathbf{Q}(\sqrt{3})$. Thus,

$$[\mathbf{Q}(\sqrt{3}, \sqrt{7}) : \mathbf{Q}(\sqrt{3})] = \deg(x^2 - 7) = 2.$$

By Theorem 21.1.14, $\{1, \sqrt{7}\}$ is a basis for $\mathbf{Q}(\sqrt{3}, \sqrt{7})/\mathbf{Q}(\sqrt{3})$. Since $x^2 - 3$ is irreducible over $\mathbf{Q}$, $[\mathbf{Q}(\sqrt{3}) : \mathbf{Q}] = 2$ and $\{1, \sqrt{3}\}$ is a basis for $\mathbf{Q}(\sqrt{3})/\mathbf{Q}$. Thus,

$$[\mathbf{Q}(\sqrt{3}, \sqrt{7}) : \mathbf{Q}] = [\mathbf{Q}(\sqrt{3}, \sqrt{7}) : \mathbf{Q}(\sqrt{3})][\mathbf{Q}(\sqrt{3}) : \mathbf{Q}] = 2 \cdot 2 = 4.$$

By Theorem 21.1.20, $\{1, \sqrt{3}, \sqrt{7}, \sqrt{21}\}$ is a basis of $\mathbf{Q}(\sqrt{3}, \sqrt{7})/\mathbf{Q}$.

$\Diamond$ **Exercise 3** Find an element $u \in \mathbf{R}$ such that $\mathbf{Q}(\sqrt{2}, \sqrt[3]{7}) = \mathbf{Q}(u)$.

Solution: We claim that $u = \sqrt{2}\sqrt[3]{7}$. Since $u = \sqrt{2}\sqrt[3]{7} \in \mathbf{Q}(\sqrt{2}, \sqrt[3]{7})$, $\mathbf{Q}(u) \subseteq \mathbf{Q}(\sqrt{2}, \sqrt[3]{7})$. Now $\sqrt{2}\sqrt[3]{7} \in \mathbf{Q}(u)$ implies that $14\sqrt{2} = (\sqrt{2}\sqrt[3]{7})^3 \in \mathbf{Q}(u)$. Hence, $\sqrt{2} \in \mathbf{Q}(u)$. Since $\sqrt{2}, \sqrt{2}\sqrt[3]{7} \in \mathbf{Q}(u)$, $\sqrt[3]{7} \in \mathbf{Q}(u)$. Therefore, $\mathbf{Q}(\sqrt{2}, \sqrt[3]{7}) \subseteq \mathbf{Q}(u)$. Thus, $\mathbf{Q}(\sqrt{2}, \sqrt[3]{7}) = \mathbf{Q}(u)$.

$\Diamond$ **Exercise 4** (i) Let F be a field and a, b be members of a field containing F. Suppose that a and b are algebraic of degree m and n over F, respectively. Suppose m and n are relatively prime. Show that $[F(a, b) : F] = mn$.

(ii) Show that the result in (i) need not be true if m and n are not relatively prime.

Solution: (i) Let $f(x) \in F[x]$ be the minimal polynomial of a of degree m. Now $f(x) \in F[x] \subseteq F(b)[x]$. Thus, a satisfies a polynomial of degree m over $F(b)$. Hence, $[F(b)(a) : F(b)] \leq m$. Since $F(b)(a) = F(a, b)$, $[F(a, b) : F(b)] \leq m$. Now $[F(a, b) : F] = [F(a, b) : F(b)][F(b) : F] \leq mn$. Also,

$$[F(a, b) : F] = [F(a, b) : F(b)][F(b) : F] = [F(a, b) : F(b)]n.$$

Thus, $n | [F(a, b) : F]$. Similarly, $m | [F(a, b) : F]$. Since m and n are relatively prime, $mn | [F(a, b) : F]$. Therefore, $[F(a, b) : F] \geq mn$. Consequently, $[F(a, b) : F] = mn$.

(ii) Let $F = \mathbf{Q}$, $a = 2^{\frac{1}{6}}$, and $b = 2^{\frac{1}{4}}$. Then a is algebraic over F of degree 6 and b is algebraic over F of degree 4. We claim that $F(a, b) = F(2^{\frac{1}{12}})$. Now $b = (2^{\frac{1}{12}})^3 \in F(2^{\frac{1}{12}})$ and $a = (2^{\frac{1}{12}})^2 \in F(2^{\frac{1}{12}})$. Thus, $F(a, b) \subseteq F(2^{\frac{1}{12}})$. Now $2^{\frac{1}{12}} = 2^{\frac{1}{4} - \frac{1}{6}} = 2^{\frac{1}{4}}(2^{\frac{1}{6}})^{-1} \in F(a, b)$. Hence, $F(a, b) = F(2^{\frac{1}{12}})$. Since $x^{12} - 2$ is the minimal polynomial of $2^{\frac{1}{12}}$, $[F(2^{\frac{1}{12}}) : F] = 12 \neq 24 = 4 \cdot 6$.

◇ **Exercise 5** Consider the unique factorization domain $F[t]$, where F is a field and t is transcendental over F. Show that the polynomial $x^2 + tx + t \in F(t)[x]$ is irreducible over $F(t)$. Also, show that $x^2 + tx + t \in F(x)[t]$ is irreducible over $F(x)$.

Solution: Now $t \nmid 1, t|t$, but $t^2 \nmid t$. Note t is prime in $F[t]$. Thus, $x^2 + tx + t \in F(t)[x]$ is irreducible over $F(t)$ by Eisenstein's criterion. If we consider $x^2 + tx + t$ as a polynomial in t over $F(x)$, then $x^2 + tx + t = (x + 1)t + x^2$. It follows that Eisenstein's criterion does not apply. However, since $(x + 1)t + x^2$ is of degree 1 in t, it is irreducible over $F(x)$.

Exercise 6 Let $K[u, v]$ denote the polynomial ring in two algebraic indepen-dent indeterminates u, v over the field K. Let F denote the field of quo-tients $K(u, v)$ of $K[u, v]$. Prove that the polynomial $x^2 + vx + u$ is irre-ducible over F.

Solution: Suppose $x^2 + vx + u$ is reducible over F. Then

$$x^2 + vx + u = \left(x + \frac{p(u, v)}{q(u, v)}\right)\left(x + \frac{f(u, v)}{g(u, v)}\right),$$

where $p(u, v), q(u, v), f(u, v), g(u, v) \in K[u, v]$. We may assume that $p(u, v)$ and $q(u, v)$ are relatively prime in $K[u, v]$ and also $f(u, v)$ and $g(u, v)$ are relatively prime in $K[u, v]$. Now

$$uq(u, v)g(u, v) = p(u, v)f(u, v). \tag{21.3}$$

Hence, $g(u, v)$ divides $p(u, v)$, $p(u, v)$ divides $ug(u, v)$, $q(u, v)$ divides $f(u, v)$, and $f(u, v)$ divides $uq(u, v)$. Also,

$$v = \frac{p(u, v)}{q(u, v)} + \frac{f(u, v)}{g(u, v)}.$$

Consequently,

$$vq(u, v)g(u, v) = p(u, v)g(u, v) + q(u, v)f(u, v). \tag{21.4}$$

Therefore, $g(u, v)$ divides $q(u, v)$ and $q(u, v)$ divides $g(u, v)$. Thus,

$$g(u, v) = kq(u, v)$$

for some $k \in K$. Hence, $g(u, v)$ and $p(u, v)$ are relatively prime. Similarly, $q(u, v)$ and $f(u, v)$ are relatively prime. Thus, $p(u, v)$ divides u and $f(u, v)$ divides u by Eq. (21.3). Hence,

$$\text{either } p(u, v) = k_1 u \text{ or } p(u, v) = k_1, \tag{21.5}$$

$$\text{either } f(u, v) = k_2 u \text{ or } f(u, v) = k_2 \tag{21.6}$$

for some $k_1, k_2 \in K$. Suppose that $p(u, v) = k_1 u$ and $f(u, v) = k_2 u$. Then substituting into Eq. (21.4) we obtain

$$vq(u, v)g(u, v) = k_1 ug(u, v) + k_2 uq(u, v).$$

Thus,

$$vq(u, v)g(u, v) = k_1 ukq(u, v) + k_2 uq(u, v).$$

Hence, $vg(u, v) = (k_1 k + k_2)u$. However, this contradicts the algebraic independence of u, v over K. Substituting the remaining possibilities in Eqs. (21.5) and (21.6) into Eq. (21.4), we also obtain a contradiction of the algebraic independence of u, v over K. Thus, $x^2 + vx + u$ is irreducible over F.

Exercise 7 Let $F = K(x, y)$, where K is a field and x, y are algebraically independent indeterminates over K. Show that $F \neq K(x)K(y)$, where

$$K(x)K(y) = \{\textstyle\sum_i (p_i(x)/q_i(x))(u_i(y)/v_i(y)) \mid p_i(x), q_i(x) \in K[x],$$
$$u_i(y), v_i(y) \in K[y], q_i(x) \neq 0, v_i(y) \neq 0\}.$$

Solution: Now $\frac{1}{x+y} \notin K(x)K(y)$ else $\frac{1}{x+y} = (\sum_i (f_i(x)g_i(y)))/h(x)k(y)$, after obtaining a common denominator. Thus,

$$h(x)k(y) = (x + y)(\sum_i (f_i(x)g_i(y))).$$

This implies that $x+y$ divides $h(x)k(y)$. Hence, $x+y$ divides $h(x)$ or $k(y)$ since $x+y$ is prime in the UFD $K[x, y]$, a contradiction of the algebraic independence of x, y over K.

21.1.2   Exercises

1. Show that $\mathbf{Q}(\sqrt{3}, -\sqrt{3}) = \mathbf{Q}(\sqrt{3})$.

2. Let F/K be a field extension. Show that $[F : K] = 1$ if and only if $F = K$.

3. Consider the field extension $\mathbf{R}/\mathbf{Q}$.

   (i) Show that π^2 is transcendental over $\mathbf{Q}$.

   (ii) Show that $\sqrt{\pi}$ is transcendental over $\mathbf{Q}$.

4. Consider the field extension $\mathbf{R}/\mathbf{Q}$. Show that $\pi - 3$ is transcendental over $\mathbf{Q}$.

5. Consider the field extension $\mathbf{R}/\mathbf{Q}$. Show that π is transcendental over $\mathbf{Q}(\sqrt{2})$.

6. Consider the field extension $\mathbf{R}/\mathbf{Q}$. Show that $\pi + \sqrt{2}$ is transcendental over $\mathbf{Q}$.

7. Let F/K be a field extension such that $[F : K] < \infty$. Let $p(x)$ be an irreducible polynomial in $K[x]$. Suppose $p(c) = 0$ for some $c \in F$. Prove that $\deg p(x)$ divides $[F : K]$.

8. Find $[\mathbf{Q}(\sqrt[3]{5}) : \mathbf{Q}]$.

9. Show that $\mathbf{Q}(\sqrt{3} - \sqrt{5}) = \mathbf{Q}(\sqrt{3}, \sqrt{5})$. Find $[\mathbf{Q}(\sqrt{3} - \sqrt{5}) : \mathbf{Q}]$.

10. Show that the polynomial $x^2 - 5$ is irreducible over $\mathbf{Q}(\sqrt{2})$.

11. Find the minimal polynomial of $\sqrt{2 + \sqrt{5}}$ over $\mathbf{Q}$.

12. Let $c = \sqrt[5]{3}$. Show that $\mathbf{Q}(c) = \mathbf{Q}(c^2)$.

13. Find $[\mathbf{Q}(\sqrt{2}, \sqrt{5}) : \mathbf{Q}(\sqrt{2})]$, $[\mathbf{Q}(\sqrt{2}, \sqrt{5}) : \mathbf{Q}]$, a basis for $\mathbf{Q}(\sqrt{2}, \sqrt{5})/\mathbf{Q}(\sqrt{2})$, and a basis for $\mathbf{Q}(\sqrt{2}, \sqrt{5})/\mathbf{Q}$.

14. Let F/K be a field extension and $c \in F$ be algebraic over K. Let $f(x) \in K[x]$. Show that $f(c)$ is algebraic over K.

15. Prove that if $[F : K] = p$, p a prime, then F/K has no proper intermediate fields.

16. Let L and M be intermediate fields of the field extension F/K. Suppose that $[L : K]$ is a prime. Prove that either $L \cap M = K$ or $L \subseteq M$.

17. Let F/K be a field extension, $f(x)$ be a nonzero polynomial in $K[x]$, and $c \in F$. If $f(x)$ is algebraic over K, prove that c is algebraic over K.

18. Let F/K be a field extension such that $[F : K] = p$, p a prime. Prove that if $c \in F$, $c \notin K$, then $F = K(c)$.

19. Let F/K be a field extension and $a, b \in F$ be algebraic over K. If a has degree m over K and $b \neq 0$ has degree n over K, prove that the elements $a + b$, ab, $a - b$, ab^{-1} have degree at most mn over K.

20. Prove that $\sqrt{2} + \sqrt{3}$, $\sqrt{2} - \sqrt{3}$ have degree 4 over $\mathbf{Q}$ and that $\sqrt{2}\sqrt{3}$, $\sqrt{2}/\sqrt{3}$ have degree 2 over $\mathbf{Q}$. Find the minimal polynomials of these elements over $\mathbf{Q}$.

21. Let F/K be a field extension and R be a ring such that $K \subseteq R \subseteq F$. Prove that if every element of R is algebraic over K, then R is a field.

22. Let F/K be a field extension and $u, v \in F$.

    (i) Prove that $K(u, u + v) = K(u, v)$.

    (ii) If u and $u + v$ are algebraic over K, prove that $[K(u, v) : K]$ is finite and v is algebraic over K.

23. Answer the following statements true or false. If the statement is true, prove it. If it is false, give a counterexample.

    (i) Let F/K be a field extension and L be an intermediate field of F/K. Let V be a basis of F/L such that $1 \in V$ and U be a basis of L/K such that $1 \in U$. Then $U \cup V$ is linearly independent over K.

    (ii) Let F/K be a field extension and L be an intermediate field of F/K. Let V be a basis of F/L and U be a basis of L/K. Then $U \cup V$ is a basis of F/K.

    (iii) Let F/K be a field extension and $c, d \in F$. If $K(c, d) = K(c)$, then $d = f(c)$ for some polynomial $f(x) \in K[x]$.

21.2 Splitting Fields

Here we give some results concerning the existence of field extensions which are generated by roots of polynomials. These results are basic to Galois theory.

Consider the polynomial ring $K[x]$ over the field K. Let $f(x) \in K[x]$. In the quotient ring $K[x]/\langle f(x) \rangle$, we let $\overline{g(x)}$ denote the coset $g(x) + \langle f(x) \rangle$. Thus, if $g(x) = \sum_{i=0}^{n} k_i x^i$, then by the definition of addition and multiplication of cosets, we have that $\overline{g(x)} = \sum_{i=0}^{n} \overline{k_i} \overline{x}^i$.

Theorem 21.2.1 (Kronecker) *Let K be a field. If $f(x)$ is a nonconstant polynomial in $K[x]$, then there exists a field extension F/K such that F contains a root of $f(x)$.*

Proof. Since $K[x]$ is a unique factorization domain, there exist irreducible polynomials $f_1(x), \ldots, f_n(x) \in K[x]$ such that $f(x) = f_1(x) \cdots f_n(x)$. Thus, a root of any $f_i(x)$, $i = 1, 2, \ldots, n$, is a root of $f(x)$. Hence, it suffices to prove the theorem for $f(x)$ irreducible in $K[x]$. The ideal $\langle f(x) \rangle$ is maximal in $K[x]$ and so $F = K[x]/\langle f(x) \rangle$ is a field. Let α be the natural homomorphism of $K[x]$ onto $K[x]/\langle f(x) \rangle$. Since $K \cap \langle f(x) \rangle = \{0\}$, α maps K one-one into F. Thus, say, $K \subseteq F$, that is, we identify $k \in K$ with $\overline{k}$ in F. Hence, $\alpha(f(x)) = \overline{f(x)} = f(\overline{x})$, where $\overline{f(x)} = f(x) + \langle f(x) \rangle$ and $\overline{x} = x + \langle f(x) \rangle$. Now $\alpha(f(x)) = \overline{0}$ and so $f(\overline{x}) = \overline{0}$. Therefore, $\overline{x}$ is a root of $f(x)$. ∎

The field extension F/K in Theorem 21.2.1 has some interesting properties. Consider the subring $K[\bar{x}]$ of F. Then $\alpha(\sum_{i=0}^{m} k_i x^i) = \sum_{i=0}^{m} k_i \bar{x}^i$ for all $\sum_{i=0}^{m} k_i x^i \in K[x]$ and so α maps $K[x]$ onto $K[\bar{x}]$. Since α also maps $K[x]$ onto F, we have $F = K[\bar{x}] = K(\bar{x})$. Thus, for $f(x)$ irreducible in $K[x]$, we have by Theorem 21.1.14 that $[F : K] = n$ and $\{1, \bar{x}, \ldots, \bar{x}^{n-1}\}$ is a basis of F/K, where $n = \deg f(x)$.

Example 21.2.2 $x^2 + 1$ *is irreducible in* $\mathbf{R}[x]$. *Now* $\mathcal{C} = \mathbf{R}/\langle x^2 + 1 \rangle = \mathbf{R}[\bar{x}] = \{a + b\bar{x} \mid a,\, b \in \mathbf{R}\}$ *is a field, where* $\bar{x} = x + \langle x^2 + 1 \rangle$. *Since* $\bar{x}^2 = -1$, *we may call* $\mathcal{C}$ *the field of complex numbers. We may think of* $\bar{x}$ *as* i.

Example 21.2.3 *Consider the polynomial* $x^4 - 3 \in \mathbf{Q}[x]$. *By Eisenstein's criterion,* $x^4 - 3$ *is irreducible in* $\mathbf{Q}[x]$. *Set* $\lambda = x + \langle x^4 - 3 \rangle$ *in the field* $\mathbf{Q}[x]/\langle x^4 - 3 \rangle$. *Then*

$$\mathbf{Q}[x]/\left\langle x^4 - 3 \right\rangle = \mathbf{Q}(\lambda) = \{a + b\lambda + c\lambda^2 + d\lambda^3 \mid a, b, c, d \in \mathbf{Q}\}$$

and $\{1, \lambda, \lambda^2, \lambda^3\}$ *is a basis of* $\mathbf{Q}(\lambda)$ *over* $\mathbf{Q}$. *Let us multiply two elements of* $\mathbf{Q}(\lambda)$ *and determine the form* $a + b\lambda + c\lambda^2 + d\lambda^3$ *for their product. Consider* $(1 + \lambda + \lambda^3)$ *and* $(1 + \lambda^2)$. *Then*

$$(1 + \lambda + \lambda^3)(1 + \lambda^2) = 1 + \lambda + \lambda^2 + 2\lambda^3 + \lambda^5.$$

Now

$$1 + x + x^2 + 2x^3 + x^5 = x(x^4 - 3) + 1 + 4x + x^2 + 2x^3$$

using the division algorithm. Thus,

$$
\begin{aligned}
1 + \lambda + \lambda^2 + 2\lambda^3 + \lambda^5 &= \lambda(\lambda^4 - 3) + 1 + 4\lambda + \lambda^2 + 2\lambda^3 \\
&= \lambda \cdot 0 + 1 + 4\lambda + \lambda^2 + 2\lambda^3.
\end{aligned}
$$

Hence,

$$(1 + \lambda + \lambda^3)(1 + \lambda^2) = 1 + 4\lambda + \lambda^2 + 2\lambda^3.$$

Let us find $(1 + \lambda + \lambda^3)^{-1}$. *Since* $x^4 - 3$ *is irreducible over* $\mathbf{Q}$, *the gcd of* $x^4 - 3$ *and* $x^3 + x + 1$ *is 1. Therefore, there exist* $s(x), t(x) \in \mathbf{Q}[x]$ *such that*

$$1 = s(x)(x^4 - 3) + t(x)(1 + x + x^3).$$

Thus,

$$
\begin{aligned}
1 &= s(\lambda)(\lambda^4 - 3) + t(\lambda)(1 + \lambda + \lambda^3) \\
1 &= 0 + t(\lambda)(1 + \lambda + \lambda^3).
\end{aligned}
$$

Hence, $t(\lambda) = (1 + \lambda + \lambda^3)^{-1}$. *We have not really calculated* $t(\lambda)$, *however. To do this calculation, we must know the exact form of* $s(x)$ *and* $t(x)$. *The method for finding s(x) and t(x) is described below. Now by repeated use of the division algorithm, we have*

$$x^4 - 3 = x(x^3 + x + 1) + (-x^2 - x - 3)$$

$$\begin{aligned}
x^3 + x + 1 &= (-x + 1)(-x^2 - x - 3) + (-x + 4) \\
-x^2 - x - 3 &= (x + 5)(-x + 4) + (-23) \\
-x + 4 &= (\tfrac{1}{23}x - \tfrac{4}{23})(-23) + 0.
\end{aligned}$$

Thus, by back substitution, we obtain

$$\begin{aligned}
-23 &= -x^2 - x - 3 - (x + 5)(-x + 4) \\
-23 &= -x^2 - x - 3 - (x + 5)[x^3 + x + 1 - (-x + 1)(-x^2 - x - 3)] \\
&= (-x^2 - 4x + 6)(-x^2 - x - 3) - (x + 5)(x^3 + x + 1) \\
&= (-x^2 - 4x + 6)[x^4 - 3 - x(x^3 + x + 1)] - (x + 5)(x^3 + x + 1) \\
&= (-x^2 - 4x + 6)(x^4 - 3) + (x^3 + 4x^2 - 7x - 5)(x^3 + x + 1).
\end{aligned}$$

This implies that

$$1 = -\frac{1}{23}(-x^2 - 4x + 6)(x^4 - 3) + (-\frac{1}{23})(x^3 + 4x^2 - 7x - 5)(x^3 + x + 1).$$

Therefore,

$$t(x) = -\frac{1}{23}x^3 - \frac{4}{23}x^2 + \frac{7}{23}x + \frac{5}{23}.$$

Consequently,

$$(1 + \lambda + \lambda^3)^{-1} = \frac{5}{23} + \frac{7}{23}\lambda - \frac{4}{23}\lambda^2 - \frac{1}{23}\lambda^3.$$

Since λ is a root of $x^4 - 3$ in $\mathbf{Q}(\lambda)$, we know by Corollary 14.1.10 that $x - \lambda$ divides $x^4 - 3$ over $\mathbf{Q}(\lambda)$. In fact, $x^4 - 3 = (x - \lambda)(x^3 + \lambda x^2 + \lambda^2 x + \lambda^3)$. We know there exists a field $\mathbf{Q}(\lambda)(\lambda_2)$, where λ_2 is a root of $x^3 + \lambda x^2 + \lambda^2 x + \lambda^3$ over $\mathbf{Q}(\lambda)$ by Theorem 21.2.1. Over the field $\mathbf{Q}(\lambda)(\lambda_2)$, $x^3 + \lambda x^2 + \lambda^2 x + \lambda^3$ factors into $(x - \lambda_2)q(x)$, where $q(x)$ has degree 2. There exists a field $\mathbf{Q}(\lambda)(\lambda_2)(\lambda_3)$, where λ_3 is a root of $q(x)$, and over the field $\mathbf{Q}(\lambda)(\lambda_2(\lambda_3)$, $q(x)$ factors into $(x - \lambda_3)(x - \lambda_4)$. Thus,

$$x^4 - 3 = (x - \lambda)(x - \lambda_2)(x - \lambda_3)(x - \lambda_4)$$

over $\mathbf{Q}(\lambda)(\lambda_2)(\lambda_3)(\lambda_4)$. In this particular example, we can take $\lambda_2 = -\lambda$ and so $\mathbf{Q}(\lambda) = \mathbf{Q}(\lambda)(\lambda_2)$. Hence,

$$\mathbf{Q}(\lambda, \lambda_2, \lambda_3, \lambda_4) = \mathbf{Q}(\lambda, \lambda_3).$$

Now over $\mathbf{Q}(\lambda)$,

$$x^4 - 3 = (x - \lambda)(x + \lambda)(x^2 + \lambda^2).$$

Also, $x^2 + \lambda^2$ is irreducible over $\mathbf{Q}(\lambda)$, a fact we leave as an exercise. Thus, $[\mathbf{Q}(\lambda) : \mathbf{Q}] = 4$ and $[\mathbf{Q}(\lambda)(\lambda_3) : \mathbf{Q}(\lambda)] = 2$. Hence, $[\mathbf{Q}(\lambda)(\lambda_3) : \mathbf{Q}] = 8$.

Example 21.2.3 leads us to believe that given any polynomial $f(x)$ in a polynomial ring $K[x]$ over a field K, there exists a field extension F/K such that $f(x)$ factors completely into linear factors. This is indeed the case, as we will presently show.

Definition 21.2.4 *Let K be a field. A polynomial $f(x)$ in $K[x]$ is said to* ***split*** *over a field $S \supseteq K$ if $f(x)$ can be factored as a product of linear factors in $S[x]$. A field S containing K is said to be a* ***splitting field*** *for $f(x)$ over K if $f(x)$ splits over S, but over no proper intermediate field of S/K.*

Example 21.2.5 *The field of complex numbers $\mathbf{C}$ is a splitting field for the polynomial $x^2 + 1$ over $\mathbf{R}$. This follows since $x^2 + 1 = (x+i)(x-i)$ in $\mathbf{C}[x]$ and $\mathbf{C}/\mathbf{R}$ has no proper intermediate fields because $[\mathbf{C}:\mathbf{R}] = 2$. (If $\mathbf{C} \supseteq L \supseteq \mathbf{R}$, where L is an intermediate field of $\mathbf{C}/\mathbf{R}$, then $2 = [\mathbf{C}:L][L:\mathbf{R}]$ and so either $[\mathbf{C}:L] = 1$ or $[L:\mathbf{R}] = 1$. Thus, either $\mathbf{C} = L$ or $L = \mathbf{R}$.) Note that $\mathbf{C}$ is not the splitting field of $x^2 + 1$ over $\mathbf{Q}$ since $x^2 + 1$ splits over $\mathbf{Q}(i) \subset \mathbf{C}$.*

Theorem 21.2.6 *Let K be a field and $f(x)$ be a polynomial in $K[x]$ of degree n. Let F/K be a field extension. If*

$$f(x) = c(x - c_1)(x - c_2) \cdots (x - c_n) \text{ in } F[x],$$

then $K(c_1, c_2, \ldots, c_n)$ is a splitting field for $f(x)$ over K.

Proof.   Since $c_1, c_2, \ldots, c_n$ are the roots of $f(x)$, $f(x)$ splits over $K(c_1, c_2, \ldots, c_n)$. Let L be an intermediate field of $K(c_1, c_2, \ldots, c_n)/K$ such that $f(x)$ splits over L. Since $K[x]$ is a UFD, there is only one way $f(x)$ can split over L, namely, $f(x) = c(x - c_1)(x - c_2) \cdots (x - c_n)$. Thus, $c_1, c_2, \ldots, c_n \in L$, whence $L \supseteq K(c_1, c_2, \ldots, c_n)$. Hence, $K(c_1, c_2, \ldots, c_n)$ is the smallest intermediate field over which $f(x)$ splits. ■

The field $\mathbf{Q}(\lambda, \lambda_3)$ of Example 21.2.3 is a splitting field for $x^4 - 3$ over $\mathbf{Q}$. We now prove the existence of splitting fields.

Theorem 21.2.7 *Let K be a field and $f(x)$ be a nonconstant polynomial over K. Then there is a splitting field for $f(x)$ over K.*

Proof.   If $\deg f(x) = 1$, then K is a splitting field for $f(x)$ over K. Assume the theorem is true for all polynomials of degree $n-1$ (≥ 1). Suppose $\deg f(x) = n$. There exists a field $K_1 \supseteq K$ such that K_1 contains a root c_1 of $f(x)$ by Theorem 21.2.1. Thus, $f(x) = (x - c_1)f_1(x)$ in $K_1[x]$ and $\deg f_1(x) = n - 1$. By the induction hypothesis, there exists a field extension E/K_1 such that $f_1(x)$ splits in $E[x]$. Thus, $f(x)$ splits in $E[x]$, say,

$$f(x) = c(x - c_1)(x - c_2) \cdots (x - c_n).$$

By Theorem 21.2.6, the intermediate field $K(c_1, c_2, \ldots, c_n)$ of E/K is a splitting field for $f(x)$ over K. ∎

The intermediate field $\mathbf{Q}(\sqrt[4]{3}, i\sqrt[4]{3})$ of $\mathbf{C}/\mathbf{Q}$ is a splitting field for $x^4 - 3$ over $\mathbf{Q}$. The field $\mathbf{Q}(\lambda, \lambda_3)$ of Example 21.2.3 is also a splitting field for $x^4 - 3$ over $\mathbf{Q}$. However, we cannot conclude that $\mathbf{Q}(\sqrt[4]{3}, i\sqrt[4]{3}) = \mathbf{Q}(\lambda, \lambda_3)$. Hence, splitting fields for a given polynomial over a field are not unique. We will show, however, that they are unique up to isomorphism.

Theorem 21.2.8 *Let α be an isomorphism of the field K onto the field K'. Let $p(x) = k_0 + k_1 x + k_2 x^2 + \cdots + k_n x^n$ be an irreducible polynomial in $K[x]$ of degree n, c be a root of $p(x)$ in some field extension of K, and $p'(y) = \alpha(k_0) + \alpha(k_1)y + \alpha(k_2)y^2 + \cdots + \alpha(k_n)y^n$ be the corresponding polynomial in $K'[y]$. Then $p'(y)$ is irreducible in $K'[y]$. If c' is a root of $p'(y)$ in some field extension of K', then α can be extended to an isomorphism α' of $K(c)$ onto $K'(c')$ with $\alpha'(c) = c'$. α' is the only extension of α such that $\alpha'(c) = c'$.*

Proof.   By an argument similar to the one used in the proof of Theorem 14.1.14, α can be uniquely extended to an isomorphism $\bar{\alpha}$ of $K[x]$ onto $K'[y]$ so that for every polynomial $b_0 + b_1 x + b_2 x^2 + \cdots + b_m x^m \in K[x]$,

$$\bar{\alpha}(b_0 + b_1 x + b_2 x^2 + \cdots + b_m x^m) = \alpha(b_0) + \alpha(b_1)y + \alpha(b_2)y^2 + \cdots + \alpha(b_m)y^m.$$

We leave to the reader the verification that $p'(y)$ is irreducible in $K'[y]$. Let β be the natural homomorphisms of $K[x]$ onto $K[x]/\langle p(x)\rangle$ and β' be the natural homomorphisms of $K'[y]$ onto $K'[y]/\langle p'(y)\rangle$. Then $\operatorname{Ker} \beta = \operatorname{Ker} \beta' \circ \bar{\alpha}$. Hence, there exists an isomorphism α^* of $K[x]/\langle p(x)\rangle$ onto $K'[y]/\langle p'(y)\rangle$ such that $\beta' \circ \bar{\alpha} = \alpha^* \circ \beta$. By Theorem 21.1.11 and Corollary 21.1.12, there exist isomorphisms γ and γ' of $K[x]/\langle p(x)\rangle$ onto $K(c)$ and $K'[y]/\langle p'(y)\rangle$ onto $K'(c')$, respectively. Thus, α' is the map $\gamma' \circ \alpha^* \circ \gamma^{-1}$. The situation is described by the following diagram:

$$
\begin{array}{ccc}
K[x] & \xrightarrow{\;\bar{\alpha}\;} & K'[y] \\
\downarrow{\scriptstyle \beta} & & \downarrow{\scriptstyle \beta'} \\
\dfrac{K[x]}{\langle p(x)\rangle} & \xrightarrow{\;\alpha^*\;} & \dfrac{K'[y]}{\langle p'(y)\rangle} \\
\downarrow{\scriptstyle \gamma} & & \downarrow{\scriptstyle \gamma'} \\
K(c) & \xrightarrow{\;\alpha'\;} & K'(c')
\end{array}
$$

Let α'' be any other extension of α to an isomorphism of $K(c)$ onto $K'(c')$ such that $\alpha''(c) = c'$. Now $\{1, c, \ldots, c^{n-1}\}$ is a basis for $K(c)/K$ and $\{1, c', \ldots, c'^{n-1}\}$ is a basis for $K'(c')/K'$. We have that

$$\alpha''\left(\sum_{i=0}^{n-1} k_i c^i\right) = \sum_{i=0}^{n-1} \alpha''(k_i)\alpha''(c^i) = \sum_{i=0}^{n-1} \alpha(k_i)c'^i = \alpha'\left(\sum_{i=0}^{n-1} k_i c^i\right).$$

Hence, $\alpha'' = \alpha'$. ∎

Corollary 21.2.9 *Let E/K be a field extension and $p(x)$ be an irreducible polynomial in $K[x]$. If a, $b \in E$ are roots of $p(x)$, then $K(a) \simeq K(b)$.*

Proof. Let $K = K'$ and α be the identity map. ∎

From Corollary 21.2.9, we have $\mathbf{Q}(\sqrt[4]{3}) \simeq \mathbf{Q}(i\sqrt[4]{3})$ in Example 21.2.3.

Theorem 21.2.10 *Let α be an isomorphism from the field K onto the field K'. Let*

$$f(x) = k_0 + k_1 x + k_2 x^2 + \cdots + k_n x^n$$

be a polynomial in $K[x]$ and

$$f'(y) = \alpha(k_0) + \alpha(k_1)y + \alpha(k_2)y^2 + \cdots + \alpha(k_n)y^n$$

be the corresponding polynomial in $K'[y]$.
   If S is a splitting field for $f(x)$ over K and S' is a splitting field for $f'(y)$ over K', then α can be extended to an isomorphism α' of S onto S'.

Proof. The proof is by induction on $\deg f(x)$. If $\deg f(x) = 1$, then $K = S$ and $K' = S'$. In this case, we can take $\alpha' = \alpha$. Assume the theorem is true for all polynomials of degree less than n (the induction hypothesis). Suppose $\deg f(x) = n$. Extend α to an isomorphism $\bar{\alpha}$ of $K[x]$ onto $K'[y]$ as in Theorem 21.2.8. Let $p(x)$ be an irreducible factor of $f(x)$ and $c_1 \in S$ be a root of $p(x)$. Let $c_1' \in S'$ be a root of $\bar{\alpha}(p(x)) = p(y)$. Then by Theorem 21.2.8, α can be extended to an isomorphism α_1 of $K(c_1)$ onto $K'(c_1')$. Extend α_1 to an isomorphism $\overline{\alpha_1}$ of $K(c_1)[x]$ onto $K'(c_1')[y]$. Now $f(x) = (x - c_1)f_1(x)$ in $K(c_1)[x]$ and $f'(y) = (y-c_1')f_1'(y)$ in $K(c_1')[y]$, where $f_1'(y) = \overline{\alpha_1}(f_1(x))$. Clearly S is a splitting field for $f_1(x)$ over $K(c_1)$ and S' is a splitting field for $f_1'(y)$ over $K'(c_1')$. Since $\deg f_1(x) = n - 1 = \deg f_1'(y)$, α_1 can be extended to an isomorphism of S onto S' by the induction hypothesis. ∎

Corollary 21.2.11 *Let $f(x) \in K[x]$. Any two splitting fields for $f(x)$ over K are isomorphic.*

Proof. Let S and S' be two splitting fields for $f(x)$ over K. In Theorem 21.2.10, take $K = K'$ and α the identity mapping on K. ∎

Definition 21.2.12 *Let F/K be a field extension and a, $b \in F$. Then a and b are called* **conjugates** *if a and b are roots of the same irreducible polynomial over K.*

We ask the reader to prove that the notion of conjugates defines an equivalence relation on F.

Example 21.2.13 *Consider the field extension* $\mathbf{C}/\mathbf{R}$. *Let* $a, b \in \mathbf{R}$. *Then* $a + bi$ *and its complex conjugate* $a - bi$ *are conjugates in the sense of Definition 21.2.12. This is obvious if* $b = 0$. *Suppose* $b \neq 0$. *Then* $a + bi \notin \mathbf{R}$. *Let* $f(x) = x^2 + 2ax + (a^2 + b^2)$. *Since* $a + bi \notin \mathbf{R}$, $[\mathbf{R}(a + bi) : \mathbf{R}] = 2$. *Now* $a + bi$ *is a root of* $f(x)$ *and* $f(x)$ *must be irreducible over* $\mathbf{R}$. $a - bi$ *is also a root of* $f(x)$.

In certain cases, the following theorem is useful in determining the irreducibility of a polynomial.

Theorem 21.2.14 *Let* F *be a field. Let* p *be a prime in* $\mathbf{Z}$ *and* $a \in F$. *Then the polynomial* $x^p - a$ *is reducible over* F *if and only if* $x^p - a$ *has a root in* F.

Proof. Suppose $f(x) = x^p - a \in F[x]$ is reducible. Let $f(x) = g(x)h(x)$ for some $g(x), h(x) \in F[x]$, $\deg g(x) = m$, $0 < m < p$, and $0 < \deg h(x) < p$. Since $f(x)$ is monic, we can take $g(x)$ to be monic. By factoring $g(x)$ as a product of linear factors in a splitting field of $g(x)$ over F, we see that the constant term of $g(x)$ is $(-1)^m d$ for some $d \in F$. Since $\gcd(m, p) = 1$, there exist integers s and t such that $1 = sm + tp$. By Theorem 21.2.1, there is a field extension of F which contains a root of $f(x)$. Let b be such a root of $f(x)$.
  Case 1: Suppose the characteristic of F is p. Since b is a root of $f(x)$, $b^p = a$. Thus,

$$(x - b)^p = x^p - b^p = x^p - a$$

and all the roots of $f(x)$ equal b. Now every root of $g(x)$ is also a root of $f(x)$. Thus, all the m roots of $g(x)$ are equal to b. Hence, $b^m = d$. Now

$$d^s = b^{ms} = b^{1-pt} = bb^{-pt} = ba^{-t}.$$

Hence, $b = d^s a^t \in F$ and so $f(x)$ has a root in F.
  Case 2: Suppose that F has characteristic 0. Let c be any other root of $f(x)$. Then

$$c^p = a = b^p.$$

Hence, $c = bu$, where $u = c^{-p+1}b^{p-1}$ and $u^p = 1$. From this, it follows that the roots of $f(x)$ are of the form

$$b, bu_1, \ldots, bu_{p-1},$$

where $u_i^p = 1$. As in case 1, we have that the product of the roots of $g(x)$ is

$$d = b^m u_1 u_2 \cdots u_{m-1} = b^m v,$$

where $v^p = 1$. Now $1 = sm + tp$ implies that

$$b^{sm} = v^{-s}d^s = b^{1-tp} = ba^{-t}.$$

Therefore, $b = v^{-s}d^s a^t$. It then follows that

$$a = b^p = (v^{-s}d^s a^t)^p = v^{-sp}(d^s a^t)^p = (d^s a^t)^p.$$

Thus, $d^s a^t \in F$ is a root of $f(x)$.

The converse follows from Corollary 14.1.10. ∎

21.2.1 Worked-Out Exercises

◇ **Exercise 1** Find a splitting field S of $x^4 - 10x^2 + 21$ over $\mathbf{Q}$. Find $[S : \mathbf{Q}]$ and a basis for $S/\mathbf{Q}$.

Solution: Note that $x^4 - 10x^2 + 21 = (x^2 - 3)(x^2 - 7)$ over $\mathbf{Q}$. Therefore, a splitting field S of $x^4 - 10x^2 + 21$ over $\mathbf{Q}$ is $\mathbf{Q}(\sqrt{3}, \sqrt{7})$. Hence, $[S : \mathbf{Q}] = 4$ and $\{1, \sqrt{3}, \sqrt{7}, \sqrt{21}\}$ is a basis for $S/\mathbf{Q}$, as can be seen from Worked-Out Exercise 2 (page 455).

◇ **Exercise 2** Show that the splitting field of $x^p - 1$ over $\mathbf{Q}$ is of degree $p - 1$, where p is a prime.

Solution: Let $f(x) = x^p - 1 \in \mathbf{Q}[x]$. Now $f(x) = (x - 1)g(x)$, where $g(x) = x^{p-1} + x^{p-2} + \cdots + x + 1$. Also,

$$g(x) = \frac{x^p - 1}{x - 1}.$$

Hence,

$$g(x + 1) = \frac{(x + 1)^p - 1}{x} = x^{p-1} + \binom{p}{1}x^{p-2} + \cdots + \binom{p}{p-1}.$$

Now since p is prime, $p | \binom{p}{r}$ for all $1 \le r \le p - 1$. Also, p^2 does not divide $\binom{p}{p-1}$. Therefore, by Eisenstein's criterion, $g(x + 1)$ is irreducible over $\mathbf{Q}$. Thus, $g(x)$ is irreducible over $\mathbf{Q}$. Let $\xi = e^{\frac{2\pi i}{p}}$, where $i^2 = -1$. Then the roots of $f(x)$ are $1, \xi, \xi^2, \ldots, \xi^{p-1}$ and the roots of $g(x)$ are $\xi, \xi^2, \ldots, \xi^{p-1}$. Now the splitting field of $f(x)$ is $S = \mathbf{Q}(1, \xi, \xi^2, \ldots, \xi^{p-1}) = \mathbf{Q}(\xi)$. Also, $g(x)$ is the minimal polynomial of ξ over $\mathbf{Q}$. Hence, $[S : \mathbf{Q}] = p - 1$.

◇ **Exercise 3** Find the splitting field of the following polynomials over $\mathbf{Q}$.

(i) $x^4 + 1$.

(ii) $x^6 + x^3 + 1$.

Solution: (i) Let $f(x) = x^4 + 1$. Then $f(x) = (x^2 + \sqrt{2}x + 1)(x^2 - \sqrt{2}x + 1)$ over $\mathbf{Q}(\sqrt{2})$. Therefore, the roots of $f(x)$ are

$$\frac{-\sqrt{2} \pm i\sqrt{2}}{2}, \quad \frac{\sqrt{2} \pm i\sqrt{2}}{2}.$$

Let S be the splitting field of $f(x)$ over $\mathbf{Q}$. We claim that $S = \mathbf{Q}(\sqrt{2}, i)$. Now

$$\sqrt{2} = \frac{\sqrt{2} + i\sqrt{2}}{2} + \frac{\sqrt{2} - i\sqrt{2}}{2} \in S$$

and

$$\sqrt{2}i = \frac{\sqrt{2} + i\sqrt{2}}{2} - \frac{\sqrt{2} - i\sqrt{2}}{2} \in S.$$

This implies that $i = \frac{i\sqrt{2}}{\sqrt{2}} \in S$. It now follows that $\mathbf{Q}(\sqrt{2}, i) \subseteq S$. Clearly $S \subseteq \mathbf{Q}(\sqrt{2}, i)$. Consequently, $S = \mathbf{Q}(\sqrt{2}, i)$. Now $x^2 - 2$ is the minimal polynomial of $\sqrt{2}$ over $\mathbf{Q}$ and $x^2 + 1$ is the minimal polynomial of i over $\mathbf{Q}$. In fact, $x^2 + 1$ is the minimal polynomial of i over $\mathbf{Q}(\sqrt{2})$. Thus, $[S : \mathbf{Q}] = [S : \mathbf{Q}(\sqrt{2})][\mathbf{Q}(\sqrt{2}) : \mathbf{Q}] = 2 \cdot 2 = 4$.

(ii) Let $f(x) = x^6 + x^3 + 1$. Now $(x^9 - 1) = (x^3 - 1)(x^6 + x^3 + 1)$. The roots of $(x^9 - 1)$ are $1, \xi, \xi^2, \ldots, \xi^8$ and $1, \xi^3, \xi^6$ are the roots of $(x^3 - 1)$, where $\xi = e^{\frac{2\pi i}{9}}$. Hence, $\xi, \xi^2, \xi^4, \xi^5, \xi^7, \xi^8$ are the roots of $x^6 + x^3 + 1$. Therefore, $S = \mathbf{Q}(\xi, \xi^2, \xi^4, \xi^5, \xi^7, \xi^8) = \mathbf{Q}(\xi)$ is the splitting field of $x^6 + x^3 + 1$ over $\mathbf{Q}$. Since $x^6 + x^3 + 1$ is irreducible over $\mathbf{Q}$, $[S : \mathbf{Q}] = 6$.

21.2.2 Exercises

1. Prove that the polynomial $p'(y)$ in Theorem 21.2.8 is irreducible in $K'[y]$.

2. Let F/K be an algebraic field extension. Define $\sim$ on F by for all a, $b \in F$, $a \sim b$ if and only if a and b are conjugates. Prove that $\sim$ is an equivalence relation.

3. (i) Show that the polynomials $x^2 - 2x - 1$ and $x^2 - 2$ have the same splitting field over $\mathbf{Q}$.

   (ii) Find a pair of polynomials in $\mathbf{Q}[x]$, other than the pair given in (i), which have the same splitting field over $\mathbf{Q}$.

4. Find a splitting field S of the polynomial $x^3 - 3$ over $\mathbf{Q}$. Find $[S : \mathbf{Q}]$ and a basis for $S/\mathbf{Q}$.

5. Find a splitting field S of the polynomial $x^2 + x + [1]$ over $\mathbf{Z}_5$. Find $[S : \mathbf{Z}_5]$ and a basis for $S/\mathbf{Z}_5$.

6. Find a splitting field S of the polynomial $x^2 + [1]$ over $\mathbf{Z}_2$. Find $[S : \mathbf{Z}_2]$ and a basis for $S/\mathbf{Z}_2$.

7. Find a splitting field S of the polynomial $x^4 - 7x^2 + 10$ over $\mathbf{Q}$. Find $[S : \mathbf{Q}]$ and a basis for $S/\mathbf{Q}$.

8. Prove that $\mathbf{Q}(-\frac{1}{2} + \frac{\sqrt{3}}{2}i)$ is a splitting field of the polynomial $x^4 + x^2 + 1$ over $\mathbf{Q}$. Find $[\mathbf{Q}(-\frac{1}{2} + \frac{\sqrt{3}}{2}i) : \mathbf{Q}]$.

9. Let $f(x) \in K[x]$, a polynomial ring over the field K. Let S be a splitting field for $f(x)$ over K. Prove that for any field L, $S \supseteq L \supseteq K$, S is a splitting field of $f(x)$ over L.

10. Let $f(x), g(x)$, and $h(x) \in K[x]$, a polynomial ring over the field K. Suppose that S is a splitting field of $f(x)$ over K and $f(x) = g(x)h(x)$. Prove that S contains a splitting field of $g(x)$ over K.

11. Let $f(x), g(x) \in K[x]$, a polynomial ring over the field K. Suppose that $g(x) = f(ax + b)$, where $0 \neq a, b \in K$. Prove that $f(x)$ and $g(x)$ have equal splitting fields over K.

12. Prove that if $f(x)$ is a polynomial in $K[x]$ of degree n, then $[S : K] \leq n!$, where S is a splitting field of $f(x)$ over K.

13. Let K be a field and $f_1(x), f_2(x), \ldots, f_n(x) \in K[x]$ be such that $\deg f_i(x) \geq 1$, $1 \leq i \leq n$. Show that there exists a field extension F/K such that each $f_i(x)$ has a root in F.

14. Let F be a field of prime characteristic p and $a \in F$. Prove that $x^p - x - a$ is reducible over F if and only if $x^p - x - a$ has a root in F.

15. Answer the following statements, true or false. If the statement is true, prove it. If it is false, give a counter example.

    (i) Let $f(x)$ be an irreducible polynomial of degree n over a field K of characteristic 0. Let $S = K(c_1, c_2, \ldots, c_n)$ be a splitting field of $f(x)$ over K, where $c_1, c_2, \ldots, c_n$ are the roots of $f(x)$. Then $K(c_2, \ldots, c_n) \subset S$.

    (ii) The polynomial $f(x) = x^5 - x - 30$ is reducible over $\mathbf{Q}$.

    (iii) $\mathbf{C}$ is a splitting field of some polynomial over $\mathbf{Q}$.

21.3 Algebraically Closed Fields

The most important result in Steinitz's work in 1910 was his proof of the existence and uniqueness of an algebraic closure of a field. In this section[1], we present these results.

---

[1]This section may be skipped without any discontinuity. The only place this section is needed is in Exercise 4 (Section 24.1).

Definition 21.3.1 *A field K is called* **algebraically closed** *if for all $f(x) \in K[x]$ with $\deg f(x) \geq 1$, $f(x)$ has a root in K.*

Theorem 21.3.2 *Let K be a field. The following conditions are equivalent.*
  (i) K is algebraically closed.
  (ii) Every irreducible polynomial in $K[x]$ is of degree 1.
  (iii) Let $f(x) \in K[x]$, $\deg f(x) \geq 1$. Then $f(x)$ splits as a product of linear factors over K.
  (iv) If F/K is an algebraic field extension, then $F = K$.

Proof.   (i)$\Rightarrow$(ii) Let $p(x) \in K[x]$ and $p(x)$ be irreducible. By (i), there exists $a \in K$ such that $p(a) = 0$. Then $p(x) = (x - a)g(x)$ for some $g(x) \in K[x]$. Since $p(x)$ is irreducible, $g(x) \in K$. Hence, $\deg p(x) = 1$.

(ii)$\Rightarrow$(iii) Let $f(x) \in K[x]$ and $\deg f(x) \geq 1$. Let $f(x) = p_1(x) \cdots p_s(x)$, where $p_i(x) \in K[x]$ is irreducible, $1 \leq i \leq s$. Then $\deg p_i(x) = 1$, $1 \leq i \leq s$. We may write $p_i(x) = k_i(x - a_i)$, where $k_i, a_i \in K$, $1 \leq i \leq s$. Let $k = k_1 \cdots k_s$. Then $f(x) = k(x - a_1) \cdots (x - a_s)$. Thus, $f(x)$ splits as a product of linear factors over K.

(iii)$\Rightarrow$(iv) Let F/K be an algebraic field extension. Let $c \in F$ and let $p(x) \in K[x]$ be the minimal polynomial of c over K. Since $p(x)$ is irreducible, $\deg p(x) = 1$ by (iii). Therefore, $p(x) = ax+b \in K[x]$. Since $p(c) = 0$, $ac+b = 0$. Thus, $c = -a^{-1}b \in K$. Hence, $K = F$.

(iv)$\Rightarrow$(i) Let $f(x) \in K[x]$, $\deg f(x) \geq 1$. There exists a field extension F/K such that F has a root of $f(x)$, say, a. Then $K(a)/K$ is an algebraic field extension. Therefore, $K(a) = K$ and so $a \in K$. Thus, K is algebraically closed. ∎

We now prove the existence of an algebraically closed field. The following proof is due to Artin.

Theorem 21.3.3 *Let K be a field. Then there exists an algebraically closed field F such that K is a subfield of F.*

Proof.   We first construct an extension F_1/K such that if $f(x) \in K[x]$ and $\deg f(x) \geq 1$, then $f(x)$ has a root in F_1. Let $\mathcal{K}$ be the set of all polynomials in $K[x]$ of degree ≥ 1. Let S be a set which is in one-one correspondence with $\mathcal{K}$. For $f(x) \in \mathcal{K}$, let x_f be the corresponding element in S.

Consider the polynomial ring $K[S]$. Let I be the ideal of $K[S]$ generated by all polynomials $f(x_f)$ in $K[S]$. We claim that $I \neq K[S]$. Suppose that $I = K[S]$. Then there exists $g_i \in K[S]$ such that

$$g_1 f_1(x_{f_1}) + g_2 f_2(x_{f_2}) + \cdots + g_n f_n(x_{f_n}) = 1. \tag{21.7}$$

Write $x_i = x_{f_i}$, $1 \le i \le n$. Since the polynomials g_i, $1 \le i \le n$, involve only a finite number of indeterminates, say, $x_1, x_2, \ldots, x_m$, with $m \ge n$, we may write Eq. (21.7) as

$$\sum_{i=1}^{n} g_i(x_1, x_2, \ldots, x_m) f_i(x_i) = 1. \tag{21.8}$$

By Exercise 13 (page 468), there exists a finite extension L/K such that each polynomial f_i, $1 \le i \le n$, has a root in L. Let c_i be a root of f_i in L, $1 \le i \le n$. Let $c_i = 0$ for $n < i \le m$. Substituting c_i for x_i, $1 \le i \le n$, in Eq. (21.8), we get $0 = 1$, a contradiction. Hence, $I \ne K[S]$.

Let M be a maximal ideal of $K[S]$ such that $I \subseteq M$. Let $F_1 = K[S]/M$. Then F_1 is a field containing an isomorphic copy $(K + M)/M$ of K. Thus, F_1 can be regarded as a field extension of K. Also, if $f \in K[x]$ and $\deg f(x) \ge 1$, then $x_f + M$ is a root of f in F_1.

By induction, we can form a chain of fields

$$F_1 \subseteq F_2 \subseteq \cdots \subseteq F_n \subseteq \cdots$$

such that every polynomial of degree ≥ 1 in F_n has a root in F_{n+1}. Let $F = \cup_{i=1}^{\infty} F_n$. Then F is a field. Let $f \in F[x]$. Then $f \in F_n[x]$ for some positive integer n. Thus f has a root in $F_{n+1} \subseteq F$. Hence, F is algebraically closed. ∎

Corollary 21.3.4 *Let K be a field. Then there exists an algebraic field extension F/K such that F is algebraically closed.*

Proof. By Theorem 21.3.3, there exists a field extension E/K such that E is algebraically closed. Let $F = \{a \in E \mid a$ is algebraic over $K\}$. Then F/K is an algebraic extension. Let $f(x) \in F[x]$ and $\deg f(x) \ge 1$. Then $f(x)$ has a root c in E. Thus, c is algebraic over F. Since F/K is an algebraic extension, c is algebraic over K. Hence, $c \in F$ and so F is algebraically closed. ∎

Definition 21.3.5 *Let K be a field. A field $F \supseteq K$ is called an **algebraic closure** of K if*
(i) F/K is algebraic and
(ii) F is algebraically closed.

For any field K, Corollary 21.3.4 guarantees the existence of an algebraic closure of K.

Lemma 21.3.6 *Let F and L be fields with L algebraically closed. Let $\sigma : F \to L$ be an isomorphism of F into L. Let a be an algebraic element over F in some field extension of F. Let $f(x) \in F[x]$ be the minimal polynomial of a. Then σ can be extended to an isomorphism η of $F(a)$ into L and the number of such extensions is equal to the number of distinct roots of $f(x)$.*

Proof. Let $f(x) = a_0 + a_1 x + \cdots + a_n x^n \in F[x]$ and $f^\sigma(x) = \sigma(a_0) + \sigma(a_1)x + \cdots + \sigma(a_n)x^n \in L[x]$. Since L is algebraically closed there exists a root b of $f^\sigma(x)$ in L. Since a is algebraic over F, $F(a) = F[a]$ by Corollary 21.1.12. Thus, if $u \in F(a)$, then $u = c_0 + c_1 a + \cdots + c_k a^k \in F[a]$. Define $\eta : F(a) \to L$ by

$$\eta(c_0 + c_1 a + \cdots + c_k a^k) = \sigma(c_0) + \sigma(c_1)b + \cdots + \sigma(c_k)b^k$$

for all $c_0 + c_1 a + \cdots + c_k a^k \in F(a)$. Suppose $c_0 + c_1 a + \cdots + c_k a^k = d_0 + d_1 a + \cdots + d_s a^s$. Let $\gamma(x) = c_0 + c_1 x + \cdots + c_k x^k$ and $\gamma'(x) = d_0 + d_1 x + \cdots + d_s x^s$. Then $(\gamma - \gamma')(a) = 0$. Hence, $f(x)$ divides $(\gamma - \gamma')(x)$. Thus, $f^\sigma(x)$ divides $(\gamma^\sigma - \gamma'^\sigma)(x)$. Consequently, $(\gamma^\sigma - \gamma'^\sigma)(b) = 0$ and so $\sigma(c_0) + \sigma(c_1)b + \cdots + \sigma(c_k)b^k = \sigma(d_0) + \sigma(d_1)b + \cdots + \sigma(d_s)b^s$. Thus, η is well defined. Clearly η is an isomorphism. The number of distinct roots of $f(x)$ in the algebraic closure of F is equal to the number of distinct roots of $f^\sigma(x)$ in L. For any extension $\xi : F(a) \to L, \xi(a)$ is a root of $f^\sigma(x)$. Therefore, the number of such extensions is equal to the number of distinct roots of $f(x)$. ∎

We close this section by showing that the algebraic closure of a field is unique up to isomorphism. Our proof uses Zorn's lemma while Steinitz's original proof used the equivalent concept of the axiom of choice.

Theorem 21.3.7 *Let F/K be an algebraic field extension. Let L be an algebraically closed field and σ be an isomorphism of K into L. Then there exists an isomorphism η of F into L such that $\eta|_K = \sigma$.*

Proof. Let $S = \{(E, \lambda) \mid E$ is a subfield of F, $K \subseteq E$ and $\lambda : E \to L$ is an isomorphism such that $\lambda|_K = \sigma\}$. Since $(K, \sigma) \in S$, $S \neq \phi$. Let (E, λ), $(E', \lambda') \in S$. Define a relation $\leq$ on S by $(E, \lambda) \leq (E', \lambda')$ if $E \subseteq E'$ and $\lambda'|_E = \lambda$. Then $(S, \leq)$ is a poset. Let $\{(E_i, \lambda_i)\}_{i \in \Lambda}$ be a chain in S. Let $E = \cup_{i \in \Lambda} E_i$. Then E is a field and $K \subseteq E$. Define $\lambda : E \to L$ as follows: Let $a \in E$. Then $a \in E_n$ for some n. Define $\lambda(a) = \lambda_n(a)$. Since $\{(E_i, \lambda_i)\}_{i \in \Lambda}$ is a chain, λ is an isomorphism of E into L. Hence, $(E, \lambda) \in S$ and (E, λ) is an upper bound of $\{(E_i, \lambda_i)\}_{i \in \Lambda}$. Hence, by Zorn's lemma, S has a maximal element, say, (T, η). Suppose $T \neq F$. Let $a \in F \backslash T$. By Lemma 21.3.6, there exists an isomorphism $\beta : T(a) \to L$ such that $\beta|_T = \eta$. From this, it follows that $(T(a), \beta) \in S$, a contradiction of the maximality of (T, η). Thus, $F = T$. ∎

Theorem 21.3.8 *Let K be a field. Let F and F' be two algebraic closures of K. Then there exists an isomorphism λ of F onto F' such that $\lambda(a) = a$ for all $a \in K$.*

Proof. Let $\sigma : K \to F'$ be such that $\sigma(a) = a$ for all $a \in K$. Then σ is an isomorphism of K into F'. By Theorem 21.3.7, there exists an isomorphism

$\lambda : F \to F'$ such that $\lambda|_K = \sigma$. Now $\lambda(F) \simeq F$. Thus, $\lambda(F)$ is algebraically closed and $K \subseteq \lambda(F)$. Now $K \subseteq \lambda(F) \subseteq F'$. Since F'/K is algebraic, $F'/\lambda(F)$ is algebraic. Thus, $F' = \lambda(F)$. Hence, $F \simeq F'$. ∎

21.3.1 Exercises

1. If F is a field with a finite number of elements, prove that F is not algebraically closed.

Ernst Steinitz (1871–1928) was born on June 13, 1871, in Laurahütte, Silesia, Germany. In 1890, he started his studies in mathematics at the University of Breslau (now Worclaw, Poland). In 1894, he received his Ph.D. He started teaching at the Technical College in Berlin-Charlottenberg. In 1920, he was appointed professor at the University of Kiel, where he remained until his death. He died on September 29, 1928.

In 1910, he published *"Algebraische Theorie der Körper"* in which he gave an abstract definition of a "field." He also introduced the notion of a prime field, separable element, perfect field, and degree of transcendence of an extension. With the help of the axiom of choice, he proved that for any field K there exists a field extension F/K such that every polynomial over K splits into linear factors over F and the smallest such field is unique up to isomorphism. He called such field, algebraically closed. His work on field theory was influenced by Weber and Kronecker.

Steinitz also worked on the theory of polyhedra.

Chapter 22

Multiplicity of Roots

22.1 Multiplicity of Roots

In some cases, an irreducible polynomial $p(x)$ of degree n over a field K does not have n distinct roots in a splitting field of $p(x)$ over K. In this chapter, we examine this situation.

If $f(x)$ is a polynomial over K and c is a root of $f(x)$ in some field F containing K, then the **multiplicity** of c is the largest positive integer m such that $(x-c)^m$ divides $f(x)$ over F.

Definition 22.1.1 *Let K be a field and $p(x)$ be an irreducible polynomial in $K[x]$ of degree n. Then $p(x)$ is called **separable** if it has n distinct roots in a splitting field S of $p(x)$ over K; otherwise $p(x)$ is called **inseparable** over K. An arbitrary polynomial in $K[x]$ is called **separable** if each of its irreducible factors in $K[x]$ is separable; otherwise it is called **inseparable**.*

Definition 22.1.2 *Let F/K be a field extension and c be an element of F which is algebraic over K. Then c is called **separable** (or **separable algebraic**) over K if its minimal polynomial over K is separable; otherwise c is called **inseparable** over K. If F/K is an algebraic extension, then F/K is called **separable** (or **separable algebraic**) if every element of F is separable over K; otherwise F/K is called **inseparable**.*

Let F/K be a field extension and L be an intermediate field of F/K. Let $c \in F$ and suppose c is separable over K. Then c must be separable over L. This follows since if $f(x)$ and $p(x)$ are the minimal polynomials of c over K and L, respectively, then $p(x) | f(x)$. Hence, c cannot be a multiple root of $p(x)$ since it is not one of $f(x)$.

Example 22.1.3 *Consider the field $K(t)$, where K is a field of prime characteristic p and t is transcendental over K. It follows that the polynomial $x^p - t^p$ is*

irreducible over $K(t^p)$ *by Eisenstein's criterion since* t^p *is irreducible in* $K[t^p]$.
Now $x^p - t^p$ *factors into*

$$\underbrace{(x - t)(x - t) \cdots (x - t)}_{p \ times} = (x - t)^p$$

over $K(t)$. *Thus,* $K(t)$ *is a splitting field for* $x^p - t^p$ *over* $K(t^p)$ *and we see that* $x^p - t^p$ *has only one root in* $K(t)$, *namely,* t. *(Since* $t \notin K(t^p)$, *we can also use Theorem 21.2.14 to deduce that* $x^p - t^p$ *is irreducible over* $K(t^p)$.*) Thus,* $x^p - t^p$, t, *and* $K(t)$ *are inseparable over* $K(t^p)$. *Note that* t *has multiplicity* p *over* $K(t^p)$.

Let K be a field and

$$f(x) = k_0 + k_1 x + \cdots + k_n x^n$$

be a polynomial in $K[x]$. Then by the **formal derivative**, $f'(x)$, of $f(x)$ we mean the polynomial

$$f'(x) = k_1 + \cdots + i k_i x^{i-1} + \cdots + n k_n x^{n-1} \in K[x].$$

Let K be a field and $f(x)$, $g(x) \in K[x]$. The following properties of formal derivatives are easily verified:

$$\begin{aligned}
(f(x) + g(x))' &= f'(x) + g'(x), \\
(f(x)g(x))' &= f(x)g'(x) + f'(x)g(x), \\
(kf(x))' &= kf'(x) \text{ for all } k \in K
\end{aligned}$$

and if $f(x) = x$, then $f'(x) = 1$.

Theorem 22.1.4 *Let* K *be a field and* $f(x) \in K[x]$, $f(x) \neq 0$. *Let* a *be a root of* $f(x)$ *in some extension field* F *of* K. *Then* a *is a multiple root of* $f(x)$ *if and only if* $f'(a) = 0$.

Proof. Suppose a is a multiple root of $f(x)$. Then $(x - a)^2$ divides $f(x)$. Hence,

$$f(x) = (x - a)^2 g(x)$$

for some $g(x) \in F[x]$. Now $f'(x) = (x - a)\{(x - a)g'(x) + 2g(x)\}$. Therefore, $f'(a) = 0$. Conversely, suppose $f'(a) = 0$. Then deg $f(x) \geq 2$. By the division algorithm,

$$f(x) = (x - a)^2 q(x) + h(x)$$

for some $q(x), h(x) \in F[x]$, where either $h(x) = 0$ or deg $h(x) \leq 1$. Suppose $h(x) \neq 0$. Since $f(a) = 0$, $h(a) = 0$. Thus, deg $h(x) = 1$ and a is a root of $h(x)$. Hence, $h(x) = b(x - a)$ for some $0 \neq b \in K$. This implies that

$$f(x) = (x - a)^2 q(x) + b(x - a)$$

and so
$$f'(x) = (x - a)\{(x - a)q'(x) + 2q(x)\} + b$$

Therefore,
$$0 = f'(a) = b,$$

a contradiction. Hence, $h(x) = 0$ and so $f(x) = (x - a)^2 q(x)$. Consequently, a is a multiple root of $f(x)$. ∎

Theorem 22.1.5 *For any field K, an irreducible polynomial $p(x)$ in $K[x]$ is separable if and only if $p(x)$ and its formal derivative $p'(x)$ are relatively prime.*

Proof. Let $d(x)$ denote the gcd of $p(x)$ and $p'(x)$. Suppose $p(x)$ is separable. Let c be a root of $p(x)$ in some field containing K. Then $p(x) = (x - c)f(x)$ for some $f(x) \in K(c)[x]$. Since $p(x)$ is irreducible, $f(c) \neq 0$. Now $p'(x) = f(x) + (x - c)f'(x)$ and so $p'(c) = f(c) + 0 \neq 0$. Hence, c is not a root of $d(x)$. But every root of $d(x)$ must be a root of $p(x)$ since $d(x)|p(x)$. Thus, since we have just seen that $d(x)$ and $p(x)$ have no common roots, $d(x)$ has no roots. Therefore, $d(x) = 1$.

Conversely, suppose that $d(x) = 1$. Let c be any root of $p(x)$. Let m denote the multiplicity of c. Then
$$p(x) = (x - c)^m f(x)$$

over $K(c)$ and c is not a root of $f(x)$. Now
$$\begin{aligned} p'(x) &= m(x - c)^{m-1}f(x) + (x - c)^m f'(x) \\ &= (x - c)^{m-1}[mf(x) + (x - c)f'(x)]. \end{aligned}$$

Thus, $(x - c)^{m-1}$ is a common divisor of $p'(x)$ and $p(x)$. Hence,
$$(x - c)^{m-1}|d(x).$$

Since $d(x) = 1$, $m = 1$. Consequently, $p(x)$ has no repeated roots. ∎

Theorem 22.1.6 *For any field K, an irreducible polynomial $p(x)$ in $K[x]$ is separable if and only if $p'(x) \neq 0$.*

Proof. Let $d(x)$ denote the gcd of $p(x)$ and $p'(x)$. Suppose $p(x)$ is separable. If $p'(x) = 0$, then $d(x) = p(x) \neq 1$, a contradiction of Theorem 22.1.5. Conversely, suppose $p'(x) \neq 0$. Since $p(x)$ is irreducible, the only common divisors of $p(x)$ and $p'(x)$ are 1 and $p(x)$. Since $1 \leq \deg p'(x) < \deg p(x)$, 1 is the only common divisor of $p'(x)$ and $p(x)$. Hence, $d(x) = 1$. Thus, $p(x)$ is separable by Theorem 22.1.5. ∎

Corollary 22.1.7 *Let K be a field of characteristic 0. Then every nonconstant polynomial in $K[x]$ is separable.*

Proof.   Let $f(x)$ be any nonconstant polynomial in $K[x]$ and $p(x) = k_0$ $+k_1 x + k_2 x^2 + \cdots + k_n x^n$ be any irreducible factor of $f(x)$, where $n \geq 1$. Then there exists $i > 0$ such that $k_i \neq 0$. Hence, $ik_i \neq 0$ since K has characteristic 0. Thus, $p'(x) \neq 0$ and so $p(x)$ is separable by Theorem 22.1.6. Hence, $f(x)$ is separable. $\blacksquare$

Example 22.1.8 *Consider the irreducible polynomial $p(x) = x^p - t^p$ over $K(t^p)$ of Example 22.1.3. Then $p'(x) = px^{p-1} = 0$. Thus, $x^p - t^p$ is inseparable over $K(t^p)$.*

Theorem 22.1.9 *Let K be a field of characteristic $p > 0$. Then an irreducible polynomial $p(x) = k_0 + k_1 x + k_2 x^2 + \cdots + k_n x^n$ over K is inseparable if and only if $p(x) = q(x^p)$ for some $q(x^p) \in K[x^p]$.*

Proof.   Clearly $p'(x) = 0$ if and only if $ik_i = 0$ for all $i = 1, 2, \ldots, n$. Thus, $p'(x) = 0$ if and only if $p|i$ for those i such that $k_i \neq 0$. Hence, $p'(x) = 0$ if and only if $p(x) = q(x^p)$ for some $q(x^p) \in K[x^p]$. The conclusion now follows from Theorem 22.1.6. $\blacksquare$

Let K be a field of characteristic $p > 0$. Let $K^p = \{a^p \mid a \in K\}$. The reader is asked to verify in Exercise 7 (page 490) that K^p is a subfield of K.

Definition 22.1.10 *Let K be a field. Then K is called **perfect** if every algebraic extension of K is separable.*

Example 22.1.11 *By Corollary 22.1.7, every field of characteristic 0 is perfect.*

The following theorem gives a necessary and sufficient condition for a field to be perfect.

Theorem 22.1.12 *Let K be a field of characteristic $p > 0$. Then K is perfect if and only if $K = K^p$.*

Proof.   Suppose K is perfect. Let $a \in K$ and F be a splitting field of $x^p - a \in K[x]$. Then F/K is a separable extension. Let $b \in F$ be a root of $x^p - a$. Then

$$x^p - a = (x - b)^p.$$

Let $p(x) \in K[x]$ be the minimal polynomial of b. Then $p(x)$ has distinct roots. If $\deg p(x) > 1$, then since $p(x)|(x - b)^p$, $p(x)$ has multiple roots, a contradiction. Hence, $\deg p(x) = 1$. This implies that $b \in K$. Hence, $a = b^p \in K^p$. Thus, $K = K^p$.

Conversely, suppose $K = K^p$. Let F/K be an algebraic field extension. Let $a \in F$ and $f(x) \in K[x]$ be the minimal polynomial of a. Suppose $f(x)$ is not separable. Then by Theorem 22.1.9, $f(x) = g(x^p)$ for some $g(x) \in K[x]$. Hence,

$$f(x) = a_0 + a_1 x^p + \cdots + a_k x^{pk},$$

$a_i \in K$, $1 \leq i \leq k$. Since $K = K^p$, $a_i = b_i^p$ for some $b_i \in K$, $1 \leq i \leq k$. Therefore,

$$f(x) = (b_0 + b_1 x + \cdots + b_k x^k)^p,$$

a contradiction, since $f(x)$ is irreducible over K. Hence, $f(x)$ is separable. Thus, F/K is a separable extension. Consequently, K is perfect. ∎

Example 22.1.13 *Let K be a finite field of characteristic p. Define $\sigma : K \to K^p$ by $\sigma(a) = a^p$. Then σ is a homomorphism. Suppose that $\sigma(a) = \sigma(b)$. Then $a^p = b^p$ and so $(a - b)^p = 0$. Since K is a field, K has no nonzero nilpotent elements. Thus, $a = b$ and so σ is one-one. Hence, $|K| = |\sigma(K)| \leq |K^p| \leq |K|$ and so $|K| = |K^p|$. Since K^p is a subfield of K and K is finite, $K = K^p$. Hence, K is perfect. We have thus shown that every finite field is perfect.*

If $p(x) = k_0 + k_1 x + k_2 x^2 + \cdots + k_n x^n$ is irreducible and inseparable over K in Theorem 22.1.9, then $p(x) = k_0 + k_p x^p + \cdots + k_{pm}(x^p)^m = q(x^p)$. It may be the case that $p(x) = q(x^p) = s(x^{p^2})$ in $K[x^{p^2}]$. However, there exists a largest positive integer e such that $p(x) = t(x^{p^e})$ for some $t(x^{p^e}) \in K[x^{p^e}]$. If $n = \deg p(x)$, then $p^e | n$.

Definition 22.1.14 *Let K be a field of characteristic $p > 0$ and $p(x)$ be an irreducible polynomial in $K[x]$. Let e be the largest nonnegative integer such that $p(x) = q(x^{p^e})$ for some $q(x^{p^e}) \in K[x^{p^e}]$. Then e is called the **exponent of inseparability** of $p(x)$ and p^e is called the **degree of inseparability** of $p(x)$. If n denotes the degree of $p(x)$, then $n_0 = \frac{n}{p^e}$ is called the **degree of separability** or **reduced degree** of $p(x)$ over K.*

By Theorem 22.1.9, $p(x)$ in Definition 22.1.14 is separable if and only if $e = 0$.

Theorem 22.1.15 *Let K be a field of characteristic $p > 0$ and*

$$p(x) = k_{n_0}(x^{p^e})^{n_0} + \cdots + k_1 x^{p^e} + k_0$$

be an irreducible polynomial in $K[x]$, where e is the exponent of inseparability of $p(x)$. Then the polynomial

$$s(y) = k_{n_0} y^{n_0} + \cdots + k_1 y + k_0 \in K[y]$$

is irreducible and separable over K.

Proof. If $s(y) = f(y)g(y) \in K[y]$, then $p(x) = f(x^{p^e})g(x^{p^e})$, contrary to the fact that $p(x)$ is irreducible in $K[x]$. Thus, $s(y)$ is irreducible in $K[y]$. If $s(y) = q(y^p)$ for some $q(y^p) \in K[y^p]$, then $p(x) = q(x^{p^{e+1}})$, contrary to the maximality of e. Hence, $s(y)$ is separable. ∎

Example 22.1.16 *Consider the polynomial $p(x) = x^{2p} + tx^p + t$ over the field $K(t)$, where K is a field of characteristic $p > 0$ and t is transcendental over K. By Eisenstein's criterion, $p(x)$ is irreducible over $K(t)$. Now $p(x) = (x^p)^2 + tx^p + t \in K(t)[x]$ and so $p(x)$ is inseparable over $K(t)$. The inseparability exponent e of $p(x)$ equals 1. Thus, $x^2 + tx + t$ is separable over $K(t)$.*

Definition 22.1.17 *Let F/K be a field extension. F is called a **simple** extension if $F = K(a)$ for some $a \in K$. Such an element a is called a **primitive** element.*

Theorem 22.1.18 *Let K be an infinite field and $K(a, b)/K$ be a field extension with a algebraic over K and b separable algebraic over K. Then there exists an element $c \in K(a, b)$ such that $K(a, b) = K(c)$, i.e., $K(a, b)/K$ is a simple extension.*

Proof. Let $f(x)$ and $g(x)$ be the minimal polynomials of a and b over K with degrees n and m and roots $a = a_1, a_2, \ldots, a_n$, and $b = b_1, b_2, \ldots, b_m$, respectively, in some extension field of K. Since b is separable, all b_i's are distinct. Also, since K is infinite, there exists $s \in K$ such that $a + sb \neq a_i + sb_j$, i.e.,

$$s \neq \frac{a_i - a}{b - b_j}$$

for all $1 \leq i \leq n$, $1 < j \leq m$. Let $c = a + sb$. Then $c - sb_j \neq a_i$ for all $1 \leq i \leq n$, $1 < j \leq m$. Also, $K(c) \subseteq K(a, b)$. Let $h(x) = f(c - sx) \in K(c)[x]$. Now

$$h(b) = f(c - sb) = f(a) = 0.$$

Thus, $g(x)$ and $h(x)$ have the common root b of multiplicity 1 in the field $K(a, b)$. Now

$$h(b_j) = f(c - sb_j) \neq 0$$

for all $1 < j \leq m$. Thus, $g(x)$ and $h(x)$ have only root b in common. Let $d(x) \in K(c)[x]$ be the greatest common divisor of $g(x)$ and $h(x)$. Then b is a root of $d(x)$. Every root of $d(x)$ is also a root of $g(x)$ and $h(x)$. Since $g(x)$ and $h(x)$ have no roots other than b in common in any field and b is of multiplicity 1, $d(x)$ is of degree 1. Hence, $d(x) = x - b$. But then $b \in K(c)$. Thus, $a = c - sb \in K(c)$. Therefore, $K(a, b) \subseteq K(c) \subseteq K(a, b)$ and so $K(c) = K(a, b)$. ∎

Corollary 22.1.19 *Let K be an infinite field. Let $a_1, a_2, \ldots, a_n$ be elements in some field containing K. Suppose that a_1 is algebraic and $a_2, \ldots, a_n$ are separable algebraic over K. Then there exists an element $c \in K(a_1, \ldots, a_n)$ such that $K(c) = K(a_1, \ldots, a_n)$, i.e., $K(a_1, \ldots, a_n)/K$ is a simple extension.*

Proof. The result follows by induction on n and Theorem 22.1.18. ∎

Corollary 22.1.20 *Let F/K be a field extension and the characteristic of K be 0. Let $a_1, a_2, \ldots, a_n \in F$ be algebraic over K. Then $K(a_1, \ldots, a_n)/K$ is a simple extension.*

Proof. The proof follows by Corollaries 22.1.7 and 22.1.19. ∎

Example 22.1.21 *Consider $\mathbf{Q}(\sqrt{2}, i)$. Now $1 \neq \frac{-\sqrt{2} - (\sqrt{2})}{i - (-i)} = -\frac{\sqrt{2}}{i}$. Thus, $\mathbf{Q}(\sqrt{2}, i) = \mathbf{Q}(\sqrt{2} + i)$ by the proof of Theorem 22.1.18, with $s = 1$ there.*

Theorem 22.1.22 (Artin) *Let K be an infinite field. Let F/K be a finite field extension. Then F/K is a simple extension if and only if there are only a finite number of intermediate fields of F/K.*

Proof. Suppose F/K is a simple extension. Let $F = K(a)$ for some $a \in F$. Let L be an intermediate field of F/K and $f(x)$ be the minimal polynomial of a over L. Let L' be the field generated by K and the coefficients of $f(x)$. Then $L' \subseteq L$ and $f(x)$ is also the minimal polynomial of a over L'. Hence,

$$[F : L] = \deg f(x) = [F : L'].$$

Thus, $[L : L'] = 1$ and so $L = L'$. Let $g(x)$ be the minimal polynomial of a over K. Then $f(x)$ divides $g(x)$. Now $g(x)$ has only a finite number of distinct monic factors. Hence, the number of intermediate fields is finite.

Conversely, suppose there are only a finite number of intermediate fields of F/K. Let $a, b \in F$. We first show that $K(a, b)/K$ is a simple extension. Let $c \in K$ and $F_c = K(a + cb)$. Then for all $c \in K$, F_c is an intermediate field of $K(a, b)/K$. Since the number of intermediate fields is finite and K is infinite, there exists $c, d \in K$, $c \neq d$ such that $F_c = F_d$. Then

$$b = (c - d)^{-1}(a + cb - a - db) \in F_c.$$

Hence, $a = a + cb - cb \in F_c$. Thus, $K(a, b) = F_c = K(a + cb)$, i.e., $K(a, b)/K$ is a simple extension. Now for all $a \in F$, $K(a)$ is an intermediate field of F/K. Since $[F : K]$ is finite, $[K(a) : K]$ is finite. Let

$$A = \{[K(a) : K] \mid a \in F\}.$$

Then A is a finite subset of **Z**. Let $a \in F$ be such that the maximum of $A = [K(a) : K]$. Suppose $F \neq K(a)$. Let $b \in F$ be such that $b \notin K(a)$. Then $K(a) \subset K(a,b)$. There exists $c \in F$ such that $K(a,b) = K(c)$. Therefore, $K(a) \subset K(c)$. Hence, $[K(c) : K] > [K(a) : K]$, a contradiction to the maximality of $[K(a) : K]$. Consequently, $F = K(a)$, i.e., F/K is a simple extension. ∎

Let F/K be a field extension. In the next chapter, we show that every finite extension of a finite field is a simple extension (Corollary 23.1.8, page 494). Hence, from this and Theorem 22.1.22, it follows that F/K is a simple extension if and only if there are only a finite number of intermediate fields of F/K.

We now focus our attention on the study of **separable algebraic and purely inseparable extensions**.[1]

Theorem 22.1.23 *Let K be a field of characteristic $p > 0$ and $f(x) = x^{p^e} - k$ be a polynomial over K, where e is a positive integer. Then $f(x)$ is irreducible over K if and only if $k \notin K^p$.*

Proof. Suppose $f(x)$ is irreducible over K. If $k = k'^p \in K^p$ for some $k' \in K$, then $f(x) = (x^{p^{e-1}} - k')^p$, contrary to the fact that $f(x)$ is irreducible over K. Hence, $k \notin K^p$. Conversely, suppose $k \notin K^p$. Let $p(x)$ be a nonconstant monic irreducible factor of $f(x)$ in $K[x]$ and c be a root of $p(x)$. Then c is a root of $f(x)$ and so $c^{p^e} = k$ and $f(x) = (x - c)^{p^e}$ over $K(c)$. Since $K(c)[x]$ is a unique factorization domain, it follows that $p(x)$ is some power of $(x - c)$, say, $p(x) = (x - c)^m$. Thus, $mn = p^e$ for some n so that $m = p^r$ and $n = p^s$ for nonnegative integers r and s. Therefore, $p(x) = x^{p^r} - c^{p^r}$ in $K[x]$. If $s > 0$, then $k = c^{p^e} = (c^{p^r})^{p^s} \in K^{p^s} \subseteq K^p$, which is contrary to the assumption $k \notin K^p$. Thus, $s = 0$ and so $r = e$. Hence, $p(x) = f(x)$, i.e., $f(x)$ is irreducible. ∎

Definition 22.1.24 *Let F/K be a field extension of characteristic $p > 0$. Let $c \in F$ be a root of the irreducible polynomial $p(x)$ in $K[x]$. If the degree of separability n_0 of $p(x)$ equals 1, then c is said to be **purely inseparable** over K. If every element of F is purely inseparable over K, then F/K is called a **purely inseparable extension**.*

In Theorem 22.1.15, let c be a root of $p(x)$. Then c^{p^e} is a root of $s(y)$. We have $K(c) \supseteq K(c^{p^e}) \supseteq K$ and c is a root of the polynomial $x^{p^e} - c^{p^e}$ over $K(c^{p^e})$. It follows that $x^{p^e} - c^{p^e}$ is irreducible over $K(c^{p^e})$, $K(c)/K(c^{p^e})$ is purely inseparable, and $K(c^{p^e})/K$ is separable.

---

[1]The remainder of this section may be skipped without any discontinuity. The only place this material is needed is in Example 24.2.8.

Theorem 22.1.25 *Let F/K be a field extension of characteristic $p > 0$ and c be an element of F. Then c is purely inseparable over K if and only if $c^{p^m} \in K$ for some nonnegative integer m.*

Proof.   Let c be purely inseparable over K. Then the degree of separability n_0 of the minimal polynomial $p(x)$ of c equals 1. Thus, $p(x) = x^{p^e} + k$ in $K[x]$, where e is the exponent of inseparability of $p(x)$ over K. Therefore, $c^{p^e} + k = 0$ or $c^{p^e} = -k \in K$. Hence, we can take $m = e$. Conversely, suppose $c^{p^m} \in K$. Let e be the smallest nonnegative integer such that $c^{p^e} \in K$. Then c is a root of the polynomial $x^{p^e} - k$ over K, where $k = c^{p^e}$. If $x^{p^e} - k$ is not irreducible over K, then $e > 0$ and $k = k'^p$ for some $k' \in K$ by Theorem 22.1.23. In this case, $x^{p^e} - k = (x^{p^{e-1}} - k')^p$. Thus, $(c^{p^{e-1}} - k')^p = 0$ and since a field has no nonzero nilpotent elements, $c^{p^{e-1}} - k' = 0$ or $c^{p^{e-1}} = k' \in K$. However, this contradicts the minimality of e. Thus, $x^{p^e} - k$ is irreducible over K. Clearly the degree of separability of $x^{p^e} - k$ is 1. Therefore, c is purely inseparable over K. ∎

Corollary 22.1.26 *Let F/K be a field extension of characteristic $p > 0$ and $c \in F$.*

   (i) If c is algebraic over K, then c is purely inseparable over K if and only if the minimal polynomial of c over K is $x^{p^e} - c^{p^e}$, where e is the smallest nonnegative integer such that $c^{p^e} \in K$.

   (ii) If c is purely inseparable over K, then $[K(c) : K] = p^e$ for some nonnegative integer e.

   (iii) If c is purely inseparable and separable algebraic over K, then $c \in K$.

Proof.   The proof of (i) follows from Theorem 22.1.25. Statement (ii) is an immediate consequence of statement (i). For the proof of statement (iii), we see that since c is purely inseparable over K the minimal polynomial of c over K has the form $x^{p^e} - k$. Since c is separable algebraic over K, the exponent of inseparability of $x^{p^e} - k$ is 0, i.e., $e = 0$. Thus, $x - k$ is the minimal polynomial of c over K, whence $c = k \in K$. ∎

Corollary 22.1.27 *Let F/K be a field extension of characteristic $p > 0$.*

   (i) If $F = K(M)$ for some subset M of F such that every element of M is purely inseparable over K, then F/K is a purely inseparable extension.

   (ii) Let L be an intermediate field of F/K. Then F/K is purely inseparable if and only if F/L and L/K are purely inseparable.

   (iii) The set of all elements of F which are purely inseparable over K is an intermediate field of F/K.

Proof.   (i) Let c be an element of F. Then there exists a finite subset $\{m_1, m_2, \ldots, m_s\}$ of M such that

$$c = \sum_{i_1, \ldots, i_s} k_{i_1 \ldots i_s} m_1^{i_1} \cdots m_s^{i_s},$$

where here we are using the fact that $F = K[M]$ since F/K is necessarily an algebraic extension. Let $e = \max\{e_1, \ldots, e_s\}$, where e_i is a nonnegative integer such that $m_i^{p^{e_i}} \in K$, $i = 1, \ldots, s$. Then

$$c^{p^e} = \sum_{i_1, \ldots, i_s} k_{i_1 \ldots i_s}^{p^e} (m_1^{p^e})^{i_1} \cdots (m_s^{p^e})^{i_s} \in K.$$

Hence, c is purely inseparable over K.

(ii) Suppose that F/K is purely inseparable. Let $c \in F$. Then there exists a nonnegative integer e such that $c^{p^e} \in K$ and so $c^{p^e} \in L$. Thus, F/L is purely inseparable. L/K is purely inseparable since every element of L is an element of F. Conversely, suppose F/L and L/K are purely inseparable. Let $c \in F$. Then there exists a nonnegative integer m such that $c^{p^m} \in L$. Since L/K is purely inseparable, there exists a nonnegative integer n such that $(c^{p^m})^{p^n} \in K$. Therefore, $c^{p^{m+n}} \in K$ so that c is purely inseparable over K.

(iii) Let J denote the set of all elements of F which are purely inseparable over K. Then $K \subseteq J$ and so $J \neq \phi$. Let $c, d \in J$. Then $c^{p^e} \in K$ and $d^{p^f} \in K$ for some nonnegative integers e and f. Let $n = \max\{e, f\}$. Then $(c - d)^{p^n} = c^{p^n} - d^{p^n} \in K$. Hence, $c - d \in J$. If $d \neq 0$, then $(cd^{-1})^{p^n} = c^{p^n}(d^{p^n})^{-1} \in K$. Thus, $cd^{-1} \in J$. Hence, J is an intermediate field of F/K. ∎

Theorem 22.1.25 and Corollary 22.1.27(i) make it quite easy to construct examples of purely inseparable field extensions.

Example 22.1.28 *Let J be any field of characteristic $p > 0$; e.g., $J = \mathbf{Z}_p$. Let $F = J(x, y, z)$, where x, y, z are algebraically independent over J. Set $K = J(x^p, y^{p^2}, z^{p^3})$. Then F/K is purely inseparable since x, y, z are purely inseparable over K. It can be shown that $[F : K] = p^6$ since x, y, z are algebraically independent over J. Since x^p, y^{p^2}, $z^{p^3} \in K$, we have $F^{p^3} \subseteq K$.*

For any field F of prime characteristic p, F/F^{p^e} is a purely inseparable field extension for any nonnegative integer e.

The following example is essentially the same as that in Example 22.1.28.

Example 22.1.29 *Let J be any field of characteristic $p > 0$. Let $K = J(x, y, z)$, where x, y, z are algebraically independent over J. Let $F = J(a, b, c)$, where a is a root of the polynomial $t^p - x$ over K, b is a root of the polynomial $t^{p^2} - y$ over $K(a)$, and c is a root of the polynomial $t^{p^3} - z$ over $K(a, b)$. Then F/K is purely inseparable, $[F : K] = p^6$, and $F^{p^3} \subseteq K$. One often writes $a = x^{p^{-1}}$, $b = y^{p^{-2}}$, and $c = z^{p^{-3}}$.*

Example 22.1.30 *Let J be any field of characteristic $p > 0$. Let $K = J(t)$, where t is transcendental over J. Let $F = K(t^{p^{-1}}, t^{p^{-2}}, t^{p^{-3}}, \ldots)$. Then F/K is purely inseparable by Corollary 22.1.27. Since $[K(t^{p^{-1}}, t^{p^{-2}}, t^{p^{-3}}, \ldots, t^{p^{-n}}) : K(t^{p^{-1}}, t^{p^{-2}}, t^{p^{-3}}, \ldots, t^{p^{-n+1}})] = p$ for all positive integers n, $[F : K] = \infty$. There does not exist a positive integer e such that $F^{p^e} \subseteq K$.*

Example 22.1.31 *Let J be any field of characteristic $p > 0$. Let $K = J(x_1, x_2, x_3, \ldots)$, where $x_1, x_2, x_3, \ldots$ are algebraically independent over J. Let $F_0 = K(x_1^{p^{-1}}, x_2^{p^{-2}}, x_3^{p^{-3}}, \ldots)$. Then F_0/K is purely inseparable and $[F_0 : K] = \infty$. Let $F_1 = K(x_1^{p^{-2}}, x_2^{p^{-2}}, x_3^{p^{-2}}, \ldots)$. Then F_1/K is purely inseparable, $[F_1 : K] = \infty$, and $F_1^{p^2} \subseteq K$.*

We now turn our attention to separable extensions.

Theorem 22.1.32 *Let F/K be a field extension of characteristic $p > 0$. If F/K is separable algebraic, then $F = K(F^p)$. If $[F : K] < \infty$ and $F = K(F^p)$, then F/K is separable algebraic.*

Proof. Suppose F/K is separable algebraic. Now every element of F is purely inseparable over F^p and thus purely inseparable over $K(F^p)$. Every element c of F is separable algebraic over K and thus separable algebraic over $K(F^p)$. Thus, every element c of F is in $K(F^p)$ by Corollary 22.1.26(iii). Hence, $F \subseteq K(F^p)$, so that $F = K(F^p)$. Conversely, suppose $[F : K] < \infty$ and $F = K(F^p)$. Let a be any element of F. Since $[F : K] < \infty$, a is algebraic over K. If a is not separable over K, then the minimal polynomial of a over K has the form

$$(x^p)^n + \cdots + k_1 x^p + k_0.$$

Therefore, $0 = a^{np} + \cdots + k_1 a^p + k_0 \cdot 1$ with not all the $k_i = 0$. Hence, $1, a^p, \ldots, a^{np}$ are linearly dependent over K. By Theorem 21.1.14, $1, a, a^2, \ldots, a^n, \ldots, a^{np-1}$ are linearly independent over K, whence $1, a, a^2, \ldots, a^n$ are linearly independent over K.

We now show that this is impossible by showing that whenever n elements $b_1, \ldots, b_n$ of F are linearly independent over K, then the elements $b_1^p, \ldots, b_n^p$ are linearly independent over K. We can assume that $b_1, \ldots, b_n$ is a basis of F/K since any linearly independent set over K can be extended to a basis of F/K, in particular, the linearly independent set $\{1, a, \ldots, a^n\}$. By Exercise 7 (page 490), the mapping $\alpha : F \to F^p$ defined by $\alpha(c) = c^p$ for $c \in F$ is an isomorphism, which maps K onto K^p. Thus, since $b_1, \ldots, b_n$ is a basis of F/K, $b_1^p, \ldots, b_n^p$ is a basis of F^p/K^p. Hence, $b_1^p, \ldots, b_n^p$ spans F^p over K^p. Consequently, $b_1^p, \ldots, b_n^p$ spans $K(F^p)$ over K; i.e., F over K. Since F has dimension n over K and the n elements $b_1^p, \ldots, b_n^p$ span F over K, the elements $b_1^p, \ldots, b_n^p$

must be a basis for F over K. ∎

The field extension F/K of Example 22.1.30 shows that the finiteness condition $[F : K] < \infty$ cannot be dropped in the above theorem. We have $F = K(F^p)$, F/K is not separable algebraic, in fact, F/K is purely inseparable.

Corollary 22.1.33 *Let F/K be a field extension of characteristic $p > 0$.*

(i) Let a be an element of F. Then $K(a) = K(a^p)$ if and only if $K(a)/K$ is separable algebraic.

(ii) Let $a_1, a_2, \ldots, a_n$ be elements of F. Then $K(a_1, \ldots, a_n)/K$ is separable algebraic if and only if a_1 is separable algebraic over K and a_i is separable algebraic over $K(a_1, \ldots, a_{i-1})$, $i = 2, 3, \ldots, n$.

Proof. (i) If $K(a) = K(a^p)$, then a cannot be transcendental over K and so a must be algebraic over K. By Theorem 22.1.32, $K(a) = K(K(a)^p)$ if and only if $K(a)/K$ is separable algebraic. We thus have the desired result since $K(K(a)^p) = K(a^p)$.

(ii) Suppose $K(a_1, \ldots, a_n)/K$ is separable algebraic. Then $a_1, \ldots, a_n$ are separable algebraic over K. By the discussion following Definition 22.1.2, a_i is clearly separable algebraic over $K(a_1, \ldots, a_{i-1})$, $i = 2, 3, \ldots, n$. Conversely, suppose a_1 is separable algebraic over K and a_i is separable algebraic over $K(a_1, \ldots, a_{i-1})$, $i = 2, 3, \ldots, n$. Then $K(a_1) = K(a_1^p), \ldots, K(a_1, \ldots, a_{i-1})(a_i) = K(a_1, \ldots, a_{i-1})(a_i^p)$, $i = 2, 3, \ldots, n$. Thus, $K(a_1, \ldots, a_n) = K(a_1^p, \ldots, a_n^p) = K([K(a_1, \ldots, a_{i-1})]^p)$. The conclusion now holds from Theorem 22.1.32. ∎

Corollary 22.1.34 *Let F/K be a field extension of characteristic $p > 0$.*

(i) If $F = K(M)$ for some subset M of F such that every element of M is separable algebraic over K, then F/K is separable algebraic.

(ii) Let L be an intermediate field of F/K. Then F/K is separable algebraic if and only if F/L and L/K are separable algebraic.

(iii) The set of all elements of F which are separable algebraic over K is an intermediate field of F/K.

Proof. (i) Let $a \in F$. There exists a finite subset $\{m_1, \ldots, m_s\}$ of M such that $a \in K(m_1, \ldots, m_s)$. Since each m_i is separable algebraic over K, we have by Corollary 22.1.33(ii) that $K(m_1, \ldots, m_s)/K$ is separable algebraic. Hence, a and thus F/K is separable algebraic.

(ii) Suppose F/K is separable algebraic. Then F/L is separable algebraic by the discussion following Definition 22.1.2. L/K is separable algebraic since every element of L is an element of F. Suppose F/L and L/K are separable algebraic. Let $a \in F$. Let $c_0, c_1, \ldots, c_n \in L$ be the coefficients of the minimal polynomial $p(x)$ of a over L. Since a is separable algebraic over L, a is separable algebraic over $K(c_0, c_1, \ldots, c_n)$. ($p(x)$ is also the minimal polynomial of a over $K(c_0, c_1, \ldots, c_n)$.) Since $c_0, c_1, \ldots, c_n \in L$ and L/K is separable algebraic, $K(c_0, c_1, \ldots, c_n)/K$ is separable algebraic by Corollary 22.1.33(ii). Thus, a and so F is separable algebraic over K.

(iii) Let S denote the set of elements of F which are separable algebraic over K. Then $S \supseteq K$. Let $a, b \in S$. Then by Corollary 22.1.33(ii), $K(a, b)/K$ is separable algebraic. Since $a - b \in K(a, b)$ and (for $b \neq 0$) $ab^{-1} \in K(a, b)$, $a - b$, and ab^{-1} ($b \neq 0$) are separable algebraic over K and thus are members of S. Hence, S is a field. ∎

Definition 22.1.35 *Let F/K be an algebraic field extension of characteristic $p > 0$. Then the intermediate field of F/K consisting of all elements of F which are separable algebraic over K is called the **separable closure** of K in F or the **maximal separable intermediate** field of F/K. We denote this field by K_s.*

Theorem 22.1.36 *Let F/K be an algebraic field extension of characteristic $p > 0$. Then F/K_s is purely inseparable, where K_s is the separable closure of F/K.*

Proof. If $F = K_s$ the theorem is immediate. Suppose $F \supset K_s$. Let $a \in F, a \notin K_s$. Let

$$p(x) = k_0 + k_1 x^{p^e} + \cdots + (x^{p^e})^{n_0}$$

be the minimal polynomial of F/K_s, where e is the exponent of inseparability and n_0 is the reduced degree of $p(x)$ over K_s. Now by Theorem 22.1.15, $k_0 + k_1 y + \cdots + y^{n_0}$ is the minimal polynomial of a^{p^e} over K_s and this polynomial is separable over K_s. Hence, a^{p^e} is separable over K_s. Thus, $K_s(a^{p^e})/K_s$ is separable algebraic and so $K_s(a^{p^e})/K$ is separable algebraic. By the definition of K_s, we have $a^{p^e} \in K_s$. Therefore, a is purely inseparable over K_s. ∎

We can think of field theory as being separated into two parts, namely, that in which the fields are of characteristic 0 and that in which the fields are of prime characteristic p. It can be shown that for any field extension F/K, there exists a subset X of F which is algebraically independent over K and which also has the property that $F/K(X)$ is algebraic. The above theorem shows that the study of algebraic field extensions of characteristic $p > 0$ can be separated into two parts, the separable part and the purely inseparable part. Separable algebraic field extensions of characteristic $p > 0$ often act entirely similar to field extensions of characteristic 0. Purely inseparable field extensions have their own distinctive behavior.

Definition 22.1.37 *Let F/K be an algebraic field extension of characteristic $p > 0$. Then the degree $[K_s : K]$ is called the **degree of separability** of*

F/K and is denoted by $[F : K]_s$. The degree $[F : K_s]$ is called the **degree of inseparability** of F/K and is denoted by $[F : K]_i$.

Theorem 22.1.38 *Let K be a field of characteristic $p > 0$ and $p(x)$ an irreducible polynomial in $K[x]$. Let $K(a)$ be an extension of K obtained by adjoining a root a of $p(x)$ to K. Then*

$$[K(a) : K]_s = n_0,$$

$$[K(a) : K]_i = p^e,$$

where n_0 is the reduced degree of $p(x)$ over K and p^e is the degree of inseparability of $p(x)$ over K.

Proof.   Let $b \in K(a)$. Then $b = \sum_{i=0}^{n-1} k_i a^i$, where n is the degree of $p(x)$ over K and each $k_i \in K$. Therefore,

$$b^{p^e} = \sum_{i=0}^{n-1} k_i^{p^e} (a^{p^e})^i \in K(a^{p^e}).$$

Thus, b is purely inseparable over $K(a^{p^e})$. Hence, $K(a)/K(a^{p^e})$ is purely inseparable. By the definition of the degree of inseparability of $p(x)$ over K, $K(a^{p^e})/K$ is separable algebraic. Now $K_s \supseteq K(a^{p^e})$. Let $b \in K_s$. We have just seen that b is purely inseparable over $K(a^{p^e})$. But b is also separable algebraic over $K(a^{p^e})$. Therefore, $b \in K(a^{p^e})$ so that $K_s = K(a^{p^e})$. By Theorem 22.1.15, the minimal polynomial of a^{p^e} over K is of degree n_0 and so $[K(a) : K]_s = [K(a^{p^e}) : K] = n_0$. Thus, $n_0 p^e = [K(a) : K] = [K(a) : K(a^{p^e})][K(a^{p^e}) : K] = [K(a) : K(a^{p^e})]n_0$. Consequently, $p^e = [K(a) : K(a^{p^e})] = [K(a) : K]_i$. ∎

Example 22.1.39 *Let K denote the field $\mathbf{Z}_p(u, v)$, where u and v are algebraically independent over $\mathbf{Z}_p$. Let a be a root of the polynomial $x^{2p} + vx^p + u$ over K. By use of Worked-Out Exercise 6 (page 456), one can deduce that $x^{2p} + vx^p + u$ is irreducible over K. Let F be the field $K(a)$. We ask the reader to verify the following properties of the field extension F/K. $K_s = K(a^p)$, $[F : K]_i = p$, and $[F : K]_s = 2$. Also, the extension F/K has no elements which are purely inseparable over K (except those elements which are already in K). Thus, if J is the intermediate field of F/K consisting of all the elements of F purely inseparable over K, then $J = K$. Hence, F/J is not separable algebraic.*

22.1.1   Worked-Out Exercises

◇ **Exercise 1** Determine if the following polynomials are separable or inseparable over the given fields.

(i) $x^2 - 6x + 9$ over $\mathbf{Q}$;

(ii) $x^4 + x^2 + [1]$ over $\mathbf{Z}_2$.

Solution: (i) $x^2 - 6x + 9 = (x-3)^2$ over $\mathbf{Q}$. Now $x - 3$ is irreducible over $\mathbf{Q}$. Since $x - 3$ is separable over $\mathbf{Q}$, $x^2 - 6x + 9$ is separable over $\mathbf{Q}$.

(ii) $x^4 + x^2 + [1] = (x^2 + x + [1])^2$ over $\mathbf{Z}_2$. Now $x^2 + x + [1]$ has no roots in $\mathbf{Z}_2$. Hence, $x^2 + x + [1]$ is irreducible over $\mathbf{Z}_2$. Now $D_x(x^2 + x + [1]) = [2]x + [1] = [1] \neq [0]$. Thus, $x^2 + x + [1]$ and so $x^4 + x^2 + [1]$ is separable over $\mathbf{Z}_2$.

◇ **Exercise 2** Prove that the following polynomials are irreducible over $\mathbf{Z}_3(t)$, where t is transcendental over $\mathbf{Z}_3$. Find the exponent of inseparability and the degree of separability of the polynomials over $\mathbf{Z}_3(t)$.

(i) $p(x) = x^{36} + tx^{18} + t$.

(ii) $q(x) = x^{24} + tx^{18} + t$.

(iii) $r(x) = x^{20} + tx^{18} + t$.

(iv) $s(x) = x^9 + t$.

Solution: Since $t|t$, $t|0$, $t \nmid 1$, $t^2 \nmid t$, the polynomials $p(x)$, $q(x)$, $r(x)$, $s(x)$ are irreducible over $\mathbf{Z}_3(t)$.

(i) $p(x) = x^{4 \cdot 3^2} + tx^{2 \cdot 3^2} + t$ and so the exponent of inseparability $e = 2$ and the degree of separability $n_0 = 4$.

(ii) $q(x) = x^{8 \cdot 3} + tx^{6 \cdot 3} + t$ and so the exponent of inseparability $e = 1$ and the degree of separability $n_0 = 8$.

(iii) Since $3 \nmid 20$, $e = 0$ and $n_0 = 20$.

(iv) Here $e = 2$ and $n_0 = 1$.

◇ **Exercise 3** Let $f(x)$ and $g(x)$ be polynomials over the field K.

(i) Does $f(c) = g(c)$ for all $c \in K$ imply that $f(x) = g(x)$?

(ii) Does $f(c) = 0$ for all $c \in K$ imply that $f(x) = 0$?

Solution: (i) Let $f(x) = [3]x^5 - [4]x^2 \in \mathbf{Z}_5[x]$ and $g(x) = x^2 + [3]x \in \mathbf{Z}_5[x]$. Now $f([0]) = [0] = g([0])$, $f([1]) = [4] = g([1])$, $f([2]) = [0] = g([2])$, $f([3]) = [3] = g([3])$, $f([4]) = [3] = g([4])$. Hence, $f(c) = g(c)$ for all $c \in \mathbf{Z}_5$. However, $f(x) \neq g(x)$.

(ii) Let $f(x) = x^2 + x \in \mathbf{Z}_2[x]$. Then $f(c) = 0$ for all $c \in \mathbf{Z}_2$, but $f(x) \neq 0$.

Exercise 4 Let $K = P(x, y, z)$ and $F = K(z^{p^{-2}}, z^{p^{-2}}x^{p^{-1}} + y^{p^{-1}})$, where P is a perfect field of characteristic $p > 0$ and x, y, z are algebraically independent indeterminates over P. Prove that $K^{p^{-1}} \cap F = K(z^{p^{-1}})$, where $K^{p^{-1}} = \{k^{p^{-1}} \mid k \in K\}$.

Solution: Clearly $F \supset K^{p^{-1}} \cap F \supseteq K(z^{p^{-1}})$. Now $[F : K] = p^3$. Suppose that $K^{p^{-1}} \cap F \supset K(z^{p^{-1}})$. Then $F = (K^{p^{-1}} \cap F)(z^{p^{-2}})$ since $z^{p^{-2}} \notin K^{p^{-1}} \cap F$ and $[K^{p^{-1}} \cap F : K]$ must be p^2. Thus, $[F : K^{p^{-1}} \cap F] = p$. Since $[K^{p^{-1}}(F) : K^{p^{-1}}] = p$, any basis of $F/(K^{p^{-1}} \cap F)$ remains a basis of $K^{p^{-1}}(F)/K^{p^{-1}}$. Now $Z = \{1, z^{p^{-2}}, \ldots, (z^{p^{-2}})^{p-1}\}$ is a basis of $F/(K^{p^{-1}} \cap F)$. Also,

$$z^{p^{-2}} x^{p^{-1}} + y^{p^{-1}} = \sum_{i=0}^{p-1} k_i (z^{p^{-2}})^i,$$

where $k_i \in K^{p^{-1}} \cap F$, $i = 0, 1, \ldots, p-1$. Since Z remains linearly independent over $K^{p^{-1}}$, $y^{p^{-1}} = k_0 \in K^{p^{-1}} \cap F$ and $x^{p^{-1}} = k_1 \in K^{p^{-1}} \cap F$. Therefore, $x^{p^{-1}}, y^{p^{-1}} \in F$. Thus, $[F : K] = p^4$, a contradiction. Hence, $K^{p^{-1}} \cap F = K(z^{p^{-1}})$.

22.1.2   Exercises

1. Let $f(x) \in K[x]$, a polynomial ring over a field K and $c \in F$, where F is an extension field of K. Prove that $(x - c)^2 | f(x)$ if and only $(x - c)|f(x)$ and $(x - c)|f'(x)$.

2. Let $f(x) \in K[x]$, a polynomial ring over a field K. Use Exercise 1 to prove that $f(x)$ has no repeated roots in any extension field of K if and only if $f(x)$ and $f'(x)$ are relatively prime.

3. Let $f(x) = x^n - x \in K[x]$, a polynomial ring over a field K. Suppose that $n \geq 2$ and that either K has characteristic 0 or a prime p such that p does not divide $n - 1$. Prove that $f(x)$ has no repeated roots in any extension field F of K.

4. Let $f(x) = x^p - k \in K[x]$, a polynomial ring over a field K of characteristic $p > 0$. Prove that either $f(x)$ is irreducible over K or that $f(x)$ is a power of a linear polynomial in $K[x]$.

5. Determine if the following polynomials are separable or inseparable over the given field.

   (i) $x^2 - 4x + 4$ over $\mathbf{Q}$.

   (ii) $x^5 + tx + t$ over $\mathbf{Z}_5(t)$, where t is transcendental over $\mathbf{Z}_5$.

6. Prove that the following polynomials are irreducible over $\mathbf{Z}_5(u)$, where u is transcendental over $\mathbf{Z}_5$. Find the exponent of inseparability and the degree of separability of the polynomials over $\mathbf{Z}_5(u)$.

   (i) $p(x) = x^{250} + ux^{125} + u$.

   (ii) $g(x) = x^{128} + ux^{125} + u$.

   (iii) $s(x) = x^{125} + u$.

7. Let F be a field of characteristic $p > 0$. Prove that for any nonnegative integer e, F^{p^e} is a subfield of F. Prove also that the mapping $\alpha : F \to F^{p^e}$ defined by $\alpha(a) = a^{p^e}$ is an isomorphism.

8. Prove that a root of the polynomials in Examples 22.1.16 and 22.1.39 is neither purely inseparable nor separable algebraic over $K(t)$ and K, respectively.

9. Let $K(a)/K$ be a field extension of characteristic $p > 0$. Prove that $(K(a))^p = K^p(a^p)$.

10. Let F/K be a finite field extension of characteristic $p > 0$. If $[F : K]$ is not divisible by p, prove that F/K is separable.

11. Let F/K be an algebraic field extension and S be an intermediate field of F/K such that F/S is purely inseparable and S/K is separable algebraic. Prove that $S = K_s$.

12. Let P be a perfect field of characteristic $p > 0$. Let $P(a)/P$ be an algebraic field extension. Prove that $P(a)/P$ is separable and that $P(a)$ is perfect.

13. Let K be any field of characteristic $p > 0$. Prove that $\mathbf{Z}_p$ is the smallest subfield of K which is perfect and $\cap_{i=0}^{\infty} K^{p^i}$ is the largest subfield of K which is perfect.

14. Verify the properties of the field extension F/K of Example 22.1.39.

15. Answer the following statements, true or false. If the statement is true, prove it. If it is false, give a counterexample.

    (i) Let F be a field of characteristic $p > 0$. Since $F \simeq F^p$ and $F^p \subseteq F$, it follows that $F^p = F$.

    (ii) Let F/K be a field extension of characteristic $p > 0$. Let $c \in F \backslash K$. Then it is impossible for c to be both separable and purely inseparable over K.

    (iii) Let F/K be a field extension of characteristic $p > 0$. Let $c \in F$. Then it is impossible for c to be both separable and inseparable over K.

Emil Artin (1898–1962) was born on March 3, 1898, in Vienna, Austria. In 1916, he passed his school certification and after one semester of university work, he was called for military service. In January 1919, he resumed his studies at the University of Leipzig, where he was awarded the Ph.D. degree in 1921.

Artin was appointed lecturer in 1923, became extraordinary professor in 1925, and became ordinary professor in 1926 at the University of Hamburg. In 1937, along with his family Artin emigrated to the United

States. He taught for a year at the University of Notre Dame and from 1938 to 1946 at Indiana University. In 1946, he joined Princeton University, and in 1958 he returned to the University of Hamburg, where he remained teaching until his death in 1962.

In 1927, Artin proved the general law of reciprocity, which included all the previous known laws of reciprocity until the time of Gauss. It has become the main theorem of class field theory. In 1961, he published, with John Tate, *Class Field Theory*.

In 1926, in collaboration with Otto Schreier, Artin developed the theory of real-closed fields. The following year, with the help of the theorem on real-closed fields, he proved the Hilbert problem of definite functions. Also in 1927 he expanded the theory of algebras of associative rings. In 1928, Artin introduced the notion of rings with minimum condition. In his honor, these are called Artinian rings.

During the 1930s Artin started to reformulate Galois theory, using techniques of linear equations. In 1942, he published *Galois Theory*, reformulating it in an abstract setting as a relationship of field extensions and the subgroups of its automorphism— the we see it today—away from the classical approach as permutations of roots of an equation. He was fascinated by Galois theory, and in a 1950 lecture he said,

"Since my mathematical youth I have been under the spell of the classical theory of Galois. This charm has forced me to return to it again and again, and to try to find new ways to prove its fundamental theorem."

Artin contributed to various areas of mathematics, including number theory, group theory, ring theory, field theory, geometric algebra, and algebraic topology. He was awarded the American Mathematical Society's Cole Prize in number theory. He died on December 20, 1962.

Chapter 23

Finite Fields

The theory of finite fields has come to the fore in the last 60 years due to newfound applications. The applications of finite fields are in coding theory, combinatorics, switching circuits, statistics via finite geometries, and certain areas of computer science.

23.1   Finite Fields

A **finite field** (or **Galois field**) is a field with a finite number of elements. If F is a finite field, then F has prime characteristic p and contains a subfield isomorphic to $\mathbf{Z}_p$. Since F has only a finite number of elements, $[F : \mathbf{Z}_p] < \infty$.

We denote a finite field of n elements by $\mathrm{GF}(n)$. We will show in the next result that n must be a power of p. The result is due to E.H. Moore (1862–1932). The United States is indebted to Moore for its beginnings in abstract algebra and for its initial international recognition in research.

Theorem 23.1.1 *If F is a finite field of characteristic p and $n = [F : \mathbf{Z}_p]$, then F contains p^n elements.*

Proof.   Since $[F : \mathbf{Z}_p] = n$, $F/\mathbf{Z}_p$ has a basis of n elements, say, $b_1, b_2, \ldots, b_n$. Every element a of F is a linear combination of $b_1, b_2, \ldots, b_n$, i.e., $a = a_1 b_1 + a_2 b_2 + \cdots + a_n b_n$, where $a_i \in \mathbf{Z}_p$, $i = 1, 2, \ldots, n$. Now $\mathbf{Z}_p$ has p elements. Hence, F has at most p^n elements. Since $\{b_1, b_2, \ldots, b_n\}$ is linearly independent over $\mathbf{Z}_p$, $a_1 b_1 + a_2 b_2 + \cdots + a_n b_n$ is distinct for every choice of $a_1, a_2, \ldots, a_n$. Thus, F has exactly p^n elements. ■

Theorem 23.1.2 *Every element of a finite field F of characteristic p and of p^n elements is a root of the polynomial $x^{p^n} - x \in \mathbf{Z}_p[x]$. Moreover, F is a splitting field of $x^{p^n} - x$ over $\mathbf{Z}_p$.*

Proof.   First note that $(F\backslash\{0\}, \cdot)$ is a commutative group of order $p^n - 1$. Thus, for all $a \in F\backslash\{0\}$, $a^{p^n-1} = 1$, whence $a^{p^n} = a$. Clearly $0^{p^n} = 0$. Since F contains all the roots of $x^{p^n} - x$, F contains a splitting field S of $x^{p^n} - x$ over $\mathbf{Z}_p$. However, F is exactly the set of all the roots of $x^{p^n} - x$ and so $F = S$. ∎

In the following result, we once again use a positive integer and the concept of an isomorphism to completely characterize an algebraic structure.

Corollary 23.1.3 *Any two finite fields of p^n elements are isomorphic, where p is a prime and n is a positive integer.*

Proof.   If F and F' are finite fields with p^n elements, then they are splitting fields of the polynomial $x^{p^n} - x$ over $\mathbf{Z}_p$. Hence, $F \simeq F'$. ∎

The next theorem can be used to show that there exists an irreducible polynomial of arbitrary degree n over $\mathbf{Z}_p$. (See Exercise 8, page 497.) Even though its proof is not constructive in nature, it is informative for certain applications. Exercises 5 and 6 can be used to actually count the irreducible polynomials of a given degree. There is an algorithm which can be used to test the irreducibility of a polynomial over a finite field—namely, Berlekamp's algorithm. This algorithm is discussed in Isaacs.

Theorem 23.1.4 *For any prime p, there exists a field extension $F/\mathbf{Z}_p$ of arbitrary finite degree n.*

Proof.   Let S be the splitting field of the polynomial $f(x) = x^{p^n} - x$ over $\mathbf{Z}_p$. Let $a \in S$ be a root of $f(x)$ of multiplicity m. Then

$$f(x) = (x - a)^m g(x),$$

where a is not a root of $g(x)$. Now

$$-1 = f'(x) = (x - a)^{m-1}[mg(x) + (x - a)g'(x)].$$

This implies that $(x - a)^{m-1}$ divides -1, whence $m - 1 = 0$. Thus, every root of $f(x)$ in S has multiplicity 1. Hence, $f(x)$ has p^n distinct roots in S. Let F denote the subset of S, which consists of all roots of $f(x)$. Let $a, b \in F$. Then $(a - b)^{p^n} = a^{p^n} - b^{p^n} = a - b$. Therefore, $a - b \in F$. For $b \neq 0$,

$$(ab^{-1})^{p^n} = a^{p^n}(b^{p^n})^{-1} = ab^{-1} \in F.$$

Thus, F is a subfield of S. Since F contains all the roots of $f(x)$ and S is generated by the roots of $f(x)$ over $\mathbf{Z}_p$, $F = S$. By Exercise 6 (page 497), $[F : \mathbf{Z}_p] = n$. ∎

Theorem 23.1.5 *Let F be a field and G be a finite subgroup of the multiplicative group $F^* = F\backslash\{0\}$. Then G is cyclic.*

Proof.   Since G is a finite Abelian group, G is a direct product of cyclic subgroups $C_1, C_2, \ldots, C_k$, where $|C_i| = n_i$, $n_1 > 1$, and $n_i | n_{i+1}$, $1 \leq i < k$, by Theorem 9.1.7. From this it follows that $g^{n_k} = 1$ for all $g \in G$. Thus, every element of G is a root of $x^{n_k} - 1 \in F[x]$. Since $x^{n_k} - 1$ has at most n_k distinct roots in F, $|G| \leq n_k$. Now C_k is a subgroup of G and $|C_k| = n_k$. Hence, $G = C_k$ and so G is cyclic. ■

The following corollary is an immediate consequence of Theorem 23.1.5.

Corollary 23.1.6 *The multiplicative group of a finite field is cyclic.* ■

Theorem 23.1.7 *Let F be a finite field and $F(a, b)/F$ a field extension with a, b algebraic over F. Then there exists $c \in F(a, b)$ such that $F(a, b) = F(c)$, i.e., $F(a, b)$ is a simple extension.*

Proof.   Since $F(a, b)/F$ is algebraic, $[F(a,b) : F] < \infty$. Thus, $F(a, b)$ is a finite field since F is a finite field. Since $F(a, b) \backslash \{0\}$ is a cyclic group with some generator, say, c by Theorem 23.1.5, it follows that $F(a, b) = F(c)$. ■

Corollary 23.1.8 *Every finite extension of a finite field is simple.* ■

23.1.1   Worked-Out Exercises

◇ **Exercise 1** Prove that $x^3 + x + [1]$ is irreducible in $\mathbf{Z}_2[x]$. Write out the addition and multiplication tables for the field

$$\mathbf{Z}_2[x] / \left\langle x^3 + x + [1] \right\rangle.$$

Find a splitting field S_1 for $x^3 + x + [1]$ over $\mathbf{Z}_2$. Find a basis for $S_1/\mathbf{Z}_2$ and $[S_1 : \mathbf{Z}_2]$.

Solution: $x^3 + x + [1]$ is irreducible over $\mathbf{Z}_2$ if and only if $\mathbf{Z}_2$ contains no root of $x^3 + x + [1]$. Since $[0]^3 + [0] + [1] \neq [0]$ and $[1]^3 + [1] + [1] \neq [0]$ in $\mathbf{Z}_2$, $\mathbf{Z}_2$ contains no roots of $x^3 + x + [1]$ over $\mathbf{Z}_2$. Hence, $x^3 + x + [1]$ is irreducible over $\mathbf{Z}_2$. By Theorem 21.1.11,

$$\mathbf{Z}_2[x] / \left\langle x^3 + x + [1] \right\rangle = \mathbf{Z}_2(\lambda),$$

where λ denotes the coset $x + \left\langle x^3 + x + [1] \right\rangle$. By Theorem 21.1.14,

$$\mathbf{Z}_2(\lambda) = \{[0], [1], \lambda, \lambda^2, [1] + \lambda, [1] + \lambda^2, \lambda + \lambda^2, [1] + \lambda + \lambda^2\}.$$

Now

$$x^3 + x + [1] = (x + \lambda)(x^2 + \lambda x + [1] + \lambda^2)$$

and λ^2 and $\lambda+\lambda^2$ are the roots of $x^2+\lambda x+[1]+\lambda^2$. Since $\lambda^2, \lambda+\lambda^2 \in \mathbf{Z}_2(\lambda)$, $\mathbf{Z}_2(\lambda)$ is a splitting field of $x^3+x+[1]$ over $\mathbf{Z}_2$. Let $S_1 = \mathbf{Z}_2(\lambda)$. Then $\{[1], \lambda, \lambda^2\}$ is a basis for $S_1/\mathbf{Z}_2$ and $[S_1 : \mathbf{Z}_2] = 3$. Let α denote $[1]+\lambda+\lambda^2$. The addition table for $\mathbf{Z}_2(\lambda)$ is given below.

+	[0]	[1]	λ	λ^2	$[1]+\lambda$	$[1]+\lambda^2$	$\lambda+\lambda^2$	α
[0]	[0]	[1]	λ	λ^2	$[1]+\lambda$	$[1]+\lambda^2$	$\lambda+\lambda^2$	α
[1]	[1]	[0]	$[1]+\lambda$	$[1]+\lambda^2$	λ	λ^2	α	$\lambda+\lambda^2$
λ	λ	$[1]+\lambda$	[0]	$\lambda+\lambda^2$	[1]	α	λ^2	$[1]+\lambda^2$
λ^2	λ^2	$[1]+\lambda^2$	$\lambda+\lambda^2$	[0]	α	[1]	λ	$[1]+\lambda$
$[1]+\lambda$	$[1]+\lambda$	λ	[1]	α	[0]	$\lambda+\lambda^2$	$[1]+\lambda^2$	λ^2
$[1]+\lambda^2$	$[1]+\lambda^2$	λ^2	α	[1]	$\lambda+\lambda^2$	[0]	$[1]+\lambda$	λ
$\lambda+\lambda^2$	$\lambda+\lambda^2$	α	λ^2	λ	$[1]+\lambda^2$	$[1]+\lambda$	[0]	[1]
α	α	$\lambda+\lambda^2$	$[1]+\lambda^2$	$[1]+\lambda$	λ^2	λ	[1]	[0]

For the multiplication table, we make a few entries, such as $([1]+\lambda)([1]+\lambda) = [1]+\lambda^2$ and $([1]+\lambda+\lambda^2)([1]+\lambda^2) = [1]+\lambda+\lambda^3+\lambda^4$. We now reduce $[1]+\lambda+\lambda^3+\lambda^4$ to the form $a+b\lambda+c\lambda^2$, where $a, b, c \in \mathbf{Z}_2$. We divide $x^4+x^3+x+[1]$ by $x^3+x+[1]$ to obtain $x^4+x^3+x+[1] = (x+[1])(x^3+x+[1])+x^2+x$. Thus, $\lambda^4+\lambda^3+\lambda+[1] = (\lambda+[1])(\lambda^3+\lambda+[1])+\lambda^2+\lambda = [0]+\lambda^2+\lambda$. Hence, $([1]+\lambda+\lambda^2)([1]+\lambda^2) = \lambda+\lambda^2$.

$\diamondsuit$ **Exercise 2** Prove that $x^3+x^2+[1]$ is irreducible in $\mathbf{Z}_2[x]$. Write out the addition and multiplication tables for the field

$$\mathbf{Z}_2[x]/\left\langle x^3+x^2+[1]\right\rangle.$$

Find a splitting field S_2 for $x^3+x+[1]$ over $\mathbf{Z}_2$. Find a basis for $S_2/\mathbf{Z}_2$ and $[S_2 : \mathbf{Z}_2]$. Compare your results with those in Worked-Out Exercise 1.

Solution: Since $[0]^3+[0]^2+[1] \neq [0]$ and $[1]^3+[1]^2+[1] \neq [0]$ in $\mathbf{Z}_2$, $\mathbf{Z}_2$ contains no roots of $x^3+x^2+[1]$ over $\mathbf{Z}_2$. Hence, $x^3+x^2+[1]$ is irreducible over $\mathbf{Z}_2$. By Theorem 21.1.11,

$$\mathbf{Z}_2[x]/\left\langle x^3+x^2+[1]\right\rangle = \mathbf{Z}_2(\mu),$$

where μ denotes the coset $x+\left\langle x^3+x^2+[1]\right\rangle$. By Theorem 21.1.14,

$$\mathbf{Z}_2(\mu) = \{[0], [1], \mu, \mu^2, [1]+\mu, [1]+\mu^2, \mu+\mu^2, [1]+\mu+\mu^2\}.$$

Now $x^3+x^2+[1] = (x+\mu)(x^2+([1]+\mu)x+\mu+\mu^2)$ and μ^2 and $[1]+\mu+\mu^2$ are the roots of $x^2+([1]+\mu)x+\mu+\mu^2$. Since $\mu^2, [1]+\mu+\mu^2 \in \mathbf{Z}_2(\mu)$, $\mathbf{Z}_2(\mu)$ is a splitting field of $x^3+x^2+[1]$ over $\mathbf{Z}_2$. Let $S_2 = \mathbf{Z}_2(\mu)$. Then $\{[1], \mu, \mu^2\}$ is a basis for $S_2/\mathbf{Z}_2$ and $[S_2 : \mathbf{Z}_2] = 3$. The addition table for $\mathbf{Z}_2(\mu)$ is determined in a manner similar to that in Exercise 1. In fact, one may obtain the addition table by substituting μ for λ in the addition table of $\mathbf{Z}_2(\lambda)$.

We now consider multiplication. We note that $([1] + \mu)([1] + \mu) = [1] + \mu^2$. However, $([1] + \mu + \mu^2)([1] + \mu^2) = [1] + \mu + \mu^3 + \mu^4 = [1]$. Hence, we note the first algebraic difference between $\mathbf{Z}_2(\lambda)$ and $\mathbf{Z}_2(\mu)$.

◊ **Exercise 3** Show that there exists an isomorphism f of $\mathbf{Z}_2(\lambda)$ onto $\mathbf{Z}_2(\mu)$ considered as vector spaces over $\mathbf{Z}_2$ such that f is the identity on $\mathbf{Z}_2$ and $f(\lambda) = \mu$, $f(\lambda^2) = \mu^2$, where λ and μ are as defined in Worked-Out Exercises 1 and 2, respectively.

Solution: $\{[1], \lambda, \lambda^2\}$ is a basis for $\mathbf{Z}_2(\lambda)$ over $\mathbf{Z}_2$ and $\{[1], \mu, \mu^2\}$ is a basis for $\mathbf{Z}_2(\mu)$ over $\mathbf{Z}_2$. Hence, there exists a unique linear transformation f of $\mathbf{Z}_2(\lambda)$ onto $\mathbf{Z}_2(\mu)$ such that $f([1]) = [1]$, $f(\lambda) = \mu$, and $f(\lambda^2) = \mu^2$. This linear transformation is given by

$$f(a[1] + b\lambda + c\lambda^2) = a[1] + b\mu + c\mu^2,$$

where $a, b, c \in \mathbf{Z}_2$. Since $\{[1], \mu, \mu^2\}$ is linearly independent, f is one-one.

◊ **Exercise 4** Show that $\mathbf{Z}_2(\lambda)$ and $\mathbf{Z}_2(\mu)$ are isomorphic as fields, where λ and μ are as defined in Worked-Out Exercises 1 and 2, respectively.

Solution: Since $|\mathbf{Z}_2(\lambda)| = 2^3 = |\mathbf{Z}_2(\mu)|$, $\mathbf{Z}_2(\lambda)$ and $\mathbf{Z}_2(\mu)$ are splitting fields of $x^8 - x$ over $\mathbf{Z}_2$ and thus are isomorphic.

◊ **Exercise 5** Factor the polynomial $x^8 - x$ over $\mathbf{Z}_2$.

Solution: $x^8 - x = x(x + [1])(x^6 + x^5 + x^4 + x^3 + x^2 + x + [1])$. Now $x^2 + x + [1]$ is the only irreducible quadratic polynomial over $\mathbf{Z}_2$. But $x^2 + x + [1]$ does not divide $x^6 + x^5 + x^4 + x^3 + x^2 + x + [1]$. We have that $x^3 + x + [1]$ and $x^3 + x^2 + [1]$ are irreducible polynomials over $\mathbf{Z}_2$ and $x^6 + x^5 + x^4 + x^3 + x^2 + x + [1] = (x^3 + x + [1])(x^3 + x^2 + [1])$. Hence, $x^8 - x = x(x + [1])(x^3 + x + [1])(x^3 + x^2 + [1])$.

◊ **Exercise 6** Find the roots of $x^3 + x^2 + [1]$ in $\mathbf{Z}_2(\lambda)$, where λ is as defined in Worked-Out Exercise 1.

Solution: $[0]$ is a root of x, $[1]$ is a root of $x + [1]$, and λ, λ^2, $\lambda + \lambda^2$ are roots of $x^3 + x + [1]$. Hence, $[1] + \lambda$, $[1] + \lambda^2$, and $[1] + \lambda + \lambda^2$ are roots of $x^3 + x^2 + [1]$.

◊ **Exercise 7** Find the roots of $x^3 + x + [1]$ in $\mathbf{Z}_2(\mu)$, where μ is as defined in Worked-Out Exercise 2.

Solution: $[0]$ is a root of x, $[1]$ is a root of $x + [1]$, and μ, μ^2, $[1] + \mu + \mu^2$ are roots of $x^3 + x^2 + [1]$. Hence, $[1] + \mu$, $[1] + \mu^2$, and $\mu + \mu^2$ are roots of $x^3 + x + [1]$.

◇ **Exercise 8** Show that there exists an isomorphism g of $\mathbf{Z}_2(\lambda)$ onto $\mathbf{Z}_2([1]+ \mu)$ such that $g(\lambda) = [1] + \mu$, where λ and μ are as defined in Worked-Out Exercises 1 and 2, respectively.

Solution: The result here follows immediately by Corollary 21.2.9.

◇ **Exercise 9** Show that there does not exist an isomorphism h of $\mathbf{Z}_2(\lambda)$ onto $\mathbf{Z}_2(\mu)$ such that $h(\lambda) = \mu$, where λ and μ are as defined in Worked-Out Exercises 1 and 2, respectively.

Solution: Suppose there exists an isomorphism h of $\mathbf{Z}_2(\lambda)$ onto $\mathbf{Z}_2(\mu)$ such that $h(\lambda) = \mu$. Then $[0] = h([0]) = h(\lambda^3 + \lambda + [1]) = \mu^3 + \mu + [1]$. Also, $[0] = \mu^3 + \mu^2 + [1]$. Hence, $\mu^3 + \mu + [1] = \mu^3 + \mu^2 + [1]$. Thus, $\mu^2 = \mu$. Therefore, $\mu = [1]$, a contradiction.

23.1.2 Exercises

1. Let F be a finite field. A generator for $F^* = F \backslash \{0\}$ is called a **primitive element** for F. Find a primitive element for the following fields.

   (i) $\mathbf{Z}_7$.

   (ii) $\mathbf{Z}_{11}$.

   (iii) F, where $F \supseteq \mathbf{Z}_2$ and $[F : \mathbf{Z}_2] = 8$.

2. Construct a field with 9 elements.

3. Construct a field with 27 elements.

4. Suppose that F is a finite field of characteristic p. If c is a primitive element of F, prove that c^p is a primitive element of F.

5. Let F be a finite field of characteristic p. If $n = [F : \mathbf{Z}_p]$, prove that there exists $c \in F$ such that c is algebraic of degree n over $\mathbf{Z}_p$ and $F = \mathbf{Z}_p(c)$.

6. If F is a finite field of p^n elements, p a prime and n a positive integer, prove that $[F : \mathbf{Z}_p] = n$.

7. Describe the splitting field of $x^{3^2} - x$ over $\mathbf{Z}_3$.

8. Prove that there exists an irreducible polynomial of arbitrary degree n over $\mathbf{Z}_p$.

9. If F is a subfield of $GF(p^n)$, prove that $F \simeq GF(p^m)$, where $m | n$.

10. Show that if m and n are positive integers such that $m | n$, then $GF(p^n)$ contains a unique subfield $GF(p^m)$, $p^m - 1$ divides $p^n - 1$, whence $x^{p^m - 1} - 1$ divides $x^{p^n - 1} - 1$ and so $x^{p^m} - x$ divides $x^{p^n} - x$.

11. Let F be a field containing $\mathbf{Z}_p$ and $f(x)$ be a polynomial over $\mathbf{Z}_p$. If $c \in F$ is a root of $f(x)$, prove that c^p is also root of $f(x)$.

12. Let $f(x) = x^p - x - [1] \in \mathbf{Z}_p[x]$. Show that a splitting field of $f(x)$ over $\mathbf{Z}_p$ is $\mathbf{Z}_p(c)$, where c is a root of $f(x)$.

13. Let F be a field and G and H be subgroups of F^*. If G and H have order n, prove that $G = H$.

14. If F is a field such that F^* is cyclic, prove that F is finite.

Evariste Galois (1811–1832) was bo-rn on October 25, 1811 in Bourg-la-Reine, near Paris, France, into a well-educated family. Galois received his early education from his mother. His father was director of a school. He read Legendre's *Géométrie* at a very young age and mastered it in one reading. He then read Lagrange's work, acquiring a solid background. In 1828, he started reading recent works on the theory of equations and the theory of elliptic functions.

Galois twice failed the entrance examination for the École Polytechnique. In 1829, he took the entrance examination for the École Normale Supérieure, which trained future secondary school teachers. There he learned about Abel's recent death and Abel's last published memoir, which contained a number of results which Galois himself had obtained and presented to the Academy. Cauchy was assigned to report on Galois's work. Cauchy advised him to revise his work, taking into account Abel's results. (It was for this reason that Cauchy did not present a report on Galois's memoir.) Galois then wrote a new text and submitted it to the Academy in February 1830. Fourier was assigned to report on it, but Fourier died before reading it and the memoir was lost.

In June 1830, Galois published a short note on the resolutions of numerical equations and a much more important article, "Sur la théorie des nombres," containing a remarkable theory of "Galois imaginaries."

On January 17, 1831, Galois presented to the Academy a new version of his memoir. Poisson reviewed it and declared much of it incomprehensible.

It was a time of great political unrest in France. Galois joined the National Guard, a republican party. He was in and out of prison. Arrested during a republican demonstration on July 14, 1931, he was placed in detention. There he revised his memoir on equations and worked on the application of his theory of elliptic functions. Later he was transferred to a nursing home because of a cholera epidemic. There he resumed his work and wrote several essays on the philosophy of mathematics. He also became involved in a love affair. He was challenged to a duel. Badly wounded, he died on May 30, 1832. On May 29, the day before his death, he wrote a letter to his friend Auguste Chevalier, sketching his principal results. He scribbled comments on the margin of his documents such as, "I have no time," and asking Jacobi and Gauss's opinion "not as to the truth, but as to the importance of these theorems."

In 1843, Louiville prepared Galois's manuscript for publication and announced to the Academy that Galois had solved the problem considered by Abel. The manuscript was finally published in the October–November 1846 issue of the *Journal des mathématiques pures et appliquées*.

Galois's terse style and the great originality of his ideas contributed to the delay in the publication of his papers.

Chapter 24

Galois Theory and Applications

The approach used today to present Galois theory is due to Artin. Artin reformulated the theory as an abstract relationship between a field extension and its group of automorphisms. He succeeded in disassociating Galois theory from the solvability of algebraic equations.

24.1 Normal Extensions

Definition 24.1.1 *Let F/K be an algebraic field extension. F/K is called a **normal extension** if every irreducible polynomial $f(x) \in K[x]$ such that $f(x)$ has a root in F, splits into linear factors in $F[x]$.*

An example of a normal extension, which comes quickly to mind is F/K, where F is an algebraic closure of K. A more trivial example of a normal extension is F/K, where $F = K$. The field extension $\mathbf{Q}(\sqrt[3]{2})/\mathbf{Q}$ is not a normal extension since the minimal polynomial of $\sqrt[3]{2}$ over $\mathbf{Q}$ is $x^3 - 2$ has two complex roots and $\mathbf{Q}(\sqrt[3]{2})$ does not contain these roots. (Example 24.2.7 to follow.)

Let F/K be a field extension and $\mathcal{F}$ be a subset of the polynomial ring $K[x]$. Then F is called a **splitting field** for $\mathcal{F}$ if for all $f(x) \in \mathcal{F}$, $f(x)$ splits into linear factors over F and for all proper intermediate fields L of F/K, there exists $f(x) \in \mathcal{F}$, which does not split over L. If $\mathcal{F}$ consists of a single polynomial $g(x)$ and F is a splitting field for $\mathcal{F}$, we say that F is a splitting field for $g(x)$.

Lemma 24.1.2 *Let F/K be a finite field extension and $c \in F$. Then there exists a field $L \supseteq F$ and a polynomial $g(x) \in K[x]$ such that the following properties hold.*

(i) L is a splitting field for $g(x)$ over K.
(ii) Every irreducible factor of $g(x)$ in $K[x]$ has a root in F.
(iii) c is a root of $g(x)$.

Proof.   Let $\{v_1, v_2, \ldots, v_n\}$ be a basis of F over K. Let $g(x)$ be the product of the minimal polynomials of $c, v_1, v_2, \ldots, v_n$ over K. Then property (ii) and (iii) hold. Let L be a splitting field of $g(x)$ over F. Then $L = F(r_1, r_2, \ldots, r_m)$, where $r_1, r_2, \ldots, r_m$ are the roots of $g(x)$. Since $v_1, v_2, \ldots, v_n \in \{r_1, r_2, \ldots, r_m\}$ and $F = K(v_1, v_2, \ldots, v_n)$, L is a splitting field of $g(x)$ over K. ∎

Theorem 24.1.3 *Let F/K be a finite field extension. Then the following conditions are equivalent.*

(i) F is normal over K.

(ii) F is a splitting field over K for some polynomial $g(x) \in K[x]$.

(iii) For every field $L \supseteq F$, all K-isomorphisms from F into L map F onto F, i.e., are K-automorphisms of F.

Proof.   Suppose that statement (i) holds. By Lemma 24.1.2, there is a polynomial $g(x) \in K[x]$ and a field $L \supseteq F$ such that L is a splitting field for $g(x)$ over K and every irreducible factor of $g(x)$ has a root in F. Since F is normal over K, each of these irreducible factors of $g(x)$ splits over K. Hence, $g(x)$ splits over K. Thus, $F = L$ and so (ii) holds.

Suppose that statement (ii) holds. Then F is a splitting field over K for some polynomial $g(x) \in K[x]$. Let L be a field containing F and α be a K-isomorphism of F into L. Then $\alpha(F)$ is a splitting field for $\alpha(g(x)) = g(x)$ over $\alpha(K) = K$. Since $g(x)$ has a unique splitting field over K and contained in L, $\alpha(F) = F$. Hence, (iii) holds.

Suppose that statement (iii) holds. Let $c \in F$ and $f(x)$ be the minimal polynomial of c over K. By Lemma 24.1.2, there is a field $L \supseteq F$ and a polynomial $g(x) \in K[x]$ such that L is a splitting field for $g(x)$ over K and c is a root of $g(x)$. Thus, $f(x)|g(x)$ and so $f(x)$ splits over L. For each root b of $f(x)$ in L, there exists a K-isomorphism α of $K(c)$ onto $K(b)$ such that $\alpha(c) = b$ by Theorem 21.2.8. By Theorem 21.2.10, α can be extended to a K-automorphism σ of L such that $\sigma(c) = b$. Since σ maps F onto F by hypothesis, $b \in F$. Hence, all the roots of $f(x)$ in L lie in F. Since $f(x)$ splits over L, it must split over F. Therefore, (iii) holds. ∎

24.1.1   Worked-Out Exercises

◇ **Exercise 1** Let F/K be a field extension. Suppose that $[F : K] = 2$. Show that F is a normal extension of K.

Solution: Let $a \in F$ be such that $a \notin K$. Since $[F : K(a)] \cdot [K(a) : K] = [F : K] = 2$ and $a \notin K$, $[K(a) : K] = 2$. Let $p(x)$ be the minimal polynomial of a over K. Then $[K(a) : K] = \deg p(x) = 2$. Now $p(x) = (x - a)h(x)$ for some $h(x) \in F[x]$. Thus, $\deg h(x) = 1$. Suppose $h(x) = cx + d$ for some $c, d \in F$, $c \neq 0$. Then $-c^{-1}d \in F$ and $-c^{-1}d$ is a root of $h(x)$. Therefore, $-c^{-1}d$ is a root of $p(x)$. Hence, both the roots of $p(x)$ are in F. Thus, F is the splitting field of $p(x)$ over K. Consequently, F is a normal extension of K.

◇ **Exercise 2** Let $F = \mathbf{Q}(\sqrt[4]{2})$ and $L = \mathbf{Q}(\sqrt[2]{2})$. Show that F is a normal extension of L, L is a normal extension of $\mathbf{Q}$, but F is not a normal extension of $\mathbf{Q}$.

  Solution: Now $[F : L] = 2 = [L : \mathbf{Q}]$. Hence, F is a normal extension of L, L is a normal extension of $\mathbf{Q}$ by Worked-Out Exercise 1. Now $x^4 - 2 \in \mathbf{Q}[x]$ is irreducible over $\mathbf{Q}$ and $\sqrt[4]{2}$ is a root of $x^4 - 2$. Thus, $x^4 - 2$ is the minimal polynomial of $\sqrt[4]{2}$. Now the roots of $x^4 - 2$ are $\pm \sqrt[4]{2}$ and $\pm i\sqrt[4]{2}$. Since $\pm i\sqrt[4]{2} \notin F$, F is not the splitting field of $x^4 - 2$. Therefore, F is not a normal extension of $\mathbf{Q}$.

◇ **Exercise 3** Let K be a field of characteristic 0. Let F/K be a finite normal extension. Let $g(x) \in K[x]$ and E be a splitting field of $g(x)$ over F. Then E/K is a normal extension.

  Solution: By Corollary 22.1.20, $F = K(a)$ for some $a \in F$. Let $h(x)$ be the minimal polynomial of a over K. Now $h(x)$ splits over F. Let $f(x) = g(x)h(x)$. Then $K \subseteq F \subseteq E$ and $f(x)$ splits over E. Let L be the splitting field of $f(x)$ over K in E. Then $K \subseteq L \subseteq E$. Now $a \in L$ and hence $K \subseteq F \subseteq L$. Thus, L is the splitting field of $g(x)$ over F. Hence, $E = L$. Consequently, E/K is normal, by Theorem 24.1.3.

24.1.2   Exercises

1. (i) Show that $\mathbf{C}$ is a normal extension of $\mathbf{R}$.

   (ii) Is $\mathbf{R}$ a normal extension of $\mathbf{Q}$?

2. Let $K \subseteq L \subseteq F$ be a chain of fields. Suppose that F/K is a normal extension.

   (i) Show that F/L is a normal extension.

   (ii) Is L/K a normal extension? Justify your answer.

3. Let $K \subseteq L_1, L_2 \subseteq F$ be fields. Suppose that L_1/K and L_2/K are normal extensions. Show that $(L_1 \cap L_2)/K$ is a normal extension.

4. [1]Let F/K be an algebraic field extension. Let $\overline{K}$ be the algebraic closure of K such that $F \subseteq \overline{K}$. Prove that the following are equivalent.

   (i) F/K is a normal extension.

---
[1]This exercise requires Section 21.3.

(ii) If $\sigma : F \rightarrow \overline{K}$ is a K-homomorphism, then σ is an automorphism of F.

(iii) F is the splitting field of a family of polynomials in $K[x]$.

24.2 Galois Theory

We have now reached the point where we can begin our study of Galois theory. Roughly speaking, this theory relates the roots of a polynomial to certain permutations of these roots. More specifically, if F is a splitting field for some polynomial $f(x)$ over a field K such that F/K is separable, then this theory sets up a one-one inclusion reversing correspondence between the intermediate fields of F/K and the subgroups of a particular group of automorphisms of F/K. These results can be applied to the solution by "radicals" of the equation $f(x) = 0$. This application will be discussed in Section 24.4.

Theorem 24.2.1 *Let F be a field and $\alpha_1, \ldots, \alpha_n$ be distinct automorphisms of F. Then for all $a \in F$ and for all $a_1, \ldots, a_n \in F$,*

$$a_1\alpha_1(a) + \cdots + a_n\alpha_n(a) = 0$$

implies that $a_1 = \cdots = a_n = 0$.

Proof. The proof is by induction on n. If $n = 1$ and $a_1\alpha_1(a) = 0$ for all $a \in F$, then $a_1 = 0$ since $\alpha_1(1) \neq 0$. Assume the theorem is valid for any m distinct automorphisms, where $1 \leq m < n$. Suppose

$$a_1\alpha_1(a) + \cdots + a_n\alpha_n(a) = 0 \text{ for all } a \in F \qquad (24.1)$$

and for some $a_1, \ldots, a_n \in F$, not all zero, say, $a_1 \neq 0$. Since the automorphisms $\alpha_1, \ldots, \alpha_n$ are distinct, there exists $b \in F$ such that $\alpha_1(b) \neq \alpha_n(b)$. Since Eq. (24.1) is valid for every element of F, we have $a_1\alpha_1(ab) + \cdots + a_n\alpha_n(ab) = 0$ or

$$a_1\alpha_1(a)\alpha_1(b) + \cdots + a_n\alpha_n(a)\alpha_n(b) = 0 \text{ for all } a \in F. \qquad (24.2)$$

Multiplying Eq. (24.1) by $\alpha_n(b)$ and subtracting this result from Eq. (24.2), we obtain

$$a_1(\alpha_1(b) - \alpha_n(b))\alpha_1(a) + \cdots + a_{n-1}(\alpha_{n-1}(b) - \alpha_n(b))\alpha_{n-1}(a) = 0$$

for all $a \in F$. Since $\alpha_1(b) \neq \alpha_n(b)$, $a_1(\alpha_1(b) - \alpha_n(b)) \neq 0$. However, this contradicts the induction hypothesis. Hence, the theorem is valid for all positive integers n. ∎

Definition 24.2.2 *Let G be a group of automorphisms of the field F. An element $a \in F$ is called **fixed** by G if $\alpha(a) = a$ for all $\alpha \in G$. We denote by F_G the set of all $a \in F$ such that a is fixed by G.*

Theorem 24.2.3 *Let G be a group of automorphisms of the field F. Then F_G is a subfield of F, called the **fixed field** of F for G.*

Proof.   Note that $F_G \neq \phi$ since $0, 1 \in F_G$. Let $a, b \in F_G$. Then for all $\alpha \in G$, $\alpha(a - b) = \alpha(a) - \alpha(b) = a - b$ so that $a - b \in F_G$. If $b \neq 0$, then $\alpha(ab^{-1}) = \alpha(a)\alpha(b^{-1}) = \alpha(a)\alpha(b)^{-1} = ab^{-1}$ so that $ab^{-1} \in F_G$. Thus, F_G is a subfield of F. ∎

Definition 24.2.4 *Let F/K be a field extension. Let $G(F/K)$ denote the set of all K-automorphisms of F.*

Theorem 24.2.5 *Let F/K be a field extension. Then $G(F/K)$ is a subgroup of the group of all automorphisms of F and is called the **group of automorphisms** of F **relative** to K.*

Proof.   Clearly the identity map is in $G(F/K)$ so that $G(F/K) \neq \phi$. Let $\alpha, \beta \in G(F/K)$. Then for all $k \in K, (\alpha \circ \beta^{-1})(k) = \alpha(\beta^{-1}(k)) = \alpha(k) = k$. Thus, $\alpha \circ \beta^{-1} \in G(F/K)$ so that $G(F/K)$ is a group. ∎

We ask the reader to verify that any automorphism of F fixes the prime subfield of F.

Theorem 24.2.6 *Let H be a finite set of automorphisms of the field F. Then*
(i) $|H| \leq [F : F_H]$ and
(ii) if H is a group, then $|H| = [F : F_H]$.

Proof.   (i) Suppose $|H| > [F : F_H]$. Then $[F : F_H] = n < \infty$ for some n. Let $b_1, \ldots, b_n$ be a basis for F/F_H. There exist $n + 1$ distinct automorphisms $\alpha_1, \ldots, \alpha_{n+1}$ in $G(F/F_H)$. Then the system of n homogeneous linear equations in the $n + 1$ unknowns $x_1, \ldots, x_{n+1}$,

$$\alpha_1(b_i)x_1 + \cdots + \alpha_{n+1}(b_i)x_{n+1} = 0, \quad i = 1, 2, \ldots, n$$

has a nontrivial solution $x_1 = a_1, \ldots, x_{n+1} = a_{n+1}$ in F. Thus,

$$\alpha_1(b_i)a_1 + \cdots + \alpha_{n+1}(b_i)a_{n+1} = 0, \quad i = 1, 2, \ldots, n. \tag{24.3}$$

Now every element $a \in F$ has the form $a = \sum_{i=1}^{n} k_i b_i, k_i \in F_H$ and so

$$a_j \alpha_j(a) = \sum_{i=1}^{n} k_i a_j \alpha_j(b_i), \quad j = 1, 2, \ldots, n + 1.$$

Then using Eq. (24.3) and the fact that each α_i fixes $k_1, \ldots, k_n$ we obtain

$$
\begin{aligned}
a_1\alpha_1(a) + \cdots + a_{n+1}\alpha_{n+1}(a) &= \sum_{i=1}^{n} k_i(a_1\alpha_1(b_i) + \cdots + a_{n+1}\alpha_{n+1}(b_i)) \\
&= 0
\end{aligned}
$$

for all $a \in F$. However, this contradicts Theorem 24.2.1. Hence, $|H| \leq [F : F_H]$.

(ii) By (i), $|H| \leq [F : F_H]$. We now show that $|H| \geq [F : F_H]$. Suppose $|H| < [F : F_H]$. Set $|H| = n$. Then there are elements $b_1, \ldots, b_{n+1}$ of F which are linearly independent over F_H. There exists a nontrivial solution $a_1, \ldots, a_{n+1}$ in F satisfying the system of n homogeneous linear equations in the $n + 1$ unknowns $x_1, \ldots, x_{n+1}$,

$$
x_1\alpha_i(b_1) + \cdots + x_{n+1}\alpha_i(b_{n+1}) = 0, \quad i = 1, 2, \ldots, n, \tag{24.4}
$$

where $\alpha_i \in H$. From all such nontrivial solutions of Eq. (24.4), choose one having the smallest number, say, m of nonzero members. We have $m > 1$ else $a_1\alpha_1(b_1) = 0$ and hence $a_1 = 0$. (Note that $\alpha_1(b_1) \neq 0$ since α_1 is one-one and $b_1 \neq 0$.) Upon reordering we have

$$
a_1\alpha_i(b_1) + \cdots + a_m\alpha_i(b_m) = 0, \quad i = 1, 2, \ldots, n \tag{24.5}
$$

and no $a_i = 0$. Let α_1 be the identity map. Then

$$
a_1 b_1 + \cdots + a_m b_m = 0,
$$

where we take $a_m = 1$. (If $a_m \neq 1$, then multiply through by a_m^{-1}.) Since $b_1, \ldots, b_m$ are linearly independent over F_H, not all $a_1, \ldots, a_m$ are in F_H, say, $a_1 \notin F_H$. Thus, for some α_j, $\alpha_j(a_1) \neq a_1$. Apply α_j to Eq. (24.5). Then

$$
\alpha_j(a_1\alpha_i(b_1)) + \cdots + \alpha_j(a_m\alpha_i(b_m)) = 0, \quad i = 1, 2, \ldots, n
$$

or

$$
\alpha_j(a_1)\alpha_{ij}(b_1) + \cdots + \alpha_j(a_m)\alpha_{ij}(b_m) = 0, \quad i = 1, 2, \ldots, n, \tag{24.6}
$$

where $\alpha_{ij} = \alpha_j \circ \alpha_i$. Since H is a group, $\{\alpha_{1j}, \ldots, \alpha_{nj}\} = H$. If we relabel the Eqs. in (24.6) and then subtract Eq. (24.6) from Eq. (24.5), we obtain

$$
(a_1 - \alpha_j(a_1))\alpha_i(b_1) + \cdots + (a_{m-1} - \alpha_j(a_{m-1}))\alpha_i(b_{m-1}) = 0, \ i = 1, 2, \ldots, n.
$$

Since $a_1 - \alpha_j(a_1) \neq 0$, $a_1 - \alpha_j(a_1)$, $\ldots$, $a_{m-1} - \alpha_m(a_{m-1})$, $0, \ldots, 0$ is a nontrivial solution of Eq. (24.4) having fewer than m nonzero members. This contradiction thus shows that the assumption $[F : F_H] > |H|$ is false. Hence, $|H| \geq [F : F_H]$ so that $[F : F_H] = |H|$. ∎

Example 24.2.7 *Consider the field extension* $\mathbf{Q}(\sqrt[3]{2})/\mathbf{Q}$. *Let* α *be any automorphism of* $\mathbf{Q}(\sqrt[3]{2})$. *Then* α *fixes every element of* $\mathbf{Q}$. *We have*

$$(\alpha(\sqrt[3]{2}))^3 = \alpha((\sqrt[3]{2})^3) = \alpha(2) = 2.$$

Hence, $\alpha(\sqrt[3]{2})$ *is a root of* $x^3 - 2$. *Thus,* $\alpha(\sqrt[3]{2}) = \sqrt[3]{2}$ *because the other two cube roots of* $x^3 - 2$ *are complex numbers, namely,* $\sqrt[3]{2}(-\frac{1}{2} + i\frac{\sqrt{3}}{2})$ *and* $\sqrt[3]{2}(-\frac{1}{2} - i\frac{\sqrt{3}}{2})$ *and so are not members of* $\mathbf{Q}(\sqrt[3]{2})$. *Hence,* α *is the identity map on* $\mathbf{Q}(\sqrt[3]{2})$. *Thus,* $G(\mathbf{Q}(\sqrt[3]{2})/\mathbf{Q}) = \{e\}$. *But* $\mathbf{Q}(\sqrt[3]{2})_{G(\mathbf{Q}(\sqrt[3]{2})/\mathbf{Q})} = \mathbf{Q}(\sqrt[3]{2}) \supset \mathbf{Q}$. *We note that* $\mathbf{Q}(\sqrt[3]{2})$ *is not the splitting field of the polynomial* $x^3 - 2$ *over* $\mathbf{Q}$. *Now* $1 = |G(\mathbf{Q}(\sqrt[3]{2})/\mathbf{Q})| < [\mathbf{Q}(\sqrt[3]{2}) : \mathbf{Q}]$. *If the other two roots of* $x^3 - 2$ *were present, then we would have found an* α *such that* $\alpha(\sqrt[3]{2}) \neq \sqrt[3]{2}$.

Example 24.2.8 *Let* F/K *be any field extension of characteristic* $p > 0$ *such that there exists* $a \in F$, $a \notin K$, *and* a *is purely inseparable over* K. *Let* α *be any automorphism of* F, *which fixes every element of* K. *Let* e *be a positive integer such that* $a^{p^e} = k \in K$. *Then* $(\alpha(a) - a)^{p^e} = \alpha(a)^{p^e} - a^{p^e} = \alpha(a^{p^e}) - a^{p^e} = \alpha(k) - k = k - k = 0$. *Since a field has no nonzero nilpotent elements,* $\alpha(a) - a = 0$ *or* $\alpha(a) = a$. *Hence,* $F_{G(F/K)} \supseteq K(a) \supset K$. *Here we note that because of the presence of* a, F/K *would not be separable even if it were algebraic.*

For a field extension F/K, we will want $F_{G(F/K)} = K$. The above two examples point out difficulties we must overcome.

Theorem 24.2.9 *Let* H *be a finite group of automorphisms on the field* F. *Then* $H = G(F/F_H)$.

Proof. Clearly $H \subseteq G(F/F_H)$. By an argument similar to that of the first part of the proof of Theorem 24.2.6, $G(F/F_H)$ is a finite group. Hence, by Theorem 24.2.6, $|G(F/F_H)| = [F : F_{G(F/F_H)}]$. Now $F_H = F_{G(F/F_H)}$ since $F_H \supseteq F_{G(F/F_H)}$ and if $a \in F_H$, then for all $\alpha \in G(F/F_H)$, $\alpha(a) = a$ so that $a \in F_{G(F/F_H)}$. Therefore, $|H| = [F : F_H] = [F : F_{G(F/F_H)}] = |G(F/F_H)|$. Since $H \subseteq G(F/F_H)$ and $G(F/F_H)$ is finite, we have $H = G(F/F_H)$. ■

Let us pause to see what we have so far. Let F/K be a finite field extension. We desire a one-one inclusion reversing correspondence between all the intermediate fields of F/K and all the subgroups of $G(F/K)$. From Examples 24.2.7 and 24.2.8, we have seen that it is possible for an intermediate field L of F/K to be strictly contained in $F_{G(F/L)}$. Hence, a mapping

$$L \to G(F/L) \tag{24.7}$$

need not be one-one since

$$F_{G(F/L)} \to G(F/F_{G(F/L)}) = G(F/L),$$

but $L \subset F_{G(F/L)}$ is possible. Note that the mapping

$$F_H \leftarrow H \qquad (24.8)$$

is one-one since by Theorem 24.2.9, $H = G(F/F_H)$. The mapping of (24.8) is the "inverse" of (24.7), but the mapping in (24.8) does not map onto all the intermediate fields of F/K. We can thus see that we need some sort of condition on F/K to force every $L = F_{G(F/L)}$. Examples 24.2.7 and 24.2.8 suggest the condition should be that F/K be separable and be the splitting field of some polynomial over K. A similar difficulty is not encountered with $G(F/K)$ since $H = G(F/F_H)$ by Theorem 24.2.9.

Definition 24.2.10 *Let F/K be a finite field extension. If $F_{G(F/K)} = K$, then $G(F/K)$ is called the **Galois group** of F/K and F/K is called a **Galois extension**.*

Theorem 24.2.11 *Let F/K be a finite extension. The following conditions are equivalent.*
   (i) $G(F/K)$ is the Galois group of F/K.
   (ii) F/K is normal and separable.
   (iii) F is the splitting field of a separable polynomial in $K[x]$.

Proof.   Suppose $[F : K] = n$. Let H be a subgroup of $G(F/K)$. Then by Theorem 24.2.6, $|H| = [F : F_H] \leq [F : K] = n$.
   (i)$\Rightarrow$(ii) Suppose $G(F/K)$ is the Galois group of F/K. Then

$$|G(F/K)| = [F : K] = n.$$

Since F/K is finite, F/K is an algebraic extension and $F = K(u_1, u_2, \ldots, u_n)$ for some $u_i \in F$, $1 \leq i \leq n$. Let $G(F/K) = \{e = \alpha_1, \alpha_2, \ldots, \alpha_n\}$. Let $a \in F$ and $a = a_1, a_2, \ldots, a_m$ be distinct elements of the set $\{\alpha_i(a) \mid i = 1, \ldots, n\}$. Now $\alpha_j \circ \alpha_i \in G(F/K)$ for all i and j. Let $a_i = \alpha_i(a)$, $i = 1, 2, \ldots, n$. Then $\alpha_j(a_i) = \alpha_j(\alpha_i(a)) = \alpha_j \circ \alpha_i(a) = \alpha_r(a) = a_r$ for some r, $1 \leq r \leq m$. Since α_k is an automorphism of F, $\alpha_k(a_i) = \alpha_k(a_j)$ if and only if $a_i = a_j$. Thus, for all k, $1 \leq k \leq n$, $\alpha_k(a_1), \alpha_k(a_2), \ldots, \alpha_k(a_m)$ are distinct elements. Let $f_a(x) = (x-a_1)(x-a_2)\cdots(x-a_m)$. Then all roots of $f_a(x)$ are distinct and lie in F. Also, the factors of $f_a(x)$ are merely permuted by any α_i of $G(F/K)$. Thus, the coefficients of $f_a(x)$ remain unaltered by any $\alpha_i \in G(F/K)$. Therefore, $f_a(x) \in K[x]$ since $K = F_{G(F/K)}$. Hence, $a = a_1$ is a root of a separable polynomial $f_a(x)$ in $K[x]$ and $f_a(x)$ splits over F. From this, it also follows that for all i, $1 \le i \le n$, u_i is a root of a separable polynomial $f_{u_i}(x)$ in $K[x]$ and $f_{u_i}(x)$ splits over F. Thus, all the roots of the polynomial $f(x) = f_{u_1}(x)f_{u_2}(x)\cdots f_{u_n}(x) \in K[x]$ are in F. Since $F = K(u_1, u_2, \ldots, u_n)$ and each u_i is a root of $f(x)$, F is the splitting field of $f(x)$ and so F/K is normal. Since each u_i is a root of a separable polynomial over K, it follows that F/K is separable. Consequently, F/K is normal and separable.

(ii)$\Rightarrow$(iii) Since F/K is a finite separable extension, there exists $a \in F$ such that $F = K(a)$. Now a is a root of a separable irreducible polynomial $f(x) \in K[x]$. Since F/K is normal, $f(x)$ splits over F. Thus, F contains all roots of $f(x)$. Hence, F is the splitting field of a separable polynomial $f(x) \in K[x]$.

(iii)$\Rightarrow$(i) Suppose F is a splitting field of a separable polynomial $f(x) \in K[x]$. Let m be the number of distinct roots of $f(x)$ in F, but not in K. We prove the result by induction on m. If $m = 0$, then $F = K$ and $G(F/K) = \{e\}$, where e is the identity automorphism of F. Hence, $K = F = F_{G(F/K)}$. Assume that the result holds for all field extensions S/T such that S is a splitting field of a separable polynomial $g(x) \in T[x]$ with $g(x)$ having fewer than $m \ge 1$ roots outside of T.

Let $f(x) = p_1(x)\cdots p_k(x)$, where each $p_i(x)$ is irreducible and separable in $K[x]$. Since $m \ge 1$, $\deg p_i(x) > 1$ for some i. By renumbering if necessary, we may assume that $i = 1$, i.e., $\deg p_1(x) = t > 1$. Let a be a root of $p_1(x)$. Then $[K(a) : K] = t$. Since $p_1(x)$ is irreducible and separable, its roots $a = a_1, a_2, \ldots, a_t$ are all distinct. Thus, there exist isomorphisms $\alpha_1', \alpha_2', \ldots, \alpha_t'$ such that $\alpha_i' : K(a) \to K(a_i)$ with $\alpha_i'(a) = a_i$ and the elements of K are fixed by α_i'. Since F is a splitting field of $f(x)$ over both $K(a)$ and $K(a_i)$, the isomorphism α_i' can be extended to an automorphism α_i of F, which maps a onto a_i and fixes the elements of K, $i = 1, 2, \ldots, t$.

Suppose now that $c \in F_{G(F/K)}$. Since $f(x)$ has fewer than m roots outside $K(a)$, $K(a) = F_{G(F/K(a))}$ by our induction hypothesis. Since $G(F/K(a)) \subseteq G(F/K)$, $c \in F_{G(F/K(a))} = K(a)$. Hence,

$$c = k_0 + k_1 a + \cdots + k_{t-1}a^{t-1}, \quad k_i \in K, \ i = 0, 1, \ldots, t - 1.$$

Thus,

$$\alpha_i(c) = c = k_0 + k_1 a_i + \cdots + k_{t-1}a_i^{t-1}, \quad i = 1, \ldots, t.$$

Therefore,

$$g(x) = (k_0 - c) + k_1 x + \cdots + k_{t-1}x^{t-1}$$

has t distinct roots $a_1, a_2, \ldots, a_t$ in F. Since $\deg g(x) < t$, $g(x)$ must be the zero polynomial. Hence, $k_0 - c = 0$ or $c = k_0 \in K$. Consequently, $K = F_{G(F/K)}$. ∎

Corollary 24.2.12 *Let F/K be a finite extension. The following conditions are equivalent.*

(i) $|G(F/K)| = [F : K]$.

(ii) F/K is normal and separable.

Proof. Write $G = G(F/K)$.

(i)$\Rightarrow$(ii)Now $K \subseteq F_G \subseteq F$ and $[F : K] = [F : F_G][F_G : K]$. Also, $[F : K] = |G(F/K)| = [F : F_G]$. Hence, $[F_G : K] = 1$ and so $F_G = K$. Thus, F/K is normal and separable by Theorem 24.2.11.

(ii)$\Rightarrow$(i) Since F/K is normal and separable by Theorem 24.2.11, $K = F_{G(F/K)}$. Hence, $|G(F/K)| = [F : F_G] = [F : K]$. ■

We are now ready to present the one-one inclusion reversing correspondence between the intermediate fields of a Galois extension and the subgroups of its Galois group.

Theorem 24.2.13 (The Fundamental Theorem of Galois Theory) *Let F/K be a finite normal and separable field extension. Let $G = G(F/K)$, $\mathcal{F} = \{L \mid L$ is an intermediate field of $F/K\}$, and $\mathcal{S}(G)$ be the set of all subgroups of G. Then the following properties hold.*

(i) $K = F_G$.

(ii) The mapping $\Psi : \mathcal{F} \to \mathcal{S}(G)$ defined by $\Psi(L) = G(F/L)$ for all $L \in \mathcal{F}$ is a one–one correspondence. The mapping $\Phi : \mathcal{S}(G) \to \mathcal{F}$ defined by $\Phi(H) = F_H$ for all $H \in \mathcal{S}(G)$ is the inverse of Ψ. Also, for all $L \in \mathcal{F}$, $[F : L] = |G(F/L)|$ and $[L : K] = [G : G(F/L)]$.

(iii) Let L, $L' \in \mathcal{F}$. Then $L' \subseteq L$ if and only if $G(F/L') \supseteq G(F/L)$. In this case, $[L : L'] = [G(F/L') : G(F/L)]$.

(iv) Let L, $L' \in \mathcal{F}$. Let $\Psi(L) = H$ and $\Psi(L') = H'$. Then there exists $\alpha \in G$ such that $\alpha(L) = L'$ if and only if $\alpha H \alpha^{-1} = H'$.

(v) Let $L \in \mathcal{F}$. Then L/K is a normal extension if and only if $G(F/L)$ is a normal subgroup of G. In this case,

$$G(L/K) \simeq G(F/K)/G(F/L).$$

Proof. (i) Immediate from Theorem 24.2.11.

(ii) Clearly Ψ is well defined. By Theorem 24.2.9, the mapping Ψ is onto. Suppose $G(F/L) = G(F/L')$. Then $F_{G(F/L)} = F_{G(F/L')}$. Since F/K is finite, normal, and separable, so is F/L for every intermediate field L of F/K. By (i), we have $L = F_{G(F/L)} = F_{G(F/L')} = L'$. Hence, the mapping Ψ is one-one. From Theorems 24.2.9 and 24.2.11, it follows that Φ is the inverse of Ψ. By Theorem 24.2.6, $[F : L] = |G(F/L)|$. That $[L : K] = [G : G(F/L)]$ follows easily by Lagrange's theorem and Theorem 21.1.20.

(iii) Clearly $L \supseteq L'$ if and only if $G(F/L') \supseteq G(F/L)$. That $[L : L'] = [G(F/L') : G(F/L)]$ follows by (ii) since F/L' is normal. (Since Ψ is one-one and onto $\mathcal{S}(G)$, we have $L \supset L'$ if and only if $G(F/L) \subset G(F/L')$.)

(iv) Suppose $\alpha(L) = L'$. For any $a' \in L'$, we have $\alpha(a) = a'$ for some $a \in L$. Now for all $\beta \in H$, $\beta(a) = a$. Therefore, $\alpha(\beta(\alpha^{-1}(a'))) = \alpha(\beta(a)) = \alpha(a) = a'$. Thus, $\alpha \circ \beta \circ \alpha^{-1} \in H'$ so that $\alpha H \alpha^{-1} \subseteq H'$. Now $|H'| = [F : L'] = [F : L] =$

$|H| = |\alpha H \alpha^{-1}|$. Hence, $\alpha H \alpha^{-1} = H'$. Conversely, suppose $\alpha H \alpha^{-1} = H'$. Then for all $a \in L$ and for all $\beta \in H$, $\alpha(\beta(\alpha^{-1}(\alpha(a)))) = \alpha(\beta(a)) = \alpha(a)$. Thus, $\alpha(L) \subseteq F_{H'} = L'$ Now $|H'| = |H|$. Therefore, $[F : L] = [F : L']$, whence $[\alpha(L) : K] = [L : K] = [L' : K]$. Consequently, $\alpha(L) = L'$.

(v) Since F/K is separable, L/K is separable and so by Corollary 24.2.12, we have L/K normal if and only if $|G(L/K)| = [L : K]$. We now show that $|G(L/K)| = [L : K]$ if and only if every isomorphism of L leaving K fixed is an automorphism of L/K.

For any $\alpha \in G$, α determines an isomorphism of L leaving K fixed. On the other hand, if β is an isomorphism of L leaving K fixed, then since L/K is normal, β can be extended to an automorphism of F leaving K fixed by Theorem 21.2.10.

Write $H = G(F/L)$ and set $m = [G : H]$. Now by (ii), $m = [L : K]$. Let $H = \alpha_1 H, \alpha_2 H, \ldots, \alpha_m H$ be the distinct cosets of H in G. For $a \in L$ and $\beta \in H$, $(\alpha_i \circ \beta)(a) = \alpha_i(a)$ for each i since $L = F_H$. Thus, the elements of G in the same coset of H determine the same isomorphism of L. Conversely, if $\alpha(a) = \alpha'(a)$ for all $a \in L$, then $a = (\alpha^{-1} \circ \alpha')(a)$ or $\alpha^{-1} \circ \alpha' \in H$ so that α, α' determine the same coset of H in G. Therefore, the number of distinct isomorphisms of L fixing the elements of K is $m = [G : H]$. If $|G(L/K)| = m$, then every isomorphism of L fixing the elements of K must be an automorphism of L/K since every automorphism of L is an isomorphism of L. Conversely, if every isomorphism of L fixing the elements of K is an automorphism of L, then $|G(L/K)|$ is the number m of these isomorphisms. Hence, $|G(L/K)| = [L : K]$ if and only if every isomorphism of L leaving the elements of K fixed is an automorphism of L/K, or L/K is normal if and only if every isomorphism of L leaving the elements of K fixed is an automorphism of L leaving the elements of K fixed.

Now, every isomorphism of L leaving the elements of K fixed is an automorphism if and only if $\alpha(L) = L$ for all $\alpha \in G$. By (iv), $\alpha(L) = L$ for all $\alpha \in G$ if and only if $H = \alpha H \alpha^{-1}$, i.e., if and only if H is normal in G.

If L/K is normal, then the distinct automorphisms of L fixing the elements of K correspond uniquely to the cosets of H in G. This one-one correspondence is clearly an isomorphism of $G(L/K)$ and G/H since for $\alpha, \alpha' \in G$, we have that $\alpha \circ \alpha'$ corresponds to $(\alpha H)(\alpha' H) = \alpha \circ \alpha' H$. ∎

Let F/K be a finite normal separable field extension and L be an intermediate field of F/K. We have seen that F/L is a normal extension, but L/K is not necessarily normal. The above result tells us when L/K is normal.

Example 24.2.14 *Let S be the splitting field of the irreducible polynomial*

$x^3 - 2$ over $\mathbf{Q}$ such that $S \subseteq \mathbf{C}$. Now

$$\begin{aligned}
x^3 - 2 &= (x - \sqrt[3]{2})(x^2 + \sqrt[3]{2}x + \sqrt[3]{4}) \\
&= (x - \sqrt[3]{2})(x - \tfrac{\sqrt[3]{2}}{2}(-1 + \sqrt{3}i))(x - \tfrac{\sqrt[3]{2}}{2}(-1 - \sqrt{3}i)).
\end{aligned}$$

Thus, $S = \mathbf{Q}(\sqrt[3]{2}, \tfrac{\sqrt[3]{2}}{2}(-1 + \sqrt{3}i), \tfrac{\sqrt[3]{2}}{2}(-1 - \sqrt{3}i)) = \mathbf{Q}(\sqrt[3]{2}, \sqrt{3}i)$ Now $S/\mathbf{Q}$ is normal and since $(x^2 + \sqrt[3]{2}x + \sqrt[3]{4})$ is irreducible over $\mathbf{Q}(\sqrt[3]{2})$, $[S : \mathbf{Q}] = 6$. Hence, $|G(S/Q)| = 6$.

The automorphisms of $G(S/\mathbf{Q})$ are completely determined by where they map $\sqrt[3]{2}$ and $\sqrt{3}i$. The following table defines the group $G(S/\mathbf{Q})$. Set $r_2 = \tfrac{\sqrt[3]{2}}{2}(-1 + \sqrt{3}i)$ and $r_3 = \tfrac{\sqrt[3]{2}}{2}(-1 - \sqrt{3}i)$.

	e	α	β	$\alpha\beta$	$\beta\alpha$	$\alpha\beta\alpha$
$\sqrt[3]{2}$	$\sqrt[3]{2}$	$\sqrt[3]{2}$	r_2	r_3	r_2	r_3
r_2	r_2	r_3	$\sqrt[3]{2}$	$\sqrt[3]{2}$	r_3	r_2
r_3	r_3	r_2	r_3	r_2	$\sqrt[3]{2}$	$\sqrt[3]{2}$

	e	α	β	$\alpha\beta$	$\beta\alpha$	$\alpha\beta\alpha$
$\sqrt[3]{2}$	$\sqrt[3]{2}$	$\sqrt[3]{2}$	r_2	r_3	r_2	r_3
$\sqrt{3}i$	$\sqrt{3}i$	$-\sqrt{3}i$	$-\sqrt{3}i$	$\sqrt{3}i$	$\sqrt{3}i$	$-\sqrt{3}i$

The subgroups of $G(S/\mathbf{Q})$ are

$$H_1 = \{e, \alpha\}, H_2 = \{e, \beta\}, H_3 = \{e, \alpha\beta\alpha\}, H_4 = \{e, \alpha\beta, \beta\alpha\}.$$

The corresponding intermediate fields are

$$L_1 = \mathbf{Q}(\sqrt[3]{2}), L_2 = \mathbf{Q}(r_3), L_3 = \mathbf{Q}(r_2), L_4 = \mathbf{Q}(\sqrt{3}i).$$

By Example 4.3.2, H_i, $i = 1, 2, 3$, is not normal in $G(S/\mathbf{Q})$ so $L_i/\mathbf{Q}$ is not normal, $i = 1, 2, 3$. Now H_4 is normal in $G(S/\mathbf{Q})$ and so L_4/K is normal.

Let S be a splitting field over the field K for a polynomial $f(x)$ in $K[x]$. Then we call $G(S/K)$ the **Galois group of the equation** $f(x) = 0$ or the **Galois group of the polynomial** $f(x)$. For any $\alpha \in G(S/K)$ and for any root a of $f(x)$ in S, $0 = \alpha(f(a)) = f(\alpha(a))$. Thus, $\alpha(a)$ is a root of $f(x)$ in S. Since α is a K-automorphism of S, distinct roots of $f(x)$ map onto distinct roots. Hence, α acts like a permutation on the roots of $G(S/K)$. Let π_α denote the permutation of the distinct roots of $f(x)$ induced by α. Then the mapping $\alpha \to \pi_\alpha$ is an isomorphism of $G(S/K)$ into S_n, where $f(x)$ has n distinct roots. Example 24.2.14 is one, where $G(S/K) \simeq S_3$.

Let K be a field of characteristic $\neq 3$. Consider a cubic polynomial $f(x) = x^3 + ax^2 + bx + c$, where $a, b, c \in K$. We eliminate the quadratic term by substituting $u - \frac{a}{3}$ for x. Then

$$
\begin{aligned}
g(u) &= (u - \frac{a}{3})^3 + a(u - \frac{a}{3})^2 + b(u - \frac{a}{3}) + c \\
&= u^3 - au^2 + \frac{a^2}{3}u - \frac{a^3}{27} + au^2 - \frac{2}{3}a^2u + \frac{a^3}{9} + bu - \frac{ab}{3} + c \\
&= u^3 + (b - \frac{a^2}{3})u - \frac{2a^3}{27} - \frac{ab}{3} + c.
\end{aligned}
$$

Hence, r is a root of $g(u)$ if and only if $r - \frac{a}{3}$ is a root of $f(x)$.

Now let $f(x) = x^3 + bx + c \in K[x]$. Then $f(x)$ is irreducible over K if and only if $f(x)$ has no roots in K. Over a splitting field S of $f(x)$ over K, we have

$$
f(x) = (x - a_1)(x - a_2)(x - a_3),
$$

where $a_1, a_2, a_3 \in S$. Thus,

$$
\begin{aligned}
a_1 + a_2 + a_3 &= 0, \\
a_1a_2 + a_1a_3 + a_2a_3 &= b, \\
-a_1a_2a_3 &= c.
\end{aligned}
$$

Define the discriminant D of $f(x)$ as follows:

$$
D = [(a_2 - a_1)(a_3 - a_1)(a_3 - a_2)]^2.
$$

Let $d = (a_2 - a_1)(a_3 - a_1)(a_3 - a_2)$. Then any K-automorphism α of $S = K(a_1, a_2, a_3)$ leaves D fixed, i.e., $\alpha(D) = D$ since $\alpha(d)$ is either d or $-d$. An easy calculation shows that

$$
D = -4b^3 - 27c^2.
$$

Theorem 24.2.15 *Let* $f(x) = x^3 + bx + c$ *be an irreducible and separable polynomial over the field* K. *Let* S *be a splitting field of* $f(x)$ *over* K *and* G *be the Galois group of* $f(x)$ *over* K. *Then* $G \simeq S_3$ *if and only if* D *is not a square in* K. *If* D *is a square in* K, *then* $[S : K] = 3$.

Proof. By the above discussion, $D \in K$. Suppose $d \in K$. Then $\alpha(d) = d$ for all $\alpha \in G$. Thus, no α can be an odd permutation. Hence, each α is in the alternating group A_3. Conversely, if $\alpha \in A_3$, then $\alpha(d) = d$. Since $f(x)$ is separable and irreducible, the roots of $f(x)$ are distinct. Therefore, $G \neq \{e\}$. Thus, the above argument shows that $G = A_3$ if and only if $d \in K$. Consequently, $G = S_3$ if and only if $d \notin K$. If $d \in K$, then $G = A_3$ and $|G| = 3$ and so $[S : K] = 3$ by the fundamental theorem of Galois theory. ∎

Theorem 24.2.16 *Let* $f(x) = x^3 + bx + c$ *be an irreducible and separable polynomial over the field* K. *Let* S *be the splitting field of* $f(x)$ *over* K. *Then* $S = K(\sqrt{D}, r)$ *for any root* r *of* $f(x)$.

Proof. Now $[K(r) : K] = 3$. If $S = K(r)$, then $S = K(\sqrt{D}, r)$. Suppose $S \supset K(r)$. Then $[S : K] = 6$ and $[S : K(r)] = 2$. Since $[S : K] = 6$, $G = S_3$, where G is the Galois group of $f(x)$ over K and so $d \notin K$. Since d is a root of $x^2 - D$ over K, $x^2 - D$ is irreducible over K. Since 2 and 3 are relatively prime, $x^2 - D$ is irreducible over $K(r)$. Thus, $S = K(\sqrt{D}, r)$. ∎

Example 24.2.17 *Consider the polynomial $x^3 - 4x + 2 \in \mathbf{Q}$. Then $x^3 - 4x + 2$ is irreducible over $\mathbf{Q}$ by Eisenstein's criterion. Now $D = -4b^3 - 27c^2 = -4(-4)^3 - 27(2)^2 = 148$. Thus, D is not a square in $\mathbf{Q}$. Hence, the Galois group of $x^3 - 4x + 2$ over $\mathbf{Q}$ is isomorphic to S_3. $S = \mathbf{Q}(\sqrt{148}, r)$, where r is any root of $x^3 - 4x + 2$.*

24.2.1 Worked-Out Exercises

◇ **Exercise 1** Let $f(x) = x^n - 1 \in \mathbf{Q}[x]$. Show that the Galois group of $f(x)$ over $\mathbf{Q}$ is commutative.

Solution: Let $\xi = e^{\frac{2\pi i}{n}}$, where $i^2 = -1$. Then the roots of $f(x)$ are 1, ξ, $\xi^2, \ldots, \xi^{n-1}$. Clearly $K = \mathbf{Q}(\xi)$ is a splitting field of $f(x)$. Let $\alpha, \beta \in G(K/\mathbf{Q})$. Now $\alpha(\xi)$ and $\beta(\xi)$ are roots of $f(x)$. Hence, $\alpha(\xi) = \xi^k$ and $\beta(\xi) = \xi^j$ for some k, j; $1 \le k, j \le n-1$. Now $(\alpha \circ \beta)(\xi) = \xi^{kj} = (\beta \circ \alpha)(\xi)$. Let $y \in K$. Then $y = \sum_{l=0}^{n-1} a_l \xi^l$ for some $a_l \in \mathbf{Q}$, $1 \le l \le n$. Now $(\alpha \circ \beta)(y) = (\alpha \circ \beta)(\sum_{l=0}^{n-1} a_l \xi^l) = \sum_{l=0}^{n-1} (\alpha \circ \beta)(a_l \xi^l) = \sum_{l=0}^{n-1} a_l (\alpha \circ \beta)(\xi^l) = \sum_{l=0}^{n-1} a_l \xi^{kjl}$. Similarly, $(\beta \circ \alpha)(y) = \sum_{l=0}^{n-1} a_l \xi^{jkl}$. Therefore, $\alpha \circ \beta = \beta \circ \alpha$. Consequently, $G(K/\mathbf{Q})$ is commutative.

◇ **Exercise 2** (i) Find a primitive element for the extension $\mathbf{Q}(\sqrt{2}, \sqrt{3})$ of $\mathbf{Q}$.

(ii) Find $[\mathbf{Q}(\sqrt{2}, \sqrt{3}) : \mathbf{Q}]$.

(iii) Show that $\mathbf{Q}(\sqrt{2}, \sqrt{3})$ is a splitting field of some polynomial $f(x)$ over $\mathbf{Q}$.

(iv) Prove that $\mathbf{Q}(\sqrt{2}, \sqrt{3})$ is a normal extension of $\mathbf{Q}$.

(v) If $F = \mathbf{Q}(\sqrt{2}, \sqrt{3})$, find the group $G(F/\mathbf{Q})$.

Solution: (i) $u = \sqrt{2} + \sqrt{3} \in \mathbf{Q}(\sqrt{2}, \sqrt{3})$. Thus, $\mathbf{Q}(\sqrt{2} + \sqrt{3}) \subseteq \mathbf{Q}(\sqrt{2}, \sqrt{3})$. Now $\sqrt{2} + \sqrt{3} \in \mathbf{Q}(\sqrt{2} + \sqrt{3})$. Therefore, $\frac{1}{\sqrt{2} + \sqrt{3}} \in \mathbf{Q}(\sqrt{2} + \sqrt{3})$ and so $\sqrt{2} - \sqrt{3} \in \mathbf{Q}(\sqrt{2} + \sqrt{3})$. Since $\sqrt{2} = \frac{1}{2}(2\sqrt{2}) = \frac{1}{2}((\sqrt{2} + \sqrt{3}) + (\sqrt{2} - \sqrt{3}))$, it follows that $\sqrt{2} \in \mathbf{Q}(\sqrt{2} + \sqrt{3})$. Again $\sqrt{3} = \frac{1}{2}((\sqrt{2} + \sqrt{3}) - (\sqrt{2} - \sqrt{3}))$ shows that $\sqrt{3} \in \mathbf{Q}(\sqrt{2} + \sqrt{3})$. Thus, $\mathbf{Q}(\sqrt{2}, \sqrt{3}) \subseteq \mathbf{Q}(\sqrt{2} + \sqrt{3})$. Hence, $\mathbf{Q}(\sqrt{2} + \sqrt{3}) = \mathbf{Q}(\sqrt{2}, \sqrt{3})$.

(ii) $[\mathbf{Q}(\sqrt{2}, \sqrt{3}) : \mathbf{Q}] = [\mathbf{Q}(\sqrt{2}, \sqrt{3}) : \mathbf{Q}(\sqrt{2})][\mathbf{Q}(\sqrt{2}) : \mathbf{Q}]$. Now $x^2 - 2$ is the minimal polynomial of $\mathbf{Q}(\sqrt{2})$ over $\mathbf{Q}$. Also, $x^2 - 3$ is the minimal polynomial of $\mathbf{Q}(\sqrt{2}, \sqrt{3})$ over $\mathbf{Q}(\sqrt{2})$ by Example 21.1.21. Hence, $[\mathbf{Q}(\sqrt{2}, \sqrt{3}) : \mathbf{Q}] = 2 \cdot 2 = 4$.

(iii) Let $f(x) = (x^2 - 2)(x^2 - 3) = x^4 - 5x^2 + 6 \in \mathbf{Q}[x]$. Since $f(x) = (x + \sqrt{2})(x - \sqrt{2})(x + \sqrt{3})(x - \sqrt{3})$, $f(x)$ splits over $\mathbf{Q}(\sqrt{2}, \sqrt{3})$. Thus, $\mathbf{Q}(\sqrt{2}, \sqrt{3})$ is a splitting field of $f(x)$ over $\mathbf{Q}$.

(iv) $f(x) = (x^2 - 2)(x^2 - 3)$ is a separable polynomial over $\mathbf{Q}$. Since $\mathbf{Q}(\sqrt{2}, \sqrt{3})$ is the splitting field of $f(x)$ by Theorem 24.2.11, it follows that $\mathbf{Q}(\sqrt{2}, \sqrt{3})$ is a normal extension of $\mathbf{Q}$.

(v) By the fundamental theorem of Galois theory 24.2.13(i), we find that $|G(F/\mathbf{Q})| = [F : \mathbf{Q}] = 4$. Now we know that $\mathbf{Z}_4$ (the cyclic group of order 4) and $\mathbf{Z}_2 \times \mathbf{Z}_2$ (the Klein 4-group) are the only (up to isomorphism) groups of order 4. Hence, either $G(F/\mathbf{Q}) \simeq \mathbf{Z}_4$ or $G(F/\mathbf{Q}) \simeq \mathbf{Z}_2 \times \mathbf{Z}_2$. If $G(F/\mathbf{Q}) \simeq \mathbf{Z}_4$, then $G(F/\mathbf{Q})$ has only one subgroup of order 2. Thus, by the fundamental theorem of Galois theory, there exists only one intermediate field L of $F/\mathbf{Q}$ such that $[L : \mathbf{Q}] = 2$. But $\mathbf{Q}(\sqrt{2})$ and $\mathbf{Q}(\sqrt{3})$ are intermediate fields of $F/\mathbf{Q}$ such that $[\mathbf{Q}(\sqrt{2}) : \mathbf{Q}] = 2$ and $[\mathbf{Q}(\sqrt{3}) : \mathbf{Q}] = 2$. Hence, $G(F/\mathbf{Q}) \not\simeq \mathbf{Z}_4$. Consequently, $G(F/\mathbf{Q}) \simeq \mathbf{Z}_2 \times \mathbf{Z}_2$.

◇ **Exercise 3** Let u be a complex number such that $u \neq 1$ and u is a root of the polynomial $x^5 - 1 \in \mathbf{Q}[x]$. Show that $G(\mathbf{Q}(u)/\mathbf{Q}) \simeq \mathbf{Z}_4$.

Solution: $x^5 - 1 = (x - 1)(x^4 + x^3 + x^2 + x + 1)$. Hence, u is a root of $f(x) = x^4 + x^3 + x^2 + x + 1$. By Worked-Out Exercise 1 (page 379), we find that $f(x)$ is irreducible in $\mathbf{Q}[x]$. From Theorem 24.3.3, $\mathbf{Q}(u)$ is a splitting field of $f(x)$. Since all roots of $f(x)$ are distinct, $f(x)$ is a separable polynomial. Hence, $\mathbf{Q}(u)$ is a normal extension of $\mathbf{Q}$. By Corollary 24.2.12,

$$|G(\mathbf{Q}(u)/\mathbf{Q})| = [\mathbf{Q}(u) : \mathbf{Q}] = 4.$$

Now u, u^2, u^3, u^4 are the four distinct roots of $f(x)$ and $\mathbf{Q}(u) = \mathbf{Q}(u^2) = \mathbf{Q}(u^3) = \mathbf{Q}(u^4)$. Hence, there exists $\sigma \in G(\mathbf{Q}(u)/\mathbf{Q})$ such that $\sigma(u) = u^2$. Thus,

$$\sigma^2(u) = \sigma(\sigma(u)) = \sigma(u^2) = \sigma(u)\sigma(u) = u^4,$$

$$\sigma^3(u) = \sigma(\sigma^2(u)) = \sigma(u^4) = u^8 = u^3,$$

and

$$\sigma^4(u) = \sigma(u^3) = u^6 = u.$$

So we find that $\sigma, \sigma^2, \sigma^3$, and σ^4 are distinct and $\sigma, \sigma^2, \sigma^3, \sigma^4 \in G(\mathbf{Q}(u)/\mathbf{Q})$. Therefore, $G(\mathbf{Q}(u)/\mathbf{Q})$ is a cyclic group of order 4. Consequently, $G(\mathbf{Q}(u)/\mathbf{Q}) \simeq \mathbf{Z}_4$.

◇ **Exercise 4** Show that the Galois group of the polynomial $f(x) = x^3 - 5$ over $\mathbf{Q}$ is isomorphic to S_3.

Solution: Let $\omega = \frac{-1+\sqrt{-3}}{2}$. Then $\omega^2 = \frac{-1-\sqrt{-3}}{2}$. Then $u = \sqrt[3]{5}$, $u\omega$ and $u\omega^2$ are the three distinct roots of $f(x)$. Thus, the splitting field of $f(x)$ over $\mathbf{Q}$ is $\mathbf{Q}(u, u\omega, u\omega^2) = \mathbf{Q}(u, w)$. $\mathbf{Q}$ is of characteristic 0. Hence, $\mathbf{Q}(u, w)$ is a normal extension of $\mathbf{Q}$. Therefore,

$$|G(\mathbf{Q}(u, \omega)/\mathbf{Q})| = [\mathbf{Q}(u, w) : \mathbf{Q}].$$

Now

$$[\mathbf{Q}(u, w) : \mathbf{Q}] = [\mathbf{Q}(u, w) : \mathbf{Q}(u)][\mathbf{Q}(u) : \mathbf{Q}].$$

The minimal polynomial of ω over $\mathbf{Q}(u)$ is x^2+x+1 and the minimal polynomial of u over $\mathbf{Q}$ is $x^3 - 5$. Consequently, $[\mathbf{Q}(u, w) : \mathbf{Q}] = 2 \cdot 3 = 6$. Thus, we find that $G(\mathbf{Q}(u, \omega)/\mathbf{Q})$ is a group of order 6 which is (up to isomorphism) either $\mathbf{Z}_6$ or S_3. If $G(\mathbf{Q}(u, \omega)/\mathbf{Q}) \simeq \mathbf{Z}_6$, then $G(\mathbf{Q}(u, \omega)/\mathbf{Q})$ has only one subgroup of order 2, i.e., $G(\mathbf{Q}(u, \omega)/\mathbf{Q})$ has only one subgroup of index 3. But $\mathbf{Q}(u, w)$ contains three distinct subfields $\mathbf{Q}(u)$, $\mathbf{Q}(u\omega)$, $\mathbf{Q}(u\omega^2)$,

$$[\mathbf{Q}(u) : \mathbf{Q}] = [\mathbf{Q}(u\omega) : \mathbf{Q}] = [\mathbf{Q}(u\omega^2) : \mathbf{Q}] = 3.$$

Hence, $G(\mathbf{Q}(u, \omega)/\mathbf{Q}) \not\simeq \mathbf{Z}_6$. Consequently, $G(\mathbf{Q}(u, \omega)/\mathbf{Q}) \simeq S_3$.

$\diamond$ **Exercise 5** Let p be a prime integer and m be a positive integer. Find the Galois group of the polynomial $f(x) = x^{p^m} - x$ over $\mathbf{Z}_p$.

Solution: The roots of $f(x)$ over $\mathbf{Z}_p$ form the Galois field, say, F with p^m elements. Now $[F : \mathbf{Z}_p] = m$ and F is the splitting field of $x^{p^m} - x$ over $\mathbf{Z}_p$ (Theorem 23.1.2). By Theorem 22.1.12, $\mathbf{Z}_p$ is perfect. Thus, $F/\mathbf{Z}_p$ is a separable extension. Also, F is a normal extension of $\mathbf{Z}_p$. Hence, by Corollary 24.2.12, we find that $|G(F/\mathbf{Z}_p)| = m$. Define $\sigma : F \to F$ by $\sigma(a) = a^p$. Let a, b be two distinct elements of F. Then $\sigma(a) - \sigma(b) = a^p - b^p = (a - b)^p \neq 0$. Thus, σ is one-one. Also, F consists of a finite number of elements. Hence, σ is also onto F. Now

$$\sigma(a + b) = (a + b)^p = a^p + b^p = \sigma(a) + \sigma(b)$$

and

$$\sigma(ab) = (ab)^p = a^p b^p = \sigma(a)\sigma(b)$$

for all $a, b \in F$. Therefore, σ is an automorphism of F. If $a \in \mathbf{Z}_p$, then $a^p = a$ and hence $\sigma(a) = a$. Thus, it follows that $\sigma \in G(F/\mathbf{Z}_p)$. For any positive integer k, $\sigma^k \in G(F/\mathbf{Z}_p)$ and $\sigma^k(a) = a^{p^k}$ for all $a \in F$. Since every element of F is a root of $x^{p^m} - x$, $\sigma^m(a) = a^{p^m} = a$ for all $a \in F$. Hence, σ^m is the identity element of $G(F/\mathbf{Z}_p)$. Suppose for some r, $1 \leq r < m$, $\sigma^r = e$. Then $a^{p^r} = a$ for all $a \in F$. Thus, every element of F is a root of $x^{p^r} - x$ over $\mathbf{Z}_p$. Since $x^{p^r} - x$ has at most p^r roots, $|F| \leq p^r < p^m$, a contradiction. Consequently, $\circ(\sigma) = m$ and so $G = \langle \sigma \rangle$.

◇ **Exercise 6** Find the Galois group of the polynomial $x^4 - 2$ over $\mathbf{Q}$.

Solution: From Eisenstein's criterion, it follows that $x^4 - 2$ is irreducible over $\mathbf{Q}$. Now $u = \sqrt[4]{2}$ is a root of $x^4 - 2$. Also,

$$x^4 - 2 = (x - \sqrt[4]{2})(x + \sqrt[4]{2})(x + i\sqrt[4]{2})(x - i\sqrt[4]{2}).$$

Hence, the splitting field of $x^4 - 2$ is $\mathbf{Q}(\sqrt[4]{2}, -\sqrt[4]{2}, i\sqrt[4]{2}, -i\sqrt[4]{2}) = \mathbf{Q}(\sqrt[4]{2}, i\sqrt[4]{2}) = \mathbf{Q}(\sqrt[4]{2}, i) = \mathbf{Q}(u, i)$. Now

$$[\mathbf{Q}(u, i) : \mathbf{Q}] = [\mathbf{Q}(u, i) : \mathbf{Q}(u)][\mathbf{Q}(u) : \mathbf{Q}].$$

The minimal polynomial of u over $\mathbf{Q}$ is $x^4 - 2$ and the minimal polynomial of i over $\mathbf{Q}(u)$ is $x^2 + 1$. Thus,

$$[\mathbf{Q}(u, i) : \mathbf{Q}] = 2 \cdot 4 = 8.$$

Also, $\mathbf{Q}(u, i)$ is the splitting field of the separable polynomial $x^4 - 2$. Hence, $\mathbf{Q}(u, i)$ is a normal extension of $\mathbf{Q}$. Therefore, by the fundamental theorem of Galois theory, it follows that $|G(\mathbf{Q}(u, i)/\mathbf{Q})| = 8$. Now $\{1, \sqrt[4]{2}, (\sqrt[4]{2})^2, (\sqrt[4]{2})^3, i, i\sqrt[4]{2}, i(\sqrt[4]{2})^2, i(\sqrt[4]{2})^3\}$ is a basis of $(\mathbf{Q}(u, i)$ over $\mathbf{Q}$. Let $a \in \mathbf{Q}(u, i)$. Then there exist $a_0, a_1, a_2, a_3, a_4, a_5, a_6$, and a_7 in $\mathbf{Q}$ such that

$$a = a_0 + a_1\sqrt[4]{2} + a_2(\sqrt[4]{2})^2 + a_3(\sqrt[4]{2})^3 + a_4 i + a_5 i\sqrt[4]{2} + a_6 i(\sqrt[4]{2})^2 + a_7 i(\sqrt[4]{2})^3.$$

If $\alpha \in G(\mathbf{Q}(u, i)/\mathbf{Q})$, then

$$\alpha(a) = a_0 + a_1\alpha(\sqrt[4]{2}) + a_2\alpha(\sqrt[4]{2})^2 + a_3\alpha(\sqrt[4]{2})^3 + a_4\alpha(i) +$$
$$a_5\alpha(i)\alpha(\sqrt[4]{2}) + a_6\alpha(i)\alpha(\sqrt[4]{2})^2 + a_7\alpha(i)\alpha(\sqrt[4]{2})^3.$$

Thus, $\alpha(a)$ will be known if we determine $\alpha(\sqrt[4]{2})$ and $\alpha(i)$. Since the minimal polynomial of $\sqrt[4]{2}$ is $x^4 - 2 \in \mathbf{Q}[x]$ and the minimal polynomial of i is $x^2 + 1 \in \mathbf{Q}[x]$, $\alpha(\sqrt[4]{2})$ is a root of $x^4 - 2$ and $\alpha(i)$ is a root of $x^2 + 1$. Hence, $\alpha(\sqrt[4]{2})$ is one of $\sqrt[4]{2}, -\sqrt[4]{2}, i\sqrt[4]{2}, -i\sqrt[4]{2}$ and $\alpha(i)$ is one of i and $-i$. It now follows that $G(\mathbf{Q}(u, i)/\mathbf{Q})$ has eight elements. The eight elements of $G(\mathbf{Q}(u, i)/\mathbf{Q})$ are given by the following table

	α_0	α_1	α_2	α_3	α_4	α_5	α_6	α_7
$\sqrt[4]{2}$	$\sqrt[4]{2}$	$-\sqrt[4]{2}$	$i\sqrt[4]{2}$	$-i\sqrt[4]{2}$	$\sqrt[4]{2}$	$-\sqrt[4]{2}$	$i\sqrt[4]{2}$	$-i\sqrt[4]{2}$
i	i	i	i	i	$-i$	$-i$	$-i$	$-i$

Now

$$(\alpha_2 \circ \alpha_6)(u) = \alpha_2(iu) = \alpha_2(i)\alpha_2(u) = i(iu) = -u$$

and

$$(\alpha_6 \circ \alpha_2)(u) = \alpha_6(iu) = \alpha_6(i)\alpha_6(u) = -i(iu) = u.$$

Consequently, $\alpha_2 \circ \alpha_6 \neq \alpha_6 \circ \alpha_2$. Therefore, we find that $G(\mathbf{Q}(u, i)/\mathbf{Q})$ is a non-commutative group of order 8. Hence, $G(\mathbf{Q}(u, i)/\mathbf{Q}) \simeq D_4$ or $G(\mathbf{Q}(u, i)/\mathbf{Q}) \simeq Q_8$. Now Q_8 has only one subgroup of index 4, but there are more than one intermediate field of $\mathbf{Q}(u, i)/\mathbf{Q}$ of dimension 4 over $\mathbf{Q}$, namely, $\mathbf{Q}(u)$ and $\mathbf{Q}(iu)$. Thus, $G(\mathbf{Q}(u, i)/\mathbf{Q} \simeq D_4$.

$\Diamond$ **Exercise 7** Find all proper subfields of $\mathbf{Q}(\sqrt{2}, \sqrt{3})$.

Solution: Let $F = \mathbf{Q}(\sqrt{2}, \sqrt{3})$. Then from Worked-Out Exercise 2 (page 513), $G(F/\mathbf{Q}) \simeq \mathbf{Z}_2 \times \mathbf{Z}_2$. Now $\mathbf{Z}_2 \times \mathbf{Z}_2$ has only three nontrivial subgroups. Each of these subgroups is of index 2. Since $\mathbf{Q}(\sqrt{2})$, $\mathbf{Q}(\sqrt{3})$, $\mathbf{Q}(\sqrt{6})$ are intermediate fields of $\mathbf{Q}(\sqrt{2}, \sqrt{3})/\mathbf{Q}$ and $[\mathbf{Q}(\sqrt{2}) : \mathbf{Q}] = [\mathbf{Q}(\sqrt{3}) : \mathbf{Q}] = [\mathbf{Q}(\sqrt{6}) : \mathbf{Q}] = 2$, it follows that $\mathbf{Q}(\sqrt{2})$, $\mathbf{Q}(\sqrt{3})$, and $\mathbf{Q}(\sqrt{6})$ are the only intermediate fields of $\mathbf{Q}(\sqrt{2}, \sqrt{3})/\mathbf{Q}$. Again $\mathbf{Q}$ is a subfield of F and $\mathbf{Q}$ has no proper subfields. Hence, $\mathbf{Q}$, $\mathbf{Q}(\sqrt{2})$, $\mathbf{Q}(\sqrt{3})$, and $\mathbf{Q}(\sqrt{6})$ are the only proper subfields of F.

$\Diamond$ **Exercise 8** Find the Galois group of the field extension

$$\mathbf{Q}(\sqrt[3]{5}, \frac{-1 + i\sqrt{3}}{2})/\mathbf{Q}.$$

Find all subgroups of this group and find all corresponding intermediate fields in the above extension according to the fundamental theorem of Galois theory.

Solution: Let $F = \mathbf{Q}(\sqrt[3]{5}, \omega)$, where $\omega = \frac{-1+i\sqrt{3}}{2}$. From Worked-Out Exercise 4 (page 514), we find that $G(F/\mathbf{Q}) \simeq S_3$. S_3 has four nontrivial subgroups $H_1 = \{e, (1\,2)\}$, $H_2 = \{e, (1\,3)\}$, $H_3 = \{e, (2\,3)\}$, and $H_4 = \{e, (1\,2\,3), (1\,3\,2)\}$. The index of H_4 is 2. Hence, the corresponding subfield of H_4 is $\mathbf{Q}(\omega)$.

Again $[S_3 : H_1] = [S_3 : H_2] = [S_3 : H_3] = 3$ and $[\mathbf{Q}(u) : \mathbf{Q}] = [\mathbf{Q}(uw) : \mathbf{Q}] = [\mathbf{Q}(uw^2) : \mathbf{Q}] = 3$, where $u = \sqrt[3]{5}$. Let $a_1 = u$, $a_2 = uw$, and $a_3 = uw^2$ and $1 \leftrightarrow a_1$, $2 \leftrightarrow a_2$, and $3 \leftrightarrow a_3$. Now

$$
\begin{aligned}
(1\,2) \quad : \quad & a_1 \to a_2 \\
& a_2 \to a_1 \\
& a_3 \to a_3.
\end{aligned}
$$

Thus, the intermediate field corresponding to H_1 is $\mathbf{Q}(uw^2)$. Similarly, the intermediate field corresponding to H_2 is $\mathbf{Q}(uw)$ and the intermediate field corresponding to H_3 is $\mathbf{Q}(u)$.

24.2.2 Exercises

1. Find the Galois group of $\mathbf{C}/\mathbf{R}$. Illustrate the Galois correspondence.

2. Find the degree of the following field extension F over $\mathbf{Q}$, the smallest extension N of F normal over $\mathbf{Q}$, and the Galois group of $N/\mathbf{Q}$.

   (i) $F = \mathbf{Q}(\sqrt{2}, \sqrt[3]{2})$.

   (ii) $F = \mathbf{Q}(\sqrt{2} + \sqrt[3]{2})$.

3. Show that the Galois group of the polynomial $(x^2 - 2)(x^2 - 3)$ over $\mathbf{Q}$ is isomorphic to $\mathbf{Z}_2 \times \mathbf{Z}_2$.

4. What are the possible degrees over $\mathbf{Q}$ of the splitting field of $x^3 + ax^2 + bx + c \in \mathbf{Q}[x]$? For each such degree, find an $f(x)$ of degree 3 in $\mathbf{Q}[x]$ whose splitting field has this degree over $\mathbf{Q}$. Can a field normal over $\mathbf{Q}$ be found in each case?

5. Find the Galois group G of the polynomial $x^3 - x - 1$ over $\mathbf{Q}$. Determine all subgroups of G and find all corresponding subfields of the splitting field. Let a_1, a_2, a_3 denote the roots of $x^3 - x - 1$. Determine $\mathbf{Q}(d)$, where $d = (a_2 - a_1)(a_3 - a_1)(a_3 - a_2)$.

6. Find the Galois group G of the following polynomials over $\mathbf{Q}$.

   (i) $(x^2 - 3x + 1)^2(x^3 - 2)$.

   (ii) $x^4 + x^2 + 1$.

7. Show that the Galois group of the polynomial $(x^2 - 2)(x^3 - 3)$ over $\mathbf{Q}$ is isomorphic to $S_3 \times \mathbf{Z}_2$. Find all subfields of the splitting field over $\mathbf{Q}$.

8. Let F be a splitting field of a polynomial $f(x)$ over a field K. Prove that the group $G(F/K)$ is isomorphic to a group of permutations of the distinct roots of $f(x)$.

9. Find the Galois group of $f(x) = 0$ over the field $\mathbf{Q}$, where $f(x) = x^3 - 7$.

10. Find all intermediate fields of $\mathbf{Q}(i, \sqrt{7})/\mathbf{Q}$.

11. Show that the Galois group of the polynomial equation $x^3 - 2 = 0$ is isomorphic to that of $x^3 - 3 = 0$ over $\mathbf{Q}$.

12. Let $F = \mathbf{Q}(\sqrt{2}, \sqrt{5}, \sqrt{7})$. Find the order of $G(F/\mathbf{Q})$.

13. Let $F = \mathbf{Q}(\sqrt{3}, \sqrt{11})$. Find the subgroups of the group $G(F/\mathbf{Q})$. Find the corresponding intermediate fields. Find all normal extensions of $\mathbf{Q}$ in F.

14. Let F be a finite field of characteristic p and $[F : \mathbf{Z}_p] = n$. Show that $F/\mathbf{Z}_p$ is a Galois extension and $G(F/\mathbf{Z}_p)$ is a cyclic group of order n.

15. Let F be a finite field of characteristic p. Let $[F : \mathbf{Z}_p] = n$. Show that for every positive divisor m of n, F has a unique subfield S of p^m elements. Also, show that F/S is a Galois extension and $G(F/S)$ is a cyclic group of order $\frac{n}{m}$.

24.3 Roots of Unity and Cyclotomic Polynomials

In Gauss's epoch-making work *Disquistiones Arithmeticae*, Gauss showed that the cyclotomic equation $x^n - 1 = 0$ is solvable for every n in the sense that the solutions are expressible in terms of radicals. He not only gave a method for finding these expressions, but also determined the values of n for which the solutions are expressible in quadratic radicals and in so doing he determined the values of n for which it is possible to construct a regular n-gon by means of ruler and compass.

Definition 24.3.1 *Let F be any field and n be a positive integer. Let $\omega \in F$. Then ω is called an nth **root of unity** if $\omega^n = 1$. ω is called a **primitive** nth root of unity if $\omega^n = 1$ and $\omega^m \neq 1$ for all m, $1 \leq m < n$.*

Let F be a field and n be a positive integer. Let $\omega \in F$ be an nth root of unity. Suppose the characteristic of F is $p > 0$ and $p|n$. Then $n = p^k m$ for some positive integer k and m such that $\gcd(p, m) = 1$. Thus, $(\omega^m - 1)^{p^k} = \omega^{p^k m} - 1 = \omega^n - 1 = 0$. Hence, $\omega^m - 1 = 0$ and so ω is also an mth root of unity.

Theorem 24.3.2 *Let K be a field and n be a positive integer. Suppose the characteristic of K does not divide n. Let G be the set of all nth roots of unity in K. Then G is a cyclic group and $|G|$ divides n. If $x^n - 1$ splits into linear factors in $K[x]$, then $|G| = n$.*

Proof.   Since $1 \in G$, $G \neq \phi$. Let $a, b \in G$. Then $(ab^{-1})^n = a^n (b^{-1})^n = 1$. Therefore, $ab^{-1} \in G$. Hence, G is a subgroup of the multiplicative group $K^* = K \backslash \{0\}$. Since $f(x) = x^n - 1 \in K[x]$ has at most n roots in K, G is finite. Thus, by Theorem 23.1.5, G is cyclic. Let F be the splitting field of $f(x)$ over K. Since the characteristic of K does not divide n, $f'(x) = nx^{n-1} \neq 0$. Consequently, all roots of $f(x)$ are simple by Theorem 22.1.4. Thus, $f(x)$ has n distinct roots in F. Let T be the set of all roots of $f(x)$ in F. Clearly T is a group, $G \subseteq T \subseteq F \backslash \{0\}$, G is a subgroup of T, and $|T| = n$. Since G is a subgroup of T, $|G|$ divides $|T| = n$. Suppose $f(x)$ splits into linear factors in

$K[x]$. Then $F = K$ and so $G = T$. Hence, $|G| = n$. ∎

Let G, K, and n be as in Theorem 24.3.2. Let $G = \langle \omega \rangle$ with $|G| = n$. Then $o(\omega) = n$. Hence, ω is a primitive nth root of unity. Conversely, if ω is a primitive nth root of unity, then $\omega \in G$, $\omega^n = 1$, and $\omega^m \neq 1$ for all m, $1 \leq m < n$. Hence, ω is of order n and so $G = \langle \omega \rangle$. Thus, ω is a primitive nth root of unity if and only if $G = \langle \omega \rangle$ if and only if $o(\omega) = n$.

Theorem 24.3.3 *Let n be any positive integer and K be a field.*

(i) There exists a finite field extension F/K such that F contains a primitive nth root of unity if and only if the characteristic of K does not divide n. (Zero is not a divisor of n.)

(ii) Suppose the characteristic of K does not divide n. Let ω be a primitive nth root of unity over K. Then $K(\omega)$ is the splitting field of $f(x) = x^n - 1 \in K[x]$, $f(x)$ has n distinct roots in $K(\omega)$, and the roots of $f(x)$ form a multiplicative cyclic group H such that H is generated by any primitive nth root of unity in $K(\omega)$.

Proof. (i) Suppose the characteristic of K does not divide n. Let $f(x) = x^n - 1 \in K[x]$. Then $f'(x) = nx^{n-1} \neq 0$. Hence, all roots of $f(x)$ are simple by Theorem 22.1.4. Thus, $f(x)$ has n distinct roots in some splitting field. Let F be the splitting field of $f(x)$ over K. Then F/K is a finite extension. Let H be the set of all nth roots of unity in F. Then by Theorem 24.3.2, H is a cyclic group of order n. Let $H = \langle \omega \rangle$. Then $\omega \in F$ and $o(\omega) = n$. Therefore, ω is a primitive nth root of unity in F.

Conversely, let ω be a primitive nth root of unity in a finite field extension F/K. Then $1, \omega, \omega^2, \ldots, \omega^{n-1} \in F$ and these are all n distinct roots of $f(x)$. Since $\deg f(x) = n$, $f(x)$ has at most n roots. Thus, all roots of $f(x)$ are simple. Hence, $f'(x) = nx^{n-1} \neq 0$. Consequently, the characteristic of K does not divide n.

(ii) By (i), there exists a finite field extension F/K such that F contains a primitive nth root of unity, say, ω. Since ω is a primitive nth root of unity, $1, \omega, \omega^2, \ldots, \omega^{n-1}$ are all distinct elements and are roots of $f(x) = x^n - 1 \in K[x]$. Thus, $f(x)$ has n distinct roots in $K(\omega)$. Hence, $K(\omega)$ is a splitting field of $f(x)$. By Theorem 24.3.2, the roots of $f(x)$ form a multiplicative cyclic group H of order n. Since the multiplicative order of a primitive nth root of unity is n, H is generated by any primitive nth root of unity. ∎

Definition 24.3.4 *Let n be a positive integer and K be a field whose characteristic does not divide n. Let $\{\omega_1, \omega_2, \ldots, \omega_m\}$ be the set of all primitive nth roots of unity in the splitting field F of $x^n - 1$ over K. The polynomial*

$$\Phi_n(x) = (x - \omega_1) \cdots (x - \omega_m) \in F[x]$$

*is called the nth **cyclotomic polynomial** over K and F/K is called the nth* **cyclotomic extension.**

In the following theorem, we describe some important properties of cyclotomic polynomials.

Theorem 24.3.5 *Let n be a positive integer and K be a field such that the characteristic of K does not divide n. Let $\Phi_n(x)$ be the nth cyclotomic polynomial over K. Then the following assertions hold.*

(i) $x^n - 1 = \prod_{d|n,\ d>0} \Phi_d(x)$.

(ii) If P is the prime subfield of K, then $\Phi_n(x) \in P[x]$.

(iii) $\deg \Phi_n(x) = \phi(n)$.

Proof.   (i) Let ω be a primitive nth root of unity over K. Then $K(\omega)$ is the splitting field of $x^n - 1 \in K[x]$ and all nth roots of unity form a multiplicative cyclic group G of order n. Let d be a positive integer such that $d|n$. Let $G_d = \{a \in G \mid o(a) = d\}$. Then $\{G_d \mid d > 0 \text{ and } d|n\}$ forms a partition of G. Clearly for any positive divisor d of n, G contains all dth roots of unity and G_d contains all primitive dth roots of unity. Hence,

$$
\begin{aligned}
x^n - 1 &= \prod_{\omega \in G}(x - \omega) \\
&= \prod_{d|n,\ d>0} \prod_{\omega \in G_d}(x - \omega) \\
&= \prod_{d|n,\ d>0} \Phi_d(x).
\end{aligned}
$$

(ii) Now $\Phi_n(x) = \prod_{\omega \in G_n}(x - \omega)$, where G_n is as defined in (i). We prove the result by induction on n. If $n = 1$, then $\Phi_1(x) = x - 1 \in P[x]$. Suppose the result is true for all positive integers k, $1 \le k < n$. Then for all $1 \le d < n$, $d|n$, $\Phi_d(x) \in P[x]$. Hence,

$$
f(x) = \prod_{d|n,\ 1 \le d < n} \Phi_d(x) \in P[x].
$$

By (i),

$$
x^n - 1 = \prod_{d|n,\ d>0} \Phi_d(x) = f(x)\Phi_n(x) \in K[x].
$$

Now $x^n - 1 \in P[x]$ and $f(x)$ is monic. By the division algorithm, there exist $q(x), r(x) \in P[x] \subseteq K[x]$ such that

$$
x^n - 1 = q(x)f(x) + r(x),
$$

where either $r(x) = 0$ or $\deg r(x) < \deg f(x)$. Hence, by the uniqueness of quotients and remainder in $K[x]$, $r(x) = 0$ and $\Phi_n(x) = q(x) \in P[x]$.

(iii)

$$
\begin{aligned}
\deg \Phi_n(x) &= \text{number of distinct primitive } n\text{th roots of unity}\\
&= \text{number of distinct elements of } G \text{ of order } n\\
&= \text{number of generators of } G\\
&= \phi(n). \ \blacksquare
\end{aligned}
$$

We now examine cyclotomic polynomials over $\mathbf{Q}$. Suppose that w is a complex root of unity. Then for some positive integer n, $|w|^n = |w^n| = 1$. Hence, $|w| = 1$ and so w lies on the unit circle in the complex plane. Also, w must be of the form $e^{\frac{2\pi k}{n}i} = \cos \frac{2\pi k}{n} + i \sin \frac{2\pi k}{n}$ for some integer k, $0 \le k < n$. Thus, there are exactly n nth roots of unity. These roots of unity divide the unit circle into n equal arcs, from which we get the word "cyclotomy."

Theorem 24.3.6 *Let $w \in \mathbf{C}$ be a primitive nth root of unity over $\mathbf{Q}$. Let $\Phi_n(x)$ be the nth cyclotomic polynomial over $\mathbf{Q}$. Then the following assertions hold.*
(i) $\Phi_n(x) \in \mathbf{Z}[x]$.
(ii) $\Phi_n(x)$ is irreducible over $\mathbf{Q}$.
(iii) $[\mathbf{Q}(w) : \mathbf{Q}] = \phi(n)$.
(iv) $G(\mathbf{Q}(w)/\mathbf{Q}) \simeq U_n$.

Proof.   (i) We prove the result by induction on n. If $n = 1$, then $\Phi_1(x) = x - 1 \in \mathbf{Z}[x]$. Suppose the result is true for all positive integers k, $1 \le k < n$. Then for all $1 \le d < n$, $d|n$, $\Phi_d(x) \in \mathbf{Z}[x]$. Hence,

$$
f(x) = \prod_{d|n,\ 1 \le d < n} \Phi_d(x) \in \mathbf{Z}[x].
$$

By Theorem 24.3.5(i),

$$
x^n - 1 = \prod_{d|n,\ d>0} \Phi_d(x) = f(x)\Phi_n(x) \in \mathbf{Q}[x].
$$

Now $x^n - 1 \in \mathbf{Z}[x]$ and $f(x)$ is monic. By the division algorithm, there exist $q(x), r(x) \in \mathbf{Z}[x] \subseteq \mathbf{Q}[x]$ such that

$$
x^n - 1 = q(x)f(x) + r(x),
$$

where either $r(x) = 0$ or $\deg r(x) < \deg f(x)$. Hence, by the uniqueness of quotients and remainder in $\mathbf{Q}[x]$, $r(x) = 0$ and $\Phi_n(x) = q(x) \in \mathbf{Z}[x]$.
   (ii) By Lemma 16.2.8, it is sufficient to show that $\Phi_n(x)$ is irreducible over $\mathbf{Z}$. Suppose $f(x) \in \mathbf{Z}[x]$ is an irreducible factor of $\Phi_n(x)$. Let $\Phi_n(x) = f(x)h(x)$ for some $h(x) \in \mathbf{Z}[x]$. Since $\Phi_n(x)$ is monic, both $f(x)$ and $h(x)$ can be taken to be monic. Let w be a root of $f(x)$. Then w is also a root of $\Phi_n(x)$ and hence w is a primitive nth root of unity. Let p be a prime such that p does

not divide n. Then $\gcd(p, n) = 1$. Hence, ω^p is also a generator of G, where G is the multiplicative cyclic group of all nth roots of unity. Thus, ω^p is also a primitive nth root of unity. We now claim that ω^p is also a root of $f(x)$.

Suppose ω^p is not a root of $f(x)$. Since ω^p is a root of $\Phi_n(x)$, ω^p is a root of $h(x)$. Therefore, ω is a root of $h(x^p)$. Since $f(x)$ is irreducible over $\mathbf{Z}$ and hence over $\mathbf{Q}$ and ω is a root of $f(x)$, $f(x) | h(x^p)$ by Corollary 21.1.9. Hence, $h(x^p) = f(x)g(x)$ for some $g(x) \in \mathbf{Q}[x]$. Since $f(x), h(x^p) \in \mathbf{Z}[x]$, we can conclude that $g(x) \in \mathbf{Z}[x]$ by using the division algorithm (as in (i)). For $t(x) \in \mathbf{Z}[x]$, let $\overline{t(x)}$ be the corresponding polynomial in $\mathbf{Z}_p[x]$, i.e., if $a \in \mathbf{Z}$ is a coefficient of $t(x)$, then $[a] \in \mathbf{Z}_p$ is a corresponding coefficient of $\overline{t(x)}$. Since the characteristic of $\mathbf{Z}_p$ is p, $\overline{h(x^p)} = (\overline{h(x)})^p$. Thus,

$$(\overline{h(x)})^p = \overline{h(x^p)} = \overline{f(x)}\ \overline{g(x)}.$$

Hence, $\overline{f(x)}$ and $\overline{h(x)}$ have a common irreducible factor. Now

$$\overline{\Phi_n(x)} = \overline{f(x)}\ \overline{h(x)}$$

and $\overline{\Phi_n(x)} | (x^n - 1)$. Therefore, $x^n - [1] \in \mathbf{Z}_p[x]$ has a multiple root. Let a be a multiple root of $t(x) = x^n - [1]$. Then $t'(a) = \overline{n}a^{n-1} = \overline{0}$. Since p does not divide n, $[n]a^{n-1} = \overline{0}$ implies that $a^{n-1} = [0]$ and so $a = [0]$. But $[0]$ is not a root of $x^n - [1]$, which gives the desired contradiction. Thus, ω^p is also a root of $f(x)$. By induction, we can show that ω^{p^r} is also a root for any positive integer r. By induction, we can also show that $\omega^{p_1^{r_1} \cdots p_s^{r_s}}$ is also a root of $f(x)$, where the p_i's are distinct primes such that p_i does not divide n and the r_i are positive integers. From this, it follows that for all k, $1 \leq k < n$, $\gcd(k, n) = 1$, ω^k is a root of $f(x)$. Since

$$\{\omega^k \mid 1 \leq k < n,\ \gcd(k, n) = 1\}$$

is the set of all primitive nth roots of unity, every primitive nth root of unity is a root of $f(x)$. Hence, $\Phi_n(x) = f(x)$ and so $\Phi_n(x)$ is irreducible over $\mathbf{Z}$.

(iii) Clearly $\mathbf{Q}(\omega)/\mathbf{Q}$ is a finite normal separable extension. Thus, by Corollary 24.2.12 and Theorem 24.3.5,

$$|G(\mathbf{Q}(\omega)/\mathbf{Q})| = [\mathbf{Q}(\omega) : \mathbf{Q}] = \phi(n).$$

(iv) Now for any $\sigma \in G(\mathbf{Q}(\omega)/\mathbf{Q})$, $\sigma(\omega)$ is a primitive nth root of unity. Hence, $\sigma(\omega) = \omega^d$ for some d, $1 \leq d < n$, and $\gcd(d, n) = 1$. Also, σ is determined if $\sigma(\omega)$ is determined. We denote this σ by σ_d. It can be easily verified that if c, d are integers such that $1 \leq c, d < n$, $\gcd(c, n) = 1$, and $\gcd(d, n) = 1$, then $\sigma_{cd} = \sigma_c \circ \sigma_d$. Define

$$\Psi : U_n \to G(\mathbf{Q}(\omega)/\mathbf{Q})$$

by $\Psi([d]) = \sigma_d$. Then Ψ is one-one function from U_n onto $G(\mathbf{Q}(\omega)/\mathbf{Q})$. Let $[c], [d] \in U_n$. Then $cd = qn + r$ for some integers q and r, $0 \leq r < n$. Then $[cd] = [r]$ and $\sigma_{cd}(\omega) = \omega^{cd} = \omega^{qn+r} = \omega^r = \sigma_r(\omega)$. Therefore, $\sigma_{cd} = \sigma_r$. Thus, $\Psi([c][d]) = \Psi([cd]) = \Psi([r]) = \sigma_r = \sigma_{cd} = \sigma_c \circ \sigma_d = \Psi([c]) \circ \Psi([d])$. Hence, Ψ is a homomorphism. Consequently,

$$G(\mathbf{Q}(\omega)/\mathbf{Q}) \simeq U_n. \ \blacksquare$$

Corollary 24.3.7 *Let n be a positive integer. Then for every positive divisor m of n,*

$$\frac{x^n - 1}{x^m - 1} \in \mathbf{Z}[x].$$

Proof.   By Theorem 24.3.5,

$$x^n - 1 = \prod_{d|n,\ d>0} \Phi_d(x),$$

where $\Phi_d(x)$ is the dth cyclotomic polynomial over $\mathbf{Q}$. Let m be a positive divisor of n. Then

$$
\begin{aligned}
x^n - 1 &= \textstyle\prod_{d|n,\ d>0} \Phi_d(x) \\
&= \textstyle\prod_{d|n,\ d>m} \Phi_d(x) \cdot \prod_{s|m,\ s>0} \Phi_s(x) \\
&= (x^m - 1) \textstyle\prod_{d|n,\ d>m} \Phi_d(x).
\end{aligned}
$$

Hence,

$$\frac{x^n - 1}{x^m - 1} = \prod_{d|n,\ d>m} \Phi_d(x).$$

By Theorem 24.3.6, $\Phi_d(x) \in \mathbf{Z}[x]$ for every positive integer d. Thus,

$$\frac{x^n - 1}{x^m - 1} = \prod_{d|n,\ d>m} \Phi_d(x) \in \mathbf{Z}[x]. \ \blacksquare$$

Corollary 24.3.8 *Let n be a positive integer. Then for every proper positive divisor m of n, $\Phi_n(x)$ divides $\frac{x^n-1}{x^m-1}$, where $\Phi_n(x)$ is the nth cyclotomic polynomial over $\mathbf{Q}$.*

Proof.   As in the proof of Corollary 24.3.7,

$$\frac{x^n - 1}{x^m - 1} = \prod_{d|n,\ d>m} \Phi_d(x).$$

Hence,

$$\frac{x^n - 1}{x^m - 1} = \Phi_n(x) \prod_{d|n,\ n>d>m} \Phi_d(x).$$

Thus, $\Phi_n(x)$ divides $\frac{x^n-1}{x^m-1}$. ∎

We now remove the assumption of commutativity in Theorem 10.1.23 as promised.

Theorem 24.3.9 (Wedderburn) *A nontrivial finite ring D without zero divisors is a field.*

Proof. We have already seen in Corollary 10.1.24 that a finite integral domain is a field. Hence, it suffices to prove that D is a commutative ring. Since D is finite, D has prime characteristic p and contains $\mathbf{Z}_p$. Set $F = \{a \mid a \in D, ad = da$ for all $d \in D\}$. Now $0, 1 \in F$ so that $F \neq \phi$. Let $a, b \in F$. Then $(a-b)d = ad - bd = da - db = d(a-b)$ for all $d \in F$. Thus, $a - b \in F$. For $b \neq 0$, $(ab^{-1})d = a(b^{-1}d) = a(db^{-1}) = d(ab^{-1})$ for all $d \in F$ since from $bd = db$, we can obtain $db^{-1} = b^{-1}d$ by multiplying on the left and right by b^{-1}. Hence, $ab^{-1} \in F$ so that since F is clearly commutative, F is a field in D. Now D is a vector space over F of finite dimension, say, n. Let q denote the number of elements in F. Then D has q^n elements and the multiplicative group G of D has $q^n - 1$ elements.

Suppose $n > 1$. We shall obtain a contradiction. For any $g \in G, g \notin F$, we set $D_g = \{d \mid d \in D, dg = gd\}$. Then as above D_g is a division ring and clearly $D_g \supseteq F$. Since D is also a vector space over D_g, we have that D_g contains q^d elements for some positive integer d, which must divide n. Thus, the multiplicative group G_g of D_g has order $q^d - 1$. Now G_g is the normalizer of g in G and hence the number of conjugates of g in G is the index $\frac{q^n-1}{q^d-1}$ of G_g in G. Decomposing G into conjugacy classes, we thus obtain

$$q^n - 1 = (q-1) + \sum \frac{q^n - 1}{q^{d_i} - 1},$$

where the sum is taken over a finite set of proper divisors d_i of n. Let $\Phi_n(x)$ be the nth cyclotomic polynomial over $\mathbf{Q}$. By Corollary 24.3.8, $\Phi_n(x)$ divides $\frac{x^n-1}{x^{d_i}-1}$. Also, by Corollary 24.3.7, $\frac{x^n-1}{x^{d_i}-1} \in \mathbf{Z}[x]$. Thus, $\Phi_n(q)$ is an integer dividing $q^n - 1$ and all the $\frac{q^n-1}{q^{d_i}-1}$ and so also dividing $q - 1$. But $\Phi_n(q) = \prod(q - \omega_j)$ and so we obtain

$$|\Phi_n(q)| = \prod |(q - \omega_j)| > q - 1$$

since $|q - \omega_j| > q - 1 \geq 1$ for all j and since $q \geq 2$. But this is contrary to the statement that $\Phi_n(q)$ divides $q - 1$. Hence, $n = 1$ and so $D = F$. ∎

24.3.1 Worked-Out Exercises

◇ **Exercise 1** Let ω be a primitive eighth root of unity over $\mathbf{Q}$. Describe $\Phi_8(x)$.

Solution: By Theorem 24.3.5,

$$x^8 - 1 = \Phi_1(x)\Phi_2(x)\Phi_4(x)\Phi_8(x).$$

Thus,

$$\Phi_8(x) = \frac{(x^8 - 1)}{\Phi_1(x)\Phi_2(x)\Phi_4(x)}.$$

Now $\Phi_1(x) = x - 1$, $\Phi_2(x) = x + 1$ and $\Phi_4(x) = x^2 + 1$. Hence,

$$\Phi_8(x) = \frac{(x^8 - 1)}{(x - 1)(x + 1)(x^2 + 1)} = x^4 + 1.$$

◇ **Exercise 2** Let n be a positive integer and ω be a primitive nth root of unity over $\mathbf{Q}$. Show that

$$[\mathbf{Q}(\omega + \frac{1}{\omega}) : \mathbf{Q}] = \frac{\phi(n)}{2}.$$

Solution: By Theorem 24.3.6, $[\mathbf{Q}(\omega) : \mathbf{Q}] = \phi(n)$. Now $\mathbf{Q} \subseteq \mathbf{Q}(\omega + \frac{1}{\omega}) \subseteq \mathbf{Q}(\omega)$. Therefore,

$$[\mathbf{Q}(\omega) : \mathbf{Q}] = [\mathbf{Q}(\omega) : \mathbf{Q}(\omega + \frac{1}{\omega})][\mathbf{Q}(\omega + \frac{1}{\omega}) : \mathbf{Q}]. \qquad (24.9)$$

By Corollary 24.2.12, $\left|G(\mathbf{Q}(\omega)/\mathbf{Q}(\omega + \frac{1}{\omega}))\right| = [\mathbf{Q}(\omega) : \mathbf{Q}(\omega + \frac{1}{\omega})]$. Now

$$G(\mathbf{Q}(\omega)/\mathbf{Q}(\omega + \frac{1}{\omega})) \subseteq G(\mathbf{Q}(\omega)/\mathbf{Q}).$$

Let $\sigma \in G(\mathbf{Q}(\omega)/\mathbf{Q})$. Since $\sigma(\omega)$ is a primitive nth root of unity, $\sigma(\omega) = \omega^d$, where $1 \le d < n$ and $\gcd(d, n) = 1$. If $d = 1$, then σ is the identity automorphism. Suppose $d \ne 1$. Also, suppose $\sigma \in G(\mathbf{Q}(\omega)/\mathbf{Q}(\omega + \frac{1}{\omega}))$. Then $\sigma(\omega + \frac{1}{\omega}) = \omega + \frac{1}{\omega}$. Hence, $\omega^d + \frac{1}{\omega^d} = \sigma(\omega + \frac{1}{\omega}) = \omega + \frac{1}{\omega}$. From this, it follows that $\omega^d - \omega = \frac{1}{\omega} - \frac{1}{\omega^d}$, i.e., $\omega^d - \omega = \frac{\omega^{d-1}-1}{\omega^d}$. Thus, $\omega(\omega^{d-1} - 1) = \frac{\omega^{d-1}-1}{\omega^d}$. Since $\omega^{d-1} - 1 \ne 0$, $\omega^{d+1} = 1$. Hence, $n = d + 1$ since $o(\omega) = n$. Thus, $d = n - 1$. Therefore, the only elements of $G(\mathbf{Q}(\omega)/\mathbf{Q})$ which fix each element of $\mathbf{Q}(\omega + \frac{1}{\omega})$ are the identity automorphism and the automorphism σ given by $\sigma(\omega) = \omega^{n-1}$. Consequently,

$$[\mathbf{Q}(\omega) : \mathbf{Q}(\omega + \frac{1}{\omega})] = \left|G(\mathbf{Q}(\omega)/\mathbf{Q}(\omega + \frac{1}{\omega}))\right| = 2.$$

Now $[\mathbf{Q}(\omega) : \mathbf{Q}] = \phi(n)$. Hence, from Eq. (24.9), it now follows that

$$[\mathbf{Q}(\omega + \frac{1}{\omega}) : \mathbf{Q}] = \frac{\phi(n)}{2}.$$

◇ **Exercise 3** Let K be a field of characteristic 0 and n be a positive integer. Let ω be a primitive nth root of unity in some field extension of K. Show that

(i) $K(\omega)/K$ is a normal extension and

(ii) $G(K(\omega)/K)$ is commutative.

Solution: (i) By Theorem 24.3.3, $K(\omega)$ is the splitting field of $x^n - 1 \in K[x]$. Hence, by Theorem 24.1.3, $K(\omega)/K$ is a normal extension.

(ii) Since the characteristic of K is 0, $K(\omega)/K$ is separable. Since $K(\omega)/K$ is also a normal extension, it follows that $K(\omega)/K$ is a Galois extension. Let $\alpha, \beta \in G(K(\omega)/K)$. Now $\alpha(\omega)$ and $\beta(\omega)$ are roots of $x^n - 1$. Thus, $\alpha(\omega) = \omega^i$ and $\beta(\omega) = \omega^j$ for some i and j. Clearly $(\alpha \circ \beta)(\omega) = (\beta \circ \alpha)(\omega)$. From this, it follows that $\alpha \circ \beta = \beta \circ \alpha$. Therefore, $G(K(\omega)/K)$ is commutative.

24.3.2 Exercises

1. Find the Galois group of $f(x) = x^2 - x + 1$ over $\mathbf{Q}$.

2. Show that the Galois groups of $x^4 - 1$ and $x^2 - x + 1$ over $\mathbf{Q}$ are isomorphic.

3. Let p be a prime and $\Phi_p(x)$ be the pth cyclotomic polynomial over $\mathbf{Q}$. Show that
$$\Phi_p(x) = 1 + x + \cdots + x^{p-1}.$$

4. Let n be a positive prime. Show that $\Phi_{2n}(x) = \Phi_n(-x)$, where $\Phi_n(x)$ is the nth cyclotomic polynomial over $\mathbf{Q}$.

5. Let n be a positive integer. Let p be a prime such that p does not divide n. Show that
$$\Phi_{pn}(x) = \frac{\Phi_n(x^p)}{\Phi_n(x)},$$
where $\Phi_{pn}(x)$ and $\Phi_n(x)$ are the pnth and nth cyclotomic polynomials over $\mathbf{Q}$, respectively.

6. Find a polynomial irreducible over $GF(3)$ having a primitive eighth root of unity as one of its roots in $GF(9)$.

7. Let K be a field of characteristic 0. Let $0 \neq a \in K$ and $f(x) = x^n - a$, where n is a positive integer. Let F/K be a field extension such that $f(x)$ splits over F. Show that F contains a primitive nth root of unity.

8. Let m and n be relatively prime positive integers.

(i) Show that the splitting field of $x^{mn} - 1$ over $\mathbf{Q}$ is the same as the splitting field of $(x^m - 1)(x^n - 1)$ over $\mathbf{Q}$.

(ii) From (i), deduce that $\phi(mn) = \phi(m)\phi(n)$.

9. Let m and n be relatively prime positive integers. Let w_m and w_n be the primitive mth and nth roots of unity, respectively. Show that $\mathbf{Q}(w_m) \cap \mathbf{Q}(w_n) = \mathbf{Q}$.

10. Let K be a field with characteristic not dividing n and F be the splitting field of $x^n - 1$ over K. Prove that F contains exactly $\phi(n)$ primitive nth roots of unity, where ϕ is the Euler ϕ-function.

11. Let n be a positive integer, K be a field containing all nth roots of unity, and $0 \neq a \in K$. Let F be the splitting field of $f(x) = x^n - a \in K[x]$ and b be a root of $f(x)$.

    (i) Show that $F = K(b)$.

    (ii) Show that the Galois group $G(F/K)$ is commutative.

24.4   Solvability of Polynomials by Radicals

The reader is familiar with the quadratic formula, which says that the roots of the polynomial $x^2 + bx + c$ are

$$\frac{-b \pm \sqrt{b^2 - 4c}}{2}.$$

The only restriction is that the field of which b and c are elements is not of characteristic 2.

By choosing cube roots correctly, the roots of the cubic polynomial $x^3 + bx^2 + cx + d$ are

$$s + t - \tfrac{b}{3},$$
$$ws + w^2 t - \tfrac{b}{3},$$
$$w^2 s + wt - \tfrac{b}{3},$$

where $w \neq 1$ is a cube root of 1,

$$s = \sqrt[3]{\tfrac{-q}{2} + \sqrt{\tfrac{p^3}{27} + \tfrac{q^2}{4}}},$$
$$t = \sqrt[3]{\tfrac{-q}{2} - \sqrt{\tfrac{p^3}{27} + \tfrac{q^2}{4}}},$$
$$p = c - \tfrac{b^2}{3},$$
$$q = \tfrac{2b^3}{27} - \tfrac{bc}{3} + d.$$

The field containing b, c, d is not of characteristic 2 or 3.

In a similar manner, there exists a formula for the roots of a quartic polynomial. This formula is also given in terms of combinations of radicals of rational functions of the coefficients. Abel showed that no such general formula can be given for the roots of fifth degree or higher degree polynomials. This does not mean that no such formula exists for certain polynomials of degree 5 or larger. Evariste Galois determined exactly for which polynomials such a formula exists. Galois's theory, polished by Emil Artin, is considered to be one of the most profound and beautiful works in the history of mathematics.

Consider the cubic polynomial $x^3 + bx^2 + cx + d$ over a field K of characteristic not equal to 2 or 3 and consider the chain of fields

$$K \subseteq K(u) \subseteq K(u, \sqrt[3]{-\frac{q}{2} + u}) \subseteq K(u, \sqrt[3]{-\frac{q}{2} + u}, \sqrt[3]{-\frac{q}{2} - u}) \subseteq F,$$

where $u = \sqrt{\frac{p^3}{27} + \frac{q^2}{4}}$ and $F = K(u, \sqrt[3]{-\frac{q}{2} + u}, \sqrt[3]{-\frac{q}{2} - u}), w)$. Then F contains the roots of the polynomial $x^3 + bx^2 + cx + d$. Also, $\pm u$ are roots of the polynomial $x^2 - u^2$, $\sqrt[3]{-\frac{q}{2} + u}$ is a root of $x^3 - (-\frac{q}{2} + u)$, $\sqrt[3]{-\frac{q}{2} - u}$ is a root of $x^3 - (-\frac{q}{2} - u)$, and w is a root of $x^3 - 1$. That is, F contains the splitting field of $x^3 + bx^2 + cx + d$ over K and F is obtained by successive adjunction of roots of a polynomial of the form $x^n - a$. In this sense, we mean that $x^3 + bx^2 + cx + d$ is solvable by radicals.

Definition 24.4.1 *A finite field extension F/K is called an* **extension by radicals** *(or* **radical extension**) *if there exists a finite chain of fields*

$$K = K_0 \subseteq K_1 \subseteq \cdots \subseteq K_m = F \tag{24.10}$$

such that $K_i = K_{i-1}(r_i)$, where r_i is a root of $x^{n_i} - a_i$, $a_i \in K_{i-1}$, for some positive integer n_i $(i = 1, 2, \ldots, m)$. The polynomial $f(x) \in K[x]$ (or the equation $f(x) = 0$) is called **solvable by radicals** *if its splitting field is contained in an extension by radicals of K.*

A chain of fields like that in (24.10) is called a **root tower.**

A question immediately comes to mind. If a polynomial is solvable by radicals, is its splitting field automatically a radical extension? The answer to this question is "no." Let $f(x) = x^3 - 4x + 2$. Since $\deg f(x) = 3$, $f(x)$ is solvable by radicals over $\mathbf{Q}$. Now $f(0) > 0$ and $f(1) < 0$. Hence, the graph of $f(x)$ must cross the x-axis three times. Thus, $f(x)$ has three real roots. Hence, a splitting field F of $f(x)$ over $\mathbf{Q}$ lies in $\mathbf{R}$. We will not show it here, but F is not a radical extension since $[F : \mathbf{Q}]$ is not a power of 2. The interested reader may find the details worked out in Isaacs.

The following is immediate from the definition of an extension by radicals.

Lemma 24.4.2 *Let $K \subseteq L \subseteq F$ be a chain of fields such that L/K and F/L are radical extensions. Then F/K is a radical extension.* ■

Theorem 24.4.3 *Let K be a field of characteristic 0 and F/K be an extension by radicals. Let $K = K_0 \subseteq K_1 \subseteq \cdots \subseteq K_m = F$ be the chain of intermediate* *fields such that $K_i = K_{i-1}(r_i)$, where r_i is a root of $x^{n_i} - a_i$, $a_i \in K_{i-1}$, for some positive integer n_i $(i = 1, 2, \ldots, m)$. Then there exists a finite chain of fields*

$$K = F_0 \subseteq F_1 \subseteq \cdots \subseteq F_m = E$$

such that F_i/K is a normal radical extension, $x^{n_i} - a_i$ splits over F_i, and $K_i \subseteq F_i$ for all i, $1 \leq i \leq m$.

Proof. Let $F_0 = K$. Suppose we have constructed a chain of fields

$$K = F_0 \subseteq F_1 \subseteq \cdots \subseteq F_i$$

such that F_j/K is a normal radical extension, $x^{n_j} - a_j$ splits over F_j, $K_j \subseteq F_j$ for all j, $1 \leq j \leq i$. Let $G = G(F_i/K) = \{e = \sigma_1, \sigma_2, \ldots, \sigma_s\}$. Now r_{i+1} is a root of $x^{n_{i+1}} - a_{i+1} \in K_i[x] \subseteq F_i[x]$. Hence, $r_{i+1}^{n_{i+1}} = a_{i+1} \in F_i$. Consider the polynomial

$$g(x) = (x^{n_{i+1}} - \sigma_1(a_{i+1}))(x^{n_{i+1}} - \sigma_2(a_{i+1})) \cdots (x^{n_{i+1}} - \sigma_s(a_{i+1})) \in F_i[x].$$

Now $-(\sigma_1(a_{i+1}) + \cdots + \sigma_s(a_{i+1}))$,

$$(\sigma_1(a_{i+1}))(\sigma_2(a_{i+1})) + (\sigma_1(a_{i+1}))(\sigma_3(a_{i+1})) + \cdots + (\sigma_{s-1}(a_{i+1}))(\sigma_s(a_{i+1})), \ldots,$$

$(-1)^k (\sigma_1(a_{i+1}))(\sigma_2(a_{i+1})) \cdots (\sigma_s(a_{i+1}))$ are the coefficients of $g(x)$, each of which is fixed under $\sigma_1, \ldots, \sigma_s$. Since K is the fixed field of $G(F_i/K)$, $g(x) \in K[x]$. Let F_{i+1} be a splitting field of $g(x)$ over F_i. Then by Worked-Out Exercise 3 (page 502), F_{i+1}/K is a normal extension. Consider the polynomial $x^{n_{i+1}} - \sigma_1(a_{i+1}) \in F_i[x]$. Let $c_1, c_2, \ldots, c_{n_{i+1}}$ be the roots of $x^{n_{i+1}} - \sigma_1(a_{i+1})$. Then $c_j^{n_{i+1}} \in F_i$, $1 \leq j \leq n_{i+1}$. Thus, we have a chain of fields

$$F_i \subseteq F_i(c_1) \subseteq F_i(c_1, c_2) \subseteq \cdots \subseteq F_i(c_1, c_2, \ldots, c_{n_{i+1}}) = F_{i1}.$$

Clearly F_{i1} is a radical extension of F_i. Similarly, we can obtain a radical extension F_{i2}/F_{i1} by adjoining the roots of the polynomial

$$(x^{n_{i+1}} - \sigma_2(a_{i+1})) \in F_i[x] \subseteq F_{i1}[x].$$

Continuing like this, we obtain a chain of fields

$$F_i \subseteq F_{i1} \subseteq F_{i2} \subseteq \cdots \subseteq F_{is} = F_{i+1},$$

such that $F_{i\,t+1}$ is a radical extension of F_{it} obtained by adjoining roots of the polynomial $(x^{n_{i+1}} - \sigma_{t+1}(a_{i+1}))$. By Lemma 24.4.2, F_{i+1} is a radical extension of F_i and hence of K. Since r_{i+1} is a root of $g(x)$, $r_{i+1} \in F_{i+1}$, and hence $K_{i+1} = K_i(r_{i+1}) \subseteq F_i(r_{i+1}) \subseteq F_{i+1}$. Therefore, we have a chain of fields

$$K = F_0 \subseteq F_1 \subseteq \cdots \subseteq F_i \subseteq F_{i+1}$$

such that F_j/K is a normal radical extension, $x^{n_j} - a_j$ splits over F_j, $K_j \subseteq F_j$ for all j, $1 \le j \le i+1$. Proceeding as above we obtain a finite chain of fields

$$K = F_0 \subseteq F_1 \subseteq \cdots \subseteq F_m = E$$

such that F_i/K is a normal radical extension, $x^{n_i} - a_i$ splits over F_i, and $K_i \subseteq F_i$ for all i, $1 \le i \le m$. ∎

The following corollary is immediate from Theorem 24.4.3

Corollary 24.4.4 *Let K be a field of characteristic 0. Let F/K be an extension by radicals. Let $K = K_0 \subseteq K_1 \subseteq \cdots \subseteq K_m = F$ be the chain of intermediate fields such that $K_i = K_{i-1}(r_i)$, where r_i is a root of $x^{n_i} - a_i$, $a_i \in K_{i-1}$ for some positive integer n_i $(i = 1, 2, \ldots, m)$. Then there exists a root tower*

$$K = F_0 \subseteq F_1 \subseteq \cdots \subseteq F_m = E$$

such that $K \subseteq F \subseteq E$ and E/K is a normal extension. ∎

Theorem 24.4.5 *Let K be a field of characteristic 0. Let F/K be a normal radical extension with root tower*

$$K = K_0 \subseteq K_1 \subseteq \cdots \subseteq K_m = F$$

such that $K_i = K_{i-1}(r_i)$, where r_i is a root of $x^{n_i} - a_i$, $a_i \in K_{i-1}$ for some n_i $(i = 1, 2, \ldots, m)$. Let $n = n_1 n_2 \cdots n_m$. Suppose K contains all nth roots of unity. Then $G(F/K)$ is a solvable group.

Proof. Now for all i, $1 \le i \le m$, K_i contains all n_{i+1}th roots of unity. Let $1 = \omega_1, \omega_2, \ldots, \omega_{n_{i+1}}$ be the distinct n_{i+1}th roots of unity. Then $r_{i+1} = r_{i+1}\omega_1$, $r_{i+1}\omega_2, \ldots, r_{i+1}\omega_{n_{i+1}}$ are the distinct n_{i+1} roots of $x^{n_{i+1}} - a_{i+1} \in K_i[x]$ and clearly all these roots are in $K_{i+1} = K_i(r_{i+1})$. Hence, K_{i+1} is the splitting field in F of $x^{n_{i+1}} - a_{i+1}$ over K_i. Thus, $G(K_{i+1}/K_i)$ is a commutative group by Exercise 11 (page 528). Let $G_i = G(F/K_i)$. Then each G_i is a subgroup of G_0 and we have the chain of subgroups

$$G_0 \supseteq G_1 \supseteq \cdots \supseteq G_m = \{e\}.$$

By the fundamental theorem of Galois theory,

$$G(K_{i+1}/K_i) \simeq G(F/K_i)/G(F/K_{i+1}) = G_i/G_{i+1}.$$

Thus, G_i/G_{i+1} is a commutative group. Hence, $G_0 = G(F/K)$ is solvable. ∎

Theorem 24.4.6 *Let K be a field of characteristic 0 and $f(x)$ be a polynomial in $K[x]$. If $f(x)$ is solvable by radicals, then the Galois group of $f(x)$ over K is solvable.*

Proof.   Let E be the splitting field of $f(x)$ over K. Let F/K be a radical extension with root tower

$$K = K_0 \subseteq K_1 \subseteq \cdots \subseteq K_m = F$$

such that $K_i = K_{i-1}(r_i)$, where r_i is a root of $x^{n_i} - a_i$, $a_i \in K_{i-1}$ for some positive integer n_i $(i = 1, 2, \ldots, m)$ and $E \subseteq F$. By Corollary 24.4.4, we may assume that F/K is a normal extension. Let $n = n_1 n_2 \cdots n_m$.

Suppose K contains all nth roots of unity. Then $G(F/K)$ is solvable by Theorem 24.4.5. Clearly E/K is a normal separable extension. Hence, by the fundamental theorem of Galois theory, $G(F/E)$ is a normal subgroup and

$$G(E/K) \simeq G(F/K)/G(F/E).$$

Thus, $G(E/K)$ is a homomorphic image of a solvable group. Hence, $G(E/K)$ is solvable.

Now suppose K does not contain all nth roots of unity. Let ω be a primitive nth root of unity over K. Let $K' = K(\omega)$. Then K'/K is a normal extension and K' contains all nth roots of unity and $G(K'/K)$ is commutative by Worked-Out Exercise 3 (page 527). Thus, $G(K'/K)$ is solvable. Suppose $\omega \notin F$. Let $F' = F(\omega)$. Then F'/F is a normal extension and F' is a splitting field of $x^n - 1 \in K[x]$ over F. Hence, by Worked-Out Exercise 3 (page 502), F'/K is a normal extension. Clearly

$$K = K_0 \subseteq K_1 \subseteq \cdots \subseteq K_m = F \subseteq F'$$

is a root tower and so F'/K is a radical extension. Also, $E \subseteq F \subseteq F'$. Therefore, we may assume that $\omega \in F$. Now F/K' is a normal extension since F/K is a normal extension. Also,

$$K' = K_0' \subseteq K_1' \subseteq \cdots \subseteq K_m' = F$$

is a root tower such that $K_i' = K_{i-1}'(r_i)$, where r_i is a root of $x^{n_i} - a_i$, $a_i \in K_{i-1} \subseteq K_{i-1}'$ for some n_i $(i = 1, 2, \ldots, m)$. Consequently, by Theorem 24.4.5, $G(F/K')$ is solvable. By the fundamental theorem of Galois theory,

$$G(K'/K) \simeq G(F/K)/G(F/K').$$

Hence, $G(F/K)$ is solvable. As in the previous case, $G(E/K)$ is solvable. ∎

To obtain the result of Abel that the general polynomial of degree $n \geq 5$ is not solvable by radicals, it suffices to find a polynomial of degree n whose Galois group is S_n because S_n is not solvable for $n \geq 5$.

We proceed to find such a polynomial. Consider the polynomial ring $F[x_1, \ldots, x_n]$ and its field of quotients $F(x_1, \ldots, x_n)$. Let S_n be the symmetric group acting on $\{1, 2, \ldots, n\}$. We can consider S_n as a group of permutations acting on $F(x_1, \ldots, x_n)$ in the following manner: For $\alpha \in S_n$ and a rational function $f(x_1, \ldots, x_n) \in F(x_1, \ldots, x_n)$, define the mapping

$$f(x_1, \ldots, x_n) \rightarrow f(x_{\alpha(1)}, \ldots, x_{\alpha(n)}). \tag{24.11}$$

We will call this mapping α. By Exercise 3, α is an automorphism of $F(x_1, \ldots, x_n)$. The fixed field of $F(x_1, \ldots, x_n)$ with respect to S_n is the field K, where

$$
\begin{aligned}
K = \{f(x_1, \ldots, x_n) \mid &f(x_1, \ldots, x_n) \in F(x_1, \ldots, x_n), \\
&f(x_1, \ldots, x_n) = f(x_{\alpha(1)}, \ldots, x_{\alpha(n)}) \text{ for all } \alpha \in S_n\}.
\end{aligned}
$$

The elements of K are called the **symmetric rational functions**. Set

$$
\begin{aligned}
a_1 &= x_1 + \cdots + x_n = \sum_{i=1}^{n} x_i \\
a_2 &= \sum_{i<j} x_i x_j \\
a_3 &= \sum_{i<j<k} x_i x_j x_k \\
&\;\;\vdots \\
a_n &= x_1 x_2 \cdots x_n
\end{aligned}
\tag{24.12}
$$

These functions are known as the **elementary symmetric functions** and they are symmetric functions. Note that for $n = 2$, x_1 and x_2 are roots of the polynomial $t^2 - a_1 t + a_2$; for $n = 3$, x_1, x_2, and x_3 are roots of $t^3 - a_1 t^2 + a_2 t - a_3$; and when $n = 4$, x_1, x_2, x_3, and x_4 are roots of $t^4 - a_1 t^3 + a_2 t^2 - a_3 t + a_4$. Since $a_1, \ldots, a_n \in K$, $F(a_1, \ldots, a_n) \subseteq K$.

Theorem 24.4.7 *Using the above notation, we have*
  (i) $[F(x_1, \ldots, x_n) : K] = n!$,
  (ii) $K = F(a_1, \ldots, a_n)$,
  (iii) $S_n = G(F(x_1, \ldots, x_n)/K)$.

Proof.   Since S_n is a group of automorphisms of $F(x_1, \ldots, x_n)$ leaving K fixed, $S_n \subseteq G(F(x_1, \ldots, x_n)/K)$. Thus, by Theorem 24.2.6,

$$[F(x_1, \ldots, x_n) : K] \geq |G(F(x_1, \ldots, x_n)/K)| \geq |S_n| = n!$$

The polynomial $p(t) = t^n - a_1 t^{n-1} + a_2 t^{n-2} + \cdots + (-1)^n a_n$ over $F(a_1, \ldots, a_n)$ has roots $x_1, \ldots, x_n$ and factors over $F(x_1, \ldots, x_n)$ into $(t - x_1) \cdots (t - x_n)$. Thus, it follows that $F(x_1, \ldots, x_n)$ is the splitting field of $p(t)$ over $F(a_1, \ldots, a_n)$. Since $p(t)$ is of degree n,

$$[F(x_1, \ldots, x_n) : F(a_1, \ldots, a_n)] \leq n!$$

Thus, since

$$F(a_1, \ldots, a_n) \subseteq K \subseteq F(x_1, \ldots, x_n),$$

$[F(x_1,\ldots,x_n) : F(a_1,\ldots,a_n)]$ is both greater than or equal to $n!$ and less than or equal to $n!$ Hence, we have

$$[F(x_1,\ldots,x_n) : F(a_1,\ldots,a_n)] = n!$$

and $K = F(a_1,\ldots,a_n)$, proving (i) and (ii). By Theorem 24.2.13,

$$|G(F(x_1,\ldots,x_n)/K)| = n!$$

and since $S_n \subseteq G(F(x_1,\ldots,x_n)/K)$, we have $S_n = G(F(x_1,\ldots,x_n)/K)$, proving (iii). ■

We have now established our goal. The Galois group of the polynomial $p(t)$ over $F(a_1,\ldots,a_n)$ is S_n and S_n is not solvable for $n \geq 5$. Hence, $p(t)$ is not solvable by radicals for $n \geq 5$.

Theorem 24.4.8 *Let G be a subgroup of S_p, where p is a prime. If G contains a p-cycle and a transposition, then $G = S_p$.*

Proof.   If $p = 2$, then $|S_p| = 2$ and the result is immediate. Suppose $p = 3$. Let $(a\ b)$ and $(x\ y\ z) \in G$, where $\{a, b\} \subseteq \{x, y, z\}$. Then it is easy to show that

$$G = \{e, (a\ b), (x\ y\ z), (x\ z\ y), (a\ b) \circ (x\ y\ z), (a\ b) \circ (x\ z\ y)\}.$$

Suppose $p = 5$. Let $\alpha = (a\ b)$ and $\beta = (x\ y\ z\ u\ v)$, where $\{a, b\} \subseteq \{x, y, z, u, v\}$. Then there exists a positive integer n such that $\beta^n = (a\ b\ c\ d\ e)$, where $\{a, b, c, d, e\} = \{x, y, z, u, v\}$. It is easily verified that

$$\begin{aligned}
\beta^n \circ \alpha \circ \beta^{-n} &= (b\ c) \\
\beta^{2n} \circ \alpha \circ \beta^{-2n} &= (c\ d) \\
\beta^{3n} \circ \alpha \circ \beta^{-3n} &= (d\ e).
\end{aligned}$$

Hence, $(a\ b), (b\ c), (c\ d), (d\ e) \in G$. Thus,

$$\begin{aligned}
(b\ c) \circ (a\ b) \circ (b\ c) &= (a\ c) \in G \\
(c\ d) \circ (a\ c) \circ (c\ d) &= (a\ d) \in G \\
(d\ e) \circ (a\ d) \circ (d\ e) &= (a\ e) \in G \\
(c\ d) \circ (b\ c) \circ (c\ d) &= (b\ d) \in G \\
(d\ e) \circ (b\ d) \circ (d\ e) &= (b\ e) \in G \\
(d\ e) \circ (c\ d) \circ (d\ e) &= (c\ e) \in G.
\end{aligned}$$

Hence, G contains the above 10 transpositions. However, these are all the transpositions of S_5 since $\binom{5}{2} = 10$. Since every permutation is a product of disjoint cycles and every cycle is a product of transpositions, G contains all the permutations of $\{x, y, z, u, v\}$. Hence, $G = S_5$. (We ask the reader to consider the theorem for arbitrary p.) ■

Theorem 24.4.9 *Let $f(x)$ be an irreducible polynomial in $\mathbf{Q}[x]$. Suppose that $\deg f(x) = p$, where p is a prime. If $f(x)$ has exactly $p - 2$ real roots and two complex roots, then the Galois group of $f(x)$ over $\mathbf{Q}$ is S_p.*

Proof.   Let S be a splitting field of $f(x)$ over $\mathbf{Q}$ such that $\mathbf{Q} \subseteq S \subseteq \mathbf{C}$. Let G denote the Galois group of $S/\mathbf{Q}$. Now $p \mid [S : \mathbf{Q}]$ and $[S : \mathbf{Q}] = |G|$. We see by viewing G as a group of permutations on the roots $\{r_1, r_2, \ldots, r_p\}$ that G must contain an element of order p, which is necessarily a p-cycle. Let $r_1 = a + bi$ and $r_2 = a - bi$. Then the automorphism α of $\mathbf{C}$, which maps every complex number to its conjugate must map S onto S since α is the identity on $\mathbf{R}$ and $\alpha(r_1) = r_2$, $\alpha(r_2) = r_1$. Hence, we see that α^2 is the identity and so is a transposition. By the previous theorem, $G = S_p$. ∎

Although Galois and Abel are most noted for their work involving the existence of formulas for finding the roots of polynomials, their approach to solving mathematical problems along with that of British algebraists marks the birth of modern algebra. Their work resulted in abstract and widely inclusive theories. Actually, Lagrange's work on algebraic equations and especially on analytic mechanics anticipated the awakening of the strength of the abstract and general approach. It was Hilbert's work on the foundations of geometry (1899) which finalized the abstract approach.

24.4.1   Worked-Out Exercises

◇ **Exercise 1** Show that the Galois group of the polynomial $f(x) = 2x^5 - 10x + 5$ over $\mathbf{Q}$ is S_5. Conclude that the equation $f(x) = 0$ is not solvable by radicals.

Solution: We have that $f(x)$ is irreducible over $\mathbf{Q}$ by Eisenstein's criterion. Now

$$f'(x) = 10(x^4 - 1).$$

Hence, $f'(x)$ has two real roots, namely, 1 and -1. Since $f(-1) > 0$ and $f(1) < 0$, it follows that $f(x)$ has three real roots, say, r_1, r_2, r_3 such that $r_1 < -1 < r_2 < 1 < r_3$. The other two roots of $f(x)$ are complex numbers. Thus, from Theorem 24.4.9, the Galois group of $f(x)$ is S_5. Hence, by Theorem 24.4.6, the equation $f(x) = 0$ is not solvable by radicals.

24.4.2   Exercises

1. Find the roots of the polynomial $2x^3 + 9x + 6$ by using the formula for the root of a cubic.

2. In $F[x_1, x_2, \ldots, x_n]$, $x_1, x_2, \ldots, x_n$ are roots of

$$p(x) = x^n - a_1 x^{n-1} + a_2 x^{n-2} + \cdots + (-1)^n a_n,$$

where the a_i's are defined on page 533. Demonstrate this result for $n = 2$ and $n = 3$.

3. Prove that α, given on page 533, is an automorphism of $F(x_1, x_2, \ldots, x_n)$ and that α fixes $F(a_1, a_2, \ldots, a_n)$.

4. It can be shown that a symmetric polynomial is a polynomial in the elementary symmetric functions in $x_1, x_2, \ldots, x_n$. Express the following as polynomials in the elementary symmetric functions in x_1, x_2, x_3.

(i) $x_1^2 + x_2^2 + x_3^2$,

(ii) $(x_1 - x_2)^2 (x_1 - x_3)^2 (x_2 - x_3)^2$.

5. Show that for every finite group G, there is a field K and a polynomial $f(x) \in K[x]$ such that the Galois group of $f(x)$ over K is isomorphic to G.

6. Find the Galois group of the polynomial $x^3 - 3x + 1$ over $\mathbf{Q}$. Solve the equation $x^3 - 3x + 1 = 0$ by radicals.

7. Show that the Galois group of the polynomial $f(x) = x^5 - 10x^4 + 2x^3 - 24x^2 + 2$ over $\mathbf{Q}$ is S_5. Is the equation $f(x) = 0$ solvable by radicals?

Joseph Henry MacLagan Wedderburn (1882–1948) was born on February 26, 1882, in Forfar, Scotland, the tenth of 14 children. His father was a physician. In 1898, he matriculated at the University of Edinburgh. In 1903, he received an M.A. degree with first-class honors in mathematics. The following year he went to Leipzig and Berlin because of the influence on him of Frobenius and Schur's work. He received his doctorate of science in 1908.

In 1909, Wedderburn was appointed professor at Princeton University. During World War I, he fought for the British army, returning to Princeton after the war and remaining there until 1945. Besides being editor of the *Annals of Mathematics* from 1912 to 1928, Wedderburn published 38 papers, and in 1934, published a textbook, *Lectures on Matrices*.

Wedderburn is most noted for the two famous theorems which bear his name. He proved both theorems between 1905 and 1908. The structure of algebras over real and complex fields had been determined by Cartan and others. Wedderburn was interested in determining the structure of algebras over arbitrary fields. He showed that a semisimple algebra is a direct sum of simple algebras. Later, in his paper "On hypercomplex numbers," he proved that every simple algebra is a matrix algebra over a division algebra. In the second theorem, he proved that every finite division ring is a field. His theorem on finite algebras gave a structure of all projective geometries with a finite number of points. Wedderburn died on October 9, 1948, in New Jersey.

Chapter 25

Geometric Constructions

25.1   Geometric Constructions

In this chapter, we consider some problems from geometry. We are concerned with constructions in the Euclidean plane that can be made by straightedge (unmarked ruler) and compass only. We identify the Euclidean plane with $\mathbf{R} \times \mathbf{R}$. We assume that we are given some length, which we take as our unit length, and two points O and X which we label $(0,0)$ and $(1,0)$, respectively.

Using straightedge and compass, we can do the following in the Euclidean plane:

(i) Draw a line through two given points.

(ii) Draw a line parallel to a given line and passing through a given point.

(iii) Draw a line perpendicular to a given line and passing through a given point.

(iv) Draw a circle with a given center and passing through another given point.

We draw a line through O and X and call it the x-$axis$. Now we draw a line perpendicular to the x-$axis$ and passing through O and call this the y-$axis$. Thus, we are able to coordinatize the plane. Hence, we have the x-$axis$, y-$axis$, origin $O = (0,0)$, and the point $X = (1,0)$.

Given line segments of lengths a and b, using straightedge and compass, we can construct line segments of lengths $a + b$, $a - b$, ab, and ab^{-1} (for $b \neq 0$). Since we have a unit length, using straightedge and compass, we can draw a line segment of any integer length in a finite number of steps. Thus, using straightedge and compass, we can draw a line segment of any rational length in a finite number of steps. We leave these facts as an exercise for the interested reader.

For the construction of a line segment of length ab^{-1}, we first draw two lines through a point P. From P, mark off a point Q on one line of length b and then mark off a point U on the same line and same direction of length a

from Q. On the other line, mark off a point S from P of length 1. Construct a line UV parallel to QS with V on line PS. An argument using similar triangles shows SV is of length ab^{-1}.

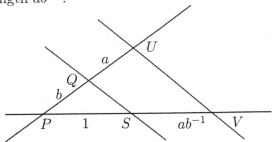

Definition 25.1.1 *Let $P \subseteq \mathbf{R} \times \mathbf{R}$. Let l be a line and C be a circle in $\mathbf{R} \times \mathbf{R}$.*
*(i) (a, b) is a **point** in P if $(a, b) \in P$.*
*(ii) l is a **line** in P if l passes through two distinct points in P.*
*(iii) C is a **circle** in P if the center of C is in P and C passes through another point in P.*

If C is a circle in $P \subseteq \mathbf{R} \times \mathbf{R}$, then the radius of C is the distance between two points in P, where the distance is the Euclidean distance.

Definition 25.1.2 *Let F be any subfield of $\mathbf{R}$. The set*

$$P_F = \{(x, y) \mid x, y \in F\}$$

*is called the **plane** of F.*

Let $p_1 = (x_1, y_1)$, $p_2 = (x_2, y_2) \in P_F$. Let l be the line passing through p_1 and p_2. If $x_1 = x_2$, then l has the form $x = x_1$. If $x_1 \neq x_2$, then l has the equation

$$y - y_1 = \frac{y_2 - y_1}{x_2 - x_1}(x - x_1)$$

which can be reduced to the form $ax + by + c = 0$ for some $a, b, c \in F$. Hence, a line in P_F is of the form

$$ax + by + c = 0$$

for some $a, b, c \in F$. Such a line is said to be a **line** in F.

Let C be a circle with center at p_1 and passing through p_2. Then the equation of C is

$$(x - x_1)^2 + (y - y_1)^2 = r^2,$$

where $r \in F$ is the radius of C. This equation of C can be put in the form $x^2 + y^2 + ax + by + c = 0$ for some $a, b, c \in F$. Hence, a circle in P_F has an equation of the form

$$x^2 + y^2 + ax + by + c = 0$$

for some $a, b, c \in F$. Such a circle is said to be a **circle** in F.

Let P_{1_F} be the set of all points of P_F and the points obtained by
(i) intersecting two lines in P_F,
(ii) intersecting two circles in P_F, and
(iii) intersecting a line and a circle in P_F.
It is easy to verify that two lines in P_F intersect in a point in P_F.
Let C_1 and C_2 be two circles in P_F with equations

$$x^2 + y^2 + a_1 x + b_1 y + c_1 = 0 \tag{25.1}$$

and

$$x^2 + y^2 + a_2 x + b_2 y + c_2 = 0 \tag{25.2}$$

for some a_1, a_2, b_1, b_2, c_1, $c_2 \in F$. Subtract Eq. (25.2) from Eq. (25.1) to obtain

$$(a_1 - a_2)x + (b_1 - b_2)y + (c_1 - c_2) = 0. \tag{25.3}$$

Thus, the points of intersection of C_1 and C_2 are the same as the points of intersection of either Eq. (25.1) or Eq. (25.2) with Eq. (25.3). Hence, case (ii) reduces to case (iii).

Let l be a line and C be a circle in P_F with equations

$$a_1 x + b_1 y + c_1 = 0 \tag{25.4}$$

and

$$x^2 + y^2 + a_2 x + b_2 y + c_2 = 0 \tag{25.5}$$

for some a_1, a_2, b_1, b_2, c_1, $c_2 \in F$. Eliminate y from Eqs. (25.4) and (25.5) to obtain an equation of the form

$$ax^2 + bx + c = 0$$

for some $a, b, c \in F$. Using our knowledge of the quadratic formula or the fact that $ax^2 + bx + c = 0$ is solvable by radicals, we have that the polynomial $ax^2 + bx + c$ has roots in $F(\sqrt{r})$ for some positive $r \in F$. (r can be taken to be positive since we have assumed the line and the circle intersect.)

Inductively, we can construct a sequence

$$P_F = P_{0_F} \subseteq P_{1_F} \subseteq P_{2_F} \subseteq \cdots \subseteq P_{i_F} \subseteq \cdots,$$

where P_{i_F} is the set of all points of P_{i-1_F} and the points obtained by
(i) intersecting two lines in P_{i-1_F},
(ii) intersecting two circles in P_{i-1_F}, and
(iii) intersecting a line and a circle in P_{i-1_F}.

Lemma 25.1.3 *Let F be a subfield of* **R**. *Let $a \in$* **R**. *The following conditions are equivalent.*
  (*i*) $(a, 0) \in P_{n_F}$ *for some $n \geq 0$.*
  (*ii*) $(a, a) \in P_{m_F}$ *for some $m \geq 0$.*
  (*iii*) $(0, a) \in P_{t_F}$ *for some $t \geq 0$.*

Proof.   (i)$\Rightarrow$(ii) Let l be the line $x = y$ and C be the circle $(x-a)^2+y^2 = a^2$. Clearly l is a line in P_{n_F}. The center of C is $(a,0) \in P_{n_F}$ and C passes through another point $(0,0) \in P_{n_F}$. Hence, C is a circle in P_{n_F}. Now l and C intersect at (a,a). Hence, $(a,a) \in P_{n+1_F}$. Let $m = n+1$. Then $(a,a) \in P_{m_F}$.

(ii)$\Rightarrow$(i) Let l be the line $y = -x$ and C be the circle $x^2 + y^2 = 2a^2$. Then l is a line in P_{m_F} and C is a circle in P_{m_F}. Now l and C intersect at $(a,-a) \in P_{m+1_F}$. Let l' be the line $x = a$ and l'' be the line $y = 0$. Then l' and l'' are lines in P_{m+1_F}. Now l' and l'' intersect at $(a,0) \in P_{m+2_F}$. Let $n = m+2$. Then $(a,0) \in P_{n_F}$.

Similarly, (ii)$\Leftrightarrow$(iii). ∎

Theorem 25.1.4 *Let F be a subfield of* **R**. *Let $a,b \in$* **R** *and $(a,0),(b,0) \in P_{n_F}$ for some $n \geq 0$. Then the following assertions hold.*

(i) $(a,b) \in P_{m_F}$ for some $m \geq 0$.
(ii) $(a \pm b, 0) \in P_{m_F}$ for some $m \geq 0$.
(iii) $(ab,0) \in P_{m_F}$ for some $m \geq 0$.
(iv) Let $b \neq 0$. Then $(\frac{a}{b},0) \in P_{m_F}$ for some $m \geq 0$.
(v) Let $a \geq 0$. Then $(\sqrt{a},0) \in P_{m_F}$ for some $m \geq 0$.

Proof.   (i) By Lemma 25.1.3, $(a,0),(a,a),(0,b),(b,b) \in P_{t_F}$ for some $t \geq 0$. Thus $x = a$ and $y = b$ are lines in P_{t_F}. These lines intersect at (a,b) and so $(a,b) \in P_{t+1_F}$. Let $m = t+1$. Then $m > 0$ and $(a,b) \in P_{m_F}$.

(ii) Let l be the line $y = 0$ and C be the circle $(x-a)^2 + y^2 = b^2$. Then l is a line in $P_{n_F} \subseteq P_{n+1_F}$. The center of C is $(a,0) \in P_{n_F} \subseteq P_{n+1_F}$ and C passes through $(a,b) \in P_{n+1_F}$. Therefore, C is a circle in P_{n+1_F}. Now l and C intersect at $(a \pm b, 0) \in P_{n+2_F}$. Let $m = n+2$. Then $(a \pm b, 0) \in P_{m_F}$.

(iii) By (i), (ii), and Lemma 25.1.3, $(a,b-1),(0,b) \in P_{k_F}$ for some $k \geq 0$. Then $ay = -x+ab$ is a line in P_{k_F} since it passes through $(0,b)$ and $(a,b-1)$. Also, $y = 0$ is a line in P_{k_F}. Both these lines intersect at $(ab,0) \in P_{k+1_F}$. Let $m = k+1$. Then $(ab,0) \in P_{m_F}$.

(iv) If $a = 0$, then the result is trivially true. Let $a \neq 0$. Now by (i), (ii), (iii), and Lemma 25.1.3, $(0,a),(a,a(1-b)) \in P_{k_F}$ for some $k \geq 0$. Then $bx = a - y$ and $y = 0$ are lines in P_{k_F}. These lines intersect at $(\frac{a}{b},0) \in P_{k+1_F}$. Let $m = k+1 \geq 0$. Then $(\frac{a}{b},0) \in P_{m_F}$.

(v) Let l be the line $y = 1$ and C be the circle

$$x^2 + (y - \frac{1+a}{2})^2 = (\frac{1+a}{2})^2.$$

The center of C is $(0, \frac{1+a}{2}) \in P_{k_F}$ for some $k \geq 0$. Also, C passes through $(0,0) \in P_{k_F}$. Thus, l is a line and C is a circle in P_{k_F}. Now l and C intersect at $(\sqrt{a},1) \in P_{k+1_F}$. Let C' be the circle

$$(x - \sqrt{a})^2 + (y - 1)^2 = a + 1.$$

The center of C' is $(\sqrt{a}, 1) \in P_{k+1_F}$ and it passes through $(2\sqrt{a}, 2) \in P_{t_F}$ for some $t \geq k + 1$. Hence, C' is a circle in P_{t_F}. Also, $y = 0$ is a line in P_{t_F}. Now C' and $y = 0$ intersect at $(2\sqrt{a}, 0) \in P_{t+1_F}$. Thus, $(\sqrt{a}, 0) \in P_{m_F}$ for some $m \geq t + 1 \geq 0$. ■

Definition 25.1.5 *Let F be a subfield of $\mathbf{R}$. A real number r is said to be **constructible from** F if $(r, 0) \in P_{n_F}$ for some $n \geq 0$.*

The following theorem is immediate from Theorem 25.1.4.

Theorem 25.1.6 *Let F be a subfield of $\mathbf{R}$. Let S be the set of all constructible real numbers from F. Then S is a subfield of $\mathbf{R}$ and $F \subseteq S$. Moreover, if $a \in S$ and $a \geq 0$, then $\sqrt{a} \in S$.*

Lemma 25.1.7 *Let F be a subfield of $\mathbf{R}$. Let a and b be real numbers. Suppose there exist real numbers $a_1, a_2, \ldots, a_n$ and $b_1, b_2, \ldots, b_m$ such that*
(i) $a_1^2 \in F$,
(ii) $a_i^2 \in F(a_1, a_2, \ldots, a_{i-1})$, $2 \leq i \leq n$, and $a \in F(a_1, a_2, \ldots, a_n)$,
(iii) $b_1^2 \in F$,
(iv) $b_j^2 \in F(b_1, b_2, \ldots, b_{j-1})$, $2 \leq j \leq m$, and $b \in F(b_1, b_2, \ldots, b_m)$.
Then there exist real numbers $s_1, s_2, \ldots, s_k$ such that
(v) $s_1^2 \in F$,
(vi) $s_i^2 \in F(s_1, s_2, \ldots, s_{i-1})$, $2 \leq i \leq k$, and $a, b \in F(s_1, s_2, \ldots, s_k)$.

Proof. Let $F_1 = F$ and $F_i = F(a_1, a_2, \ldots, a_{i-1})$, $2 \leq i \leq n + 1$. Let $k = n + m$, $s_i = a_i$, $1 \leq i \leq n$ and $s_{n+j} = b_j$, $1 \leq j \leq m$. Then
(a) $s_1^2 = a_1^2 \in F$, proving (v),
(b) $s_i^2 = a_i^2 \in F(a_1, a_2, \ldots, a_{i-1}) = F(s_1, s_2, \ldots, s_{i-1})$, $2 \leq i \leq n$,
(c) $s_{n+1}^2 = b_1^2 \in F \subseteq F(s_1, s_2, \ldots, s_n)$,
(d) $s_{n+j}^2 = b_j^2 \in F(b_1, b_2, \ldots, b_{j-1}) \subseteq F_{n+1}(b_1, b_2, \ldots, b_{j-1}) = F(a_1, a_2, \ldots, a_n, b_1, b_2, \ldots, b_{j-1}) = F(s_1, s_2, \ldots, s_{n+j-1})$, $2 \leq j \leq m$. Also, $a \in F(a_1, a_2, \ldots, a_n) \subseteq F(s_1, s_2, \ldots, s_k)$ and $b \in F(b_1, b_2, \ldots, b_m) \subseteq F(s_1, s_2, \ldots, s_k)$, proving (vi). ■

Lemma 25.1.8 *Let F be a subfield of $\mathbf{R}$. Let $r \in \mathbf{R}$ be such that $r^2 \in F$. Then*
(i) for all $a \in F(r)$, $(a, 0)$, $(0, a) \in P_{1_F}$ and
(ii) for all $a, b \in F(r)$, $(a, b) \in P_{3_F}$, i.e., $P_{F(r)} \subseteq P_{3_F}$.

Proof. (i) If $r \in F$, then the result is trivially true. Suppose $r \notin F$. Let $a \in F(r)$. Then $a = b + cr$ for some $b, c \in F$. Let l be the line

$$y = 0$$

and C be the circle

$$(x - b)^2 + (y - \frac{c^2 r^2 - 1}{2})^2 = (\frac{c^2 r^2 + 1}{2})^2.$$

Then l is a line in F. The center of C is $(b, \frac{c^2 r^2 - 1}{2}) \in P_F$ and it passes through another point $(b, c^2 r^2) \in P_F$. Hence, C is a circle in P_F. Now l and C intersect at a point $(b + cr, 0) = (a, 0)$. Thus, by definition, $(a, 0) \in P_{1_F}$. By a similar argument, we can show that $(0, a) \in P_{1_F}$.

(ii) Let $a, b \in F(r)$. Then by (i), $(a, 0), (0, b) \in P_{1_F}$. As in the proof of (i)$\Rightarrow$(ii) in Lemma 25.1.3, $(a, a), (b, b) \in P_{2_F}$. Hence, $x = a$ and $y = b$ are lines in P_{2_F} which intersect at the point (a, b). Therefore, by definition, $(a, b) \in P_{3_F}$. ∎

Lemma 25.1.9 *Let F be a subfield of $\mathbf{R}$. Let $x, y \in \mathbf{R}$. Suppose there exist real numbers $s_1, s_2, \ldots, s_k$ such that*
(i) $s_1^2 \in F$,
(ii) $s_i^2 \in F(s_1, s_2, \ldots, s_{i-1})$, $2 \leq i \leq k$.
Let $T = F(s_1, s_2, \ldots, s_k)$. Then $P_T \subseteq P_{n_F}$ for some $n \geq 1$.

Proof. If $k = 1$, then the result holds by Lemma 25.1.8. Suppose the result is true for all i, $1 \leq i < k$. If $s_k \in F(s_1, s_2, \ldots, s_{k-1})$, then the result holds by the induction hypothesis. Suppose $s_k \notin F(s_1, s_2, \ldots, s_{k-1})$. Let $K = F(s_1, s_2, \ldots, s_{k-1})$. By the induction hypothesis, $P_K \subseteq P_{m_F}$ for some $m \geq 1$. Clearly, $P_{1_K} \subseteq P_{m+1_F}$ and $P_{3_K} \subseteq P_{m+3_F}$. Let $a, b \in T = K(s_k)$. Then by Lemma 25.1.8, $(a, b) \in P_{3_K} \subseteq P_{m+3_F}$. Let $n = m + 3 \geq 1$. Then $P_T \subseteq P_{n_F}$. ∎

Theorem 25.1.10 *Let F be a subfield of $\mathbf{R}$. Let $x, y \in \mathbf{R}$. Then $(x, y) \in P_{n_F}$ for some $n \geq 1$ if and only if there exists a sequence of real numbers $s_1, s_2, \ldots, s_k$ such that*
(i) $s_1^2 \in F$,
(ii) $s_i^2 \in F(s_1, s_2, \ldots, s_{i-1})$, $2 \leq i \leq k$, and $x, y \in F(s_1, s_2, \ldots, s_k)$.

Proof. Let $(x, y) \in P_{n_F}$ for some $n \geq 1$. We prove the result by induction on n. Let $n = 1$ and $(x, y) \in P_{1_F}$. If $x, y \in F$, then the result is trivial. Suppose $x, y \notin F$. Then (x, y) is obtained by either intersecting a line and a circle or two circles in F. Then, as shown before, $x, y \in F(\sqrt{r})$ for some $r \in F$. Let $s_1 = r^2$. Then $s_1 \in F$ and $x, y \in F(s_1)$. Hence, the result is true for $n = 1$. Suppose the result is true for all P_{k_F} such that $1 \leq k < n$. Let $(x, y) \in P_{n_F}$. If $(x, y) \in P_{n-1_F}$, then the result holds by induction. Suppose $(x, y) \notin P_{n-1_F}$. Then (x, y) is obtained by intersecting two lines or two circles or a line and a circle in P_{n-1_F}. Suppose (x, y) is obtained by intersecting a line L and a circle C in P_{n-1_F}. Then L passes through two distinct points $(a, b), (c, d) \in P_{n-1_F}$ and C has its center $(u, v) \in P_{n-1_F}$ and the radius r of C is the distance between

two points in P_{n-1_F}. Now by induction hypothesis and by Lemma 25.1.7, it follows that there exists a sequence of real numbers $s_1, s_2, \ldots, s_{k-1}$ such that
(i) $s_1^2 \in F$,
(ii) $s_i^2 \in F(s_1, s_2, \ldots, s_{i-1})$, $2 \leq i \leq k-1$, and $a, b, c, d, u, v, r \in F(s_1, s_2, \ldots, s_{k-1})$.

Thus, (x, y) is obtained by intersecting a line and a circle in $F(s_1, s_2, \ldots, s_{k-1})$. Hence, there exists a real number s_k such that $s_k^2 \in F(s_1, s_2, \ldots, s_{k-1})$ and $x, y \in F(s_1, s_2, \ldots, s_{k-1})(s_k) = F(s_1, s_2, \ldots, s_k)$. The other cases are similar.

The converse follows by Lemma 25.1.9. ■

The following theorem is immediate from Theorem 25.1.10.

Theorem 25.1.11 *Let F be a subfield of $\mathbf{R}$. A real number r is constructible from F if and only if there exist real numbers $r_1, r_2, \ldots, r_n$ such that*
(i) $r_1^2 \in F$,
(ii) $r_i^2 \in F(r_1, r_2, \ldots, r_{i-1})$, $2 \leq i \leq n$, and $r \in F(r_1, r_2, \ldots, r_n)$. ■

Definition 25.1.12 *A real number a is **constructible** if it is constructible from $\mathbf{Q}$.*

Definition 25.1.13 *(i) A point (a, b) is **constructible** (or **located**) in the Euclidean plane if a and b are constructible real numbers.*

*(ii) A **line segment** is **constructible** in the Euclidean plane if its end points are constructible.*

*(iii) A **line** is **constructible** in the Euclidean plane if it passes through two distinct constructible points.*

*(iv) A **circle** is **constructible** in the Euclidean plane if its center is constructible and it passes through another constructible point.*

Theorem 25.1.14 *Let S be the set of all constructible numbers in $\mathbf{R}$. Then S is a subfield of $\mathbf{R}$ and $\mathbf{Q} \subseteq S$. Moreover if $a \in S$ and $a \geq 0$, then $\sqrt{a} \in S$.*

Proof.   The proof follows from Theorem 25.1.6. ■

Theorem 25.1.15 *The real number r is constructible if and only if there exists a finite number of real numbers $s_1, \ldots, s_n$ such that*
(i) $s_1^2 \in \mathbf{Q}$,
(ii) $s_i^2 \in \mathbf{Q}(s_1, \ldots, s_{i-1})$ for $i = 2, \ldots, n$ such that $r \in \mathbf{Q}(s_1, \ldots, s_n)$. ■

Corollary 25.1.16 *If the real number r is constructible, then r lies in some extension of $\mathbf{Q}$ of degree a power of 2.*

Proof.   If r is constructible, then there exist real numbers $s_1, \ldots, s_n$ satisfying conditions (i) and (ii) of Theorem 25.1.15 such that $r \in \mathbf{Q}(s_1, \ldots, s_n)$. Now $[\mathbf{Q}(s_1, \ldots, s_n) : \mathbf{Q}] = [\mathbf{Q}(s_1, \ldots, s_n) : \mathbf{Q}(s_1, \ldots, s_{n-1})] \, [\mathbf{Q}(s_1, \ldots, s_{n-1}) : \mathbf{Q}(s_1, \ldots, s_{n-2})] \cdots [\mathbf{Q}(s_1) : \mathbf{Q}]$, which is clearly a power of 2. ∎

Corollary 25.1.17 *If the real number r is a root of an irreducible polynomial over $\mathbf{Q}$ of degree k, where k is not a power of 2, then r is not constructible.* ∎

Theorem 25.1.18 *A real number a is constructible if and only if using straightedge and compass we can construct a line segment of length $|a|$ in the Euclidean plane.*

Proof.   Suppose using straightedge and compass that we can construct a line segment PQ of length $|a|$ in the Euclidean plane. We may assume that $a > 0$. Let $P = (x_1, y_1)$ and $Q = (x_2, y_2)$. Then P and Q are constructible points in the Euclidean plane and hence x_1, y_1, x_2, y_2 are constructible real numbers. Hence, by Theorem 25.1.15 and Lemma 25.1.7, there exist real numbers $s_1, \ldots, s_n$ such that

(i) $s_1^2 \in \mathbf{Q}$,

(ii) $s_i^2 \in \mathbf{Q}(s_1, \ldots, s_{i-1})$ for $i = 2, \ldots, n$ such that x_1, y_1, x_2, $y_2 \in \mathbf{Q}(s_1, \ldots, s_n)$. Now $a^2 = (x_1 - x_2)^2 + (y_1 - y_2)^2 \in \mathbf{Q}(s_1, \ldots, s_n)$. Let $s_{n+1} = \sqrt{a}$. Then $s_{n+1}^2 \in \mathbf{Q}(s_1, \ldots, s_n)$ and $a \in \mathbf{Q}(s_1, \ldots, s_{n+1})$. Hence, a is constructible from $\mathbf{Q}$.

Conversely, suppose a is constructible from $\mathbf{Q}$. Then $A = (a, 0) \in P_{n\mathbf{Q}}$ for some $n \geq 0$, where $P_{n\mathbf{Q}}$ is defined as above (here the arbitrary field $F = \mathbf{Q}$). Let $B = (0, 0) \in P_{n\mathbf{Q}}$. Then A and B are two constructible points in the Euclidean plane. Hence, we can construct the line segment AB in the Euclidean plane in a finite number of steps, and AB is of length $|a|$. ∎

We have now laid enough groundwork to answer by algebraic methods some ancient questions of geometry.

Theorem 25.1.19 *It is impossible to trisect an angle of $60°$ by means of straightedge and compass alone.*

Proof.   Suppose that an angle of $60°$ can be trisected by straightedge and compass. Then the real number $r = \cos 20°$ is constructible. From the trigonometric formula $\cos 3\theta = 4\cos^3 \theta - 3\cos\theta$ and by setting $\theta = 20°$, we obtain $\frac{1}{2} = 4r^3 - 3r$ or $8r^3 - 6r - 1 = 0$. Thus, r is a root of the polynomial $8x^3 - 6x - 1 = 0$ over $\mathbf{Q}$. The possible linear factors of $8x^3 - 6x - 1$ over $\mathbf{Z}$ are $(x \pm 1)$, $(2x \pm 1)$, $(4x \pm 1)$, and $(8x \pm 1)$. However, it is easily verified that ± 1, $\pm\frac{1}{2}$, $\pm\frac{1}{4}$, $\pm\frac{1}{8}$ are not roots of $8x^3 - 6x - 1$. Therefore, $8x^3 - 6x - 1$ is irreducible over $\mathbf{Z}$ and thus over $\mathbf{Q}$. Thus, by Corollary 25.1.17, r is not constructible. Consequently, it is impossible to trisect an angle of 60°. ■

There are some angles which can be trisected by means of straightedge and compass alone; for example, angles of 90° and 72°. We ask the reader to verify this fact.

Another ancient problem is that of "squaring the circle," that is, constructing a square whose area is equal to that of a given circle. Since the area of a circle is πr^2, where r is the radius of the circle, this problem is equivalent to the constructibility of $\sqrt{\pi}$. However, it can be shown that π, whence $\sqrt{\pi}$, is not even algebraic over $\mathbf{Q}$, let alone a root of a quadratic polynomial. Hence, it is impossible to square the circle. Thus, we have the following result.

Theorem 25.1.20 *It is impossible to square the circle by straightedge and compass alone.* ■

We now consider the problem of "duplicating the cube," that is, constructing a cube whose volume is twice that of a given cube. If the original cube is the unit cube, then the problem reduces to the construction of a real number r such that $r^3 = 2$. Since the polynomial $x^3 - 2$ is irreducible over $\mathbf{Q}$, we have by Corollary 25.1.17 that it is impossible to duplicate a cube.

Theorem 25.1.21 *It is impossible to duplicate the cube by straightedge and compass alone.* ■

Example 25.1.22 *Consider a triangle of sides of length $1, 1, r$, where the side of length r is opposite an angle of 36°. Then the other two angles are 72° each. Draw a bisector from one of the 72° angles to the opposite side. Similar triangles are obtained. The ratios of the corresponding sides yield $\frac{r}{1-r} = \frac{1}{r}$. Thus, $r^2 + r - 1 = 0$. Hence,*

$$r = \frac{-1 + \sqrt{5}}{2}.$$

Thus, r is constructible and so an angle of 36° is constructible.

Theorem 25.1.23 *Let $\theta \in \mathbf{R}$. Then the following conditions are equivalent.*
 (i) The angle θ is constructible.
 (ii) The number $\cos \theta$ is constructible.
 (iii) The number $\sin \theta$ is constructible.

Proof. (i)$\Rightarrow$(ii): There exist constructible points p and q such that the radian measure of the angle $p(0,0)q$ is θ. Without loss of generality, we may assume that q lies on the x-*axis*. The unit circle then intersects the line containing q and p at the point $r = (\cos \theta, \sin \theta)$. Thus, $\cos \theta$ and $\sin \theta$ are constructible since r is constructible.

(ii)$\Rightarrow$(iii): Since $\cos\theta$ is constructible, the point $q = (\cos\theta, 0)$ is constructible. We may construct a line containing q and perpendicular to the x-axis. This line intersects the unit circle at the point $(\cos\theta, \sin\theta)$. Hence, $\sin\theta$ is constructible since $\cos\theta$ is constructible.

(iii)$\Rightarrow$(i): Since $\sin\theta$ is constructible, $\cos\theta$ is constructible by an argument similar to that of (ii) implies (iii). The line through $(0, \sin\theta)$ parallel to the x-axis intersects the unit circle at the point $p = (\cos\theta, \sin\theta)$. Therefore, p and so $q = (\cos\theta, 0)$ are constructible. Consequently, the angle θ is constructible since the angle $p(0,0)q$ has radian measure θ. ∎

25.1.1 Worked-Out Exercises

◇ **Exercise 1** Let $n \in \mathbf{N}$. Let θ_n denote an angle with radian measure $\frac{2\pi}{n}$. Show that a regular polygon with n sides is constructible if and only if the angle θ_n is constructible.

   Solution: The desired result follows by noting that we may inscribe a regular polygon in the unit circle.

◇ **Exercise 2** Let $\theta, \phi \in \mathbf{R}$ and $m, n \in \mathbf{Z}$. If θ and ϕ are constructible, show that the angle with radian measure $m\theta + n\phi$ is constructible.

   Solution: The numbers $\cos\theta, \sin\theta, \cos\phi$, and $\sin\phi$ are constructible. Now $\sin(m\theta + n\phi)$ is equal to an algebraic expression involving $\cos\theta, \sin\theta, \cos\phi$, and $\sin\phi$. Since the set of constructible numbers is a field, $\sin(m\theta + n\phi)$ is constructible and so the desired result follows by Theorem 25.1.23.

◇ **Exercise 3** Let $m, n \in \mathbf{N}$. Let θ_n denote an angle with radian measure $\frac{2\pi}{n}$.

   (i) Show that if θ_{mn} is constructible, then θ_m and θ_n are constructible.

   (ii) Show that if θ_m and θ_n are constructible, where m and n are relatively prime, then θ_{mn} is constructible.

   Solution: (i) We note that

$$m\theta_{mn} = m\frac{2\pi}{mn} = \frac{2\pi}{n} = \theta_n$$

and similarly $n\theta_{mn} = \theta_m$. Hence, the result follows from Worked-Out Exercise 2 (page 547).

   (ii) Since m and n are relatively prime, there exist integers s and t such that $1 = sm + tn$. Thus,

$$\theta_{mn} = \frac{2\pi}{mn} = \frac{2\pi ms + 2\pi tn}{mn} = s\theta_n + t\theta_m.$$

Hence, θ_{mn} is constructible by Worked-Out Exercise 2 (page 547).

Exercise 4 Show that the regular 9-gon is not constructible.

Solution: Suppose that a regular 9-gon is constructible. Then an angle of $40°$ $(= 360°/9)$ could be constructed. However, an angle of $20°$ could then be constructed by bisecting the $40°$ angle. But this is impossible by Theorem 25.1.19 since it is shown there that it is impossible to construct an angle of $20°$.

Exercise 5 Show that it is possible to construct an angle of $30°$.

Solution: Since $\sqrt{3}$ is constructible, $\sqrt{3}/2$ is constructible. Thus, $\cos 30°$ is constructible and so $30°$ is constructible.

Exercise 6 Show that the regular 20-gon is constructible.

Solution: By Example 25.1.22, an angle of $36°$ can be constructed. Hence, an angle of $18°$ can be constructed by bisecting the angle of $36°$. Since $\frac{360}{20} = 18$, the regular 20-gon is constructible.

25.1.2   Exercises

1. Given line segments of length a and b, show that it is possible by straightedge and compass to construct line segments of length $a \pm b$, ab.

2. Prove that it is impossible to construct a cube whose volume equals that of a given sphere.

3. Prove that an angle of $40°$ cannot be constructed.

4. Prove that it is impossible to construct a regular septagon by straightedge and compass.

5. Prove that the regular pentagon and hexagon are constructible.

6. Prove that it is possible to trisect angles of $90°$ and $72°$.

7. Prove that it is impossible to construct a cube whose volume is three times the volume of a given cube.

8. Let $n \in \mathbf{N}$, $n > 1$. Let $n = p_1^{e_1} \cdots p_r^{e_r}$ be the prime factorization of n. Prove that a regular polygon with n sides is constructible if and only if a regular polygon with $p_i^{e_i}$ sides is constructible, $i = 1, 2, \ldots, r$.

Chapter 26

Coding Theory

26.1 Binary Codes

In this section, we examine techniques for transmitting information across a noisy channel. The information is often represented as a sequence of binary digits (0's and 1's). The channel may be space, as in satellite communication systems, or wires or cables, as in the telephone system, or wires as in circuits in a digital computer. Erratic currents called **noise** are always present to interfere with transmitted signals. Erratic currents can also be caused by such things as sunspots or magnetic storms. The channel noise will occasionally cause a transmitted one to be mistakenly interpreted as a zero or a transmitted zero to be mistakenly interpreted as a one. In order to reduce the effects of such errors, the transmitter may adjoin to the sequence of m (binary) message digits, s check digits.

The s check digits are selected by a method that makes them dependent on the m message digits. This is accomplished by mapping the sequence of message digits onto a sequence of $n = m + s$ digits called the **codeword**. This function is called the **encoding scheme**. The codeword is then transmitted. The receiver or decoder maps the received word, which may be different from the codeword due to channel noise, onto a sequence of m digits. This function is called the **decoding scheme.** Claude E. Shannon is credited as the originator of general coding theory.

The main aim of this section is to discuss the concepts of error detection and error correction.

Throughout this chapter, we let 0 and 1 denote the elements of the field $\mathbf{Z}_2$. For $n \geq 1$, let

$$B^n = \underbrace{\mathbf{Z}_2 \times \mathbf{Z}_2 \times \cdots \times \mathbf{Z}_2}_{n \text{ times}}.$$

Definition 26.1.1 *A binary (m, n)-code is a 4-tuple (B^m, B^n, E, D), where B^m is the set of all binary m-tuples, B^n is the set of all binary n-tuples ($n >$*

m) and $E : B^m \rightarrow B^n$ *and* $D : R \rightarrow B^m$, *where* $R \subseteq B^n$. *The functions* E *and* D *are called the* **encoding scheme** *and the* **decoding scheme**, *respectively.*

A nonempty subset of B^m is called a set of message words. Let $X \subseteq B^m$ be a set of message words. Then $E(X)$ is called a set of codewords. These codewords are transmitted across a noisy channel. Let X' be the set of received words after transmission. These received words are decoded by the decoding function D. Then $D(X')$ is the set of decoded words. We show by the following diagram the above coding and decoding process.

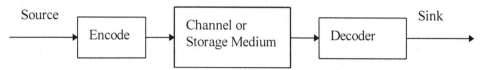

Block diagram of a general data communication or storage system.

We use the notation $\bar{c} = (c_1, \ldots, c_n) \in B^n$ for a codeword and $\bar{r} = (r_1, \ldots, r_n) \in B^n$ for a received word.

Example 26.1.2 ((m, $m + 1$)-***Parity-Check Code***) *This code is an error detecting code. The encoding function* E *is defined by*

$$E(a_1, a_2, \ldots, a_m) = (a_1, a_2, \ldots, a_m, a_{m+1}),$$

where $a_{m+1} = (a_1 + a_2 + \cdots + a_m)(\bmod\ 2)$. *Then* a_{m+1} *is 0 or 1, depending on whether the number of 1's in* $a_1, a_2, \ldots, a_m$ *is even or odd.*

For example, let us consider the $(3, 4)$-*parity-check code. Then* B^3 *is the set of message words and* $C = \{(0,\ 0,\ 0,\ 0),\ (0,\ 0,\ 1,\ 1),\ (0,\ 1,\ 0,\ 1),\ (0,\ 1,\ 1,\ 0),\ (1,\ 0,\ 0,\ 1),\ (1,\ 0,\ 1,\ 0),\ (1,\ 1,\ 0,\ 0),\ (1,\ 1,\ 1,\ 1)\}$ *is the set of codewords. Any odd number of errors can be detected, but the code fails to detect an even number of errors.*

Example 26.1.3 ((m, $3m$)-***Repetition Code***) *In this code,* $E : B^m \rightarrow B^{3m}$ *is defined as*

$$E(a_1, a_2, \ldots, a_m) = (a_1, a_2, \ldots, a_m, a_1, a_2, \ldots, a_m, a_1, a_2, \ldots, a_m).$$

Let x, y, $z \in B^m$. *Then* xyz *denotes the word* $w \in B^{3m}$ *such that the first* m *letters of* w *are those of* x, *the next* m *letters of* w *are those of* y, *and the last* m *letters of* w *are those of* z. *Define the decoding function* $D : B^{3m} \rightarrow B^m$ *as follows: The ith digit of* $D(w)$, $w \in B^{3m}$, *is the member that appears as the ith digit in at least two of the words,* x, y, z, *where* x, y, $z \in B^m$ *and* $w = xyz$.

For example, if $m = 3$ *and* $a = (1,\ 0,\ 1) \in B^m$, *then* $E(a) = (1,\ 0,\ 1,\ 1,\ 0,\ 1,\ 1,\ 0,\ 1)$. *Now* $w = aaa$ *is a codeword. Suppose that the transmission makes an error in the sixth digit. Then the received word, say,* v, *is* $(1,\ 0,\ 1,\ 1,\ 0,\ 0,$

1, 0, 1). *Let* $x = (1, 0, 1)$, $y = (1, 0, 0)$, $z = (1, 0, 1)$. *Now the first digit of* $D(v)$ *is 1 since 1 is the first digit of* x, y, *and* z. *The second digit of* $D(v)$ *is 0 since 0 is the second digit of* x, y, *and* z. *The third digit of* $D(v)$ *is 1 since 1 is the third digit of* x *and* z. *Hence,* $D(v) = (1, 0, 1)$.

We find that this code can detect a single error and can also correct the error. It follows that this code can also detect two errors, but it can correct only one error.

Example 26.1.4 (Repetition Code) *Let* m *be an integer and* d *an even integer. Let* $s = dm$. *Then* $n = (d + 1)m$. *Define* E *and* D *as follows: For all* $\bar{a} = (a_1, \ldots, a_m) \in B^m$, $E(\bar{a}) = \bar{c}$, *where for* $j = 0, 1, \ldots, d$, $c_{jm+i} = a_i$, $i = 1, \ldots, m$. *That is,* $\bar{a}$ *is encoded by breaking it into m-character blocks, each of which is transmitted* $(d+1)m$ *times. For all* $\bar{r} \in B^n$,

$$D(\bar{r}) = \begin{cases} 0 & \text{if more than half the } r_{jm+i} \text{ are zero, } j = 0, 1, \ldots, d, \\ 1 & \text{if more than half the } r_{jm+i} \text{ are one, } j = 0, 1, \ldots, d. \end{cases}$$

If more than half the digits in a fixed position of a codeword $\bar{c}$ of Example 26.1.4 are altered by channel noise, then the decoder will commit a **decoding error.**

If s were allowed to be odd in Example 26.1.4, then it would be possible for the number of zero r_i making up $\bar{r}$ to be equal to the number of nonzero r_i. In such a case, the decoder may decide not to decode the received word $\bar{r}$. This is an example of a **decoding failure.**

If a decoding algorithm decodes all received words, the algorithm is called **complete**; otherwise it is called **incomplete**. A decoding algorithm is complete if and only if $R = B^n$ in Definition 26.1.1.

Assume now that errors in transmitting successive digits occur independently. Thus, if p is the probability that a given digit will be received correctly, then the random variable counting the number of errors in a received word has a binomial distribution. That is, the probability p_k of exactly k errors in an n-digit received word is $p_k = \binom{n}{k}p^{n-k}q^k$, $k = 0, 1, \ldots, n$, where $q = 1 - p$. This simplified mathematical model for the channel is called the **binary symmetric channel.**

Encoding by Matrix Multiplication

We now describe a technique for encoding binary words by matrix multiplication.

Definition 26.1.5 *An* $m \times n$ *matrix* $(m < n)$ *with entries from* $\mathbf{Z}_2$ *is called a **generator matrix** if the submatrix consisting of the first m columns of this matrix has rank m.*

An $m \times n$ generator matrix M can be written as a partitioned matrix $M = (M'' \; M')$, where M'' is an $m \times m$ invertible submatrix of M and M' is the submatrix of M consisting of the last $n - m$ columns of M.

Let M be an $m \times n$ $(m < n)$ generator matrix. Define the corresponding coding function

$$E_M : B^m \to B^n$$

by $E_M(\bar{a}) = \bar{a}M$ for all $\bar{a} \in B^m$. Here we regard an element $\bar{a}$ of B^m as a $1 \times m$ matrix over $\mathbf{Z}_2$, $\bar{a}M = \bar{c} \in B^n$, where $\bar{c} = (c_1, \dots, c_n)$, $M = (e_{ij})$ and

$$c_j = \sum_{i=1}^{n} a_i e_{ij} \pmod 2, \; j = 1, 2, \dots, n. \tag{26.1}$$

Let $M'' = I_m$, the $m \times m$ identity matrix. From Eq. (26.1), we find that $c_j = a_j$ for $j = 1, 2, \dots, m$ and the submatrix M' of M consisting of the last $n - m$ columns of m determines the check digits to be adjoined to the message word $\bar{a} = (a_1, \dots, a_m)$.

Example 26.1.6 *Let*

$$M = \begin{bmatrix} 1 & 0 & 0 & 0 & 1 & 1 \\ 0 & 1 & 0 & 1 & 0 & 1 \\ 0 & 0 & 1 & 1 & 1 & 0 \end{bmatrix}$$

be a generator matrix over $\mathbf{Z}_2$. *This defines an encoding function* $E_M : B^3 \to B^6$. *For example, let* $\bar{a} = (1 \;\; 0 \;\; 1) \in B^3$. *Then*

$$E_M(\bar{a}) = (1\;0\;1) \begin{bmatrix} 1 & 0 & 0 & 0 & 1 & 1 \\ 0 & 1 & 0 & 1 & 0 & 1 \\ 0 & 0 & 1 & 1 & 1 & 0 \end{bmatrix} = (1\;0\;1\;1\;0\;1) \in B^6.$$

Hence, 101 is a message word and the corresponding codeword is 101101. 1, 0, 1 are the check digits which are adjoined to the message word 101.

Definition 26.1.7 *If* $M = (M'' \; M')$ *is an* $m \times n$ *generator matrix* $(m < n)$, *then the* $n \times (n - m)$ *matrix*

$$K = \begin{pmatrix} M''^{-1} M' \\ I_{n-m} \end{pmatrix},$$

where I_{n-m} *is the* $(n-m) \times (n-m)$ *identity matrix, is called the corresponding* ***parity check matrix*** *of* M.

We ask the reader to verify that $\bar{c}$ is a codeword if and only if $\bar{c}K = \bar{0}$, where K is the parity check matrix.

Example 26.1.8 *The parity check matrix for the generator matrix of Example 26.1.6 is*

$$K = \begin{bmatrix} 0 & 1 & 1 \\ 1 & 0 & 1 \\ 1 & 1 & 0 \\ 1 & 0 & 0 \\ 0 & 1 & 0 \\ 0 & 0 & 1 \end{bmatrix}.$$

We now turn our attention to an important class of codes discovered by Richard W. Hamming.

Definition 26.1.9 *An (m,n) code (B^m, B^n, E, D) is called a **Hamming** (m,n) code if E is defined by a generator matrix M such that the rows of the corresponding parity check matrix K contain all the $2^t - 1$ nonzero vectors of B^t, where $t = n - m$.*

Example 26.1.10 *Let*

$$M = \begin{bmatrix} 1 & 0 & 0 & 0 & 1 & 1 & 1 \\ 0 & 1 & 0 & 0 & 1 & 1 & 0 \\ 0 & 0 & 1 & 0 & 1 & 0 & 1 \\ 0 & 0 & 0 & 1 & 0 & 1 & 1 \end{bmatrix}.$$

Then M is a 4×7 matrix. The corresponding parity check matrix is

$$K = \begin{bmatrix} 1 & 1 & 1 \\ 1 & 1 & 0 \\ 1 & 0 & 1 \\ 0 & 1 & 1 \\ 1 & 0 & 0 \\ 0 & 1 & 0 \\ 0 & 0 & 1 \end{bmatrix}.$$

K contains all the nonzero elements of B^3. Hence, the encoding function defined by the generator matrix M defines a $(7,4)$ Hamming code.

The (m,n)-Hamming code corrects every single error pattern. No other errors and no other $(2^t - 1 - t,\ 2^t - 1)$ code can be constructed, which will correct more than all single errors. Any received word with two or more errors will be decoded as if it had one error. In this case, a decoding error is made. This follows since if the received word $\bar{r}$ has a single error in the ith component, then $\bar{r}K$ is just the ith row of K. Thus, every received word with a single error can be corrected because K has n rows, which are nonzero and distinct.

Weight and Distance

Definition 26.1.11 *Let $a \in B^n$. The **weight** of a, denoted by $wt(a)$, is the number of 1's in a.*

Example 26.1.12 *For $01100 \in B^5$, $wt(01100) = 2$.*

Definition 26.1.13 *Let $a, b \in B^n$. The **distance between** a and b, denoted by $d(a, b)$, is defined by*

$$d(a, b) = wt(a + b).$$

Example 26.1.14 *For $a = 0101$ and $b = 1110$, $a + b = 1011$. Then $wt(a+b) = 3$. Hence, $d(a, b) = 3$.*

We leave the proofs of the next two results for the exercises.

Theorem 26.1.15 *Let $a, b, c \in B^n$. Then*
   (i) $d(a, b) =$ the number of locations i with $a_i \neq b_i$.
   (ii) $d(a, 0) = wt(a)$.
   (iii) $d(a, b) = 0$ if and only if $a = b$.
   (iv) $d(a, b) = d(b, a)$.
   (v) $d(a, b) + d(b, c) \geq d(a, c)$. ∎

Theorem 26.1.16 *(i) A code (B^m, B^n, E, D) can detect all sets of k or fewer errors if and only if the minimum distance between any two distinct codewords is at least $k + 1$.*
   (ii) For a code to correct all sets of k or fewer errors, it is necessary that the minimum distance between any two distinct codewords be at least $2k + 1$. ∎

Example 26.1.17 *Consider the following set C of codewords in B^6.*

$C = \{000000,\ 001110,\ 010101,\ 100011,\ 011011,\ 101101,\ 110110,\ 111000\}.$

The minimum distance between two distinct codewords is 3. Hence, this code can detect two or fewer errors.

Group Codes

Now $(B^m, +)$ and $(B^n, +)$ are commutative groups, where for both, $+$ is defined by componentwise addition (mod 2) of vectors. Clearly $|B^n| = 2^n$. Let C be the subset of B^n consisting of all codewords. That is, $C = \{\bar{c} \mid \bar{c} \in B^n, \bar{c}K = \bar{0}\}$. Clearly C is a subgroup of B^n and $|C| = 2^m$.

We know that the cosets of C in B^n partition B^n and that the difference between any two vectors in the same coset is a codeword. Also, the sum of a vector $\bar{x}$ in B^n and any codeword gives another vector in the same coset as $\bar{x}$.

After the message is encoded into the full codeword, the codeword $\bar{c}$ is transmitted across the noisy channel. The channel adds to $\bar{c}$ the **error** or **noise word** $\bar{e} = (e_1, \ldots, e_n)$, where

$$
e_i = \begin{cases} 0 & \text{if the channel does not change the } i\text{th digit,} \\ 1 & \text{if the channel does change the } i\text{th digit.} \end{cases}
$$

The received word $\bar{r}$ is equal to the codeword plus the error word, i.e., $\bar{r} = \bar{c} + \bar{e}$. The set of possible error patterns must be exactly the coset of C in B^n determined by $\bar{r}$. This follows because $\bar{e} = \bar{r} - \bar{c} \in \bar{r} + C$. Hence, the decoder may immediately exclude all error patterns which do not lie in the same coset as that determined by $\bar{r}$. However, all error patterns in this coset are possible. Those error patterns with a smaller number of ones are more probable than those with many ones since channel errors are relatively infrequent. An element in a coset with the fewest number of ones is called a **coset leader**.

Definition 26.1.18 *Let* (B^m, B^n, E, D) *be a code. If* $E(B^m)$, *the image of* B^m *under* E, *is a subgroup of* B^n, *then this code is called a **linear** or **group** code.*

Theorem 26.1.19 *Let* (B^m, B^n, E, D) *be a group code. Then the minimum distance between any two distinct codewords is the least weight of a nonzero codeword.*

Proof. Let d be the minimum distance between two distinct codewords. Then there exist two distinct codewords u and v in B^n such that $d(u, v) = d$. Let x be a nonzero codeword such that $\text{wt}(x) \leq \text{wt}(y)$ for all nonzero codewords y. Now $d \leq d(x, 0) = \text{wt}(x)$. Again $d = d(u, v) = \text{wt}(u + v) \geq \text{wt}(x)$. Hence, $d = \text{wt}(x)$. ∎

Let M be an $m \times n$ generator matrix. We now show that the encoding function

$$
E = E_M : B^m \to B^n
$$

defined by $E(\bar{a}) = \bar{a}M$ defines a group code. For this, let us prove that $E(B^m)$ is a subgroup of B^n. Let $\bar{b}, \bar{c} \in E(B^m)$. There exist $\bar{a}, \bar{d} \in B^m$ such that $\bar{b} = E(\bar{a}) = \bar{a}M$ and $\bar{c} = E(\bar{d}) = \bar{d}M$. Therefore, $\bar{b} + \bar{c} = \bar{a}M + \bar{d}M = (\bar{a} + \bar{d})M = E(\bar{a} + \bar{d})$. Thus, $\bar{b} + \bar{c} \in E(B^m)$. Now in B^n, $\bar{c} = -\bar{c}$. Hence, $E(B^m)$ is a subgroup of B^n, proving that E defines a group code.

A group code can be decoded by the following procedure, known as the tabular procedure. We explain this procedure with the help of an example.

Consider the code (B^3, B^6, E, D) with generator matrix

$$M = \begin{bmatrix} 1 & 0 & 0 & 1 & 1 & 1 \\ 0 & 1 & 0 & 0 & 1 & 1 \\ 0 & 0 & 1 & 1 & 0 & 1 \end{bmatrix}.$$

We have

Elements of B^3	The Set of Codewords $E_M(B^3)$
000	$000M = 000000$
001	$001M = 001101$
010	$010M = 010011$
011	$011M = 011110$
100	$100M = 100111$
101	$101M = 101010$
110	$110M = 110100$
111	$111M = 111001.$

Let $C = E_M(B^3)$. Then C is a subgroup of B^6.

Step 1: List the codewords in a row with 000000 first.

000000 001101 010011 011110 100111 101010 110100 111001

Step 2: Choose a word x in B^6 of least weight among those not in the previously chosen cosets. Then list the elements of the left coset $x + C$ as the next row appearing below a for every $a \in C$. Let us take $x = 100000$.

C :	000000	001101	010011	011110	100111	101010	110100	111001
$x + C$:	100000	101101	110011	111110	000111	001010	010100	011001

Step 3: Repeat step 2 until all elements of B^6 are exhausted.

Step 4: Decode each received word as the codeword of the column in which the received word appears.

The table obtained by the above process is called the **decoding table**. The decoding table for the above code is shown below:

C :	000000	001101	010011	011110	100111	101010	110100	111001
$100000 + C$:	100000	101101	110011	111110	000111	001010	010100	011001
$010000 + C$:	010000	011101	000011	001110	110111	111010	100100	101001
$001000 + C$:	001000	000101	011011	010110	101111	100010	111100	110001
$000100 + C$:	000100	001001	010111	011010	100011	101110	110000	111101
$000010 + C$:	000010	001111	010001	011100	100101	101000	110110	111011
$000001 + C$:	000001	001100	010010	011111	100110	101011	110101	111000
$100001 + C$:	100001	101100	110010	111111	000110	001011	010101	011000

The x's chosen are the coset leaders in the coset $x + C$. Suppose the received word is 110011. In the above table, it appears in the third column. Hence, the decoder decodes the received word as 010011. We note that a decoding error is possible; for if the error pattern was actually 001010, then the codeword transmitted was 111001.

26.1.1 Worked-Out Exercises

Exercise 1 Let $\bar{c} \in B^n$. Let q be an integer such that $0 \le q \le n$. Prove that there exist $\binom{n}{q}$ elements $w \in B^n$ such that $d(w, c) = q$.

Solution: We have that $d(w, c) = q$ if and only if w and c differ in exactly q bits. There are exactly $\binom{n}{q}$ ways to change q bits of c.

Exercise 2 Let C be a set of codewords in B^n. Prove that if C can correct k errors, then

$$|C| \le \frac{2^n}{\binom{n}{0} + \binom{n}{1} + \cdots + \binom{n}{k}}.$$

Solution: Let $N = \binom{n}{0} + \binom{n}{1} + \cdots + \binom{n}{k}$. For all $c \in C$, let $S_k(c) = \{w \in B^n \mid d(w, c) \le k\}$. Now the $S_k(c)$ are pairwise disjoint and contain N elements. Since there are $|C|$ distinct $S_k(c)$, $N |C| \le |B^n| = 2^n$. Thus, $|C| \le \frac{2^n}{N}$.

26.1.2 Exercises

1. Find the weight of each word below:

   (i) 11011010,

   (ii) 11000110.

2. Find the distance between the following pairs of words:

   (i) 11011011 and 10001010,

   (ii) 11000100 and 00111011.

3. Let M be an $m \times n$ matrix whose submatrix consisting of the first m columns is the identity matrix. Let M' be the submatrix of M consisting of the last $m - n$ columns of M. Set

   $$K = \begin{pmatrix} M' \\ I_{n-m} \end{pmatrix}.$$

   Prove that $\bar{c}$ is a codeword if and only if $\bar{c}K = \bar{0}$.

4. For the matrices M and K of Exercise 3, prove that C is a subgroup of B^n, where

   $$C = \{\bar{c} \mid \bar{c} \in B^n, \ \bar{c}K = \bar{0}\}.$$

5. Let M be an $m \times n$ encoding matrix whose submatrix M'' consisting of the first m columns of M is invertible. Let M' be as defined in Exercise 3 and set

   $$K = \begin{pmatrix} M''^{-1}M' \\ I_{n-m} \end{pmatrix}.$$

   Prove that $\bar{c}$ is a codeword if and only if $\bar{c}K = \bar{0}$.

6. Find the set of codewords of Example 26.1.10.

7. For each of the following generator matrices, find how many errors the corresponding code can detect and how many errors it can correct.

(i) $\begin{bmatrix} 1 & 0 & 0 & 1 & 1 \\ 0 & 1 & 0 & 0 & 1 \\ 0 & 0 & 1 & 1 & 0 \end{bmatrix}$

(ii) $\begin{bmatrix} 1 & 0 & 0 & 0 & 1 & 0 & 0 & 1 \\ 0 & 1 & 0 & 0 & 1 & 1 & 0 & 1 \\ 0 & 0 & 1 & 0 & 0 & 1 & 1 & 1 \\ 0 & 0 & 0 & 1 & 1 & 0 & 1 & 0 \end{bmatrix}$.

8. Write the complete coset decoding table for the code given by the generator matrix
$$\begin{bmatrix} 1 & 0 & 0 & 1 & 1 & 0 \\ 0 & 1 & 0 & 1 & 0 & 1 \\ 0 & 0 & 1 & 0 & 1 & 1 \end{bmatrix}.$$

From the table, decode the following received words:

$$001111, \ 101010, \ 011110.$$

9. Let C be an (m, n) code. Suppose that each word $b \in B^n$ with $\text{wt}(b) \leq t$ is the coset leader of $b + C$. Prove that C corrects t or fewer errors.

10. (i) Show that no $(2, 4)$ code can correct single errors.

(ii) Show that no $(3, 6)$ code can correct two errors.

11. (i) Construct a $(2, 5)$ code that corrects a single error.

(ii) Construct a $(3, 6)$ code that corrects a single error.

12. Let $s = 4$. Construct the $(11, 15)$-Hamming code.

13. Prove Theorem 26.1.15.

14. Prove Theorem 26.1.16.

26.2 Polynomial and Cyclic Codes

In this section, we describe a technique, which encodes m-digit message words into n-digit codewords by polynomial multiplication.

Let $\bar{a} = (a_0, a_1, \ldots, a_{m-1}) \in B^m$. Then the correspondence

$$\bar{a} \rightarrow a_0 + a_1 x + \cdots + a_{m-1} x^{m-1} = a(x) \qquad (26.2)$$

is a one-one mapping of B^n into the polynomial ring $\mathbf{Z}_2[x]$. Set $s = n - m$ and let

$$g(x) = b_0 + b_1 x + \cdots + b_s x^s$$

be a polynomial of degree s over $\mathbf{Z}_2$ such that $b_0 \neq 0$. Then

$$c(x) = c_0 + c_1 x + \cdots + c_{n-1} x^{n-1} = a(x)g(x)$$

is a polynomial of degree $\leq n - 1$. Considering the correspondence,

$$\bar{c} = (c_0, c_1, \ldots, c_{n-1}) \leftrightarrow c(x)$$

leads us to the following definition.

Definition 26.2.1 *Let n, m be positive integers such that $n > m$. Let $g(x) = b_0 + b_1 x + \cdots + b_s x^s$ be any fixed polynomial of $\mathbf{Z}_2[x]$ of degree $s = n - m$ such that $b_0 \neq 0$. The encoding polynomial $g(x)$ encodes each message word $\bar{a} \in B^m$, which corresponds to $a(x)$ in (26.2) into the codeword $\bar{c}$, which corresponds to the code polynomial $c(x) = a(x)g(x)$.*

Example 26.2.2 *Let $m = 3, n = 7$, and the encoding polynomial be $g(x) = 1 + x + x^4$. By considering the product $c(x) = a(x)g(x)$, we obtain the following encoding scheme:*

$000 \to 0000000$		$100 \to 1100100$	
$001 \to 0011001$		$101 \to 1111101$	
$010 \to 0110010$		$110 \to 1010110$	
$011 \to 0101011$		$111 \to 1001111$	

Define the matrix G by

$$G = \begin{bmatrix} b_0 & b_1 & b_2 & b_3 & b_4 & 0 & 0 \\ 0 & b_0 & b_1 & b_2 & b_3 & b_4 & 0 \\ 0 & 0 & b_0 & b_1 & b_2 & b_3 & b_4 \end{bmatrix}$$

$$= \begin{bmatrix} 1 & 1 & 0 & 0 & 1 & 0 & 0 \\ 0 & 1 & 1 & 0 & 0 & 1 & 0 \\ 0 & 0 & 1 & 1 & 0 & 0 & 1 \end{bmatrix}$$

Then the above encoding scheme can be obtained by the matrix multiplication $\bar{a}G$.

Example 26.2.3 *Let $m = 3, n = 4$, and $g(x) = 1 + x$. Then the encoding polynomial $g(x)$ encodes message words $\bar{a} \in B^3$ by the following encoding scheme:*

$000 \to 0000$		$100 \to 1100$	
$001 \to 0011$		$101 \to 1111$	
$010 \to 0110$		$110 \to 1010$	
$011 \to 0101$		$111 \to 1001$	

The encoding matrix is

$$G = \begin{bmatrix} 1 & 1 & 0 & 0 \\ 0 & 1 & 1 & 0 \\ 0 & 0 & 1 & 1 \end{bmatrix}.$$

The code satisfies an even parity check.

Example 26.2.4 Let $m = 11, n = 15$, and $g(x) = 1+x+x^4$. Then the encoding matrix corresponding to the encoding polynomial $g(x)$ is given by

$$G = \begin{bmatrix}
1 & 1 & 0 & 0 & 1 & 0 & 0 & 0 & 0 & 0 & 0 & 0 & 0 & 0 & 0 \\
0 & 1 & 1 & 0 & 0 & 1 & 0 & 0 & 0 & 0 & 0 & 0 & 0 & 0 & 0 \\
0 & 0 & 1 & 1 & 0 & 0 & 1 & 0 & 0 & 0 & 0 & 0 & 0 & 0 & 0 \\
0 & 0 & 0 & 1 & 1 & 0 & 0 & 1 & 0 & 0 & 0 & 0 & 0 & 0 & 0 \\
0 & 0 & 0 & 0 & 1 & 1 & 0 & 0 & 1 & 0 & 0 & 0 & 0 & 0 & 0 \\
0 & 0 & 0 & 0 & 0 & 1 & 1 & 0 & 0 & 1 & 0 & 0 & 0 & 0 & 0 \\
0 & 0 & 0 & 0 & 0 & 0 & 1 & 1 & 0 & 0 & 1 & 0 & 0 & 0 & 0 \\
0 & 0 & 0 & 0 & 0 & 0 & 0 & 1 & 1 & 0 & 0 & 1 & 0 & 0 & 0 \\
0 & 0 & 0 & 0 & 0 & 0 & 0 & 0 & 1 & 1 & 0 & 0 & 1 & 0 & 0 \\
0 & 0 & 0 & 0 & 0 & 0 & 0 & 0 & 0 & 1 & 1 & 0 & 0 & 1 & 0 \\
0 & 0 & 0 & 0 & 0 & 0 & 0 & 0 & 0 & 0 & 1 & 1 & 0 & 0 & 1
\end{bmatrix}.$$

We have for the matrix M'' defined following Definition 26.1.5 that

$$M''^{-1} = \begin{bmatrix}
1 & 1 & 1 & 1 & 0 & 1 & 0 & 1 & 1 & 0 & 0 \\
0 & 1 & 1 & 1 & 1 & 0 & 1 & 0 & 1 & 1 & 0 \\
0 & 0 & 1 & 1 & 1 & 1 & 0 & 1 & 0 & 1 & 1 \\
0 & 0 & 0 & 1 & 1 & 1 & 1 & 0 & 1 & 0 & 1 \\
0 & 0 & 0 & 0 & 1 & 1 & 1 & 1 & 0 & 1 & 0 \\
0 & 0 & 0 & 0 & 0 & 1 & 1 & 1 & 1 & 0 & 1 \\
0 & 0 & 0 & 0 & 0 & 0 & 1 & 1 & 1 & 1 & 0 \\
0 & 0 & 0 & 0 & 0 & 0 & 0 & 1 & 1 & 1 & 1 \\
0 & 0 & 0 & 0 & 0 & 0 & 0 & 0 & 1 & 1 & 1 \\
0 & 0 & 0 & 0 & 0 & 0 & 0 & 0 & 0 & 1 & 1 \\
0 & 0 & 0 & 0 & 0 & 0 & 0 & 0 & 0 & 0 & 1
\end{bmatrix}.$$

Thus, the parity-check matrix K of Definition 26.1.7 is given by

$$K = \begin{bmatrix} 1 & 1 & 0 & 0 \\ 0 & 1 & 1 & 0 \\ 0 & 0 & 1 & 1 \\ 1 & 1 & 0 & 1 \\ 1 & 0 & 1 & 0 \\ 0 & 1 & 0 & 1 \\ 1 & 1 & 1 & 0 \\ 0 & 1 & 1 & 1 \\ 1 & 1 & 1 & 1 \\ 1 & 0 & 1 & 1 \\ 1 & 0 & 0 & 1 \\ 1 & 0 & 0 & 0 \\ 0 & 1 & 0 & 0 \\ 0 & 0 & 1 & 0 \\ 0 & 0 & 0 & 1 \end{bmatrix}.$$

The matrix G is the encoding matrix for the $(11, 15)$-Hamming code.

Theorem 26.2.5 *The error polynomial associated with any undetected error vector $\bar{e} = (e_0, e_1, \ldots, e_{n-1})$ of a polynomial (m, n)-code with generator $g(x)$ must be a nontrivial multiple of $g(x)$.*

Proof. If the received word $\bar{r} = (r_0, r_1, \ldots, r_{n-1})$ with the corresponding polynomial $r(x) = r_0 + r_1 x + \cdots + r_{n-1} x^{n-1}$ is erroneous, but undetected. Then $\bar{r}$ is a codeword and so the error $\bar{e} = \bar{r} - \bar{c}$ is a nonzero codeword. Hence $\bar{e}$ must correspond to a nonzero codeword $q(x)g(x)$. ∎

The division algorithm is very convenient in the detection of errors. Consider a polynomial (m, n)-code with generator $g(x)$. Suppose $r(x)$ is the polynomial corresponding to the received word $\bar{r}$. By the division algorithm, there exist polynomials $q(x)$ and $t(x)$ such that $r(x) = q(x)g(x) + t(x)$, where either $t(x) = 0$ or $\deg t(x) < \deg g(x)$. If $t(x) \neq 0$, an error has occurred and we deduce that $\bar{r}$ was not a codeword.

We now begin our discussion of cyclic codes.

The **cyclic shift** of the n-tuple $(c_0, c_1, \ldots, c_{n-1})$ is defined to be the n-tuple $(c_{n-1}, c_0, c_1, \ldots, c_{n-2})$. A linear code is said to be **cyclic** if the cyclic shift of every codeword is again a codeword. Once again we identify the n-tuple $(c_0, c_1, \ldots, c_{n-1})$ with the polynomial $c(x) = c_0 + c_1 x + \cdots + c_{n-1} x^{n-1}$.

Theorem 26.2.6 *Every linear cyclic (m, n)-code has a unique monic codeword $g(x)$ of degree $s = n - m$ and $g(x)$ divides $x^n - 1$. Let $c(x)$ be of degree $n - 1$*

or less. Then $c(x)$ is a codeword if and only if $c(x)$ is a multiple of $g(x)$. Also, every monic polynomial of degree s, which divides $x^n - 1$, is the generator polynomial of a linear cyclic (m, n)-code with $s = n - m$.

Proof. Consider an arbitrary linear cyclic code and let $g(x)$ of degree s be the minimum degree polynomial among the codewords. ($g(x)$ is necessarily monic since the coefficients of $g(x)$ lie in $\mathbb{Z}_2$.) Now $xg(x), x^2g(x), \ldots, x^{n-s-1}g(x)$ are just cyclic shifts of $g(x)$ and must be codewords. Since the code is linear, all combinations of $g(x), xg(x), x^2g(x), \ldots, x^{n-s-1}g(x)$, i.e., the 2^{n-s} polynomials $a(x)g(x)$, where the degree of $a(x)$ is less than $n - s = m$, must be codewords. But this is all the codewords since if $c(x)$ is any codeword, the division algorithm gives $c(x) = a(x)g(x) + r(x)$, $r(x) = 0$ or $\deg r(x) < s$, $a(x) = 0$ or $\deg a(x) < n - s$.

Thus, $r(x) = c(x) - a(x)g(x)$, which by the linearity of the code shows that $r(x)$ must be a codeword. Since $r(x) = 0$ or $\deg r(x) < s$, we have by the minimality of $\deg g(x)$ that $r(x) = 0$. Hence, $c(x)$ is a multiple of $g(x)$. Therefore, $g(x)$ is the generator polynomial of the cyclic code. Finally, we note that $x^{n-s}g(x) - (x^n - 1)$ is the cyclic shift of $x^{n-s-1}g(x)$ and hence a codeword and thus a multiple of $g(x)$, i.e., there is a polynomial $a(x)$ of degree less than $n - s$ such that

$$x^{n-s}g(x) - (x^n - 1) = a(x)g(x)$$

or

$$x^n - 1 = h(x)g(x),$$

where $h(x) = x^{n-s} - a(x)$ is a monic polynomial of degree $n - s$. Thus, $g(x)$ divides $x^n - 1$. There are 2^{n-s} codewords, so $m = n - s$.

Conversely, let $g(x)$ be a polynomial of degree s which divides $x^n - 1$. If the polynomial

$$c(x) = c_0 + c_1x + \cdots + c_{n-1}x^{n-1}$$

is multiplied by $x \bmod(x^n - 1)$, the result is $c_{n-1} + c_0x + \cdots + c_{n-2}x^{n-1}$. The codeword represented by the polynomial $xc(x)\bmod(x^n-1)$ is seen to be a cyclic shift of the codeword represented by the polynomial $c(x)$. Since every cyclic shift of a codeword therefore gives another codeword, the code is a cyclic code. Hence, the set of multiples of $g(x)\bmod(x^n - 1)$ forms a linear cyclic code. ∎

Example 26.2.7 *Consider the linear binary cyclic code with the codeword set*

$$\{000, 011, 101, 110\}.$$

The codeword 110 corresponds to the polynomial $1 + x$, which is the minimum degree codeword polynomial. Hence, this cyclic code is the length $n = 3$ cyclic code generated by $g(x) = 1 + x$.

Example 26.2.8 *Choose the generator polynomial of a binary cyclic code to be $g(x) = 1 + x + x^3$. It is readily checked that $g(x)$ divides $x^7 - 1$ so that we may choose $n = 7$. The number of information digits in this code is $m = 7 - 3 = 4$. The $2^m = 2^4$ codewords are the 7-tuples corresponding to the polynomials*

$$a(x)g(x) = (a_0 + a_1 x + a_2 x^2 + a_3 x^3)(1 + x + x^3).$$

This code is the second in the class of Hamming single-error-correcting codes. (There is a code in this class with $n = 2^i - 1$ and $m = 2^i - i - 1$ for $i = 2, 3, 4, \ldots$.)

Example 26.2.9 *Choose the generator polynomial of a binary cyclic code to be $g(x) = 1 + x + x^4$. It is readily verified that $g(x)$ divides $x^{15} - 1$ so that we may choose $n = 15$. The number of information digits in this code is $m = 15 - 4 = 11$. This code is the third in the class of Hamming single-error-correcting cyclic codes.*

The representation of codewords by the polynomials modulo $x^n - 1$ in the proof of Theorem 26.2.6 suggests that we could have introduced cyclic codes by means of ideals of commutative rings. Consider the polynomial ring $\mathbf{Z}_2[x]$, the ideal generated by $\langle x^n - 1 \rangle$, and the quotient ring

$$B_n = \mathbf{Z}_2[x] / \langle x^n - 1 \rangle.$$

Then

$$B_n = \{a_0 + a_1 \bar{x} + \cdots + a_{n-1} \bar{x}^{n-1} \mid a_i \in \mathbf{Z}_2, \, i = 0, 1, \ldots, n - 1\},$$

where $\bar{x}$ denotes the coset $x + \langle x^n - 1 \rangle$ in B_n. Let $C \subseteq B_n$ be a cyclic code and $f(\bar{x}) = a_0 + a_1 \bar{x} + \cdots + a_{n-1} \bar{x}^{n-1} \in C$. Since $\bar{x}^n = 1$ in B_n, we see that $\bar{x} f(\bar{x}) = a_{n-1} + a_0 \bar{x} + a_1 \bar{x}^2 + \cdots + a_{n-2} \bar{x}^{n-1}$ is the cyclic shift of $f(\bar{x})$. Since C is cyclic, $\bar{x} f(\bar{x}) \in C$. Therefore, $g(\bar{x}) f(\bar{x}) \in C$ for any $f(\bar{x}) \in C$ and any $g(\bar{x}) \in B_n$. Hence, C is an ideal of B_n. Clearly if C is an ideal of B_n, then $\bar{x} f(\bar{x}) \in C$ for all $f(\bar{x}) \in C$. Thus, we have shown the following result.

Theorem 26.2.10 *Let $C \subseteq B_n$ be a linear code. Then the following conditions are equivalent.*
   (i) C is cyclic.
   (ii) $\bar{x} C \subseteq C$.
   (iii) C is an ideal of B_n. ■

Now B_n is the homomorphic image of the principal ideal ring $\mathbf{Z}_2[x]$ and so B_n is a principal ideal ring. Thus, if C is an ideal of B_n, there exists $g(x) \in \mathbf{Z}_2[x]$ such that $\langle g(\bar{x}) \rangle = C$. The polynomial $g(x)$ has special properties which we describe in the exercises, and which can be seen by Theorem 26.2.6.

We now determine the dimension of a cyclic code C, where C is considered to be a subspace of the vector space B_n over $\mathbf{Z}_2$. Let $C = \langle g(\overline{x}) \rangle$, where $g(x)$ divides $x^n - 1$. We recall that if R is a commutative ring and $a \in R$, then the annihilator of a, $\mathrm{ann}(a) = \{ r \in R \mid ra = 0 \}$, is an ideal of R.

Lemma 26.2.11 *Let $g(x) \in \mathbf{Z}_2[x]$ divide $x^n - 1$. Then in B_n,*

$$\mathrm{ann}(g(\overline{x})) = \langle h(\overline{x}) \rangle,$$

where $x^n - 1 = h(x)g(x)$.

Proof. Since $0 = \overline{x}^n - 1 = h(\overline{x})g(\overline{x})$, $h(\overline{x}) \in \mathrm{ann}(g(\overline{x}))$. Thus,

$$\langle h(\overline{x}) \rangle \subseteq \mathrm{ann}(g(\overline{x})).$$

Let $f(\overline{x}) \in \mathrm{ann}(g(\overline{x}))$. Then $f(\overline{x})g(\overline{x}) = 0$. Therefore, $f(x)g(x) \in \mathrm{Ker}\, \eta = \langle x^n - 1 \rangle$, where η is the natural homomorphism of $\mathbf{Z}_2[x]$ onto $\mathbf{Z}_2[x]/\langle x^n - 1 \rangle = B_n$. Thus, there exists $q(x) \in \mathbf{Z}_2[x]$ such that $f(x)g(x) = q(x)(x^n - 1)$. Therefore, $f(x)g(x) = q(x)h(x)g(x)$ and so $f(x) = q(x)h(x)$. Thus,

$$f(\overline{x}) = q(\overline{x})h(\overline{x}) \in \langle h(\overline{x}) \rangle.$$

Hence, $\mathrm{ann}(g(\overline{x})) \subseteq \langle h(\overline{x}) \rangle$. Consequently, $\mathrm{ann}(g(\overline{x})) = \langle h(\overline{x}) \rangle$. ∎

Theorem 26.2.12 *Let $g(x) \in \mathbf{Z}_2[x]$ divide $x^n - 1$. Let $C = \langle g(\overline{x}) \rangle$ be a cyclic code in B_n. Let $s = \deg g(x)$ and $m = n - s$. Then $X = \{g(\overline{x}), \overline{x}g(\overline{x}), \ldots, \overline{x}^{m-1}g(\overline{x})\}$ is a basis of C over $\mathbf{Z}_2[x]$.*

Proof. Since $g(x)$ divides $x^n - 1$, there exists $h(x) \in \mathbf{Z}_2[x]$ such that $x^n - 1 = h(x)g(x)$. We show that X spans C. Let $f(\overline{x}) \in C$. Then $f(\overline{x}) = k(\overline{x})g(\overline{x})$ for some $k(x) \in \mathbf{Z}_2[x]$. By the division algorithm, there exists $q(x), r(x) \in \mathbf{Z}_2[x]$ such that $k(x) = q(x)h(x) + r(x)$, where either $r(x) = 0$ or $\deg r(x) < \deg h(x) = m$. By Lemma 26.2.11, $h(\overline{x})g(\overline{x}) = 0$. Thus,

$$f(\overline{x}) = (q(\overline{x})h(\overline{x}) + r(\overline{x}))g(\overline{x}) = r(\overline{x})g(\overline{x})$$

and $\deg r(x) \leq m - 1$. Hence, $f(\overline{x})$ is a linear combination over $\mathbf{Z}_2$ of the elements of X. Therefore, X spans C. Suppose $0 = a_0 g(\overline{x}) + a_1 \overline{x} g(\overline{x}) + \cdots + a_{m-1} \overline{x}^{m-1} g(\overline{x})$, where $a_i \in \mathbf{Z}_2$, $i = 0, 1, \ldots, m - 1$. Let $f(x) = a_0 + a_1 x + \cdots + a_{m-1} x^{m-1}$. Then $0 = f(\overline{x})g(\overline{x})$. Hence, $f(\overline{x}) \in \mathrm{ann}(g(\overline{x})) = \langle h(\overline{x}) \rangle$. Thus, $\langle f(\overline{x}) \rangle \subseteq \langle h(\overline{x}) \rangle$ and so $h(x)|f(x)$ by Exercise 5. Now $\deg f(x) \leq m - 1 < m = \deg h(x)$. Consequently, $f(x) = 0$. Hence, $a_0 = a_1 = \cdots = a_{m-1} = 0$. Therefore, X is linearly independent over $\mathbf{Z}_2$. ∎

Corollary 26.2.13 *Let $g(x) \in \mathbf{Z}_2[x]$ divide $x^n - 1$. Let $C = \langle g(\overline{x}) \rangle$ be a cyclic code in B_n. Then C has dimension m over $\mathbf{Z}_2$. Furthermore, C is an (n, m)-code.* ■

We recall that if $C = \langle g(\overline{x}) \rangle$ is a cyclic code in B_n, then

$$
G = \begin{bmatrix}
g_0 & g_1 & g_2 & \cdots & g_s & 0 & \cdots & 0 \\
0 & g_0 & g_1 & \cdots & g_{s-1} & g_s & \cdots & 0 \\
\vdots & \vdots & \vdots & \vdots & \vdots & \vdots & \vdots & \vdots \\
0 & 0 & \cdots & g_0 & g_1 & g_2 & \cdots & g_s
\end{bmatrix}_{m \times n}
\tag{26.3}
$$

is a generator matrix for C, where $g(x) = g_0 + g_1 x + \cdots + g_s x^s$. We now determine a parity-check matrix H for C. We have that $x^n - 1 = h(x)g(x)$ for some $h(x) \in \mathbf{Z}_2[x]$, where $\deg h(x) = n - s = m$. Write $h(x) = h_0 + h_1 x + \cdots + h_m x^m$. Define H to be the $n \times s$ matrix,

$$
H = \begin{bmatrix}
0 & 0 & \cdots & h_m \\
\vdots & \vdots & \cdots & \vdots \\
0 & h_m & \cdots & h_{m-(s-2)} \\
h_m & h_{m-1} & \cdots & h_{m-(s-1)} \\
\vdots & \vdots & \cdots & \vdots \\
h_{s-1} & h_{s-2} & \cdots & h_0 \\
\vdots & \vdots & \cdots & \vdots \\
h_2 & h_1 & \cdots & 0 \\
h_1 & h_0 & \cdots & 0 \\
h_0 & 0 & \cdots & 0
\end{bmatrix}.
\tag{26.4}
$$

Lemma 26.2.14 *Let G and H be defined as above. Then $GH = 0$.*

Proof. The (i, j)th component of the matrix GH is given by

$$
\sum_{k=0}^{n-1} g_{n-i+k+1} h_{n-j-k},
\tag{26.5}
$$

where $g_{s+1} = \cdots = g_{n-1} = 0$ and $h_{m+1} = \cdots = h_{n-1} = 0$ and where the subscripts of the $g_{n-i+k+1}$ and the h_{n-j-k} are each taken modulo n. Since $g(\overline{x})h(\overline{x}) = 0$, we have

$$
g_0 h_t + \cdots + g_k h_{t-k} + \cdots + g_t h_0 = 0
$$

for $k = 0, 1, \ldots, t$; $t = 0, 1, \ldots, n - 1$. Hence, if we take the subscripts in Eq. (26.5),

$$
g_i h_{j+n-1} + \cdots + g_{i+k} h_{j+n-k-1} + \cdots + g_{i+n-1} h_j,
$$

modulo n, we have

$$\sum_{k=0}^{m-1} g_{i+k}h_{j+n-k-1} = 0.$$

Thus, $GH = 0$. ∎

It seems advisable to give an example illustrating Lemma 26.2.14 before stating our next result.

Example 26.2.15 *Let $n = 7$ and $g(x) = 1+x+x^3$. Then $h(x) = 1+x+x^2+x^4$. Thus, $s = 3$ and $m = 4$. We have that*

$$G = \begin{bmatrix} 1 & 1 & 0 & 1 & 0 & 0 & 0 \\ 0 & 1 & 1 & 0 & 1 & 0 & 0 \\ 0 & 0 & 1 & 1 & 0 & 1 & 0 \\ 0 & 0 & 0 & 1 & 1 & 0 & 1 \end{bmatrix}$$

and

$$H = \begin{bmatrix} 0 & 0 & 1 \\ 0 & 1 & 0 \\ 1 & 0 & 0 \\ 0 & 1 & 1 \\ 1 & 1 & 1 \\ 1 & 1 & 0 \\ 1 & 0 & 0 \end{bmatrix}.$$

Now GH is a 4×3 matrix and the $(3,2)$ entry of GH is

$$g_5h_5 + g_6h_4 + g_7h_3 + g_8h_2 + g_9h_1 + g_{10}h_0 + g_{11}h_{-1}$$
$$= g_5h_5 + g_6h_4 + g_0h_3 + g_1h_2 + g_2h_1 + g_3h_0 + g_4h_6$$
$$(\textit{taking subscripts modulo } 7)$$
$$= g_0h_3 + g_1h_2 + g_2h_1 + g_3h_0$$

since $g_4 = g_5 = g_6 = 0 = h_5 = h_6$. Now $g_0h_3 + g_1h_2 + g_2h_1 + g_3h_0 = 0$ since $g(\bar{x})h(\bar{x}) = 0$.

In the following theorem, we show that the matrix H given by Eq. (26.4) is a parity check matrix of the cyclic code which is generated by the matrix G of Eq. (26.3).

Theorem 26.2.16 *Let $C = \langle g(\bar{x}) \rangle$ be a cyclic code in B_n, where $g(x)h(x) = x^n - 1$ and $\deg g(x) = s \geq 1$. Let G and H be the matrices given in Eqs. (26.3) and (26.4). Let $D = \{r(\bar{x}) \in B_n \mid \bar{r}H = 0\}$, where $r(\bar{x}) = \sum_{i=0}^{n-1} r_i\bar{x}^i$ and $\bar{r} = r_0r_1\cdots r_{n-1}$ for $r_i \in \mathbf{Z}_2$, $i = 0, 1, \ldots, n-1$. Then $D = C$.*

Proof. By Lemma 26.2.14, $D \subseteq C$. By Theorem 26.2.12, $\dim C = m$. Hence, it suffices to show that $\dim D = m$. Since $g(x)h(x) = x^n - 1$, $h_m = 1$, where $h(x) = \sum_{i=0}^m h_i x^i$. Thus, the s columns of H are linearly independent. Therefore, if we let D_0 denote the subspace of B_n spanned by these columns, then $\dim D_0 = s$. From linear algebra, we recall that the orthogonal complement $D_0^\perp = \{r(\overline{x}) \in B_n \mid \overline{r} \circ \overline{t} = 0, t(\overline{x}) \in D_0\}$ has dimension $m = n - s$. However, $D_0^\perp = D$ and so $\dim D = m$, the desired conclusion. ■

Example 26.2.17 *Let $n = 4$. It follows that $1 - x^4 = (1+x)(1+x+x^2+x^3)$. Let $g(\overline{x}) = 1+\overline{x}$. Then $s = 1$ and $m = 4 - 1 = 3$. Hence, $h(\overline{x}) = 1+\overline{x}+\overline{x}^2+\overline{x}^3$. Thus,*

$$G = \begin{bmatrix} 1 & 1 & 0 & 0 \\ 0 & 1 & 1 & 0 \\ 0 & 0 & 1 & 1 \end{bmatrix}$$

and

$$H = \begin{bmatrix} 1 \\ 1 \\ 1 \\ 1 \end{bmatrix}.$$

We recall that this is the $(4,3)$ code in Example 26.2.7.

We now examine error detection and correction for cyclic codes. We know from Theorem 26.1.16 that for a code to detect all sets of k or fewer errors, it is necessary and sufficient that the minimum distance between any two codewords be at least $k + 1$, and for a code to correct all sets of k or fewer errors, it is necessary that the minimum distance between codewords be at least $2k + 1$. Theorem 26.1.19 says that the minimum distance between any two distinct codewords is the least weight of a nonzero codeword.

Let F be a field containing $\mathbf{Z}_2$ and $a \in F$ be a primitive nth root of unity over $\mathbf{Z}_2$, that is, a has order n in the group $(F\backslash\{0\}, \cdot)$. Thus, $1, a, \ldots, a^{n-1}$ are distinct. Hence, n is odd else $n = 2k$ for some k and so $(a^k - 1)^2 = a^n - 1 = 0$. In this case, $a^k - 1 = 0$, which contradicts the fact that $1, a, \ldots, a^{n-1}$ are distinct. Since $a \in F$, $a^2, \ldots, a^{n-1} \in F$. Therefore,

$$x^n - 1 = (x-1)(x-a)(x-a^2)\cdots(x-a^{n-1}) \text{ over } F.$$

Let $g(x) \in \mathbf{Z}_2[x]$ be any polynomial which divides $x^n - 1$. Then the set of all roots of $g(x)$ is a subset of $\{1, a, \ldots, a^{n-1}\}$.

Theorem 26.2.18 *Let a be a primitive nth root of unity over $\mathbf{Z}_2$, $n \geq 1$. Let*

$$C = \langle g(\overline{x}) \rangle$$

be a cyclic code in B_n. Let u, v be integers such that $1 \leq u \leq v \leq n - 1$, $a^u, a^{u+1}, \ldots, a^v$ are roots of $g(x)$. Then $d \geq v - u + 2$, where d is the minimum distance of the code C.

Proof. Let $t = v - u + 1$ and $f(\overline{x}) = a_0 + a_1\overline{x} + \cdots + a_{n-1}\overline{x}^{n-1} \in C$. It suffices to show that $\text{wt}(\overline{a}) \geq t + 1$, where $\overline{a} = a_0 a_1 \cdots a_{n-1}$. There exists $q(x) \in \mathbf{Z}_2[x]$ such that $f(x) = q(x)g(x)$. Thus, $f(a^i) = q(a^i)g(a^i) = 0$ for $i = u, u+1, \ldots, v$. Consequently,

$$[a_0 a_1 \cdots a_{n-1}] \begin{bmatrix} 1 & 1 & \cdots & 1 \\ a^u & a^{u+1} & \cdots & a^{u+t-1} \\ a^{2u} & a^{2(u+1)} & \cdots & a^{2(u+t-1)} \\ \vdots & \vdots & \vdots & \vdots \\ a^{(n-1)u} & a^{(n-1)(u+1)} & \cdots & a^{(n-1)(u+t-1)} \end{bmatrix} = [00\cdots 0].$$

Suppose that $\text{wt}(\overline{a}) = s < t + 1$. Then exactly s of the $a_i = 1$, say, $a_{i_1} = \cdots = a_{i_s} = 1$, where $i_1 < \cdots < i_s$. Now

$$[a_{i_1} \; a_{i_2} \; \cdots \; a_{i_s}] \begin{bmatrix} a^{i_1 u} & a^{i_1(u+1)} & \cdots & a^{i_1(u+s-1)} \\ a^{i_2 u} & a^{i_2(u+1)} & \cdots & a^{i_2(u+s-1)} \\ \vdots & \vdots & \vdots & \vdots \\ a^{i_s u} & a^{i_s(u+1)} & \cdots & a^{i_s(u+s-1)} \end{bmatrix} = [0 \; 0 \cdots 0].$$

Thus,

$$0 = a^{i_1 u + i_2 u + \cdots + i_s u} \det \begin{bmatrix} 1 & a^{i_1} & \cdots & (a^{i_1})^{s-1} \\ 1 & a^{i_2} & \cdots & (a^{i_2})^{s-1} \\ \vdots & \vdots & \vdots & \vdots \\ 1 & a^{i_s} & \cdots & (a^{i_s})^{s-1} \end{bmatrix}.$$

However, this is impossible since the $a^{i_1}, a^{i_2}, \ldots, a^{i_s}$ are distinct and the determinant is a Vandermonde determinate. ∎

We now consider only binary cyclic codes of odd length n. Then $x^n - 1$ has distinct factors. The factors are generators of binary cyclic codes. However, it can be difficult to find the factors. Fortunately, an ideal can have more that one generator. We will be interested in certain kinds of generators since there is a method of determining them.

A generator $e(\overline{x})$ of an ideal in B_n is called an **idempotent generator** if it is an idempotent, that is, if $e^2(\overline{x}) = e(\overline{x})$. Note that if $a(\overline{x})$ is in $\langle e(\overline{x}) \rangle$, then $a(\overline{x}) = b(\overline{x})e(\overline{x})$ for some $b(\overline{x})$ and so $a(\overline{x})e(\overline{x}) = b(\overline{x})e^2(\overline{x}) = b(\overline{x})e(\overline{x}) = a(\overline{x})$. That is, $e(\overline{x})$ is an identity of $\langle e(\overline{x}) \rangle$. Conversely, an idempotent that is an identity for an ideal I, generates I. This follows from the following argument.

Suppose $e(\overline{x})$ is an identity for an ideal I in B_n. Then for all $a(\overline{x}) \in I$, $a(\overline{x}) = a(\overline{x})e(\overline{x})$. Thus, $I \subseteq \langle e(\overline{x}) \rangle$. Since $e(\overline{x}) \in I$, $\langle e(\overline{x}) \rangle \subseteq I$. Hence, $I = \langle e(\overline{x}) \rangle$.

In order to determine how we find idempotent generators, we introduce the notion of a cyclotomic coset.

Consider any integer s such that $0 \le s < p^m - 1$, where p is a prime and let r be the smallest nonnegative integer such that $p^{r+1}s \equiv s(\mathrm{mod}(p^m - 1))$. The cyclotomic coset containing s consists of $\{s, ps, p^2s, \ldots, p^rs\}$, where each p^is is reduced $\mathrm{mod}(p^m - 1)$. If $\gcd(s, p^m - 1) = 1$, then $r = m - 1$, but if $\gcd(s, p^m - 1) \ne 1$, then r varies with s. Note $(p^m - 1)|(p^{r+1}s - s)$ and so $(p^m - 1)|(p^{r+1} - 1)s$. Even though cyclotomic cosets are not cosets of a group, they partition the integers mod $(p^m - 1)$, that is, each integer mod $(p^m - 1)$ is in exactly one coset. If u is the smallest element in its cyclotomic coset, we denote the coset by C_u.

Let $m = 4$ and $p = 2$. Let $s = 0$. Then $(2^4 - 1)|(2^{0+1}0 - 0)$ and so $r = 0$. Hence, $C_0 = \{0\}$. Let $s = 1$. Then $(2^4 - 1)|(2^{r+1}1 - 1)$ and so $r = 3$. Thus, $C_1 = \{1, 2, 4, 8\}$. Let $s = 3$. Then $(2^4 - 1)|(2^{r+1}3 - 3)$, i.e., $(2^4 - 1)|(2^{r+1} - 1)3$. Now $(2^4 - 1) \nmid (2^{0+1}1 - 1)3$, $(2^4 - 1) \nmid (2^{1+1}1 - 1)3$, $(2^4 - 1)$ does not divide $(2^{2+1}1 - 1)3$, but $(2^4 - 1)|(2^{3+1}1 - 1)3$. Therefore, $r = 3$. Consequently, $C_3 = \{3, 6, 12, 24\} = \{3, 6, 12, 9\}$. Let $s = 5$. Then $(2^4 - 1)|(2^{r+1}5 - 5)$. Now $(2^4 - 1)$ does not divide $(2^{0+1}1 - 1)5$, but $(2^4 - 1)|(2^{1+1}1 - 1)5$. Thus, $r = 1$. Hence, $C_5 = \{5, 10\}$. Note that $C_0 = \{0\}$, $C_1 = \{1, 2, 4, 8\}$, $C_3 = \{3, 6, 12, 9\}$, $C_5 = \{5, 10\}$, and $C_7 = \{7, 14, 13, 11\}$ partition $\{0, 1, 2, \ldots, 14\}$.

We now illustrate how we determine idempotent generators. Let $n = 7$. Suppose we have $e(\bar{x}) = a_0 + a_1\bar{x} + \cdots + a_6\bar{x}^6$. Now the coefficient of $e^2(\bar{x})$ for $\bar{x}^i$ is

$$\sum_{j=0}^{i} a_j a_{i-j},$$

where $i = 0, 1, \ldots, 12$ and we have $a_i = 0$ for $i = 7, \ldots, 12$. Hence,

$$\sum_{j=0}^{i} a_j a_{i-j} = \begin{cases} 0 & \text{if } i \text{ is odd} \\ a_{i/2}^2 & \text{if } i \text{ is even.} \end{cases}$$

For example, $a_0a_0 = a_0^2$, $a_0a_1 + a_1a_0 = 0$, $a_0a_2 + a_1a_1 + a_2a_0 = a_1^2$, and $a_0a_3 + a_1a_2 + a_2a_1 + a_3a_0 = 0$. Now 0 is the coefficient of $\bar{x}^7$, $\bar{x}^9$, $\bar{x}^{11}$ and $\bar{x}^8 = \bar{x}$, $\bar{x}^{10} = \bar{x}^3$, and $\bar{x}^{12} = \bar{x}^5$. Thus,

$$e^2(\bar{x}) = a_0^2 + a_4^2\bar{x} + a_1^2\bar{x}^2 + a_5^2\bar{x}^3 + a_2^2\bar{x}^4 + a_6^2\bar{x}^5 + a_3^2\bar{x}^6 = e(\bar{x}).$$

Hence, $a_0 = a_0$, $a_1 = a_4$, $a_2 = a_1$, $a_3 = a_5$, $a_4 = a_2$, $a_5 = a_6$, and $a_6 = a_3$ or $a_0 = a_0$, $a_1 = a_2 = a_4$, $a_3 = a_5 = a_6$. Note that modulo 7, we have $2 \cdot 0 = 0$, $2 \cdot 1 = 2$, $2 \cdot 2 = 4$, $2 \cdot 4 = 1$, $2 \cdot 3 = 6$, $2 \cdot 6 = 5$. Thus, we see that this can only happen if S is a union of cyclotomic cosets for $n = 7$, where S is the set of powers of $\bar{x}$ that occurs with nonzero coefficients in $e(\bar{x})$. The proof of the following result follows by a similar argument.

Lemma 26.2.19 *Let $f(\bar{x}) \in B_n$. Then $f(\bar{x})$ is an idempotent in B_n if and only if the set S of powers of $\bar{x}$ that occur with nonzero coefficients in $f(\bar{x})$ is a union of cyclotomic cosets.*

We now determine the cyclotomic cosets for $n = 7 = 2^3 - 1$, i.e., $m = 3$ and $p = 2$. Let $s = 0$, then $7|(2^{0+0}0 - 0)$ and so $r = 0$. Thus, $C_0 = \{0\}$. Let $s = 1$. Then $7|(2^{2+1}1 - 1)$ and so $r = 2$. Hence, $C_1 = \{1, 2, 4\}$. Let $s = 3$. Then $7|(2^{2+1}3 - 3)$ and so $r = 2$. Thus, $C_3 = \{3, 6, 12\} = \{3, 6, 5\}$. The cyclotomic cosets yield the eight idempotents. We list them below.

Idempotent generator

$$e_1(x) = 1 + x + x^2 + x^3 + x^4 + x^5 + x^6; \qquad C_0 \cup C_1 \cup C_3$$
$$e_2(x) = 1 + x^3 + x^5 + x^6; \qquad C_0 \cup C_3$$
$$e_3(x) = 1 + x + x^2 + x^4; \qquad C_0 \cup C_1$$
$$e_2(x) + e_3(x) = x + x^2 + x^3 + x^4 + x^5 + x^6; \quad C_1 \cup C_3$$
$$e_1(x) + e_2(x) = x + x^2 + x^4; \qquad C_1$$
$$e_1(x) + e_3(x) = x^3 + x^5 + x^6; \qquad C_3.$$

The two remaining idempotents are 0 and 1. Now 1 generates the whole space, while 0 generates the zero space. We see every code of length 7 has an idempotent generator.

Theorem 26.2.20 *Every cyclic code has an idempotent generator.*

Proof. Let $g(\bar{x})$ be the generator polynomial of the cyclic code C. Then $x^n - 1 = g(x)h(x)$ for some $h(x)$. Since $x^n - 1$ has distinct factors, $g(x)$ and $h(x)$ are relatively prime. Thus, there exist $s(x)$, $t(x) \in \mathbf{Z}_2[x]$ such that

$$1 = s(x)g(x) + t(x)h(x). \tag{26.6}$$

Let $e(x) = s(x)g(x)$. Then $e(\bar{x}) \in C$. We multiply Eq. (26.6) by $s(x)g(x)$ to obtain $s(x)g(x) = s^2(x)g^2(x) + s(x)g(x)t(x)h(x) = s^2(x)g^2(x) + s(x)t(x)(x^n - 1)$. Hence, $e(\bar{x}) = e^2(\bar{x}) + 0$ in B_n. Thus, $e(\bar{x})$ is idempotent. Let $c(\bar{x}) \in C$. Then $c(\bar{x}) = r(\bar{x})g(\bar{x})$ for some $r(\bar{x})$ in B_n. Now multiply Eq. (26.6) by $c(x)$. We obtain $c(x) = s(x)g(x)c(x) + t(x)c(x)h(x)$. Thus, $c(\bar{x}) = s(\bar{x})g(\bar{x})c(\bar{x}) = e(\bar{x})c(\bar{x})$. Therefore, $e(\bar{x})$ is an identity for C. Consequently, $e(\bar{x})$ generates C. ∎

26.2.1 Exercises

1. Determine the number of cyclic codes of the following lengths:

   (i) length 6,

   (ii) length 7,

   (iii) length 10.

2. Let $m = 3$, $n = 7$ and the generator polynomial be $g(x) = 1 + x + x^2 + x^3 + x^4$. Determine the corresponding $(3, 7)$-code and give the encoding matrix G.

3. In Exercise 2, find the parity-check matrix K.

4. Show that the codes in Examples 26.2.8 and 26.2.9 are Hamming codes.

5. Let $f(x)$ and $h(x) \in \mathbf{Z}_2[x]$ be such that $h(x)$ divides $x^n - 1$ in $\mathbf{Z}_2[x]$. Prove that $\langle f(\overline{x}) \rangle \subseteq \langle h(\overline{x}) \rangle$ if and only if $h(x)$ divides $f(x)$ in $\mathbf{Z}_2[x]$.

6. Let C be an ideal of B_n. Prove the following assertions.

   (i) There exists an ideal I of $\mathbf{Z}_2[x]$ such that $C = \eta(I)$ and $I \supseteq \mathrm{Ker}\ \eta$, where η is the natural homomorphism of $\mathbf{Z}_2[x]$ onto B_n.

   (ii) If $C = \langle g(\overline{x}) \rangle$, then $I = \langle g(x) \rangle$, where I is the ideal in (i).

   (iii) If $g(x) \in \mathbf{Z}_2[x]$ is such that $I = \langle g(x) \rangle$, then $C = \langle g(\overline{x}) \rangle$ and $g(x)|(x^n - 1)$.

7. Let C be a cyclic code in B_n. Let $g(x) \in \mathbf{Z}_2[x]$ be the smallest degree polynomial such that $g(\overline{x}) \in C$. Prove that $\langle g(\overline{x}) \rangle = C$ and that $g(x)$ is unique.

26.3 Bose-Chauduri-Hocquenghem Codes

In this section, we take a very brief look at Bose-Chauduri-Hocquenghen codes (BCH codes). For codewords of length several thousand, these codes preform very well. BCH codes are multiple-error-correcting codes. The number of check digits is a function of the number of errors to be detected or corrected. In the following, we give a systematic way to construct binary BCH codes of any length.

Since we only consider binary BCH codes here, our symbols are once again from $\mathbf{Z}_2$. Two words are said to have **distance** d if they differ in d places. We wish to construct a code with minimum distance d, i.e., the distance between two codewords is at least d.

Definition 26.3.1 *Let a be a primitive nth root of unity over $\mathbf{Z}_2$. Let $m_i(x)$ be the minimum polynomial of a^i over $\mathbf{Z}_2$, $i = 1, \ldots, n - 1$. Let d and u be integers, where $d \geq 2$ and $u \geq 0$. If*

$$g(x) = \mathrm{lcm}(m_u(x), m_{u+1}(x) \ldots, m_{u+d-2}(x)),$$

*then the cyclic code $\langle g(\overline{x}) \rangle$ in B_n is called a **binary BCH code of length** n and **distance** d.*

Since the polynomials $m_u(x), m_{u+1}(x) \ldots, m_{u+d-2}(x)$ are irreducible over $\mathbf{Z}_2$, $g(x)$ is the product of the distinct $m_i(x)$, $i = u, u + 1, \ldots, u + d - 2$.

Theorem 26.3.2 *Let $C = \langle g(\overline{x}) \rangle$ be the binary BCH code of length n and distance d. Then the following assertions hold.*

(i) The minimum distance of C is at least d.

(ii) $f(\overline{x}) \in C$ if and only if $f(a^i) = 0$ for $i = u, u + 1, \ldots, u + d - 2$.

(iii) A parity-check matrix for C is given by

$$H = \begin{bmatrix} 1 & 1 & \cdots & 1 \\ a^u & a^{u+1} & \cdots & a^{u+d-2} \\ a^{2u} & a^{2(u+1)} & \cdots & a^{2(u+d-2)} \\ \vdots & \vdots & \vdots & \vdots \\ a^{(n-1)u} & a^{(n-1)(u+1)} & \cdots & a^{(n-1)(u+d-2)} \end{bmatrix}.$$

Proof. (i) Since $m_i(x)$ divides $g(x)$, a^i is a root of $g(x)$, for $i = u, u + 1, \ldots, u + d - 2$. Thus, the desired result follows from Theorem 26.2.18.

(ii) Let $f(\overline{x}) \in C$. Since $g(x)$ divides $f(x)$ and each a^i is a root of $g(x)$ by (i), each a^i is a root of $f(x)$. Conversely, suppose that a^i is a root of $f(x)$ for $i = u, u + 1, \ldots, u + d - 2$. Then $m_i(x)$ divides $f(x)$ since $m_i(x)$ is the minimal polynomial of a^i, $i = u, u + 1, \ldots, u + d - 2$. Therefore, $g(x)$ divides $f(x)$ since $g(x)$ is the product of the distinct $m_i(x)$, which are relatively prime. Thus, $f(\overline{x}) \in C$.

(iii) Let $f(\overline{x}) = a_0 + a_1\overline{x} + \cdots + a_{n-1}\overline{x}^{n-1} \in B_n$. Then

$$\overline{a}H = f(a^u)f(a^{u+1}) \cdots f(a^{u+d-2}),$$

where $\overline{a} = a_0 a_1 \cdots a_{n-1}$. Thus, $\overline{a}H = 0$ if and only if $f(a^u) = f(a^{u+1}) = \cdots = f(a^{u+d-2}) = 0$ if and only if $f(\overline{x}) \in C$ by (ii). ∎

Example 26.3.3 *We construct a binary BCH code with codeword length $n = 15$, which has minimum distance $d = 5$. Let a be a root of the primitive polynomial $1 + x + x^4$. Consider the successive powers of a*

$$
\begin{aligned}
a^2 & & & & a^9 &= a^3 + a \\
a^3 & & & & a^{10} &= a^2 + a + 1 \\
a^4 &= a + 1 & & & a^{11} &= a^3 + a^2 + a \\
a^5 &= a^2 + a & & & a^{12} &= a^3 + a^2 + a + 1 \\
a^6 &= a^3 + a^2 & & & a^{13} &= a^3 + a^2 + 1 \\
a^7 &= a^3 + a + 1 & & & a^{14} &= a^3 + 1 \\
a^8 &= a^2 + 1 & & & a^{15} &= 1
\end{aligned}
$$

Let $m_i(x)$ denote the minimum polynomial of a^i, $i = 1, 2, 3, 4$. It is easily verified that a, a^2, a^4, a^8 are the roots of $1 + x + x^4$ and that a^3, a^6, a^9, a^{12} are the roots of $1 + x + x^2 + x^3 + x^4$. Thus,

$$m_1(x) = m_2(x) = m_4(x) = 1 + x + x^4$$

and
$$m_3(x) = 1 + x + x^2 + x^3 + x^4.$$

Hence,
$$lcm(m_1(x), m_2(x), m_3(x), m_4(x)) = lcm(m_1(x), m_3(x)).$$

Since $m_1(x)$ and $m_3(x)$ have no common roots, $lcm(m_1(x), m_3(x))$ is of degree at least 8. Hence,

$$lcm(m_1(x), m_3(x)) = m_1(x)m_3(x) = 1 + x^4 + x^6 + x^7 + x^8.$$

Since $\deg g(x) = 8$, this BCH code has $15 - 8 = 7$ information digits. This code detects all sets of $5 - 1 = 4$ or fewer errors and corrects all sets of $(5-1)/2 = 2$ or fewer errors.

26.3.1 Exercises

1. Show that BCH codes are cyclic.

2. Show that the polynomial $1 + x + x^4$ is irreducible over $\mathbf{Z}_2$.

3. Let F be a finite field such that $F \supseteq \mathbf{Z}_2$. Let $m_i(x)$ be the minimal polynomial of c^i over $\mathbf{Z}_2$, where $F^* = \langle c \rangle$. Prove that $m_i(x)$ divides $x^{2^n-1} - 1$, where $2^n = |F|$.

Chapter 27

Gröbner Bases

27.1   Affine Varieties

This chapter is concerned with the geometry dealing with **affine varieties**. An affine variety is defined by polynomial equations. These polynomial equations may define, for example, curves and surfaces. Throughout this chapter, we let $K[x_1, \ldots, x_n]$ denote the polynomial ring in the algebraically independent indeterminates $x_1, \ldots, x_n$ over the field K.

Definition 27.1.1 *A product of the form* $x_1^{\alpha_1} \cdots x_n^{\alpha_n}$, *where* $\alpha_1, \ldots, \alpha_n$ *are nonnegative integers, is called a* **monomial** *in* $x_1, \ldots, x_n$. *The sum* $\alpha_1 + \cdots + \alpha_n$ *is called the* **total degree** *of the monomial.*

Let $x_1^{\alpha_1} \cdots x_n^{\alpha_n}$ be a monomial. Then we simply write x^α for $x_1^{\alpha_1} \cdots x_n^{\alpha_n}$, where we let $\alpha = (\alpha_1, \ldots, \alpha_n)$. If $\alpha = 0$, then $x^\alpha = 1$. We sometimes write $|\alpha|$ for $\alpha_1 + \cdots + \alpha_n$.

Definition 27.1.2 *Let* $f = \sum_\alpha a_\alpha x^\alpha \in K[x_1, \ldots, x_n]$. *Then* a_α *is called the* **coefficient** *of the monomial* x^α *and* $a_\alpha x^\alpha$ *is called a* **term** *of* f *if* $a_\alpha \neq 0$. *The* **total degree** *of* f, *denoted by* $\deg(f)$, *is the largest* $|\alpha|$ *for which* $a_\alpha \neq 0$.

Definition 27.1.3 *Let* n *be a positive integer. The set*

$$K^n = \{(a_1, \ldots, a_n) \mid a_i \in K, \ i = 1, \ldots, n\}$$

is called the **affine space** *over* K.

For $f \in K[x_1, \ldots, x_n]$, we can interpret f as a function from K^n into K as follows: For all $(a_1, \ldots, a_n) \in K^n$,

$$f((a_1, \ldots, a_n)) = \sum_\alpha a_\alpha a_1^{\alpha_1} \cdots a_n^{\alpha_n},$$

where $f = \sum_\alpha a_\alpha x_1^{\alpha_1} \cdots x_n^{\alpha_n}$.

Theorem 27.1.4 *Suppose that K is infinite. Let $f \in K[x_1, \ldots, x_n]$. Then $f = 0$ in $K[x_1, \ldots, x_n]$ if and only if $f : K^n \to K$ is the zero function.*

Proof. If $f = 0$, then clearly $f : K^n \to K$ is the zero function. Conversely, suppose that f is the zero function. The proof is by induction on n, the number of indeterminates. Suppose that $n = 1$. Then by Theorem 14.1.11, f has at most m roots, where m is the degree of f. Since $f(a) = 0$ for all $a \in K$ and since K is infinite, it follows that $f = 0$. Now assume that the converse is true for $n - 1$. Let $f \in K[x_1, \ldots, x_n]$ be such that $f(a_1, \ldots, a_n) = 0$ for all $(a_1, \ldots, a_n) \in K^n$. Now we can express f in the form

$$f = \sum_{i=0}^{q} g_i(x_1, \ldots, x_{n-1}) x_n^i,$$

where $g_i \in K[x_1, \ldots, x_{n-1}]$, $i = 0, 1, \ldots, q$. Consider any arbitrary fixed $(a_1, \ldots, a_{n-1}) \in K^{n-1}$. Then $f(a_1, \ldots, a_{n-1}, x_n)$ is a polynomial in one indeterminate. By the $n = 1$ case, $f(a_1, , \ldots, a_{n-1}, x_n)$ is the zero polynomial in $K[x_n]$ since $f(a_1, \ldots, a_n) = 0$ for all $a_n \in K$. Hence, $g_i(a_1, \ldots, a_{n-1}) = 0$ for $i = 0, 1, \ldots, q$. Since $(a_1, \ldots, a_{n-1})$ is arbitrary, it follows by the induction hypothesis that each g_i is the zero polynomial in $K[x_1, \ldots, x_{n-1}]$. Thus, f is the zero polynomial in $K[x_1, \ldots, x_n]$. ∎

Corollary 27.1.5 *Suppose that K is infinite. Let $f, g \in K[x_1, \ldots, x_n]$. Then $f = g$ if and only if $f : K^n \to K$ and $g : K^n \to K$ are the same function.*

Proof. The proof follows from Theorem 27.1.4 by considering $f - g$. ∎

Definition 27.1.6 *Let $f_1, \ldots, f_m \in K[x_1, \ldots, x_n]$. The set*

$$V(f_1, \ldots, f_m) = \{(a_1, \ldots, a_n) \in K^n \mid f_i(a_1, \ldots, a_n) = 0 \text{ for all } i = 1, \ldots, m\}$$

*is called the **affine variety** defined by $f_1, \ldots, f_m$.*

We sometimes use the notation $V(\{f_i \mid i = 1, 2, \ldots, m\})$ for $V(f_1, \ldots, f_m)$. Consider, for example, the following linear system of equations

$$
\begin{aligned}
x + 2y + z &= 2 \\
x + y - z &= 1.
\end{aligned}
$$

We replace the second equation by the second equation minus the first equation to obtain

$$
\begin{aligned}
x + 2y + z &= 2 \\
-y - 2z &= -1.
\end{aligned}
$$

We then replace the first equation by the first equation plus two times the second equation. We then have

$$\begin{aligned} x - 3z &= 0 \\ -y - 2z &= -1. \end{aligned}$$

Thus,

$$V(x + 2y + z - 2, x + y - z - 1) = \{(3t, 1 - 2t, t) \mid t \in K\}.$$

The method used to solve the above system of equations was that of elimination of variables. The equations $x + 2y + z - 2 = 0$ and $x + y - z - 1 = 0$ form what is called an **implicit representation** of V.

For an application of polynomial equations, we turn to robotics. We consider the motion of a robot's arm in the plane. We assume that we have three linked rods of lengths 6, 4, 2, respectively.

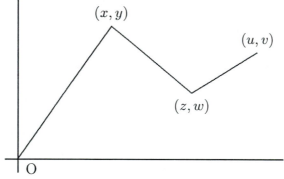

The positions or states of the arm are determined by the solution in $\mathbf{R}^6$ to the following polynomial equations.

$$\begin{aligned} x^2 + y^2 &= 36 \\ (z - x)^2 + (w - y)^2 &= 16 \\ (u - z)^2 + (v - w)^2 &= 4. \end{aligned}$$

Another application to polynomial equations is in automatic geometric theorem proving. We introduce Cartesian coordinates in the Euclidean plane. Having done this, many geometric theorems can be expressed as polynomial equations. We show, for example, how polynomial equations can be used to determine results concerning the diagonals of a square. Let A, B, C, D be vertices of a square.

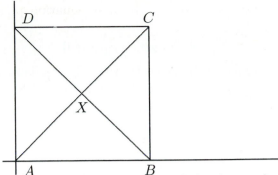

Let X denote the point of intersection of the diagonals AC and BD. We place side AB on the x-axis with vertex A at the origin. Then the Cartesian coordinates of A are $(0,0)$ and those of B are $(a,0)$, where a is arbitrary. The Cartesian coordinates of C and D are determined by B. We write (x_1, y_1) and (x_2, y_2) for the Cartesian coordinates of C and D, respectively. We use the slope formula for a line segment to translate the defining properties of a square into polynomial equations.

$$AB \perp AD : x_2 = 0$$
$$AB \| CD : 0 = (y_2 - y_1)/(x_2 - x_1)$$
$$|AB| = |DC| : a^2 = (x_2 - x_1)^2 + (y_2 - y_1)^2$$
$$|AB| = |AD| : y_2 = \pm a.$$

Thus, we obtain the polynomial equations,

$$\begin{aligned} y_1 - y_2 &= 0 \\ x_1^2 + (y_2 - y_1)^2 - a^2 &= 0. \end{aligned} \tag{27.1}$$

(Hence, $x_1 = \pm a$ and $y_1 = \pm a$.)

We also know that A, X, C are collinear, as are D, X, B. Thus, if we let (x_3, y_3) denote the Cartesian coordinates of X, we have the following equations,

A, X, C are collinear : $y_3/x_3 = (y_3 - y_1)/(x_3 - x_1)$
D, X, B are collinear : $(y_3 - y_2)/(x_3 - x_2) = (y_3 - 0)/(x_3 - a)$.

We hence obtain the polynomial equations,

$$\begin{aligned} x_1 y_3 - x_3 y_1 &= 0 \\ x_2 y_3 - x_3 y_2 - a y_3 + a y_2 &= 0. \end{aligned} \tag{27.2}$$

(Thus, $x_3 = a/2 = y_3$.)

Consider the property that the diagonals of a square intersect in right angles.

$$\overline{AX}^2 + \overline{XD}^2 = \overline{AD}^2 : (0 - x_3)^2 + (0 - y_3)^2 + (x_3 - x_2)^2 + (y_3 - y_2)^2$$
$$= (0 - x_2)^2 + (0 - y_2)^2$$

or

$$x_3^2 + y_3^2 - x_2 x_3 - y_2 y_3 = 0. \tag{27.3}$$

Therefore, the statement which says that the diagonals of a square intersect in right angles translates into the statement that Eqs. (27.1) and (27.2) imply the Eq. (27.3). There are, of course, many other conclusions we could derive.

Lemma 27.1.7 *Let $V, W \subseteq K^n$ be affine varieties. Then $V \cup W$ and $V \cap W$ are affine varieties.*

Proof.   Let $V = V(f_1, \ldots, f_m)$ and $W = V(g_1, \ldots, g_q)$ for some $f_1, \ldots, f_m$, $g_1, \ldots, g_q \in K[x_1, \ldots, x_n]$. Now $(a_1, \ldots, a_n) \in V \cap W$ if and only if $f_i(a_1, \ldots, a_n)$ $= 0$ and $g_j(a_1, \ldots, a_n) = 0$ for all $i = 1, \ldots, m$ and $j = 1, \ldots, q$ if and only if

$$(a_1, \ldots, a_n) \in V(f_1, \ldots, f_m, g_1, \ldots, g_q).$$

Thus, $V \cap W = V(f_1, \ldots, f_m, g_1, \ldots, g_q)$.

Let $(a_1, \ldots, a_n) \in V$. Then $f_i(a_1, \ldots, a_n) g_j(a_1, \ldots, a_n) = 0$ for all $i = 1, \ldots, m$ and $j = 1, \ldots, q$. Hence, $V \subseteq V(\{f_i g_j \mid i = 1, \ldots, m \; ; \; j = 1, \ldots, q\})$ and similarly $W \subseteq V(\{f_i g_j \mid 1, \ldots, m \; ; \; j = 1, \ldots, q\})$. Thus, $V \cup W \subseteq V(\{f_i g_j \mid i = 1, \ldots, m \; ; \; j = 1, \ldots, q\})$. Let $(a_1, \ldots, a_n) \in V(\{f_i g_j \mid i = 1, \ldots, m \; ; \; j = 1, \ldots, q\})$. Suppose there exists i such that $f_i(a_1, \ldots, a_n) \neq 0$. Since $f_i(a_1, \ldots, a_n) g_j(a_1, \ldots, a_n) = 0$, we have that $g_j(a_1, \ldots, a_n) = 0$ for all $j = 1, \ldots, q$. Therefore, $(a_1, \ldots, a_n) \in W$. Suppose $f_i(a_1, \ldots, a_n) = 0$ for all $i = 1, \ldots, m$. Then $(a_1, \ldots, a_n) \in V$. Thus, $V(\{f_i g_j \mid i = 1, \ldots, m \; ; \; j = 1, \ldots, q\}) \subseteq V \cup W$. Consequently, $V(\{f_i g_j \mid i = 1, \ldots, m \; ; \; j = 1, \ldots, q\}) = V \cup W$. ∎

We now wish to consider a way of describing the points of an affine variety. We can accomplish this at times by parametrizing the variety. Parametric representations of curves and surfaces are used to draw them on a computer. The implicit representation of a variety is useful in determining whether or not a point lies on the curve or surface.

Consider again the linear system of equations

$$\begin{aligned} x + 2y + z &= 2 \\ x + y - z &= 1. \end{aligned} \tag{27.4}$$

Then

$$\begin{aligned} x &= 3t \\ y &= 1 - 2t \\ z &= t. \end{aligned} \tag{27.5}$$

t is called a **parameter** and Eqs. (27.5) is called a **parametrization** of Eqs. (27.4).

For another example, consider the equation $x^2 + y^2 = 1$. Then

$$x = \cos t$$
$$y = \sin t$$

is another example of a parametrization. Another known parametrization of $x^2 + y^2 = 1$ (except for the point $(-1, 0)$) is

$$x = (1 - t^2)/(1 + t^2)$$
$$y = 2t/(1 + t^2).$$

Note that $-1 = (1 - t^2)/(1 + t^2)$ is impossible, else $-1 = 1$. Next, we show how to obtain this parametrization.

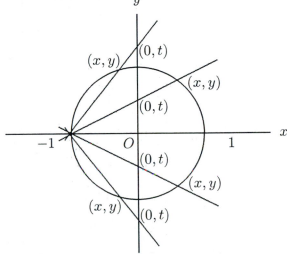

Each nonvertical line through $(-1, 0)$ will intersect the circle in a unique point other than $(-1, 0)$. As t varies from $-\infty$ to ∞, the corresponding point (x, y) traverses all of the circle except for the point $(-1, 0)$. The slope of each nonvertical line is given by

$$(t - 0)/(0 - (-1)) = (y - t)/(x - 0).$$

Thus,

$$t = y/(x + 1)$$
$$y = t(x + 1). \tag{27.6}$$

Substituting $t(x + 1)$ in for y in $x^2 + y^2 = 1$ yields

$$x^2 + t^2(x + 1)^2 = 1$$

or

$$(1 + t^2)x^2 + 2t^2 x + t^2 - 1 = 0.$$

Solving this latter equation for x in terms of t, we get the x-coordinate of where the line intersects the circle $x^2 + y^2 = 1$. One solution is $x = -1$. Hence, $x + 1$ divides $(1 + t^2)x^2 + 2t^2x + t^2 - 1$. This division yields

$$(1 + t^2)x^2 + 2t^2x + t^2 - 1 = (x + 1)((1 + t^2)x - (1 - t^2)) = 0.$$

Setting

$$(1 + t^2)x - (1 - t^2) = 0,$$

we obtain

$$x = (1 - t^2)/(1 + t^2).$$

Substituting this into Eq. (27.6) yields

$$y = 2t/(1 + t^2).$$

Definition 27.1.8 *Let $V \subseteq K^n$ be an affine variety. Let*

$$I(V) = \{f \in K[x_1, \ldots, x_n] \mid f(a_1, \ldots, a_n) = 0 \text{ for all } (a_1, \ldots, a_n) \in V\}.$$

Lemma 27.1.9 *If $V \subseteq K^n$ is an affine variety, then $I(V)$ is an ideal of $K[x_1, \ldots, x_n]$.*

Proof.   Clearly the zero polynomial is a member of $I(V)$ since $0(a_1, \ldots, a_n) = 0$ for all $(a_1, \ldots, a_n) \in K^n$. Let $f, g \in I(V)$. Then

$$(f + g)(a_1, \ldots, a_n) = f(a_1, \ldots, a_n) + g(a_1, \ldots, a_n) = 0 + 0 = 0.$$

Thus, $f + g \in I(V)$. Let $h \in K[x_1, \ldots, x_n]$. Then

$$(hf)(a_1, \ldots, a_n) = h(a_1, \ldots, a_n)f(a_1, \ldots, a_n) = h(a_1, \ldots, a_n)0 = 0.$$

Therefore, $hf \in I(V)$. Hence, $I(V)$ is an ideal of $K[x_1, \ldots, x_n]$. ∎

The ideal $I(V)$ in Lemma 27.1.9 is called the **ideal** of V.

Example 27.1.10 *Let $V = \{(0, 0)\} \subseteq K^2$. In this example, we show that*

$$I(V) = \langle x, y \rangle.$$

Let $f(x, y)x + g(x, y)y \in \langle x, y \rangle$. Then

$$f(0, 0)0 + g(0, 0)0 = 0.$$

Thus, $f(x, y)x + g(x, y)y \in I(V)$. Hence, $\langle x, y \rangle \subseteq I(V)$. Let $f(x, y) \in I(V)$. Then $f(0, 0) = 0$. Now

$$f(x, y) = \sum_{j=0}^{n} \sum_{i=0}^{m} a_{ij}x^i y^j$$

for some $a_{ij} \in K$. Therefore,

$$0 = a_{00} + a_{10}0 + a_{01}0 + \sum_{j=1}^{n} \sum_{i=1}^{m} a_{ij}0^i 0^j.$$

This implies that $a_{00} = 0$ and so

$$f(x,y) = a_{10}x + a_{01}y + \sum_{j=1}^{n} \sum_{i=1}^{m} a_{ij}x^i y^j \in \langle x, y \rangle.$$

Hence, $I(V) \subseteq \langle x, y \rangle$. Consequently, $I(V) = \langle x, y \rangle$.

Example 27.1.11 *Let $V = K^n$. Then $f \in I(K^n)$ if and only if $f(a_1, \ldots, a_n) = 0$ for all $(a_1, \ldots, a_n) \in K^n$. Hence, if K is infinite, then f is the 0 polynomial. Thus, $I(K^n) = \{0\}$ if K is infinite.*

Lemma 27.1.12 *Let $f_1, \ldots, f_s \in K[x_1, \ldots, x_n]$. Then*

$$\langle f_1, \ldots, f_s \rangle \subseteq I(V(f_1, \ldots, f_s)).$$

Proof. Let $f \in \langle f_1, \ldots, f_s \rangle$. Then $f = h_1 f_1 + \cdots + h_s f_s$ for some $h_1, \ldots, h_s \in K[x_1, \ldots, x_n]$. Thus, for all $(a_1, \ldots, a_n) \in V(f_1, \ldots, f_s)$,

$$\begin{aligned}
f(a_1, \ldots, a_n) &= h_1(a_1, \ldots, a_n)f_1(a_1, \ldots, a_n) + \cdots + \\
&\quad\, h_s(a_1, \ldots, a_n)f_s(a_1, \ldots, a_n) \\
&= 0 + \cdots + 0 \\
&= 0
\end{aligned}$$

and so $f \in I(V(f_1, \ldots, f_s))$. Hence, $\langle f_1, \ldots, f_s \rangle \subseteq I(V(f_1, \ldots, f_s))$. ∎

Proposition 27.1.13 *Let V and W be affine varieties in K^n. Then*
  (i) $V \subseteq W$ if and only if $I(V) \supseteq I(W)$,
  (ii) $V = W$ if and only if $I(V) = I(W)$.

Proof. (i) Suppose that $V \subseteq W$. Let $f \in I(W)$. Then for all $(a_1, \ldots, a_n) \in W$, $f(a_1, \ldots, a_n) = 0$. Hence, for all $(a_1, \ldots, a_n) \in V$, $f(a_1, \ldots, a_n) = 0$ and so $f \in I(V)$. Thus, $I(W) \subseteq I(V)$. Suppose that $I(W) \subseteq I(V)$. Since W is an affine variety, there exist $g_1, \ldots, g_t \in K[x_1, \ldots, x_n]$ such that $W = V(g_1, \ldots, g_t)$. Therefore, $g_1, \ldots, g_t \in I(W) \subseteq I(V)$. Consequently, $g_1(a_1, \ldots, a_n) = \cdots = g_t(a_1, \ldots, a_n) = 0$ for all $(a_1, \ldots, a_n) \in V$. Thus, $(a_1, \ldots, a_n) \in W$ for all $(a_1, \ldots, a_n) \in V$ and so $V \subseteq W$.
   (ii) Clearly, if $V = W$, then $I(V) = I(W)$. Suppose that $I(V) = I(W)$. Then by (i), $V \supseteq W$ and $V \subseteq W$. Hence, $V = W$. ∎

We have seen in this section how the generators of an ideal in a polynomial ring can be associated with a system of polynomial equations. An ideal may have different sets of generators. See for example Exercises 3 and 4. In the next section, we will be interested in determining the "best" generating set.

27.1.1   Worked-Out Exercises

◇ **Exercise 1** Let $V = V(y - x) \subseteq \mathbf{R}^2$. Show that

$$I(V) = \langle y - x \rangle \,.$$

Solution: Clearly $y - x \in I(V)$. Hence, $\langle y - x \rangle \subseteq I(V)$. Let $x^\alpha y^\beta$ be a monomial in $\mathbf{R}[x, y]$. Then α and β are nonnegative integers. By the binomial theorem

$$
\begin{aligned}
x^\alpha y^\beta &= x^\alpha (x + (y - x))^\beta \\
&= x^\alpha (x^\beta + (\textstyle\sum_{i=1}^\beta \binom{\beta}{i} x^{\beta-i}(y - x)^{i-1})(y - x)) \\
&= h_{\alpha\beta}(y - x) + x^{\alpha+\beta}
\end{aligned}
$$

for some $h_{\alpha\beta} \in \mathbf{R}[x, y]$. Thus, for all $f \in \mathbf{R}[x, y]$,

$$
\begin{aligned}
f &= \textstyle\sum_{\alpha=0}^m \sum_{\beta=0}^n r_{\alpha\beta} x^\alpha y^\beta && (r_{\alpha\beta} \in \mathbf{R}) \\
&= \textstyle\sum_{\alpha=0}^m \sum_{\beta=0}^n r_{\alpha\beta}(h_{\alpha\beta}(y - x) + x^{\alpha+\beta}) \\
&= h(y - x) + \widehat{r}
\end{aligned}
$$

for some $h \in \mathbf{R}[x, y]$, $\widehat{r} \in \mathbf{R}[x]$.

Let $f \in I(V)$. Then $f = h(y - x) + \widehat{r}$ as above. Since f vanishes on V, we obtain

$$0 = f(t, t) = 0 + \widehat{r}(t)$$

for any real number t. Therefore, $\widehat{r} = 0$ and so $f = h(y - x) \in \langle y - x \rangle$. Hence, $I(V) \subseteq \langle y - x \rangle$. Consequently,

$$I(V) = \langle y - x \rangle \,.$$

(We see that $f \in \langle y - x \rangle$ if and only if $f(t, t) = 0$.)

27.1.2   Exercises

1. Let $V = V(y - x, z - x^2, w - x^3) \subseteq \mathbf{R}^4$. Show that

$$I(V) = \left\langle y - x, z - x^2, w - x^3 \right\rangle \,.$$

2. Show by an example that equality need not hold in Lemma 27.1.12.

3. In the polynomial ring $K[x, y]$, show that

$$\langle x, y \rangle = \langle x + y, x - y \rangle = \left\langle x + xy, y + xy, x + x^2 y^2, y - x^2 y^2 \right\rangle \,.$$

4. In the polynomial ring $\mathbf{Q}[x]$, find a single generator for the ideal

$$\left\langle x^4 + 3x^3 + 2x^2, x^2 + 2x + 1 \right\rangle \,.$$

5. Show that

$$V(3x^2 + 2y^2 - 11, x^2 - y^2 + 3) = \{(1,2),\ (-1,2),\ (1,-2),\ (-1,-2)\}$$

by showing

$$\left\langle 3x^2 + 2y^2 - 11, x^2 - y^2 + 3 \right\rangle = \left\langle x^2 - 1, y^2 - 4 \right\rangle.$$

27.2   Gröbner Bases

The concept of Gröbner bases provides computational means for solving problems in mathematics, computer science, engineering, and science. The real impact of Gröbner bases is that they can be computed. Influenced by Wolfgang Gröbner, Bruno Buchberger introduced Gröbner bases in 1965. His algorithm for computing such bases is the major contribution to the theory. We will not explicitly give his algorithm here. The interested reader may pursue the subject further in Adams and Loustaunau or Becker and Weispfenning or Cox, Little and O'Shea.

In the following, we let $\mathbf{W}$ denote the set of whole numbers, i.e., the non-negative integers.

Definition 27.2.1 *Let $\succ$ be a relation on*

$$\mathbf{W}^n = \{(\alpha_1, \ldots, \alpha_n) \mid \alpha_i \in \mathbf{W}, i = 1, \ldots, n\}$$

*and $>$ be the relation on the set of monomials of $K[x_1, \ldots, x_n]$ defined by $x^\alpha > x^\beta$ if and only if $\alpha \succ \beta$. If $\succ$ satisfies properties (i), (ii), and (iii), then $>$ is called a **monomial ordering**, where*

(i) $\succ$ is a total (linear) ordering on $\mathbf{W}^n$;
(ii) for all $\alpha, \beta, \gamma \in \mathbf{W}^n$, $\alpha \succ \beta$ implies $\alpha + \gamma \succ \beta + \gamma$;
(iii) every nonempty subset of $\mathbf{W}^n$ has a smallest element relative to $\succ$.

We ask the reader to verify that a relation $\succ$ on $\mathbf{W}^n$ satisfies (iii) of Definition 27.2.1 if and only if every strictly decreasing sequence in $\mathbf{W}^n$ terminates.

Definition 27.2.2 *Define the relation $\succ_l$ on $\mathbf{W}^n$ by for all $\alpha, \beta \in \mathbf{W}^n$, $\alpha \succ_l \beta$ if the left-most nonzero entry in $\alpha - \beta \in \mathbf{Z}^n$ is positive. Define the relation $>_l$ on the set of monomials of $K[x_1, \ldots, x_n]$ by $x^\alpha >_l x^\beta$ if and only if $\alpha \succ_l \beta$. Then $\succ_l$ and $>_l$ are called **lexicographic** (or **lex**) **orders**.*

Definition 27.2.3 *Define the relation $\succ_{grl}$ on $\mathbf{W}^n$ by for all $\alpha, \beta \in \mathbf{W}^n$, $\alpha \succ_{grl} \beta$ if*

$$|\alpha| > |\beta|, \ \text{or} \ |\alpha| = |\beta| \ \text{and} \ \alpha \succ_l \beta.$$

*Define the relation $>_{grl}$ on the set of monomials of $K[x_1, \ldots, x_n]$ by $x^\alpha >_{grl} x^\beta$ if and only if $\alpha \succ_{grl} \beta$. Then $\succ_{grl}$ and $>_{grl}$ are called **graded lexicographic orders**.*

Definition 27.2.4 *Define the relation* $\succ_{grel}$ *on* $\mathbf{W}^n$ *by for all* $\alpha, \beta \in \mathbf{W}^n$, $\alpha \succ_{grel} \beta$ *if either* $|\alpha| > |\beta|$ *or* $|\alpha| = |\beta|$ *and the right-most nonzero entry in* $\alpha - \beta \in \mathbf{Z}^n$ *is negative. Define the relation* $>_{grel}$ *on the set of monomials of* $K[x_1, \ldots, x_n]$ *by* $x^\alpha >_{grel} x^\beta$ *if and only if* $\alpha \succ_{grel} \beta$. *Then* $\succ_{grel}$ *and* $>_{grel}$ *are called* **graded reverse lexicographic orders.**

$$(1, 2, 3) \succ_l (1, 1, 4) \text{ since } (1, 2, 3) - (1, 1, 4) = (0, 1, -1) \text{ and } 1 > 0.$$

$$(1, 2, 3) \succ_{grl} (1, 1, 4) \text{ since } |(1, 2, 3)| = 6 = |(1, 1, 4)|$$
$$\text{and } (1, 2, 3) \succ_l (1, 1, 4).$$

$$(1, 2, 3) \succ_{grel} (1, 1, 4) \text{ since } |(1, 2, 3)| = 6 = |(1, 1, 4)|, \text{ but}$$
$$(1, 2, 3) - (1, 1, 4) = (0, 1, -1) \text{ and } -1 < 0.$$

$$(1, 2, 3) \succ_{grl} (1, 3, 1) \text{ and } (1, 2, 3) \succ_{grel} (1, 3, 1) \text{ since}$$
$$|(1, 2, 3)| = 6 > 5 = |(1, 3, 1)|.$$

We ask the reader verify the following result.

Theorem 27.2.5 *The lex ordering on* $\mathbf{W}^n$ *is a monomial ordering.* ∎

Definition 27.2.6 *Let* $f = \sum_\alpha a_\alpha x^\alpha \in K[x_1, \ldots, x_n]$, $f \neq 0$ *and* $>$ *be a monomial ordering.*
(i) The **multidegree** *of* f, *written* multideg(f), *is defined to be*

$$\max\{\alpha \in \mathbf{W}^n \mid a_\alpha \neq 0\},$$

where the maximum is taken with respect to $>$;
(ii) the **leading coefficient** *of* f, *written* $LC(f)$, *is defined to be* $a_\mu \in K$, *where* $\mu = $ multideg(f);
(iii) the **leading monomial** *of* f, *written* $LM(f)$, *is defined to be*

$$x^\mu, \text{ where } \mu = multideg\ (f);$$

(iv) the **leading term** *of* f, *written* $LT(f)$, *is defined to be* $LC(f) \cdot LM(f)$.

Example 27.2.7 *Let* $f = 2x^4 z^2 - 3x^4 y^2 z^2 + 5xy^3 + 5y^3 z^2$ *with* $x > y > z$. *Then, with respect to lexicographic ordering we have*
multideg(f) = $(4, 2, 2)$
$LC(f) = -3$
$LM(f) = x^4 y^2 z^2$
$LT(f) = -3x^4 y^2 z^2$.

The following theorem is a generalization of Theorem 14.1.4.

Theorem 27.2.8 *Let* $f, g \in K[x_1, \ldots, x_n]$ *be nonzero. Then the following properties hold.*
(i) multideg(fg) = multideg(f)+multideg(g);
(ii) If $f+g \neq 0$, *then multideg(f+g) $\leq$ max{multideg(f),multideg(g)} with equality holding when multideg(f) $\neq$ multideg(g).* ∎

We now consider the division algorithm for polynomials of several variables. The idea is to divide a polynomial f in $K[x_1, \ldots, x_n]$ by polynomials $f_1, \ldots, f_t \in K[x_1, \ldots, x_n]$ to obtain quotients $q_1, \ldots, q_t$ and a remainder $r \in K[x_1, \ldots, x_n]$ such that $f = q_1 f_1 + \cdots + q_t f_t + r$. We illustrate the procedure with some examples before we state the actual division algorithm.

Example 27.2.9 *Let* $f = x^2 y^2 + y + 1$, $f_1 = xy + 1$, *and* $f_2 = y + 1$. *We use lex order with* $x > y$. *Since* $LT(f_1) > LT(f_2)$, *we list* f_1 *first in the following scheme:*

$$
\begin{array}{rl}
q_1: & \\
q_2: & \\
\hline
xy+1 \quad & x^2y^2 + y + 1 \\
y+1 \quad & x^2y^2 + xy.
\end{array}
$$

The leading terms, $LT(f_1) = xy$ *and* $LT(f_2) = y$, *both divide* $LT(f)$. *Hence, divide* f *by* f_1 *first. We obtain*

$$
\begin{array}{rl}
q_1: xy & \\
q_2: & \\
\hline
xy+1 \quad & x^2y^2 + y + 1 \\
y+1 \quad & x^2y^2 + xy \\
\hline
& -xy + y + 1.
\end{array}
$$

Both $LT(f_1)$ *and* $LT(f_2)$ *divide* $-xy$. *Hence, we divide* $-xy + y + 1$ *by* $LT(f_1)$. *This time we obtain*

$$
\begin{array}{rl}
q_1: xy - 1 & \\
q_2: & \\
\hline
xy+1 \quad & x^2y^2 + y + 1 \\
y+1 \quad & x^2y^2 + xy \\
\hline
& -xy + y + 1 \\
& -xy \qquad -1 \\
\hline
& y + 2.
\end{array}
$$

This time $LT(f_1)$ *does not divide* y, *but* $LT(f_2)$ *does. Hence, we divide* $y + 2$

by $LT(f_2)$ *to obtain*

$$
\begin{array}{r}
q_1 : xy - 1 \\
q_2 : 1 \\
\hline
\end{array}
$$

$$
\begin{array}{r|l}
xy + 1 & x^2y^2 + y + 1 \\
y + 1 & x^2y^2 + xy \\
\hline
 & -xy + y + 1 \\
 & -xy \quad\quad - 1 \\
\hline
 & y + 2 \\
 & y + 1 \\
\hline
 & 1.
\end{array}
$$

Thus, we have that $x^2y^2 + y + 1 = (xy - 1)(xy + 1) + 1(y + 1) + 1.$

In the following example, we illustrate a slight complication of this procedure.

Example 27.2.10 *Let* $f = x^2y + xy + y$, $f_1 = x^2 + y$, *and* $f_2 = y^2 + 1$. *We use lex order with* $x > y$. *Since* $LT(f_1) > LT(f_2)$, *we list* f_1 *first in the following scheme:*

$$
\begin{array}{r}
q_1 : y \\
q_2 : \\
\hline
\end{array}
$$

$$
\begin{array}{r|l}
x^2 + y & x^2y + xy + y \\
y^2 + 1 & x^2y + y^2 \\
\hline
 & xy - y^2 + y.
\end{array}
$$

Now neither $LT(f_1)$ *nor* $LT(f_2)$ *divides* xy. *Hence, we pull* xy *out as a remainder. We thus arrive at*

$$
\begin{array}{r}
q_1 : y \\
q_2 : \\
\hline
\end{array}
$$

$$
\begin{array}{r|l}
x^2 + y & x^2y + xy + y \\
y^2 + 1 & x^2y + y^2 \\
\hline
 & -y^2 + y \qquad r : xy.
\end{array}
$$

Now $LT(f_1)$ *does not divide* $-y^2$, *but* $LT(f_2)$ *does. Hence,*

$$
\begin{array}{r}
q_1 : y \\
q_2 : -1 \\
\hline
\end{array}
$$

$$
\begin{array}{r|l}
x^2 + y & x^2y + xy + y \\
y^2 + 1 & x^2y + y^2 \\
\hline
 & -y^2 + y \\
 & -y^2 - 1 \\
\hline
 & y + 1 \qquad r : xy.
\end{array}
$$

Now neither $LT(f_1)$ nor $LT(f_2)$ divides y. Hence, we pull out $y + 1$ as a remainder to obtain

$$
\begin{array}{r|l}
\begin{array}{l} q_1 : y \\ q_2 : -1 \end{array} & \\
\hline
\begin{array}{l} x^2 + y \\ y^2 + 1 \end{array} & \begin{array}{l} x^2y + xy + y \\ x^2y + y^2 \end{array} \\
\hline
& -y^2 + y \\
& -y^2 - 1 \\
\hline
\end{array}
$$

$$r : xy + y + 1.$$

Hence, $x^2y + xy + y = y(x^2 + y) - 1(y^2 + 1) + xy + y + 1.$

Theorem 27.2.11 (Division Algorithm) *Let $>$ be a monomial ordering on the set of monomials of $K[x_1, \ldots, x_n]$. Let $f, f_1, \ldots, f_t \in K[x_1, \ldots, x_n]$, where*

$$LT(f_1) > LT(f_2) > \cdots > LT(f_t).$$

Then there exist $q_1, \ldots, q_t, r \in K[x_1, \ldots, x_n]$ such that $f = q_1 f_1 + \cdots + q_t f_t + r$, where either $r = 0$ or r is a K-linear combination of monomials, none of which is divisible by any of $LT(f_1), \ldots, LT(f_t)$. Also, multideg$(f) \geq$ multideg$(q_i f_i)$ for those $q_i f_i \neq 0$. ∎

We ask the reader to show in Exercises 3 and 4 that if the order of $f_1, \ldots, f_t$ by which we divide f is altered, then the remainder r may also be altered.

Definition 27.2.12 *An ideal $I \subseteq K[x_1, \ldots, x_n]$ is called a **monomial ideal** if there exists $A \subseteq \mathbf{W}^n$ such that $I = \langle \{x^\alpha \mid \alpha \in A\} \rangle$.*

If $I = \langle \{x^\alpha \mid \alpha \in A\} \rangle$ is a monomial ideal, then every element of I is a finite sum of the form $\sum_{\alpha \in A} h_\alpha x^\alpha$, where $h_\alpha \in K[x_1, \ldots, x_n]$.

Lemma 27.2.13 *Let $I = \langle \{x^\alpha \mid \alpha \in A\} \rangle$ be a monomial ideal. Then a monomial $x^\beta \in I$ if and only if $x^\alpha \mid x^\beta$ for some $\alpha \in A$.*

Proof.   If $x^\alpha \mid x^\beta$, then there exists $h_\alpha \in K[x_1, \ldots, x_n]$ such that $x^\beta = h_\alpha x^\alpha \in I$. Suppose $x^\beta \in I$. Then $x^\beta = \sum_{i=1}^q h_i x^{\alpha(i)}$, where $h_i \in K[x_1, \ldots, x_n]$ and $\alpha(i) \in A$. Now $h_i = \sum_{j=1}^{m_i} k_{ij} \, x^{\beta(ij)}$, $k_{ij} \in K$ for all i, j. Thus, $x^\beta = \sum_{i=1}^q (\sum_{j=1}^{m_i} k_{ij} \, x^{\beta(ij)}) x^{\alpha(i)}$ and so $x^\beta = \sum_{i=1}^q \sum_{j=1}^{m_i} k_{ij} \, x^{\beta(ij)+\alpha(i)}$. Hence, $\beta = \beta(ij) + \alpha(i)$, $j = 1, \ldots, m_i$; $i = 1, \ldots, n$ except for those i and j which drop out when like terms are combined. Now $x^{\alpha(i)} \mid x^{\beta(ij)+\alpha(i)}$, $j = 1, \ldots, m_i$; $i = 1, \ldots, n$. Thus, $x^{\alpha(i)} \mid x^\beta$, $i = 1, \ldots, n$. ∎

The set

$$\alpha + \mathbf{W}^n = \{\alpha + \gamma \mid \gamma \in \mathbf{W}^n\}$$

consists of the exponents of all monomials divisible by x^α. If

$$I = \left\langle x^4 y^2, x^3 y^4, x^2 y^5 \right\rangle,$$

then the exponents of the monomials in I form the set

$$((4,2) + \mathbf{W}^n) \cup ((3,4) + \mathbf{W}^n) \cup ((2,5) + \mathbf{W}^n).$$

Lemma 27.2.14 *Let I be a monomial ideal and $f \in K[x_1, \ldots, x_n]$. Then the following are equivalent.*

(i) $f \in I$.
(ii) Every term of f lies in I.
(iii) $f = \sum_{\alpha=0}^{t} a_\alpha x^\alpha$ for some $a_\alpha \in K$ and $x^\alpha \in I$, $0 \le \alpha \le t$.

Proof. That (iii) $\Rightarrow$ (ii) $\Rightarrow$ (i) is immediate.

(i) $\Rightarrow$ (iii): Since $f \in I$, $f = \sum_{i=1}^{q} h_i \, x^{\alpha(i)}$ for some $h_i \in K[x_1, \ldots, x_n]$ and $\alpha(i) \in A$, where $I = \langle \{x^\alpha \mid \alpha \in A\} \rangle$. Now $h_i = \sum_{j=1}^{n_i} k_{ij} \, x^{\beta(ij)}$, $i = 1, \ldots, q, k_{ij} \in K$ for all i, j. Thus,

$$f = \sum_{i=1}^{q} \sum_{j=1}^{m_i} k_{ij} x^{\beta(ij)} x^{\alpha(i)} \tag{27.7}$$

Therefore, (iii) holds since $x^{\beta(ij)} x^{\alpha(i)}$ is a monomial in I. ■

Corollary 27.2.15 *Let I and J be monomial ideals. Then $I = J$ if and only if I and J have the same monomials.*

Proof. Let $I = \langle \{x^\alpha \mid \alpha \in A\} \rangle$ and $J = \left\langle \{x^\beta \mid \beta \in B\} \right\rangle$. Suppose I and J have the same monomials. Then $x^\beta \in \langle \{x^\alpha \mid \alpha \in A\} \rangle$ and $x^\alpha \in \left\langle \{x^\beta \mid \beta \in B\} \right\rangle$. Hence,

$$\langle \{x^\alpha \mid \alpha \in A\} \rangle = \left\langle \{x^\beta \mid \beta \in B\} \right\rangle.$$

The converse is immediate. ■

Theorem 27.2.16 (Dickson's Lemma) *Let $I = \langle \{x^\alpha \mid \alpha \in A\} \rangle \subseteq K[x_1, \ldots, x_n]$ be a monomial ideal. Then there exist $\alpha(1), \ldots, \alpha(s) \in A$ such that*

$$I = \left\langle x^{\alpha(1)}, \ldots, x^{\alpha(s)} \right\rangle.$$

Proof.   If $n = 1$, then $I = \langle \{x_1^\alpha \mid \alpha \in A\} \rangle$, where $\alpha \in A \subseteq \mathbf{W}$. Let β be the smallest element of A. Then $x_1^\beta | x_1^\alpha$ for all $\alpha \in A$ and so $I = \langle x_1^\beta \rangle$. Now assume that $n > 1$ and that the theorem is true for $K[x_1, \ldots, x_{n-1}]$. Consider $K[x_1, \ldots, x_{n-1}, y]$ so that the monomials in $K[x_1, \ldots, x_{n-1}, y]$ can be written $x^\alpha y^m$, where $\alpha = (\alpha_1, \ldots, \alpha_{n-1}) \in \mathbf{W}^{n-1}$ and $m \in \mathbf{W}$.

Suppose that $I \subseteq K[x_1, \ldots, x_{n-1}, y]$ is a monomial ideal. Let

$$J = \langle \{x^\alpha \mid x^\alpha y^m \in I \text{ for some } m\} \rangle$$

in $K[x_1, \ldots, x_{n-1}]$. Then by the induction hypothesis, there exist $x^{\alpha(1)}, \ldots, x^{\alpha(s)}$ ($x^{\alpha(i)} \in \{x^\alpha \mid x^\alpha y^m \in I \text{ for some } m\}$) such that $J = \langle x^{\alpha(1)}, \ldots, x^{\alpha(s)} \rangle$ in $K[x_1, \ldots, x_{n-1}]$. By the definition of J, $x^{\alpha(i)} y^{m_i} \in I$ for some $m_i > 0, i = 1, \ldots, s$. Let $m = \max\{m_1, \ldots, m_s\}$. Let $J_k = \langle \{x^\beta \mid x^\beta y^k \in I\} \rangle$ in $K[x_1, \ldots, x_{n-1}]$, $k = 0, \ldots, m-1$. By the induction hypothesis, there exist $x^{\alpha_k(1)}, x^{\alpha_k(2)}$, $\ldots, x^{\alpha_k(s_k)}$ such that $J_k = \langle x^{\alpha_k(1)}, \ldots, x^{\alpha_k(s_k)} \rangle$ in $K[x_1, \ldots, x_{n-1}]$, where $x^{\alpha_k(i)} \in \{x^\beta \mid x^\beta y^k \in I\}$.

We now show that I is generated from the following list of monomials

$$x^{\alpha(1)} y^m, \ldots, \ x^{\alpha(s)} y^m \text{ from } J$$
$$x^{\alpha_0(1)}, \ldots, \ x^{\alpha_0(s_0)} \text{ from } J_0$$
$$x^{\alpha_1(1)} y, \ldots, \ x^{\alpha_1(s_1)} y \text{ from } J_1$$
$$\vdots$$
$$x^{\alpha_{m-1}(1)} y^{m-1}, \ldots, \ x^{\alpha_{m-1}(s_{m-1})} y^{m-1} \text{ from } J_{m-1}.$$

Let $x^\alpha y^p \in I$. If $p \geq m$, then $x^\alpha y^p$ is divisible by some $x^{\alpha(i)} y^m$ by the construction of J and Lemma 27.2.13. If $p \leq m - 1$, then $x^\alpha y^p$ is divisible by some $x^{\alpha_p(i)} y^p$ by the construction of J_p and Lemma 27.2.13. It follows from Lemma 27.2.13 that the above monomials generate an ideal having the same monomials as I. By Corollary 27.2.15, these ideals are the same.

Now $x^{\alpha_k(i)} y^k \in I$, but we don't know $x^{\alpha_k(i)} y^k \in \{x^\alpha \mid \alpha \in A\} \subseteq K[x_1, \ldots, x_n]$, where $x_n = y$. We write $I = \langle x^{\beta(1)}, \ldots, x^{\beta(s)} \rangle$, where $x^{\beta(i)} \in I$, $i = 1, \ldots, s$. By Lemma 27.2.13, each $x^{\beta(i)}$ is divisible by some $x^{\alpha(i)} \in A$ and $I = \langle \{x^\alpha \mid \alpha \in A\} \rangle$. Thus,

$$I \supseteq \langle x^{\alpha(1)}, \ldots, x^{\alpha(s)} \rangle \supseteq \langle x^{\beta(1)}, \ldots, x^{\beta(s)} \rangle = I.$$

Hence, $I = \langle x^{\alpha(1)}, \ldots, x^{\alpha(s)} \rangle$. ∎

The ideal I in the following example already has a finite basis. However, it is merely our intention to illustrate the proof of Dickson's lemma.

Example 27.2.17 *Suppose that*

$$I = \left\langle x^5 y^2, x^2 y^3, x^2 y^5 \right\rangle.$$

Then

$$J = \left\langle x^5, x^2, x^2 \right\rangle = \left\langle x^2 \right\rangle \ \textit{in } K[x].$$

Clearly $m = 5$. *Now*

$$
\begin{aligned}
J_0 &= \left\langle \{x^\beta \mid x^\beta \in I\} \right\rangle = \langle \phi \rangle = \{0\} \\
J_1 &= \left\langle \{x^\beta \mid x^\beta y \in I\} \right\rangle = \langle \phi \rangle = \{0\} \\
J_2 &= \left\langle \{x^\beta \mid x^\beta y^2 \in I\} \right\rangle = \langle x^5 \rangle \\
J_3 &= \left\langle \{x^\beta \mid x^\beta y^3 \in I\} \right\rangle = \langle x^2 \rangle \\
J_4 &= \left\langle \{x^\beta \mid x^\beta y^4 \in I\} \right\rangle = \langle x^2 \rangle.
\end{aligned}
$$

Hence,

$$I = \left\langle x^2 y^5, x^5 y^2, x^5 y^3, x^2 y^4 \right\rangle.$$

Corollary 27.2.18 *Let $>$ be a relation on $\mathbf{W}^n$ satisfying the following properties:*

(i) $>$ is a total ordering on $\mathbf{W}^n$;

(ii) if $\alpha > \beta$ and $\gamma \in \mathbf{W}^n$, then $\alpha + \gamma > \beta + \gamma$.

Then $>$ is a well-ordering if and only if $\alpha \geq 0$ for all $\alpha \in \mathbf{W}^n$.

Proof. Assume $>$ is a well-ordering. Let α_0 be the smallest element of $\mathbf{W}^n$. Suppose that $0 > \alpha_0$. Adding $n\alpha_0$ to both sides we obtain $n\alpha_0 > (n+1)\alpha_0$, where n is a positive integer. Hence,

$$0 > \alpha_0 > 2\alpha_0 > \cdots > n\alpha_0 > (n+1)\alpha_0 > \cdots$$

is an infinite descending sequence, a contradiction.

Conversely, suppose that $\alpha \geq 0$ for all $\alpha \in \mathbf{W}^n$. Let $A \subseteq \mathbf{W}^n$ be nonempty. Since $I = \langle \{x^\alpha \mid \alpha \in A\} \rangle$ is a monomial ideal, we have by Dickson's lemma that there exists $\alpha(1), \ldots, \alpha(s) \in A$ such that $I = \left\langle x^{\alpha(1)}, \ldots, x^{\alpha(s)} \right\rangle$. There is no loss in generality in assuming $\alpha(1) < \alpha(2) < \cdots < \alpha(s)$. Let $\alpha \in A$. Then $x^\alpha \in I$ and so $x^{\alpha(i)}$ divides x^α for some $\alpha(i)$. Thus, $\alpha = \alpha(i) + \gamma$ for some $\gamma \in \mathbf{W}^n$. Therefore, $\gamma \geq 0$ and so by (ii), $\alpha = \alpha(i) + \gamma \geq \alpha(i) + 0 = \alpha(i) > \alpha(1)$. Consequently, $\alpha(1)$ is the smallest element of A. ∎

Hence, in Definition 27.2.1, (iii) can be replaced by the simpler condition that $\alpha \geq 0$ for all $\alpha \in \mathbf{W}^n$.

Definition 27.2.19 *Let I be an ideal of $K[x_1, \ldots, x_n]$, $I \neq \langle 0 \rangle$. Let*

$$LT(I) = \{cx^\alpha \mid there\ exists\ f \in I\ with\ LT(f) = cx^\alpha\}.$$

Example 27.2.20 *Let $y > x$. Let $I = \langle f_1, f_2 \rangle$, where $f_1 = y^3 - yf(x)$ and $f_2 = y^2x - xf(x) + y$, where $f \in K[x, y]$ is a nonzero polynomial in x alone. Use the grl ordering on monomials in $K[x, y]$. Then $y^2 = y(y^2x - xf(x) + y) - x(y^3 - yf(x)) \in I$. Thus, $y^2 = LT(y^2) \in \langle LT(I) \rangle$. By Lemma 27.2.13, $y^2 \notin \langle LT(f_1), LT(f_2) \rangle$ since $LT(f_1)$ does not divide y^2 and $LT(f_2)$ does not divide y^2, using grl ordering. Hence, $\langle LT(f_1), LT(f_2) \rangle \subset \langle LT(I) \rangle$.*

Theorem 27.2.21 *Let I be an ideal of $K[x_1, \ldots, x_n]$. Then*
    (i) $\langle LT(I) \rangle$ is a monomial ideal,
    (ii) there exist $g_1, \ldots, g_t \in I$ such that

$$\langle LT(I) \rangle = \langle LT(g_1), \ldots, LT(g_t) \rangle.$$

Proof.   (i) Let $g \in I \backslash \{0\}$. Recall that $LM(g) = x^m$, where $m = \text{multideg}(g) = \max\{\alpha \in \mathbf{W}^n \mid a_\alpha \neq 0\}$, where $g = \sum_\alpha a_\alpha x^\alpha$. Recall also that $LT(g) = a_\alpha LM(g)$, where α is the multidegree of g. Since $LM(g)$ and $LT(g)$ differ by a nonzero constant, $\langle \{LM(g) \mid g \in I \backslash \{0\}\} \rangle = \langle LT(I) \rangle$. Thus, $\langle LT(I) \rangle$ is a monomial ideal.

    (ii) Since $\langle LT(I) \rangle$ is generated by the monomials $LM(g)$ for $g \in I \backslash \{0\}$, $\langle LT(I) \rangle = \langle LM(g_1), \ldots, LM(g_t) \rangle$ for finitely many $g_1, \ldots, g_t \in I$ by Dickson's lemma. Since $LM(g_i)$ differs from $LT(g_i)$ by a nonzero constant, it follows that

$$\langle LT(I) \rangle = \langle LT(g_1), \ldots, LT(g_t) \rangle. \ \blacksquare$$

    In the following result, we have selected one particular monomial order to use the division algorithm and to compute leading terms.

Theorem 27.2.22 (Hilbert Basis Theorem) *Let I be an ideal of $K[x_1, \ldots, x_n]$. Then there exist $g_1, \ldots, g_t \in I$ such that $I = \langle g_1, \ldots, g_t \rangle$. That is, I has a finite generating set.*

Proof.   If $I = \{0\}$, then $I = \langle 0 \rangle$. Suppose that I contains some nonzero polynomial. By Theorem 27.2.21, there exist $g_1, \ldots, g_t \in I$ such that $\langle LT(I) \rangle = \langle LT(g_1), \ldots, LT(g_t) \rangle$. Clearly $\langle g_1, \ldots, g_t \rangle \subseteq I$. Let $f \in I$. Divide f by $g_1, \ldots, g_t$ to get an expression of the form

$$f = a_1 g_1 + \cdots + a_t g_t + r,$$

where every term in r is divisible by none of $LT(g_1), \ldots, LT(g_t)$. Now

$$r = f - (a_1 g_1 + \cdots + a_t g_t) \in I.$$

If $r \neq 0$, then $\mathrm{LT}(r) \in \langle \mathrm{LT}(I) \rangle = \langle \mathrm{LT}(g_1), \dots, \mathrm{LT}(g_t) \rangle$ and so by Lemma 27.2.13, $\mathrm{LT}(r)$ is divisible by some $\mathrm{LT}(g_i)$. This contradicts what it means to be a remainder. Hence, $r = 0$. Thus,

$$f = a_1 g_1 + \dots + a_t g_t \in \langle g_1, \dots, g_t \rangle.$$

Therefore, $I \subseteq \langle g_1, \dots, g_t \rangle$. Consequently, $I = \langle g_1, \dots, g_t \rangle$. ∎

In addition to answering the ideal description question, the basis $\{g_1, \dots, g_t\}$ used in the proof of Theorem 27.2.22 has the special property that

$$\langle \mathrm{LT}(I) \rangle = \langle \mathrm{LT}(g_1), \dots, \mathrm{LT}(g_t) \rangle.$$

Definition 27.2.23 *Fix a monomial ordering. A finite subset $G = \{g_1, \dots, g_t\}$ of an ideal I is said to be a **Gröbner basis** (or **standard basis**) if*

$$\langle LT(g_1), \dots, LT(g_t) \rangle = \langle LT(I) \rangle.$$

Corollary 27.2.24 *Fix a monomial order. Let I be a nonzero ideal of $K[x_1, x_2, \dots, x_n]$. Then I has a Gröbner basis. Furthermore, any Gröbner basis for I is a basis for I.*

Proof. The set $G = \{g_1, \dots, g_t\}$ constructed in the proof of Theorem 27.2.22 is a Gröbner basis by definition. Note also that if

$$\langle \mathrm{LT}(I) \rangle = \langle \mathrm{LT}(g_1), \dots, \mathrm{LT}(g_t) \rangle,$$

then the argument in Theorem 27.2.22 shows that $I = \langle g_1, \dots, g_t \rangle$ so that G is a basis for I. ∎

Example 27.2.25 *Let $y > x$. Consider I in Example 27.2.20. $\{y^3 - yf(x), y^2 x - xf(x) + y\}$ is a basis for I, but not a Gröbner basis since $y^2 \in \langle LT(I) \rangle$, but $y^2 \notin \langle LT(y^3 - yf(x)), LT(y^2 x - xf(x) + y) \rangle$.*

Example 27.2.26 *Let $x > y > z$. Consider the ideal $J = \langle x - z^i, y - z^j \rangle$ in $\mathbf{R}[x, y, z]$, where i and j are fixed positive integers. To show that $\{x - z^i, y - z^j\}$ is a Gröbner basis for J, it suffices to show that the leading term of every nonzero element in J lies in $\langle LT(x - z^i), LT(y - z^j) \rangle = \langle x, y \rangle$. By Lemma 27.2.13, this is equivalent to showing that the leading term of any nonzero element of J is divisible by either x or y. Consider any $f = Ag_1 + Bg_2 \in J$, where $g_1(x, y, z) = x - z^i$ and $g_2(x, y, z) = y - z^j$. Suppose that $f \neq 0$ and x and y do not divide $LT(f)$. Then by the definition of lex order, f must be a polynomial in z alone. However, f vanishes on $V = V(x - z^i, y - z^j) \subseteq \mathbf{R}^3$ since $f \in J$. Clearly, $(t^i, t^j, t) \in V$ for any real number t. The only polynomial in z alone that vanishes at all these points is the zero polynomial, a contradiction, since $f \neq 0$. Hence, $\{g_1, g_2\}$ is a Gröbner basis for J.*

Theorem 27.2.27 *Let I be an ideal of $K[x_1, \ldots, x_n]$, $G = \{g_1, \ldots, g_t\}$ be a Gröbner basis for I, and $f \in K[x_1, \ldots, x_n]$. Then there exists a unique $r \in K[x_1, \ldots, x_n]$ such that*

(i) *no term of r is divisible by one of $LT(g_1), \ldots, LT(g_t)$,*

(ii) *there exists $g \in I$ such that $f = g + r$.*

Proof. The division algorithm gives $f = a_1 g_1 + \cdots + a_t g_t + r$, where r satisfies (i). g satisfies (ii) by letting $g = a_1 g_1 + \cdots + a_t g_t$.

To prove uniqueness, suppose that $f = g_1 + r_1 = g_2 + r_2$ with (i) and (ii) holding. Then $r_2 - r_1 = g_1 - g_2 \in I$. Thus, if $r_2 - r_1 \neq 0$, then $LT(r_2 - r_1) \in LT(I) = \langle LT(g_1), \ldots, LT(g_t) \rangle$. By Lemma 27.2.13, it follows that $LT(r_2 - r_1)$ is divisible by some $LT(g_i)$. This is impossible since no term of r_1, r_2 is divisible by one of $LT(g_1), \ldots, LT(g_t)$. Therefore, $r_1 - r_2 = 0$, i.e., $r_1 = r_2$. (Every term of $r_2 - r_1$ is a term of either r_1 or r_2, except for a constant multiple.) Hence, $g_1 = g_2$ also. ∎

By Theorem 27.2.27, we can list the elements of G in any order when dividing f by G since the remainder r is unique.

Corollary 27.2.28 *Let I be an ideal of $K[x_1, \ldots, x_n]$ and $G = \{g_1, \ldots, g_t\}$ be a Gröbner basis for I. Let $f \in K[x_1, \ldots, x_n]$. Then $f \in I$ if and only if the remainder on division of f by G is zero.*

Proof. If $r = 0$, then $f = a_1 g_1 + \cdots + a_t g_t \in I$. Conversely, suppose that $f \in I$. Then $f = f + 0$ satisfies (i) and (ii) of Theorem 27.2.27. Thus, $r = 0$ by the uniqueness of r. ∎

Definition 27.2.29 *Let f^F denote the remainder on division of f by the ordered s-tuple $F = (f_1, \ldots, f_s)$.*

If F in Definition 27.2.29 is a Gröbner basis for $\langle f_1, \ldots, f_s \rangle$, then we can regard F as a set (without any particular order) by Theorem 27.2.27.

Example 27.2.30 *Let*

$$F = (x^2 y - xy, x^4 y^3 - xy)$$

and

$$f(x, y) = x^5 y^2.$$

Use lex ordering with $x > y$ to obtain

$$f^F = xy^2.$$

$$q_1 : 0$$
$$q_2 : x^3y + x^2y + xy + y$$

$$
\begin{array}{r|l}
x^4y^3 - xy & x^5y^2 \\
x^2y - xy & x^5y^2 - x^4y^2 \\
\hline
& x^4y^2 \\
& x^4y^2 - x^3y^2 \\
\hline
& x^3y^2 \\
& x^3y^2 - x^2y^2 \\
\hline
& x^2y^2 \\
& x^2y^2 - xy^2 \\
\hline
& xy^2.
\end{array}
$$

Example 27.2.31 *Let* $y > x$. *Consider the ideal* I *of Example 27.2.20. Then* $I = \langle f_1, f_2 \rangle$, *where* $f_1 = y^3 - yf(x)$ *and* $f_2 = y^2x - xf(x) + y$. *Use the grl ordering. Now* $y(y^2x - xf(x) + y) - x(y^3 - yf(x)) = y^2$. *Thus,* $y^2 \in I$. *Now* $LT(f_1) \nmid y^2$ *and* $LT(f_2) \nmid y^2$. *Hence,*

$$LT(y^2) = y^2 \notin \langle LT(f_1), LT(f_2) \rangle.$$

We see that $xLT(f_1) - yLT(f_2) = 0$. *That is, the leading terms in* $xf_1 - yf_2$ *cancel, leaving only smaller terms.* $\{f_1, f_2\}$ *is not a Gröbner basis because*

$$\langle LT(I) \rangle \neq \langle LT(f_1), LT(f_2) \rangle.$$

Note: $LT(y^2) \in LT(I)$ *and* $LT(y^2) \notin \langle LT(f_1), LT(f_2) \rangle.$

Definition 27.2.32 *Let* f, g *be nonzero polynomials in* $K[x_1, \ldots, x_n]$.
   (i) *Let* $multideg(f) = \alpha$, $multideg(g) = \beta$, *and* $\gamma_i = \max\{\alpha_i, \beta_i\}$, $i = 1, 2, \ldots, n$. *Let* $\gamma = (\gamma_1, \ldots, \gamma_n)$. *Then* x^γ *is called the **least common multiple** of* $LM(f)$ *and* $LM(g)$, *written*

$$L = x^\gamma = LCM(LM(f), LM(g)).$$

   (ii) *The polynomial*

$$S(f, g) = \frac{x^\gamma}{LT(f)} \cdot f - \frac{x^\gamma}{LT(g)} \cdot g$$

*is called the **S-polynomial** of* f *and* g.

Example 27.2.33 *Let* $y > x$. *In Example 27.2.31, let* $f(x) = x$ *so that* $f_1 = y^3 - yx$ *and* $f_2 = y^2x - x^2 + y$ *in* $\mathbf{R}[x, y]$ *with the grl order. Then* $multideg(f_1) = (3, 0)$ *and* $multideg(f_2) = (2, 1)$. *Thus,* $L = y^3x$ *and*

$$
\begin{aligned}
S(f_1, f_2) &= \frac{y^3x}{y^3}f_1 - \frac{y^3x}{y^2x}f_2 \\
&= xf_1 - yf_2 \\
&= -yx^2 + yx^2 - y^2 \\
&= -y^2.
\end{aligned}
$$

The purpose of an S-polynomial $S(f,g)$ is to produce cancellation of leading terms. The next lemma shows that S-polynomials are involved in every such cancellation.

Lemma 27.2.34 *Consider the sum $\sum_{i=1}^{t} c_i x^{\alpha(i)} g_i$, where $c_1, \ldots, c_t$ are constants and*

$$\alpha(i) + multideg(g_i) = \delta \in \mathbf{W}^n$$

whenever $c_i \neq 0$. If

$$multideg(\sum_{i=1}^{t} c_i x^{\alpha(i)} g_i) < \delta,$$

then there are constants c_{jk} such that

$$\sum_{i=1}^{t} c_i x^{\alpha(i)} g_i = \sum_{j,k} c_{jk} x^{\delta - \gamma_{jk}} S(g_j, g_k), \qquad (27.8)$$

where $x^{\gamma_{jk}} = LCM(LM(g_j), LM(g_k))$. Furthermore, each $x^{\delta - \gamma_{jk}} S(g_j, g_k)$ has multidegree $< \delta$. ∎

In Eq. (27.8), every summand $c_i x^{\alpha(i)} g_i$ on the left has multidegree δ. Thus, the cancellation occurs after the summands have been added. Each summand on the right has multidegree $< \delta$. Hence, the cancellation has already occurred. We see that the S-polynomials account for the cancellation.

Theorem 27.2.35 *Let I be an ideal in $K[x_1, \ldots, x_n]$. Then a basis $G = \{g_1, \ldots, g_t\}$ for I is a Gröbner basis for I if and only if the remainder, $S(g_i, g_j)^G$, on division of $S(g_i, g_j)$ by G (listed in some order) is zero for all i, j with $i \neq j$.* ∎

Example 27.2.36 *Consider the ideal*

$$I = \langle y - x^2, z - x^3 \rangle.$$

We show with the help of Theorem 27.2.35 that $G = \{y - x^2, z - x^3\}$ is a Gröbner basis for lex order with $y > z > x$. Consider $S(y - x^2, z - x^3)$. Now

$$S(y - x^2, z - x^3) = \frac{yz}{y}(y - x^2) - \frac{yz}{z}(z - x^3) = -zx^2 + yx^3.$$

$$
\begin{array}{ll}
q_1 : x^3 & \\
q_2 : -x^2 & \\
\hline
\end{array}
$$

$$
\begin{array}{l|l}
y - x^2 & yx^3 - zx^2 \\
z - x^3 & yx^3 - x^5 \\
\hline
 & -zx^2 + x^5 \\
 & -zx^2 + x^5 \\
\hline
 & 0.
\end{array}
$$

Thus,

$$-zx^2 + yx^3 = x^3(y - x^2) + (-x^2)(z - x^2) + 0.$$

Hence,

$$S(y - x^2, z - x^3)^G = 0.$$

Therefore, G is a Gröbner basis for I.

Now consider G with lex order $x > y > z$. Then $g_1(x, y, z) = -x^2 + y$ and $g_2(x, y, z) = -x^3 + z$. Now $multideg(g_1) = (2, 0, 0)$ and $multideg(g_2) = (3, 0, 0)$. Thus, $\gamma = (3, 0, 0)$. Hence,

$$
\begin{aligned}
S(-x^2 + y, -x^3 + z) &= \tfrac{x^3}{-x^2}(-x^2 + y) - \tfrac{x^3}{-x^3}(-x^3 + z) \\
&= x^3 - xy - x^3 + z \\
&= -xy + z.
\end{aligned}
$$

$$
\begin{array}{r l}
 & q_1 : 0 \\
 & q_2 : 0 \\
\hline
\begin{array}{r} -x^3 + z \\ -x^2 + y \end{array} & \left|\; \begin{array}{l} -xy + z \\ 0 \end{array} \right. \\
\hline
 & -xy + z
\end{array}
$$

$$-xy + z = 0(-x^3 + z) + 0(-x^2 + y) + (-xy) + z = -xy + z \neq 0.$$

Therefore, $\{-x^2 + y, -x^3 + z\}$ is not a Gröbner basis with lex order $x > y > z$.

Every ideal in $K[x_1, \ldots, x_n]$ has a Gröbner basis, but the proof of this result was nonconstructive. We will now show how to construct a Gröbner basis.

Example 27.2.37 Let $y > x$. Consider $K[x, y]$ with grl order and let $I = \langle f_1, f_2 \rangle$, where $f_1 = y^3 - yx$ and $f_2 = y^2x - x^2 + y$. By Example 27.2.31, $\{f_1, f_2\}$ is not a Gröbner basis since $LT(S(f_1, f_2)) = -y^2 \notin \langle LT(f_1), LT(f_2) \rangle$. By Example 27.2.33,

$$S(f_1, f_2) = -y^2.$$

Now

$$
\begin{array}{r l}
 & q_1 : 0 \\
 & q_2 : 0 \\
\hline
\begin{array}{r} y^3 - yx \\ y^2x - x^2 + y \end{array} & \left|\; \begin{array}{l} -y^2 \\ 0 \end{array} \right. \\
\hline
 & -y^2
\end{array}
$$

The remainder of $S(f_1, f_2)$ upon division of f_1, f_2 is not zero. Hence, we should include the remainder in our generating set. Let $F = \{f_1, f_2, f_3\}$, where $f_3 =$

$-y^2$. Then $S(f_1, f_2) = f_3$. Consequently, $S(f_1, f_2)^F = 0$. Consider f_1 and f_3. Then $multideg(f_1) = (3,0)$ and $multideg(f_3) = (2,0)$. Therefore, $L = y^3$. Thus,

$$S(f_1, f_3) = (y^3/y^3)(y^3 - yx) - (y^3/(-y^2))(-y^2) = -yx,$$

but

$$S(f_1, f_3)^F = -yx \neq 0.$$

Therefore, we must add $f_4 = -yx$ to our generating set. Let $F = \{f_1, f_2, f_3, f_4\}$. Then

$$S(f_1, f_2)^F = S(f_1, f_3)^F = 0.$$

Consider f_1 and f_4. Then $multideg(f_1) = (3,0)$ and $multideg(f_4) = (1,1)$. Hence, $L = y^3 x$. Thus,

$$S(f_1, f_4) = x(y^3 - yx) - (-y^2)(-yx) = -yx^2 = xf_4$$

and so $S(f_1, f_4)^F = 0$. Consider f_2 and f_3. Then $multideg(f_2) = (2,1)$ and $multideg(f_3) = (2,0)$. Thus, $L = y^2 x$. Hence,

$$S(f_2, f_3) = (y^2 x - x^2 + y) - (-x)(-y^2) = -x^2 + y,$$

but

$$S(f_2, f_3)^F = -x^2 + y \neq 0.$$

Thus, we must also add $f_5 = -x^2 + y$ to our generating set. Let $F = \{f_1, f_2, f_3, f_4, f_5\}$. Then one can show that $S(f_i, f_j)^F = 0$ for all $1 \leq i < j \leq 5$. Hence, by Theorem 27.2.35, $\{f_1, f_2, f_3, f_4, f_5\}$ is a Gröbner basis for I.

Lemma 27.2.38 *Let I be an ideal in $K[x_1, \ldots, x_n]$. Let G be a Gröbner basis for I. Let $g \in G$ be a polynomial such that $LT(g) \in \langle LT(G\backslash\{g\})\rangle$. Then $G\backslash\{g\}$ is also a Gröbner basis for I.*

Proof. We have that $\langle LT(G)\rangle = \langle LT(I)\rangle$. Suppose that

$$LT(g) \in \langle LT(G\backslash\{g\})\rangle.$$

Then $\langle LT(G\backslash\{g\})\rangle = \langle LT(G)\rangle$. Consequently, $\langle LT(G\backslash\{g\})\rangle = \langle LT(I)\rangle$ and so $G\backslash\{g\}$ is a Gröbner basis for I. ∎

Definition 27.2.39 *Let I be an ideal in $K[x_1, \ldots, x_n]$ and G be a Gröbner basis for I. Then G is a **minimal Gröbner basis** for I if the following conditions hold:*

(i) $LC(g) = 1$ for all $g \in G$ and
(ii) for all $g \in G$, $LT(g) \notin \langle LT(G\backslash\{g\})\rangle$.

Example 27.2.40 *Let $y > x$. It is possible to construct a minimal Gröbner basis for a nonzero ideal by applying the above procedure and then using Lemma 27.2.38 to eliminate any unneeded generators. In Example 27.2.37, using grl order, we constructed the Gröbner basis*

$$
\begin{aligned}
f_1 &= y^3 - yx \\
f_2 &= y^2x - x^2 + y \\
f_3 &= -y^2 \\
f_4 &= -yx \\
f_5 &= -x^2 + y.
\end{aligned}
$$

We multiply f_3, f_4, and f_5 by -1 to make the leading coefficients of the generators equal to 1. Now $LT(f_1) = y^3 = -y \cdot LT(f_3)$. By Lemma 27.2.38, we can eliminate f_1. Similarly, since $LT(f_2) = y^2x = -y \cdot LT(f_4)$, we can eliminate f_2. There are no more cases, where the leading term of a generator divides the leading term of another generator:

$$
y^2 \not| yx, \, y^2 \not| x^2, \, yx \not| y^2, \, yx \not| x^2, \, x^2 \not| y^2, \, x^2 \not| yx
$$

Hence,

$$
g_3 = y^2, \quad g_4 = yx, \quad g_5 = x^2 - y
$$

is a minimal Gröbner basis for I. Now $y^2 + ayx$, yx, $x^2 - y$ is also a minimal Gröbner basis for I, where $a \in K$ is any nonzero constant. Thus, for K infinite, there exists infinitely many Gröbner bases. Hence, a Gröbner basis for an ideal is not necessarily unique.

Definition 27.2.41 *Let I be an ideal in $K[x_1, \ldots, x_n]$. Let G be a Gröbner basis for I. Then G is a **reduced Gröbner basis** for I if the following conditions hold:*

(i) $LC(g) = 1$ *for all $g \in G$.*
(ii) *For all $g \in G$, no monomial of g lies in $\langle LT(G \setminus \{g\}) \rangle$.*

In the above example, y^2, yx, $x^2 - y$ is a reduced Gröbner basis for $I : y^2 \notin \langle yx, x^2 \rangle$, $yx \notin \langle y^2, x^2 \rangle$, $x^2 \notin \langle y^2, yx \rangle$, $y \notin \langle y^2, yx \rangle$.
For the minimal Gröbner basis $y^2 + ayx$, yx, $x^2 - y$, $ayx \in \langle yx, x^2 \rangle$ and so this basis is not reduced when $a \neq 0$.

Theorem 27.2.42 *Let I be an ideal of $K[x_1, \ldots, x_n]$. Then for a given monomial ordering, I has a unique reduced Gröbner basis.* ∎

We now consider the problem of solving polynomial equations.

Theorem 27.2.43 *Let I be an ideal of $K[x_1, x_2, \ldots, x_n]$. Let $f_1, f_2, \ldots, f_m \in I$ be such that $I = \langle f_1, \ldots, f_m \rangle$. Then $V(I) = V(\{f_1, \ldots, f_m\})$.* ∎

Example 27.2.44 *Consider the polynomial equations*

$$x^3 + y + z^2 = 0$$
$$x^2 + z^2 = y \qquad\qquad (27.9)$$
$$x = z.$$

Let

$$I = \left\langle x^3 + y + z^2, x^2 + z^2 - y, x - z \right\rangle \subseteq \mathbf{C}[x, y, z].$$

By Theorem 27.2.43, we can compute $V(I)$ using any basis of I. Using lex order with $x > y > z$, we obtain the following Gröbner basis:

$$g_1 = x - z$$
$$g_2 = y - 2z^2$$
$$g_3 = z^3 + 3z^2.$$

The polynomial g_3 depends on z alone, and its roots are

$$z = 0, \ -3.$$

Next, we can solve the equations $g_1 = 0$ and $g_2 = 0$ uniquely for x and y, respectively, by substituting the value of z. Thus, the solutions of $g_1 = g_2 = g_3 = 0$ are $(0,0,0)$ and $(-3, 18, -3)$. Since $V(I) = V(g_1, g_2, g_3)$, we have found all solutions to Eqs. (27.9).

In the above example, the variables are eliminated successively. Also, note that the order of elimination corresponds to the ordering of the variables. It does follow that lex order gives a Gröbner basis that successively eliminates the variables.

27.2.1   Worked-Out Exercises

◇ **Exercise 1** Let $I = \langle f_1, f_2 \rangle$, where $f_1 = xz - y^2$ and $f_2 = x^3 - z^2$ in $\mathbf{C}[x, y, z]$. Use grl order with $x > y > z$. Let $f = -4x^2y^2z^2 + y^6 + 3z^5$.

(i) Show that $\{f_1, f_2\}$ is not a Gröbner basis for I.
(ii) Find a Gröbner basis for I.
(iii) Determine if $f \in I$.
(iv) Show that $g = xy - 5z^2 + x \notin I$.

Solution: (i) multideg$(f_1) = (1, 0, 1)$ and multideg$(f_2) = (3, 0, 0)$. Thus, $\gamma = (3, 0, 1)$ and so

$$S(f_1, f_2) = \frac{x^3z}{xz}(xz - y^2) - \frac{x^3z}{x^3}(x^3 - z^2) = -x^2y^2 + z^3.$$

LT$(S(f_1, f_2)) = -x^2y^2 \in$ LT(I) since $-x^2y^2 + z^3 = x^2 f_1 - z f_2 \in I$. But

$$-x^2y^2 \notin \langle \text{LT}(f_1), \text{LT}(f_2) \rangle = \left\langle xz, x^3 \right\rangle.$$

(ii) Let $f_1 = xz - y^2$, $f_2 = x^3 - z^2$, $f_3 = x^2y^2 - z^3$, $f_4 = xy^4 - z^3$, and $f_5 = y^6 - z^5$. Then $G = \{f_1, f_2, f_3, f_4, f_5\}$ is a Gröbner basis for I. G is a reduced Gröbner basis.

(iii) Divide f by G. We obtain $f = 0f_1 + 0f_2 - 4z^2 f_3 + 0f_4 + 1f_5 + 0$. Since the remainder is zero, we have $f \in I$.

(iv) $\mathrm{LT}(g) = xy \notin \langle \mathrm{LT}(G) \rangle = \langle xz, x^3, x^2y^2, xy^4, y^6 \rangle$. Hence, $g^G \neq 0$, so that $g \notin I$.

27.2.2   Exercises

1. Let $f = 3x^2y^3z - 5x^4yz^2 + 3xy - 2x \in \mathbf{R}[x, y, z]$. Use lex, graded lex, and reverse lex orderings for the following determinations, where $x > y > z$.

   (i) Find multideg(f).

   (ii) Find LC(f).

   (iii) Find LM(f).

   (iv) Find LT(f).

2. Let $f = x^3y^3 + y^2$, $f_1 = xy^2 + x + y^2$, and $f_2 = y^2 - y - 1 \in \mathbf{Q}[x, y]$. Use lex order with $x > y$ to divide f by f_1 and f_2.

3. Let f, f_1, and f_2 be defined as in Exercise 2. Use lex order with $x > y$ to divide f by f_2 and f_1, i.e., reverse the role of f_1 and f_2. Compare the remainder with the remainder obtained in Exercise 2.

4. Let $f = x^3yz^2 - 2xyz^2$ and $g = x^2z - xy^2z + xz$. Compute $S(f, g)$ using the lex ordering with $x > y > z$.

5. Suppose that $I = \langle x^4y^2, x^3y^4, x^2y^5 \rangle$. In Dickson's lemma, determine J, m, and $J_0, J_1, \ldots, J_{m-1}$.

6. In Example 27.2.37, show that

   (i) $S(f_1, f_3)^F = -yx$ with $F = \{f_1, f_2, f_3\}$;

   (ii) $S(f_1, f_2)^F = S(f_1, f_3)^F = 0$ with $F = \{f_1, f_2, f_3, f_4\}$;

   (iii) $S(f_2, f_3)^F = y - x^2$ with $F = \{f_1, f_2, f_3, f_4\}$.

7. Show that a relation $\succ$ on $\mathbf{W}^n$ satisfies (iii) of Definition 27.2.1 if and only if every strictly decreasing sequence in $\mathbf{W}^n$ terminates.

8. Prove that lex ordering on $\mathbf{W}^n$ is a monomial ordering.

9. Prove Theorem 27.2.8.

10. Prove Theorem 27.2.43.

11. Consider the polynomial equations

$$x^2 - y = 0$$
$$x^2 z = 0.$$

Find a reduced Gröbner basis for the ideal $I = \langle x^2 - y, x^2 z \rangle$ in $\mathbf{R}[x, y, z]$. Compute $V(I)$.

Selected Bibliography

1. Adams, W. W., and Loustaunau, P. *An Introduction to Gröbner Bases, Graduate Studies in Mathematics*, Vol. 3. American Mathematical Society, 1994.

2. Aschbacher, M. The classification of finite simple groups. *Mathematics Intelligencer*, 3(2), 59–65, 1981.

3. Artin, E. *Galois Theory*. Notre Dame, Ind.: University of Notre Dame Press, 1944.

4. Barnes, W. E. *Introduction to Abstract Algebra*. Boston: D.C. Heath and Company, 1963.

5. Becker, T., and Weispfenning, V., In cooperation with Kredel, H. *Gröbner Bases, A Computational Approach to Commutative Algebra*, Graduate text in mathematics. New York: Springer Verlag, 1993.

6. Bell, E. T. *Men of Mathematics*. 2d ed. New York: Simon and Schuster, 1962.

7. Berlekamp, E. R. *Algebraic Coding Theory*. New York: McGraw-Hill, 1968.

8. Burton, D. M. *Elementary Number Theory*. Boston: Allyn & Bacon, 1980.

9. Cohn, P. M. *Algebra*, Vols. 1 and 2. New York: Wiley, 1974, 1977.

10. Cox, D., Little, J. and O'Shea, D. *Ideals, Varieties, and Algorithms*. New York: Springer Verlag, 1992.

11. Edwards, H. M. The genesis of ideal theory. *Arch. History Exact Sci.* 23, 321–378, 1980.

12. Edwards, H. M. Dedekind's invention of ideals. In *Studies in the History of Mathematics*, E.R. Phillips, ed. The Mathematical Association of America, 1987.

13. Fuchs, L. *Infinite Abelian Groups,* Vols. 1 and 2. New York: Academic Press, 1970,1973.

14. Gillispie, C. C., ed. *Dictionary of Scientific Biography,* Vols. 1-14. New York: Charles Scribner's Sons.

15. Goodearl, K. R. *Ring Theory: Nonsingular Rings and Modules.* New York: Marcel Dekker, 1976.

16. Halmos, P. R. *Naive Set Theory.* New York: Springer Verlag, 1974.

17. Hardy, G. H., and Wright, E. M. *An Introduction to the Theory of Numbers,* 4th ed. Oxford, England. Clarendon Press, 1960.

18. Herstein, I. N. *Topics in Algebra.* 2d ed. New York: Wiley, 1975.

19. Hungerford, T. W. *Algebra.* New York: Holt, Reinhart and Winston, 1974.

20. Isaacs, I. M. *Algebra.* California: Brooks/Cole, 1994.

21. Jacobson, N. *Basic Algebra,* Vols. 1 and 2. San Francisco: Freeman, 1974, 1980.

22. Karpilovsky, G. *Topics in Field Theory.* New York: North-Holland, 1989.

23. Kiernan, B. M. The development of Galois theory from Lagrange to Artin. *Arch. History Exact Sci.* 8, 40–154, 1971–1972.

24. Kleiner, I. The evolution of group theory: A brief survey. *Mathematics Magazine* 59(4), 195–215, 1986.

25. Kleiner, I. A sketch of the evolution of (noncommutative) ring theory. *L'Enseignement Mathématique,* 33, 227–267, 1987.

26. McCoy, N. H. *The Theory of Rings.* New York, Chelsea Publishing Company, 1973.

27. Pless, V. *Introduction to the Theory of Error-Correcting Codes.* New York: Wiley-Interscience Series Discrete Mathematics, 1982.

28. Rotman, J. J. *An Introduction to the Theory of Groups.* Iowa, Wm. C. Brown, 1988.

29. Rotman, J. J. *Galois Theory.* New York: Springer Verlag, 1990.

30. Van der Waerden, B. L. *A History of Algebra.* New York: Springer Verlag, 1985.

31. Zariski, O., and Samuel, P. *Commutative Algebra,* Vol. 1. New Jersey: D. Van Nostrand Co. Inc., 1960.

Answers and Hints to Selected Exercises

Exercises 1.1.2 (page 6) (Sets)

1. $A \cup B = \{x, y, z, w\}$; $A \cap B = \{y\}$; $A \backslash B = \{x, z\}$; $B \backslash A = \{w\}$; $A \times B = \{(x, y), (x, w),$ $(y, y), (y, w), (z, y), (z, w)\}$; $\mathcal{P}(A) = \{\phi, \{x\}, \{y\}, \{z\}, \{x, y\}, \{x, w\}, \{y, z\}, A\}$.

3. (i) $x \in A \cup B$ if and only if $x \in A$ or $x \in B$ if and only if $x \in B$ or $x \in A$. Thus, $A \cup B = B \cup A$. Similarly, $A \cap B = B \cap A$.

4. $|\mathcal{P}(S)| = 4096$. 4095 subsets are properly contained in S.

6. (i) $A \triangle B = \{a, d, e\}$.

7. (ii) Note that $A = (A \backslash B) \cup (A \cap B)$ and $(A \backslash B) \cap (A \cap B) = \phi$. Now use (i).

8. (i) True (ii) False (iii) False (iv) True (v) True.

Exercises 1.2.2 (page 19) (Integers)

1. $\gcd(90, 252) = 18$, $s = 3$ and $t = -1$.

2. $s = 239$ and $t = -353$.

3. $s = 22$ and $t = -15$.

4. (ii) Let $S(n)$ be the statement: $7^n - 1$ is divisible by 6 for all $n \in \mathbf{Z}^{\#}$. For $n = 0$, $7^0 - 1 = 0$, which is divisible by 6. Hence, $S(0)$ is true. Suppose $S(n)$ is true for some $n \geq 0$. Consider $S(n + 1) : 7^{n+1} - 1$ is divisible by 6. Now $7^{n+1} - 1 = 7^n \cdot 7 - 1 = 7^n(6 + 1) - 1 = 7^n \cdot 6 + 7^n - 1$. Now $7^n \cdot 6$ is divisible by 6 and by the induction hypothesis $7^n - 1$ is divisible by 6. This implies that $7^n \cdot 6 + 7^n - 1$ is divisible by 6 and so $7^{n+1} - 1$ is divisible by 6. Thus, $S(n + 1)$ is true. Hence, by induction $7^n - 1$ is divisible by 6 for all $n \in \mathbf{Z}^{\#}$.

5. (i) Suppose $a|b$. Then $b = an$ for some $n \in \mathbf{Z}$. Thus, $bc = a(cn)$ for some $n \in \mathbf{Z}$ and for all $c \in \mathbf{Z}$. Thus, $a|bc$ for all $c \in \mathbf{Z}$.

(iii) Suppose $a|b$ and $a|c$. Then $b = an$ and $c = am$ for some $n, m \in \mathbf{Z}$. Let $x, y \in \mathbf{Z}$. Now $bx + cy = a(nx + my)$. Since $n, m, x, y \in \mathbf{Z}$, $nx + my \in \mathbf{Z}$. Hence, $a|(bx + cy)$.

7. $ab \neq 0$. Now $c = at$ and $d = bs$ for some $s, t \in \mathbf{Z}$. Hence, $cd = abts$. Thus, $ab|cd$.

9. Suppose $\gcd(m, n) = c$. There exist $u, v \in \mathbf{Z}$ such that $m = uc$ and $n = vc$. Thus, $a = cdu$ and $b = cdv$ and so $cd|a$ and $cd|b$. Since $\gcd(a, b) = d$, $cd|d$. Thus, $d = cdk$ for some $k \in \mathbf{Z}$. This implies that $1 = ck$, and since c is a positive integer, $c = 1$. Consequently, $\gcd(m, n) = 1$.

11. There exist $u, v, t, s \in \mathbf{Z}$ such that $1 = xu + yv$ and $1 = xt + zs$. Thus, $zs = 1 - xt$. Now $zs = zs \cdot 1 = zs(xu + yv)$. Therefore, $1 - xt = xzus + yzvs$ and so $1 = x(t + zus) + yzvs$. Hence, by Theorem 1.2.11, $\gcd(x, yz) = 1$.

14. Let $b = a + 1$. Now $1 = (a + 1) \cdot 1 + a(-1)$. Thus, by Theorem 1.2.11, $\gcd(a, a + 1) = 1$.

20. (i) True (ii) True (iii) True (iv) True (v) True

Exercises 1.3.2 (page 28) (Relations)

1. (i) $R = \{(1,1),(1,5),(2,2),(2,6),(3,3),(3,7),(4,4),(5,1),(5,5),(6,2),(6,6),(7,3),$ $(7,7)\}$. (ii) $\mathcal{D}(R) = A$. (iii) $\mathcal{I}(R) = A$. (iv) $R^{-1} = \{1,1),(1,5),(2,2),(2,6),(3,3),(3,7),$ $(4,4),(5,1),(5,5),(6,2),(6,6),(7,3),(7,7)\}$. (v) $\mathcal{D}(R^{-1}) = A$. (vi) $\mathcal{I}(R^{-1}) = A$.

2. (i) $R = \{(1,1),(1,2),(1,3),(1,4),(1,5),(1,6),(2,1),(2,2),(2,3),(2,4),(2,5),(2,6),$ $(3,1),(3,2),(3,3),(3,4),(3,5),(3,6),(4,1),(4,2),(4,3),(4,4),(4,5),(5,1),(5,2),(5,3),$ $(5,4),(6,1),(6,2),(6,3)\}$. (ii) No (iii) Yes (iv) No.

3. (i) Yes (ii) No (iii) No (iv) No (v) Yes (vi) Yes (vii) No.

5. $[1] = \{1,4,7\}$, $[2] = \{2,5,8\}$, and $[3] = \{3,6\}$.

6. $\mathcal{D}(R) = A$ and $\mathcal{I}(R) = A$.

8. $[-1] = [5] = [11] = [23]$ and $[2] = [8]$.

10. $x - y = nk$ and $z - w = nl$ for some $k, l \in \mathbf{Z}$. Thus, $(x+z)-(y+w) = (x-y)+(z-w) = nk + nl = n(k+l)$. Thus, $x + z \equiv_n y + w$. Also, $xz - yz = nkz$ and $yz - yw = nly$. Thus, $xz - yw = (xz - yz) + (yz - yw) = nkz + nly = n(kz + ly)$ and so $xz \equiv_n yw$.

16. (i) Let $(a,b),(b,c) \in R^\infty$. Then $(a,b) \in R^n$ and $(b,c) \in R^m$ for some positive integers m and n. Let $m \geq n$. Then $(a,b),(b,c) \in R^m$. Thus, $(a,c) \in R^{m+1} \subseteq R^\infty$. Hence, R^∞ is transitive.

(ii) Let $(a,b) \in R^\infty$. Then $(a,b) \in R^n$ for some positive integer n. Thus, there exist $a_1, a_2, \ldots, a_{n-1} \in A$ such that $(a,a_1),(a_1,a_2),\ldots,(a_{n-2},a_{n-1}),(a_{n-1},b) \in R$. Hence, $(a,a_1),$ $(a_1,a_2),\ldots,(a_{n-2},a_{n-1}),(a_{n-1},b) \in T$. Since T is transitive, it follows that $(a,b) \in T$. Hence, $R^\infty \subseteq T$.

17. Let $(x,y) \in R_2 \circ R_1$. Thus, there exists $z \in S$ such that $(x,z) \in R_1$ and $(z,y) \in R_2$. Since R_1 and R_2 are symmetric, $(y,z) \in R_2$ and $(z,x) \in R_1$. Thus, $(y,x) \in R_1 \circ R_2 \subseteq R_2 \circ R_1$. Hence, $R_2 \circ R_1$ is symmetric. Now $(y,x) \in R_2 \circ R_1$. As before, $(x,y) \in R_1 \circ R_2$. Hence, $R_2 \circ R_1 \subseteq R_1 \circ R_2$. Thus, $R_1 \circ R_2 = R_2 \circ R_1$.

19. (i) True (ii) False (iii) False (iv) True (v) True.

Exercises 1.4.2 (page 38) (Partially Ordered Sets)

2. Let $A = \{1,2\}$ and $R = \{(1,2)\}$.

5. Yes.

7. (i) $a \leq a \vee (b \vee c)$ and $b \vee c \leq a \vee (b \vee c)$. Thus, $a,b,c \leq a \vee (b \vee c)$. Since $a \vee b$ exists, $a \vee b \leq a \vee (b \vee c)$. Therefore, $a \vee (b \vee c)$ is an upper bound of $\{a \vee b, c\}$. Let x be an upper bound of $\{a \vee b, c\}$. Then $a, b \leq a \vee b \leq x$ and $c \leq x$. Since $b \vee c$ exists, $b \vee c \leq x$. Again $a \vee (b \vee c)$ exists. Hence, $a \vee (b \vee c) \leq x$. Consequently, $a \vee (b \vee c)$ is the least upper bound of $\{a \vee b, c\}$ and so $a \vee (b \vee c) = (a \vee b) \vee c$.

8. (i) is not a lattice. (ii) and (iii) are lattices.

9. $4 \wedge (8 \vee 10) = 4$ and $(2 \vee (2 \wedge 8)) \vee 20 = 20$.

10. (i) $a \leq a \vee b$ and $a \leq a \vee c$. Thus, a is a lower bound of $\{(a \vee b),(a \vee c)\}$. Hence, $a \leq (a \vee b) \wedge (a \vee c)$. Now $b \wedge c \leq b \leq a \vee b$ and $b \wedge c \leq c \leq a \vee c$. Thus, $b \wedge c$ is a lower bound of $\{(a \vee b),(a \vee c)\}$. Hence, $b \wedge c \leq (a \vee b) \wedge (a \vee c)$. Thus, $(a \vee b) \wedge (a \vee c)$ is an upper bound of $\{a, b \wedge c\}$. Hence, $a \vee (b \wedge c) \leq (a \vee b) \wedge (a \vee c)$.

13. (i) False (ii) False (iii) True.

Exercises 1.5.2 (page 50) (Functions)

1. (i) f is one-one and onto. (ii) f is one-one and onto. (iii) f is neither one-one nor onto.

2. f is neither one-one nor onto.

3. $(f \circ g)(x) = \sqrt{3x + 1}$, $(g \circ f)(x) = 3\sqrt{x} + 1$, and $f \circ g \neq g \circ f$.

4. $(g \circ f)(x) = 2 + \frac{1}{x}$.

5. (i) $g(x) = x - 2$ is a left inverse of f. (ii) Let $g : \mathbf{Z} \to \mathbf{Z}$ be defined by $g(x) = \frac{x}{2}$ if x is even and $g(x) = 1$ if x is odd. Then g is a left inverse of f. (iii) f has no left inverse.

6. (i) Let $h(x) = x + 3$ for all $x \in \mathbf{Z}$. Then h is a right inverse of f. (ii) f does not have a right inverse. (iii) f has no right inverse.

7. $f_1 : 1 \to 1, 2 \to 2, 3 \to 3$; $f_2 : 1 \to 1, 2 \to 3, 3 \to 2$; $f_3 : 1 \to 2, 2 \to 1, 3 \to 3$; $f_4 : 1 \to 3, 2 \to 2, 3 \to 1$; $f_5 : 1 \to 2, 2 \to 3, 3 \to 1$; $f_6 : 1 \to 3, 2 \to 1, 3 \to 2$.

10. (i) $x \in f(A \cup B)$ if and only if $x = f(u)$ for some $u \in A \cup B$ if and only if $x = f(u)$ for some $u \in A$ or $u \in B$ if and only if $x \in f(A)$ or $x \in f(B)$ if and only if $x \in f(A) \cup f(B)$. Hence, $f(A \cup B) = f(A) \cup f(B)$.

11. (ii) $x \in f^{-1}(A \cap B)$ if and only if $f(x) \in A \cap B$ if and only if $f(x) \in A$ and $f(x) \in B$ if and only if $x \in f^{-1}(A)$ and $x \in f^{-1}(B)$ if and only if $x \in f^{-1}(A) \cap f^{-1}(B)$. Hence, $f^{-1}(A \cap B) = f^{-1}(A) \cap f^{-1}(B)$.

13. Define $f : \mathbf{Z} \to \mathbf{E}$ by $f(n) = 2n$. Then f is one-one and onto $\mathbf{E}$. Hence, $\mathbf{Z} \sim \mathbf{E}$.

15. (i) Define $f : \mathbf{Z} \to 3\mathbf{Z}$ by $f(n) = 3n$. Then f is one-one and onto $3\mathbf{Z}$. Hence, $\mathbf{Z} \sim 3\mathbf{Z}$.

19. Yes.

21. (i) False (ii) False.

Exercises 1.6.2 (page 55) (Binary Operations)
1. (i) No (ii) Yes (iii) Yes (iv) Yes (v) Yes (vi) Yes.

2. (ii) (iii), (iv), (v), (vi).

3. (ii) and (iv) have identity.

Exercises 2.1.2 (page 77) (Elementary Properties of Groups)
1. (i) $(\mathbf{N}, *)$ is a semigroup but not a group. (ii) $(\mathbf{Z}, *)$ is not a semigroup and so not a group. (iii) $(\mathbf{R}, *)$ is a semigroup but not a group. (iv) $(\mathbf{R}, *)$ is a group and so a semigroup. (v) $(\mathbf{R}, *)$ is a semigroup but not a group. (vi) $(\mathbf{Q}, *)$ is a semigroup but not a group. (vii) $(G, *)$ is a group and so a semigroup. (viii) $(G, *)$ is a group and so a semigroup.

2. $(0, 1)$ is the identity and $(-\frac{a}{b}, \frac{1}{b})$ is the inverse of (a, b).

5. $n = 7$.

6. $[b] = [8]$ and $[b] \in U_9$.

7. $n = 2$.

8. $U_6 = \{[1], [5]\}$; $U_9 = \{[1], [2], [4], [5], [7], [8]\}$; $U_{12} = \{[1], [5], [7], [11]\}$; $U_{24} = \{[1], [5], [7], [11], [13], [17], [19], [23]\}$.

9. For all $0 < a < p$, $\gcd(a, p) = 1$. Thus, for all $0 < a < p$, $[a] \in U_p$. Hence, $U_p = \mathbf{Z}_p \setminus \{[0]\}$.

12. Note that $a^2 = e$ implies that $a = a^{-1}$. By using $a * b^4 * a = b^7$, first show that $t^? = b^{16}$.

17. Suppose G is commutative. Let $a, b \in G$. Now $(a * b)^{-1} = b^{-1} * a^{-1} = a^{-1} * b^{-1}$. Conversely, suppose $(a * b)^{-1} = a^{-1} * b^{-1}$ for all $a, b \in G$. Let $a, b \in G$. Then $(a * b)^{-1} = a^{-1} * b^{-1}$, which implies that $((a * b)^{-1})^{-1} = (a^{-1} * b^{-1})^{-1}$, i.e., $a * b = b * a$. Hence, G is commutative.

19. Let $a, b \in G$. Suppose $(a * b)^i = a^i * b^i$, $(a * b)^{i+1} = a^{i+1} * b^{i+1}$, $(a * b)^{i+2} = a^{i+2} * b^{i+2}$. Now $a^{i+1} * b^{i+1} = (a * b)^{i+1} = (a * b)(a * b)^i = a * b * a^i * b^i$, which implies that $a^i * b = b * a^i$. Also, $a^{i+2} * b^{i+2} = (a * b)^{i+2} = (a * b)(a * b)^{i+1} = a * b * a^{i+1} * b^{i+1}$, which implies that $a^{i+1} * b = b * a^{i+1}$. Hence, $b * a^{i+1} = a^{i+1} * b = a * a^i * b = a * b * a^i$ and this implies that $b * a = a * b$. Hence, G is commutative.

21. x is unique.

22. Use induction on n.

23. Consider the set $\{a^n \mid n \in \mathbf{N}\} \subseteq G$. $\{a^n \mid n \in \mathbf{N}\}$ has finitely many elements.

24. Use induction on n.

26. Suppose $|G| = n$. Note that $\{e, a, a^2, \dots, a^n\} \subseteq G$ and G has n elements.

28. (ii) Suppose that $a * b = b^{2n+3} * a^{2n+1}$, where $n \geq 1$ is an integer. Then $\circ(b * a^{-1}) = \circ(b^{2n+3} * a^{2n-1}) = \circ(b^{2n+1} * a^{2n+1})$.

29. Use Theorem 2.1.28.

31. $\frac{n}{m}$.

32. (ii) Use induction and part (i).

33. Use induction and Worked Out Exercise 4 (page 74).

34. First note that $(1,0)$ is the identity element of G and $(\frac{1}{a}, -\frac{b}{a})$ is the inverse of $(a, b) \in G$. Now proceed as in Worked Out Exercise 2 (page 72).

36. Let $b \in S$. Now $\{b, b^2, \ldots\} \subseteq S$. Since S is finite, there exist integers m and n, $m > n$ such that $b^m = b^n$. Let $k = m - n$. Then $b^{n+k} = b^n$. Now $b^{2n+k} = b^n b^{n+k} = b^n b^n = b^{2n}$. By induction, $b^{ns+k} = b^{ns}$ for all positive integers s. Also, $b^{ns+2k} = b^{ns+k}b^k = b^{sn}b^k = b^{ns+k} = b^{ns}$. Hence, by induction $b^{ns+kr} = b^{ns}$ for all positive integers s and r. Thus, $b^{2nk} = b^{nk+nk} = b^{nk}$. Let $a = b^{nk}$. Then $a^2 = a$.

41. Use induction.

42. (i) False (ii) False (iii) True (iv) True (v) False (vi) False.

Exercises 3.1.2 (page 96) (Permutation Group)

1.(i) $(1\ 3\ 4) \circ (2\ 5\ 6) = (1\ 4) \circ (1\ 3) \circ (2\ 6) \circ (2\ 5)$.   (ii) $(1\ 3) \circ (4\ 5)$.

2. $(5\ 4\ 6)$.

3. $(2\ 4\ 8) \circ (3\ 5\ 6)$.

4. $(2\ 1\ 6\ 7)$.

5. $(3\ 9\ 7) \circ (5\ 4\ 1\ 2)$.

6. $(1\ 2) \circ (5\ 6) \circ (7\ 8)$.

7. Let $\alpha_1 = (1\ 6)$, $\alpha_2 = (1\ 5)$, $\alpha_3 = (1\ 4)$, $\alpha_4 = (1\ 3)$, $\alpha_5 = (1\ 2)$. Let $\beta = \alpha_1 \circ \alpha_2 \circ \alpha_3 \circ \alpha_4 \circ \alpha_5$. Show that $\beta(i) = \alpha(i)$ for all i.

8. 6.

13. $A_4 = \{e, (1\ 2\ 3), (1\ 3\ 2), (2\ 3\ 4), (2\ 4\ 3), (1\ 3\ 4), (1\ 4\ 3), (1\ 2\ 4), (1\ 4\ 2), (1\ 2) \circ (3\ 4), (1\ 4) \circ (3\ 2), (1\ 3) \circ (2\ 4)\}$.

15. Let H be the set of all odd permutations in S_n. Then $S_n = A_n \cup H$ and $A_n \cap H = \phi$. Hence, $|S_n| = |A_n| + |H|$. First show that there exists a one-one function from A_n onto H. Therefore, $|A_n| = |H|$. Now use the fact that $|S_n| = n!$ and $|S_n| = |A_n| + |H|$.

Exercises 4.1.2 (page 107) (Subgroups)

2. (i) Since $\begin{bmatrix} 1 & 0 \\ 0 & 1 \end{bmatrix} \in S, S \neq \phi$. Let $\begin{bmatrix} a & b \\ c & d \end{bmatrix}$, $\begin{bmatrix} g & h \\ u & v \end{bmatrix} \in S$. Now $\begin{bmatrix} g & h \\ u & v \end{bmatrix}^{-1} =$

$\begin{bmatrix} v & -h \\ -u & g \end{bmatrix}$. Thus,

$$\begin{bmatrix} a & b \\ c & d \end{bmatrix}\begin{bmatrix} g & h \\ u & v \end{bmatrix}^{-1} = \begin{bmatrix} a & b \\ c & d \end{bmatrix}, \begin{bmatrix} v & -h \\ -u & g \end{bmatrix} = \begin{bmatrix} av - bu & -ah + bg \\ cv - du & -ch + dg \end{bmatrix}.$$

Since $(av - bu)(-ch + dg) - (-ah + bg)(cv - du) = 1$, $\begin{bmatrix} a & b \\ c & d \end{bmatrix}\begin{bmatrix} g & h \\ u & v \end{bmatrix}^{-1} \in S$. Hence, S is a subgroup.

(iv) Since $\begin{bmatrix} 1 & 0 \\ 0 & 1 \end{bmatrix} \in S$, $S \neq \phi$. Let $\begin{bmatrix} a & b \\ 0 & d \end{bmatrix}$, $\begin{bmatrix} c & e \\ 0 & f \end{bmatrix} \in S$. Then $ad \neq 0$ and $cf \neq 0$. Now

$$\begin{bmatrix} a & b \\ 0 & d \end{bmatrix}\begin{bmatrix} c & e \\ 0 & f \end{bmatrix}^{-1} = \begin{bmatrix} a & b \\ 0 & d \end{bmatrix}\begin{bmatrix} \frac{f}{cf} & -\frac{e}{cf} \\ 0 & \frac{c}{cf} \end{bmatrix}$$

$$= \begin{bmatrix} \frac{af}{cf} & \frac{-ae+bc}{cf} \\ 0 & \frac{dc}{cf} \end{bmatrix}.$$

Now $\frac{af}{cf}\frac{dc}{cf} = \frac{adcf}{(cf)^2} \neq 0$. Hence, $\begin{bmatrix} a & b \\ 0 & d \end{bmatrix}\begin{bmatrix} c & e \\ 0 & f \end{bmatrix}^{-1}$ and so S is a subgroup.

4. Note that $(0,1)$ is the identity element and $(-\frac{a}{b}, \frac{1}{b})$ is the inverse of $(a, b) \in G$.

(ii) Since $(0,1) \in K$, $K \neq \phi$. Let (a,b), $(c,d) \in K$. Thus, $b > 0$ and $d > 0$, which implies that $\frac{b}{d} > 0$. Hence, $(a,b)(c,d)^{-1} = (a,b)(-\frac{c}{d}, \frac{1}{d}) = (a - \frac{bc}{d}, \frac{b}{d}) \in K$. Thus, K is a subgroup.

(iv) Elements of order 2 are of the form, $(a, -1)$, where $a \in \mathbf{R}$.

5. No.

6. $\langle 4, 6 \rangle = \langle 2 \rangle$.

7. $\langle 4, 5 \rangle = \mathbf{Z}$.

8. (i) Let $a = \begin{pmatrix} 1 & 2 & 3 & 4 \\ 4 & 3 & 2 & 1 \end{pmatrix}$ and $b = \begin{pmatrix} 1 & 2 & 3 & 4 \\ 2 & 1 & 4 & 3 \end{pmatrix}$. Then $\langle a, b \rangle = \{e, a, b, ab\}$.

(ii) $\langle h, v \rangle = \{r_{360}, h, v, r_{180}\}$.

9. (i) $\circ(a) = 4$ and $\circ(b) = 2$. (iii) $H = \{e, a, a^2, a^3, b, ba, ba^2, ba^3\}$. (iv) $|H| = 8$.

10. (ii) $(a^2b)^2 = a^2ba^2b = a(aba)ab = abab = (ab)^2 = e$.

13. Note that $\circ(bab^{-1}) = \circ(a)$ for all $b \in G$.

15. $ea = a = ae$. Thus, $e \in C(a)$ and hence $C(a) \neq \phi$. Let $b, c \in C(a)$. Now $ab = ba$ implies that $ab^{-1} = ba^{-1}$ and so $b^{-1} \in C(a)$. Also, $a(bc) = (ab)c = (ba)c = b(ac) = b(ca) = (bc)a$. Thus, $bc \in C(a)$. Hence, $C(a)$ is a subgroup. Let $x \in Z(G)$. Then $ax = xa$ for all $a \in G$. Thus, $x \in C(a)$ for all $a \in G$ and so $Z(G) \subseteq \cap_{a \in G} C(a)$. Conversely, let $x \in \cap_{a \in G} C(a)$. Then $x \in C(a)$ for all $a \in G$. Thus, $xa = ax$ for all $a \in G$, which implies that $x \in Z(G)$. Hence, $Z(G) = \cap_{a \in G} C(a)$.

18. $\langle H \rangle = \cap \{K \mid K$ is a subgroup of G such that $H \subseteq K\} = H$.

21. (i) Note that $(1\,2\,3 \cdots n) \circ (i\,i+1) \circ (1\,2\,3 \cdots n)^{-1} = (i+1\,i+2)$ for all $i = 1, \ldots, n-2$.

22. Suppose $\mathbf{Q} = \langle \frac{p_1}{q_1}, \frac{p_2}{q_2}, \ldots, \frac{p_n}{q_n} \rangle$, where $\gcd(p_i, q_i) = 1$ for all i. Let $q = \text{lcm}(q_1, q_2, \ldots, q_n)$. Then $q = q_i r_i$ for some $r_i \in \mathbf{Z}$, $1 \leq i \leq n$. Now $\frac{p_i}{q_i} = \frac{p_i r_i}{q_i r_i} = p_i r_i \frac{1}{q} \in \langle \frac{1}{q} \rangle$. Hence, $\mathbf{Q} = \langle \frac{1}{q} \rangle$. Now $\frac{1}{2q} \in \mathbf{Q}$. There exists $k \in \mathbf{Z}$ such that $\frac{1}{2q} = k\frac{1}{q}$, which implies that $\frac{1}{2} = k \in \mathbf{Z}$, a contradiction. Hence, $(\mathbf{Q}, +)$ is not finitely generated.

23. Note that, if $|G| = n$, then $|\mathcal{P}(G)| = 2^n$.

24. No.

25. (i) True (ii) True (iii) False (iv) False (v) False (vi) False (vii) True.

Exercises 4.2.2 (page 114) (Cyclic Groups)

1. (i) $\langle a^5 \rangle = \{e, a^5, a^{10}, a^{15}, a^{20}, a^{25}\}$. (ii) $\langle a^2 \rangle = \{e, a^2, a^4, a^6, a^8, a^{10}, a^{12}, a^{14}, a^{16}, a^{18}, a^{20}, a^{22}, a^{24}, a^{26}, a^{28}\}$.

2. Two elements of order 6 and four elements of order 5.

3. Use Worked Out Exercise 3 (page 114).

4. (i) $(\mathbf{Q}, +)$ is a subgroup of $(\mathbf{R}, +)$. Every subgroup of a cyclic group is cyclic. Thus, since $(\mathbf{Q}, +)$ is not cyclic, $(\mathbf{R}, +)$ is not cyclic.

7. Consider $\mathbf{Z}_2 \times \mathbf{Z}_2$.

11. Let G be the set of all 2×2 nonsingular matrices over $\mathbf{R}$. Let H be the cyclic subgroup generated by $\begin{bmatrix} -1 & 0 \\ 0 & -1 \end{bmatrix}$.

13. (i) True (ii) False (iii) True (iv) False (v) False.

Exercises 4.3.2 (page 125) (Lagrange's Theorem)

1. (i) The right cosets of H are H, $H(1\,2)$, and $H(1\,3)$. (ii) $\{e, (1\,2)\}$.

2. Let $H = 6\mathbf{Z}$. The right cosets of H are $H + 0$, $H + 1$, $H + 2$, $H + 3$, $H + 4$, and $H + 5$.

3. Write $a = \begin{pmatrix} 1 & 2 & 3 & 4 \\ 2 & 1 & 4 & 3 \end{pmatrix}$, $b = \begin{pmatrix} 1 & 2 & 3 & 4 \\ 3 & 4 & 1 & 2 \end{pmatrix}$, and $c = \begin{pmatrix} 1 & 2 & 3 & 4 \\ 4 & 3 & 2 & 1 \end{pmatrix}$.

Since $e \in H$, $H \neq \phi$. Now $x^2 = e$ for all $x \in H$. Thus, $x^{-1} = x \in H$ for all $x \in H$. Also, $ab = ba = c \in H$, $ac = ca = b \in H$ and $bc = cb = a \in H$. Hence, H is a subgroup of G.

4. H, $r_{90}H$, $r_{180}H$, $r_{270}H$ are the left cosets and H, Hr_{90}, Hr_{180}, Hr_{270} are the right cosets.

6. $\langle(1\ 2\ 3\ 4)\rangle = \{e, (1\ 2\ 3\ 4), (1\ 3)\circ(2\ 4), (1\ 4\ 3\ 2)\};$ $\langle(1\ 3\ 2\ 4)\rangle = \{e, (1\ 3\ 2\ 4), (1\ 2)\circ(3\ 4), (1\ 4\ 2\ 3)\};$ $\langle(1\ 3\ 4\ 2)\rangle = \{e, (1\ 3\ 4\ 2), (1\ 4)\circ(2\ 3), (1\ 2\ 4\ 3)\};$ $\{e, (1\ 2), (3\ 4), (1\ 2)\circ(3\ 4)\};$ $\{e, (1\ 4), (3\ 2), (1\ 4)\circ(3\ 2)\};$ $\{e, (1\ 3), (2\ 4), (1\ 3)\circ(2\ 4)\};$ $\{e, (1\ 2)\circ(3\ 4), (1\ 4)\circ(3\ 2), (1\ 3)\circ(2\ 4)\}.$

7.

	d	a	b	c
d	d	a	b	c
a	a	c	d	b
b	b	d	c	a
c	c	b	a	d

11. Let G be the group of symmetries of the square and, H denote the subgroup $\{r_{360}, h\}$.

14. Consider $H = \{\alpha \in S_n \mid \alpha(1) = 1\}$.

15. Consider $|HK|$.

16. Suppose that H and K are two subgroups of order p. Use previous exercise to conclude that $H = K$.

19. 175

20. Find $|AB|$.

23. (i) False (ii) False (iii) False (iv) True (v) True (vi) True.

Exercises 4.4.2 (page 136) (Normal Subgroup)

1. H is a normal subgroup.

2. H is not normal.

5. For all $h \in H$, $Hh = H$, and for all $x \notin H$, $Hx = Ha$. Thus, H is of index 2 in G. Hence, H is a normal subgroup.

7. Replace $Z(G)$ by H in Worked-Out Exercise 4.4.1 (page 136).

9. (i) $\{\mathbf{E}, 1 + \mathbf{E}\}$. (ii) $\mathbf{Q}/\mathbf{Z} = \{\frac{a}{b} + \mathbf{Z} \mid 1 < a < b\}$.

12. Show that $hkh^{-1}k^{-1} \in H \cap K$ for all $h \in H$ and $k \in K$.

14. Let $G = \{\pm 1, \pm i, \pm j, \pm k\}$, where $i^2 = j^2 = k^2 = -1$, $ij = k = -ji$, $jk = i = -kj$, $ki = j = -ik$.

17. Note that $|aHa^{-1}| = |H|$ for all $a \in G$.

19. Let H be a subgroup of order 6. Since A_4 has eight 3-cycles and $|H| = 6$, there exists a 3-cycle, α say, such that $\alpha \notin H$. Then $\alpha^2 = \alpha^{-1} \notin H$. Let $K = \{e, \alpha, \alpha^2\}$. Then K is a subgroup of A_4 such that $|K| = 3$ and $H \cap K = \{e\}$. Thus, $|HK| = \frac{|H||K|}{|H\cap K|} = 6\cdot 3 = 18 > |A_4|$, a contradiction. Hence, A_4 has no subgroup of order 6.

20. $\{e\}$, $\{e, (1\ 2)\circ(3\ 4)\}$, $\{e, (1\ 4)\circ(3\ 2)\}$, $\{e, (1\ 3)\circ(2\ 4)\}$, $\langle(1\ 2\ 3)\rangle$, $\langle(1\ 3\ 4)\rangle$, $\langle(1\ 2\ 4)\rangle$, $\langle(2\ 3\ 4)\rangle$, $\{e, (1\ 2)\circ(3\ 4), (1\ 4)\circ(3\ 2), (1\ 3)\circ(2\ 4)\}$, and A_4.

25. (i) True (ii) True (iii) False (iv) False (v) True.

Exercises 5.1.2 (page 151) (Homomorphism)

1. (i) f is a homomorphism. Ker $f = \{1\}$. (ii) f is a homomorphism. Ker $f = \{0\}$. (iii) f is a homomorphism. Ker $f = \{1, -1\}$. (iv) f is not a homomorphism. (v) f is a homomorphism. Ker $f = \{0\}$.

2. There are two homomorphisms from $\mathbf{Z}$ onto $\mathbf{Z}$. One is the identity homomorphism and the other maps 1 to -1.

3. There are two homomorphisms from $\mathbf{Z}$ onto $\mathbf{Z}_6$.

4. There are four homomorphisms from $\mathbf{Z}_8$ to $\mathbf{Z}_{12}$ and there are 10 homomorphisms from $\mathbf{Z}_{20}$ to $\mathbf{Z}_{10}$.

6. Suppose that $(\mathbf{Q}, +) \simeq (\mathbf{R}, +)$ and let $f : \mathbf{Q} \to \mathbf{R}$ be an isomorphism. Then $f(0) = 0$. Let $0 \neq \frac{p}{q} \in \mathbf{Q}$. Then $f(\frac{p}{q}) = f(\frac{1}{q} + \frac{1}{q} + \cdots + \frac{1}{q}) = f(\frac{1}{q}) + f(\frac{1}{q}) + \cdots + f(\frac{1}{q}) = pf(\frac{1}{q})$. Now $f(1) = f(\frac{p}{p}) = pf(\frac{1}{p})$. Hence, $f(\frac{1}{p}) = \frac{1}{p}f(1)$. Thus, $f(\frac{p}{q}) = \frac{p}{q}f(1)$. Now $1 \in R$. Since f is onto there exists $\frac{m}{n} \in \mathbf{Q}$ such that $1 = f(\frac{m}{n})$. If $m = 0$, then $1 = f(\frac{m}{n}) = f(0) = 0$, which is a contradiction. Hence, $m \neq 0$. This implies that $1 = f(\frac{m}{n}) = \frac{m}{n}f(1)$ and so $f(1) = \frac{n}{m} \in \mathbf{Q}$.

Hence, $f(\frac{p}{q}) \in \mathbf{Q}$ for all $\frac{p}{q} \in \mathbf{Q}$. Thus, irrational numbers have no preimage. Consequently, $(\mathbf{Q}, +) \not\simeq (\mathbf{R}, +)$.

10. Consider the function $f(a) = \log \frac{1+a}{1-a}$ for all $a \in G$.

11. (i) Ker $f = \{e, a^4\}$. (ii) Ker $f = \{e, a^2, a^4, a^6\}$.

15. Ker $f = \{e\} \times H$.

19. Define $\psi : A \to B$ by $\psi(a) = f(a)$. Show that ψ is one-one and onto B.

21. (i) True (ii) True (iii) False (iv) False (v) False (vi) True (vii) False.

Exercises 5.2.2 (page 164) (Isomorphism and Correspondence Theorems)

1. Define $f : \mathbf{R}^* \to \mathbf{R}^+$ by $f(x) = x^2$ for all $x \in \mathbf{R}^*$. Show that f is an epimorphism and Ker $f = \{1, -1\}$.

3. Define $f : 8\mathbf{Z} \to \mathbf{Z}_7$ by $f(t) = [t]$ for all $t \in 8\mathbf{Z}$. Show that f is an epimorphism and Ker $f = 56\mathbf{Z}$.

4. Let $G = (\mathbf{Z}, +)$; $A = 2\mathbf{Z}$ and $B = 4\mathbf{Z}$.

7. The correspondence is given by $Z(G) \to \{e\}$, $\{r_{360}, r_{180}, h, v\} \to \{e, b\}$, $\{r_{360}, r_{90}, r_{180}, r_{270}\} \to \{e, a\}$, $\{r_{360}, r_{180}, d_1, d_2\} \to \{e, c\}$, and $G \to G_1$.

12. The subcollections of isomorphic groups are $\{\mathbf{Z}_2, S_2\}$, $\{\mathbf{Z}_6\}$, $\{S_6\}$, $\{(\mathbf{Z}, +), (17\mathbf{Z}, +), (3\mathbf{Z}, +), (<\pi>, \cdot)\}$, $\{(\mathbf{Q}, +)\}$, $\{(\mathbf{R}, +), (\mathbf{R}^+, \cdot)\}$, $\{(\mathbf{Q}^*, \cdot)\}$, $\{(\mathbf{C}^*, \cdot)\}$, and $\{(\mathbf{R}^*, \cdot)\}$.

13. Use Worked-Out Exercise 7 (page 163).

14. $|\text{Aut}(\mathbf{Z}_6)| = 2$. One automorphism is the identity mapping and other mapping [1] onto [5].

16. Now $Z(S_3) = \{e\}$. Since $S_3/Z(S_3) \simeq \text{Inn}(S_3)$, $\text{Inn}(S_3) \simeq S_3$. Hence, $6 = |S_3| = |\text{Inn}(S_3)| \le |\text{Aut}(S_3)|$. Now $S_3 = \langle \alpha, \beta \rangle$, where $\alpha = (1\ 2)$ and $\beta = (1\ 2\ 3)$. Let $f \in \text{Aut}(S_3)$. Then f is determined if $f(\alpha)$ and $f(\beta)$ are determined. Now $o(f(\alpha)) = o(\alpha) = 2$ and $o(f(\beta)) = o(\beta) = 3$. Since S_3 has three elements of order 2 and two elements of order 3, $f(\alpha)$ has three choices and $f(\beta)$ has two choices. This shows that $\text{Aut}(S_3)$ has at most six elements. Since $6 \le |\text{Aut}(S_3)|$, $6 = |\text{Aut}(S_3)|$. Hence, $\text{Inn}(S_3) = \text{Aut}(S_3)$ and so $\text{Inn}(S_3) \simeq S_3 \simeq \text{Aut}(S_3)$.

17. $\text{Inn}(S_4) \simeq S_4 \simeq \text{Aut}(S_4)$.

20. (ii) Consider $G = \mathbf{Z}_2 \times \mathbf{Z}_2$ and $H = \{([0], [0]), ([1], [0])\}$.

22. (i) True (ii) False (iii) True (iv) False (v) True (vi) True (vii) False

Exercises 5.3.2 (page 171) (The Groups D_4 and Q_8)

1. Let $H = \{e, b, a^2, ba^2\}$ and $K = \{e, b\}$.

3. The homomorphic images are D_4, $\mathbf{Z}_0$, $\mathbf{Z}_2$, and K_4.

Exercises 5.4.2 (page 178) (Group Actions)

1. $G_1 = \{e, (2\ 3)\}$, $G_2 = \{e, (1\ 3)\}$, $G_3 = \{e, (1\ 2)\}$.

8. Use Worked-Out Exercise 5 (page 177).

9. Use Worked-Out Exercise 5 (page 177).

11. Use Worked-Out Exercise 4 (page 177).

12. Use Corollary 5.4.10.

Exercises 6.1.2 (page 187) (Direct Product of Groups)

3. Let $x \in G$. Then there exists unique $h_i \in H_i$, $1 \le i \le n$, such that $x = h_1 h_2 \cdots h_n$. Define $f : G \to \frac{H_1}{K_1} \times \frac{H_2}{K_2} \times \cdots \times \frac{H_n}{K_n}$ by $f(x) = (h_1 K_1, h_2 K_2, \ldots, h_n K_n)$. It is easy to verify that f is an epimorphism and Ker $f = K$.

5. Define $f : G \to H$ as follows: Let $x \in G$. Then there exists unique $h \in H$ and $k \in K$ such that $x = hk$. Define $f(x) = h$. Clearly f is an epimorphism and Ker$f = K$. Thus, $\frac{G}{K} \simeq H$. Similarly, $\frac{G}{H} \simeq K$.

7. Use Worked-Out Exercise 6 (page 187).

8. Show that the mapping $f : G \to \frac{H}{N} \times \frac{K}{N}$ defined by $f(a) = (Nh, Nk)$, where $a = hk$, is an epimorphism with kernel N.

11. Note that if C_4 is a cyclic group of order 4 and C_2 is a cyclic group of order 2, then C_4 has an element of order 4 while $C_2 \times C_2$ has no element of order 4.

13. Yes.

15. Suppose H and K are proper subgroups of D_4 such that $D_4 = H \times K$. Then H and K are normal subgroups of D_4 and either $|H| = 4$ and $|K| = 2$ or $|H| = 2$ and $|K| = 4$. Suppose $|H| = 4$ and $|K| = 2$. Then H and K are commutative. Also, $hk = kh$ for all $h \in H$ and $k \in K$. Now it follows that D_4 is commutative, a contradiction.

16. No.

17. Suppose $\mathbf{Z}$ is an internal direct product of its nontrivial subgroups H and K. If $H = \langle n \rangle$ and $K = \langle m \rangle$, then show that $mn \in H \cap K$.

Exercises 7.1.2 (page 195) (Conjugacy Classes)

5. H is normal in K if and only if $xHx^{-1} = H$ for all $x \in K$ if and only if $x \in N_G(H)$ for all $x \in K$ if and only if $K \subseteq N_G(H)$.

7. Let α be a 5-cycle, β be a 4-cycle, γ be a 3-cycle, σ be a 2-cycle, δ be a product of a 3-cycle and a 2-cycle and μ be a product of two 2-cycles. The conjugate classes are $Cl(e)$, $Cl(\alpha)$, $Cl(\beta)$, $Cl(\gamma)$, $Cl(\sigma)$, $Cl(\delta)$, and $Cl(\mu)$. Also, $|Cl(e)| = 1$, $|Cl(\alpha)| = 24$, $|Cl(\beta)| = 30$, $|Cl(\gamma)| = 20$, $|Cl(\sigma)| = 10$, and $|Cl(\delta)| = 20$, and $|Cl(\gamma)| = 15$. Now $[S_5 : C(a)] = |Cl(a)|$ for all $a \in S_5$. Hence,

$$
\begin{aligned}
|S_5| &= [S_5 : C(e)] + [S_3 : C(\alpha)] + [S_5 : C(\beta)] + [S_5 : C(\gamma)] + \\
&\quad + [S_5 : C(\sigma)] + [S_5 : C(\delta)] + [S_5 : C(\mu)] \\
&= |Cl(e)| + |Cl(\alpha)| + |Cl(\beta)| + |Cl(\gamma)| + |Cl(\sigma)| + |Cl(\delta)| + |Cl(\gamma)| \\
&= 1 + 24 + 30 + 20 + 10 + 20 + 15 \\
&= 120.
\end{aligned}
$$

Exercises 7.2.2 (page 200) (Cauchy's Theorem and p-groups)

2. 6.

4. Use induction on n.

5. The 2-subgroups of $\mathbf{Z}_{12}$ are $\{[0], [6]\}$ and $\{[0], [3], [6], [9]\}$. $\{[0], [4], [8]\}$ is the only 3-subgroup of $\mathbf{Z}_{12}$.

6. $\langle (1\ 2) \circ (3\ 4) \rangle$, $\langle (1\ 3) \circ (2\ 4) \rangle$, $\langle (14) \circ (23) \rangle$, and $\{e, (1\ 2) \circ (3\ 4), (1\ 3) \circ (2\ 4), (1\ 4) \circ (2\ 3)\}$.

8. Use Cauchy's theorem and Worked-Out Exercise 5 (page 177)

12. Use Worked-Out Exercise 5 (page 199).

13. (i) By Cauchy's theorem, G has a subgroup of order 11, say H. Suppose K is any other subgroup of G of order 11. Suppose $H \neq K$. It follows that $|H \cap K| = 1$. Hence, $|HK| = \frac{|H||K|}{|H \cap K|} = \frac{11 \cdot 11}{1} = 121 > |G|$, a contradiction. Hence, $H = K$ and so H is unique. Since H is the only subgroup of order 11, H is a normal subgroup of G.

(ii) Since $|H|$ is prime, H is cyclic. Let $H = \langle a \rangle$ for some $a \in H$. Let $g \in G$. Then $gag^{-1} \in H$. Thus, $gag^{-1} = a^i$ for some i, $0 \leq i < 11$. Clearly $i \neq 0$. We claim that $i = 1$. Now $g^2 a g^{-2} = g(gag^{-1})g^{-1} = ga^i g^{-1} = (gag^{-1})^i = (a^i)^i = a^{i^2}$. By induction, $g^r a g^{-r} = a^{i^r}$. Now G contains an element of order 3, say b. Then $a = b^3 a b^{-3} = a^{i^3}$. Hence, $a^{i^3 - 1} = e$. Since $o(a) = 11$, $i^3 \equiv_{11} 1$. By Fermat's theorem, $i^{10} \equiv_{11} 1$. Thus, $i \equiv_{11} 1$. Since $1 \leq i \leq 10$, we must have $i = 1$. Thus, $gag^{-1} = a$, i.e., $ga = ag$. Hence, $H \subseteq Z(G)$.

Exercises 7.3.2 (page 209) (Sylow Theorems)

2. Use induction on n.

4. First show that if $|G/Z(G)| = 91$, then $G/Z(G)$ is cyclic.

8. Let $x \in G$. Then $xPx^{-1} \subseteq xHx^{-1} = H$. Hence, xPx^{-1} is a Sylow p-subgroup of H. There exists $h \in H$ such that $hxPx^{-1}h^{-1} = P$. This implies that $hx \in N_G(P)$. Thus, $hx = y$ for some $y \in N_G(P)$, which implies that $x = h^{-1}g \in HN_G(P)$. Hence, $G = HN_G(P)$.

10. Since K is a p-subgroup there exists a Sylow p-subgroup P of G such that $K \subseteq P$. Let Q be a Sylow p-subgroup of G. Then $Q = xPx^{-1}$ for some $x \in G$. Since K is a normal subgroup, $K = xKx^{-1}$. Hence, $K = xKx^{-1} \subseteq xPx^{-1} = Q$.

12. Let K be a maximal normal subgroup of G such that $K \subseteq H$, (possibly $K = \{e\}$). Then $|K| = p^r$ such that $r < m$. Now $\left|\frac{G}{K}\right| = p^{m-r} = p^n$, $n > 0$. Then $Z(\frac{G}{K})$ is a nontrivial subgroup of $\frac{G}{K}$. Since $Z(\frac{G}{K})$ is a normal subgroup of G, $Z(\frac{G}{K}) = \frac{L}{K}$, where L is a normal subgroup of G such that $K \subseteq L$. Since $\left|Z(\frac{G}{K})\right| > 1$, $K \neq L$. Since K is a maximal normal subgroup of G such that $K \subseteq H$, $L \neq H$. Let $a \in L$ be such that $a \notin H$. We now show that $aHa^{-1} = H$. Let $b \in aHa^{-1}$. Then $b = aha^{-1}$ for some $h \in H$. Since $Ka \in L/K = Z(\frac{G}{K})$, $KaKh = KhKa$. Thus, $aha^{-1}h^{-1} = (ah)(ah)^{-1} \in K \subseteq H$. Hence, $aha^{-1} = aha^{-1}h^{-1}h \in H$. Thus, $aHa^{-1} \subseteq H$. Since $\left|aHa^{-1}\right| = |H|$, $aHa^{-1} = H$.

Exercises 7.4.2 (page 219) (Some Applications of Sylow Theorems)

1. For order 20, show that the group has a unique Sylow 5-subgroup. For order 28, show that the group has a unique Sylow 7-subgroup. For order 36, show that the group either has a unique Sylow 3-subgroup or a normal subgroup of order 3.

2. Use Sylow's first theorem and Worked-Out Exercise 5 (page 199).

6. As in Worked-Out Exercise 2 (page 217), show that either G has a normal subgroup of order 32 or a normal subgroup of order 16.

7. Use Corollary 7.4.12.

9. The number of Sylow 5-subgroups is $1 + 5k$ such that $1 + 5k | 7 \cdot 19$. Thus, $k = 0$ and hence, G has a unique Sylow 5-subgroup, H say. Thus, H is normal. Similarly, G has a unique Sylow 7-subgroup, K say, and a unique Sylow 19-subgroup, L say. Thus, K and L are normal subgroups. Clearly H, K and L are cyclic groups, $H \cap K = \{e\}$, $K \cap L = \{e\}$, and $L \cap H = \{e\}$. Let $H = \langle h \rangle$, $K = \langle k \rangle$ and $L = \langle l \rangle$. Since $hk = kh$, $hl = lh$ and $kl = lk$, $o(hkl) = o(h) \circ (k) \circ (l) = 5 \cdot 7 \cdot 19 = |G|$. Thus, $G = \langle hkl \rangle$.

15. (i) The number of Sylow 7-subgroups of G is $1 + 7k$ such that $1 + 7k | 24$. Thus, $k = 0$ or 1. If $k = 0$, then G has a unique Sylow 7-subgroup, which must be normal. This is a contradiction since G is simple. Thus, $k = 1$. Then G has eight Sylow 7-subgroups.

(iii) Let K be a subgroup of order 14 in G. Now $|K| = 14 = 2 \cdot 7$. The number of Sylow 7-subgroups of K is $1 + 7k$ such that $1 + 7k | 2$. Thus, $k = 0$. Hence, K has a unique Sylow 7-subgroup, say, P, which is normal in K. Now P is also a Sylow 7-subgroup of G. Since for all $a \in K$, $aPa^{-1} = P$, $K \subseteq N_G(P)$. This implies that $14 | |N_G(P)| = 21$, a contradiction. Hence, G has no subgroup of order 14.

18. Let G be a group of order 70. Let H be a Sylow 7-subgroup of G and K be a Sylow 5-subgroup of G. Then H and K are unique and hence normal. Also, H, K, and HK are cyclic subgroups, HK is a normal subgroup of G, $|H| = 7$, $|K| = 5$, and $|HK| = 35$. Let $HK = \langle a \rangle$. Then $o(a) = 35$. Let $b \in G$ and $o(b) = 2$. Now $b^{-1}ab \in HK$. Thus, $b^{-1}ab = a^r$ for some r, $1 \leq r \leq 34$. Now $G = HK \cup bHK$. From this, it follows that every element of G is of the form $b^s a^r$, where $s = 0$ or 1 and $0 \leq r \leq 34$. Now $b^{-1}ab = a^r$ implies that $o(a^r) = o(a) = 35$. Thus, $\gcd(r, 35) = 1$. Now $a = b(b^{-1}ab)b^{-1} = a^{r^2}$. Therefore, $35 | (r^2 - 1)$. Hence, $5 | (r^2 - 1)$ and $7 | (r^2 - 1)$, $1 \leq r \leq 34$. Now it follows that the only possible choices of r satisfying the above conditions are $r = 1, 6, 29, 34$. Thus, there are four groups of order 70.

Case 1: $r = 1$. Then $b^{-1}ab = a$, i.e., $ab = ba$. In this case, G is commutative and it is easy to verify that $G \simeq \mathbf{Z}_{70}$.

Case 2: $r = 34$. Then $b^{-1}ab = a^{34} = a^{-1}$, i.e., $ab = ba^{-1}$. Thus, $G = \langle a, b \rangle$ such that $a^{35} = e = b^2$ and $ab = ba^{-1}$. Hence, in this case $G \simeq D_{35}$.

Case 3: $r = 6$. Then $b^{-1}ab = a^6$. Thus, $ab = ba^6$. Clearly $o(a^k) \neq 2$ for all k, $0 \leq k \leq 34$. Let ba^k be an element of order 2. Then $(ba^k)^2 = e$ implies that $a^{7k} = e$. Hence, $35 | 7k$, which implies that $5|k$, $0 \le k \le 34$. Thus, $k = 0, 5, 10, 15, 20, 25, 30$. Hence, elements of order 2 in G are b, ba^5, ba^{10}, ba^{15}, ba^{20}, ba^{25}, ba^{30}. Let $u = a^5$ and $S = \langle u, b \rangle$. Then $u^7 = e = b^2$, $b^{-1}ub = ba^5b = a^{30} = u^6 = u^{-1}$. From this, it follows that $S \simeq D_7$. Now

$$a^k ba^{5t} a^{-k} = ba^{6k} a^{5t-k} = ba^{5(k+t)} \in S$$

and

$$ba^k ba^{5t} (ba^k)^{-1} = bba^{6k} a^{5t-k} b = ba^{30(k+t)} \in S.$$

Also, $a^k a^{5t} a^{-k} = a^{5t} \in S$ and $ba^k a^{5t} (ba^k)^{-1} = ba^k a^{5t-k} b = a^{30t} \in S$. Hence, S is a normal subgroup of G. Let $v = a^7$ and $T = \langle v \rangle$. Then T is a subgroup of G and $T \simeq \mathbf{Z}_5$. Now T is a normal subgroup of G (since T is a Sylow 5-subgroup of G). Clearly $S \cap T = \{e\}$ and $|G| = |ST|$. Thus, S and T are normal subgroups of G, $S \cap T = \{e\}$, and $|G| = |ST|$. Hence, $G = S \times T \simeq D_7 \times \mathbf{Z}_5$.

Case 4. $r = 29$. As in Case 3, we can show that $G \simeq D_5 \times \mathbf{Z}_7$.

21. Let G be a group of order 14. Then G is cyclic or $G \simeq D_7$.

23. If n is odd, then $Z(D_n) = \{e\}$, and if n is even, then $Z(D_n) = \{e, r^{\frac{n}{2}}\}$.

24. The conjugacy classes in D_{2n+1} are

$$\{e\}, \ \{b, ba, ba^2, \ldots, ba^{2n}\}, \ \{a^r, \ a^{-r}\}, \ 1 \le r \le n.$$

The conjugacy classes in D_{2n} are

$$\{e\}, \ \{b, ba^2, \ \ldots, \ ba^{2n}\}, \ \{ba, ba^3, \ \ldots, \ ba^{2n-1}\},$$
$$\{a^r, \ a^{-r}\}, \ 1 \le r \le n-1, \ \{a^n\}.$$

25. (ii) Suppose $n_p = q^2$. Then $pk + 1 = q^2$ for some $k \ge 0$. This implies $p|(q^2 - 1)$. Thus, either $p|(q + 1)$ or $p|(q - 1)$. Since $p > q$, $p \nmid (q - 1)$. Hence, $p|(q + 1)$. Thus, $q + 1 \ge p > q$. This implies that $p = 3$ and $q = 2$.

29. (i) False (ii) True (iii) True (iv) False.

Exercises 8.1.2 (page 237) (Solvable Groups)

1. Let $H_1 = \{r_{180}, r_{360}, h, v\}$, $H_2 = \{r_{360}, h\}$, $H_3 = \{r_{180}, r_{360}, d_1, d_2\}$ and $H_4 = \{r_{360}, d_1\}$. Note that $|G/H_1| = 2 = |H_1/H_2| = |G/H_3| = |H_3/H_4|$.

2. Let $H = \langle 66 \rangle$. The composition series is

$$\mathbf{Z}/H \supset 2\mathbf{Z}/H \supset 6\mathbf{Z}/H \supset 66\mathbf{Z}/H = \{H\},$$
$$\mathbf{Z}/H \supset 2\mathbf{Z}/H \supset 22\mathbf{Z}/H \supset 66\mathbf{Z}/H = \{H\},$$
$$\mathbf{Z}/H \supset 3\mathbf{Z}/H \supset 6\mathbf{Z}/H \supset 66\mathbf{Z}/H = \{H\},$$
$$\mathbf{Z}/H \supset 3\mathbf{Z}/H \supset 33\mathbf{Z}/H \supset 66\mathbf{Z}/H = \{H\},$$
$$\mathbf{Z}/H \supset 11\mathbf{Z}/H \supset 22\mathbf{Z}/H \supset 66\mathbf{Z}/H = \{H\},$$
$$\mathbf{Z}/H \supset 11\mathbf{Z}/H \supset 33\mathbf{Z}/H \supset 66\mathbf{Z}/H = \{H\}.$$

4. (i) $S_3 \supset A_3 \supset \{e\}$.

(ii) Let $K = \{e, (1\ 2) \circ (3\ 4), (1\ 3) \circ (2\ 4), (1\ 4) \circ (2\ 3)\}$, $H_1 = \{e, (1\ 2) \circ (3\ 4)\}$, $H_2 = \{e, (1\ 3) \circ (2\ 4)\}$ and $H_3 = \{e, (1\ 4) \circ (2\ 3)\}$. The composition series of A_4 is

$$A_4 \supset K \supset H_1 \supset \{e\},$$
$$A_4 \supset K \supset H_2 \supset \{e\},$$
$$A_4 \supset K \supset H_3 \supset \{e\}.$$

(iii) Let K, H_1, H_2 and H_3 be as in (ii). The composition series of S_4 is

$$S_4 \supset A_4 \supset K \supset H_1 \supset \{e\},$$
$$S_4 \supset A_4 \supset K \supset H_2 \supset \{e\},$$
$$S_4 \supset A_4 \supset K \supset H_3 \supset \{e\}.$$

(iv) The composition series of $\mathbf{Z}_2 \times \mathbf{Z}_2$ is

$$\mathbf{Z}_2 \times \mathbf{Z}_2 \supset \mathbf{Z}_2 \times \{[0]\} \supset \{([0],[0])\},$$
$$\mathbf{Z}_2 \times \mathbf{Z}_2 \supset \{[0]\} \times \mathbf{Z}_2 \supset \{([0],[0])\},$$
$$\mathbf{Z}_2 \times \mathbf{Z}_2 \supset \{([0],[0]),\ ([1],[1])\} \supset \{([0],[0])\}.$$

5. Let G be a finite group. Since G is finite, there exists a maximal normal subgroup G_1 of G. Thus, G/G_1 is simple. If $G_1 \neq \{e\}$, then since G_1 is finite, there exists a maximal normal subgroup G_2 of G_1. Then G_1/G_2 is simple. If $G_2 \neq \{e\}$, then continuing as before, we obtain the following series $G = G_0 \supset G_1 \supset \cdots \supset G_n \supset \cdots$ such that G_i/G_{i+1} is simple for all i. Since G is finite, there exists $n \geq 0$ such that $G_n = \{e\}$. Thus, $G = G_0 \supset G_1 \supset \cdots \supset G_n = \{e\}$ is a composition series.

12. If $p = q$, then G is a p-group and so solvable. Suppose $p \neq q$. Show that G has a unique Sylow p-subgroup or a unique Sylow q-subgroup. Let H be a Sylow p-subgroup and K be a Sylow q-subgroup. Then $|H| = p^2$ and $|K| = q$. Clearly both H and K are commutative and hence solvable. Suppose H is normal. Then $\left|\frac{G}{H}\right| = q$. This implies that $\frac{G}{H}$ is a cyclic group and hence solvable. Since H and $\frac{G}{H}$ are solvable, G is solvable. Now suppose K is normal. Then $\left|\frac{G}{K}\right| = p^2$. This implies that $\frac{G}{K}$ is a commutative group and hence solvable. Since K and $\frac{G}{K}$ are solvable, G is solvable.

14. A solvable series for $S_3 \times S_3$ is $S_3 \times S_3 \supset S_3 \times A_3 \supset A_3 \times A_3 \supset A_3 \times \{e\} \supset \{e\} \times \{e\}$.

15. If $G = \{e\}$, then the result is trivially true. Suppose $G \neq [e]$. Suppose G is not commutative. Then $G' \neq \{e\}$. Since G' is a normal subgroup of G and G is simple, $G = G'$. Thus, $G^{(n)} = G \neq \{e\}$ for all positive integers n. However, since G is solvable, there must exist a positive integer n such that $G^{(n)} = \{e\}$, a contradiction. Hence, G is commutative.

18. Suppose G is solvable and $H \neq \{e\}$ is a subgroup of G. Then H is solvable and hence $H' \neq H$ by Worked-Out Exercise 2 (page 234). Conversely, suppose that $H' \neq H$ for any subgroup $H \neq \{e\}$ of G. Then $G \neq G'$. Thus, $G \supset G'$. If $G^{(n)} \neq \{e\}$, then $G^{(n)} \neq G^{(n+1)}$, i.e., $G^{(n)} \supset G^{(n+1)}$. Hence, we have the following strictly descending chain of subgroups:

$$G \supset G' \supset \cdots \supset G^{(n)} \supset G^{(n+1)} \supset \cdots.$$

Since G is finite and $H' \neq H$ for any subgroup $H \neq \{e\}$ of G, there must exist a positive integer n such that $G^{(n)} = \{e\}$. Hence, G is solvable.

21. Note that $\frac{AB}{A} \simeq \frac{B}{A \cap B}$.

22. (i) False (ii) False (iii) True (iv) False (v) False (vi) False (vii) False.

Exercises 8.2.2 (page 244) (Nilpotent Groups)

3. $D_n = \langle a, b \rangle$, where $o(a) = n$, $o(b) = 2$, and $ab = ba^{-1}$. Suppose D_n is nilpotent. Also, suppose $n = 2^m k$, where k is an odd integer. Let $u = a^{2^m}$. Then $o(u) = k$. Now $ub = a^{2^m} b = ba^{-2^m} = bu^{-1}$. Let $H = \langle u, b \rangle$. Then $H \simeq D_k$. Now $|H| = 2k$ and H is nilpotent. Let $K = \{e, b\}$. Then K is a Sylow 2-subgroup of H. Since H is nilpotent, K is normal. But then $ubu^{-1} = bu^{-2} \in K$, a contradiction. Hence, $n = 2^m$ for some positive integer m.

5. No.

Exercises 9.1.2 (page 257) (Finite Abelian Groups)

2. Order 9 : $\mathbf{Z}_9$ and $\mathbf{Z}_3 \oplus \mathbf{Z}_3$.

Order 16 : $\mathbf{Z}_{16}$, $\mathbf{Z}_8 \oplus \mathbf{Z}_2$, $\mathbf{Z}_4 \oplus \mathbf{Z}_4$, $\mathbf{Z}_4 \oplus \mathbf{Z}_2 \oplus \mathbf{Z}_2$, and $\mathbf{Z}_2 \oplus \mathbf{Z}_2 \oplus \mathbf{Z}_2 \oplus \mathbf{Z}_2$.

Order 27 : $\mathbf{Z}_{27}$, $\mathbf{Z}_9 \oplus \mathbf{Z}_3$, and $\mathbf{Z}_3 \oplus \mathbf{Z}_3 \oplus \mathbf{Z}_3$.

Order 32 : $\mathbf{Z}_{32}$, $\mathbf{Z}_{16} \oplus \mathbf{Z}_2$, $\mathbf{Z}_8 \oplus \mathbf{Z}_4$, $\mathbf{Z}_8 \oplus \mathbf{Z}_2 \oplus \mathbf{Z}_2$, $\mathbf{Z}_4 \oplus \mathbf{Z}_4 \oplus \mathbf{Z}_2$, $\mathbf{Z}_4 \oplus \mathbf{Z}_2 \oplus \mathbf{Z}_2 \oplus \mathbf{Z}_2$, and $\mathbf{Z}_2 \oplus \mathbf{Z}_2 \oplus \mathbf{Z}_2 \oplus \mathbf{Z}_2 \oplus \mathbf{Z}_2$.

3. Order 15 : $\mathbf{Z}_{15}$.

4. Order 60 : $\mathbf{Z}_4 \oplus \mathbf{Z}_3 \oplus \mathbf{Z}_5$ and $\mathbf{Z}_2 \oplus \mathbf{Z}_2 \oplus \mathbf{Z}_3 \oplus \mathbf{Z}_5$.

Order 80 : $\mathbf{Z}_{16} \oplus \mathbf{Z}_5$, $\mathbf{Z}_8 \oplus \mathbf{Z}_2 \oplus \mathbf{Z}_5$, $\mathbf{Z}_4 \oplus \mathbf{Z}_4 \oplus \mathbf{Z}_5$, $\mathbf{Z}_4 \oplus \mathbf{Z}_2 \oplus \mathbf{Z}_2 \oplus \mathbf{Z}_5$, and $\mathbf{Z}_2 \oplus \mathbf{Z}_2 \oplus \mathbf{Z}_2 \oplus \mathbf{Z}_2 \oplus \mathbf{Z}_5$.

Order 240 : $\mathbf{Z}_{16} \oplus \mathbf{Z}_3 \oplus \mathbf{Z}_5$, $\mathbf{Z}_8 \oplus \mathbf{Z}_2 \oplus \mathbf{Z}_3 \oplus \mathbf{Z}_5$, $\mathbf{Z}_4 \oplus \mathbf{Z}_4 \oplus \mathbf{Z}_3 \oplus \mathbf{Z}_5$, $\mathbf{Z}_4 \oplus \mathbf{Z}_2 \oplus \mathbf{Z}_2 \oplus \mathbf{Z}_3 \oplus \mathbf{Z}_5$, and $\mathbf{Z}_2 \oplus \mathbf{Z}_2 \oplus \mathbf{Z}_2 \oplus \mathbf{Z}_2 \oplus \mathbf{Z}_3 \oplus \mathbf{Z}_5$.

Order 540 : $\mathbf{Z}_4 \oplus \mathbf{Z}_{27} \oplus \mathbf{Z}_5$, $\mathbf{Z}_2 \oplus \mathbf{Z}_2 \oplus \mathbf{Z}_{27} \oplus \mathbf{Z}_5$, $\mathbf{Z}_4 \oplus \mathbf{Z}_9 \oplus \mathbf{Z}_3 \oplus \mathbf{Z}_5$, $\mathbf{Z}_2 \oplus \mathbf{Z}_2 \oplus \mathbf{Z}_9 \oplus \mathbf{Z}_3 \oplus \mathbf{Z}_5$, $\mathbf{Z}_4 \oplus \mathbf{Z}_3 \oplus \mathbf{Z}_3 \oplus \mathbf{Z}_3 \oplus \mathbf{Z}_5$, and $\mathbf{Z}_2 \oplus \mathbf{Z}_2 \oplus \mathbf{Z}_3 \oplus \mathbf{Z}_3 \oplus \mathbf{Z}_3 \oplus \mathbf{Z}_5$.

6. (i) $2^2, 2^3, 2^4, 3, 3^2$. (ii) $2, 2, 2^3, 3, 3, 5, 5, 5$.

10. $\mathbf{Z}_{p^3} \oplus \mathbf{Z}_{q^2}$, $\mathbf{Z}_{p^2} \oplus \mathbf{Z}_p \oplus \mathbf{Z}_{q^2}$, $\mathbf{Z}_p \oplus \mathbf{Z}_p \oplus \mathbf{Z}_p \oplus \mathbf{Z}_{q^2}$, $\mathbf{Z}_{p^3} \oplus \mathbf{Z}_q \oplus \mathbf{Z}_q$, $\mathbf{Z}_{p^2} \oplus \mathbf{Z}_p \oplus \mathbf{Z}_q \oplus \mathbf{Z}_q$ and $\mathbf{Z}_p \oplus \mathbf{Z}_p \oplus \mathbf{Z}_p \oplus \mathbf{Z}_q \oplus \mathbf{Z}_q$.

11. $\mathbf{Z}_4 \oplus \mathbf{Z}_2 \oplus \mathbf{Z}_9$ and $\mathbf{Z}_4 \oplus \mathbf{Z}_2 \oplus \mathbf{Z}_3 \oplus \mathbf{Z}_3$.

15. Only two elements of order 3.

17. $\mathbf{Z}_{81}$ and $\mathbf{Z}_{27} \oplus \mathbf{Z}_3$.

18. (i) True (ii) False (iii) True.

Exercises 9.2.2 (page 267) (Finitely Generated Abelian Groups)

3. $|T(G)| = 252$.

4. The torsion coefficients are 30 and 60 and the betti number is 2.

5. The elementary divisors are $2, 2^4, 3, 3, 5, 11$; $d_1 = 6$ and $d_2 = 2640$.

7. (i) No (ii) No.

13. (i) False (ii) False (iii) False.

Exercises 10.1.2 (page 282) (Elementary Properties)

1. (i) [1], [3], [5], and [7] are the units of $\mathbf{Z}_8$. [1] and [5] are the units of $\mathbf{Z}_6$. (ii) [2], [4], and [6] are the nilpotent elements of $\mathbf{Z}_8$. $\mathbf{Z}_6$ has no nonzero nilpotent elements. (iii) The zero divisors in $\mathbf{Z}_8$ are [2], [4], and [6]. The zero divisors in $\mathbf{Z}_6$ are [2], [3], and [4].

3. (i) $a + (-1)a = 1 \cdot a + (-1)a = (1 + (-1))a = 0 \cdot a = 0$. Hence, $(-a) + (a + (-1)a) = 0 + (-a) = -a$. This implies that $((-a) + a) + (-1)a = -a$, i.e., $0 + (-1)a = -a$. Hence, $(-1)a = -a$. Similarly, $a(-1) = -a$. Now $(-1)(-1) = -(-1) = 1$ since 1 and -1 are additive inverses of each other.

5. Suppose R is commutative. Let $a, b \in R$. Then $(a + b)(a - b) = a(a - b) + b(a - b) = aa - ab + ba - bb = a^2 - ab + ab - b^2$ (since $ab = ba$) $= a^2 - b^2$. Conversely, suppose that $a^2 - b^2 = (a + b)(a - b)$ for all $a, b \in R$. Let $a, b \in R$. Now $(a + b)(a - b) = a^2 - ab + ab - b^2$. Hence, $a^2 - ab + ba - b^2 = a^2 - b^2$ and this implies that $-ab + ba = 0$ and so $ba = ab$.

7. Use induction on n.

8. Use induction.

9. Use Exercise 7 (page 282) and the fact that $\binom{p}{r}x = 0$ for all $x \in R$ and for all $1 \le r \le p - 1$ since the characteristic of R is p.

11. Suppose a is nilpotent. There exists a positive integer n such that $a^n = 0$. Now $(1 - a)(1 + a + \cdots + a^{n-1}) = 1 + a + \cdots + a^{n-1} - a - a^2 - \cdots - a^{n-1} - a^n = 1 - a^n = 1 - 0 = 1$.

16. Since $\gcd(m, n) = 1$, there exist integers r and s such that $1 = mr + ns$. Thus, $a = a^{mr+ns} = a^{mr}a^{ns} = (a^m)^r(a^n)^s = (b^m)^r(b^n)^s = b^{mr}b^{ns} = b^{mr+ns} = b$.

20. Note that $\begin{bmatrix} 1 & 0 \\ 0 & 1 \end{bmatrix}$ is the identity of $M_2(R)$ and for any positive integer n, $n \cdot 1 = 0$ if and only if $n \begin{bmatrix} 1 & 0 \\ 0 & 1 \end{bmatrix} = \begin{bmatrix} 0 & 0 \\ 0 & 0 \end{bmatrix}$. Hence, R and $M_2(R)$ have the same characteristic.

22. (i) False (ii) False (iii) False (iv) True (v) False (vi) True (vii) False (viii) False (ix) False (x) False.

Exercises 10.2.3 (page 287) (Some Important Rings)

3. Every nonzero nonunit is a zero divisor.

5. Let R be a regular ring with 1. Let $a \in R$ and $a \ne 0$. Suppose a is not a zero divisor. There exists $b \in R$ such that $aba = a$. This implies that $a(ba - 1) = 0$. Since a is not a zero divisor, $ba - 1 = 0$ and so $ba = 1$. Similarly, $ab = 1$. Hence, a is a unit.

Exercises 11.1.2 (page 293) (Subrings and Subfields)

1. (i) Clearly $T_1 \neq \phi$. Let $\begin{bmatrix} a & b \\ 0 & c \end{bmatrix}, \begin{bmatrix} d & e \\ 0 & f \end{bmatrix} \in T_1$. Now

$$\begin{bmatrix} a & b \\ 0 & c \end{bmatrix} - \begin{bmatrix} d & e \\ 0 & f \end{bmatrix} = \begin{bmatrix} a-d & b-e \\ 0 & c-f \end{bmatrix} \in T_1$$

and $\begin{bmatrix} a & b \\ 0 & c \end{bmatrix} \begin{bmatrix} d & e \\ 0 & f \end{bmatrix} = \begin{bmatrix} ad & ae+bf \\ 0 & cf \end{bmatrix} \in T_1$. Hence, T_1 is a subring.

2. (i) No (ii) No (iii) Yes.

4. Clearly $T \neq \phi$. Let $n1, m1 \in T$. Then $n1 - m1 = (n-m)1 \in T$ and $(n1)(m1) = nm1 \in T$. Hence, T is a subring.

7. To show that $\mathbf{Z}[\sqrt{2}]$ is not a subfield of $\mathbf{R}$. Assume $(\sqrt{2})^{-1} \in \mathbf{Z}[\sqrt{2}]$ and obtain a contradiction.

11. Since $0^p = 0$, $0 \in T$ and so $T \neq \phi$. Let $a, b \in T$. Since F is commutative, $(a-b)^p = a^p - pa^{p-1}b + \cdots + (-1)^i \binom{p}{i} a^{p-i}b^i + \cdots + (-1)^{p-1} pab^{p-1} + (-1)^p b^p$. Also, since $p \mid \binom{p}{i}$, $\binom{p}{i}x = 0$ for all $x \in F$ and for all $1 \leq i \leq p-1$. Hence, $(a-b)^p = a^p + (-1)^p b^p = a - b$ since if $p = 2$, then $1 = -1$, and if p is odd, then $(-1)^p = -1$. Hence, $a - b \in T$. Suppose $b \neq 0$. Then $(ab^{-1})^p = a^p (b^p)^{-1} = ab^{-1}$ and so $ab^{-1} \in T$. Since $0, 1 \in T$, T has at least two elements. Hence, T is a subfield of F.

14. Since $a0 = 0 = 0a$ for all $a \in R$, $0 \in C(R)$ and so $C(R) \neq \phi$. Let $a, b \in C(R)$. Then $ax = xb$ and $bx = xb$ for all $x \in R$. Now $x(a-b) = xa - xb = ax - bx = (a-b)x$ and $x(ab) = (xa)b = (ax)b = a(xb) = a(bx) = (ab)x$. Hence, $a - b, ab \in C(R)$. Thus, $C(R)$ is a subring.

16. The elements in the center of $M_2(\mathbf{R})$ are of the form $\begin{bmatrix} a & 0 \\ 0 & a \end{bmatrix}$, where $a \in \mathbf{R}$.

23. (i) False (ii) False (iii) False (iv) True (v) True (vi) True.

Exercises 11.2.2 (page 306) (Ideals and Quotient Rings)

1. (i) Clearly $I \neq \phi$. Let $\begin{bmatrix} 0 & b \\ 0 & c \end{bmatrix}, \begin{bmatrix} 0 & d \\ 0 & e \end{bmatrix} \in I$, $\begin{bmatrix} a & u \\ 0 & v \end{bmatrix} \in T_2(\mathbf{Z})$. Then $\begin{bmatrix} 0 & b \\ 0 & c \end{bmatrix} - \begin{bmatrix} 0 & d \\ 0 & e \end{bmatrix} = \begin{bmatrix} 0 & b-d \\ 0 & c-e \end{bmatrix} \in I$, $\begin{bmatrix} a & u \\ 0 & v \end{bmatrix} \begin{bmatrix} 0 & b \\ 0 & c \end{bmatrix} = \begin{bmatrix} 0 & ab+uc \\ 0 & vc \end{bmatrix} \in I$, and

$$\begin{bmatrix} 0 & b \\ 0 & c \end{bmatrix} \begin{bmatrix} a & u \\ 0 & v \end{bmatrix} = \begin{bmatrix} 0 & bv \\ 0 & cv \end{bmatrix} \in I.$$

Hence, I is an ideal. Now $\begin{bmatrix} a & u \\ 0 & v \end{bmatrix} + I = \begin{bmatrix} a & 0 \\ 0 & 0 \end{bmatrix} + I + \begin{bmatrix} 0 & u \\ 0 & v \end{bmatrix} + I = \begin{bmatrix} a & 0 \\ 0 & 0 \end{bmatrix} + I$.

Thus, $T_2(\mathbf{Z})/I = \left\{ \begin{bmatrix} a & 0 \\ 0 & 0 \end{bmatrix} + I \mid a \in \mathbf{Z} \right\}$.

2. $\mathbf{Z}_{24}/I = \{I, [1] + I, [2] + I, [3] + I, [4] + I, [5] + I, [6] + I, [7] + I\}$.

3. Since $0 = 0 + 0\sqrt{-5} \in I$, $I \neq \phi$. Let $a + b\sqrt{-5}, c + d\sqrt{-5} \in I$. Then $a - b$ and $c - d$ are even and so $(a-c) - (b-d) = (a-b) - (c-d)$ is even. Thus, $(a + b\sqrt{-5}) - (c + d\sqrt{-5}) = (a-c) + (b-d)\sqrt{-5} \in I$. Let $x + y\sqrt{-5} \in \mathbf{Z}[\sqrt{-5}]$. Then $(a + b\sqrt{-5})(x + y\sqrt{-5}) = (ax - 5by) + (ay + bx)\sqrt{-5}$. Now $(ax - 5by) - (ay + bx) = (a-b)x - (a-b)y - 6by$, which is even. Thus, $(a+b\sqrt{-5})(x+y\sqrt{-5}) \in I$. Since $\mathbf{Z}[\sqrt{-5}]$ is commutative, $(x+y\sqrt{-5})(a+b\sqrt{-5}) \in I$. Hence, I is an ideal.

5. Since $0 = 00 \in AB$, $AB \neq \phi$. Let $x, y \in AB$ and $r \in R$. Then $x = \sum_{i=1}^{n} a_i b_i$ and $y = \sum_{j=1}^{m} c_j d_j$, where $a_i, c_j \in A$ and $b_i, d_j \in B$ for all i and j. Clearly $x - y =$

$\sum_{i=1}^{n} a_i b_i - \sum_{j=1}^{m} c_j d_j \in AB$. Also, $rx = r(\sum_{i=1}^{n} a_i b_i) = \sum_{i=1}^{n} (ra_i)b_i \in AB$ since $ra_i \in A$ for all i, and $xr = (\sum_{i=1}^{n} a_i b_i)r = \sum_{i=1}^{n} a_i(b_i r) \in AB$ since $b_i r \in B$ for all i. Hence, AB is an ideal.

7. Use Theorem 11.2.11 and Exercise 6 (page 306).

11. (ii) Since $m|q$ and $n|q$, $q \in \langle m \rangle$ and $q \in \langle n \rangle$. Hence, $\langle q \rangle \subseteq \langle m \rangle \cap \langle n \rangle$. Let $a \in \langle m \rangle \cap \langle n \rangle$. Then $m|a$ and $n|a$. Since q is the lcm of m and n, $q|a$ and so $a \in \langle q \rangle$. Therefore, $\langle m \rangle \cap \langle n \rangle \subseteq \langle q \rangle$. Consequently, $\langle m \rangle \cap \langle n \rangle = \langle q \rangle$.

12. $\{0\} \times \{0\}$, $F_1 \times \{0\}$, $\{0\} \times F_2$, and $F_1 \times F_2$ are the only ideals of $F_1 \times F_2$.

13. (ii) Yes.

16. $\mathbf{Z}$ has no zero divisor; $\langle 6 \rangle$ is a nonzero ideal of $\mathbf{Z}$. Since $(2 + \langle 6 \rangle)(3 + \langle 6 \rangle) = 6 + \langle 6 \rangle = \langle 6 \rangle$, $2 + \langle 6 \rangle \neq \langle 6 \rangle$, and $3 + \langle 6 \rangle \neq \langle 6 \rangle$, it follows that $\mathbf{Z}/\langle 6 \rangle$ has zero divisors.

18. $\operatorname{ann} I = \{[0], [10]\}$.

19. $\operatorname{ann} I = \{0\}$.

21. Let I be an ideal of a regular ring R. Let $a \in I$. There exists $b \in R$ such that $a = aba$. Since I is an ideal, $x = bab \in I$. Now $axa = a(bab)a = (aba)ba = aba = a$. Hence, I is regular.

27. (i) False (ii) True (iii) False.

Exercises 11.3.2 (page 315) (Homomorphisms and Isomorphisms)

2. Let $a, b, c, d \in \mathbf{Z}$. Suppose that $a = c$ and $b = d$. Then $a + b - 1 = c + d - 1$ and $a + b - ab = c + d - cd$ since $+$ and $\cdot$ are well defined on $\mathbf{Z}$. Thus, $\oplus$ and $\odot$ are well defined on $\mathbf{Z}$. Define $f : \mathbf{Z} \to \mathbf{Z}$ by $f(a) = 1 - a$ for all $a \in \mathbf{Z}$. Clearly f is a one-one function of $\mathbf{Z}$ onto $\mathbf{Z}$. Now $f(a + b) = 1 - (a + b) = (1 - a) + (1 - b) - 1 = (1 - a) \oplus (1 - b) = f(a) \oplus f(b)$. Also, $f(ab) = 1 - ab = (1 - a) + (1 - b) - (1 - a)(1 - b) = (1 - a) \odot (1 - b) = f(a) \odot f(b)$. Thus, f preserves addition and multiplication of $(\mathbf{Z}, +, \cdot)$ onto $(\mathbf{Z}, \oplus, \odot)$. Hence, it follows that $(\mathbf{Z}, \oplus, \odot)$ is a ring isomorphic to the ring $(\mathbf{Z}, +, \cdot)$.

3. (ii) If there exists an isomorphism f of $\mathbf{R}$ onto $\mathbf{C}$, then there exists $r \in \mathbf{R}$ such that $f(r) = i$. In this case $f(r^2) = i^2 = -1$. But $f(-1) = -1$ since $f(1) = 1$ and so $r^2 = -1$. However, no such real number r exists. (iii) Yes.

4. (ii) Yes (iii) No (iv) $\operatorname{Ker} f = \left\{ \begin{bmatrix} 0 & b \\ 0 & c \end{bmatrix} \mid b, c \in \mathbf{Z} \right\}$.

5. No.

7. Suppose that there exists an isomorphism f of $2\mathbf{Z}$ onto $3\mathbf{Z}$. Then $2\mathbf{Z}$ and $3\mathbf{Z}$ are isomorphic as additive groups under f. Thus, f maps a generator of the cyclic group $2\mathbf{Z}$ onto a generator of the cyclic group $3\mathbf{Z}$. Hence, $f(2) = 3$ or $f(2) = -3$. Suppose $f(2) = 3$. Then $f(4) = f(2 + 2) = f(2) + f(2) = 3 + 3 = 6$. Since f also preserves multiplication, $f(4) = f(2 \cdot 2) = f(2)f(2) = 3 \cdot 3 = 9$. However, this is impossible.

10. No.

12. $S = \{0, 1, \ldots, n - 1\}$ is a subring of R isomorphic to $\mathbf{Z}_n$. The isomorphism is given by $f(i) = [i]$ for all $i \in S$.

13. Let f be a homomorphism of $\mathbf{R}$ into $\mathbf{R}$. Since $\mathbf{R}$ is a field, either $\operatorname{Ker} f = \{0\}$ or $\operatorname{Ker} f = \mathbf{R}$. If $\operatorname{Ker} f = \mathbf{R}$, then $f(x) = 0$ for all $x \in \mathbf{R}$. Suppose $\operatorname{Ker} f = \{0\}$. Then f is one-one. The desired result now follows as in Worked-Out Exercise 2 (page 313).

16. (i) True (ii) True (iii) False (iv) False (v) True (vi) False.

Exercises 12.1.2 (page 323) (Ring Embeddings)

3. Define $f^* : F \to F'$ by $f^*(ab^{-1}) = f(a)f(b)^{-1}$ for all $ab^{-1} \in F$. Then $ab^{-1} = cd^{-1}$ if and only if $ad = bc$ if and only if $f(ad) = f(bc)$ if and only if $f(a)f(d) = f(b)f(c)$ if and only if $f(a)f(b)^{-1} = f(c)f(d)^{-1}$ if and only if $f^*(ab^{-1}) = f^*(cd^{-1})$. Thus, f^* is a one-one mapping. Let $a'b'^{-1} \in F'$. Since f maps R onto R', there exist $a, b \in R$, $b \neq 0$ such that $f(a) = a'$ and $f(b) = b'$. Thus, $f^*(ab^{-1}) = f(a)f(b)^{-1} = a'b'^{-1}$. Hence, f^* maps F onto F'. Let $ab^{-1}, cd^{-1} \in F$. Then $f^*(ab^{-1} + cd^{-1}) = f^*((ad + bc)(bd)^{-1}) = f(ad + bc)f(bd)^{-1} = (f(a)f(d) + f(b)f(c))f(b)^{-1}f(d)^{-1} = f(a)f(b)^{-1} + f(c)f(d)^{-1} = f^*(ab^{-1}) +$

$f^*(cd^{-1})$. Also, $f^*(ab^{-1} \cdot cd^{-1}) = f^*(ac(bd)^{-1}) = f(ac)f(bd)^{-1} = f(a)f(c)f(b)^{-1}f(d)^{-1} = f(a)f(b)^{-1}f(c)f(d)^{-1} = f^*(ab^{-1}) \cdot f^*(cd^{-1})$. Suppose g^* is an isomorphism of F onto F' such that $g^* = f$ on R. Then $g^*(ab^{-1}) = g^*(a)g^*(b^{-1}) = g^*(a)g^*(b)^{-1} = f(a)f(b)^{-1} = f^*(ab^{-1})$. Thus, $f^* = g^*$. Hence, f^* is unique.

6. The field of quotients of $\mathbf{Z}[i] = \{\frac{a+bi}{c+di} \mid a,b,c,d \in \mathbf{Z},\ c+di \neq 0\} = \{p+qi \mid p,q \in \mathbf{Q}\}$.

The field of quotients of $\mathbf{Z}[\sqrt{2}] = \{\frac{a+b\sqrt{2}}{c+d\sqrt{2}} \mid a,b,c,d \in \mathbf{Z},\ c+d\sqrt{2} \neq 0\} = \{p+q\sqrt{2} \mid p,q \in \mathbf{Q}\}$.

Exercises 13.1.2 (page 334) (Direct Sum of Rings)

2. We have $A = RA = (R_1 \oplus \cdots \oplus R_n)A = R_1 A \oplus \cdots \oplus R_n A$. Let $A_i = R_i A$ for $i = 1, 2, \ldots, n$. Then A_i is an ideal of R_i since $R_i A_i = R_i(R_i A) = R_i A = A_i$. Let g be the natural homomorphism of R onto R/A. Then $R/A \simeq g(R_1) \oplus \cdots \oplus g(R_n)$ and $g(R_i) \simeq R_i/A_i$ for $i = 1, 2, \ldots, n$.

4. $\mathbf{Z}_{mn}$ is a cyclic group of order mn with generator $[1]$. Since $\gcd(m,n) = 1$, $\mathbf{Z}_m \oplus \mathbf{Z}_n$ is a cyclic group of order mn with generator $([1],[1])$. Define $g : \mathbf{Z}_{mn} \to \mathbf{Z}_m \oplus \mathbf{Z}_n$ by $g(i[1]) = i([1],[1])$, where $i \in \mathbf{Z}$. Then g is a (additive) group isomorphism of $\mathbf{Z}_{mn}$ onto $\mathbf{Z}_m \oplus \mathbf{Z}_n$. Now $g([i][j]) = g(i[1]j[1]) = g(ij[1]) = ij([1],[1]) = i([1],[1])j([1],[1]) = g(i[1])g(j[1]) = g([i])g([j])$.

Exercises 14.1.2 (page 343) (Polynomial Rings)

1. Let $\sum_{i=0}^{m} a_i x^i, \sum_{i=0}^{n} b_i x^i \in I[x]$, where $a_i, b_i \in I$. Either $m \geq n$ or $n \geq m$, say, $m \geq n$. If $m > n$, let $b_i = 0$ for $i = n+1, \ldots, m$. Then $(\sum_{i=0}^{m} a_i x^i - \sum_{i=0}^{n} b_i x^i) = \sum_{i=0}^{m}(a_i - b_i)x^i \in I[x]$. Let $\sum_{j=0}^{k} r_j x^j \in R[x]$. Then $(\sum_{j=0}^{k} r_j x^j)(\sum_{i=0}^{m} a_i x^i) = \sum_{j=0}^{k+m}(\sum_{i=0}^{j} r_i a_{j-i})x^j \in I[x]$. Similarly, $(\sum_{i=0}^{m} a_i x^i)(\sum_{j=0}^{k} r_j x^j) \in I[x]$.

3. $\langle x \rangle$ is the set of all polynomials over R with constant term 0.

4. (i) $q(x) = x^2 + x - 1$, $r(x) = x + 3$. (ii) $q(x) = x^2 + x + [4]$, $r(x) = x + [3]$.

5. $q(x) = x$, $r(x) = [4]x^3 + [3]x^2 + x + [3]$.

8. $([1] + [2]x)([1] - [2]x + [4]x^2) = [1] + [8]x^3 = [1]$.

11. The units of $\mathbf{Z}[x]$ are 1 and -1.

12. The units of $\mathbf{Z}_6[x]$ are $[1]$ and $[5]$.

14. (iii) Use Exercise 9.

15. (i) The proof is by induction on n. Suppose $n = 1$. Suppose that $\sum_{i=0}^{m} r_i(x_1)^i = 0$. Then $r_i = 0$ for $i = 0, 1, \ldots, m$ by the definition of a polynomial. Suppose the result is true for $n - 1 \geq 1$. Let $p(x_1, x_2, \ldots, x_n) = \sum_{i_n \ldots i_1} r_{i_1 \ldots i_n} x_1^{i_1} \cdots x_n^{i_n} \in R[x_1, x_2, \ldots, x_n]$. Suppose that $p(x_1, x_2, \ldots, x_n) = 0$. Now $p(x_1, x_2, \ldots, x_n) = \sum_{i_n}(\sum_{i_{n-1} \ldots i_1} r_{i_1 \ldots i_n} x_1^{i_1} \cdots x_{n-1}^{i_{n-1}})x_n^{i_n}$. Hence, $\sum_{i_{n-1} \ldots i_1} r_{i_1 \ldots i_n} x_1^{i_1} \cdots x_{n-1}^{i_{n-1}} = 0$ for all i_n. Thus, by induction $r_{i_1 \ldots i_n} = 0$ for all $i_1, \ldots, i_{n-1}$ and for each i_n.

(ii) Since $\sum_{i_n \ldots i_1} r_{i_1 \ldots i_n} x_1^{i_1} \cdots x_n^{i_n} = \sum_{i_n \ldots i_1} s_{i_1 \ldots i_n} x_1^{i_1} \cdots x_n^{i_n}$ if and only if $r_{i_1 \ldots i_n} = s_{i_1 \ldots i_n}$ for all $i_1, \ldots, i_n$, α is well defined. By Defintion 14.1.13, α maps $R[x_1, x_2, \ldots, x_n]$ onto $R[c_1, c_2, \ldots, c_n]$. Since for any two polynomials $f(x_1, x_2, \ldots, x_n), g(x_1, x_2, \ldots, x_n) \in R[x]$, $k(x_1, x_2, \ldots, x_n) = f(x_1, x_2, \ldots, x_n) + g(x_1, x_2, \ldots, x_n)$ implies $k(c_1, c_2, \ldots, c_n) = f(c_1, c_2, \ldots, c_n) + g(c_1, c_2, \ldots, c_n)$ and $h(x_1, x_2, \ldots, x_n) = f(x_1, x_2, \ldots, x_n)g(x_1, x_2, \ldots, x_n)$ implies $h(c_1, c_2, \ldots, c_n) = f(c_1, c_2, \ldots, c_n)g(c_1, c_2, \ldots, c_n)$, it follows that α preserves $+$ and $\cdot$.

16. $K[x]/\langle f(x)\rangle = \{k(x) + \langle f(x)\rangle \mid k(x) \in K[x]\}$. Let $k(x) + \langle f(x)\rangle \in K[x]/\langle f(x)\rangle$. Then there exist $q(x), r(x) \in K[x]$ such that $k(x) = q(x)f(x) + r(x)$, where $r(x) = 0$ or $\deg r(x) < n$. Thus, $k(x) + \langle f(x)\rangle = (q(x)f(x) + r(x)) + \langle f(x)\rangle = (q(x)f(x) + \langle f(x)\rangle) + (r(x) + \langle f(x)\rangle) = (0 + \langle f(x)\rangle) + (r(x) + \langle f(x)\rangle) = r(x) + \langle f(x)\rangle$.

17. (i) True (ii) False (iii) True.

Exercises 15.1.2 (page 352) (Euclidean Domains)

2. $(q_0 + q_1\sqrt{3}) = 1$, $r_0 + r_1\sqrt{3} = 8 - 2\sqrt{3}$.

3. $q_0 + q_1 i = 3 + 0i$, $r_0 + r_1 i = 0 + 1i$.

6. Let $f : \mathbf{Z} \to \mathbf{Z}_n$ be defined by $f(a) = [a]$ for all $a \in \mathbf{Z}$. Then f is an epimorphism. Use Exercise 5.

7. (i) True (ii) False (iii) False.

Exercises 15.2.2 (page 359) (Greatest Common Divisors)

1. (i) The associates of $3 - 2i$ are $3 - 2i$, $-3 + 2i$, $2 + 3i$, $-2 - 3i$. (ii) The associates of $1 + i\sqrt{5}$ are $1 + i\sqrt{5}$, $-1 - i\sqrt{5}$. (iii) The associates of $[6]$ in $\mathbf{Z}_{10}$ are $[2]$, $[4]$, $[6]$, and $[8]$. (iv) $[1]$, $[2]$, $[3]$, $[4]$ are the associates of $[4]$ in $\mathbf{Z}_5$. (v) The associates of $[2] + x$ are $[2] + [x]$ and $[1] + [2]x$.

2. ± 1.

3. The units of $\mathbf{Z}[x]$ are 1 and -1. The associates of $2 + x - 3x^2$ are $2 + x - 3x^2$ and $-2 - x + 3x^2$.

5. Units of $\mathbf{Z}_7[x]$ are the nonzero elements of $\mathbf{Z}_7$. The associates of $x^2 + [2]$ are $x^2 + [2]$, $[2]x^2 + [4]$, $[3]x^2 + [6]$, $[4]x^2 + [1]$, $[5]x^2 + [3]$, $[6]x^2 + [5]$.

7. Since a and b are associates, $a = bu$ and $b = aw$ for some units u and w. Now $v(a) = v(bu) \geq v(b)$ and $v(b) = v(aw) \geq v(a)$. Hence, $v(a) = v(b)$.

13. $\gcd(2 - 7i, 2 + 11i) = 1$; $x = -(2 + 4i)$ and $y = -3i$.

16. Let $a + b\sqrt{2}$ be a unit such that $1 < a + b\sqrt{2} < 1 + \sqrt{2}$. Then $a^2 - 2b^2 = \pm 1$. That is, $(a + b\sqrt{2})(a - b\sqrt{2}) = \pm 1$. Hence, $a - b\sqrt{2} = \frac{\pm 1}{a + b\sqrt{2}}$. This implies that $-1 < a - b\sqrt{2} < 1$. From $1 < a + b\sqrt{2} < 1 + \sqrt{2}$ and $-1 < a - b\sqrt{2} < 1$, it follows that $0 < 2a < 2 + \sqrt{2}$, i.e., $0 < a < 1 + \frac{\sqrt{2}}{2}$. Thus, since a is an integer, $a = 1$. This implies that $1 < 1 + b\sqrt{2} < 1 + \sqrt{2}$, a contradiction. Hence, there is no unit between 1 and $1 + \sqrt{2}$.

Exercises 15.3.2 (page 365) (Prime and Irreducible Elements)

1. In $\mathbf{Z}[i\sqrt{5}]$, the only units are ± 1 since $x + yi\sqrt{5} \in \mathbf{Z}[i\sqrt{5}]$ is a unit if and only if $x^2 + 5y^2 = 1$. Let $2 + i\sqrt{5} = (a + bi\sqrt{5})(c + di\sqrt{5})$ for some $a + bi\sqrt{5}, c + di\sqrt{5} \in \mathbf{Z}[i\sqrt{5}]$. Then $2 - i\sqrt{5} = (a - bi\sqrt{5})(c - di\sqrt{5})$. Hence, $9 = (2 + i\sqrt{5})((2 - i\sqrt{5}) = (a^2 + 5b^2)(c^2 + 5d^2)$. Now proceeding as in Example 15.3.11, it follows that either $a^2 + 5b^2 = 1$ or $c^2 + 5d^2 = 1$. Thus, either $a + bi\sqrt{5}$ is a unit or $c + di\sqrt{5}$ is a unit. Hence, $2 + i\sqrt{5}$ is irreducible. We now show that $2 + i\sqrt{5}$ is not prime. Since $(2 + i\sqrt{5})(2 - i\sqrt{5}) = 9 = 3 \cdot 3$, $(2 + i\sqrt{5})|3 \cdot 3$. Suppose $2 + i\sqrt{5}$ is prime. Then $(2 + i\sqrt{5})|3$ and so $3 = (2 + i\sqrt{5})(a + bi\sqrt{5})$ for some $a + bi\sqrt{5} \in \mathbf{Z}[i\sqrt{5}]$. Then $9(a^2 + 5b^2) = 9$. Thus, $a^2 + 5b^2 = 1$ and so $a + bi\sqrt{5}$ is a unit. Hence, $a + bi\sqrt{5} = \pm 1$. Suppose $a + bi\sqrt{5} = 1$. This implies that $3 = 2 + i\sqrt{5}$. Thus, $3 = 2$ and $\sqrt{5} = 0$, which is absurd. If $a + bi\sqrt{5} = -1$, then we would get $3 = -2$ and $\sqrt{5} = 0$, which is again absurd. Hence, $2 + i\sqrt{5}$ is not prime.

2. Suppose $1 + i = (a + bi)(c + di)$ for some $a, b, c, d \in \mathbf{Z}$. Thus, $1 - i = (a - bi)(c - di)$ and so $1^2 + 1^2 = (a^2 + b^2)(c^2 + d^2)$, i.e., $2 = (a^2 + b^2)(c^2 + d^2)$. Hence, $(a^2 + b^2 = 2$ and $c^2 + d^2 = 1)$ or $(a^2 + b^2 = 1$ and $c^2 + d^2 = 2)$. Suppose $a^2 + b^2 = 2$ and $c^2 + d^2 = 1$. Now $c^2 + d^2 = 1$ implies that $(c = 0, d \pm 1$ or $c = \pm 1, d = 0)$. Thus, $c + di$ is a unit. If $a^2 + b^2 = 1$ and $c^2 + d^2 = 2$, then $a + bi$ is a unit. Hence, $1 + i$ is irreducible.

3. Since $3|(2 + i\sqrt{5})(2 - i\sqrt{5})$, $3 \nmid (2 + i\sqrt{5})$, and $3 \nmid (2 - i\sqrt{5})$, 3 is not prime in $\mathbf{Z}[i\sqrt{5}]$.

5. No.

9. $[3]$ and $[6]$ are the prime elements. $[3]$ and $[6]$ are also irreducible.

12. (i) False (ii) False (iii) False (iv) True (v) True.

Exercise 16.1.2 (page 374) (Unique Factorization Domains)

1. Since $\mathbf{Z}$ is a PID, the result follows by Lemma 16.1.6.

4. Define $N : \mathbf{Z}[i\sqrt{6}] \to \mathbf{N}$ by $N(a + ib\sqrt{6}) = |a^2 + 6b^2|$. As in Example 16.1.4, show that $\mathbf{Z}[i\sqrt{6}]$ is an FD. Now $10 = (2 + i\sqrt{6})(2 - i\sqrt{6}) = 5 \cdot 2$. Show that 5 is an irreducible element, but not a prime element, and use Theorem 16.1.10.

6. (i) False (ii) False.

Exercises 16.2.2 (page 380) (Factorization of Polynomials over a UFD)

2. Suppose $f(x)$ is not irreducible in $\mathbf{Q}[x]$, but $f(x)$ is irreducible in $\mathbf{Z}[x]$. By Exercise 1 (page 380), $f(x)$ is primitive. Then by Lemma 16.2.8, $f(x)$ is irreducible in $\mathbf{Q}[x]$, a contradiction. Hence, $f(x)$ is not irreducible in $\mathbf{Z}[x]$.

3. Show that the ideal $I = \langle x, y \rangle$ is not a principal ideal in $\mathbf{Q}[x, y]$.

Exercises 16.3.2 (page 386) (Irreducibility of Polynomials)

3. In $Z_2[x]$, $x^2 + [2]x + [6] = x^2 = x \cdot x$. Hence, the polynomial $x^2 + [2]x + [6]$ is reducible in $\mathbf{Z}_2[x]$. Now $x^2 + 2x + 6$ has no roots in $\mathbf{Q}$ and so it is irreducible in $\mathbf{Q}$. Thus, $x^2 + 2x + 6$ is irreducible in $\mathbf{Z}$.   **4.** For $x^2 + 2x + 6 \in \mathbf{Z}[x]$. Let $p = 2$.

6. As in Example 16.3.6, show that $f(x)$ has no roots in $\mathbf{Q}$.

8. First show that $g(x) = 15x^2 + 5x - 6$ is irreducible in $\mathbf{Q}[x]$.

9. Consider $f(x - 1)$ and use Eisenstein's criterion.

11. Use Eisenstein's criterion.

12. $x^2 + [1]$, $x^2 + [1]x + [2]$, $x^2 + [2]x + [2]$, $[2]x^2 + [2]$, $[2]x^2 + x + [1]$, and $[2]x^2 + [2]x + [1]$.

15. (ii) $\frac{1}{2}p(p - 1)$.

Exercises 17.1.2 (page 399) (Maximal, Prime, and Primary Ideals)

1. Maximal ideals: $\{[0], [5]\}$, $\{[0], [2], [4], [6], [8]\}$. Prime ideals: $\{[0], [5]\}$, $\{[0], [2], [4], [6], [8]\}$, and $\mathbf{Z}_{10}$.

3. Only one maximal ideal: $\{[0], [p], [2p], \ldots, [(p-1)p]\}$.

4. I is a maximal ideal.

7. Let I be a nonzero proper ideal of R such that I is prime. Let J be an ideal of R such that $I \subset J$. There exists $a \in J$ such that $a \notin I$. Now $a(1 - a) = 0 \in I$ and $a \notin I$. Hence, $1 - a \in I \subset J$. Also, $a \in J$ and so $1 \in J$. Thus, $J = R$ and so I is a maximal ideal. The converse follows by Theorem 17.1.7, since every Boolean ring is commutative.

13. Let $f(x), g(x) \in K[x]$. Then $\phi_a(f(x) + g(x)) = f(a) + g(a) = \phi_a(f(x)) + \phi_a(g(x))$ and $\phi_a(f(x)g(x)) = f(a)g(a) = \phi_a(f(x))\phi_a(g(x))$. Therefore, ϕ_a is a homomorphism. Let $b \in K$. Now $f(x) = b - ax + x^2 \in K[x]$ and $\phi_a(f(x)) = b - a^2 + a^2 = b$. Thus, ϕ_a is an epimorphism. Hence, $K[x]/\text{Ker } \phi_a \simeq K$. Since K is a field, $K[x]/\text{Ker } \phi_a$ is a field and so Ker ϕ_a is a maximal ideal.

15. Clearly $\cap_\alpha I_\alpha$ is an ideal. Let $ab \in \cap_\alpha I_\alpha$. If either $a \in I_\alpha$ for all α or $b \in I_\alpha$ for all α, then either $a \in \cap_\alpha I_\alpha$ or $b \in \cap_\alpha I_\alpha$. Suppose $a, b \notin \cap_\alpha I_\alpha$. Then there exist α and β such that $a \notin I_\alpha$ and $b \notin I_\beta$. Since $\{I_\alpha\}$ is a chain, either $I_\alpha \subseteq I_\beta$ or $I_\beta \subseteq I_\alpha$. To be specific, let $I_\alpha \subseteq I_\beta$. This implies that $b \notin I_\alpha$. Since $ab \in I_\alpha$, and I_α is prime we must have either $a \in I_\alpha$ or $b \in I_\alpha$, a contradiction. Therefore, either $a \in \cap_\alpha I_\alpha$ or $b \in \cap_\alpha I_\alpha$. Consequently, $\cap_\alpha I_\alpha$ is a prime ideal.

19. (ii) $(x + 2)(x + 3) \in \langle x, 6 \rangle$, but neither $(x + 2)^n \in \langle x, 6 \rangle$ nor $(x + 3)^n \in \langle x, 6 \rangle$ for any positive integer n. Hence, $\langle x, 6 \rangle$ is not primary.

24. (i) $\mathbf{Z}_8$ has only one maximal ideal, namely, $\{[0], [2], [4], [6]\}$ and so $\mathbf{Z}_8$ is a local ring.

30. (i) True (ii) True (iii) True (iv) False (v) True (vi) False (vii) False (viii) True.

Exercises 17.2.2 (page 405) (Jacobson Semisimple Ring)

1. Suppose the ring $\mathbf{Z}_n$, $n > 1$, is J-semisimple. There exist prime integers $p_1, p_2, \ldots, p_k$ such that $n = p_1 p_2 \cdots p_k$. We claim that $p_1, p_2, \ldots, p_k$ are all distinct. Suppose there exist $1 \leq i < j \leq k$ such that $p_i = p_j = p$ (say). The mapping $\beta : \mathbf{Z} \to \mathbf{Z}_n$ defined by $\beta(m) = [m]$ is an epimorphism of rings and Ker $\beta = n\mathbf{Z}$. Now Ker $\beta \subseteq t\mathbf{Z}$, where $t \geq 1$ and t divides n. Hence, the ideals of $\mathbf{Z}_n$ are of the form $I_t = \{[0], [t], [2t], \ldots\}$, where $t \geq 1$ and t divides n. Note that $I_{t_1} \subseteq I_{t_2}$ if and only if t_2 divides t_1. Hence, the maximal ideals are given by I_{p_i} for $i = 1, 2, \ldots, k$. Now if $t = pp_1p_2 \cdots p_{i-1}p_{i+1} \cdots p_k$, then t is divisible by each p_i and $t | n$.

Therefore, $I_t \subseteq I_{p_i}$ for all $i = 1, 2, \ldots, k$. Thus, $I_t \subseteq \cap I_{p_i} = \text{rad}\mathbf{Z}_n$ and so $\text{rad}\mathbf{Z}_n \neq \{0\}$, a contradiction. Hence, n is square free. If n is square free, then from the above argument we can prove that $\mathbf{Z}_n$ is J-semisimple

2. Yes.

3. Let R be a PID together with the given properties. Each maximal ideal of a PID is generated by a prime element and each prime element of R generates a maximal ideal. Hence, $\text{rad}R = \cap_{p \text{ is a prime elment of } R} \langle p \rangle$. Let $a \in \text{rad}R$. Then $a \in \langle p \rangle$ for all prime elements p of R. Hence, a has an infinite number of nontrivial divisors. Since R is also a UFD, it follows that $a = 0$ or a is a unit. But $\text{rad}R$ does not contain any units. Hence, $a = 0$ and so R is J-semisimple.

6. (i) True (ii) True (iii) False (iv) True.

Exercises 18.1.2 (page 419) (Noetherian and Artinian Rings)

2. Consider the field $\mathbf{Q}$ of rational numbers. The polynomial ring $\mathbf{Q}[x]$ is Noetherian. Let $R = \{f(x) \in \mathbf{Q}[x] \mid \text{the constant term of } f(x) \text{ is an integer}\}$. Then R is a subring of $\mathbf{Q}[x]$. We show that R is not Noetherian. For this, let $I_r = \langle 2^r x \rangle$ for $r = 0, -1, -2, \ldots$. Then $I_0 \subset I_{-1} \subset I_{-2} \subset \cdots$ is an infinite strictly ascending chain of ideals of R. Hence, R is not Noetherian.

3. $\mathbf{Z}(p^\infty)$.

10. (i) Since $I \in \mathcal{F}_1$, $\mathcal{F}_1 \neq \phi$. Since R is right Artinian, $\mathcal{F}_1$ has a minimal element, say I_0. Now $I_0^2 \subseteq I_0$. Since I_0 is minimal, either $I_0^2 = \{0\}$ or $I_0^2 = I_0$. If $I_0^2 = \{0\}$, then I_0 is a nilpotent right ideal, a contradiction. Hence $I_0^2 = I_0$. (ii) Since $I_0^2 = I_0$, we find that $I_0 \in \mathcal{F}$ and so $\mathcal{F} \neq \phi$. Now R is right Artinian. So $\mathcal{F}$ contains a minimal element, say I_1. Then $I_1 I_0 \neq \{0\}$. This implies that there exists a nonzero element $u \in I_1$ such that uI_0 is a nonzero right ideal of R and $uI_0 \subseteq I_1 \subseteq I_0$. Also, $(uI_0)I_0 = uI_0^2 = uI_0 \neq \{0\}$ shows that $uI_0 \in \mathcal{F}$. Since I_1 is minimal in $\mathcal{F}$, it now follows that $uI_0 = I_1$. (iii) $uI_0 = I_1$ implies that $ua = u$ for some $a \in I_0$. Since $u \neq 0$, we find that $a \neq 0$. Also, $u = ua = ua^2 = ua^3 = \cdots$ shows that $a^n \neq 0$ for all $n \geq 1$. Thus, I contains nonnilpotent elements. Hence, I is not a nil right ideal. (iv) Let $T = \{r \in I_0 \mid ur = 0\}$. Then T is a right ideal of R such that $T \subseteq I_0$. Since $uI_0 = I_1 \neq \{0\}$, it follows that $T \subset I_0$. Since I_0 is a minimal element in $\mathcal{F}_1$, it follows that T is a nilpotent right ideal. Also, $u(a^2 - a) = 0$ ($a \in I_0$, see (iii)) implies that $a^2 - a \in T$. Hence, $a^2 - a$ is nilpotent. Thus, there exists a positive integer n such that $(a^2 - a)^n = 0$. This implies that $a^n = a^{n+1}f(a)$ for some polynomial $f(x)$ in $\mathbf{Z}[x]$. Thus, $a^n = a(a^n f(a)) = a^{n+2}f(a)^2$. Proceeding in this way, we obtain $a^n = a^{2n}f(a)^n$. Let $e = a^n f(a)^n$. Now $e \in I$ and $e^2 = a^{2n}f(a)^{2n} = (a^{2n}f(a)^n)f(a)^n = a^n f(a)^n = e$. Clearly $e \neq 0$, otherwise, $e = 0$ implies $a^n = 0$. Hence, I contains a nonzero idempotent.

13. (i) True (ii) False (iii) True.

Exercises 19.1.2 (page 431) (Modules and Vector Spaces)

1. (i) Yes (ii) Yes (iii) Yes (iv) No.

6. (ii) Let $x \in (A + B) \cap C$. Then $x \in C$ and $x = a + b$ for some $a \in A$ and $b \in B$. Now $b = x - a \in C + A = C$. This implies that $b \in B \cap C$. Hence, $x = a + b \in A + (B \cap C)$. Thus, $(A + B) \cap C \subseteq A + (B \cap C)$. On the other hand, let $x \in A + (B \cap C)$. Then $x = a + b$ for some $a \in A$ and $b \in B \cap C$. Then $a + b \in A + B$ and $a + b \in A + C = C$. This implies that $x \in (A + B) \cap C$. Thus, $A + (B \cap C) \subseteq (A + B) \cap C$. Consequently, $A + (B \cap C) = (A + B) \cap C$.

8. Suppose M is a simple module and let $0 \neq m \in M$. Then $Rm \neq \{0\}$ and Rm is a submodule of M. Hence, $M = Rm$ and so M is generated by m. Conversely, assume that $Rm = M$ for all nonzero elements $m \in M$. Let T be a nonzero submodule of M. Let $0 \neq a \in T$. Then $Ra \neq \{0\}$ and $Ra \subseteq T$. Thus, $M = Ra \subseteq T \subseteq M$ and so $T = M$. Hence, M is simple.

16. Let W be a subspace of V. Let $\dim V = n$. Then $\dim W = m \leq n$. Let $\{w_1, w_2, \ldots, w_m\}$ be a basis of W. Now $\{w_1, w_2, \ldots, w_m\}$ is a linearly independent set in V and hence can be extended to a basis of V, say $\{w_1, w_2, \ldots, w_m, w_{m+1}, \ldots, w_n\}$. Let U be the subspace of V generated by $\{w_{m+1}, \ldots, w_n\}$. Now $W + U \subseteq V$. Let $x \in V$. Then $x = a_1 w_1 + \cdots + a_m w_m + a_{m+1} w_{m+1} + \cdots + a_n w_n$ for some $a_i \in F$, $1 \leq i \leq n$. Since $a_1 w_1 + \cdots + a_m w_m \in W$ and $a_{m+1} w_{m+1} + \cdots + a_n w_n \in V$, $x \in W + U$. Hence, $V = W + U$. Let $x \in W \cap U$. Then $x = b_1 w_1 + \cdots + b_m w_m$ and $x = b_{m+1} w_{m+1} + \cdots + b_n w_n$ for some $b_i \in F$. Thus, $b_1 w_1 + \cdots + b_m w_m = b_{m+1} w_{m+1} + \cdots + b_n w_n$, which implies that $b_1 w_1 + \cdots + b_m w_m + (-b_{m+1}) w_{m+1} + \cdots + (-b_n) w_n = 0$. This implies that $b_i = 0$, $1 \leq i \leq n$. Thus, $x = 0$. Hence, $W \cap U = \{0\}$. Consequently, W is a direct summand of V.

Exercises 20.1.2 (page 437) (Full Matrix Rings)
1. $\{0\}$ and $M_n(R)$.

Exercises 20.2.2 (page 444) (Rings of Triangular Matrix)
2. Since $\mathbf{R}$ is of infinite dimension over $\mathbf{Q}$, $\mathbf{R}$ is not an Artinian right $\mathbf{Q}$-module. Hence,
$\begin{bmatrix} 0 & 0 \\ \mathbf{R} & 0 \end{bmatrix}$ and so $\begin{bmatrix} \mathbf{Q} & 0 \\ \mathbf{R} & \mathbf{Q} \end{bmatrix}$ is not right Artinian by Theorem 20.2.3.

Exercises 21.1.2 (page 457) (Algebraic Extensions)
3 (i). Suppose that π^2 is algebraic over $\mathbf{Q}$. Then $[\mathbf{Q}(\pi^2) : \mathbf{Q}] < \infty$. Now $[\mathbf{Q}(\pi) : \mathbf{Q}(\pi^2)] = 2$ and so $[\mathbf{Q}(\pi) : \mathbf{Q}] < \infty$. Thus, π is algebraic over $\mathbf{Q}$, a contradiction. Hence, π^2 is transcendental over $\mathbf{Q}$.

5. Suppose that π is algebraic over $\mathbf{Q}(\sqrt{2})$. Then $[\mathbf{Q}(\sqrt{2})(\pi) : \mathbf{Q}(\sqrt{2})] < \infty$ and since $[\mathbf{Q}(\sqrt{2}) : \mathbf{Q}] < \infty$, $[\mathbf{Q}(\sqrt{2})(\pi) : \mathbf{Q}] < \infty$. Hence, every element of $\mathbf{Q}(\sqrt{2})(\pi)$ is algebraic over $\mathbf{Q}$. However, this is impossible since π is transcendental over $\mathbf{Q}$.

8. $[\mathbf{Q}(\sqrt[3]{5}) : \mathbf{Q}] = 3$.

9. $[\mathbf{Q}(\sqrt{3}, \sqrt{5}) : \mathbf{Q}] = 4$.

10. Suppose that $x^2 - 5$ is not irreducible over $\mathbf{Q}(\sqrt{2})$. Then there exist polynomials $x - a, x - b \in \mathbf{Q}(\sqrt{2})$ such that $x^2 - 5 = (x - a)(x - b) = x^2 - (a + b)x + ab$. Hence, $a + b = 0$ and $ab = -5$. Thus, $a^2 = 5$. Hence, $\sqrt{5} \in \mathbf{Q}(\sqrt{2})$. Thus, there exists $p, q \in \mathbf{Q}$ such that $\sqrt{5} = q + p\sqrt{2}$. Hence, $5 = q^2 + 2pq\sqrt{2} + 2p^2$ and so $\sqrt{2} \in \mathbf{Q}$, a contradiction.

11. $x^4 - 4x^2 - 1$.

13. $[\mathbf{Q}(\sqrt{2}, \sqrt{5}) : \mathbf{Q}] = 4$. $\{1, \sqrt{2}, \sqrt{5}, \sqrt{10}\}$ is a basis of $\mathbf{Q}(\sqrt{2}, \sqrt{5})/\mathbf{Q}$.

14. By Theorem 21.1.18, every element of $K(c)/K$ is algebraic over K. Now $f(c) \in K(c)$. Thus, $f(c)$ is algebraic over K.

23. (i) True (ii) False (iii) False.

Exercises 21.2.2 (page 467) (Splitting Fields)
4. $S = \mathbf{Q}(\sqrt[3]{3}, i\sqrt{3})$, $[S : \mathbf{Q}] = 6$, and $\{1, \sqrt[3]{3}, \sqrt[3]{9}, i\sqrt{3}, i\sqrt{3}\sqrt[3]{3}, i\sqrt{3}\sqrt[3]{9}\}$ is a basis of S over $\mathbf{Q}$.

5. $[S : \mathbf{Z}_5] = 2$ and $\{1, \lambda\}$ is a basis of $S/\mathbf{Z}_5$, where λ is a root of $x^2 + x + [1]$.

6. $S = \mathbf{Z}_2$, and $\{[1]\}$ is a basis of $S/\mathbf{Z}_2$.

7. $S = \mathbf{Q}(\sqrt{2}, \sqrt{5})$, $[S : \mathbf{Q}] = 4$ and $\{1, \sqrt{2}, \sqrt{5}, \sqrt{10}\}$ is a basis for $S/\mathbf{Q}$.

8. $[\mathbf{Q}(-\frac{1}{2} + \frac{\sqrt{3}}{2}i) : \mathbf{Q}] = 2$.

12. The proof is by induction on n. Suppose that $n = 1$. Then $f(x) = ax + b$ for some $a, b \in K$, $a \neq 0$. Hence, $S = K(-\frac{b}{a})$ and $[S : K] = 1 \leq 1!$ Assume the result is true for all polynomials of degree $\leq n - 1$. Suppose $\deg f(x) = n$. Let λ be a root of $f(x)$. Then $[K(\lambda) : K] \leq n$ and $f(x) = (x - \lambda)g(x)$ over $K(\lambda)$ and $\deg g(x) = n - 1$. Let S be the splitting field of $g(x)$ over $K(\lambda)$. Then $[S : K(\lambda)] \leq (n - 1)!$ by the induction hypothesis. Now S is the splitting field of $f(x)$ over K and $[S : K] = [S : K(\lambda)][K(\lambda) : K] \leq (n - 1)! \cdot n = n!$

15. (i) False (ii) True.

Exercises 22.1.2 (page 489) (Multiplicity of Roots)

5. (i) separable (ii) separable.

6. (i) $e = 3$ and $n_0 = 2$. (ii) $e = 0$ and $n_0 = 128$. (iii) $e = 3$, $n_0 = 1$.

8. Consider $p(x)$ in Example 22.1.16. $p(x)$ is irreducible over $K(t)$ by Eisenstein's criterion. $D_x(p(x)) = 0$ and so $p(x)$ is inseparable over $K(t)$. Now $e = 1$ and $n_0 = 2$. Thus, a root of $p(x)$ is inseparable, but not purely inseparable over $K(t)$.

Let $p(x) = x^{2p} + vx^p + u$ in Example 22.1.39. By Worked-Out Exercise 6 (page 456), $p(x)$ is irreducible over K. Now $D_x(p(x)) = 0$ and so $p(x)$ is inseparable over K. Also, $e = 1$ and $n_0 = 2$. Thus, any root of $p(x)$ is inseparable, but not purely inseparable over K.

10. Since $[F : K] < \infty$, F/K is algebraic. Let $a \in F$. Let $p(x)$ be the minimal polynomial of a over K. Then $[K(a) : K]$ divides $[F : K]$. Hence, p does not divide $[K(a) : K]$. Thus, the exponent of inseparability of $p(x)$ is 0. Hence, a is separable over K. Thus, F/K is separable.

13. $\mathbf{Z}_p$ is the smallest subfield of K. Now for all $x \in \mathbf{Z}_p \backslash \{0\}$, $x^{p-1} = 1$ and so $x^p = x$. Hence, $(\mathbf{Z}_p)^p = \mathbf{Z}_p$ and so $\mathbf{Z}_p$ is perfect.

15. (i) False (ii) True (iii) True.

Exercises 23.1.2 (page 497) (Finite Fields)

2. In order to construct a field with $9 (= 3^2)$ elements, we find an irreducible polynomial of degree 2 over $\mathbf{Z}_3$. For this, consider the polynomial $x^2 + [1]$ over $\mathbf{Z}_3$. Since $[0]^2 + [1] \neq [0]$, $[1]^2 + [1] \neq [0]$, $[2]^2 + [1] \neq [0]$, $\mathbf{Z}_3$ contains no roots of $x^2 + [1]$. Hence, $x^2 + [1]$ is an irreducible polynomial of degree 2 over $\mathbf{Z}_3$. This implies that $F = \mathbf{Z}_3/\langle x^2 + [1] \rangle$ is a field. Since $[F : \mathbf{Z}_3] = 2$, F has nine elements by Theorem 23.1.1.

6. Let $[F : \mathbf{Z}_p] = m$. Then $|F| = p^m$ by Theorem 23.1.1. Hence, $n = m$.

7. Let F be the splitting field of $x^{3^2} - x$ over $\mathbf{Z}_3$. Then $|F| = 3^2$ and $[F : \mathbf{Z}_3] = 2$. Since every element of F is a root of $x^{3^2} - x$ by Theorem 23.1.2 and since $[F : \mathbf{Z}_3] = 2$, $F = \mathbf{Z}_p(c)$ for any root c of $g(x)$, where $g(x) \in \mathbf{Z}_3[x]$ is such that $x^{3^2} - x = x(x - [1])(x - [2])g(x)$.

9. By Theorem 23.1.1, $F \simeq GF(p^m)$ for some m and clearly $m \leq n$. Now $n = [GF(p^n) : \mathbf{Z}_p] = [GF(p^n) : F][F : \mathbf{Z}_p] = [GF(p^n) : F]m$. Thus, $m|n$.

Exercises 24.1.2 (page 502) (Normal Extensions)

1. (ii) $\mathbf{R}$ is not a normal extension of $\mathbf{Q}$.

2. (ii) No. Consider $L = \mathbf{Q}(\sqrt[3]{2})$ over $\mathbf{Q}$.

Exercises 24.2.2 (page 518) (Galois Theory)

1. $G(\mathbf{C}/\mathbf{R})$ consists of the identity automorphism and the automorphism α of $\mathbf{C}/\mathbf{R}$ such that $\alpha(i) = -i$.

2. (i) $[F : \mathbf{Q}] = 6$, $[N : \mathbf{Q}] = 12$, and $G(N/\mathbf{Q}) \simeq \mathbf{Z}_2 \times S_3$. (ii) $[F : \mathbf{Q}] = 6$, $[N : \mathbf{Q}] = 12$, and $G(N/\mathbf{Q}) \simeq \mathbf{Z}_2 \times S_3$.

5. $G(S/\mathbf{Q}) \simeq S_3$. $x^3 - x - 1$ has one real root, say r_1, and two complex roots, say r_2 and r_3. The following table defines $G(S/\mathbf{Q})$.

	e	α	β	$\alpha\beta$	$\beta\alpha$	$\alpha\beta\alpha$
r_1	r_1	r_1	r_2	r_3	r_2	r_3
r_2	r_2	r_3	r_1	r_1	r_3	r_2
r_3	r_3	r_2	r_3	r_2	r_1	r_1

The proper subgroups of $G(S/\mathbf{Q})$ are $H_1 = \{e, \alpha\}$, $H_2 = \{e, \beta\}$, $H_3 = \{e, \alpha\beta\alpha\}$, $H_4 = \{e, \alpha\beta, \alpha\beta\}$, and the corresponding intermediate fields of $S/\mathbf{Q}$ are $L_1 = \mathbf{Q}(r_1)$, $L_2 = \mathbf{Q}(r_3)$, $L_3 = \mathbf{Q}(r_2)$, $L_4 = \mathbf{Q}(d)$.

6. (i) $\mathbf{Z}_2 \times S_3$. (ii) $\mathbf{Z}_2$.

10. The intermediate fields of $\mathbf{Q}(i, \sqrt{7})/\mathbf{Q}$ are $\mathbf{Q}$, $\mathbf{Q}(i, \sqrt{7})$, $\mathbf{Q}(i)$, $\mathbf{Q}(\sqrt{7})$, and $\mathbf{Q}(i\sqrt{7})$.

12. $|G(F/\mathbf{Q})| = 8$.

13. The Galois group G of F over $\mathbf{Q}$ is isomorphic to $\mathbf{Z}_2 \times \mathbf{Z}_2$. All the intermediate fields of $\mathbf{Q}(\sqrt{3}, \sqrt{11})/\mathbf{Q}$ are normal extensions of $\mathbf{Q}$ in $\mathbf{Q}(\sqrt{3}, \sqrt{11})$.

Exercises 24.3.2 (page 527) (Roots of Unity and Cyclotomic Polynomials)

1. $\mathbf{Z}_2$.

2. Both are isomorphic to $\mathbf{Z}_2$.

4. Show that $\Phi_{2n}(x) = \Phi_n(-x) = \frac{x^n+1}{x+1}$.

5. Let $\Psi(x) = \frac{\Phi_n(x^p)}{\Phi_{pn}(x)}$. If ω is a primitive pnth root of unity, then ω^p is a primitive nth root of unity. Therefore, ω is a root of $\Phi_n(x^p)$. Hence, $\frac{\Phi_{pn}(x)}{\Phi_n(x^p)}$ is a monic polynomial in $\mathbf{Q}[x]$ of degree $\phi(n)$. Thus, so is $\Psi(x)$. If δ is a primitive nth root of unity, then so is δ^p since p and n are relatively prime. Therefore, δ is a root of $\Phi_n(x^p)$. Thus, δ is a root of $\Psi(x)$. Consequently, $\Psi(x) = \Phi_n(x)$.

6. $x^2 + 1$.

10. By Exercise 7 (527), F contains a primitive nth root of unity, say ω. Now ω^r is a primitive nth root of unity if and only if ω^r generates $\langle \omega \rangle$ if and only if r and n are relatively prime.

Exercises 24.4.2 (page 535) (Solvability of Polynomials by Radicals)

1. Consider $x^3 + \frac{9}{2}x + 3$. Then $b = 0$, $c = \frac{9}{2}$, $d = 3$. Now $p = \frac{9}{2}$, $q = 3$, $s = \sqrt[3]{-\frac{3}{2} + \sqrt{\frac{27}{8} + \frac{9}{4}}} = \sqrt[3]{-\frac{3}{2} + \frac{3}{2}\sqrt{\frac{5}{2}}}$, and $t = \sqrt[3]{-\frac{3}{2} - \frac{3}{2}\sqrt{\frac{5}{2}}}$. The roots of $2x^3 + 9x + 6$ are $s + t$, $\omega s + \omega^2 t$, and $\omega^2 s + \omega t$, where ω is a cube root of 1, $\omega \neq 1$.

4. (i) $x_1^2 + x_2^2 + x_3^2 = (x_1 + x_2 + x_3)^2 - 2(x_1 x_2 + x_1 x_3 + x_2 x_3)$. (ii) $(x_1 - x_2)^2 (x_1 - x_3)^2 (x_2 - x_3)^2 = (x_1 + x_2 + x_3)^2 (x_1 x_2 + x_1 x_3 + x_2 x_3)^2 - 4(x_1 x_2 + x_1 x_3 + x_2 x_3)^2 - 4(x_1 + x_2 + x_3)^3 (x_1 x_2 x_3) - 27(x_1 x_2 x_3)^2 + 18(x_1 + x_2 + x_3)(x_1 x_2 + x_1 x_3 + x_2 x_3)(x_1 x_2 x_3)$.

6. A_3.

7. Show that the Galois group of $f(x)$ is isomorphic to S_5. The equation $f(x) = 0$ is not solvable by radicals.

Exercises 25.1.2 (page 548) (Geometric Constructions)

4. Prove that $2\cos \frac{2\pi}{7}$ is a root of the polynomial $x^3 + x^2 - 2x - 1$ and that this polynomial is irreducible over $\mathbf{Q}$.

5. Use Example 25.1.21 and Worked-Out Exercises 2 (page 547) and 5 (page 548).

6. In the proof of Theorem 25.1.19, we see that an angle of $\theta°$ can be trisected if and only if the polynomial $4x^3 - 3x - \cos\theta$ is reducible over $\mathbf{Q}(\cos\theta)$. Since $\cos 90° = 0$, we have that an angle of $90°$ can be trisected.

Exercises 26.1.2 (page 557) (Binary Codes)

4. $\bar{0} \in C$ and so $C \neq \phi$. Let $\bar{c_1}, \bar{c_2} \in C$. Then $(\bar{c_1} - \bar{c_2})K = \bar{c_1}K - \bar{c_2}K = \bar{0} - \bar{0} = \bar{0}$. Hence, $\bar{c_1} - \bar{c_2} \in C$. Hence, C is a subgroup of B^n.

6.

$$
\begin{aligned}
0000M &= 0000000 \\
0001M &= 0001011 \\
0010M &= 0010101 \\
0011M &= 0011110 \\
0100M &= 0100110 \\
0101M &= 0101101 \\
0110M &= 0110011 \\
0111M &= 0111000 \\
1000M &= 1000111 \\
1001M &= 1001100 \\
1010M &= 1010010 \\
1011M &= 1011001 \\
1100M &= 1100001 \\
1101M &= 1101010 \\
1110M &= 1110100 \\
1111M &= 1111111
\end{aligned}
$$

7. (i) The code can detect one error. The code cannot correct every single error.

8.

C :	000000	001011	010101	011110	100110	101101	110011	111000
$100000 + C$:	100000	101011	110101	111110	000110	001101	010011	011000
$010000 + C$:	010000	011011	000101	001110	110110	111101	100011	101000
$001000 + C$:	001000	000011	011101	010110	101110	100101	111011	110000
$000100 + C$:	000100	001111	010001	011010	100010	101001	110111	111100
$000010 + C$:	000010	001001	010111	011100	100000	101111	110001	111010
$000001 + C$:	000001	001010	010100	011111	100111	101100	110010	111001
$100001 + C$:	100001	101010	110100	111111	000111	001100	010010	011001.

001111 is decoded as 001011; 101010 is decoded as 001011; 111110 is decoded as 011110.

12. $s = 4$. K is $(2^4 - 1) \times 4$ or 15×4.

$$
K = \begin{bmatrix}
0001 \\
0010 \\
0011 \\
0100 \\
0101 \\
0110 \\
0111 \\
1000 \\
1001 \\
1010 \\
1011 \\
1100 \\
1101 \\
1110 \\
1111
\end{bmatrix} .
$$

Exercises 26.2.1 (page 570) (Polynomial and Cyclic Codes)

2. $G = \begin{bmatrix} 1 & 1 & 1 & 1 & 1 & 0 & 0 \\ 0 & 1 & 1 & 1 & 1 & 1 & 0 \\ 0 & 0 & 1 & 1 & 1 & 1 & 1 \end{bmatrix}$. **3.** $K = \begin{bmatrix} 1 & 1 & 0 & 1 \\ 0 & 0 & 0 & 1 \\ 1 & 1 & 1 & 1 \\ 1 & 0 & 0 & 0 \\ 0 & 1 & 0 & 0 \\ 0 & 0 & 1 & 0 \\ 0 & 0 & 0 & 1 \end{bmatrix}$.

5. Suppose $\langle f(\overline{x}) \rangle \subseteq \langle h(\overline{x}) \rangle$. Then there exists $q(\overline{x})$ such that $h(\overline{x})q(\overline{x}) = f(\overline{x})$. Hence, $h(x)q(x) = f(x) + x^n - 1$. Now $h(x)|(x^n - 1)$ and so there exists $m(x) \in \mathbf{Z}_2[x]$ such that $h(x)m(x) = x^n - 1$. Thus, $h(x)q(x) = f(x) + h(x)m(x)$ and so $h(x)(q(x) - m(x)) = f(x)$. Hence, $h(x)|f(x)$.

7. The proof follows from the proof of Theorem 26.2.6 since every element of C is of the form $\overline{x}^i g(\overline{x})$ and C is closed under addition.

Exercises 26.3.1 (page 573) (Bose-Chauduri-Hocquenghem Codes)
2. $1 = 0 + 0^4 \neq 0$ and $1 + 1 + 1^4 \neq 0$. Hence, if $1 + x + x^4$ factors nontrivially over $\mathbf{Z}_2$, it must be a product of two quadratics. Suppose

$$1 + x + x^2 = (a + bx + cx^2)(d + ex + fx^2),$$

where $a, b, c, d, e, f \in \mathbf{Z}_2$. Then $1 = ab$, $1 = ae + bd$, $0 = af + be + cd$, $0 = bf + ce$, $1 = cf$. Thus, $a = d = c = f = 1$, $1 = e + b$, $0 = 1 + be + 1$, $0 = b + e$. This is impossible. Hence, $1 + x + x^4$ is irreducible over $\mathbf{Z}_2$.

Exercises 27.1.2 (page 582) (Affine Varieties)
2. Show that $\langle x^i, y^j \rangle \subset I(V(x^i, y^j))$.
4. $x + 1$.

Exercises 27.2.2 (page 600) (Gröbner Bases)
1. (i) multideg$(f) = (4, 1, 2)$. (ii) LC$(f) = -5$. (iii) LM$(f) = x^4yz^2$. (iv) LT$(f) = -5x^4yz^2$.
2. $a_1 = x^2y - xy + y$, $a_2 = -y$ and $r = -xy^3 + x^2y - xy - y$.
3. $a_1 = x^3y + x^3 + 1$, $a_2 = 0$ and $r = 2x^3y + x^3 + y + 1$.
4. $S(f, g) = -2xyz^2 + x^2y^3z^2 - x^2yz^2$.
5. $J = \langle x^2 \rangle$, $J_0 = \{0\}$, $J_1 = \{0\}$, $J_2 = \langle x^4 \rangle$, $J_3 = \langle x^4 \rangle$, $J_4 = \langle x^3 \rangle$.
11. $\{x^2 - y, yz\}$ is a reduced Gröbner basis for I. $V(I) = \{(x, x^2, 0) \mid x \in \mathbf{R}\} \cup \{(0, 0, z) \mid z \in \mathbf{R}\}$.

Index

A

Abel, Niels Henrik, 82
Abelian group, 59
    fundamental theorem of finite, 251
    fundamental theorem of finitely generated, 262
ACC, 406
action of groups, 172
affine variety, 574–575
algebraic closure, 470
algebraic element, 447
algebraic field extension, 453
algebraically closed field, 469
algebraically independent, 341
alternating group, A_n, 94
Artin, Emil, 491
Artinian ring, 412
    left, 412
    right, 412
ascending chain condition for principal ideals, 368
associate, 353
associated prime ideal, 394
automorphism
    of groups, 144
    of rings, 309

B

basis, 423
    Gröbner, 592
    minimal Gröbner, 597
    reduced Gröbner, 598
    standard, 592
betti number, 266
binary operation, 52
    associative, 52
    closed under, 52
    commutative, 52
binary symmetric channel, 551

Boolean ring, 284
Burnside theorem, 175

C

Cartesian product, 4, 47
Cauchy's theorem, 196
Cauchy, Augustin-Louis, 98
Cayley's theorem, 149
Cayley, Arthur, 180
center
    of groups, 101
    of rings, 271
centralizer, 190
chain, 32
    factors of, 223
    length of, 223
    normal, 223
    subnormal, 223
chain condition
    ascending, 406
    descending, 406
characteristic
    of a ring, 278
    subgroup, 165
Chinese remainder theorem, 30
Chinese remainder theorem for rings, 333
circle, 539–540
class equation, 192
code
    BCH, 571
    distance between, 554
    group, 555
    Hamming, 553
    polynomial, 558
    weight of, 554
codeword, 549
commutative
    group, 59